Perspectives in Insect Behaviour

M.L. Agarwal
University Professor
Department of Entomology
Rajendra Agricultural University
Pusa (Samastipur)- 848 125, India

International Book Distributing Co.
(Publishing Division)

Published by

INTERNATIONAL BOOK DISTRIBUTING CO.
(Publishing Division)
Khushnuma Complex Basement
7, Meerabai Marg (Behind Jawahar Bhawan),
Lucknow 226 001 U.P. (INDIA)
Tel. : 91-522-2209542, 2209543, 2209544, 2209545
Fax : 0522-4045308
E-Mail : ibdco@airtelmail.in

First Edition 2009

Price: MAXIMUM RETAIL PRICE INCLUSIVE OF ALL TAXES $2100.00

ISBN 81-8189-247-X

Composed & Designed at :

Panacea Computers
3rd Floor, Agarwal Sabha Bhawan, Subhash Mohal
Sadar Cantt. Lucknow-226 002
Phone : 0522-2483312, 9335927082, 9452295008
E-mail : prasgupt@rediffmail.com

Printed at:

Salasar Imaging Systems
C-7/5, Lawrence Road Industrial Area
Delhi - 110 035
Tel. : 011-27185653, 9810064311

To

Kusum
Monika
and
Sarthak

Foreword

Insects represent the zenith of evolutionary developement among animals and an overview or world of insects reveals that they are most amazing creatures on the earth. The behavioural patterns exhibited by insects have enthralled the man since long ago. Most people admire insects' activities and become fascinated by watching their interesting behaviours. Inscet behaviour is an intricate subject and envisages a multidisciplinary appraoch to investigate the answers for many questions about how and why insects do the things they do. Integration of knowledge on insect behaviour with other branches of biological sciences, like genetics taxonomy, and other has led to deal this subject in more intricate manner.

A wealth of knowledge has been generated in recent years on inscet behaviour and in this context the publication of 'Perspectives in Insect Behaviour' which includes all important aspects pertaining to insect behaviour is the need of the time. This meticulously written by book interperts many interesting facts about insects' behavioural patterns. The efforts made by Professor M.L. Agarwal for bringing out such an excellent text are appreciable and I am sure that this text will of immense interest and useful for the readers, particularly students, teachers and researchers.

Rajendra Agricultural University, Bihar
Pusa (Samastipur)

V.S. Verma
Vice-Chancellor

Preface

Insects exhibit culmination of evolutionary development and vast majority of them occupy different ecological niches. Most people are attracted to them on account of their interesting ways of life; however, many of them remain undaunted by the vigilant eyes of the watcher. Chorusing of crickets and cicadas, light production by fireflies and glowworms, path finding by ants, dances of honey bees, migration in butterflies, defence measures and many other amazing behaviours shown by insects have fascinated and entertained the men since ancient times. Their great number and diversity have led them to express diverse behavioural patterns. Though most behaviours of insects' are innate; however, many of these clearly indicate their learning ability to some extent. The subject of insect behaviour is engrossing for the readers like scientists, teachers and students. Insect behaviour is an important subject taught to the undergraduate and postgraduate students of Zoology and Entomology in all Universities and is notable that can be cherished by any student of biology. Introductory entomology texts include only a little on this intellectually interesting and practically important subject. A few books are available dealing with various aspects of insect behaviour, but the studies in stride with increasing specialization in almost all spheres of insect behaviour have traversed substantially in various directions and need thorough discussion. While teaching the subject at postgraduate level I thought for writing a manuscript on this subject and in this text an attempt has been made to discuss different sections in an easy and lucid manner.

The whole text has been divided into 21 chapters dealing with various aspects of insects' behaviour. The book starts with an introduction. Both innate and learned behavioural patterns are discussed in detail. A brief account of sensory reception in insects and neural control of behaviour has also been incorporated. The movement of insects is one of the major behavioural patterns and has been discussed in chapters, viz. orientation, locomotion, navigation and migration. The feeding, reproduction, communication, defence, and social behaviour are incorporated in other chapters. An account of insects' interactions among themselves and with plants and fungi has been added in separate chapters. The rhythmic and light production behaviours are dealt in additional chapters. Chemicals like hormones and pheromones have a great influence on behavioural patterns of various insects and such aspects are dealt in other noteworthy chapters. The subject of insect behaviour is intimately related with other branches of entomology and therefore, chapters on behavioural taxonomy, behavioural genetics, and behavioural manipulations for insect pest management have also been included in the present text.

Suggestions received from peers' will certainly encourage me and help in improvement of future editions. I shall consider this effort successful and gratifying if the book serves the purpose for which it is intended.

I am thankful to Dr V.S. Verma, Vice-Chancellor and Dr K.R. Maurya, Dean, Faculty of Agriculture, Rajendra Agricultural University, Pusa for inspiration and encouragement. I am also thankful to Dr I.P. Singh, Chairman, Department of Entomology, Dr R. Singh, Professor of Entomology, Dr M.K. Mallik, Professor of Soil Science, Dr S.K. Jain, Associate Professor, Agricultural Engineering and Shri R.C. Sharma, Assistant Librarian (Selection Grade), Rajendra Agricultural University, Pusa (Samastipur), India for their kind cooperation. Thanks are also due to Dr. M. Kumar, University Professor, Department of Biotechnology and Molecular Biology for critically reviewing the chapter on behavioural genetics. I also extend my gratitude for the cooperation received from other colleagues. I acknowledge my gratitude to various authorities whose works have been referred in the present text.

Thanks are also due to Mr Suneel Gomber (Managing Partner), International Book Distributing Co., Lucknow for his keen interest in publication of this book.

I cannot forget the wholehearted cooperation received from my family members during the course of writing the present text.

M.L. Agarwal

Acknowledgements

The author deeply appreciates the kind permission granted by following scientists/ persons for using photographs/ figures from their website/ publications for the present text.

Professor J.E. Lloyd, Professor Emeritus, Department of Entomology & Nematology, University of Florida, PO Box 110620, Gainesville, Florida 32611-0640; John Tyler (www.johntyler.co.uk) 5 Woodfield, Lacey Green H P27 OQQ, UK; Alex Wild, (www.myrmecos.net) Department of Entomology, University of Arizona, Tucson, AZ 85705; Dr John R. Conway Associate Professor of Biology, Department of Biology, University of Scranton, Scranton, Pennsylvania 18510-4625; David L. Mohn (http:\\ccs-ht.org/DM/), Light Creations, Hong Kong; John Moore (www.thaibugs.com/thaibugs.htm) 79/90 Moo. Intrachitchal Doi Saket Road, T. Talatkhwan Doi Saket, Chiang Mai, Thailand; Robin Scagell (www.galazypix.com/glowworms) 34 Fennels Way, High Wycombe HP 10 9BY, UK; Garry Maguire (www.springbrook.info/glow_worms) Springbrook Research Centre, Springbrook, Queensland, Australia; A. San Juan (www.blueboard.com/leafcutters/images.htm) 12 Mendell Ave Cranford, NJ 07016 USA; Ramana Athreya (www.cls.jhu.edu/people/zak/ramana), Englenest Biodiversity Project, Meghalaya, India. He is also grateful to his colleagues Dr I.P. Singh, Professor of Entomology; Dr R. Singh, Professor of Entomology; Dr S.K. Jain, Associate Professor of Agricultural Engineering; Dr P.P. Singh, Associate Professor of Entomology and Dr Neeraj Kumar, Assistant Professor of Entomology, Rajendra Agricultural University, Pusa (Samastipur)- 848 125 India for supplying some photographs used in the present text.

Acknowledgments

Contents

Chapter 1

Introduction

Insects comprise a highly specialized group of animals and belong to great phylum of jointed-legged animals- the Arthropoda. They represent the culmination of evolutionary development among terrestrial forms. In comparison with other groups of animals they have been observed populating almost all types of habitats. An overview of the world of insects reveals their great diversity and constancy. The vast number of over one million insect species named so far and many more yet to be named, clearly indicates their dominance on the globe. They have exploited many advantages and other specializations in their morphological, physiological, behavioural and other characteristics to become the dominion of evolutionary sequences. Such outstanding specializations are: small body size, exoskeleton, functional wings, hexapod locomotion, compound eyes, scattered sense organs, direct respiration, high fecundity, adaptability of structures, indirect development, behaviouristic adaptations and so on. It also seems to be rather difficult to study and explain such a large group having great diversity. The differences in the groups and species of insects are primarily due to differences in the environment they occupy. The strong and intricate relationships that exist among insects and vast multitude of living and non-living components of the environment form a fascinating subject for the present day entomologists. Efforts have been made to study such relationships and their responses to environmental events. In addition the responses shown by insects are governed by the genetic factors. Thus, their responses are directly or indirectly results of genes, e.g. the flight of any insect depends on presence or absence of wings, and whether it has wings or not depends largely on its genes. The genes determine the characteristics it can develop, but the development of traits depends on the interaction of genes and environment during development. The similar phenomenon holds true for developmental responses shown by the insects or their behavioural patterns. However, the development of behavioural patterns is rather complicated as compared with the development of body characteristics. The body characters develop until the full growth is achieved, whereas the behavioural patterns alter throughout the life as the interaction of genes and environment conditions continues. Presently, many zoologists have begun to accept the premise that an animal's behaviour has the evolved function of propagating the individual's genes rather than contributing to the survival of species (Williams, 1966). Dawkins (1989) pointed that animals do not act for the 'good of the species'.

This is an archaic view- in fact, it is more appropriate to think of them as acting for the good of their genetic material and to think of genes as being selfish.

1.1 Concept

The concept of behaviour, as conceived at present, is of recent origin. Broadly speaking, the study of animal behaviour is also termed 'ETHOLOGY' (*Gr.* Ethos, meaning character or habit or custom; and Logos, meaning study). Thus, ethology literally means "study of characters or habits of animals." Insidore Geoffray Saint-Hilaire (1859) used the term ethology to refer to the study of relationships between organisms and environment. He defined ethology as "the study of relations of the organisms within the family and society in the aggregate and in the community." There seems to be some controversy and this term was used by him to living organisms in ecological sense rather than behavioural. After 1930, the term ethology was widely used in the study of animal behaviour.

In simple words behaviour has been defined as what animals do or movement of animals, e.g. movement of wings during flight, legs in walking and head and mouthparts during feeding. However, in some behavioural activities no movement occurs, such as production of sound by vibration of a membrane in cicada. The sound is produced when tymbal muscles contract and relax. Here movement of musculature associated with tymbal occurs. In other activities, viz. feeding, walking, flying there occur movement of muscles. Therefore, more precisely animal behaviour may be called to consists of a series of muscular movement. Another example of motionless activity considered as a behavioural pattern is death-feigning. Many insects, e.g. stick insects, many beetles, caterpillars and others become apparently feign dead, if disturbed (Plate 1.1). Some drop from leaf or twig and lie completely stiff on the ground.

The behaviour is also the way in which an organism adjusts to and interacts with its environment. Thus, it includes a wide range of activities, which lead to better survival of the individuals of a species. The appropriate behavioural responses of insects provide better chances of living to them and their progeny, thereby results into substantial increase in survival value. The others that are not able to response appropriately are weeded out in the process of natural selection. Thus, the behavioural patterns of the individuals of a species are evolved through a long process of evolutionary development.

Survival also depends on the maintenance of suitable relations with the non-living environment. The adaptive relation between an animal and its environment is also behaviour. The affirmation that an insect maintains relations with the environment (both living and non-living), implies that an insect changes in the environment. These change that we call behaviour are not passive, they are direct actions, which promote survival. Another efficient property of living beings is their ability to respond to stimuli and is termed as 'irritability'. In multicellular animals, the course of evolution of responsiveness has been shaped by the requirements of the whole individual acting in the environment, not solely in terms of the individual unit, e.g. a tissue or an organ. Further with an increase in the complexity of body structures, there was a great need for coordination and conducting of information. To meet this need certain cells became specialized in enhancing both irritability and rapid conduction of irritability changes. With further specialization, these cells became so organized that they were able to perform specific function of conduction of information, were able to detect environmental changes (both external and internal), to integrate information and to initiate a response. Such cells became neurons, the matrix of behaviour. It follows that the behaviour of an animal is an expression principally of the capabilities of its nervous system. The study of behaviour is therefore, an analysis of the potentialities of the nervous system. Insects, with a well-defined body structure and highly evolved sense organs have a precisely defined nervous system, with well-organized brain, ganglia and nerves. The complexity of nervous system in insects has enabled them to display many complicated behavioural patterns.

People who study animal behaviour are concerned with understanding the causes, functions, development, and evolution of behaviour. The causes of behaviour include both the external stimuli that affect behaviour, and the internal hormonal and neural mechanisms that control behaviour. The functions of behaviour include its immediate effects on animals and its adaptive value in helping animals to survive or reproduce successfully in a particular environment. The development of behaviour pertains to the ways in which behaviour changes over the lifetime of an animal, and how both genes and experience affect these changes. The evolution of behaviour relates to the origins of behaviour patterns and how these change over generations. The study of behaviour involves not only what an animal does but also when, how, why and where the behaviour occurred (Lehner, 1979).

McFarland (1985) stated that the scientific study of behaviour involves a variety of approaches and it can be explained in terms of: (i) its evolutionary history, (ii) benefits it brings to the animal, (iii) psychological mechanisms, and (iv) physiological mechanisms.

The behavioural patterns seen in all animals are result of three main evolutionary causes or selective pressures:

(i) An animal's behavioural patterns will be selected for as they contribute to its evolutionary success.

(ii) The behavioural patterns must allow the animal to react and to solve immediate problem.

(iii) Reaction of animal to stimuli that affect survival and reproduction by completing appropriate behaviour. Ability of animals to discriminate between various stimuli and for ensuring that an animal completes a behaviour pattern is crucial parts of any animal's behavioural makeup.

The science of ethology forms a fascinating subject to be studied. Though modern technologies and devices have provided numerous aids to ethologists to undertake detailed observations in the laboratory, the essence of the subject is to study behaviour as it occurs in nature. For this reason most researches are still conducted in nature and observing and describing of what animals do in nature is one of the principal preoccupation of the ethologists. Recently many ethologists have joined their hands with ecologists and both these branches of biological science are integrated by studying population behaviour. In a number of cases systematists also have separated closely related taxa and sibling species on the basis of behavioural patterns and are able to determine the phyletic relationships among different groups of animals. The other branches of biological science like genetics, neurology, physiology and others are also intimately interlinked with the science of ethology.

1.2 Historical aspects

Ethology is as old in respect of its history as man himself. The early primitive tribes were able to know the fauna of their own regions and were mainly hunters. They were using tools for hunting and had some ideas of behaviour of the animals to be hunted. Behaviour has evolved by natural selection and knowledge and understanding of it has always been a practical matter for early trappers, hunters, traditional shepherds and herdsmen. Before any scientific study of animal behaviour there was great interest in it because of the practical benefits and raw human curiosity, and there was the belief that understanding nature contributed to understanding God. The insects find mention in many ancient books and sacred writings. Aristotle- a student of Plato and Greek Philosopher, Educator and Scientist may be called as founder of Entomology. He (372 B.C.) for the first time wrote excellent descriptions of animal behaviour in *Historia Animalium*. His writings contain remarkable observations on feeding, locomotion, sound production, reproduction and parental care in insects. Réaumer (1734-1742) provided first insight of scientific knowledge of insects, and the results of his detailed studies and precise observations were published in six large volumes.

During early periods of 19th century in Switzerland F. Huber, though blind but genius, was having a deep sense of scientific investigation and made outstanding observations on honey bees and ants with the help of his son and valet. His son Jean Pierre Huber further extended the work on ants started by his father and the results of his exemplary studies were published in 1810. Other notable contributions of this period need to be cited were of Kirby and Spence (1815-1828) and Burmeister (1836).

Darwin's monumental book *"The Origin of Species by Means of Natural Selection, or the Preservation of Favoured Races in the Struggle for Life"* in 1859 includes general principles of evolution, and he included a chapter on instinct in this book. His valuable treatise *"The Expression of the Emotions in Man and the Animals"* published in 1872 includes discussion on behaviour of various animals. His contributions were based on the results of vast study of nature and its belongings. Some other worthily written contributions of that period were: Hagen (1858), Muller (1873, 1874), Graber (1877-1879), Fabre (1879-1908), Lubbock (1883), Peckham and Peckham (1898) and Forel (1886-1888; 1908, 1910).

In the first half of 20th century around 1930 ethologists adopted many new techniques and efforts were made to solve complex problems of behaviour. During this period comprehensive study of behaviour began to emerge. Though the attention of ethologists and psychologists was seems to be more concentrated towards birds and mammals; however, valuable contributions also appeared on insects' behaviour. Three European biologists established the foundations for the study of the behaviour of animals in their natural environments, in mid-20th century. Konrad Lorenz, an Austrian biologist, is considered to be the founder father of modern ethology and carried out extensive studies on animal behaviour. His studies focussed on innate animal behaviour, especially imprinting of birds. He also found patterns of instinctual actions, not guided by insight or conscious reflection. Niko Tinbergen, a zoologist from Holland studied numerous animals including insects like butterflies and digger wasps, and his contributions on the concept of sign stimuli are innumerable. Karl von Frisch, a biologist from Germany conducted tedious experiments on communication and other behavioural patterns in honey bees. These three received the Nobel Prize in 1973 in recognition of their significant contributions. The notable contributions of Lorenz (1952, 1965, 1970, 1971); Tinbergen (1951, 1953, 1958, 1965, 1973), and Frisch (1954, 1967, 1971) are cited in the references.

After 1930 numerous valuable publications appeared in the form of books/ chapters on animal behaviour. Some of these generalized texts exclusively dealing with insects are: Evans (1966a), Haskell (1966), Roeder (1967), Barton Browne (1974), Markl (1974), Matthews and Matthews (1978), Atkins (1980), Hansell (1985), and Kerkut and Gilbert (1985).

Klopfer and Hailman (1967) and Richards (1973) discussed historical development of studies on insect behaviour during different periods. Romoser (1981) listed selected references of the work done on various aspects of insect behaviour in the 20th century.

Other texts which include numerous insects' examples and of interest are: Thorpe, 1963, 1979), Wigglesworth (1964a), Carthy (1958, 1965), Marler and Hamilton (1966), Hinde (1970, 1982), Eibl-Eibesfeldt (1970), Brown (1975), Gould (1982), Barnard (1983), Halliday and Slater (1983), Alcock (1984), Huntingford (1984), McFarland (1985), Dawkins (1986), Slater (1985, 1999), Manning and Dawkins (1992), and Goodenough *et al.* (1993). Some books published in India and embrace various aspects of animal behaviour are: Arora (1992), Mathur (1994), Singh (1995) and Awasthi (2001).

It is worth mentioning that most behavioural researches have been conducted in the second half of 20th century and the field of ethology is rather young and developing science. Insects form the vast majority among animal groups, rather easy to handle and study, and there is ample scope to study and solve complex problems pertaining to behaviour by taking insects as key animals in near future.

Chapter **2**

Innate and Learned Behaviour

Broadly speaking, most insects show two kinds of behavioural patterns, innate and learned. At first, behaviour consists of more or less fixed response or a series of responses and is considered to be genetically determined. Thus, the behaviour is 'stereotyped' and the animal to a large extent is stimulus bound. Since this behaviour is essentially the outcome of inherited properties of the nervous system of an animal or species, it is termed 'innate behaviour'. Later, behaviour becomes more variable and more modifiable through experiences. The adaptations of the individual animal may develop uniquely in its life through the process of learning and such behavioural patterns are called 'learned'. Obviously, such patterns are not inherited. From the biological point of view, in learning the change in behaviour must be a change in the functioning of the nervous system and if it is a permanent change, perhaps it is also a change in the structure of the nervous system (Dethier and Stellar, 1963). As the complexities of the nervous system grow, new properties of behaviour emerge. Behaviour is now not so much stimulus bound, much of it may originate on the basis of past experiences.

2.1 Innate behaviour

Innate behavioural patterns are considered to be genetically determined. Their imprints are genetically programmed into the nervous system and it is rather difficult to alter such patterns, as they also transmit from parents to offspring. It has also been named as inborn, inherent or instinctive behaviour in literature. However, the ethologists have controversies in the use of these terms as well as the concept of innate behaviour. This is partly because in many cases it is rather hard to test whether a given pattern was innate or learned. Being determined by heredity it is the result of genetic makeup of any animal and therefore, independent of animals experiences. Lorenz (1937, 1939) considered it to be hereditary, fixed and thus open to evolutionary analysis. Tinbergen (1942, 1951) suggested these acts as being highly stereotyped coordinated movements controlled by genetic constitution of an animal. He and many others altogether ceased to use the word innate and criticized Lorenz's views of genetic determination of behavioural patterns. Hebb (1953) writes, "that the term 'instinct' implying a mechanism or neural process independent of environmental factors, and distinct from the neural processes in which learning enters, is completely a misleading term." Lehrman (1953) contended that it would never be practically possible to exclude the participation of learning in early ontogenic processes in the egg or *in utero,* which are inaccessible to observation. Tinbergen (1955, 1963), Hinde (1970) and many others also proposed similar conjectures.

Lorenz (1965) critically examined the concept of innate elements of behaviour and published critiques of other ethologists' attitudes. Contemporaneously many others supported the ideas proposed by Lorenz. Thorpe (1974) mentioned that instinctive behaviour in animals have four main characteristics:

(i) It follows a recognizable and predictable pattern in almost all members of a species or all members of one sex of a species and is inheritable.

(ii) It shows a patterned sequence in time.

(iii) Its all or some consequences are of obvious value in contributing to the preservation of an individual or continuity of a species, which means it is adaptive.

(iv) It develops when all the ordinary opportunities of learning and practicing are absent, that is to say it is spontaneous or endogenous.

Alcock (1984) believed that 'innate behaviour is an action that is performed in a functional manner the first time an animal of certain age and motivational state encounters the appropriate environmental cue for the behaviour'. McFarland (1985) regarded it to be as 'behaviour that develops without obvious environmental influence'. Both Tinbergen and Frisch found that most insect behavioural patterns were innate, and indeed it must be because most insects have no parents around to learn from and must behave appropriately upon first exposure to stimuli.

Innate behaviour develops without the animal's having learning to it, so that way by which it can be tested by depriving the animal of any environmental stimuli. Such experiments are called to be as 'depriving experiments'. The behavioural patterns are similar in all individuals of a species and their performance can only change over generations as individuals with different first-time responses are for or against. A number of patterns, e.g. songs of crickets reared in isolated conditions, nest building by a sand wasp without any contact with members of previous generation, courtship displays, parental care are considered to be the release of previously stored programmes. The workers of honey bee do not learn to build the hive and thier comb construction behaviour seems to be the result of a prewritten composition.

2.1.1 Reflexes

The basic and simplest form of innate behaviour is 'reflex', in which simplest type of impulse conduction known as 'reflex action' occurs in the central nervous system (CNS). The term 'reflex action' refers to a rapid, automatic, unlearned response to a stimulus. These actions are involuntary as opposed to voluntary actions, which are governed by willful or conscious control of the animal. In reflex, an effector organ as a result of a stimulus transmitted to it along nerve fibre from a receptor organ carries out, an action at once. Thus, the stimulus is followed by the immediate response. These actions are deeply planned in the CNS that they are carried out even when the purpose can no more be achieved. The reflex action requires a threshold intensity to elicit a response and the stimulus is termed a releaser, e.g. when an insect's legs are not touching the ground, there is a flight response, but when the legs touch the ground, the wings stop beating.

The pathway of reflex action called 'reflex arc' is presented in Fig. 2.1. A simple reflex arc consists of (i) a sensory neuron in the receptor organ, (ii) a part of CNS called 'reflex centre', and (iii) a motor neuron, which transmit impulses from the reflex centre to the efferent organ. Besides, in most reflex action an associated neuron is also present between the two neurons. The reflex arcs are of two types.

(i) Somatic reflex arc involving effector organs located in (soma) body structures, e.g. muscles associated with legs, wings, mouthparts.

(ii) Visceral reflex arc involves effectors located in the visceral organs, e.g. glands.

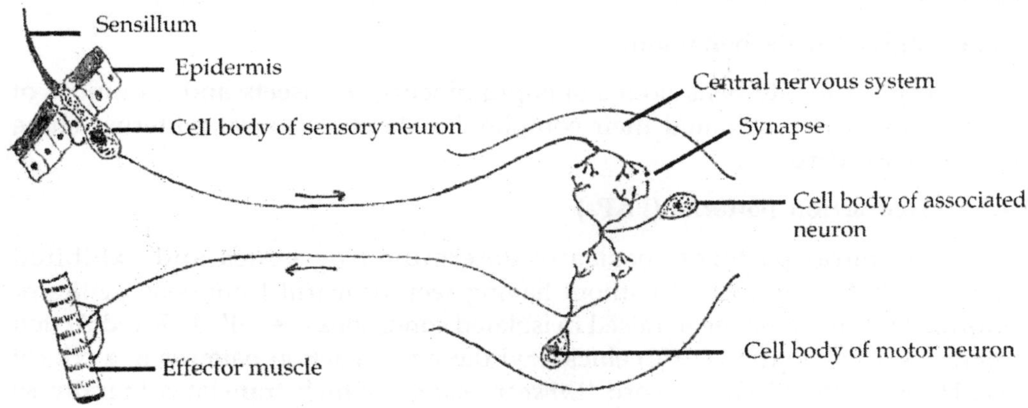

Fig. 2.1. Diagram showing a simple reflex arc.

A reflex may be 'phasic' or 'tonic'. Phasic reflexes are rapid, of short duration and involved in rapid movement of body or its parts, e.g. proboscis extension, sudden flight to a threat, running of cockroach on a touch to circus. Tonic reflexes, on the other hand, are rather slow, long-lived and are involved in maintenance of body posture, position in space, balance, muscle tone, etc.

Reflexes may be organized at various levels of nervous system and occur in varying degree of complexity. In insects, the simplest reflex involves many more neurons. They are also highly plastic, e.g. cockroach normally cleans antennae with fore legs, and if fore legs are removed it may use middle leg of opposite site for the purpose. Similarly, among the long-horned beetles one may clean antennae with legs, another with mouthparts or by using both the appendages. Once a reflex has been set going, subsequent conduction of impulses through the same channel favour the response and this may persists for a long time. The phenomenon is called 'facilitation'. However, the strong or repeated stimuli can cause overflow of impulses in other reflexes or decline in the magnitude of responses due to fatigue. A series of reflexes in a definite sequence lead to 'chain reflex' or 'cyclic reflex', e.g. walking of an insect, probing in aphids. Besides, failure of an expected reflex or a check to the chain reflex is due to inhibition from other centres of CNS.

Righting reflex occurs when an insect is placed on its back and help to restore normal body position. The reflex bleeding has also been reported in many Orthoptera, Coleoptera, Lepidoptera and other groups, in which small drops of fluid or blood are released from various points of their body, due to an increase in hydrostatic pressure. In *Dictyophorus* (Order Orthoptera) the blood is mixed with air to form brownish foam. Most meloids and aphids also exhibit similar reflexes. This behaviour of insects is of protective value and help in escaping.

The most apparent and complex reflexes involve orientation of body in space, time or in response to certain stimulus. Such patterns are being discussed in chapter 5.

2.1.2 Complex innate behaviour

The concept of innate behaviour holds prominence in insects and a number of experiments have elucidated their complex innate behavioural patterns. These are as discussed below.

2.1.2.1 Fixed action patterns (FAPs)

The behavioural patterns, basically determined by genes and exhibited automatically by any animal without having seen or learnt from conspecifics or performed by an animal even raised in isolated conditions are called "Fixed Action Patterns" (FAPs). Lorenz (1970) elucidated these fixed action patterns in a critical way. He used the German word *'Erbkoordination'* which translates literally as 'inherited co-ordination' to describe them.

The properties of FAPs are:

(i) They are inherited and similar in individuals of a species.

(ii) Cannot be influenced in performance by external stimuli.

(iii) They are conservative in evolution of a species and resistant to phylogenetic changes occurring in the course of evolution.

(iv) Elicited by simple but specific stimuli, which are termed as 'releasers' or 'sign stimuli'.

A number of behavioural acts are considered to be programmed in the CNS of female digger wasp, e.g. by the time she emerges, her parents are already dead; she finds a mate and mates; then forms a nest in the ground by performing complex activities; then go for hunting and bring other insects as food for developing larvae; lays egg and seals the nest and ultimately die. During her short life span, she has a very tight schedule, no chance to see her parents and other conspecifics. These observations were made by the great naturalist J.H. Fabre more than hundred years ago and presently considered to be the result of preprogrammed FAP. In other cases insects have been observed to exhibit the pattern of appendages cleaning, even if they are removed. Most impressive examples are found in the songs of crickets and grasshoppers, which are species-specific. They can be precisely displayed in oscillograms and spectrograms of the sound and remain unaltered even if they are reared in isolation.

The early ethologists suggested that there must be some built-in mechanism in the nervous system of animals to detect sign stimuli. The special neurosensory mechanism that releases the behaviour in response to sign stimuli is called 'Innate Releasing Mechanism' (IRM) (Lorenz, 1950; Tinbergen, 1951). The important aspects of IRM are:

(i) The mechanism is considered to be innate.

(ii) There exists an inborn neural centre called IRM that responds to a sign stimulus.

(iii) IRM is situated at the hypothesized locus in CNS on which sign stimulus acts to release fixed action pattern (FAP) and for each FAP there is a distinct IRM.

Baerends (1950) supported the conception of IRM by studying the responses of digger wasp, *Ammophila adriaansei* to the perception of a caterpillar. During hunting the wasp stung the caterpillar and bring it to the nest as food for larvae; if the caterpillar is found near the nest opening, just after the wasp has opened the nest it is drawn in; but if a caterpillar is present near nest opening when the wasp is closing it, the caterpillar is used as filling material; and lastly during nest digging if she finds a caterpillar, it is removed away. Thus, the wasp releases different responses with the same object in different conditions.

Prokopy and Economopoulos (1975) demonstrated that both laboratory-cultured and wild olive fruit flies, *Bactrocera oleae* were highly responsive to visual stimuli and had a significance preference to fluorescent yellow, over other colours. Honey bees reared in incubator, after having followed a tail-wagging dance for the first time, fly in the indicated direction and exactly to the feeding site (Lindauer, 1971).

Nowadays the term Fixed Action Pattern has been substituted by the phrases 'behaviour patterns' or 'behavioural acts' because:

(i) Behaviour is not as fixed as implied by the term Fixed Action Pattern. There are subtle variations between and within animals for example, the duration of individual components they vary around a mean value.

(ii) Fixed action patterns are not simply innate. They can be subtly modified by experience.

(iii) Behaviour is modified as a result of the animal's environment- it is not always triggered in the presence of the external stimulus.

2.1.2.2 Motivation

For the initiation of a behavioural pattern a number of factors are responsible, which affect the development of a state of 'internal readiness' or 'motivation'. These factors may be external or internal, and lower the thresholds of mechanisms that exert control over the release of a specific behaviour, and have been named as 'drive'. The basis of intrinsic motivational factors may be neural or hormonal.

Neural excitations of enteroreceptors, sensitive to internal conditions of an insect and spontaneous activity of nervous system are the internal neural factors. Endogenous activity- rhythms are responsible for regular cycles of behavioural patterns, having internal origin. Many diurnal insects are active during a particular period of day and this rhythm continues for many days even after they are constantly kept in darkness.

Many receptors react to the conditions within the body and thereby activate the CNS, e.g. spinning activity in cecropia moth, *Hyalophora cecropia* depends upon the stimuli arising from spinning glands (Kloot and Williams, 1953). The postcopulatory courtship of *Gryllus* males and the readiness to copulate in *Periplaneta* males depend on the stimuli arising from spermatophore.

The influence of hormones on behaviour has been analysed in many cases (chapter 17). The relative concentrations of juvenile hormone and ecdysone determine spinning behaviour of *Galleria* moth. A hormone of corpora allata determines readiness to mate in *Drosophila*. Besides, chemicals secreted from exocrine glands (pheromones) also have pronounced effect on behaviour (chapter 18).

External factors, viz. temperature, light, time, contact, etc. also influence motivation. Swarming of ants, flight of beetles, foraging in bees and wasps, glimmering of fireflies, singings of cicada are examples of external regulation of motivation.

2.1.2.3 Appetitive behaviour

When an insect is in a state of readiness (motivated) it may engage in activities that increase the possibility of exposure to a releasing stimulus. Such a pattern is named as 'appetitive behaviour'. Sexually aroused male cicada move randomly (appetitive behaviour), thereby increasing the probability to encounter a female. After searching the female, the copulatory behaviour (consummatory action) is released and sexual arousal decreases. Therefore, the final consummatory action has usually to be preceded by appetitive behaviour. In most searching there occurs locomotion, but always not necessarily so as in some predatory insects, waiting for prey is appetitive behaviour. In some cases, there may be an increase in appetitive behaviour, in terms of kinesis reactions, e.g. speed of movement or frequency of turning, which also result in a decrease in receiving inhibitory stimuli. Blow fly, *Phormia* that has not eaten for a long time, searches actively and does not receive inhibitory stimuli from gut receptors.

2.2 Learned behaviour

Learning is the process that brings about certain adaptive changes in the behaviour of an animal as a result of experiences. The ability to learn is striking character of most animals. However, it is more pronounced in higher animals and has been observed increasing sequentially with an increase in the complexity of animal forms. The animals perform the learned behavioural patterns only after seeing the conspecifics performing the same, through experiences and also due to interaction with the environment. Ethologists have proposed many possible definitions of learning. Thorpe (1963) proposed "learning is the process which manifests itself by adaptive changes in individual behaviour as a result of experiences." Learning also involves accumulation and storage of information (memory) and has a subsequent effect on animal's behaviour. Papaj and Lewis (1993), Abramson, (1994) texts are important compilation of works dealing with various aspects of learning.

Though most actions of insects are stamped with the rigid structure of fixed action patterns, in many cases they also exhibit learning ability. Insects with relatively less number of neurons in the brain are considered poor learners; however, they display many forms of learning, ranging from simple modifications of innate behaviour to highly complex patterns. There are differences in opinions and under noted classifications have been proposed for learning.

After Thorpe (1963)

- (i) Habituation
- (ii) Associative learning
    - (a) Classical conditioning
    - (b) Instrumental conditioning
- (iii) Latent learning
- (iv) Insight learning*
- (v) Imprinting**

After Alloway (1972)***

- (i) Classical conditioning
- (ii) Instrumental conditioning
- (iii) Shock avoidance learning
- (iv) Olfactory conditioning

After Markl (1974)

- (i) Habituation
- (ii) Associative learning
    - (a) Classical conditioning

(b) Instrumental conditioning (Pavlovian conditioning)

(c) Shock avoidance conditioning (Operant conditioning)

(iii) Latent learning or exploratory learning

(iv) Insight learning*

After others

(i) Habituation

(ii) Latent learning

(iii) Insight learning*

(iv) Imprinting**

The associative learning, in broad sense is considered to be the ability to form associations between previously meaningless stimuli and reinforcements, such as, positive (reward) and negative (punishment). This type of conditioning has been thoroughly exploited by taking insects as experimental animals. In the present discussion following learning patterns are being incorporated.

2.2.1 Habituation

It is the simplest form of learning and observed throughout the life of an animal. In habituation a repeated stimulus is given to an animal, the intensity of response gradually decreases and may disappear ultimately. By showing no response to old and unimportant stimuli animals save energy and their reactions to new stimuli increases. The phenomenon of habituation can be well understood by performing simple experiments with grasshoppers, mantids, ants, etc. They can be tamed if handled frequently enough and after repeated picking their responses and movements gradually become slow. Wigglesworth (1941) demonstrated that in human louse, *Pediculus humanus* repeated contact stimuli, not followed by any injury are no longer responded. Similar observations have been made in response to visual, chemical and acoustic stimuli. Bernays and Wrubel (1985) demonstrated that nymphs of lesser migratory grasshopper, *Melanoplus sanguinipes* learn to associate food with visual cues, e.g. light/ colour. This also improves foraging efficiency by reducing time spent in searching unsuitable areas. Fukushi (1989) discussed visual learning in Australian sheep blow fly, *Lucilia cuprina* by presenting sugar droplets on a patch of coloured paper. Gould (1986a) studied that honey bees learn to recognize flowers on the basis of their colour, odour or pattern. Harrison and Breed (1987) demonstrated time and location learning in nectar feeding tropical ant, *Paraponera clavata*. Artificial nectar sources were given to workers at fixed time and location for five consecutive days and after training most ants arrived at feeding site within 30 minutes.

---

* Only circumstantial evidences in insects

** No evidence of imprinting in insects

*** Proposed only for insects

Braun and Bicker (1992) reported that in foragers of *Apis mellifera* touching one antenna with a droplet of sugar-water elicited a proboscis extension reflex; on repeated stimulation, the response decreased and finally disappeared. Daly and Figueredo (2000) demonstrated that pheromone-mediated sexual response in the adults of tobacco budworm, *Heliothis virescens* can in fact be modulated by conditions consistent with the production of habituation. In conditions appropriate to the formation of habituation, it will develop and disrupt male sexual response to female sex pheromone. Givois and Pollack (2000) observed that when crickets are stimulated with cricket-song-like trains of sound pulses, response strength declines for successive pulses in the train, and the decrease becomes more pronounced as the stimulus intensity increases.

2.2.2 Classical conditioning

The concept was developed by Russian Physiologist Pavlov (1927) after conducting experiments on dog and therefore, also called as Pavlovian conditioning. In this type of learning when two different stimuli are repeated simultaneously for a number of times, animal develops the ability to respond to one of the stimulus to

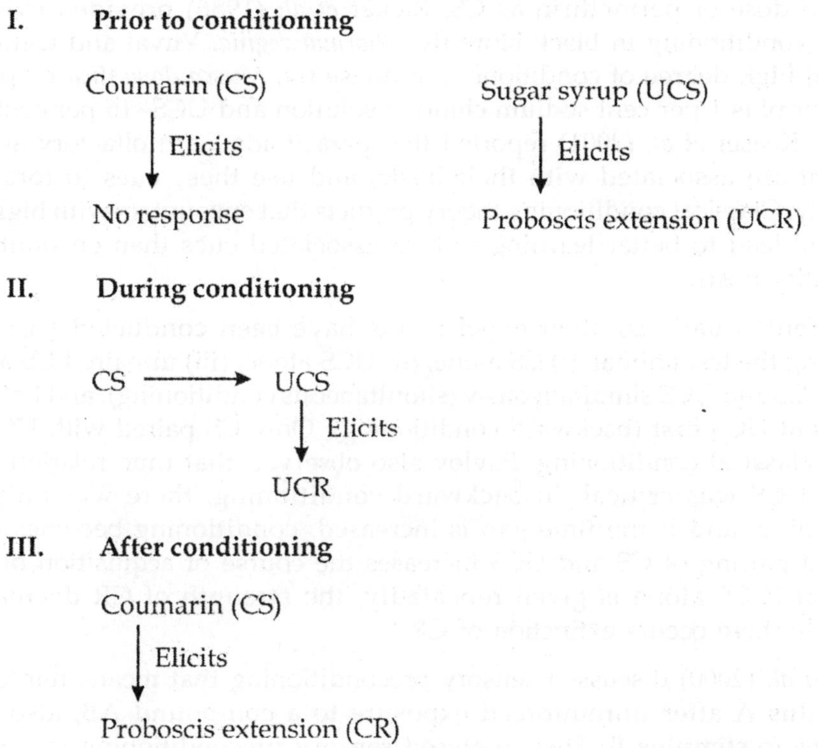

Fig. 2.2. Sequence of events and relationships among stimuli and responses for classical conditioning in honey bee, ***Apis mellifera***.

which it does not customarily elicit the response. Thus two stimuli are involved, one- the unconditioned stimulus (UCS), which elicits the unconditioned response (UCR), without any prior training. The other, conditioned stimulus (CS), must not elicit a response similar to UCR prior to conditioning. The conditioning procedure involves pairing of CS and UCS in a close succession (CS briefly preceding the UCS), and after conditioning the animal gradually elicits the response called conditional response CR (similar to UCR) also to CS (Fig. 2.2).

Kimble (1961) discussed this type of conditioning in detail. Dethier and Stellar (1963) defined it as "a process in which a previously natural stimulus (CS) is enabled to elicit a response (CR) that is never elicited before training." Honey bees afford a good example of classical conditioning among insects. Takeda (1961) used various olfactory stimuli (CS), which were initially unable to produce proboscis extension response (PER) in *Apis mellifera,* but after repeated pairing with the presentation of sugar syrup (UCS), the olfactory stimulus (odour of coumarin) also elicited proboscis extension. The workers performed the best at the age of 14-16 days old.

Mammod and Waller (1990) evaluated this conditioning in *A. mellifera* by using sublethal dose of permethrin as CS. Ricker *et al.* (1986) provided evidence of classical conditioning in black blow fly, *Phormia regina.* Yuval and Galun (1987) observed high degree of conditioning in house fly, *Musca domestica* by providing CS- water plus 1 per cent sodium chloride solution and UCS- 16 per cent sucrose solution. Keaser *et al.* (2001) reported that parasitoids learn olfactory and visual cues that are associated with their hosts, and use these cues to forage more efficiently. Classical conditioning theory predicts that encounters with high-quality hosts will lead to better learning of host-associated cues than encounters with low-quality hosts.

In different situations, other experiments have been conducted variously by presenting the test animal: (i) CS alone, (ii) UCS alone, (iii) unpaired CS and UCS, (iv) both CS and UCS simultaneously (simultaneous conditioning), and both stimuli paired, but UCS first (backward conditioning). Only CS paired with UCS comes to elicit classical conditioning. Pavlov also observed that time relation between CS and UCS was critical. In backward conditioning, there was only a little conditioning, and if the time gap is increased, conditioning becomes difficult. Repeated pairing of CS and UCS increases the course of acquisition of CR and after that if CS alone is given repeatedly, the strength of CR decreases and ultimately there occurs extinction of CR.

Muller *et al.* (2000) discussed sensory preconditioning that means reinforcement of stimulus A after unreinforced exposure to a compound AB, also leads to responses to stimulus B. They analysed sensory preconditioning in honey bee, *Apis mellifera* by using two-element odorant compounds in classical conditioning of the proboscis extension reflex, and found: (i)Sensory preconditioning is not due to stimulus generalization, (ii) paired, but not unpaired, presentation of elements supports sensory preconditioning, (iii) simultaneous, but not sequential, exposure to the elements of the compound supports sensory preconditioning, and (iv) a single presentation of the compound yields maximal sensory preconditioning.

2.2.3 Instrumental conditioning

This is more complex learning as compared with classical conditioning. It includes a number of related training procedures in which a stimulus or a series of stimulus (called reinforcers) are presented or removed contingent upon the performance of some response or series of responses. The reinforcers may be positive, like food or negative like electric shock and induce positive or negative responses, respectively. Maze learning is a good example of instrumental learning. In it the animal starts showing a variety of responses that are stored in him. In insects it has been observed that after repeated trials of learning the maze, the number of errors also decrease.

Turner (1913) was the first to use a metal maze mounted over water for cockroach. Longo (1964) trained American cockroach, *Periplaneta americana* to escape from light in a maze. Alloway (1969) used three different techniques to train adults and larvae of yellow mealworm beetle, *Tenebrio molitor* to turn either left or right consistently in T-mazes. Similar experiments were done for training *Drosophila melanogaster*. Ants, bees and wasps were also taken as test insects for conducting experiments on instrumental learning.

Instrumental leaning can be made complicated by increasing the complexities of stimuli or response possibilities. Thus, the animal is forced to a number of responses and ultimately to choose the right one for its better survival. In this type of situation, the animal is more obviously involved in a process of trial, error and accidental success. This type of conditioning has been further studied by using other methods in which an insect has to achieve a goal. Not necessarily it must always be a maze to be mastered, but the models for such tests were somewhat similar. The maze is; only a suitable experimental setup to study instrumental conditioning.

2.2.4 Shock avoidance conditioning

Horridge (1962) demonstrated the shock avoidance responses in cockroaches and locusts. Pairs of decapitated insects were suspended over a container of electrified saline solution. In one insect (test insect) a shock was received, whenever its legs fell below a predetermined level. Thus, it was trained to stay above a particular level. For another insect (control) same number and intensity of shocks were given to the corresponding leg in a random fashion. In the second phase of experimentation, two insects were separated and shocked independently, whenever their legs fell below a predetermined level. It was observed that test insect received significantly fewer shocks than control insect. Learning is demonstrated as a difference between number of shocks received by the test and control insects. Since the insects were headless, the ventral nerve cord and associated ganglia are considered to be capable of mediating learning. Pritchatt (1968) further supported this type of avoidance reaction to a weak electrical signal in cockroach. Experiments have also been conducted on cockroaches with intact head and complete nervous system. In another experiment Eisenstein and Cohen (1966) demonstrated that even a single ganglion in the ventral nerve cord of cockroach mediate learned leg flexion of a single leg.

2.2.5 Olfactory conditioning

It may be a form of habituation or latent learning. Exposure of some insects to repellent odour during a certain period of their development results in the loss of response to that odour by the adults. Thorpe and Jones (1937) investigated that ichneumon parasite, *Nemaritis canescens* develops positive olfactory response to an abnormal host, a lesser wax moth, *Meliphora grisella* when reared on that host. Thorpe (1939a) showed that ichneumon adults also indicate their positive response to both odour of wax moth and that of cedar wood oil, after a seven-day period of exposure to these odours. He (1939b) also demonstrated that adults of *Drosophila malanogaster* are normally repelled by odour of peppermint oil. However, if their larvae are reared in peppermint-scented diet, adults show a strong positive response to oil. Among possible feeding signals in bees, odour is always being the dominant cue (Lindauer, 1970). They also remember the colour and characteristics odour signals of feeding place (Koltermann, 1971). Lee *et al.* (1987) discussed associative learning for host location and selection in grasshoppers using olfactory cues.

Ding *et al.* (1989) evaluated the olfactory response of *Macrocentrus grandii* (Fam. Braconidae), a larval parasitoid of European corn borer, *Ostrinia nubilalis* to 51 plant species. The experimental females indicated strongest attraction to corn, potato or sunflower, on which they previously had found their hosts, the European corn borers. Roces (1990) observed that during recruitment workers of a leaf cutting ant, *Acromyrmex lundii* are conditioned to the odour of food fragments initially carried by the scout workers. The learned odour cue is used as a decision criterion during food collection. Laloi *et al.* (1999) reported that bumble bees, *Bombus terrestris* were able to learn a pure odorant when it was presented in paired association with a sugar reward, but not when odour and reward were presented in an explicitly unpaired procedure. Laloi *et al.* (2001) observed that proboscis extension reflex conditioning in honey bee shows variability in bees of different age.

2.2.6 Latent learning

It occurs without apparent reward or punishment. The animals learn a lot about their surroundings during the courses of their daily activities. Such information may not have apparent and immediate functional value, but may be valuable for survival any time. Even the simple forms of animals are capable of learning when the circumstances are appropriate, that is, when their ecological situation requires it. Bees and wasps learn the location of their nests through landmarks recognized and remembered from a previous orientation flight. Tinbergen and Kruyt (1938) discussed learning in digger wasp, *Philanthus triangulum* (Fig. 2.3). They kept a few pine cones in a circle surrounding the nest as physical cues. The wasp learns to associate her nest with these distinctive landmarks by making an orientation flight or survey (training condition) over the nest entrance. The pine cones were displaced, while the wasp was away for hunting and on return with prey she flew over the circle of pine cones (testing condition) even though it had been moved during her absence and expects to find her nest in the centre of ring of cones. So it had learned the geography of its nest entrance, something it could not possibly have been born knowing. A record of the position of one or more

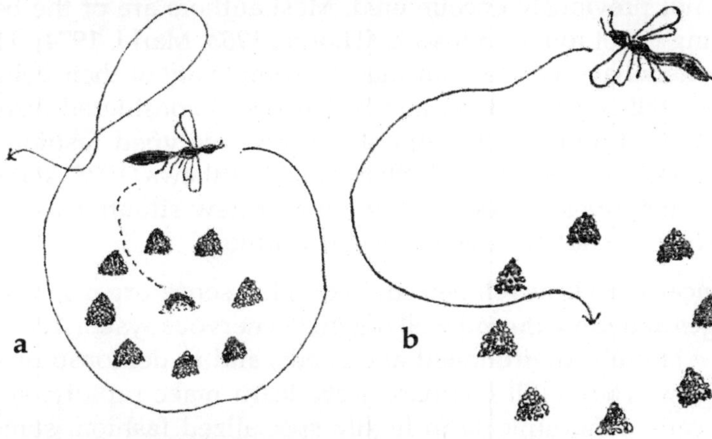

Fig. 2.3. A digger wasp finds her nest by learning to recognize visual landmarks (pine cones), (a)- The wasp makes an orientation flight over the nest entrance ringed with pine cones, (b)- The pine cones have been moved in the wasp's absence. She returns to the centre of the ring of cones where she expects to find her nest (modified after Tinbergen and Kruyt, 1938).

points of interest relative to the positions of landmarks or salient terrain features is a map. A map that resides in the brain of wasp is called a 'cognitive map'. Thus, the results of Tinbergen and Kruyt's experiment suggest that the wasp makes a cognitive map of the immediate vicinity of her nest and uses that map to choose a landing site. The predatory wasp, *Ammophila* can perform the action determined in the visit for up to 15 hours (Baerends, 1941). Bees learn to visit different feeding sites and times and memorize not only time and place, but also odour and signals (Wahl, 1932; Koltermann, 1971). Ants can learn four different routes at one time with landmarks (Jander, 1957). Similar landmark learning has been observed in honey bees (Gould, 1987). Menzel (1993) discussed two types of learning in *Apis mellifera* foragers: latent (or observatory) and associative learning and concluded that memory needs time to be established and is processed in sequential phases. The neuronal compartments of the brain involved in the chemosensory pathway appear to participate differently in the sequential memory phases. Li and Liu (2004) reported that phytophagous insects have the ability of learning in both larval and adult stages. However, larval diet and experience of feeding do not have direct influence on adult behaviour. Learning in phytophagous insects is related to dietary range, types of host stimuli, and the spatial-temporal distribution dynamics of host plants. The ability of learning may enable phytophagous insects to cope with a complex and dynamic plant environment more effectively, enhance their efficiency of host utilization, and thereby increase their fitness.

2.2.7 Insight learning

It is considered to be the highest and most complex form of learning and is somewhat similar to the term we call as 'intelligence'. Insight learning is drawing on past experience to solve a new problem. It involves the new adaptive response in circumstances not previously encountered. Most authors are of the belief that this type of learning is not found in insects (Thorpe, 1963; Markl, 1974). However, some fictitious writers have pointed out that the complexity of their behaviour is the result of their intelligence. Dethier and Stellar (1963) considered insects to be mainly stimulus-bound animals that operate in a stereotyped fashion in strict accordance with the stimuli received. Matthews and Matthews (1978) stated, "their ability to transfer memories to assist in learning of new situations is nearly or totally absent." They are not capable of insight learning.

In insects, presence of well-developed and complex sense organs, jointed and flexible appendages and intricate and well organized nervous system have enabled them to assess the stimuli, environment and needs, and to demonstrate complex behavioural patterns. They built complex nests, learn maze rapidly, show high degree of brood care, communicate in highly specialized fashion, some live in societies and can learn signals quickly, thereby indicating to possess specific abilities to learn selected things. But such patterns are not the results of their intelligence. Baerends (1941) conducted a series of experiments on *Ammophila campestris*, which makes a number of nests. These nests are constructed at different times, thereby having young ones of different stages and capable of consuming variable number of caterpillars in a day according to their growth stage. The female used to inspect all nests and their contents in the morning. He replaced these nests by artificial ones, manipulated their contents and observed that during subsequent visits the patterns of wasp's activity remained unaltered and she provided the number of caterpillars to the growing larvae as per observations made in the morning. This is a complex type of learning and is guided by sign stimuli observed in a brief examination of the nests, but cannot be considered as insight. The worker honey bees receive information about position, kind and relative abundance of food source from the dancer bees. This directs the behaviour of followers in great detail and ultimate success in learning (Frisch, 1967; Lindauer, 1970, 1971). In these cases the stimuli not only release behaviour, but also determine the course of behaviour. Thus, instructions for the performance of ensuing activity are derived from external stimuli and for this reason Markl (1974) termed these complex processes as 'instructive learning'.

Chapter **3**

Sensory Reception

The sense organs are placed on the periphery of an insect and are able to receive the forms of energy existing in the environment (stimuli), which activate the neurons for impulse conduction. All sense organs (receptors) act as transducers- converting light energy, chemical energy, or mechanical energy from the environment into electrical energy of nerve impulses in sensory neurons. Behavioural patterns of any insect are under the influence of stimuli received by a receptor cell or organ. The study of these subjects is rather mingled and before penetrating into the details of insect behaviour, it is of utmost importance to have brief discussion on structural and functional aspects of insect receptors. Besides, insects also respond to internal physiological state, but no internal receptors are known for these senses.

Most sense organs of insects are microscopic in size and are found on their body wall. The receptor cells are not uniformly distributed over the body surface, but localized in a suitable receptor, the 'sense organ or sensillum'. Most sensilla originate from integument and hence are ectodermal. All sense organs have one or more nerves leading to them. The sensilla are classified into mechano-, chemo-, thermo-, hygro-, and photoreceptors. Insects' receptors have been discussed by Snodgrass (1935), Imms (1957), Dethier (1963), Chapman (1972), Atkins (1980) and Romoser (1981). Among insects, blow fly is the best studied in terms of its sensory receptors (Dethier, 1976). Keil (1997) reviewed morphogenesis of insect sensilla.

Generally the sensillum of insect consists of two types of cells 'receptor cell' and 'accessory cell'. Receptor cells are usually bipolar neurons and conduct impulses. Some multipolar neurons also act as receptor cells. Accessory cell surrounds the receptor cell and is usually involved in the secretion of specialized cuticular structure of sense organ. Four types of sensilla, viz. hair sensilla, sensilla campaniformia, sensilla placodea and scolophorous sensilla have been recognized in insects.

The simplest types of hair sensilla are derived from setae (sensilla trichodea), spiny setae (sensilla chaetica), scales (sensilla squamiformia), and cone or peg-like structure (sensilla basiconica). Further modified sensilla consist of sunken processes in a shallow pit (sensilla coeloconica) or deep flask-shaped pit (sensilla ampullacea). The other types of sensilla do not have hairs, pegs or bristles. Sensilla campaniformia are characterized by a round or oval pit covered by a bell-shaped cuticular cap innervated by a receptor cell. In placoid sensillum the cuticular cap is surrounded by a membranous ring and innervated by many receptor cells. Scolophorous sensillum or chordotonal organ occurs in bundles of structures called scolopidia or scolophores. Each scolopidium (Fig. 3.1) consists of a bipolar neuron enclosed by a scolopale cell and an attachment cell. Scolopidia are generally stretched between two internal surfaces of integument. The complex sensilla have great modifications with respect to their structures and functions. Besides, the photoreceptors are entirely different in their structural aspects. The structural details of some common sensilla are shown in figure 3.2.

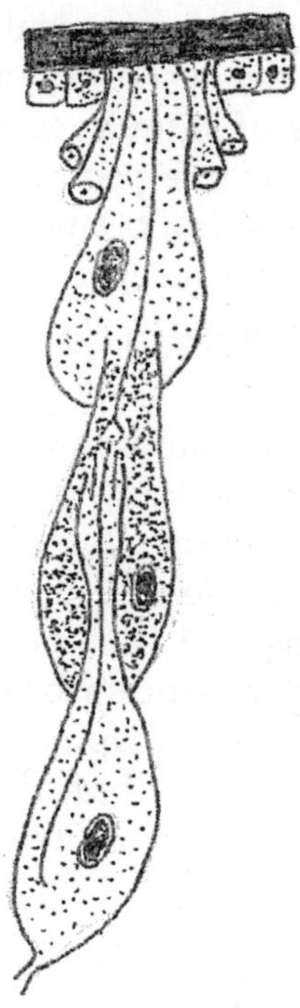

Fig. 3.1. A scolopidium.

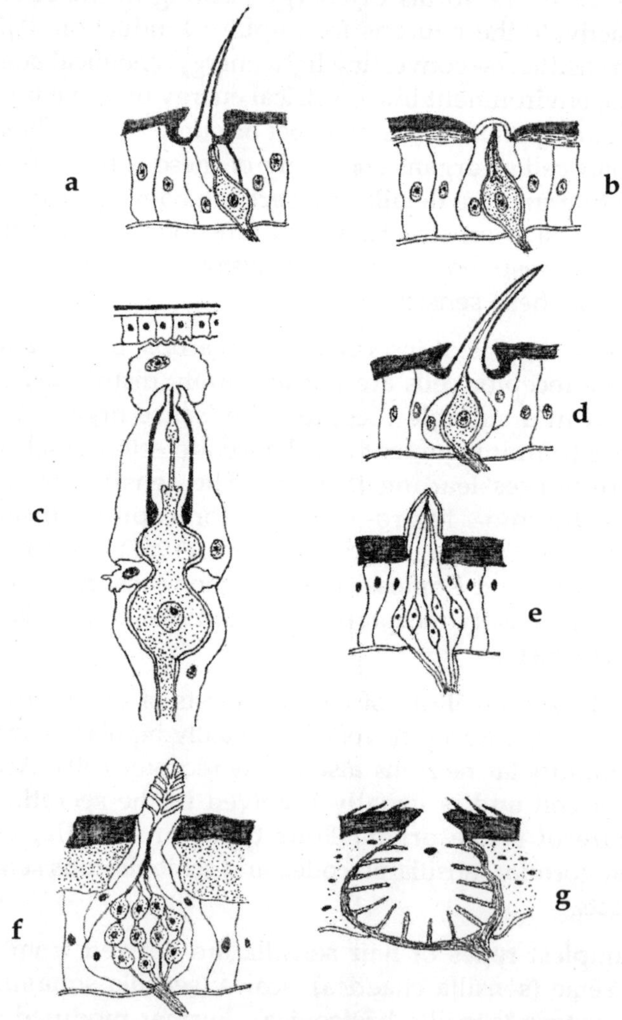

Fig. 3.2. Common Mechano- and chemoreceptors of insects. (a-c)- Mechanoreceptors, (d-g)- Chemoreceptors.

Earlier approaches in the study of insect sensilla were mainly concentrated towards the study of their morphological details and efforts were also made to know their functions by implying variable stimuli having differences in their concentrations and intensities. Recent techniques of electron microscopy (both TEM and SEM), sophisticated elctro-physiological and biochemical techniques have made possible to analyse structural and functional aspects of insect sensilla at cellular and molecular levels, and to construct suitable models that reveals the physiological basis of sensory behaviour.

The receptors are information acquisition, processing and decision making system. Behaviour can be seen as the crossroads where heredity information interacts with resource availability and sensory information, and provide specific direction to use environmental resources for replication. For achieving this insect needs not only sensory input from the external environment (ecoassessment), but also from position of body parts and internal physiological state (autoassessment) (Rice, 1975).

3. 1 Mechanoreceptors

Mechanoreceptors (Fig. 3.2a-c) are capable of receiving stimuli arising from touch, local excitation, gravity, pressure, flight, displacement of body organs, displacement of surrounding media and vibrations. In some cases, they also provide information pertaining to the state of internal body organs. Dethier (1953a, 1963), Schwartzkopff (1974), Rice (1975) and (McIver, 1975, 1985) reviewed mechanoreception in insects.

Most common types of hair-like mechanoreceptors are widely distributed over insect integument in the form of cuticular outgrowths. The sensillum is formed by an epidermal hair-forming cell 'trichogen cell' and an adjacent socket forming 'tormogen cell'. The bipolar neuron is connected with the base of the hair. Such receptors do not have holes in them. These receptors are concentrated in regions that come in direct contact of a substrate, e.g. mouthparts, antennae, legs, wings, ovipositor, and cerci. Besides, they are also located between body segments, joints and other areas. They show a high rate of discharge initially and if the stimulus persists continuously the conduction become rather slow. They have also been termed as 'velocity sensitive' and 'pressure sensitive'. The former generate impulses in presence of constantly changing stimuli, e.g. impact of velocity on an insect during flight. The other sends a steady flow of impulses to CNS, e.g. when an insect is resting on a substrate.

The different stimuli received by hair sensilla lead an insect to show variable behavioural patterns, e.g. tactile hairs on legs help in maintaining body position during locomotion, and hairs on head detect the flow of air and determine flying speed. These sensory hairs also help honey bees to orient to the earth's gravity when they are upside down on their hive. Such sensilla are widely scattered on the cerci of crickets and grasshoppers and on antennae of most insects.

The proprioceptive sensilla are present between body segments or joints of body parts and provide continuous information related to body position or body parts with respect to one another. They are also responsible for proper locomotion and orientation. Both sensory and other sensilla act as proprioceptors. In cockroach well-developed proprioceptors are found at joints of palps and legs. They are also present between head and thorax in ants, at anterior portion of prothorax in mantids, and between head and thorax and abdomen in honey bees.

The other, campaniform sensilla are located in different areas of integument and act as contact or stretch receptors. They are more concentrated in areas like wing bases, leg joints, mandibular bases, halteres and ovipositor, and responsible for regulation of their movement. Pringle (1938a,b) demonstrated that in palpi and legs of cockroach sensilla also act as proprioceptors. In fruit flies (Fam. Tephritidae) both hair and campaniform sensilla located on ovipositior are responsible for monitoring host surface texture and mechanical contact prior to penetration of host tissues (Stoffolano and Yin, 1987). Eisemann and Rice (1989) sowed that the ovipositor of Queensland fruit fly, *Bactrocera tryoni* possesses hygro-, mechano- and chemosensilla.

The hair and campaniform sensilla have been studied on the ovipositor of insects belonging to orders: Coleoptera, Lepidoptera, Hymenoptera and Diptera. These sensilla have also been studied on antennae and halteres of bluebottle blow fly, *Calliphora vicina*; cibarium of tsetse fly and blow fly and mouthparts of tsetse fly. It is worth mentioning that most sensilla bearing body parts of insects, e.g. antennae, mouthparts, tarsi, and ovipositor have both mechano- and chemoreceptors.

Chordotonal or scolophorous organs are distributed in the pedicel of antennae, wing bases, halteres, legs and abdominal segments. They are concerned with orientation in dragonflies during flight, equilibrium and body movements in flies. In blow fly five sets of campaniform sensilla on surface and two sets of chordotonal sensilla within are located on halteres and proven to be responsible for orientation feedback during flight (Smith, 1969). In some cases they have also been observed to be associated with trachea, pulsatile structures and haemocoel.

The complex form of chordotonal sensilla is 'Johnston's organ', which lies in pedicel of antenna. Its structure varies in complexity and greatest development has been observed in male mosquitoes and midges. In these insects it performs the function of sound reception, while in others it acts as proprioceptor. In aquatic bugs it helps in proper orientation of body during swimming, and in whirligig beetle, *Gyrinus* it enables the insect to perceive vibrations of water current. It also acts as a speed indicator in *Calliphora* and during flight respond to the movement of the flagellum.

The sound sensitive structures or tympanic organ consists of a thin area of integument called 'tympanum' and associated chordotonal sensilla (Fig. 3.3). Their location also varies: as in tettigoniid grasshoppers present on fore legs, in acridid grasshoppers on first abdominal segment, in metathorax of noctuid moths and in abdomen of geometrid, pyralid moths and cicada. Autrum (1963) and Michelsen and Larsen (1985) discussed the sound receptors and hearing mechanism in insects.

The subgenual organs are also modified chordotonal sensilla located in basal portion of tibia. They are well-developed in Orthoptera, Coleoptera, and Lepidoptera, and considered to be responsible for perception of high frequencies. Bats make sound that will echo from a flying insect, and the bats use this echo-location to catch their food. Some moths have a tympanum and can hear the clicking sounds of the bats and take evasive action by dropping in the air or changing their flight path to avoid being caught (Roeder, 1965).

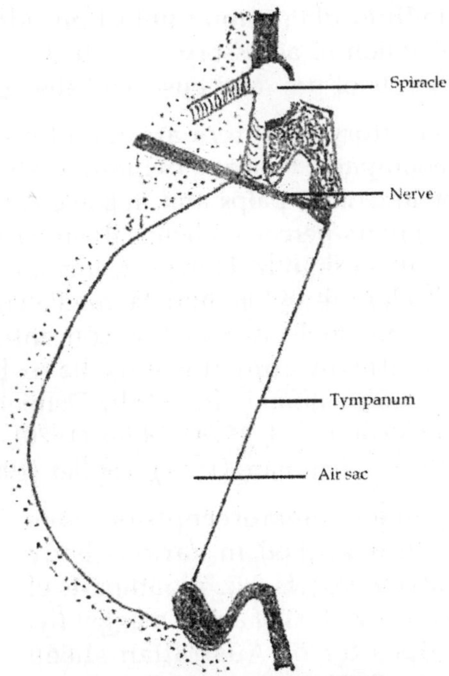

Fig. 3.3. Tympanum- specialized mechanoreceptor that is found in insects on legs, abdomen, thorax, and even wings. It assists in avoiding predators such as bats and for finding mates.

3.2 Chemoreceptors

The chemical receptors (Fig. 3.2d-g) of insects are of prime significance for their successful survival. They provide a means to identify wide variety of stimuli and hence been studied in detail in many insects. Three broad categories of receptors are involved in the process of chemoreception: (i) internal chemoreceptors, which determine internal factors; however, many details of such receptors are not known, (ii) olfactory receptors responding to stimuli of smell, and (iii) gustatory receptors for receiving stimuli of taste. Dethier (1953b) considered that insects also have another type of 'common or general chemical sense', which need a high concentration of stimulant or so-called irritating compounds, viz. ammonia, chlorine, or essential oils, and the response is always in the nature of avoiding reaction. Insects can detect these compounds even when all known chemoreceptors have been covered or destroyed. The irritants evidently trigger a generalized response from other types of sensory neurons.

Chemical receptors facilitate to identify host, to monitor host suitability and quality, taste, food procurement, habitat selection, to detect con- and interspecifics, mate finding, pheromone detection, identification of noxious stimuli, to monitor secretion of accessory reproductive glands, to locate oviposition site, numerous aspects of pre- and eusocial behaviour, etc.

Gustatory receptors are found on all body parts and more concentrated on mouthparts, tarsi and ovipositor while olfactory receptors are mainly concentrated on antennae, palps and in some cases on ovipositor. Broadly speaking, olfactory receptors perceive chemicals in gaseous state, while gustatory receptors are for solids or liquids. However, this must not be taken always as an assertion. Some olfactory receptors may be mediated from a far distance by showing response to gaseous molecules of low concentration and are highly sensitive and specific. Significant contributions have been published on chemoreceptors and chemoreception in insects by Dethier and Chadwick (1948), Dethier (1953b, 1963), Hodgson (1958, 1974), Slifer (1970), Kaissling (1971), Seabrook (1977), Zacharuk (1980), Chapman, (1982), Städler (1984), and Chapman *et al.* (1987).

Contact chemoreceptors have been described in various body parts of insects, viz. mouthparts of tsetse fly; tarsi of apple maggot fly; ovipositor of Australian sheep blow fly, locust and potato tuber moth. Olfactory receptors have been described on ovipositor of blow fly, antenna of face fly and horn fly, and flagellum of grasshopper.

Trichodea, basiconica, placodea and coelonica sensilla are involved in chemoreception. They are specialized in having pores at their extreme tips, through which chemical stimuli reach to neurons (Fig. 3.4). Gustatory receptors usually have one distally located pore, while olfactory ones are characterized by many pores scattered over surface through which airborne molecules diffuse. A single sense organ on the antenna of the polyphemus moth, *Antheraea polyphemus* can have 18,000 pores for the chemicals to enter. Dendrites of sensory

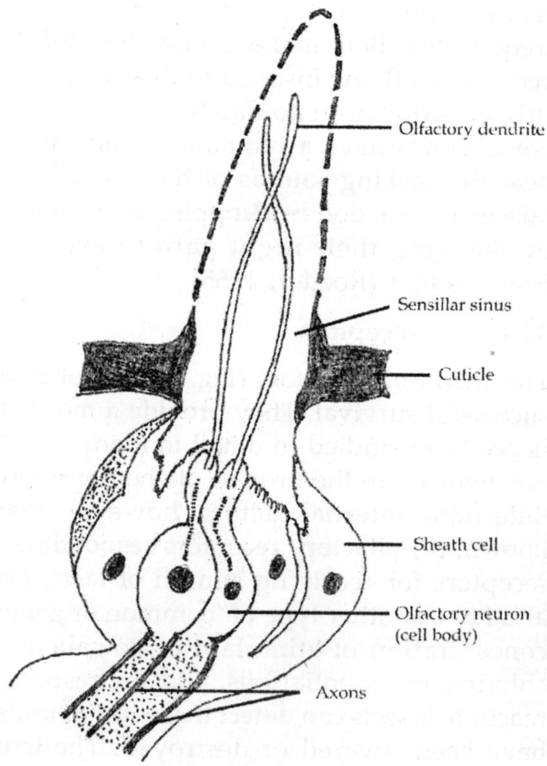

Fig. 3.4. A sensillum trichodeum with pores at their extreme tips, through which chemical stimuli reach to neurons.

neurons branch profusely within these pores and may respond to very low concentrations of detectable compounds (e.g. sex pheromones). The number of receptor cells associated with a chemoreceptor also varies. Some receptors respond to a wide range of substances while others are highly specific. These receptors are also found in buccal cavity of cockroach and mosquito and foregut of house fly. Many insects have them on tarsi and they extend their proboscis, if the tarsi come in contact with sugar solution. Olfactory cues play an important role in chemical communication and sexual behaviour of many insects. The male of the lesser emperor moth can smell the chemical pheromone of the female at a distance of more than six miles. These receptors also have prime significance in host selection and location of oviposition site. Wasps and crickets know where to lay their eggs because they have receptors on their ovipositor.

3.3 Thermo- and hygroreceptors

Insects are poikilothermic and have no precise mechanism to regulate their body temperature. There is a fairly well-defined favourable range of temperature for every insect species within which it is able to survive. They are able to detect rather small changes in temperature and respond accordingly. All insects are sensitive to high temperature and avoid hot places, but some like chironomid larvae can flourish in hot springs at 55°C. Other like snow-beetle can survive well at temperature around 0°C. Honey bees, locusts and moths can rise their body temperature as well as temperature of their surrounding. In other social insects like *Formica, Vespa* chemical regulation of body temperature is also known. In fruit flies, *Bactrocera dorsalis, B. zonata* and many others all activities are ceased at 3°C, but they again become active with a rise in temperature. In the blood-sucking insects, e.g. mosquito, lice, bed bug perception of warmth is important in host selection. Some insects have a tendency to aggregate in a particular zone of temperature gradient, e.g. in *Schistocerca* normal range extends from 35-45°C, with a peak at 40-41°C (Chapman, 1965). Temperature also has a pronounced effect on migration, feeding, locomotion, diapause, growth, fecundity, development, survival, metabolism and other activities.

Temperature receptors are located on antennae, mouthparts and tarsi in the form of small pegs without any pore (basiconic sensilla). They also act as hygroreceptors. Altner and Prinninger (1980); Altner and Loftus (1985) discussed the ultrastructure of such receptors in insects. These receptors are with 3-4 neurons, of which one or two respond to temperature and other two to humidity. A thermo-hygroreceptor may contain wet, dry, warm and cold receptor. Hallberg *et al.* (1984) demonstrated the presence of such a sensillum with wet, dry and cold receptor in a pit on the antenna of olive fruit fly, *Bactrocera oleae*. In blood-sucking bug, *Rhodnius* thick-walled setae on antennae are considered to be thermoreceptors, which can detect small change in air temperature. In grasshopper the entire body is sensitive to rise in temperature, but the sensitivity is most marked in the proximal half of antenna, pulvilli and tarsi. Similarly in cockroach there is a temperature sensitive region in arolium and pulvilli. *Melanophila* beetle has sensory pits on underside of mesothorax, which are sensitive to infrared radiations (Evans, 1964).

Some insects respond to stimuli of moisture very effectively. However, much details of sensory mechanism involved are not known, but there are ample evidences to support it. They also exhibit the phenomenon of 'humidity preferendum' and congregate within a favourable range. Some have moisture-conserving devices and some others can absorb moisture from the environment. The antennae of human louse, *Pediculus humanus* bear tuft of hairs called 'tuft organ' that absorb water from the atmosphere (Wigglesworth, 1972). Collembolans and other soil-dwelling insects are highly sensitive to moisture content of soil and air. They are attracted towards the region of high humidity. On the other hand, earwigs avoid areas of high moisture content.

In most cases thin walled trichoid or basoconic hygroreceptors are located on antennae. In yellow fever mosquito, *Aedes aegypti* such receptors are located on palpi and tarsi while in Queensland fruit fly, *Bactrocera tryoni* on ovipositor. In cockroach hygroreceptors are located on antennae. The guard hairs of the spiracles in tsetse fly are sensitive to moisture.

3.4 Photoreceptors

The insects are able to perceive light in the visible or near visible (near ultraviolet) range of electromagnetic spectrum. They can detect form, pattern, distance, light versus dark, movement, polarization plane, intensity, and colours. Their photoreceptors are well equipped with light absorbing pigment and producing a nerve impulse. Compound eyes are the chief photoreceptors present in insects. Besides, some also have 2 or 3 dorsally placed ocelli, and larvae of holometabolous insects have lateral ocelli or 'stemmata'. Although both types of ocelli are similar in structure, they are believed to have separate phylogenetic and embryological origin.

Some insects exhibit response to light even if their photoreceptors are removed or covered with opaque materials, e.g. decapitated meal moth larvae continue to avoid light and blind cockroach prefers dark situations. The general body surface (dermal light sense) appears to be photosensitive in these cases. Trumen (1976) discussed such extraretinal photoreceptors in insects.

The maggots of certain higher Diptera have no externally visible photoreceptors. However, they have a marked set of photosensitive behavioural patterns. It has been considered that they have some photoreceptor cells associated with the cephalopharyngeal skeleton (Fig. 3.5) (Bolwig, 1946).

Most adult insects and nymphs have a pair of compound eyes. In some others like larvae of Holometabola, cave-dwelling, soil insects and parasites, compound eyes are absent or secondarily lost.

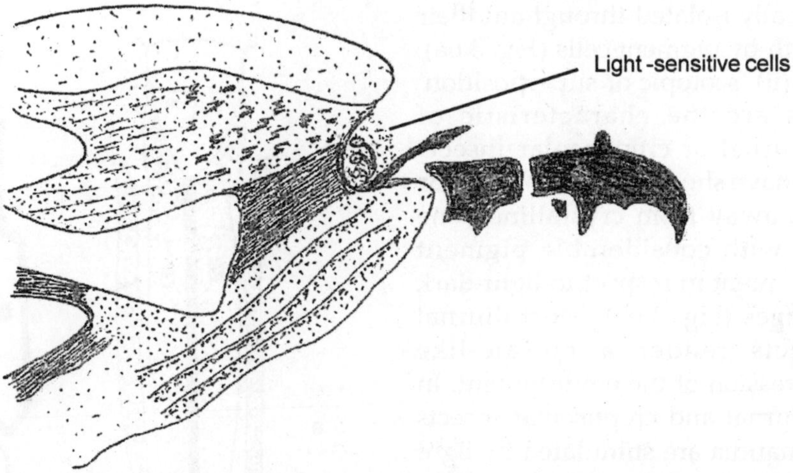

Fig. 3.5. Cephalopharyngeal skeleton of a fruit fly, ***Bactrocera*** sp. larva showing position of light-sensitive cells.

The structural organization of compound eye and mechanism of vision in insects have been discussed and reviewed by Dethier (1953c, 1963), Kupier (1962), Burtt and Catton (1966), Ruck (1964), Bernhard (1966), Mazokhin-Porschnyakov (1969), Wehner (1971), Trujillo-Cenoz (1972), Goldsmith and Bernard (1974), Horridge (1975), Carlosn and Chi (1979), Land (1985), and Järvilehto (1985).

A pair of laterally placed compound eyes located on either side of head gives a wide field of vision in all directions. Each compound eye consists of many similar, closely-packed facets or optic units 'ommatidia' which are the structural and functional units of vision. The number of ommatidia varies considerably from species to species: worker ants, *Ponera* species have only one while some dragonflies may have 28,000. Externally, each ommatidium is marked by a biconvex thickening of transparent cuticle, the 'cuticular lens' or 'corneal lens' or 'cornea'. Beneath the lens, there is often a crystalline cone secreted by a pair of semper cells. Together, the lens and the crystalline cone form a 'dioptric apparatus' that refracts incoming light down into a receptor region containing visual pigment. The light-sensitive part of an ommatidium is called the rhabdom. It is a rod-like structure, secreted by an array of 6-8 specialized neurons (retinula cells), and centred on the optical axis just below the crystalline cone. The rhabdom contains an array of closely packed microtubules where light-sensitive pigments, e.g. rhodopsin and metarhodopsin are stored (White, 1985). These pigments absorb certain wavelengths of incident light and generate nerve impulses.

Two basic types of compound eyes have been recognized in insects, (i) 'photopic or apposition', which is present in diurnal insects and characterized by long, thin rhabdom extending from basement membrane to crystalline cone, retinular cells immediately beneath crystalline cone, and individual ommatidia are completely optically isolated throughout their length by pigment cells (Fig. 3.6a), and (ii) 'scotopic or superposition' eyes are the characteristic of nocturnal or crepuscular insects and have short rhabdom, retinular cells away from crystalline cone and with considerable pigment movement in respect to light-dark changes (Fig. 3.6b). Most diurnal insects render a mosaic-like impression of the environment. In nocturnal and crepuscular insects ommatidia are stimulated by light from larger fields of view. This produces a brighter but theoretically less distinct mosaic image.

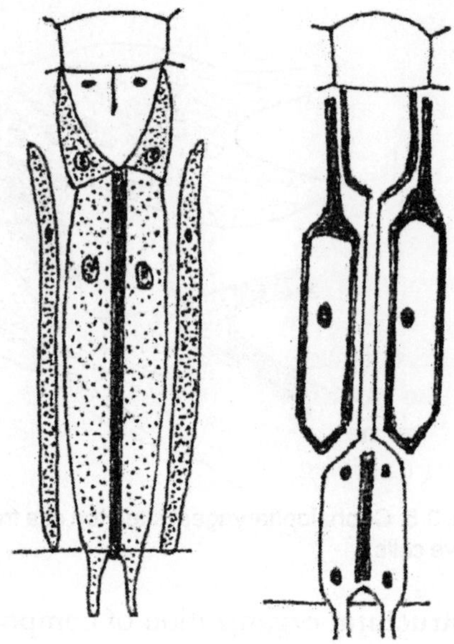

Fig. 3.6. Ommatidium, (a)- From apposition eye, (b)- From superposition eye.

The classical 'mosaic theory of vision' was proposed by Johannes Müller (1826) and further elaborated by Exner (1891). According to this theory each ommatidium is sensitive only to light entering at a small angle to its axis. Oblique rays are absorbed by pigments. Thus, each ommatidium can see a small portion of any object and the entire eye forms an erect image from the mosaic of reports from ommatidia.

Insects respond to spectral range from about 240 (ultraviolet) to 700 (yellow-orange) nm. However, there occurs variation in this range with respect to different species. They are more sensitive to ultraviolet and blue-green regions of the

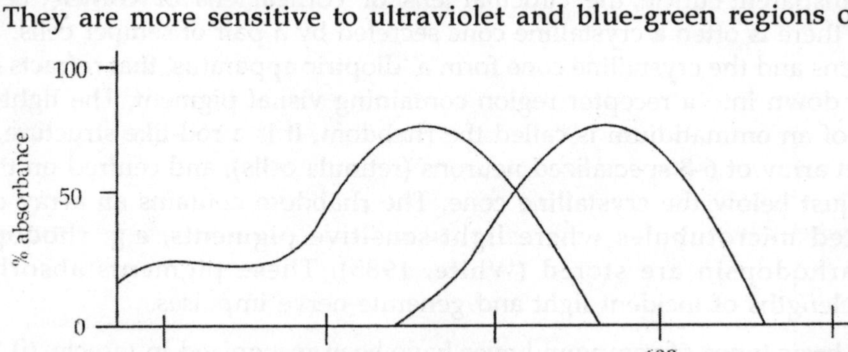

Fig. 3.7. Bichromatic insects have two types of visual pigment receptors, one pigment absorbs green and yellow light and the other absorbs blue and ultraviolet light.

spectrum. Blow flies have three peaks of sensitivity: 350, 480 and 550 nm. In the honey bee, four of the visual cells in each ommatidium respond best to yellow-green light (544 nm), two respond maximally to blue light (436 nm), and the remaining two respond best to ultraviolet light (344 nm). Bichromatic insects (those with only two types of colour pigment receptors) are often unable to discriminate pure colours from mixtures of colours (Fig. 3.7). Using a special ultraviolet-transmitting lens, Eisner *et al.* (1969) demonstrated that many insect-pollinated flowers appear to the honey bee quite different from the way they appear to us (Fig. 3.8a,b). The sharp contrasts between flowers that appear similar to us partly explains the efficiency with which honey bees secure nectar from only one species of flower at a time even when other species are also in bloom. Some like nocturnal stick insect, *Dixippus* sp. is unable to distinguish colours. In water boatman, *Notonecta* only dorsal part of compound eye is colour sensitive, while in *Calliphora* converse is true.

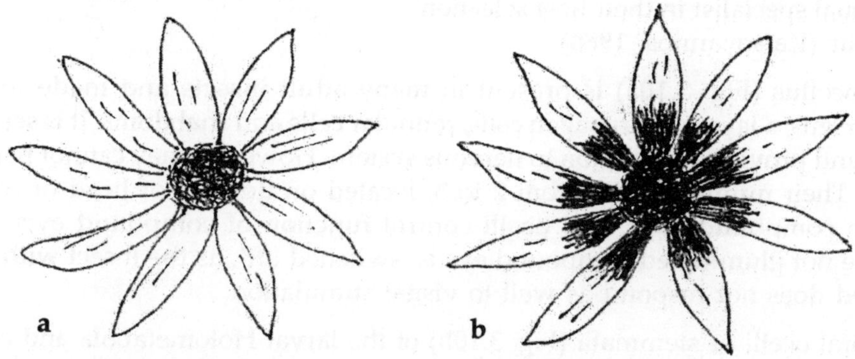

Fig. 3.8 a,b. The figures showing a flower in, (a)- Visible light, (b)- Under ultraviolet light. It shows us the flower the way a honey bee would see it under ultraviolet light.

Bees and ants have polarization analyser in their compound eyes and are able to detect the polarization pattern of the sky, which varies with the position of the sun and enables them to determine the direction to find hive or nest after foraging.

The compound eyes are generally 'near eyes' that permit detection of movement and distance of objects. The combined use of both the eyes enables perception of distance in many insects. Various species of insects can detect movement of objects at different distances. Such perceptions are markedly evolved in predatory insects. Honey bee can detect movement of an object at a minimum distance of 40-60 cm, house fly 40-70 cm and cicindelid beetle 1-1.5 m. Insects like cockroaches and dragonflies have large compound eyes and a wide field of view of 360°. The compound eye is excellent at detecting motion. As an object moves across the visual field, ommatidia are progressively turned on and off. Because of the resulting 'flicker effect', insects respond far better to moving objects than stationary ones. Honey bees, for example, will visit wind-blown flowers more readily than still ones.

Behavioural responses of insects also indicate that they are capable of simple perception of pattern and form. Honey bees respond to different shapes and can be trained to come on a particular one. Zerrahn (1933) demonstrated that the honey bees do not apparently distinguish figures of various shapes. But they can readily differentiate the figures that are markedly broken up into black and white areas (Fig. 3.9). In fruit flies (Fam. Tephritidae) visual responses have been studied to fruit models of different colours, shapes and sizes, and they are considered to be visual specialist in their host selection behaviour (Katsoyannos, 1988).

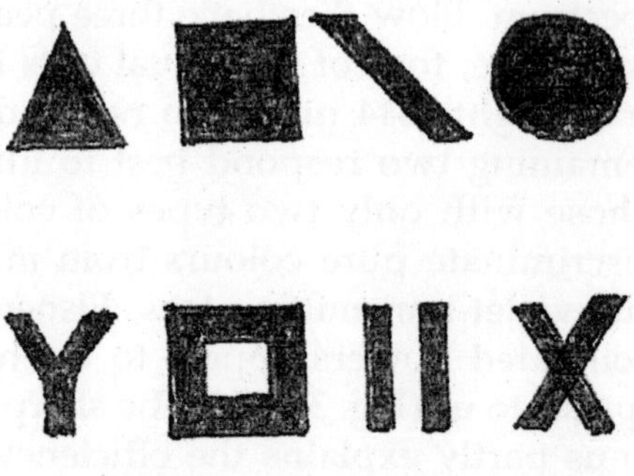

Fig. 3.9. Honey bees can not distinguish the figures in upper row, but they can readily distinguish the figures of lower row from those of the upper row.

Dorsal ocellus (Fig. 3.10a) is present in many adult insects and made up of a cuticular lens, a layer of corneagen cells, retinular cells and rhabdoms. It is sensitive to light and provide stimulation to nervous system. However, they cannot perceive images. Their number varies from 2 to 3, located on dorsum of head or on face between compound eyes. The ocelli control function of compound eye- when ocelli are not illuminated compound eye is 'switched off', as the insect with ocelli darkened does not respond as well to visual stimulation.

The lateral ocelli or stemmata (Fig. 3.10b) of the larval Holometabola and certain

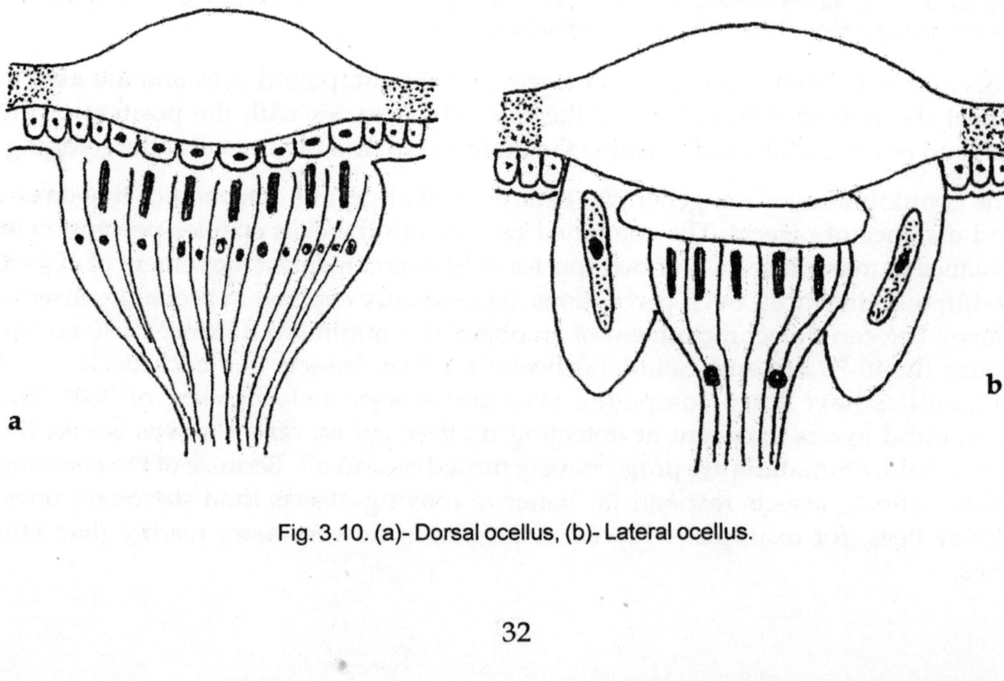

Fig. 3.10. (a)- Dorsal ocellus, (b)- Lateral ocellus.

adults (*e.g.* Collembola, Thysanura, Siphonaptera, and Strepsiptera) are similar to an individual ommatidium, e.g. in caterpillar each consists of a cornea, crystalline cone, retinular cells that form a rhabdom. In other insects, a cornea lies over a group of retinular cells and many rhabdoms. They are considered to be transient representatives of compound eye and probably not image forming but sensitive to direction of light, movement and boundaries as each stemma has its own narrow field of view.

Chapter 4

Neural Control of Behaviour

One of the foremost characters of living being is to respond to stimuli and is termed as 'irritability'. The existence of individuals of a species depends on the effectiveness of response and maintenance of suitable reactions with environment. In multicellular animals certain ectodermal cells become differentiated and specialized to perform the function of information conduction. These cells (neurons) along with other type of supportive cells (neuroglia) form the nervous system. The nervous system is involved in the behaviour of insects from the point of regulation of intracellular activity, coordination of movement, conduction of information, receiving and responding to stimuli, learning and memory.

Bullock and Horridge (1965) published a general review of insect nervous system; Smith and Treherne (1963) gave fine structure of nervous system; Huber (1974) reviewed neural integration; Schneirla (1953a) and Thorpe (1963) discussed neurological basis of learning. Diethier and Stellar (1963), Hoyle (1964, 1970, 1974), Roeder (1967, 1970, 1974), Camhi (1974), Miller (1974a,b), Howse (1974, 1975) and Young (1989) discussed various aspects of neurological basis of insect behaviour.

4.1 Neurons

Neuron is the functional unit of nervous system and has highly evolved properties of sensitivity and conductivity. Typically, a neuron consists of a cell body or soma having many processes called dendrites, and a long process called axon or nerve fibre. From sides of axon third type of branches called collaterals may also arise. The axon endings also have branched terminal arborization. Neuron may be uni-, bi-, and multipolar types. Most insect neurons may also be afferent, efferent and interneurons. Mostly bipolar and some multipolar peripheral neurons receive the stimuli and conduct impulses towards central nervous system (CNS) and are called afferent or sensory neurons. The other unipolar neurons conduct stimuli from CNS to efferent organ and hence called efferent or motor neurons. The afferent and efferent neuron may synapse with each other directly, but frequently interlinked by interneurons.

The bundles of nerve fibres form the nerve trunks, which may contain only sensory, only motor or both neurons. The aggregated cell bodies constitute a ganglion. Both trunks and ganglia are enclosed in a sheath of connective tissue, which regulate the ionic composition of extraneuronal fluid and help in proper conduction of impulses through the neuron.

4.2 Nervous system

The CNS is bilaterally symmetrical with segmentally arranged ganglia. The ganglia are laterally connected by commissures and longitudinally by connectives. The brain or supraoesophageal ganglion consists of ganglia of first three head segments, viz. proto-, deuto-, and tritocerebrum. The suboesophageal ganglion is also formed by the fusion of next three ganglia of head segments. In thorax pro-, meso-, and metathoracic ganglia are present, while the number of abdominal ganglia varies. In higher insects due to fusion of various ganglia, there occurs reduction in total number of ganglia.

The nerve impulses are conducted in the form of a self-propagating wave of electrical negativity along the surface of axon's plasma membrane and during impulse conduction the membrane become depolarized. The transmission of impulse through synapse involves a chemical transmitter, viz. acetylcholine, cholinesterase or choline acetylase.

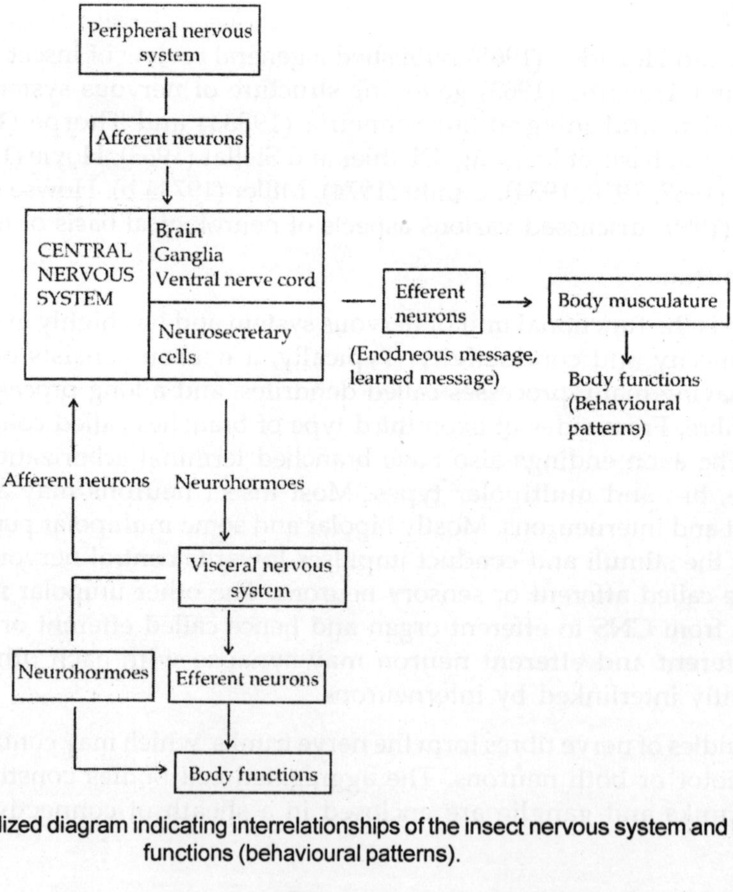

Fig. 4.1. A generalized diagram indicating interrelationships of the insect nervous system and body functions (behavioural patterns).

The other type of nervous systems, namely (i) visceral or sympathetic nervous system, which control movement of alimentary canal and is also responsible for hormonal secretion, and (ii) peripheral nervous system comprising of nerves arising from CNS also have a pronounced influence on behavioural patterns of insects. Neurosecretory cells in the brain and other structures associated with visceral nervous system secrete the hormones. The role of hormones in behaviour of insects has been discussed in chapter 17. The generalized functioning of nervous system is depicted in figure 4.1.

4.3 Neural basis of behaviour

How the nervous system of insect functions and coordinates various activities and behavioural patterns? It is rather a persuadable problem with severe intricacy and needs careful discussion. A number of references are available carrying discussion on the same and still in many cases the problem remains to be unanswered. Neurons are able to receive and process the stimuli, can store the effects of previous stimuli, organize and regulate simple and complex outputs, and are responsible for elicitation of behavioural patterns to a great extent. To analyse the problem a number of neuroanatomical and neurophysiological methods have been applied to the insect nervous system.

The neural basis of insect behaviour can be investigated by studying the input-output reactions of a neuron in quantitative terms, i.e. by analysing the transfer functions in neuronal circuits and by determining their behavioural significance in the whole system. The combination of information theory (information moves to and from CNS via nerve fibres in a coded form) and control-system analysis will lead to more adequate and complete interpretation of integrative activities within the CNS underlying behaviour (Huber, 1974).

Many behavioural patterns of an insect can occur without being triggered by sensory output. The spontaneous patterns are the result of endogenous activity of the nervous system and thus have intrinsic origin. Such activities have been observed by conducting experiments in isolated nerve cord, ganglion, brain and its parts. Besides, in other types of ablation experiments behavioural patterns have been studied before and after removal of some part of nervous system, some body part or whole head. After taking into consideration, the results of these experiments one may conclude:

(i) There exists preprogrammed 'neural-pattern generators' in CNS, which regulate stereotypes sequence of behaviour.

(ii) The endogenous activities of a part of CNS are regulated by excitatory and inhibitory stimuli from other centres.

(iii) Most behaviour patterns are the result of cooperation of many ganglia or tracts, and there occur pleurisegmental integration to govern pleurisegmental behaviour.

(iv) They have some sort of 'sensory and motor tapes' and the CNS select and reduce sensory input into small series or sets of information. Thus, there occurs a filtering device (stimulus filtering) to derive motor output.

(v) Activities of some motor mechanisms also depend upon the stimuli received.

(vi) Learning, on the other hand is the result of addition of new tapes or modification of already existing ones. Certain centres are considered to be involved in learning; however, most details are yet to be known.

(vii) There is no single behaviour for which all the insights of neuronal machinery are exactly known and still there are many gaps.

The brain and suboesophageal ganglion have been observed to exert excitatory and inhibitory influences on endogenously controlled motor patterns. In *Drosophila* an excited male exhibit copulatory responses in presence of unconscious female. This pattern appears to be preprogrammed in CNS and their sequences are rather fixed. Many motor patterns, e.g. locomotion, flight, stridulation and others are at least partially organized by nervous generators. A decapitated *Mantis* turns continuously and performs copulatory movements. Its sexual activities are also increased. Most insects walk or even fly after decapitation, therefore coordinating centres for locomotion are supposed to be present in thorax. In *Mantis* walking is ceased if the ventral nerve cord is cut in the thorax, but in *Periplaneta* walking is observed even after similar cut. Without brain *Mantis* cannot move backward and exhibit continuous cleaning movement once started. The decapitation of female of *Tipula* shows immediate oviposition after removal of head thus, indicates inhibitory effect of brain on oviposition. Similarly, decapitated female of silkworm, *Bombyx mori* could lay eggs normally, but not selectively on mulberry leaves, when thorax was damaged eggs were not laid in definite rows and after the neural connections of abdomen were cut, the eggs were laid. Varalakshmi and Mohan (2000) demonstrated that lesioning of one or both lobes of the brain led to inability of the silkworm, *B. mori* to spin the cocoon as well as to pupate. Frye and Dickinson (2001) reported that flies exhibit a repertoire of aerial acrobatics unmatched in robustness and aerodynamic sophistication. The exquisite control of this complex behaviour emerges from encoding intricate patterns of optic flow, and the translation of these visual signals into the mechanical language of the motor system.

Corpora pedunculata or mushroom bodies of brain also have a general inhibitory effect on activities like running. In grasshoppers lesions made in these bodies cause significant increase in locomotor activity, resulting into continuous jumping and running. In *Gomphocerus* (Order Orthoptera) removal of mushroom bodies make them unresponsive to female sound and their excitation induces continuous chirping. Honey bees are unable to distinguish colours after giving lesions to these bodies and an injury to one half of the brain in bees resulted in asymmetrical movements.

The suboesophageal ganglion also has a marked influence on many activities, e.g. walking, crawling and copulatory movements. Such activities increased after all sensory connections of this ganglion are cut, indicating its inhibitory influence on these activities. When a reflex occurs through a single ganglion, other regions of the nervous system may also influence its activity. In *Cloeon* nymph centres for the movement of gills are segmentally arranged, but the excitatory centre is present in sixth abdominal and inhibitory in second thoracic ganglia. In honey bees insertion and withdrawal of sting; discharge of faecal pallets by caterpillars, and oviposition by female silkworm are under the control of last abdominal ganglion. Kumashiro and Sakai (2001) demonstrated that motor neurons innervating the muscles in genitalia from nerve bundles of the terminal abdominal ganglion control reproductive behaviour and spermatophore formation in male cricket, *Gryllus bimaculatus*. Some reflex responses do not have fixed centres in the nervous system, e.g. water beetle uses hind legs for swimming and after detachment of hind legs, and it uses middle legs for the purpose. Similarly dragonfly nymphs without mask seize the food in mandibles.

In *Dytiscus* and *Periplaneta* swallowing is affected by frontal ganglion of visceral nervous system. The destruction of brain or suboesophageal ganglia is not affecting the swallowing and it is abolished if frontal ganglion is removed.

Many insects are able to discriminate colours, shapes, sizes and chemicals. If similar sensory inputs are repeatedly exerted on an insect, it is able to restore its abstracts. This ability has been exemplified in many Hymenoptera, but details of learning at neuronal level are not yet exactly known. The corpora pedunculata is considered to be involved in learning. It is highly developed in worker honey bees and occupies 13.5 per cent of brain, which is rather high when compared with queen (9.2 per cent) and drone (5.6 per cent) (Chapman, 1972). The ants loose the ability of instrumental learning through maze if the tracts from corpora pedunculata are cut. On the other hand, headless locust, *Schistocerca* can learn to keep leg flexed to avoid electric shocks, thereby indicating no involvement of corpora pedunculata in the shock avoidance learning.

Chapter **5**

Orientation

Orientation refers to the mechanism by which an animal shows its response to various environmental stimuli. It is a self-controlled and preprogrammed innate behavioural pattern, which assists in position maintenance and postural adjustments by an animal. The orientation ability of an insect depends on the receptor organs that detect the stimuli and both afferent and efferent conductions are capable to find optimum localized conditions for survival. By orienting themselves the insects tend to avoid unfavourable conditions and find a suitable environment. The orientation also involves random or actively controlled spatial movement. In many cases these responses are also time dependent and the temporal aspects are also taken into consideration. The phenomenon of orientation is based on simple principles, but may involve complex mechanisms.

Early discussions on animal orientation were of Mast (1911), Loeb (1918) and Kühn (1919). Jacques Loeb worked on the assumption that all motions of an animal were determined by internal or external forces and proposed that:

> *"Normally the responses including locomotion are equal in both halves of the CNS, and the tension of the symmetrical muscles being equal, the animal moves as in a straight line. If, however, the velocity of chemical reactions in one side of the body, e.g. in one eye of an insect, is increased, the physiological symmetry of both sides of the brain and, as a consequence, the equality of tension of the symmetrical muscles no longer exists. The animal will thus be compelled to change the direction of its motion..... The conduct of animals consists of forced movements."*

The concept of 'forced movement' proposed by Loeb has long since been abandoned and his theories and terminologies have undergone many changes. He used the term 'tropism' for such movement of animals, which is commonly used to refer responses of plants in response to various stimuli. Taking into consideration that most plants are stationary, while most animals move actively, the term tropism was replaced by 'taxis' (*pl.* taxes) when used with reference to animals. Mast (1938) concluded that Loeb's ideas of sensory organ-muscle-scheme could not be fitted to all types of orientation and different animals have different systems of orientation.

Gottfried Fraenkel and Donald Gunn (1961) summarized the work on animal orientation and discussed various kinds of orienting reactions. Other useful discussions on the subject are given by Schneirla (1953b), Carthy (1958), Birukow (1966), Jander (1963, 1975), Adler (1971), Wigglesworth (1972), Markl (1974) and Matthews and Matthews (1978).

A number of definitions of orientation appeared in literature and to mention some of them are by Mittelstaedt (1954) as "activity of receptors, centres and effectors by which the spatial relations of one part of an organism to another and to external reference systems is established and controlled.' Jander (1963) defined orientation as "the capacity and activity of controlling location and attitude in space and time with the help of external and internal references, i.e. stimuli". Markl (1974) as "orientation in a broader sense, makes direction of body axes and speed of movements dependent on internal and/or external stimuli, and occasionally dependent upon time."

The process of orientation exhibits complex behavioural patterns in many cases and is independent of locomotion. In some orientation reactions, movement of the body does not occur and in others the orientation sets the direction. If the information for a directional choice comes from the environment, the orientation is said to be as 'exokinetic' or 'extrinsic orientation' whereas, if regulated by an internal, genetically controlled programme, the orientation is 'endokinetic' or 'intrinsic orientation'. Görner (1973) suggested 'allothetic orientation' in which the body's position in space is controlled with respect to external factors, and 'idiothetic orientation' in which position is controlled with respect to signals produced by the animal, while maintaining or changing its position in space. The orientation reactions may also vary with respect to various factors, e.g. an insect showing a positive response towards a stimulus at a particular temperature may exhibit negative response with a change in temperature. Sometimes, more than one type of stimuli are involved in orientation, as in many cases olfactory stimulus releases locomotion, but the insect orients toward visual stimulus. In some cases a particular orientation reaction may be antagonistic to other, e.g. an insect eliciting positive response to light may show negative response to gravity simultaneously.

5.1 Primary and secondary orientation

Frankel and Gunn (1961) recognized two basic types of orientation, viz. primary and secondary orientation.

I. Primary orientation. It is the assumption and maintenance of the basic body position in space. The insects tend to maintain their body position in space while at rest, by keeping all legs in contact with the ground. Most flying and some swimming insects keep the longitudinal axis of the body parts in relation to one another. Insects like dragonflies have mechanoreceptors on the neck and flies at halteres, which assist in body balance during flight. They also show responses towards the changes in visual surroundings by optomotor reactions. The optomotor control assists in flying, swimming, landing and walking by maintaining a position in space.

II. Secondary orientation. It is positioning of the locomotor axis in response to various stimuli. In secondary orientation both direction and range of movements require accurate control. It may be important to a predator to determine the distance and size of the prey, in escape movements, in construction of nest and case, to avoid unfavourable environmental conditions, etc.

Frankel and Gunn (1961) proposed another classification on the basis of mechanism of orientation and suggested 'kinesis', 'taxis' and 'transverse orientation' as three basic kinds of orientation.

5.2 Kineses

Kineses are the simple and unlearned forms of spatial orientation. These are essentially random or undirected locomotory responses, in which the intensity of response varies with the intensity of stimulus. The kineses can be further distinguished into 'orthokinesis' and 'klinokinesis'.

5.2.1 Orthokinesis

In orthokinesis the speed of locomotion depends on intensity of stimulus. The stimulus induces movement in insects and when it ceases to act the insects come at rest and tend to aggregate. Thus, an insect is inactive in dark starts showing its response at low light intensity. When the light intensity reaches threshold level the insect become active and its intensity increases further within the threshold range, and beyond which the activity ceases.

5.2.2 Klinokinesis

In klinokinesis the frequency of directional change or turning depends on the intensity of stimulus. When an insect is locating a host or food it moves as long as the strength of stimulus remains constant or increases, and a decrease in the same leads to random turns. If the stimulus is quite strong the insect turns aside instantly giving an 'avoidance reaction'. The avoidance of high temperature by insects is a klinokinetic response. Some insects that normally fly in a straight line perform rapid turning movements with an unfavourable stimulus, e.g. change in temperature, humidity, wind, light, etc. Wigglesworth (1941) demonstrated that human louse, *Pediculus humanus* orients klinokinetically to different type of stimuli of temperature, humidity and odour. While moving towards host it also makes less number of random turns on approaching closer to the host. Havukkala (1980) redefined klinokinesis as "a change in circular variance of turning angles per unit time which does not change the mean direction (or directions in case of multimodal distribution) of the circular distribution."

Schneirla (1953b) suggested that orientation might combine both generalized and specific types. The different stimulus may induce random movements; however, as gradient of stimulus increases movement may be more specific and directional. Thus, a kinesis type of orientation may introduce a taxis type. Thus, the two effects can also be studied independently.

The kinesis have also been categorized as: photo-, hygro-, thermo-, stereo-, geo-, and chemokinesis in respect to the insects' reactions in response to stimuli of light, moisture, temperature, contact, gravity and chemical. The locusts begin to move in the morning as soon as they get light. Their response is photokinetic to start with, although the influence of temperature soon becomes overriding (Chapman, 1959). If locusts are provided with choice between moist and dry situations, they prefer dry areas. The wireworms, *Agriotes* larvae are inactive in saturated air, but become active with a decrease in relative humidity. Wigglesworth (1972) suggested that inhibiting effect of contact is termed as 'stereokinesis' or 'thigmotaxis'. The loss of contact between tarsi and ground immediately after induces flight in many insects. Many insects are active on smooth surfaces but stop moving (akinesis) or rough surfaces. The bed bug, earwig may come to rest if their contact receptors are sufficiently stimulated. Anderson (1985) observed that some riparian, *Bembidion* prefer to hide under stones, while others under grasses and withered foliage. The kinetic responses to contact stimuli seem to be a key factor in habitat selection by these beetles. Orientation of many insects in response to weak chemical stimuli is chemokinetic.

5.3 Taxes

Taxes are directional movements relative to the source of stimuli. The taxis involves movement of whole body and the long axis of the body takes on a definite orientation in response to the source of stimulus. It provides an efficient means of orientation and requires complex system of sense organs. If the orientation and movement are toward the source of stimulation we call of it 'positive taxis' and if away from it, a 'negative taxis'. Jander (1963) used the terms 'protaxis' (taxis without angular constancy), and 'metataxis' (taxis with angular constancy) for angular orientation. Depending on situation taxes are switched, e.g. an ant going to forage leave nest (negatively geotactic; positively phototactic), move away from the nest using astrotaxis, find the nest again by switching astrotaxis 180°, enter the nest (negative phototaxis; positive geotaxis).

5.3.1 Taxes and behavioural response

On the basis of behavioural responses taxes have been divided into:

5.3.1.1 Klinotaxis

Klinotaxis is orientation reaction towards or away from a source of stimulation in which insect compares the intensity of stimulation by alternate right and left side. Thus, the stimulus gradient is successively scanned. The fly maggot after completion of feeding phase and before going into pupal stage tends to move away from the source of light. As it crawls, it moves the head alternatively left and right and compares light intensities by using receptors located in cephalopharyngeal skeleton. The differences in the intensities between successive light stimuli determine the direction of movement (Fig. 5.1) (Fraenkel and Gunn, 1961). Klinotactic orientation in response to odour gradients has also been observed in a number of insect species. Havukkala (1980) defined klinotaxis as "a change in the mean direction(s) of the circular distribution of angular velocities."

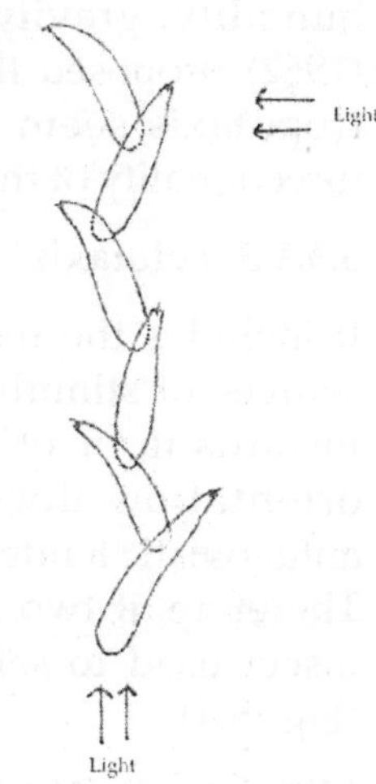

Fig. 5.1 Klinotaxis in a maggot in response to light. The arrow indicates direction of light (modified after Mast, 1911).

5.3.1.2 Tropotaxis

It is characterized by a straight path directly towards or away from the source of stimulation. The paired receptors on each side of the body compare the intensity of stimulation. The movement may be positive or negative and is performed without any trial.

If the light source is kept in front of a photopositive insect, it moves directly towards the source and upon shifting the light source, the insect takes a turn toward the light. The photonegative larvae of Mediterranean flour moth, *Ephestia kuehniella* are placed between two light sources of variable intensities; they move away and show a divergence from high intensity stimulation (Fig. 5.2) (Brandt, 1934). If a photopositive insect with one eye covered is released in a uniformly lighted field, it performs circus movement toward the open eye (Fig. 5.3). Contrarily, the photonegative *Pediculus* makes similar movements to the covered side if one eye is blackened. Tropotactic orientation has also been studied and discussed in response to temperature,

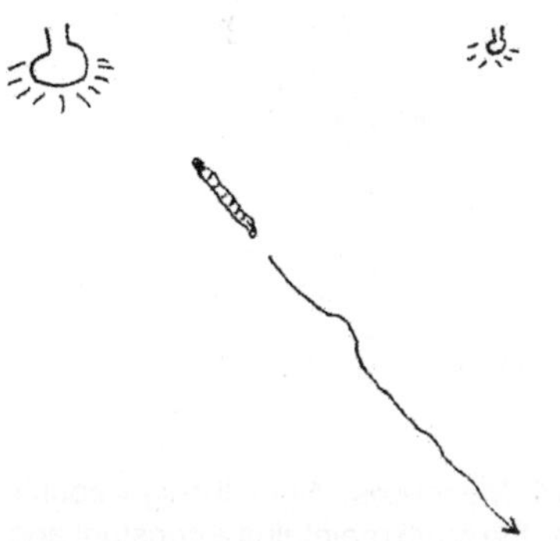

Fig. 5.2. Tropotaxis orientation of larva of Mediterranean flour moth, *Ephestia kuehniella*.

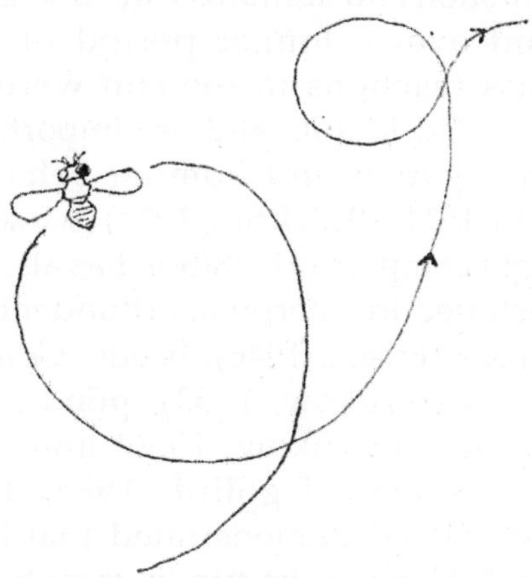

Fig. 5.3. Phototropotaxis as shown by a photopositive fly. If a fly with one eye blackened is released in a uniformly lighted field, it performs circus movement toward the open eye.

humidity, gravity, chemical, and other stimuli. Markl (1962) proposed that the differences between klino- and tropotaxis seem to be gradual, depending upon the directionality of the receiving system and/or the stimulus.

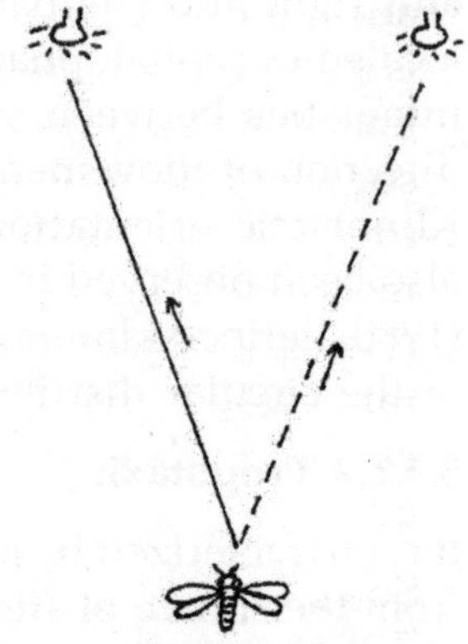

Fig. 5.4. Telotactic orientation. Two equal sources of light are provided to a photopositive insect it orients toward one.

5.3.1.3 Telotaxis

It includes the response in which an insect fixates one sources of stimulus with its sense organs and advances towards it, or orients itself in respect to stimulus. The orientation does not depend upon a balance of antagonistic tendencies controlled by the paired receptors. Therefore, if two stimuli are offered simultaneously, the insect used to select one and orient towards the same (Fig. 5.4).

Mittelstaedt (1962) discussed visual telotaxis process of praying mantis. It fixes the prey in its fixative range by a complex control mechanism involving compound eyes and hair plates. The eyes tend to bring the object into the fixative position and hair plate assists in keeping head position constant. Honey bees also show telotactic orientation and have been observed selecting one of the two stimuli, rather than going between them.

5.3.1.4 Menotaxis

It has also been termed as 'compass orientation' (Buddenbrock, 1917). In menotaxis the angle between the direction of locomotion and stimulus direction is kept constant over a longer period of time. Compass reactions to the sun were first demonstrated in ants and are important in their navigation and homing behaviour (Santschi, 1911, 1923; Brun, 1914) (see section 7.1). Light compass orientation has also been demonstrated in caterpillars (Buddenbrock, 1917); bees (Frisch, 1946); beetle, *Geotrupes stercorosus* (Birukow, 1953); pond-skater, *Velia currens* (Birukow, 1956) and wasp, *Polistes gallicus* (Ugolini, 1985, 1986). Kennedy (1939) demonstrated that locust hoppers basking in the sun fix their body in the angle of incidence of light. If they are shaded with a blanket and light is reflected on them by using a mirror, they change the

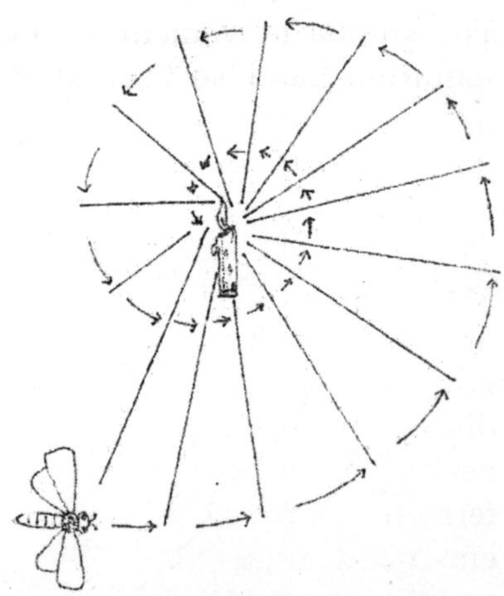

Fig. 5.5. Menotaxis. When the light source is close, the moth maintains a constant angle with the light rays that causes it to circle toward the source (modified after Buddenbrock, 1917).

angle of body in response to the source of light. In menotaxis if the source of light is distant, rays striking the insect are approximately parallel, it maintains a straight line. When the source of light is closed, the insect turns continually toward the source to keep the constant incidence of light on its retina. Buddenbrock (1917) explained the orientation of moth toward a near light source on these lines. The moth maintains a constant angle and flies into the light along a logarithmic spiral (Fig. 5.5). The caterpillar of small tortoise shell butterfly, *Aglais urticae* moves randomly in dark, but when a single light is introduced, they first move in a curved path with the light on its left side and then shift for a different light adjustment (Buddenbrock and Schulz, 1933).

Some insects like bees, ants and others also use polarized pattern of the sky (Fig. 7.3) as orientation signal without seeing the sun. Honey bees perform orientation flights at the age of about 13 days. During their first outward flight they gradually acquire knowledge of appearance and location of their hives in accordance with optical and olfactory cues. The celestial cues, colour of the hive and landmarks assist in proper compass reaction. Of these, the sun and polarized light of the sky mainly constitute the compasses that guide the bees. For their own orientation and for exchange of information they use them in such a multitude of way as no other animal does. The waggle dance of honey bees is monotactic orientation to gravity (Frisch, 1946). The 'astrotaxis' is a menotaxis with a continuously changing orientation in order to compensate for the apparent movement of a celestial body, mostly the sun (Jander, 1963).

5.3.2 Taxes in response to stimuli

The taxes have also been classified on the basis of stimuli by adding prefixes denoting the sensory modality to taxis, e.g. geotaxis is directed movement towards or away with respect to gravity. In many cases insects show responses to more than one type of stimuli simultaneously.

5.3.2.1 Chemotaxis (Osmotaxis)

Many insects use chemical stimuli to locate hosts, food materials, mates and oviposition sites. The orientation reactions of monophagous insects are mainly governed by host's odour. Behavioural effects of various chemicals have been discussed by Dethier (1957) and Finch (1980). The insects may be positively or negatively oriented from different chemicals. *Drosophila melanogaster* orients toward fermenting fruits. Adults of many *Bactrocera* species are attracted toward ammonia emanating compounds, oils, fruit juices and synthetic attractants. The extracts of vegetative plant tissues of many plants attract a number of other tephritid fruit flies (Agarwal and Kapoor, 1986). The bluebottle blow fly, *Calliphora vicina* orients toward meat. The dung beetles, *Scarabaeus* and *Geotrupes* find their food exclusively through odour. The honey bees are able to distinguish various odours and orient toward a specific one while foraging. The cabbage root flies, *Delia radicum* are allured by iso-thiocynates and nitrites produced by cruciferous plants (Finch,

1978), and the onion fly, *Delia antiqua* is attracted by sulphur compounds in onion. The sweetclover weevil, *Sitona cylindricollis* is attracted to sweet clover by the substance coumarin, which is the source of characteristic odour of the plants (Hans and Thorsteinson, 1961). Nordlander *et al.* (1986) discussed orientation of large pine weevil, *Hylobius abietis* towards conifer roots by utilizing host volatiles diffusing through soil. Torr (1989) demonstrated that tsetse flies, *Glossina pallidipes* and *G. morsitans* use both host odour and visual stimuli in orientation. The human louse, *Pediculus humanus* aggregate on filter paper impregnated with aqueous extracts of their faeces. Of the various components of faeces only ammonium salts caused marked orientation (Mumcuoglu *et al.*, 1986). Orientation by odour also involves following scent trail in ants and attraction towards pheromones (see section 18.4.4).

Negative chemotaxis enables insects to avoid injurious and other substances not liked by them. A number of chemicals are known to be acting as insect repellents. Orphanidis and Kalmoukos (1970) studied reaction of olive fruit fly, *Bactrocera oleae* to 27 essentials oils and other aromatic substances and observed that these substances not only entirely lack any attractive action but also act as strong agents of negative chemotaxis.

5.3.2.2 Phototaxis

Phototaxis is a convenient measure of visual capacities in animals, particularly in insects, because this stereotyped behaviour is easily elicited and relatively stable. The stimulus parameters influencing the response are mainly light intensity wavelength or spectral composition. Directed locomotory movement towards and away from a source of light is referred as phototaxis (Fig. 5.6). The insects are able to perceive the direction and intensity of light in a more accurate manner with their compound eyes. A number of experiments have been conducted to study phototactic behaviour of insects by covering compound eyes fully or partially and the results indicate that if one eye is painted with black, the muscular tone of other side is

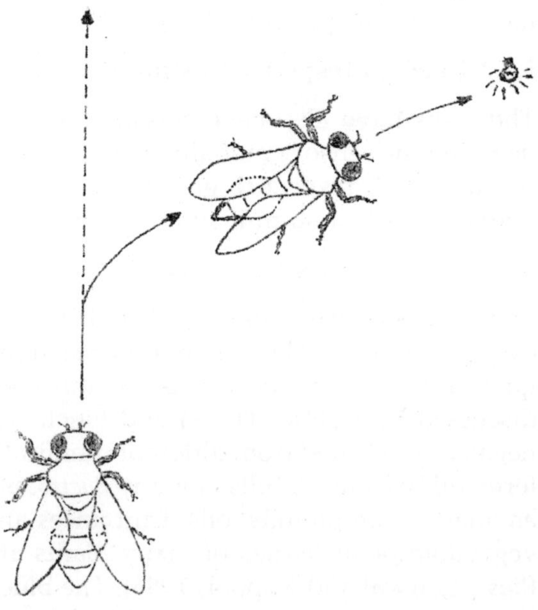

Fig. 5.6. A positive phototactic fly is deflected by light from the straight path. Jacques Loeb referred it as forced movement.

relatively increased (Fig. 5.3). Larvae of *Vanessa* are positively phototactic and if blinded on both sides they still move toward the light source probably by cuticular stimulation.

The orientation is greatly influenced by wavelength and intensity of light. Many insects show their positive response toward green foliage for deposition of eggs, which later become food for hatched larvae. The adults of *Nymphula* respond selectively to different regions of visible spectrum. The green-yellow region (510-570 nm) was most suitable and orange-red (590-640 nm) was extremely inhibitory (Narain, 1979). Male and female cabbage flea beetles, *Phyllotreta striolata* showed identical phototactic behaviours and were most sensitive to light with peak wavelengths between 350 nm and 430 nm in the morning. In the afternoon and evening, the sensitivity to wavelengths shorter than 430 nm decreased, but they remained most sensitive to 430 nm. Thus, the blue wavelengths were more attractive than others (Yang *et al.*, 2003). El-Helaly *et al.* (1981) observed that the adults of tobacco whitefly, *Bemisia tabaci* at 35.6 fc light intensity and the median attracting exposure period (AP_{50}) of 17.5 minutes show marked preference to yellow colour than others. Kokkinn and William (1989) observed that chironomid midges, *Tanytarsus barbitarsis* shows positive phototactic response to artificial lights immediately after emergence. The honey bee's compound eye contains three spectral types of receptors (UV-, blue and green receptors) which are assembled in each ommatidium. Phototaxis in honey bees seems to be elicited by all three spectral receptor types, but dominated by the UV-receptors.

Factors like temperature may also have pronounced influence on phototactic behaviour, e.g. adults of syrphid fly, *Eristalis* are attracted to light at temperatures between 10-30°C, but move away at higher temperature (Dolley and White, 1951), and *Apis* at temperatures below 16°C is negatively phototactic, while at higher temperature exhibits positive response (Chapman, 1972). The tsetse flies, *Glossina austeni* and *G. morsitans* were positively phototactic below 30°C and negative response was observed above it. They also show preference to humidity gradient and aggregate in damp situations (Popham and Vickere, 1979). The paddy leafhopper, *Nephotettix* is attracted to light on a hot and humid evening but is indifferent to it during dry weather (Atwal and Bains, 1974).

5.3.2.3 Thermotaxis

Insects being poikilothermic have no mechanism for regulating their body temperature. There is fairly a well-defined favourable range of temperature called 'temperature preferendum' within which an insect is able to survive well. They tend to orient in the preferred range and avoid places, which are unduly hot. Parasites of warm-blooded animals are attracted by heat and odour. Mosquitoes and bed bugs are positively thermotactic at temperatures above 98°F, which is mammalian blood temperature.

The blood-sucking bug, *Rhodnius* exhibits klinotaxis towards a test tube containing warm water with a temperature of 37°C (Wigglesworth and Gillett, 1934). The human louse, *Pediculus humanus* enclosed in an arena with one-half above the preferred temperature and other below, shows orientation along the boundary (Wigglesworth, 1941). The *Musca* larvae during feeding stage show their klinotactic response to temperature and humidity (Hafez, 1950).

5.3.2.4 Hygrotaxis

The insects may directly orient and aggregate in zone of preferred humidity or avoid such zones of high or low humidity. Thus they exhibit phenomenon of 'humidity preferendum' also like that of temperature (Yazdani and Agarwal, 1997). Termites usually move towards a zone of high humidity when subjected to slightest desiccation. This orientation response was greater with a high hygropositive response when *Microcerotermes beesoni* was subjected to preconditioning in dry air (Sen-Sarma and Chatterjee, 1966).

In addition, these responses are also influenced by temperature, degree of hunger, water balance in body, age and sex. The flour beetle, *Tribolium* normally prefer dry conditions; however, starved and desiccated beetles exhibit preference to high humidity (Willis and Roth, 1950). Many carabids normally preferring dry air and reverse their response after desiccation (Perttunen, 1951). The orientation response of spruce budworm, *Choristoneura fumiferana* vary with humidity gradient and greatly influenced by temperature and evaporating power of air (Wellington, 1949). Gunn and Cosway (1938) observed that the oriental cockroach, *Blatta orientalis* prefers lower humidities in optimum water balance condition, but exhibits hygropositive reaction when desiccated. Contrary responses have been observed for the beetles, *Ptinus tectus* and *Tenebrio molitor*.

5.3.2.5 Phonotaxis

Many insects produce various kind of sounds and even in some cases the wavelength of sound lie outside the range of human ear. Some sounds have no evident significance, but in others may induce orientation. The responses of different insects vary with sound frequencies. The gyrinid beetles are able to perceive ripples on water surface due to displacement of flagellum and avoid collision with other insects (Chapman, 1972). The sound signals are also perceived by conspecifics for orientation of insects of opposite sex. Papov and Shuvalov (1977) discussed properties of phonotactic reactions of five species of crickets to calling songs of males. The females of *Gryllus campestris* and *G. bimaculatus* show phonotactic response to calling songs of conspecifics males. Bailey and Cunninghum (1990) observed that female of bush cricket, *Requena verticalis* in a two male field trail shows preference to higher frequency component of calls spectrum. Tauber *et al.* (2001) reported that in phaneropterine bush cricket, *Phaneroptera nana* both males and females sing, producing a duet. A short female 'tick' answers the male's song. A male that receives a response in a defined interval after the end of his chirp usually initiates phonotaxis towards the female. The male chirps vary considerably in several temporal song characters. Females exhibited equivalent response rates to solitary singing males regardless of their song characters. However, when simultaneously presented with two singing males, females showed consistent preferences and responded preferentially to relatively longer chirps produced by larger males. Only males receiving a high female response rate actually approach and mate with the female.

5.3.2.6 Geotaxis

The positive response of a number of insects has been observed with respect to gravity. Soil-dwelling insects and most borers are mainly positive geotactic. In many cases after hatching the young ones burrow in the soil. Both photo- and geotaxis are interlinked with each other and exhibited simultaneously (Fig. 5.7). Positive geotaxis is combined with negative phototaxis and vice versa. The larvae of dacine fruit flies after maturation become positively geotactic and leave the host tissues to go deep into the soil for pupation, while adults show their negative response towards gravity. Larvae of many aquatic *Culex* species normally rest with their heads downward.

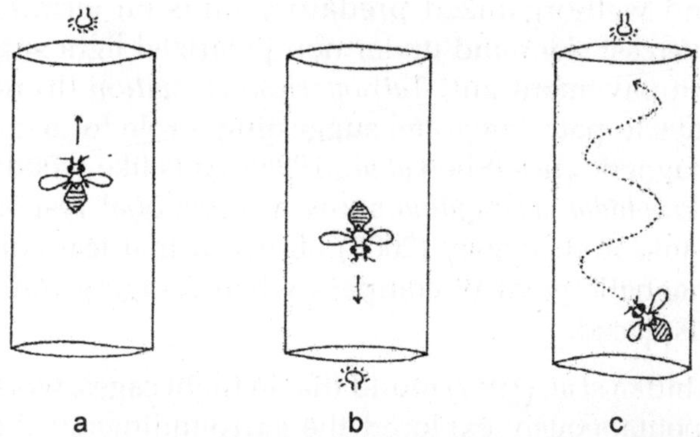

Fig. 5.7. *Drosophila melanogaster* is geonegative and photopositive. (a)- Move upwards in a dark tube towards light. (b)- If the light source is provided at bottom of tube orientation is downward- phototaxis dominates over geotaxis. (c)- If one eye blinded moves in a spiral.

Crocker and Hamon (1983) observed that citrus whiteflies, *Dialeurodes citri* congregate almost exclusively on the bottom sides of leaves and show response to light and gravity. Ng (1986) demonstrated that larvae of some aphidophagous coccinellids show negative response to gravity. Under uniform light they search randomly, but under non-uniform light the response was influenced by stage and hunger. Larvae those were satiated, hungry, about to die exhibited negative geotaxis. Subramanyan and Krishnamoorthy (1990) discussed gravity orientation in silkworm, *Bombyx mori*. The third, fourth and fifth instar larvae show positive responses, which also depend on feeding status, disease condition, dark-light cycles and larval age. The negative orientation towards gravity was observed at the termination of cocoon construction.

5.3.2.7 Magnetotaxis

The magnetic orientation has been studied in insects; however, the number of insect species observed is still small. In some social insects such as ants, bees, wasps and termites magnetic measurements have been performed to characterize biomineralized magnetic material as a magnetoreceptor sensor. However, the mechanisms that transduce the magnetic information into useful sensory signals are still poorly understood. The most popular hypothesis, named the ferromagnetic hypothesis, considers that tiny magnetic particles, such as magnetite (a natural magnet), can play the role of magnetic sensors. In fact, magnetite has been found in *Apis mellifera* bees and in ants of the species *Pachycondyla marginata* and *Solenopsis* sp. *P. marginata* is an interesting ant to study magnetoreception for its migratory and well-organized predatory raids on termite. The ability to return home in overcast skies and under non-polarized light suggests that at least some ants, e.g. the pavement ant, *Tetramorium caespitum* do not require celestial or landmark cues to path integrate, suggesting a role for a combination of proprioceptive and magnetic cues (Shen *et al.*, 1998). Ants like wood ants, *Formica rufa* and tailor ants, *Oecophylla smaragdina* show a directional response to shifted magnetic polarity. Banks and Srygley (2003) observed that leaf cutting ants, *Atta colombica* also use magnetic polarity compass when foraging trails degrade or when ants become displaced.

Chittka *et al.* (1999) found that in flight cages, worker bumble bees, *Bombus impatiens* spontaneously explored the surroundings of their nest and foraged in complete darkness by walking instead of flying from feeders up to 150 cm away from the nest. This behaviour was wholly unexpected in these classically visual foragers. The finding provides a controlled system for dissecting possible non-visual components of navigation. The bees also laid odour marks and when such odour cues were eliminated, bees maintained correct directionality suggesting a magnetic compass. They were also able to assess travel distance correctly, using an internal, non-visual, measure of path length.

5.3.2.8 Skototaxis

It involves directed movement in response to darkness. A large number of nocturnal, soil dwelling, borers, others found beneath stones, filth and those live in crevices prefer dark areas. Insects like *Aedes, Blatta, Forficula, Lasius, Lepisma, Pediculus* and *Periplaneta* and many others show such response. If the earwig, *Forficula* if provided with a dark screen and a light source, it may go directly away from the light (negative phototaxis) or towards the dark screen. Atkins *et al.* (1987) discussed skototactic responses of adults of female cricket, *Acheta* toward various shaped targets in an orientation arena and on a compensatory treadmill. The crickets oriented toward dark targets if they had horizontal visual axis $\geq 30°$, and if the vertical dimensions of the target was $\leq$ its horizontal dimension. Taylor (1988) observed that adults of three species of flat bugs, *Neuroctenus simplex, Mezira granulata* and *M. sayi* show positive responses to dark and the same is important in habitat selection.

5.3.2.9 Rheotaxis

The ability to respond in respect of water currents is known as rheotaxis. The gaseous and nutrient contents also influence insect's orientation in water. Many insects adjust themselves to the water currents in streams. The Simuliidae, Blepharoceridae, Deuterophlebiidae larvae (Fig. 5.8) are found in swift water streams and attach themselves to objects by means of suckers to maintain their position. The orientation due to currents has also been explained as due to differences in pressure produced by the currents in different body parts. Many aquatic insects while swimming maintain their fixed position by using visual stimuli. *Notonecta* maintains a fixed position in a running stream, it swims against water current to maintain a constant visual field, and thus a constant position is also maintained during swimming.

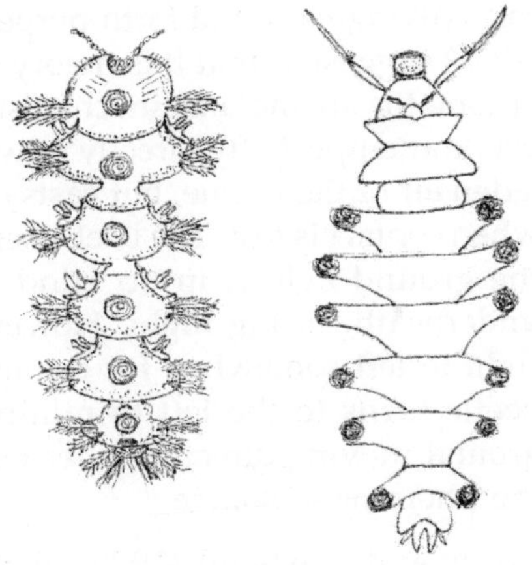

Fig. 5.8. (a)- *Blepharocera* larva, (b)- *Deuterophlebia* larva.

5.3.2.10 Anemotaxis

The taxis response with respect to air currents is called anemotaxis. Such responses are mainly elicited by flying insects, and mostly they tend to fly with the wind. The locusts are activated by odour of grass and elicit positive anemotaxis (Kennedy and Moorehouse, 1969). Their cerci bear two types of hair sensilla that act as anemotaxis receptors. Bell and Kramer (1979) discussed anemotaxis behaviour of three cockroaches in still air and in wind. The odour of fully-grown potato plant elicits a true odour-conditioned positive anemotaxis in Colorado potato beetle, *Leptinotarsa decemlineata* (Thiery and Visser, 1986). Kennedy and Marsh (1974) reported that certain male moths flying upwind toward a scent-producing female appear to be guided anemotactically by optomotor reaction to the ground pattern. Loss of odour stimulus changes the anemotactic angle from into wind to across wind with left-right reversals. They gave experimental evidences on optomotor anemotaxis in male of Indian meal moth, *Plodia interpunctella* stimulated by wind-borne sex pheromone.

An insect on a substrate can use mechanoreceptors to gauge wind direction, but once airborne, mechanoreceptive input is no longer a useful cue. Instead, flying insects use visual feedback (drift or side-slip) to determine an upwind heading, termed 'optomotor anemotaxis'. The optomotor anemotaxis mechanism for orientating to pheromone sources is proposed for insects, especially moths (Kennedy, 1983; Baker, 1989). In this theory, an insect attempts to fly directly upwind when in contact with a packet of pheromone-laden air of the plume, but casts (flying back and forth perpendicular to the wind) when contact is lost. Byers (1995) suggested that this theory is also applicable for the bark beetles, which use optomotor anemotaxis mechanism for orientating to pheromone sources. A bark beetle attempts to fly directly upwind when in contact with a packet of pheromone-laden air of the plume, but casts (flies from side to side with respect to the source) when contact is lost. The beetle senses the wind direction while flying by observing the ground below; in no wind, or head-on wind, the ground moves directly underneath during flight. However, if the visual ground field also moves from right to left somewhat, for example, then wind is coming from the left, and the beetle turns to the left to minimize the transverse ground shift and keep the ground moving directly underneath so that the insect heads upwind and toward the pheromone source.

Meats and Hartland (1999) observed anemotaxis movements of mature male Queensland fruit fly, *Bactrocera tryoni* individually in a wind tunnel under conditions of 'cue-lure with wind', 'cue-lure with no wind', 'wind only' and 'no wind or cue-lure'. Patterns of walking or flying were essentially the same in all treatments except that in the 'cue-lure with wind' treatment.

5.3.2.11 Thigmotaxis

Orientation by loss of contact stimuli has been termed as thigmotaxis or stereokinesis (Wigglesworth, 1972). Antennae, cerci and tarsi have most tactile sense organs and receive the stimuli of contact. Insects like earwig, cockroach, beetle if released in a large glass container; they tend to move in a circle along the sides of the container (Fig. 5.9). If an insect is released between two long and straight pieces its movement is affected by contact stimuli and it moves in the direction as indicated in the figure 5.10a-c. Thus, insects tend to maintain a contact with the surface and when a break is reached turn reflexly according to previous contact.

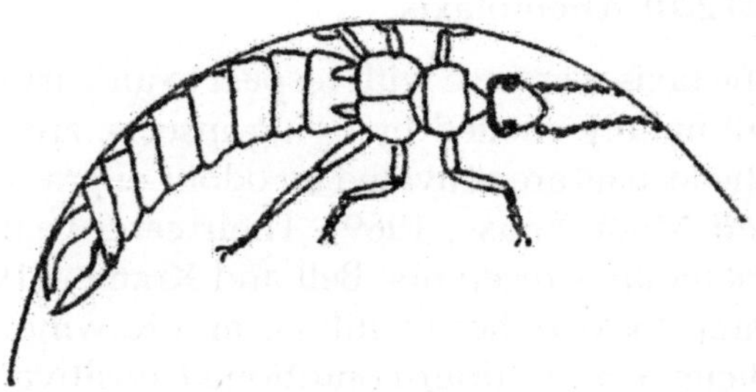

Fig. 5.9. Earwig at rest in a Petri dish- body remains as much as possible in contact with glass.

5.3.2.12 Astrotaxis

It is compass orientation with respect to sun and also to moon and stars in some cases (Jander, 1963). Some authors have also used the term 'heliotaxis' for taxis response with respect to the sun. It has been demonstrated in ants (Santschi, 1911), honey bees (Wolf, 1927) and other insects. Honey bees are also able to use a different kind of astrotaxis, i.e. polarized pattern of the sky as the degree and plane of polarization vary with position of the sun at every point of the sky (see section 7.2).

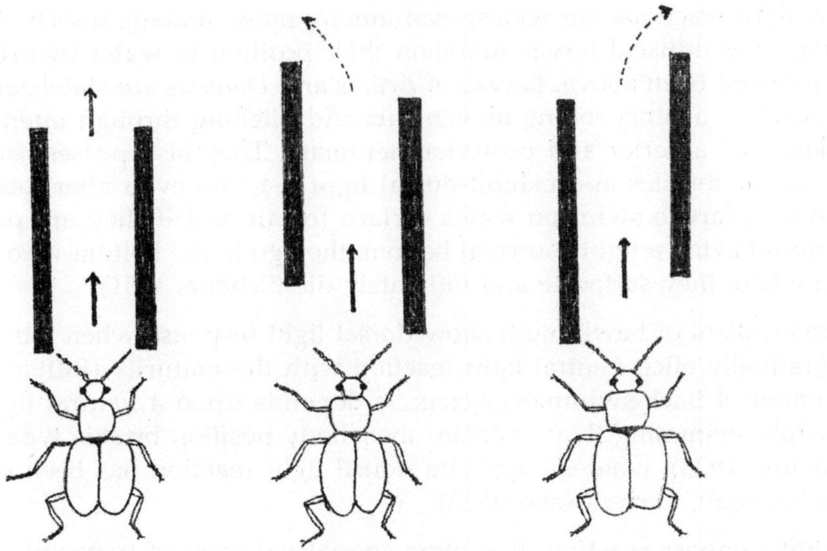

Fig. 5.10a-c. If an insect is released between two long and straight pieces its movement is affected by contact stimuli and it moves in the direction as indicated by the arrow.

5.4 Transverse orientation

In transverse orientation the body is oriented at a fixed angle relative to the direction of the source of the stimulus and hence it is also a type of 'positional orientation'. In it locomotion is not necessarily involved. Fraenkel and Gunn (1961) discussed its two types, viz. 'dorsal light reaction' and 'light compass reaction'. In case of transverse orientation, Lambin (1987) proposed a method to locate nearby cues by an animal. By studying the value constancy of the angles formed by a moving animal and the direction of all spatial points, he set up cartography of the neighbouring space of the animal during a path.

I. Dorsal light reaction. In dorsal (or ventral) light reaction the direction of light in which the insect maintains posture by keeping the dorsal side of the body toward the light. In this type of reaction locomotion is not usually involved, and such reactions are different from phototaxis because orientation is transverse to the light rays (parallel to the light rays in phototaxis). When insects are sunning themselves, they often take characteristic postures as many butterflies and locusts do. The dragonfly, *Anax imperator* keeps the body in a position so that sun's rays are perpendicular to their backs. The compound eyes are mainly responsible; however, there is some disturbance if ocelli are covered. When light is presented from ventral side the dragonfly spirals downward and crashes (Mittelstaedt, 1950). Many Odonata and Orthoptera during flight align the head in a position so that the dorsal ommatidia receive maximum light. Goodman (1965) described that in *Schistocerca gregaria* reaction to the position of horizon line in the visual field helps to stabilize in the rolling plane.

Dorsal light reactions are widespread among many aquatic insects. Nymphs of mayflies and dytiscid larvae maintain their position in water by orientation to light received from above. Larvae of *Acilius* and *Dytiscus* are stabilized by dorsal light reaction against rolling movements and pitching through interaction right and left, and anterior and posterior stemmata. They also possess dermal light sense in the tergites and exhibit dorsal light reaction even after total blinding. The *Acilius* larvae swim on water surface for air and if they are placed in an aquarium having a light source at bottom, they go to the bottom with their backs down where they suffocate and ultimately die (Schöne, 1951).

The caterpillars of hawk moth show dorsal light response when they are young and gradually elicit ventral light reaction with the maturity (Ruiter, 1956). The movement of back-swimmer, *Notonecta* depends upon a ventral light reaction and while swimming they maintain their body position by this type of reaction (Lehmann, 1923), whereas opposite dorsal light reaction has been observed in water boatman, *Corixa* (Rabe, 1953).

II. Light compass reaction. It is more specialized form of transverse orientation involving locomotion. It also assists in navigation in many insects as they orientate obliquely or at right angles to the source of light and maintain a constant angle with the direction of light rays from the sun. Moon compass orientation has been demonstrated in some cases and the use of starlight by insects cannot be neglected. This type of orientation has been observed in many insects. For details see section 7.1.

5.5 Temporal aspects of orientation

Majority of insects exhibit daily cycles of activity and perform them at a particular time of the day. These phenomena are responses to environmental changes and also manifestations of an endogenous periodicity. Buttel-Reepen (1900) concluded that insects possess a 'time sense' (*Zeitsinn*), and Forel (1910) proposed that the bees have a 'time memory' (*Zeitgedächtnis*). Beling (1929) demonstrated that bees tend to come back at the same time each day and could be trained accordingly. Wahl (1932) observed that bees returned to the feeding site at a particular time even in absence of cosmic radiations, thereby indicating an endogenous periodicity. He succeeded in training bees to come as many as five times a day and also trained them to collect food in the morning at one site in the afternoon at another. They can also correlate the location of a food source with other factors, e.g. its odour and colour with the time of the day (Koltermann, 1971). Lindauer (1960) concluded that the time-compensated orientation mechanism of honey bees was innate, however they had to learn the direction of movement of sun and its speed, besides they can memorize simultaneously up to four different feeding places with for different times. Experiments concluded to change bees' metabolism by using thyroxine (to increase metabolism) or euchinin (to slow down metabolism) showed that these chemicals has no apparent effect on time sense (Renner, 1957), while deep-CO_2-necrosis for long duration and chilling for 6 hours at 4-5°C caused delay (Renner, 1957). Other experiments with constant temperature, light, relative humidity and electrical conductivity of air indicated that time sense of bees is not influenced by these factors (Beling, 1929; Wahl, 1932; Renner, 1957).

Many other insects are also known to orientate themselves by maintaining a fixed angle to the sun, artificial light or pattern of polarized light from a blue sky. The experiments on ants conducted by Santschi (1911, 1913) and Brun (1914) indicated light compass reaction without time-compensation (see section 7.1). Similar light compass orientation without time-compensation was demonstrated in honey bees. Contrary to these claims time-compensated sun orientation for the changing azimuth of the sun has also been described in other insects, e.g. pond-skater, *Velia currens* (Birukow, 1956); beetle, *Geotrupes stercorosus* (Birukow, 1953, 1960); white grub cockchafer, *Melolontha melolontha* (Cuoturier and Robert, 1955); beetle, *Phaleria provincialis* (Pardi, 1955); and bumble bee, *Bombus terrestris* (Jacobs-Jessen, 1959).

5.6 Ecological aspects of orientation

Orientation in ecological context can be considered as a form of adaptive behaviour. The environment includes both positive (resources needed for sustaining life) and negative (stress sources) factors. The ability of an organism to maintain its distance from resources (food, shelter, etc.) and maximize its distance from stress sources is called 'orientation fitness'. Taking these facts into consideration a meaningful classification of orientation was proposed by Jander (1975) that includes six categories.

Positional orientation. Most animals maintain their position in space by compensating various disturbances. While at rest or moving they compensate for displacement by air or water currents. Transverse orientation to light and gravity are referred to be positional orientation.

Object orientation. Universally exhibited by all motile animals, as they are able to search and approach resources or avoid stress sources. Animals have either precise information about the spatial location of resources or stresses (objects) in the environment and the spatial orientation related to them is termed object orientation, e.g. straight-line search towards or away from the light.

Strato orientation. The objects are extensive in at least one direction and their vertical distribution is called stratification (layering). Movement of animals across the vertical layers of a habitat is strato orientation.

Zonal orientation. In contrast with the stratification, horizontal distribution of objects is called zonation and orientation with respect to these gradients is called zonal orientation, e.g. time-compensated sun compass orientation.

Topographic orientation. This is an elaborated approach phase of orientation involving components like distance, direction and learning. In topographic or home range orientation an animal retraces its path to the point where it started from, and distance is controlled by an intrinsic mechanism while direction by using celestial cues and landmark learning.

Geographical orientation. Many motile animals exhibit a tendency to disperse or migrate over considerable distances. Dispersal can be considered as a rudimentary form of geographical orientation and further lengthening of search phase leads to true geographic orientation.

Chapter **6**

Locomotion

The ability to move from place to place at some stage in their life history is an important attribute of all animals and locomotor abilities are well-developed in insects. Insects are able to move with the help of three pairs of jointed legs and the same play a vital role in their successful survival. They evolved from a terrestrial ancestor and subsequently adopted aquatic and aerial modes of life. It is thought that the ancestors of insects had more than six legs but as the legs became longer to facilitate faster speeds, they became fewer to reduce problems of coordination and six is the minimum number for great stability over a wide range of speeds. Functional efficiency is increased by having the bases of the legs close together and near the head. With the development of wings the insect thorax was evolved as locomotor centre. The hexapod condition provides a stable equilibrium during all stages of terrestrial locomotion and assists in feeding, dispersal, escape, mate finding, adjustment in response to climatic factors, etc. The terrestrial locomotion can be studied from various points; however, the present discussion includes mainly behavioural aspects.

The subject has been discussed in the publications: Hughes (1952), Wilson (1966), Wendler (1966), Hughes and Mill (1974), Hoyle (1976), Bowerman (1977), Delcomyn (1981, 1985a,b,c, 2004), Cruse and Graham (1984), Graham (1985), Gewecke and Wendler (1985) and Alexander (2003).

6.1 Walking

Most terrestrial insects move by walking, running, jumping and crawling. During walking or running the legs move in a rather fixed sequence, which varies with the speed so that stability is maintained. Additional support may also be provided by abdominal tip by making contact with the ground. The chitinous exoskeleton and leg segmentation support the body in a fashion that it hangs down from the legs (Fig. 6.1), and thus resulting into a low centre of gravity. The legs are inserted with strong cross-striated extrinsic and intrinsic muscles capable

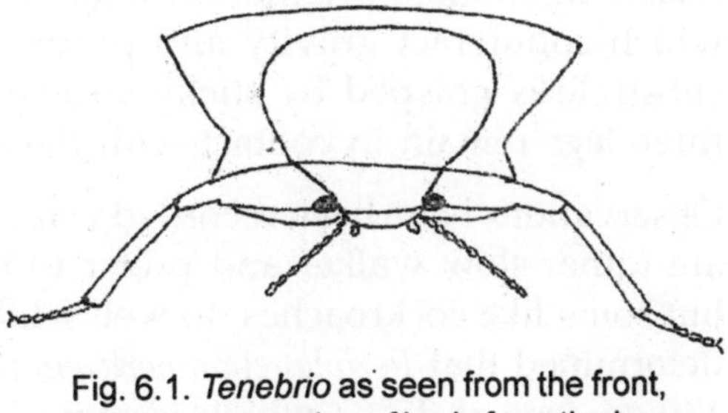

Fig. 6.1. *Tenebrio* as seen from the front, showing suspension of body from the legs.

of rapid phasic contraction and expansion. These muscles are innervated by neurons that regulate the coordination of legs movements through central and segmental control mechanisms. In addition, legs are also provided with sensory hairs, plates, campaniform sensilla and chordotonal organs. These sensilla also act as proprioceptors and stretch receptors to provide information to position of body parts in respect to one another and are responsible for movement and orientation. There are three basic systems required for effective locomotion: (i) a sensory nervous system reporting on conditions in, on and around the body, (ii) a central nervous system to identify, interpret and coordinate that information, and (iii) a system of muscles, controlled by the nervous system and capable of producing the required movements.

Legs serve the purpose to support the body off the ground and to assist in walking by exerting propulsive forces on the body. Gray (1968) considered insect leg in its capacity as a strut and as a lever. The legs act as struts propping up the weight of the body and also as levers magnifying the movements, which the thoracic muscles apply to their bases. One of the most significant attributes of insect leg is to be a multisegmental structure, which facilitate in balancing the body and in propulsion. The movements of legs are determined by the nature of articulation of coxa with thorax and structures of legs. In different insects distinct differences occur between all three pairs of legs. The hind legs are usually longer and exert an axial thrust against the body and ground, which propels the body forward. Most studies on walking have been done by taking American cockroach, *Periplaneta americana*; stick insect, *Carausius morosus* and European mantid, *Mantis religiosa* as test animals.

The structure and modifications of tarsal segments is another important feature to increase the friction, as most adults walk faster on rough surface and on smooth surface tend to take shorter steps. Insect like house fly is able to walk easily on smooth surface, because of the presence of various adhesive structures like pulvilli and pads. These structures are covered with tufts of minute hairs, which remain wet with an adhesive secretion from glands located at the base of hairs. Such hairs are ultimately responsible for clinging ability and are called as 'tenent hairs'. When an insect walks upright, the leg system maintains equilibrium in space, and assists in suitable orientation relative to substrate. The legs provide the forces, which counteract gravity and propel the insect in desired direction, while the substrate is grasped by sticky structures of the legs. Further, at a time at least three legs remain in contact with the substrate to hold the body.

Observations have been recorded concerning speed of various insects. Most insects are rather slow walker and prefer to take-off wings for covering long distances, but some like cockroaches do well while running. McConnell and Richards (1955) determined that *Periplaneta americana* may go at a fast speed of 2.9 miles/hour or 130 cm/sec at 35°C, which is remarkably fast in comparison to its body size. Temperature also has a pronounced influence on insects speed and they move fast in a favourable range. The speed of some other insects has been mentioned in table 6.1.

Table 6.1. Walking/running speed of some insects.

Common name	Scientific name	Speed cm/sec.	Reference
American cockroach	*Periplaneta americana*	70-80	Delcomyn, 1971a
German cockroach	*Blattella germanica*	18-29	Wille, 1920
Desert locust	*Schistocerca gregaria*	8-15	Burns, 1973
Stick insect	*Carausius morosus*	8-9	Graham, 1972
European earwig	*Forficula auricularia*	10	Manton, 1972
Bug	*Gastrodes grossipes*	5-7	Manton, 1972
Beetle	*Cicindela hybrida*	58	Evans, 1977
Common black ant	*Lasius niger*	1.6	Barnes and Kohn, 1932
Bristle-tail	*Petrobius brevistylus*	21	Manton, 1972

6.1.1 Walking patterns

Wilson (1966) described walking as a cooperative interaction between individual leg oscillators. He proposed a model relating to different six-legged gaits. The sequences of movement of various legs during walking are rather fixed at a given speed for an insect. Usually all legs are used for walking, but some may walk efficiently on four legs. The most common pattern is called as 'alternation of triangles of support' (Fig. 6.2), in which each triangle consists of fore- and hind legs of one side and mid leg of other side (Fig.6.3a,b). During walking three legs in a triangular fashion give support to the body and remaining three moves forward. The alternate cycles are repeated and result into continuous

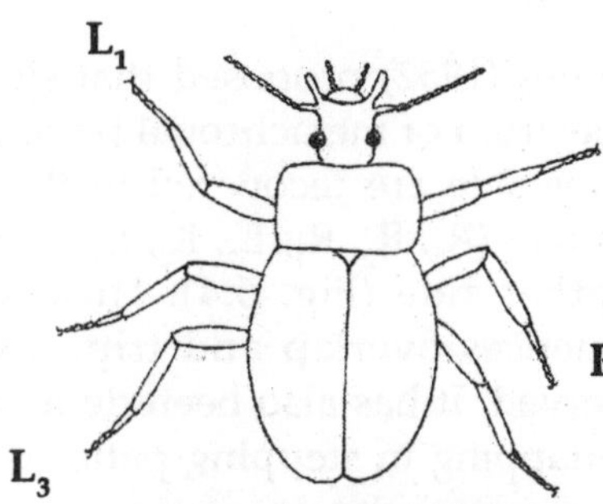

Fig. 6.2. The sequence of leg movement in a ground beetle. When moving it stands on a tripod consisting of three legs.

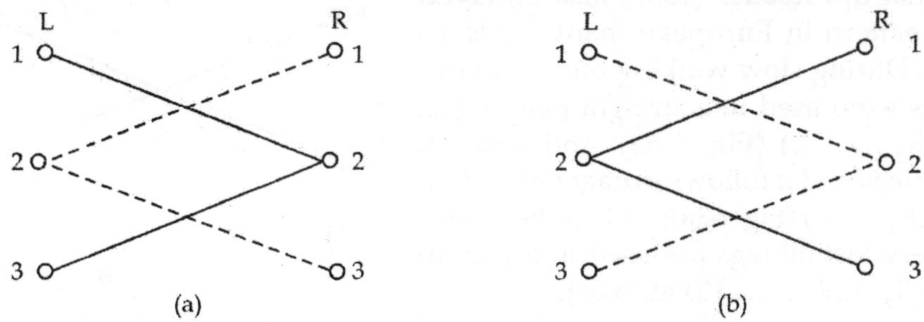

Fig. 6.3a-b. Alteration of triangles of support (L- left; R-right and 1,2,3- fore-, mid-, and hind legs).

zigzag movement. The walking pattern can be indicated as: L_1 R_2 L_3, R_1 $L_2 R_3$ or
L_1 R_1 L_1 R_1 L_1.........
$R_2 L_2$ $R_2 L_2 R_2$.........
$L_3 R_3$ L_3 R_3 L_3.........

The other patterns may also be followed by insects and have been studied by using cinematography and videography. Two general principles (i and ii) were proposed by Hughes (1952) for nearly all types of walking sequences. Wilson (1966) further added three more (iii-v) and these are:

(i) No leg is raised until the leg behind it in a supporting position.

(ii) Movement of the two legs of a segment alternate.

(iii) Protection time remains constant at all stepping frequencies.

(iv) Retraction time decrease with increase in stepping frequencies.

(v) Stepping interval between legs of two consecutive segments remain constant at all frequencies, whereas decrease between fore- and hind leg with an increase in frequency.

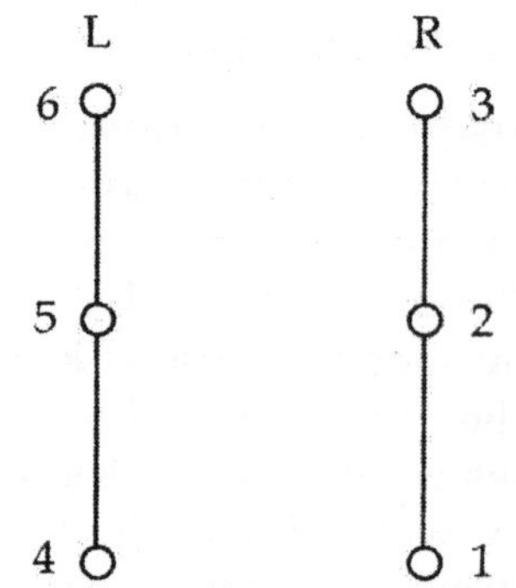

Fig. 6.4. The sequence of leg lifting in American cockroach, *Periplaneta americana* at very slow speed.

Hughes (1952) proposed that slow insects fit within a spectrum of metachronal patterns in which the legs of one side are recovered in the order of hind, mid and fore (R_3, R_2, R_1, L_3, L_2, L_1 ....), followed by legs of other side (Fig. 6.4). However, at more speed sequences overlap and tripod stepping pattern is observed. It has also been demonstrated that further overlapping in stepping patterns may also occur.

In grasshopper, *Rhipipteryx* only anterior four legs are used for walking and are lifted in a diagonal pattern (L_1, R_2, R_1, L_2 ....) (Fig. 6.5). In this case only two legs support the body at one time and further support is provided by abdominal tip. Roeder (1937) also observed similar pattern in European mantid, *Mantis religiosa*. During slow walking only posterior four legs were used in a straight pattern (L_3, L_2, R_3, R_2 ....) (Fig. 6.6a), and with an increase in speed it follows a diagonal pattern (L_3R_2, L_2R_3 ....) (Fig. 6.6b). When the *Mantis* move very fast all legs are used in a pattern (L_1R_3, L_3R_2, L_2R_1 ....) (Fig. 6.6c).

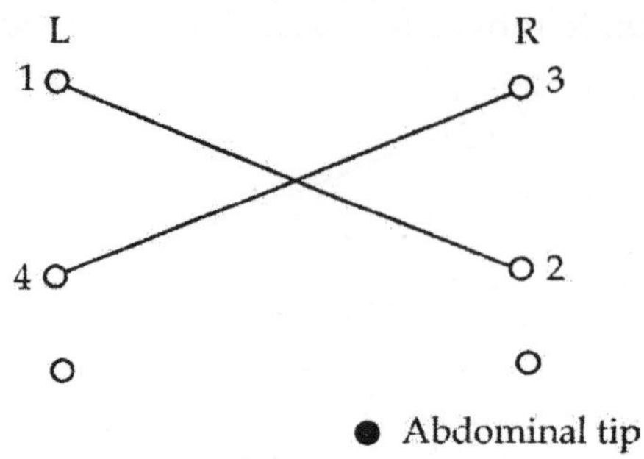

Fig. 6.5. Sequence of leg lifting in grasshopper, *Rhipipteryx* sp.

La Greca (1947) observed that in acridid,

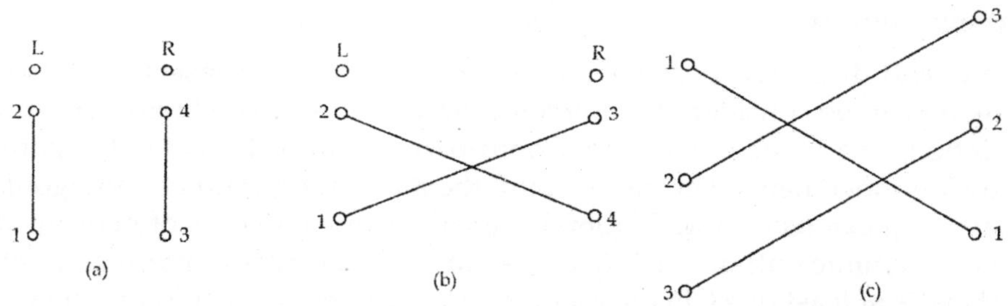

Fig. 6.6a-c. Orders of legs lifting in *Mantis religiosa* at three different speeds.

Tropidopola cylindrica at fast speed pattern was L_1R_1, L_2R_2 .... (Fig. 6.7), and tip of abdomen provide further support. However, at slow speed it follows a tetrapod diagonal pattern or other patterns were also observed.

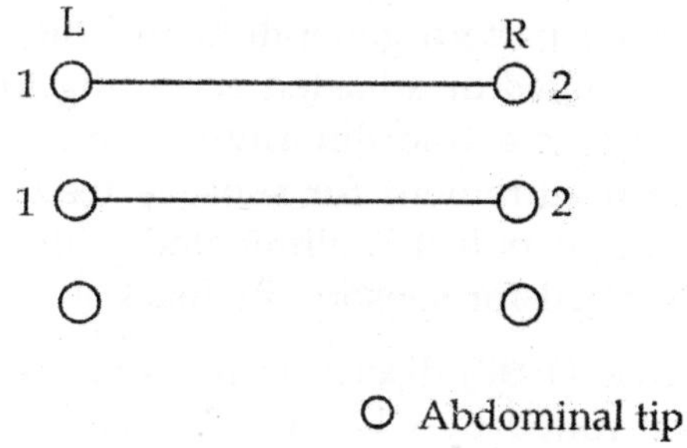

Fig. 6.7. Sequence of leg lifting in *Tropidopola cylindrica* at fast speed.

The pattern of leg movements and number of legs on the ground at a given time also depend on the relative times of protraction (when legs are in air) and retraction (when legs are on the ground). A change in the relative periods of protraction and retraction resulted into a different walking pattern. Wilson (1966) proposed that at low speed retraction time is more in comparison with protraction time, hence the protraction/retraction will be low, e.g. in *Blatta* it comes to be 0.31 and with an increase in the speed retraction time becomes less and protraction/retraction approaches to 1.

Some species show two different gaits. In first instar of stick insect, *Carausius morosus* two different gaits, gait I and II were described by Graham (1972), which were found to occupy same frequency range. Other species like *Blatta orientalis* and *Periplaneta americana* also show similar patterns. Pringle (1940) suggested distinction between running and walking based upon the use of either fast or slow fibres in the leg muscles. Delcomyn and Usherwood (1973) found significant differences in the number of impulses and their mean interval in different gaits in cockroach. Kozocik (1981) also reported that in *Periplaneta americana* the frequency of occurrence of a given step is bimodal.

The insects are also able to change the direction of a walk, which is brought into by a change in step frequency on right and left sides. This turning behaviour can be induced by various stimuli, e.g. food, visual, gravity, mate, etc.

6.1.2 Nervous control

Delcomyn (1999) discussed nervous control of locomotion in insects. Control of insect walking can be considered hierarchical and modular. The brain determines onset, direction, and speed of walking. Coordination is done locally in the ganglia that control leg movements. During walking the alternating action of antagonistic muscle pairs (leg extensors and flexors) is generated by networks of neurons that can produce rhythmic output. Such networks are called central pattern generators (CPGs). There is at least one CPG per leg; or may even be one per joint. The CPGs are coupled together via interneurons so that the activity of one can influence the activity of another. The sensory feedback from the moving legs is also in most cases critical for the proper coordination of the legs during walking. This peripheral feedback provides information about leg load, position, velocity, and acceleration, as well as information about joint angles and foot contact. In addition, both the central pattern generators and the sensory information that feeds them may be modulated or adjusted according to circumstances. Consequently, locomotion in insects is extraordinarily robust and adaptable. In stick insect, *Carausius morosus* the requirement for sensory feedback is so strong that coordination virtually disappears if it is eliminated while in American cockroach, *Periplaneta americana* the need for sensory feedback may be reduced during fast walking.

Cruse (1990) discussed neuronal control of walking system. Leg movements are controlled by CNS and by feedback from the proprioceptors in the legs. The head ganglia and other peripheral sense organs have a pronounced influence on walking. The reflexes also have a role in the control of leg movements. Each pair of legs is associated with thoracic ganglion of its corresponding segment by bilateral pair of connectives. The head and thoracic ganglia are also connected by connectives. Brain has an inhibitory effect on walking and many insects walk continuously after decapitation. If the connectives between suboesophageal and prothoracic ganglia are cut, stick insect showed unorderly walking (Roeder, 1937). Hughes (1957) demonstrated that at least one connective is necessary between each thoracic ganglion for coordinated walking.

In American cockroach, *Periplaneta americana* stimulation of anal cerci resulted into running and it has been considered that the impulses are being transmitted from last abdominal ganglia to the thoracic locomotory centres by the giant nerve fibres (Roeder, 1967).

6.1.3 Leg amputation and walking

The removal of one or more legs causes quantifiable changes in the temporal and spatial patterning of movement of the remaining legs (Hughes, 1957). Removal of only one leg has only relatively slight effect, but removal of two legs resulted in a tetrapod gait and the remaining legs move in a diagonal pattern (L_1, R_3, R_1, L_3 ....) (Buddenbrock, 1921) (Fig. 6.8a). The amputee does not appear to require any learning period to adopt the new coordination. Removal of tibia and tarsi of one or both mesothoracic legs has no effect on gait (Delcomyn, 1971b). Hughes (1957) demonstrated that in some insects' removal of mesothoracic legs resulted into a very unstable gait with a pattern (R_3R_1, L_3L_1 .... (Fig. 6.8b), but usually the pattern followed is either (R_3, R_1, L_3, L_1 .....) (Fig. 6.8c) or a diagonal pattern (R_1L_3, L_1R_3 ....) (Fig. 6.8d). Wendler (1966) showed that the metathoracic coxae of amputees of *Carausius* also move synchronously with fore legs, but if mid legs are cut in the middle of femur and then attached with artificial legs, normal coordination is again restored.

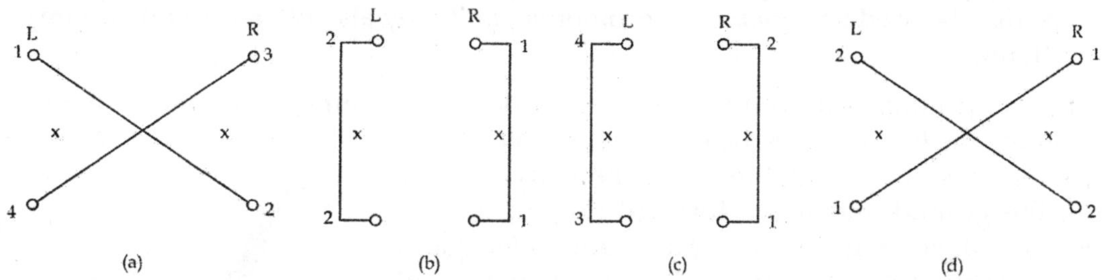

Fig. 6.8a-d. Walking patterns of insects without metathoracic legs.

6.1.4 Turning

The insects rarely walk straight lines for any significant length of time and used to take turns. Ants nicely demonstrate the turning while they are attempting to find food. They make random turns and continue these until they smell some food. However, in order to actually cover any real distance while randomly exploring, the ant must make sure that the random turns are not too large. If they are so, then the ant could just spend its time zigzagging the same ground over and over again while making little progress. This type of behaviour is pretty normal for a number of different insects and is also used to explore the environment in an effective manner.

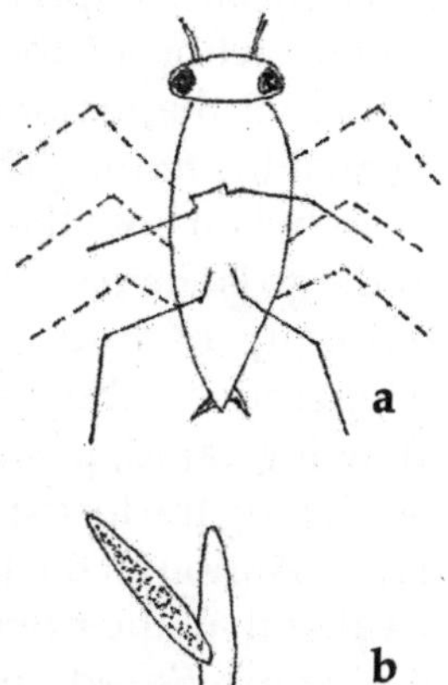

Fig. 6.9. (a)- Insect leg extension showing that how an insect extends its legs in order to turn its body. (b)- The original and new orientation of insect.

Cockroach uses extension of its legs to turn its body (Fig. 6.9). If one of the legs extends then it is pushing the insect in the opposite direction with a force equivalent to the amount of extension. Extending the right rear leg during walking will turn the insect clockwise. Extending the right front leg will turn the insect counter-clockwise. The motor neurons directly control the extension of the front legs and degree of turn.

6.2 Jumping

The jumping is an important strategy for many insects whose normal speed of locomotion is relatively slow. Different insects utilize a remarkable variety of mechanisms to generate a jump. Insects like grasshoppers, fleas, flea-beetle and certain hemipterans exhibit jumping behaviour by using their modified hind legs, while some others like springtails (Collembola) and click beetles have different mechanisms. In most cases, except grasshoppers jumping is undirected and probably serves as an escape mechanism. Most studies on jumping behaviour of insects have been done by taking locust hoppers and adults. In these insects short hops may be used as a means of locomotion, and they also often assist their jump by flying.

The locust jump has been investigated in detail in a number of studies. Before jumping the locust raises the body and bends its legs. As the knee joint straightens out the body is lifted off the ground and move forward (Fig. 6.10). The jump is dependent on the storage and subsequent rapid release of energy. In the locust and grasshoppers, the back (metathoracic) legs (Fig. 6.11) show a number of specializations for optimizing jumping, including their great length. The jump mechanism comprises of three stages. The first stage is contraction of the flexor tibiae muscle producing flexion of the femoro-tibial joint to bring the tibia up against the femur. This is followed by a cocontraction phase when both the flexor and extensor tibiae muscle undergo isometric contraction. This phase allows the build up of potential energy. The jump is then triggered by the relaxation of the flexor muscles allowing rapid, powerful extension of the tibia, thus propelling the locust into the air (Heitler and Burrows, 1977). Although the flexor muscle is much smaller and weaker than the extensor muscle, it is able to maintain the leg in a flexed position. This is due to the presence of a lump of cuticle on the inner surface of the distal end of the femur, which acts as a catch to hold the flexor apodeme. When the flexor muscle relaxes, the apodeme slips off the lump, allowing extension of the leg (Heitler, 1974).

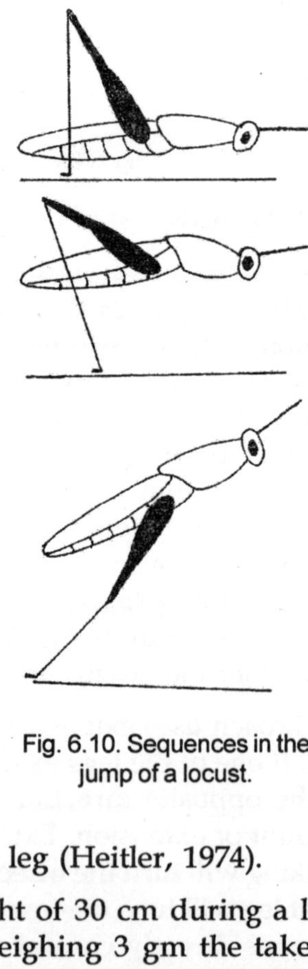

Fig. 6.10. Sequences in the jump of a locust.

The fifth instar nymph of *Locusta* may jump to a height of 30 cm during a long jump of 70 cm at 27°C (Hoyle, 1955). For a locust weighing 3 gm the take-off velocity has been calculated to be 340 cm/sec. The take-off angle is usually 60° and height of jump increases with the take-off angle and with the distance through which legs are extended. However, with an increase in insect's weight the height of jump decreases. The height can be calculated by the formula:

$$\frac{1}{2}mV_h^2 = mgh$$

$$\therefore h = V_h^2 2g$$

where:

h - height

m - mass of insect

V_h - vertical take-off velocity

g - acceleration due to gravity

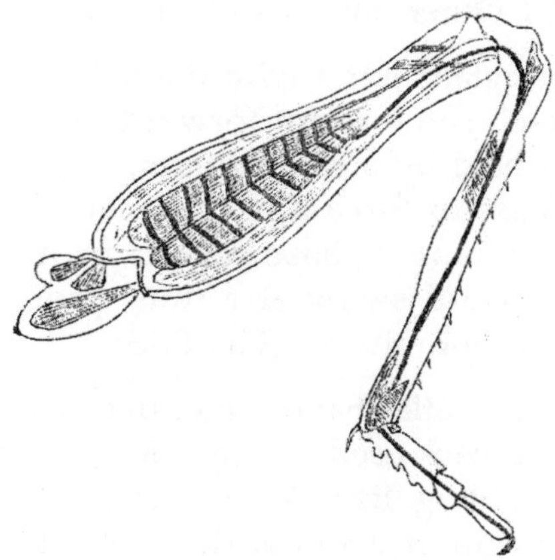

Fig. 6.11. Hind leg of a grasshopper with musculature.

Bennet-Clark and Lucey (1967) discussed the jumping mechanism of the rabbit flea, *Spilopsyllus cuniculi* and oriental rat flea, *Xenopsylla cheopis*. The flea raises mesothoracic femora and brings the trochanters in contact with the substratum. The muscular contraction exerts a force into the patch of resilin in the thoracic wall. The thorax is shortened by contraction of longitudinal muscles, due to which the thoracic segments are held in new position by cuticular catches (pegs and sockets on the exoskeleton). The jump is initiated by relaxation of muscles that raised the femora and shortened the thorax. Consequently the femora descend and thrust the tibiae hard against the substratum, and the release of catches brings the thorax in natural shape. In addition the energy stored in the resilin pads amplifies the acceleration. The sequences of the jump of a flea are depicted in figure 6.12.

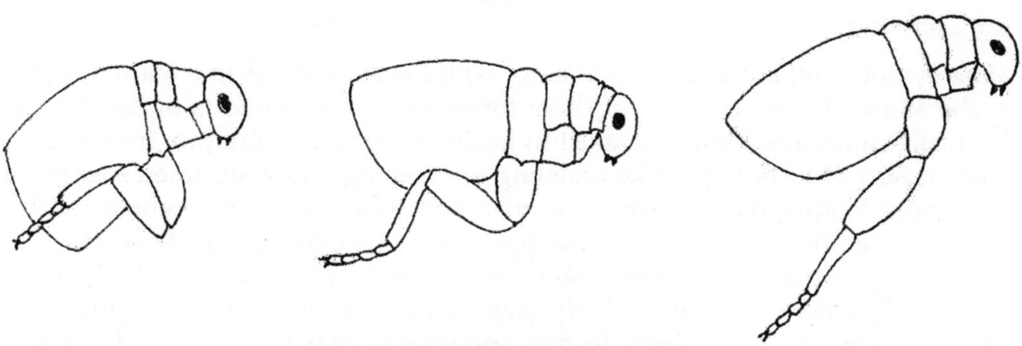

Fig. 6.12. Sequences of the jump of a flea.

Certain hemipterans belonging to families Cercopidae, Jassidae, Psyllidae and Membracidae jump by a rotation of the leg on the coxo-trochanteral joint. In psyllids the coxa is fused with thorax and thus femora lies parallel with the trunk. Powerful muscles originating from thorax are inserted into legs and assist in jumping.

6.2.1 Other patterns of jumping

Springtails jump with the help of modified abdominal appendages called furca, which can be bent forward and its tips are held by a retinaculum. The sudden release of furca due to development of tension at its base results in its quick backward swing and thus, propels the insect into the air (Fig. 6.13).

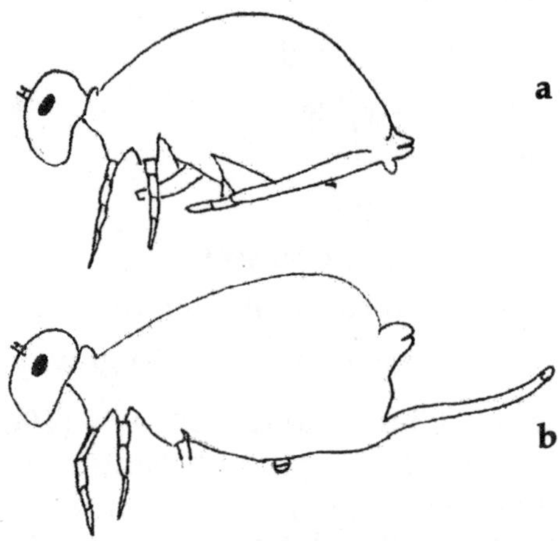

Fig. 6.13. Jumping in a springtail. (a)- Furca in flexed position, (b)- Furca in extended position.

Click beetle (Fam. Elateridae) when lies on its back used to jumps for the purpose of righting. Its jackknife jump may reach a height of 0.3 m with a velocity of 2.4 m/sec. The beetle arches its back between pro- and mesothorax and then contracts ventral muscles to straighten the body due to which an elongated prosternal spine is released suddenly resulting into jerk in the air.

The last instar of cheese maggot, *Piophila* (Order Diptera) jump by bending the head and holding the last abdominal segment between mouthhooks. A muscular tension is built up within the body and abdomen is suddenly released giving a jerk to the maggot. Mature fruit fly (Fam. Tephritidae) larvae of some genera have the ability to hop in a similar way, which assists in leaving the host, and also provide defence against the soil predators, particularly ants.

6.3 Crawling

Most nymphs and larvae of insects move with the help of legs in a manner similar to the adults; however, some of these forms are either apodous or do not have relatively rigid exoskeleton to assist in locomotion. Such immature insects adopt other means of locomotion like crawling or creeping. They are often assisted by additional modifications of the abdomen and its musculature, which resulted into a peristaltic movement for the purpose of propulsion. In these cases the haemolymph act as a hydrostatic skeleton and the muscles keep the body turgid. The muscle actions in different body regions are antagonistic to each other and a change in one part of the body lead to an opposite change in some other part.

The lepidopteran caterpillars and some hymenopterous larvae have additional prolegs or accessory legs on some abdominal segments. Prolegs are hollow outgrowths of integument and their lumen is filled with haemolymph. The apical tip of proleg bears a tuft of curved hooks called crochets (Fig. 6.14a,b). The prolegs are supplied with two types of muscles, one the 'turgor muscles' that maintains a steady internal haemolymph pressure and other 'locomotor muscles' involved in movement of prolegs. Crawling patterns with the help of prolegs have been described by Hughes and Mill (1974). The movement is resulted by serial contractions of the longitudinal muscles along with leg movements. Usually only a single wave appears along the body at a time, which starts from posterior end and is followed by legs movements. Two legs of a segment move together and the crawling rhythms progress from posterior to anterior end.

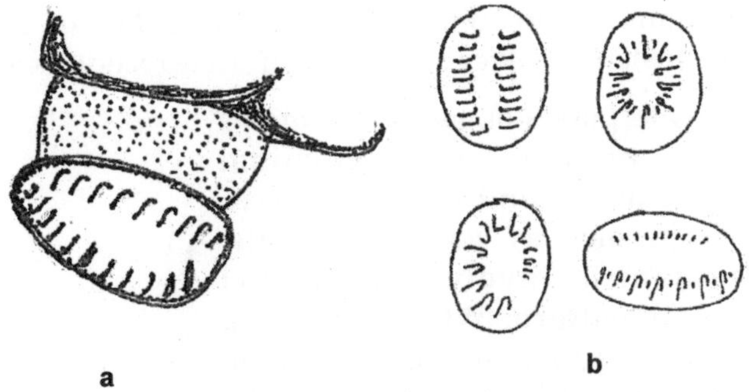

Fig. 6.14. (a)- Proleg of a lepidopteran caterpillar showing crochets, (b)- Arrangements of crochets.

During crawling the contraction of dorsal longitudinal, dorsal intersegmental and transverse muscles resulted into shortening of the segment dorsally. Consequently the segment behind is lifted from the ground and the prolegs are released. Then the ventral longitudinal muscles contract so as to bring the segment down toward the ground and the prolegs become fixed a little ahead. It starts from the last segment and the terminal prolegs are lifted first and placed on the ground at varying distances forward. The wave of peristalsis involves simultaneously at least three body segments, which are in different stages of contraction at one time (Fig. 6.15).

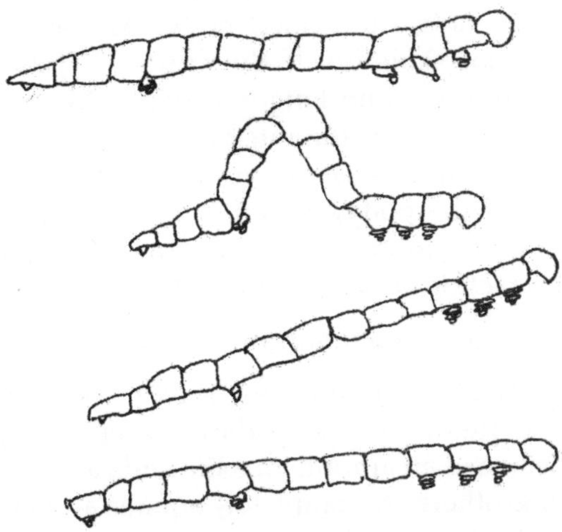
Fig. 6.15. Stages in the locomotion of caterpillar of geometrid moth. The caterpillar repeatedly gives a very rapid looping action and each time this result into forward movement of the body.

Some caterpillars are able to move backward at high speed. Brackenbury (1999) described the suite of reverse gaits available to caterpillars of *Pleuroptya ruralis*, *Cacoecimorpha pronubana*, *Cucullia verbasci* and *Tyria jacobaeae*. These reverse forms of locomotion are produced primarily in response to threat, involve bilateral activation of the intersegmental muscles and are relatively simple in terms of neural control.

The crawling involves a high degree of coordination and the control is largely through CNS and if the ventral nerve cord is cut, the coordination between anterior and posterior halves ceases. The muscle receptors are also involved in this response.

The caterpillars of geometrid moths have prolegs on abdominal segments six and ten. They have been termed as 'loopers' and the mechanism involves somewhat similar pattern. While moving a large portion of the abdomen remains out of contact with the ground giving the shape of Greek character Omega (Ω) (Fig. 6.16).

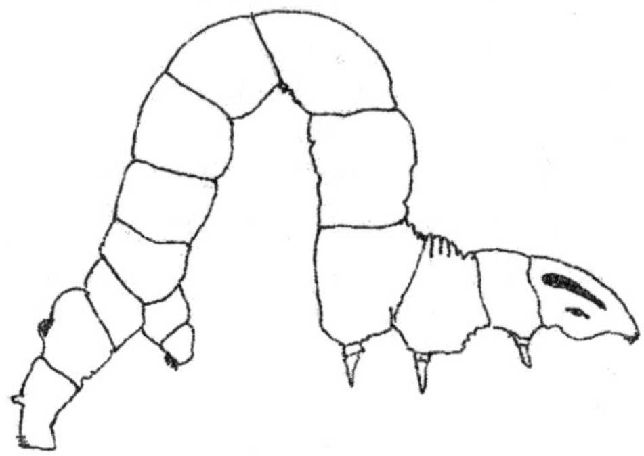

Fig. 6.16. A looper caterpillar with looped body.

The movement of apodous larvae like maggots of flies entirely depends on peristaltic changes in the body shape as result of muscular action against the body fluids. This movement is achieved by the action of muscles on the soft, fluid-filled body. Although the cuticle is soft, it is essential as an external skeleton. If the cuticle is punctured, the internal fluid escapes and locomotion is no longer possible. Many have minute spines or other structures associated with the integument, which increase friction between their body and the substratum. The wave of peristaltic contraction passes from posterior to anterior end of the body, which is the direction of progression. The reverse has been observed in soil dwelling tipulid and bibionid larvae in which the wave of contraction proceeds opposite to that of progression.

6.4 Aquatic locomotion

A number of adult insects particularly belonging to the orders Hemiptera and Coleoptera have secondarily adopted aquatic mode of life (Fig. 6.17). Some of them lead an amphibious life and can survive both on land and in water; however, some others are completely aquatic and may be surface or bottom dwellers. Among the bottom dwelling insects a few remain fixed with some object, while others are able to walk or swim by using their legs or by other means. Many insects pass a part of their life cycle in water and usually their immature stages are aquatic. Some in water are able to control their buoyancy and can remain suspended at different depth. Many aquatic forms have also retained their ability to fly.

The bugs of families: Belostomatidae (giant water bugs), Corixidae (water boatman), Gerridae (pond-skaters), Hebridae, Hydrometridae, Nepidae (water scorpions), Notonectidae (back-swimmers), Mesoveliidae, Veliidae and among beetles Dytiscidae (diving beetles), Gyrinidae (whirligigs), Hydrophilidae (water scavengers) and a few others are distinguished inhabitants of water. Nymphs of

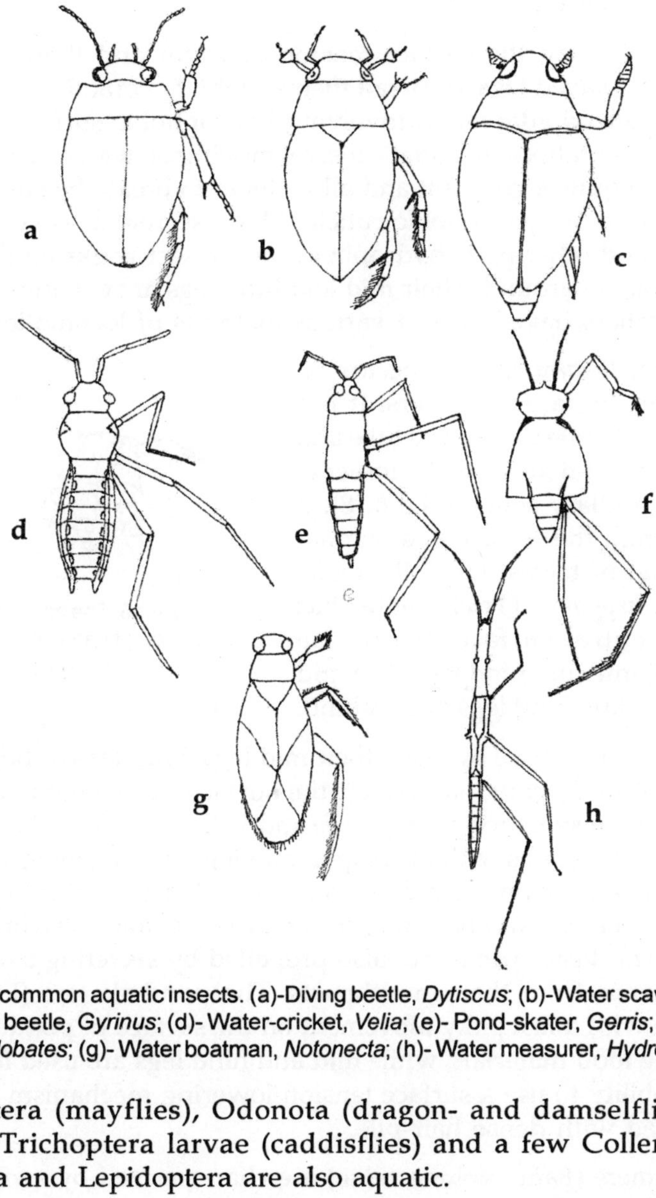

Fig. 6.17. Some common aquatic insects. (a)-Diving beetle, *Dytiscus*; (b)-Water scavenger, *Hydrophilus*; (c)- Whirligig beetle, *Gyrinus*; (d)- Water-cricket, *Velia*; (e)- Pond-skater, *Gerris*; (f)-Water-skater, *Halobates*; (g)- Water boatman, *Notonecta*; (h)- Water measurer, *Hydrometra*.

Ephemeroptera (mayflies), Odonota (dragon- and damselflies), Plecoptera (stoneflies), Trichoptera larvae (caddisflies) and a few Collembola, Diptera, Hymenoptera and Lepidoptera are also aquatic.

The major problems of aquatic forms are those of respiration and locomotion. The surface dwellers use oxygen of the air, some come to the surface to renew their air supply, while others have highly specialized adaptations for underwater breathing. The locomotory modifications are relatively simple and their body and legs are modified for the purpose. The subject of aquatic locomotion in insect has been reviewed by Nachtigall (1974b, 1985).

6.4.1 Surface locomotion

Neuston insects live on the surface tension of water and therefore tend to be lightly built with splayed legs to spread their weight. Locomotion on water surface is a highly specialized adaptation involving hydrophobic surface, modified legs and respiratory adaptations. In aquatic insects moving on water surface, the surface tension forces are quite significant and allow them to break the surface film. Such insects usually have a hydrophobic cuticle. Many secrete a waxy substance that makes the tarsi and other parts hydrophobic and insect walks on the surface in a similar manner as on ground. Their mid and hind legs are expanded and oar-like. The surface dwellers have adopted various methods of locomotion.

The collembolan, *Podura aquatica* occurs on the water surface in large numbers. It has hydrofuge cuticle, while the ventral tube and claws are wettable and anchors the insect to the surface by capillary action (Fig. 6.18). It can also jump from the surface in a manner similar to as that of terrestrial Collembola. The imago of *Phryganea* (Trichoptera) has hydrophobic ventral surface. It runs on surface by using mid and hind legs. The mid legs make rowing and hind legs drag behind.

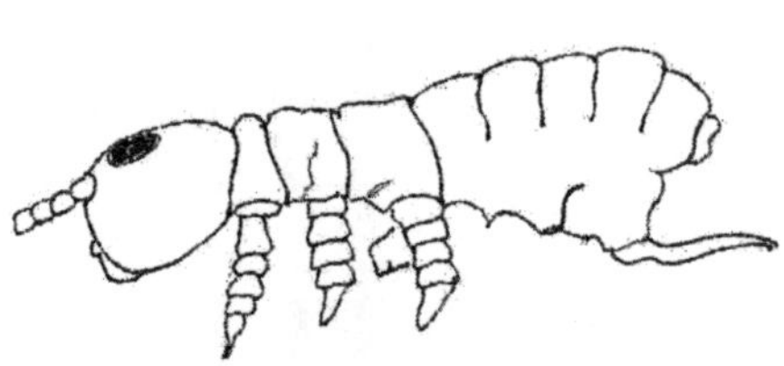

Fig. 6.18. The collembolan, *Podura aquatica*. It can walk on water surface rapidly as on land.

Surface dwelling bugs move by using their mid legs, hind legs or both. *Hydrometra* (Fam. Hydrometridae) crawl slowly on the surface of stagnant water by using legs as oar beating in synchrony. *Rhagovelia* and *Velia* (Fam. Veliidae) swim against the current of swift streams with the help of mid legs. The water strider, *Rhagovelia* has highly specialized swimming fans on last tarsal segments of mid legs, which spreads out and act as a webbed foot, so the insect is able to swim on surface of rapid streams. The *Velia* species are also propelled by secreting a surface tension reducing substance from their mouthparts. The pond-skaters (Fam. Gerridae) (Plate 6.1) glide with a fast jerky movement on the surface of pond. They use fore legs for grasping food materials, while mid and hind legs are used for locomotion. They have an ability to use a surface tension lowering mechanism as the body is ventrally covered with dense hair-pile.

The back-swimmers (Fam. Notonectidae) are characterized by swimming on the back, which is shaped like the bottom of a boat. They are usually observed floating on the surface with long oar-like legs outstretched. When alarmed, they dive readily carrying a supply of air beneath wings. They can also leap into the air to fly.

The whirligig beetles (Fam. Gyrinidae) have divided compound eyes for seeing above in air and below in water. Their fore legs are long and prehensile, while hind legs are broad, greatly flattened and highly adapted for swimming

(Fig. 6.19). Vuline (1987) demonstrated that their pygidial glands secrete a substance, which contain a surface-active component and may help to propel the beetle across the water.

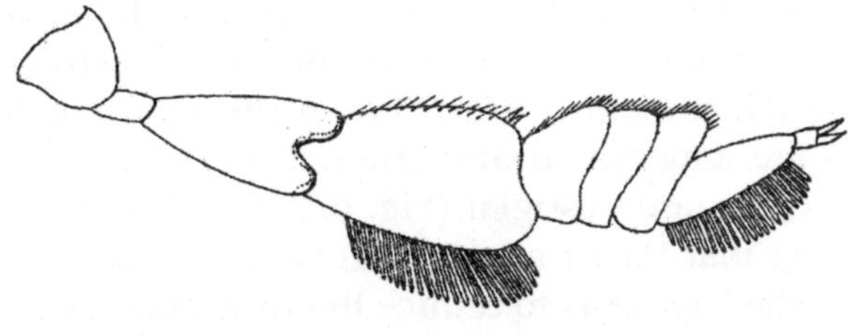

Fig. 6.19. Hind leg of whirligig beetle, *Gyrinus* with swimming blades.

The rove beetle, *Stenus* found on stems of small plants bordering streams and often fall into the water, and walk slowly on the surface. When alarmed, it releases a secretion from pygidial glands, which reduce the surface tension behind the insect and causes more rapid locomotion (Billard and Bruyant, 1905). It shoots forward on its fore legs, and steers itself by flexing its abdomen. This tiny beetle can travel nearly one metre per second this way and avoids predators.

In water strider, *Halobates* many specializations in the thoracic skeleton and musculature, leg structure and surface fine structure of body and legs are adaptations towards a life in the water-air interface and especially for locomotion on the water surface (Andersen, 1976). The surface structures consist of two kinds of macro-hairs inserted in sockets and there is a velvety undercoat of hook-like microtrichia. The hair-layers are to protect the insect from being wetted when occasionally submerged. When submerged in water it carries a large bubble of gas entangled in the hair-layers of its body and appendages. This gas store makes the insect strongly buoyant and it will rise to the surface rapidly. However, the hydrofuge property of the hair-layers is not permanent and upon prolonged exposure to water the hairs become wetted and the submerged insect faces difficulties in regaining an above-water position. When resting only the distal segments of the legs are in contact with the surface film. The short and stout front legs support the body or grasps and hold the prey during feeding or the female during copulation. The long and slender middle legs propel the body like oars beating in synchrony, and the hind legs are chiefly used for steering and supporting the body.

6.4.2 Subsurface locomotion

A number of insect species thrive well beneath the water surface. They may be bottom dwellers or swim at different depths. Certain dwellers are also able to go deep in the water. Many insects have air stores of different types, some carry air while going in and others have modified structures like gills, respiratory siphons, etc. Some come out to renew their air supply. Their air stores increase buoyancy and help in maintenance of hydrostatic balance. The body has hydrophobic surfaces, hard cuticle and shape is also modified to minimize friction during movement in water. Their legs are modified and function in steering, propulsion and also provided with dense hairs, which assist in swimming.

The whirligig beetle, *Gyrinus* is also able to move below the surface with flat hind legs. The tibiae and tarsi form paddle-like structure and are further supported by hairs (Nachtigall, 1962a) (Fig. 6.19). In water beetles, *Dytiscus* and *Acilius* hind legs are lengthened and fringed with hairs. Their tibiae and tarsi are flattened and claws are vestigial (Fig. 6.20a,b). The stiff hairs greatly increase the area of the leg that thrust against the water on power stroke, while on return they fold back the legs so as to reduce the resistance. The larvae of *Acilius sulcatus* move all legs according to basic pattern of insect locomotion and its swimming mechanism

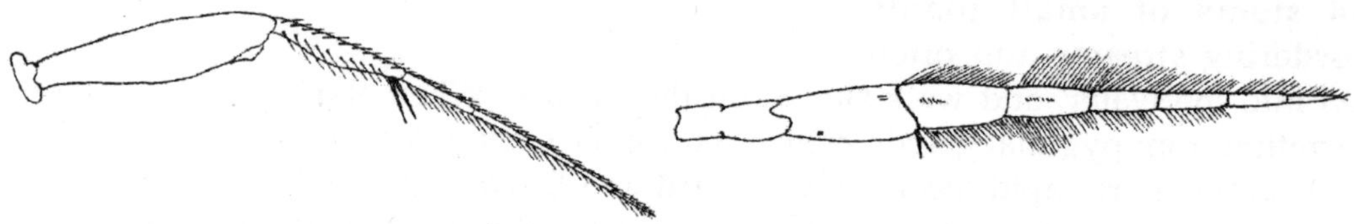

Fig. 6.20. (a)- Hind leg of diving beetle, *Dytiscus*, (b)- Hind leg of water beetle, *Acilius sulcatus*.

has been described by Nachtigall (1962b). The hydrophilid beetles have a dorsal air reservoir beneath the elytra and ventral hairy tracts, which also serve to retain an air film. They row by using all legs, which move in a counteracting alternation of phases. The swimming patterns in *Dytiscus* and *Hydrophilus* were discussed by Hughes (1958).

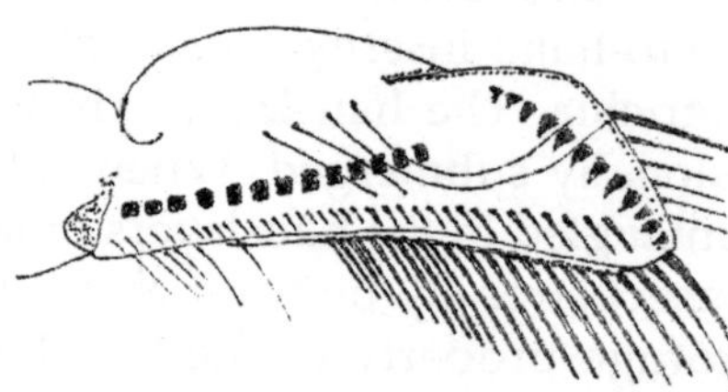

Fig. 6.21. Terminal segment of hind leg of water boatman, *Cenocorixa bifida* that propel the body by synchronous rowing action.

The water boatman, *Cenocorixa bifida* is propelled by synchronous rowing action of hind legs (Fig. 6.21). The propulsive cycle consists of a power stroke and a recovery stroke (Blake, 1986). The back-swimmers can also swim beneath the surface with ventral side up and head directed downward. The water scorpions, *Nepa* and *Ranatra* (Plate 6.2) swim by using meso- and metathoracic legs. In *Nepa* hairs on legs are absent, while in *Ranatra* only a few hairs are present. The giant water bug, *Belostoma* has dorsoventrally flattened body and legs are fringed with hairs (Plate 6.3).

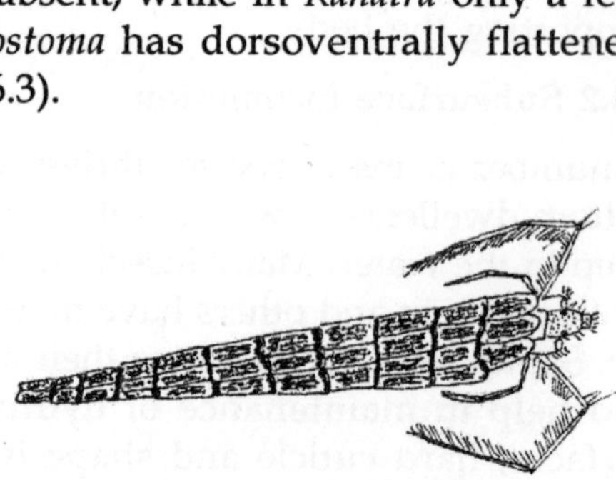

Fig. 6.22. *Triaenodes* larva.

The bottom-dwelling *Triaenodes* (Trichoptera) larva lives in a spiral case, of which the anterior end of the body projects out and it move in a rotary motion (Fig. 6.22). It has swimming hairs on long hind legs. During thrust the leg stretches and during recovery it is bent. The larvae of *Leptocerus tineiformis* (Trichoptera) move the water by hopping motion of strongly hirsute legs. Larvae of *Sialis lutaria* (Megaloptera) have hairs on legs and abdominal appendages, and are very good swimmers. Larvae of dragon- and damselflies also walk on substrata with legs similar to terrestrial forms.

6.4.3 Aquatic locomotion by other means

The apodous aquatic larvae and pupae of certain Diptera move by undulation of their bodies. The larvae of *Ceratopogon, Chironomus, Culex, Aedes, Corethra* and others move due to curling or eel-like undulations of their elongated bodies (Fig. 6.23). The phenomenon has been described by Gray (1953), Nachtigall (1963, 1964). Some of them also have additional structures, e.g. larvae of *Aedes* and *Culex* have swimming fans consisting of dense hairs on last abdominal segment. *Corethra* larvae also have similar fan but with less number of hairs. Brackenbury (2000) discussed locomotory kinematics of larvae and pupae of *Chironomus plumosus.* Larvae display three independent modes of motile activity: swimming, crawling and whole-body respiratory undulation. Swimming and respiratory undulation involve the use of metachronal waves of body bending which travel in a head-to-tail direction. Crawling appears to result from an independent programme of muscle activation. The pupa displays two swimming modes, somersaulting and eel-like whole-body undulation, and the former being principally an escape manoeuvre, the latter being a faster form of locomotion.

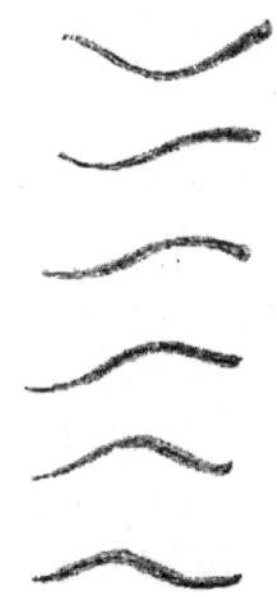

Fig. 6.23. Undulating movements of legless aquatic *Ceratopogon* larva (modified after Nachtigall, 1963).

In larvae of mayflies entire abdomen undergo undulating movement and is further assisted by abdominal gills and caudal filaments. But they do not move continuously and movement is interrupted by pauses. The nymphs of damselflies have three anal gills and move by undulations of these gills.

The nymphs of mayfly, *Cloeon* and dragonfly fill their hindgut with water and push it out suddenly, resulting into forward movement. As the abdomen of dragonfly nymph contract, the legs are retracted along the body sides, so as to offering a minimum resistance in the forward movement (Hughes, 1958). Mosquitoes, *Clunio* and *Pontomyia* propel by means of their wings on surface.

Insects that inhabit lotic waters are usually small and show various morphological adaptations suiting them to this environment. In *Drunella doddsi* dorsoventral flattening of the body allows the insect to exploit the 'boundary layer effect' that is a thin layer of water directly above the insect's body. In this region the current becomes greatly reduced and therefore the drag exerted on the insect within this zone is considerably reduced. This allows safer movement over the substrate and less energy expenditure to remain stationary.

6.5 Flight (Aerial locomotion)

The ability to fly is one of the most significant attributes of insects and has made them dominant among all animal groups. They have an advantage over their allies and are able to explore different modes of life and great diversity of habitats inaccessible to other animals. Besides the need to search for food or a resting habitat, there are other purposes for insects to remain in flight. With wings, insects were able to spread over the globe; if conditions became unfavourable in one place, they simply took to the air to find another. Flight has given them an advantage over land bound animals in being able to forage widely for food, make good their escape from enemies and search actively for mates. The drones (males of the honey bees) fly only to fertilize the queen. Many swarms of social insects are mating swarms. Insect flight evolved about 330 million years ago. There is genetic evidence that wings evolved from articulated gill plates on the limbs of aquatic ancestors, rather than being novel outgrowths from the body wall (Carroll *et al.*, 1995; Averof and Cohen, 1997). It is evident that some stonefly species can stand on their hind limbs and flap their wings as they sail. Sailing like this may have had sufficient potential to drive the evolution of insect wings in the past (Kramer and Marden, 1997). Marden *et al.* (2000) supported hypothesis for the evolutionary origin of insect wings is that they evolved from articulated, leg-derived respiratory structures of aquatic ancestors. The surface-skimming locomotion in stoneflies and mayflies offer a plausible solution by showing how rudimentary wings and muscle power can be used to accomplish two-dimensional aerodynamic locomotion on the surface of water. Distinct forms of wing-flapping surface skimming include: (i) stoneflies that flap their wings weakly while maintaining their body in contact with the water and undulating their abdomen laterally, (ii) stoneflies that skim while elevating their body above the water and maintaining all six legs on the surface, (iii) stoneflies and mayflies that skim with only four legs on the water surface, (iv) stoneflies that skim with only their two hind legs on the surface, and (v) stoneflies that use their hind legs to jump from the water into the air to initiate flapping flight. These forms of skimming show that wing-beat amplitude, horizontal velocity, and the verticality of aerodynamic force production increase as the body orientation becomes more upright and contact with the water is minimized. These behaviours illustrate a mechanical pathway by which flying insects could have evolved from swimming ancestors.

Some of the most remarkable contributions published on various aspects of insect flight are: Chadwick (1953a,b), Weis-Fogh (1964), Wilson (1968, 1971), Nachtigall (1974a, 1981), Pringle, (1957, 1968, 1974), Rainey (1974, 1976), Dalton (1975), Alexander (1981), Kammer (1985), Wilkin (1985), Brodsky (1994) and Dudley (2000).

Flight is a characteristic of all wing-bearing insects (pterygotes); however, aptery may occur in some of these insects as a result of adaptations to special circumstances. In the powerfully flying insects, the wings are most beautifully adapted for the stresses and aerodynamics of flight. The veins are thicker, stronger, and closer together towards the front edge (leading edge) and thinner yet flexible towards the rear edge (trailing edge). This makes the insect wing an excellently constructed 'airfoil' capable of exerting both propulsion and lift while minimizing drag. All pterygotes except primitive Ephemeroptera and Odonata are able to flex their wings over abdomen, but the Odonata though primitive are amongst the fastest fliers. The wing movement is facilitated by both direct and indirect striated muscles (Fig. 6.24). The direct muscles have their origin on ventral exoskeleton and are directly inserted on wing base, while indirect muscles are hinged between thoracic tergum and pleuron. Direct muscles are predominantly operative in Dictyoptera, Odonata, Orthoptera and Coleoptera. In other groups flight is the result of action of mainly indirect muscles; however, direct muscles also participate in forward and backward movements and rotation of wings. Wigglesworth (1973) proposed a direct correlation between flight muscles weight in comparison with body weight and flying capability of insects.

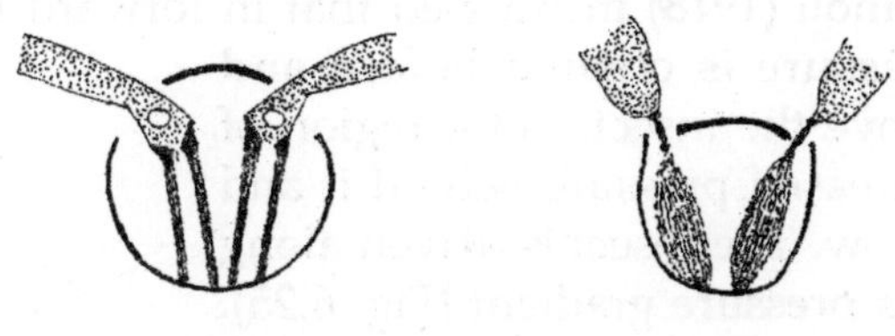

Fig. 6.24. Flight muscles of insect wing, (a)- Direct, (b)- Indirect.

The flight mechanism involves wings, their articulation with body, muscles, energy requirements, oxygen consumption and circulation. The flight is the result of wing movements, which is a complex process and involves elevation, depression, flexion, rotation, promotion, re-motion and change in shape. In primitive insects fore and hind wings move independently, whereas in complex forms during flight both are united by various coupling mechanisms to make a functional unit. In Orthoptera and some Neuroptera hind wings work in a region of turbulence produced by movement of fore wings. In Odonata hind wings beat first, and in *Schistocerca* hind wings beat with large amplitude to avoid the turbulence produced by fore wings. The elytra of beetles are not moved much during flight, but they function effectively as fixed lifting surfaces.

A number of sensilla located in the wings and other body parts are important to control the flight. They tend to correct any forces that disrupt the stability of insect during flight. In primitive insect orders flight is initiated by synchronous volley of nerve impulses to the muscles, whereas, in other groups the insects have evolved such a system that does not require synchronous nervous control for all actions. The impulses in motor nerves supplying the flight muscles are asynchronous with the wing-beats and the rhythm is generated by the muscles themselves (Pringle, 1974).

Chadwick (1953a) pointed out three forces for propelling an insect in a direction and at a velocity.

(i) Reaction to the thrust generated by the wings.

(ii) Resistance offered by the air to the passage of the insect through it.

(iii) The downward pull of gravity.

Demoll (1918) mentioned that in forward horizontal flight a region of decreased pressure is created before and above the insect and a region of increased pressure behind it and below. The insect is driven along this pressure gradient (Fig. 6.25).

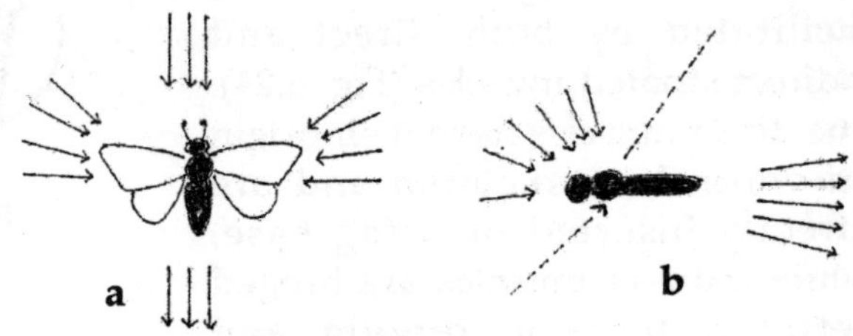

Fig. 6.25. Diagram showing the air currents which lift the insect and push it forward, (a)- Dorsal view, (b)- Side view. The broken arrows indicate the resultant thrust on the insect (modified after Demoll, 1918).

For a flight an insect must beat its wings at a frequency and with stroke amplitude to generate the necessary thrust and lift. The thoracic and other muscles associated with wings are mainly responsible for up and down stroke of the wings. Such strokes operate on a 'click mechanism'. The elastic stiffness of thorax, wing position, attachment of wing base to the sclerite and muscles action keep the wing in a stable position of fully up or fully down. Any position in between will be unstable and the wing will tend to return to any one of the stable positions. This mechanism is greatly helpful in accelerating wing motion with little input of energy.

Muscles inserted on wing sclerites cause the forward and backward movement and partial rotation. The rotation of wing during motion is because of differences in texture of its anterior and posterior parts. Thus, the rotary movement of wing is the result of anterior and posterior deflections along with the effect of changing air pressure on its heavier anterior and membranous posterior surfaces (Dickinson *et al.*, 1999). Such rotations are best exhibited by swift fliers, e.g. wasps, bees, flies, dragonflies, etc.

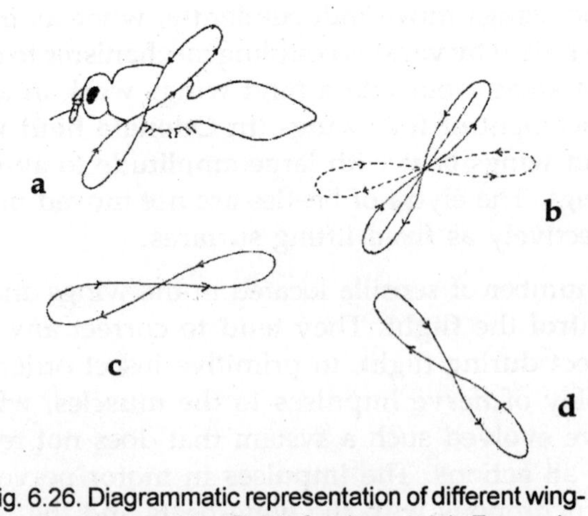

Fig. 6.26. Diagrammatic representation of different wing-beat patterns during flight of an insect, (a)- Forward flight, (b)- Turning to right, (c)- During hovering, (d)- During backward flight.

The detailed studies on the movement of wings during flight have been conducted on *Schistocerca, Calliphora* and dragonfly, *Sympetrum frequens*. In general, the tips of wings during forward flight trace in air an elongated figure of eight '8' (Fig. 6.26a). Steering during flight is effected by differential action of wings on the two sides. The Orthoptera are usually incapable of steering and therefore dash in a straight line. Hymenoptera and Lepidoptera often steer effectively by abdominal movement. Coleoptera use their elytra as rudders. By rotating the wing axis many flies achieve quick steering when in flight. Dipterous flies have only one pair of wings and round shaped organs on thorax called halteres. The halteres beat like a normal wing during flight but do not produce any lift like wings do. The halteres move as the flight direction changes, and prevent the insect from losing its direction. Some authors have considered halteres of Diptera as steering organs, while most others consider them to organs of balance. Turning is induced by a change in vibration of the wings on two sides (Fig. 6.26b). Some insects like hover flies and dragonflies remain at a fixed position in air during hovering, and vibration caused the creation of a low-pressure zone above the insect, which is just sufficient to balance its body weight against gravity. They change inclination of the figure '8' relative to body (Fig. 6.26c). During hovering in front of insect no low pressure zone is produced resulting into no forward drive. Some insects are able to revert their flight suddenly, which is effected by varying the amplitude of vibrations on the two sides (Fig. 6.26d). In dragonfly two wings are cross-placed on a body. No matter at what speed or direction it is already moving, it can immediately stop and start flying in the opposite direction. It can accelerate up to a speed of 40km/h that is quite surprising for an insect.

The wings of some insects are vibrated according to the electric signals conducted by the nerves, e.g. in a grasshopper each one of these nerve signals results in contraction of the muscle that in turn moves the wing. Two opposing muscle groups, known as "lifters" and "sinkers", enable the wings to move up and down by pulling in opposite directions. Grasshoppers flap their wings twelve to fifteen times a second but smaller insects need a higher rate in order to fly. Honey bees, wasps and flies flap their wings 200 to 400 times per second this rate goes up to 1000 in sand flies. The wing-beat frequency and speed varies tremendously in different insects. Various external and internal factors also have an influence on these. In some Diptera and Hymenoptera wing-beat exceeds considerably, as in males of the fly, *Forcipomyia* (Fam. Ceratopogonidae) the frequency is 988-1046/ sec. (Sotavalta, 1953). The values of wing-beats per second and flight velocity of some insects are given in the table 6.2. The beating provides both lifting force and propelling power. In honey bees wing-beats are not only responsible for flight but also for stationary condition, hive ventilation and for the dissemination of the odour.

Table 6.2. Insect Wing-beat frequency and flight speed.

Insect	Wing-beat/ second	Flight speed km/hr
Dragonfly	20-28	25
Beetles	46-90	5
Butterflies	9-12	9
Hawk moth	70-85	18
Mosquito	300-550	32
Honey bee	200	22
Wasp	110	9
House fly	250	7
Bumble bee	130	12.3
Hornet	100	20.6
Horse fly	96	14

6.5.1 Temperature regulation in flight

The overall thermoregulatory strategy in insects involves both behavioural and physiological components. As insects are poikilothermic, pre-flight warm-up behaviour is also important in many groups and they increase thoracic temperature during warm-up. Insect flight is supported by a set of biochemical machinery that must be maintained within a relatively narrow temperature range for efficient operation. The ability to maintain the particular body temperature permits flight in a rather wide range of environmental temperatures. Honey bees forage at ambient temperatures ranging from 7 to 46°C while maintaining a thoracic temperature of about 31 to 32°C. Similarly, some butterflies fly at low temperatures, but maintain thoracic temperatures near 37 to 38°C. The bumble bees while perching on a flower maintain the thorax at a temperature of 32°C by shivering, but do not waste energy in heating the abdomen, which is not involved in powered flights (Heinrich, 1979).

During flight, the rate of heat production is necessarily high because of the intense and continuous muscular work. The rate of heat production depends upon the work performed in flight. In tobacco hornworm, *Manduca sexta* during flight at low ambient temperature, thoracic temperature may be 20°C higher than the abdominal temperature, and 23°C higher than the ambient temperature. But at high ambient temperatures, the abdomen can become almost as warm as the thorax and effectively operate as a thermal radiator. At higher temperatures, thoracic heat is transferred to the abdomen by movements of the haemolymph. Besides, the thorax is highly insulated by a thick coat of scales and is thermally isolated from the abdomen by an insulating air sac. At low temperature the heart and ventral diaphragm beat at fairly low rates, but at high thoracic temperatures their action increase leading to fast flow of haemolymph between thorax and abdomen. From uninsulated abdomen the heat is radiated to the environment. The dragonfly, *Anax* appears to rely on a similar mechanism. When the dragonfly's dorsal heart is tied off, heat transfer from the thorax is abolished.

6.5.2 Energy supply in flight

Flight also involves rate of energy consumption and rapid supply of oxygen. The flight fuels for insects are : sugars (mainly glucose), lipids (diglycerides) and amino acid (proline). Most insects are probably capable of using more than one energy substrate during several of their activities. In some insects during fight energy consumption is very little and they are carried to great heights and over considerable distance by natural air movements (Rainey, 1958). The rapid metabolism of flight muscles and modifications of the tracheal system are valuable for constant supply of oxygen and energy during flight.

The fat body is the major site of energy storage. During flight the fuels are transported mainly from the fat body, found mostly in the abdomen, to the thoracic muscles. The lipids when used as flight fuel are capable of making long-range flights, whereas those that utilize sugars are capable of only relatively short periods of continuous flight. However, sugars offers advantages as the energy from glucose can be realized much faster than the energy from a lipid. It is generally true that the most rapidly metabolizing insects oxidize sugar. Some insects of the orders Orthoptera and Lepidoptera begin using carbohydrates within flight muscles, then switch to lipids and mobilized reserves from the fat body.

Chapter 7

Navigation

The movement of insects is one of the most significant characteristics and ensures their survival. Most insects are active within the so-called 'home range' and move for performing various life activities in a given area. Such insects usually cover short distances. The term homing indicates a return to a place where an animal lives and has its shelter, den nest. Navigation occurs when an animal has to rely on local cues to calculate the goal direction, which is selected by means of a compass, no matter whether it is over a familiar area or not (Papi, 1992). The movements of insects may be trivial or migratory. In migratory movement insects tend to reach a particular region or area to occupy a new habitat and mostly cover long distances. Some of them also show return movement and come back at original site from where they have started their movement. Both trivial and migratory movements require some degree of orientation and some ability to navigate. The phenomenon of navigation in insects is rather complex but fascinating because they do many things without any mechanical aids and with little or no learning. However, the ability requires time and energy. The insects must develop such a physiological and behavioural mechanism that makes the process possible.

Although much has been learned about how insects find their way around, we are still far from detailed understandings of this remarkable collection of behavioural patterns. It is rather easy to study the orientation and navigation in insects covering short distances between their nests and feeding sites. However, the problem is more complex in long-range migrants and in some cases it remains a complete mystery. Most studies on these aspects had been conducted by taking hymenopterans particularly bees, wasps and ants as test insects and the literature on the subject is widely scattered.

Within the home range some species particularly those that construct some sort of nests may need to navigate with considerable precision, whereas others need only to recognize the boundaries of their home range. Taking this into consideration Lindauer (1976) suggested two types of navigation, viz. 'short-range navigation' in which insect locates and identifies the goal itself, and 'long-range navigation' by which insect reaches the neighbourhood of the goal.

A number of means are used for maintaining an appropriate course during navigation and various cues, viz. celestial (sun, stars, moon, and polarized light), landmarks, electromagnetic field, chemicals and others are valuable. To be sure that a particular cue is used as a compass, it is necessary to show that the insect can detect the phenomenon and that it can use it for orientation under natural conditions. The mechanism varies form species to species and many possess more than one compass system.

7.1 Light compass reaction

The homing of ants involves orientation at an angle to the direction of stimulus as indicated in the light compass reaction. The ants are guided, in part by the direction of the sun. If the apparent direction of the sun is changed by means of a mirror, the ants also change the course of their movement accordingly (Santschi, 1911). Brun (1914) observed that if the common black ant, *Lasius niger* was kept in a dark box for a certain period in the middle of their homeward journey, it travelled in a new direction from its original path and maintained the same angle to the sun when released (Fig. 7.1). Jander (1957) demonstrated that the ants compensated for the movement of the sun and continued in the same direction upon being released. He suggested that ants could utilize the polarization pattern of the blue sky. They can remember a direction in relation to a light source for at least five days. They also have an ability to calculate the time and can adjust their movement in response to the direction of the sun and time. The result, therefore, was contrary to that of Brun (1914). Similar time-compensated compass reactions have also been observed in the beetle, *Geotrupes stercorosus* (Birukow, 1953) and pond-skater *Velia currens* (Birukow, 1956). Collett *et al.* (1999) suggested that desert ants, *Cataglyphis* sp. monitor their position relative to the nest using a form of dead reckoning known as 'path integration' (PI). They do this with a sun compass and an odometer to update an accumulator that records their current position. Ants can use PI to return to the nest. They can also use PI to approach a previously visited food source.

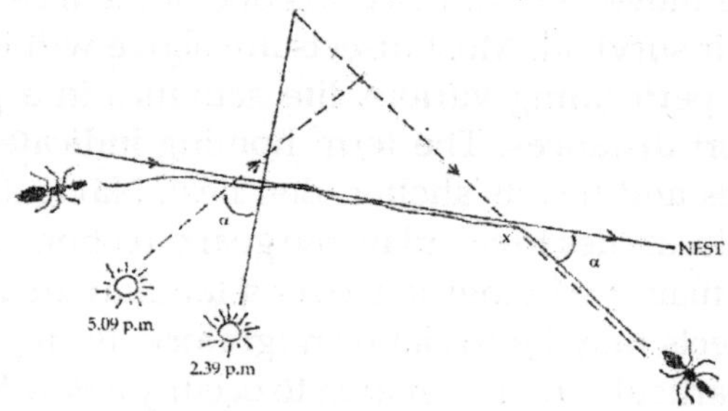

Fig. 7.1. Light-compass response in the ant, *Lasius niger* returning to her nest. On reaching x the ant was kept in dark for 2.5 hours. It was released again and observed deviating from its earlier path by the same angle as that through which the sun had moved during the period (modified after Brun, 1914).

The honey bees are skilled navigators as demonstrated by their ability to return to hive after an outward journey in search of food. Frisch and Lindauer (1954) trained honey bees to visit a food source 180 m northwest of the hive. The colony was then shifted overnight to a new territory and bees were provided four feeding choices in different directions. They observed that despite a change in sun's position and in absence of any familiar landmark most bees came to the feeding site in the training direction (Fig. 7.2). Thus, the bees are able to calculate the position of the sun according to the time of the day and orient themselves accordingly (Meder, 1958). A number of other insects show their response to

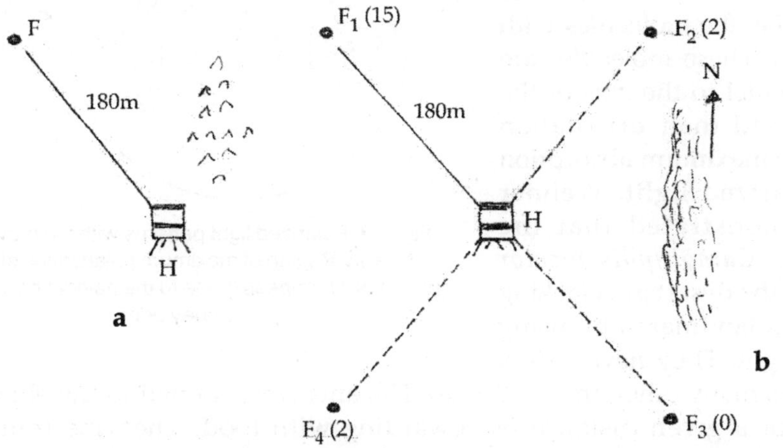

Fig. 7.2. Time-compensated sun orientation in honey bee. (a)- Bees were trained for food (F) at a distance of 180 m, (b)- During night the hive (H) was displaced at other locality and in the morning bees were provided four food sources at the same distance in four directions. Most bees were observed to move in the direction for which they were trained. Numbers in parentheses indicate number of bees visited the feeding sites (modified after Frisch and Lindauer, 1954).

light and the same may be used for navigational purposes.

Ugolini (1985, 1986) discussed homing ability in workers of wasp, *Polistes gallicus*. The workers were displaced from their nests in plexiglass test tubes, which allow sight of sun and sky and then released at various distances. They were able to correct initial orientation towards home. Beugnon (1987) studied time-compensated celestial orientation in swimming cricket, *Pteronemobius lineolatus* by varying the time differences between capture and release. He observed that crickets were able to compensate time for the apparent azimuthal change of sun's position.

7.2 Polarized light

The polarized light pattern of the sky provides remarkable navigational cues to a number of insect species. This pattern varies with position of the sun or with the orientation of the plane of a triangle formed by the sun, the observer and the point observed. The polarized pattern shifts around the celestial hemisphere as the sun moves across the sky (Fig. 7.3), and provides a compass guide to some navigational insects. Karl von Frisch (1949) demonstrated for the first time the ability of honey bees to utilize polarized light. Wehner *et al.* (1975) determined the mechanism of polarized light perception. In the compound eyes of insects

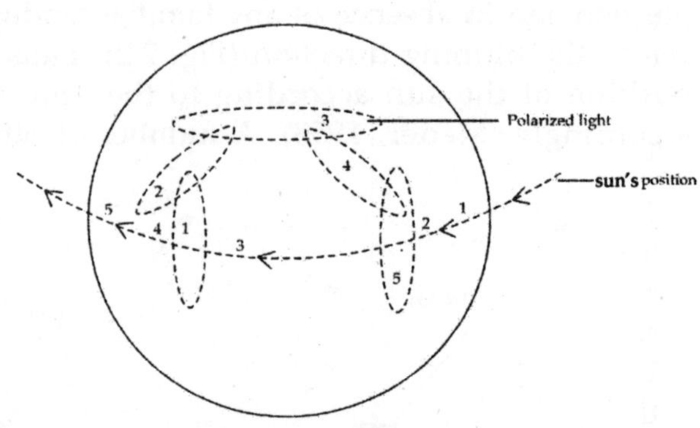

Fig. 7.3. Polarized light patterns with the movement of the sun. Region of maximum polarization shifts and provides a compass guide to the navigating insects like honey bees.

there are no polarized crystals; however, the visual pigment rhodopsin is present in the microvillar membrane of the visual cells in the form of dipolar molecules, i.e. the molecules with distinct axis. These molecules are aligned parallel to the axis of the microvilli and their orientation result in the maximum absorption of the polarized light. Wehner (1976) demonstrated that the desert ant, *Cataglyphis bicolor* navigate in the desert area lacking conspicuous landmarks by using polarized light. They never use a scent trail as many other ant species do. The ants were trained to travel in a given direction for a given distance by rewarding with food. Then the trained ants were released in other area and it was observed that they would set off in the home direction covering a distance to which they were trained. The ants lost their ability to detect polarized light at wavelengths more than 410 nm, i.e. the range visible to man. He concluded that the polarized pattern is least affected by atmospheric disturbances and is more stable, and the process of natural selection has enabled them to use such pattern as an orientation signal. Fent (1986) suggested that ants also use additional cues like degree of polarization, colour or intensity to find its way to home when sun is obscured.

Other insects like *Drosophila* and *Sarcophaga* also respond to the plane of polarization of the light and change their direction of movement if the polarization plane is rotated (Frisch *et al.*, 1960). In the flesh fly, *Sarcophaga* the dorsal ocelli and in larval *Neodiprion* the lateral ocelli are able to determine the plane of polarization (Wellington, 1953). Wellington (1974) noted that if the workers of bumble bee, *Bombus terricola* are provided with polarized light overhead they fly directly back to their nest in a straight line, and when the polarized light failed they follow a zigzag course dependent on landmarks. The dorsal ocelli were important in navigation when the light intensity was very low.

Rossel and Wehner (1982) demonstrated how the bee's celestial map could be derived from *e*-vector patterns in the sky. They proposed that the bees rely on a generalized celestial map that is used invariably throughout the day. Wehner and Rossel (1985), Rossel and Wehner (1986) discussed the use of celestial pattern of polarization as a compass by honey bees. The mechanism involves the transformation of polarization information into modulations of perceived brightness, while the bees scan the sky by rotating its field of view. Using this technique the bee needs only a simple strategy to read compass information from the polarization pattern in the sky. Schwind (1989) shown that many insects can orient to water by reflected polarized light.

7.3 Moon and stars

The moon and stars have also been reported providing navigational cues to some nocturnal insects. But they are not as prominent as the sun for diurnal species. A number of night-flying insects show their positive response to the light. Moon compass orientation has been observed in ant, *Monomorium salomonis* (Santschi, 1923) and wood ant, *Formica rufa* (Jander, 1957). The wood ant workers transport their broods into the nest during full moon. Similarly, the water boatman, *Trichocorixa reticulata* also shows moon compass orientation. The common yellow underwing moth, *Noctua pronuba* uses moon's azimuth and does not compensate for shift in the moon's position during night (Sotthibandhu and Baker, 1979). Mellor and Hamilton (2003) reported that males and females of sand flies, *Lutzomyia longipalpis* can discriminate between different wavelengths at different intensities and thus have true colour vision. They may be able to navigate at dusk or under moonlight or starlight conditions using light in the blue-green-yellow part of the spectrum.

Fig. 7.4a,b. Dung roller beetle rolling a dung ball. After locating a source of fresh droppings, it forms a ball of dung and rolls it off at high speed to escape competition at and around the dung pile. The beetle pushes the ball and on reaching the nest site he places the ball into it.

Dacke *et al.* (2003) demonstrated that African dung beetle, *Scarabaeus zambesianus* begins to forage for a fresh pile of dung around sunset. The receptors used to detect the skylight are found in the ommatidia of the dorsal rim area of the eye, and are well-adapted for polarized light detection at low light levels. After locating a source of fresh droppings, the beetle forms a ball of dung and rolls it off at high speed to escape competition at and around the dung pile (Fig. 7.4a,b). The beetle pushes the ball along with its back legs while walking with its front legs and face down to the ground (Plate 7.1). At sunset, the beetle is able to orientate itself using the polarization pattern formed around the setting sun, but when the sun is 18° beyond the horizon, known as astronomical twilight, the pattern is lost. Then the moon creates a new polarized path in the sky. It lasts for as long as the moon is in the sky. The beetles are able to use the polarized light of the moon to navigate. On nights when there was a moon, the beetles continued to forage and roll their dung balls in a straight line. On moonless or cloudy nights the beetles could not maintain a straight path. To determine if this continued ability to forage after astronomical twilight is a result of the polarization of moonlight or of the moon itself, they placed a polarizing light filter over a ball-rolling beetle feeding inside a ten-foot diameter arena. With a change in the pattern of polarized moonlight by 90°, the beetles also changed their course by 90°. The results indicated that indeed the beetles were using the polarized light of the moon to navigate.

7.4 Magnetic fields

Magnetic fields seem to provide navigational cues to some insects. The insects using such cues navigate efficiently during day and night. Their compasses are based upon the features of the geophysical environment, e.g. magnetic field of the earth. Lindauer and Martin (1968) demonstrated that honey bees respond to earth's electromagnetic field. Gould *et al.* (1978) discovered the presence of magnetic material in abdomen of honey bee.

Baker and Mather (1982) demonstrated that the orientation of common yellow underwing moth, *Noctua pronuba* is effected by moon, stars and magnetic field of the earth, and if moths are placed in a cage with reverse field there occur corresponding reversal in the orientation. The heart and dart moth, *Agrotis exclamationis* also use the geomagnetic field to calibrate a moon compass (Baker, 1987). Etheredge *et al.* (1999) tested fall migratory monarch butterflies, *Danaus plexippus* for their directional responses to magnetic cues under three conditions, magnetic, normal, and reversed magnetic fields. The butterflies showed three distinct patterns. In the absence of a magnetic field, monarchs lacked directionality as a group. In the normal magnetic field, monarchs oriented to the southwest with a group pattern typical for migrants. When the horizontal component of the magnetic field was reversed, the butterflies oriented to the northeast. In contrast, nonmigratory monarchs lacked directionality in the normal magnetic field. The results were direct demonstration of magnetic compass orientation in monarch butterflies. Banks and Srygley (2003) reported that leaf cutting ants, *Atta colombica* use trail following to travel between foraging sites and the home nest. However, this combination of pheromone and visual cues is likely to be complemented by a directional reference system such as a compass, used not only when foraging but also during colony formation, where foraging trails degrade or where ants become displaced. One candidate system is the magnetic polarity compass. They tested the orientation of leaf cutting ants under a magnetic field of reversed-polarity and observed that when the sun's disc was unobstructed by clouds, magnetic cues were not used to orient. However, when the sky was overcast, the ants shifted their mean orientation. The results demonstrate that fungus-growing ant, *Atta* respond to magnetic reversal in the absence of sunlight cues, and suggest a role for magnetic cues in determining direction during orientation.

7.5 Landmarks

A large number of insects use various landmarks as navigational cues. Simple landmarks like bushes, trees, rocks are more conspicuous for ground insects, whereas, others like road, contour, river, forest edge, railway line, telephone wires, etc. provide significant directional cues to long distance fliers. Many nest-dwelling ants and wasps identify the objects in their terrain and memorize their location. The classical experiment on homing in digger wasp, *Philanthus* as studied by Tinbergen and Kruyt (1938) provides a good example (section 2.2.6). The wasp located the nest entrance by reference to nearby landmarks, not by the sight or smell, etc. She notes the position of her nest relative to these landmarks before her departure, and in fact, the departing wasp makes a special survey flight for that purpose. Moreover, it implies that when she returns she uses the remembered position of her nest relative to those landmarks to guide her flight path to a landing site. A record of the position of one or more points of interest relative to the positions of landmarks or salient terrain features is a 'map'. A map that resides in the brain of wasp is called a 'cognitive map'. Thus, the results of Tinbergen and Kruyt's experiment suggest that the wasp makes a cognitive map of the immediate vicinity of her nest and uses that map to choose a landing site. Tinbergen and Kruyt experiment did not however, shed light on how the wasp navigated over the large distances covered by her foraging. They did not show how she found her way back from the place where she captured her prey to the vicinity of her nest. Thorpe (1950) did some experiments that suggested that digger wasps, *Ammophila pubescens* relied on a cognitive map of their large-scale environment to find their way to the vicinity of their nest from wherever they happened to capture prey. He worked with a species of digger wasp that preys on caterpillars too large to be carried home in flight. The female of this species drags her prey home across the gravely ground. He placed obstacles in her path that forced her to deviate from her course. As soon as she had cleared each obstacle, she resumed her course. This proved that the wasp was not marching blindly and when forced to deviate from her course, she corrected for the deviation she had made. The ability to use currently perceived terrain features to set a course for an area generally requires the use of a map (also called a chart), which specifies the position of destination relative to the positions of the landmarks. It also implies the ability to do navigational computations utilizing the data on the map. Thus, these results suggest the brain of an insect has computational capabilities analogous to those found in a 'Geographic Position Sensor' (GPS).

Gaul (1951) gave detailed description of orientation flight in workers of wasp, *Vespula*. When the wasp leaves the nest it makes about 20 side-sweeping flights around the nest. Again it returns to the nest and then starts a second orientation flight and the same may also be repeated for the third time. After second or third flight it flies abruptly away from the nest for foraging. Upon returning to the nest the wasp generally makes several sweeping flights to locate the nest entrance, after 15-20 such return flights the nest locality is learnt and wasp thereafter flies directly to the nest entrance. Heran and Lindauer (1963) observed that honey bees never diverted away from their flight-route by crosswind and compensate the wind by keeping the body oblique to the wind. The bees were trained to fly across a lake to a food source, and they fail to this allowance for drift due to absence of visual cues. When they were provided with visual markers floating over water, the bee tried to follow these markers.

The migratory cabbage white butterfly, *Ascia monuste* uses the visual guides when on flight. It makes a few rectilinear flights along the coast of groups of flowers and then finally flies straight (Neilsen, 1961). Hölldobler (1980) observed that in ant, *Paltothyreus tarsatus* celestial cues have not detectable effect in precise homing, but canopy patterns have a major influence in the home range orientation. Oliveira and Hölldobler (1989) described that the giant tropical ant, *Odontomachus bauri* employs canopy orientation during foraging and homing. It can also utilize horizontal visual cues and chemical marks in localizing the nest entrance. Workers of the ant, *O. bauri* use trail pheromones for orientation and recruitment of nestmates. However, chemical markings may not always be sufficient for successful navigation in complex three-dimensional terrain, and additional orientation cues may be required. Ants, which were prevented from seeing the canopy, were unable to navigate back to the nest, even though trail pheromones were still present. In contrast, foragers found their way back to the nest after their trail pheromones had been abolished but their visual scenes remained unchanged. Baader (1996) emphasize the important role of visual landmarks (Baader, 1996) in homing of the foragers of the ant, *Paraponera clavata*. Fukushi and Wehner (2004) reported that wood ants, *Formica japonica* could steer their outbound (foraging) and inbound (homing) courses without using celestial compass information, by relying exclusively on landmark cues.

Gould (1986b) discussed landmark learning in honey bees. He proposed that the bees are able to construct a 'mental map' of their home area and use this knowledge of the spatial relation between landmarks to navigate along novel routes. When a bee leaves her nest for the first time, she takes short flight and memorizes landmarks around the hive. She does not collect any nectar or pollen and a few such flights are enough to learn all the features of nearby area. If several hives are present in a locality the bee recognizes its own hive by colour, shape, odour and location. Honey bees, *Apis mellifera* forage from a fixed location, the hive. This type of foraging, often termed 'central-place foraging' imposes a requirement for good navigation abilities.

Information used in home-based navigation can be divided into two categories:

(i) **Egocentric** (self-centred) information, which is generated internally by the insect and is independent of its immediate surroundings. Internal calculations of the distance and direction travelled, which are used in path integration, are examples of egocentric information. The positional information is obtained by some kind of route integration (dead reckoning), which provides the insect with a continuous representation of its spatial position relative to its starting point (home).

(ii) **Geocentric** (terrain-centred) information, which includes landmarks and any map information. The insect gains positional information from the relative location of its start within the environmental framework of its home-range area.

Honey bees primarily use path integration (PI) in making their way to and from foraging sites (Wehner *et al.*, 1996). Dance information provides outgoing bees with a distance and direction to be travelled. Flight direction is determined by a sun compass orientation and distance of flight by an internal 'odometer' that measures the rate at which visual images pass through the eyes. Other inputs, such as odours, provide supplementary information. Once a route is learned, bees incorporate visual landmarks when they make repeat trips to a foraging site. How do bees remember landmarks? For this two models have been proposed. The simplest 'snapshot model' calls for the bee to remember a series of visual images (snapshots) of the landscape as it passes. The bee also remembers images of particularly prominent landscape features. These images can then be compared with the actual landscape surrounding the bee at any given moment. Ultimately, the bee may be able to use the landmark snapshot information in an allocentric manner, projecting its position when it is displaced to an unfamiliar location but still in view of an array of familiar landmarks (Dyer *et al.* 1993). More complex is the 'cognitive map model', which requires the bee to construct a relatively complete neural representation of the landscape based on its experiences while flying.

Menzel *et al.* (1998) proposed that navigation in honey bees arises from a rich repertoire of behavioural routines that are intimately connected by the motivational state during foraging, incidental and associative learning, and selective memory retrieval. The results of their study provide evidence for:

(i) The association of home vector memories with specific locations of feeding sites.

(ii) The dependence of memory retrieval on motivation.

(iii) The selection of a novel flight direction towards the hive, if two home vector memories are simultaneously activated.

Menzel *et al.* (2000) showed that honey bees also possess an allocentric form of spatial memory, which allows localization of multiple places relative to the intended goal, the hive. The egocentric route memory, which is also called the specialized route memory (SRM), initially dominates navigation when an animal is first trained to a feeding site and then released at an unexpected site. However, the SRM can be replaced by an allocentric spatial memory called the general landscape memory (GLM). The GLM is directly accessible to the honey bee if no SRM exists, for example, if bees were not trained along a route before testing. Under these conditions bees return to the hive from all directions around the hive at a speed comparable to that of an equally long flight along a trained route. The flexible use of the GLM indicates that bees may store relational information on places; connections between landmarks and the hive and/or views of landmarks from different directions and thus, the GLM may have a graph structure, at least with respect to the hive.

Collett *et al.* (2002) suggested that bees seem to use landmarks to segment familiar routes. They can associate, with a landmark, a memory that encodes the direction and distance of the path segment between that landmark and the nest. The expression of the memory results in the performance of a local vector matching the distance and direction of the path segment. The memories of path segments appear to be 'chained' together, so that the performance of one local vector is sometimes sufficient to elicit the subsequent local vector, even in the absence of the associated landmark.

7.6 Chemical cues

Many insects use chemical cues in orientation to food source, their return movement and communication. They make scent trails, which provide effective means of navigation. For good many years it has been known that chemical sign spots guide the individuals. Many terrestrial forms form a trail in continuous or intermittent lines on a solid substrate, while some others moving on wings also make trails.

A number of ant species lay trails between nests and foraging sites with certain glandular secretions from Dufour's gland, hindgut, poison gland, Pavan's gland or metathoracic legs, which can be utilized by themselves and their mates as aid to orientation. The sensilla with which ants perceive trail substances are mainly located on antennae. Attygalle and Morgan (1985) suggested that a foraging ant worker returning to the nest lays a narrow band of chemical on the substratum 'recruitment trail' which may excite other workers to follow the same. Trail chemical also facilitate migration of the colony to a new site 'emigration trail'. Some long-lasting trails may serve as chemical cues in home range orientation in marking home territories. The subject has been discussed in reviews published by Carthy (1952), Hölldobler (1978), Wilson (1963), Parry and Morgan (1979) and Attygalle and Morgan (1985).

The worker of imported fire ant, *Solenopsis saevissima* after finding a food source leaves minute streaks of trail pheromone from Dufour's gland, which activates and guides other workers (Wilson, 1962). Bossert and Wilson (1963); Wilson and Bossert (1963) investigated the odour field of fire ant. The trail substance is rather volatile and diffuses within two minutes and the workers can follow it only for the distance they can travel in this time. It provides two advantages, first the old trails do not confuse the hunting workers, and second the intensity of trail laid by many workers gives a sensitive index of the amount of food at a given site and the rate of its depletion. Similar observations have also been made in tropical fire ant, *Solenopsis geminata* (Hangartner, 1969b). The Florida harvester ant,

Pogonomyrmex badius also secrete a trail making pheromone from Dufour's gland (Hölldobler, 1971b). On the contrary, the trail formed by the ant, *Tetramorium bicarinatum* last for days (Blum and Ross, 1965). The subterranean ant, *Acanthomyops interjectus* emits the chemical through the anal opening by pressing abdominal tip intermittently against the substrate (Hangartner, 1969a), and workers of *Crematogaster ashmeadi* lay trail from a gland located in hind tibia (Leuthold, 1968). Jaffe *et al.* (1990) suggested that ant species use both visual and chemical cues for homing.

Lindauer (1971) observed that workers of the bee, *Trigona postica* on returning from a food source to the hive land intermittently on stones, etc. and mark them with secretion from their mandibular glands. Perhaps the worker bees follow the trail in a way that is analogous to terrestrial trail. Similarly, workers of the bee, *Trigona subterranea* also produce a trail-following pheromone from mandibular glands (Blum *et al.*, 1970). Butler *et al.* (1969) observed that workers of honey bees secrete a pheromone that stimulates the foragers to enter the hive; and in common wasp, *Vespula vulgaris* another pheromone attracts homecoming wasps.

In addition the scent from other sources may also be helpful in navigation and homing. Lindauer (1976) discussed that the honey bees are also able to recognize particular flower scent by olfactory receptors of the plate organ of the antennae. The source of odour can be detected either klinotactically, i.e. by the successive comparison of different concentrations emanating from the centre or tropotactically- by simultaneous comparison of different concentrations with two antennae. The bees may be helped by tropotactic orientation to find the actual nectar source.

Chapter **8**

Migration

Most insects are highly mobile creatures and have tendency to move from one place to another. They are able to walk quickly as in case of their ancestors; however, the additional ability to fly is perhaps their most significant attribute. Their mobility has enabled them to occupy a great diversity of habitats and various modes of life.

When we consider movement of insects in a broad sense or in terms of population, there may be passive or active movements. The movement permits a population to colonize new sites, to explore new food resources, and to avoid overcrowding and unfavourable conditions. In addition, it also provides a mean to exchange genes, promotes chances of better survival and plays an important role in their geographical distribution. The insects move in response to environmental changes, population pressure and due to some physiological and behavioural reasons. In ecological sense the individuals may move into or out of the population and such movements have been categorized into three types :

(i) *Immigration*. It is inward movement, which lead to rise in population level.

(ii) *Emigration*. It is outward movement resulting in depopulation.

(iii) *Migration*. It involves mass two-way movement of entire population.

However, there are controversies when we apply this definition of migration in insects, as all these three processes are closely interlinked with each other and play a significant role in the determination of their numbers.

8.1 Dispersal

Usually the dispersal of insects occurs either passively or under their own volition, but in some cases other factors may also be responsible. The passive transport occurs by wind, water, human agencies, phoresy and locomotion. One-way travel has been described as dispersal by many investigators. However, the direction and other activities during movement are also important. Schneider (1962) defined dispersal in a spatial sense as the "lengthening of the mean distance between neighbouring individuals"; Crumpacker and Williams (1973) considered it as "movement in a randomly selected direction in the absence of specific geographical information."

Passive dispersal by wind is of considerable importance in the distribution patterns of nearby all insects. For delicate insects light wind is an effective means of wide dispersal, but even heavy insects are also carried to a great distance by strong winds. Man has also played a valuable role in the insects' dispersal either through accidental introducer or through deliberate transport of species. *Periplaneta americana, Cimex lectularius, Sitophilus oryzae, S. granarius, Leptinotarsa decemlineata, Carpophilus hemipterus, Ephestia kuehniella, Acanthoscelides obtectus, Bruchus pisorum, Lasioderma serricorne,* etc. are some common insect species which have become widely dispersed throughout the world by human agencies.

Stinner *et al.* (1983) discussed dispersion of insect pests. If the pest insects are introduced in previously pest free area dispersal may be followed by damage. Chemical control operations of a pest may also lead to an unbalanced dispersal. In spruce budworm *Choristoneura fumiferana* population outbreaks do not occur in discontinuous forests, because of high dispersal losses suffered by early instars (Morris *et al.*, 1956). Fletcher (1974) released marked adults of Queensland fruit fly, *Bactrocera tryoni* in N.S.W., Australia and they were again trapped at a distance of 24 km from the release site within three weeks. The fruit flies, *Bactrocera cucurbitae, B. dorsalis* and *Ceratitis capitata* disperse ranging from 24-45 miles (including distances up to 40 miles over water). Mosquitoes move from their breeding places to hosts and come back to suitable pools for oviposition. The dispersion of many mosquitoes to human habitations seems to be directed by topography and wind carrying human odour.

Many flightless insects are dispersed by other organisms and the phenomenon is called 'phoresy'. A wide variety of small insects ride upon other insects. The parasites disperse with their hosts. The bird lice attach themselves to the bodies of hippoboscids parasitizing birds, thus being carried from one host to another (Fig. 12.3). In parasitic wasp, *Dimorphothynnus haemorrhoidalis* the male carry small wingless female and inseminate her while flying. It has been termed as 'phoretic copulation' and provides effective dispersal of female into new areas (Evans, 1969). The human warble fly, *Dermatobia* oviposits on other insects particularly mosquitoes, ticks which transport eggs to the hosts.

8.2 Dispersion versus migration

A number of insects travel by active means, in which movement is determined by insect itself and leads to a change in habitat. Such movements are usually seasonal or periodical. When we consider both dispersion and migration together, it is rather very difficult to define the exact boundaries of these two terms. There occur movement in both cases and the views regarding these have changed markedly after 1960. Since the publication of Southwood's (1962) review, and Johnson's (1966, 1969) review and monograph, the study of such movement has received increasing attention. Johnson (1966) considered insect dispersal as a confused subject, which lacks attention. This confusion is not entirely the result of diversity of insects but also due to ill-defined concept of dispersal. General assumptions about dispersal (and migration, with which it is linked) based on the result drawn by studying only a few insect species are completely specialized rather than general. It is extremely difficult to measure how much dispersion occurs accidentally, incidentally or adaptively. In his 1966 monograph he stated that the dispersal of plant seeds is involuntary and the dispersal of even the feeblest insect depends on a specialized behaviour pattern. They too, can therefore be characterized as migrants, not only ecologically but also behaviourally. Kennedy (1975) also argued in favour of Johnson's thoughts. He was rather reluctant to use the term dispersal. Evans (1984) stated that dispersal and migration should not be considered as different phenomenon. Dispersal is a general term for all forms of displacement to new habitats. Migration is usually considered a special form of dispersal. In practice also, it sometimes proves arbitrary as to whether a particular insect is said to be dispersing or migrating. Johnson, Kennedy and Southwood do not distinguish between the two, except to consider migration as a set of adaptation to ensure that dispersal occurs under specific conditions.

In earlier literature a distinction was made between migration and other forms of movements. But in recent years it has become increasingly accepted that this distinction is artificial and that dispersal and migration are one in the same thing (Baker, 1981). Dispersal is a term that is inclusive of migratory and trivial flight (insects can disperse by either behaviour or by random locomotion). Dispersal can be accomplished by populations and not individuals (Southwood, 1981).

Vegetative movements of insects differ from migration in that they are directed toward exploitation of resources, particularly required for growth and reproduction and these movements are readily interrupted by an encounter with the resource item. They often tend to be 'station-keeping' and retain the insect within a locality (Dingle, 1996). If the required resource is not available, an individual may switch to a different mode of behaviour which may allow it to explore a wider area than its local habitat patch. Such movements are sometimes called 'ranging' and tend to cease if the resource is found.

Thus, the term dispersal is now considered as a synonym for migration, and its use should be discouraged because it confuses individual behaviour with the consequences for the population. Nevertheless, some authors have preferred to retain the term dispersal, partly because it is not always possible to know the behavioural state of an individual insect during a particular movement. Dispersal is perhaps best reserved to describe the increase in separation between members of a population. A more neutral term such as 'population redistribution' has also been used to describe the ecological consequences of movement, because some so-called 'dispersals' may well result in aggregation of population, rather than spatial spread, e.g. the bringing together of African armyworm moths, *Spodoptera exempta* and other airborne migrants by wind convergence (Gatehouse, 1997). At present, there is no unanimity on the definitions of insect movement, but attempts to attain this should identify, exceptions which prove the rule and thus improve the conceptual framework of movement behaviour (Kennedy, 1985; Dingle, 1996).

8.3 Trivial and migratory flight

Flight is an essential component of insect movement and can be broadly classified into two categories, viz. 'trivial or appetitive flight' and 'migratory flight'.

I. **Trivial or appetitive flight.** It involves local flight of different duration and is without any specific direction. Such flights are concerned with feeding, mating, oviposition, escape, foraging and other such activities. During flight insects may show a variety of responses to various stimuli resulting into numerous pauses, and usually long distance flight may not occur; however, in some cases they may also lead to some dispersion.

II. **Migratory flight.** It involves a behavioural pattern, which appears to be motivated by the inner drive. Other activities like feeding, mating are suppressed and their stimuli are ignored by migrants. It also leads them to occupy new habitat. In most cases these are usually post-teneral flights of different duration and often followed by muscle autolysis (which may be reversible).

8.4 Migration

In migration insects travel by active means seasonally or periodically. When the migrants are in passage it is called 'trans-migration'. Earlier authors considered two directional movements as an essential part of migration. However, only a few insect species fulfil this criterion and indicate return movements also. The course of migration in case of birds, which involves periodic to and fro movements in two different habitats- one breeding and other feeding may not be applicable to all insects, because of their small size, short life span, great population fluctuations and break down of flight muscles after the migratory flight. Therefore, the return to the original home was not considered to be an essential component of insect migration.

The problem of migration in insects seems to be rather complicated and many questions are related with the subject like: the kind of insect migrate, time, route, season, direction, causes, etc. Williams (1957) summarized the evidence of migration in insects as:

(i) Steady flight of large numbers of insects in one direction.
(ii) Sudden appearance of winged insects in large number in an area, with no evidence of their local breeding or emergence, followed by sudden disappearance.
(iii) Presence of large numbers on oceanic islands, high altitudes, far out at sea.
(iv) Seasonal occurrence.
(v) Evidence as indicated by marked recaptured studies.

The earlier mentions of migrant insects are available in old literature and in many cases persons have mentioned regarding immense flocks of insects. In the beginning of past century J.W. Tutt (1902) compiles his articles on migration in the form of a small booklet. The significant contributions dealing with insect migration are: Fraenkel (1932), Williams (1930, 1957, 1958), Schneider (1962), Southwood (1962), Johnson (1963, 1969, 1974, 1976), Dingle (1972, 1978, 1985, 1996), Kennedy (1961, 1975, 1985), Rainey (1976), Baker (1978), Aidley (1981), Danthanarayana (1986), Drake and Gatehouse (1995), Gatehouse (1997) and Woiwod *et al.* (2001).

A number of definitions appeared concerning migration in insects and to mention some of them are: Williams (1958) employed it for "continued movement in a more or less or less defined direction, in which both movement and direction are under the control of animal concerned"; Kennedy (1961) considered the term refers "a persistent straightened-out movement with some internal inhibition of response that will eventually arrest it"; Schneider (1962) defined as "prolonged escape movement with a tendency to maintain the direction and to abandon definitely or periodically a habitat."

Kennedy (1985) defined it as: "migratory behaviour is persistent and straightened-out movement effected by the animal's own locomotory exertions on or by its active embarkation on a vehicle. It depends on some temporary inhibition of station-keeping responses, but promotes their eventual disinhibition and recurrence." Important characteristics of this definition are:

(i) Persistence- which means the activity of the insect is tenacious so that it carries the migrant beyond its original habitat to a new one where it gathers previously unavailable resources.

(ii) Dispersal is straightened out- which means there is directionality to the dispersal and a minimum of turning and backtracking.

(iii) Dispersal is undistracted- which means the insect ignores vegetative cues (cues to resources that could promote growth and maintenance) and keys on other cues, such as sky light that might be used for activities such as reproduction, and are employed to support migration.

Johnson (1960, 1966) considered migration as an integral part of the life-strategy of those insects that do it, whether or not they travel regularly to and fro between geographically distinct areas and regardless of the manner of their displacement or the life-stages at which it occurs. In behavioural sense it is an accentuation of locomotor function with a depression of vegetative function. Thus, the migratory insects exhibit persistent movement and usually do not respond to vegetative stimuli, e.g. food, sexual signals, etc. He also enumerated that migration is very often an exodus flight made from the original breeding site by new adults motivated to get away from the place of birth. Mass migration is the result of mass emergence and a maximum of physiologically determined flight capacity. However, there are exceptions to this rule. Kennedy (1975) mentioned three interconnected diagnostic features of migration- persistent locomotion, straightening out of movements, and the depression of vegetative responses, but there have been considerable reservations about them. He used the term 'adaptive travelling' as it may be viewed as an active mass movement adapted to displace population. He also considered it as an intense 'locomotory drive' that sets the insects off on a relatively prolonged and undistracted flight compared with other flight.

The airborne migration is one of the most common adaptations for surviving and exploiting habitat variability. Studies on migration have established that the phenomenon affects population processes to a great extent. The migratory insects tend to '*r* strategists' having high reproductive value. They are polarized towards '*r*' end of MacArthur's '*r-k*' spectrum (MacArthur, 1962).

The process of migration and reproduction are closely interlinked. In females of many species migration occurs prior to egg development and Johnson (1966) called it 'oogenesis-flight syndrome'. During this period wing muscles develop and reproduction is suppressed. However, the same was not applicable to males and hence Dingle and Arora (1973) referred it as 'reproductive-flight syndrome'. Thus, it is post-teneral and pre-reproductive movement characterized by immature ovaries, large fat bodies and usually positive phototaxis response.

Migration and diapause are also linked with each other as both involve escape form unfavourable conditions and temporary suppression of reproductive function. The two shares certain common features and influenced by same hormones. Both usually occur before reproduction and migration is being regarded as escape in space and diapause as escape in time.

Migratory movements have been mainly observed in many insects belonging to orders: Odonata, Hemiptera, Coleoptera, Lepidoptera, Diptera and Hymenoptera. In big insects it is very spectacular and can be studied easily. However, there occur problems when insects are small, fast and migrate at night. Insect migration has been studied by using many recent techniques, e.g. airborne suction traps, aircraft, aerial balloons, radars, remote sensing, radio-telemetry, video observations of movement in laboratory and field insects, by marked-recapture method, and biochemical and molecular techniques. Movements of locust swarms have been studied successfully from airplanes. Insects are being marked with labels, signs, paints, radioisotopes, and their migratory movements have been studied. Use of radars has contributed significantly to our knowledge of insect migration.

Flight is not necessarily an essential part of migration. The army ants, *Eciton hamatum* migrate on foot and its migration has been termed as 'pedestrian migration' (Schneirla, 1957a). These ants make simple nest, from which they make two different types of excursion. One is simple foraging activity and in other cases the whole colony move and the excursion ends with the setting up of a new bivouac nest at a new site. Kennedy (1975) discussed migration of aphid, *Pemphigus* sp. The migratory stage is minute first instar nymph, while the other instars and apterous adults live and feed in the roots of sea-aster plants found on mud banks in tidal salt marshes. The first instar nymphs become active, photopositive, come out and drifted by rising tide. The migration is waterborne and the nymphs colonize in new aster roots. The nymphs of desert locust had been observed migrating on foot under crowded conditions.

8.4.1 Types of migration

Johnson (1966) proposed three types of migration on the basis of adult life span.

I. Short-lived adults that emigrate and die within a season- insects fly from their birth place, oviposit elsewhere and die without returning, e.g. locusts, many butterflies.

II. Short-lived adults that emigrate and return- the adults leave their birth place, fly to feeding sites, where their eggs mature, then they return to old breeding site or nearby areas to oviposit, e.g. dragonflies; and white grub cockchafer, *Melolontha melolontha.*

III. Long-lived adults that hibernate or aestivate- adults fly from their birth place, to a resting site, where they aggregate. The return migration occurs to their original breeding site in the next season, e.g. *Danaus plexippus, Agrotis infusa,* etc.

Taylor (1986) described four kinds of migration, two of which he applied to the movements of insects.

I. Dynamic migration. It is directed movement controlled by tides or wind, with navigation abilities not essential. The desert locust, *Schistocerca gregaria* found in Africa is a good example of this type of migration. The majority of migratory insects fall into this category.

II. Homeostatic migration. In these migrations two-way movements occur in which migrants or offspring returning to breeding areas, hence they need some navigational abilities. The monarch butterfly, *Danaus plexippus* is a good example of this type of migration. The monarch butterflies migrate to overwintering sites and then migrate back toward their summer range taking several generations. Almost all the homeostatic migrating insects are lepidopterans.

8.4.2 Migration within and outside the boundary layer

I. Migration within the boundary layer. Some insects migrate within a boundary layer and fly only at a height of few feet above the ground as that of normal flight. They have considerable control over where they are going and if they go against wind, they go at lower height. Some syrphids, chafers and many butterflies have been observed flying at shorter height. The cabbage white butterflies, *Ascia monuste* in Florida is a strong flier and fly only 1-4 m above the ground. It can fly against the wind of 10 km/h and its movement is not much affected by the wind direction.

II. Migration outside the boundary layer. Many insects fly at a greater height during migration. *Ascia monuste* which usually fly inside the boundary layer have also been observed flying up to a height of 3,000 ft in Argentina. Locusts usually fly at a greater height, so that sometimes the swarm looks like faint brownish cloud. The insects crossing the mountains in the course of migration may go very high. The Alps dead silver-Y moth and cabbage white butterflies have been found at a height of 12,000 ft. Many other insects of the orders Odonata, Coleoptera and Diptera go beyond boundary layer.

8.5 Some important considerations

Start of migration. A number of factors seem to be responsible for the start of flight. The stimuli for this may be spontaneous as well as from outside. Their strength varies according to physiological state and external factors. In some cases it has been suggested that adverse conditions may not be fully responsible to initiate migration, as the monarch butterflies begin to move before the onset of cold and locusts leave their habitats while food supplies are still abundant. The teneral period between emergence and exodus flight is also very important. It determines the start of flight as during this period cuticular hardening takes place and may last for a few hours to a few days. Aphids may take-off as soon as they are mechanically or physiologically strong and the teneral period is of few hours. Dragonflies may fly, while their cuticular parts are still soft. Internal changes like growth of fat bodies and more feeding may also occur during this period. At the end of teneral period even the weak stimuli may induce flight.

Other migrants also stimulate resting locusts and butterflies. Temperature has a significant role in take-off flight and most small insects have ambient temperature thresholds for start of flight; however, larger insects are able to rise their body temperature by increasing muscular activity. Other factors like light, wind, humidity, crowding, etc. have also been reported to influence the migratory take-off.

Time of migration. The flight may occur during day (locusts and butterflies) or by night (moths) but it is not always true. Butterflies usually land by dusk and rarely fly even during night. However, they continually fly over the sea. Moths like humming bird hawk moth and silver-Y moth migrate during day and night with equal dexterity.

Orientation. It is one of most distinguished characteristics of migratory insects, i.e. while flying they have the ability to follow straight path over long distances and long periods of time. Individuals of different species migrating simultaneously in the same locality have been observed flying in different directions. In most cases the migratory flight end up after reaching to a desirable ecological goal, but in some cases they deviate from their well-defined routes possibly because of effects of wind. At the beginning it is greatly influenced by light and most diurnal and some crepuscular migrants, e.g. aphids, mosquitoes, chafers take-off with a strong positive phototaxis and are attracted by short waves of spectrum to the sky. After reaching to a particular height they move in a particular direction or blown by the wind. Chafers orient themselves toward the highest point in the horizon or to silhouettes of thickets or woods (Couturier and Robert, 1958). The cabbage white butterfly, *Ascia monuste* makes a few rectilinear feeding flights along the coast of groups of flowers and then finally a straight flight is established (Neilsen, 1961). Thus, visual guides keep the butterflies on flight. Fraenkel and Gunn (1961) suggested that many butterflies move ahead by optically fixing an object, then shift to another one in the same line and continue their flight. Locusts usually take-off heading into wind and their sensilla on body parts assist in the same, while swarming they exhibit remarkable accurate turning power and show complex orientation. Other factors like earth's magnetic field, humidity, moonlight, scent, position of the sun, contours, etc. have also been considered but need detailed investgations.

Reppert *et al.* (2004) reported that during their spectacular migratory journey in the fall, North American monarch butterflies, *Danaus plexippus* use a time-compensated sun compass, which help them in navigating to their overwintering sites in central Mexico. They suggested that the angle of polarized skylight (the *e*-vector) is a relevant orientation parameter. Monarch butterflies possess the anatomical structure needed for polarized skylight detection, as rhabdoms in the dorsal-most row of photoreceptor cells in monarch eye show the organization characteristic of polarized light receptors. The existence of polarized light detection could allow migrants to accurately navigate under a variety of atmospheric conditions and reveals a critical input pathway into the sun compass mechanism.

Speed. In some cases speed of insects during migration has been accurately measured. However, it is also governed by the speed and direction of wind. The speed is usually more soon after take-off and fall subsequently to a steady level. Locusts have maximum air speed of about 10 miles/hour and yellow butterflies, *Catopsilia florella* fly at a speed of 12-16 miles/hour (Williams, 1958). Insects like locusts and *Ascia* butterflies are able to regulate their air speed to a fairly constant level (Kennedy, 1951; Neilsen, 1961).

Distance. The distances covered by migratory insects may be small or long. In insects, which migrate upward the hills from plains, lesser distances are covered as in Himalayas some butterflies move up hill, but details of distances they cover are not yet known. Locusts have been observed covering more than 1,000 miles or more. Desert locust, *Schistocerca gregaria* a dynamic migrator, migrated westward across the Atlantic Ocean and performed the longest recorded migration. The records of *S. gregaria* found in many Caribbean islands and parts of the east coast of South America during October 1988 indicate that they flew for a distance of about 4500 km (Tipping, 1995). The longest migration with return movement has been reported in monarch butterflies, *Danaus plexippus*. The approximate distances travelled by some important migrants are shown in the table 8.1.

Table 8.1. Approximate distances covered and conditions of migration for some migrant insects.

Insect	Approximate distance covered (km)	Conditions
LocustaSchistocerca	Up to 8×10^3	Crowded
Aphids	10^{-1} to 10^2	Crowded or lack of hosts
Danaus plexippus	10^2 to 4×10^3	To and from winter hibernation site
Cynthia cardui	$10^2 \times 10^3$	Northward in spring, southward in autumn
Melolontha melolontha	10^{-1} to 10	To feeding site and back to oviposition site

Duration. The duration of migration has been correlated with the preoviposition period, and the delayed preoviposition period usually resulted in extended flight. In *Danaus plexippus* the early-matured females are not able to reach their hibernation site, but oviposit and die in the way (Urquhart, 1960). On the contrary, aphids travel for a long distance on the wind.

Number. The number of individuals participating in migratory flight varies from little to large swarms consisting of millions of individuals covering several hundred kilometres. Great swarms of locusts and butterflies have been reported. In California a flight of painted lady, *Cynthia cardui* was believed to contain about 3,000 million butterflies, while in contrast only a few individuals of the same species were also observed migrating (Williams, 1958). Huge swarms of locusts also contain good many millions of individuals.

Sexes involved. The migration of different sexes depends on mating of insects, but it always includes females. In desert locust, *Schistocerca gregaria* both sexes migrate and mate at the end of journey. Sexually immature individuals of sunn pest, *Eurygaster integriceps* migrate to hibernation site and mate after hibernation. In *Danaus plexippus* males mature soon after beginning to migrate and accompany immature females hundred of miles. Males of *Aedes taeniorhynchus* mate at exodus and fly along with females for a short distance. In *Rhyacionia* only fertilized females migrate and in aphids parthenogenetic females need no males.

Species involved. Most migratory flights include individuals of only one species; however, sometimes more than one species are involved, e.g. *Pieris brassicae* and *P. rapae* have been observed migrating together. Syrphid flies and ladybird beetles have been noted flying together in Britain. Moths and grasshoppers were observed travelling in opposite directions in the lower and upper winds, respectively.

Fuel for migration. The insects can fly as long as the fuel is available to drive the wing muscles. In long distance migrants like locusts, noctuids and monarch butterflies reserve fat is utilized for energy in the course of migration. Flight duration is also correlated with fat content of insects. Small insects like aphids with small stored fat have a short migratory period, while monarch butterflies may fly for many days on initial food reserves. Locusts at rest feed intermittently and thus replenishing the fuel required for further flight. The consumption of fuel is faster if they fly against the wind. Some insects also use carbohydrates as source of energy. Locusts and aphids use carbohydrates at the initiation of flight, but subsequently burn fat.

Termination. It is usually associated with the consumption of fuel and wing muscles degeneration, thereby preventing further flight. In many insects further migration is suppressed after feeding. In aphids flights occur repeatedly until they find suitable hosts. However, the role of environmental factors and physiological state of migratory individuals in the termination of flight cannot be ignored.

Return movement. A number of insects are known to indicate return movements. In some cases the return flight is made by the individuals of next generation. The best-studied example of a return movement is that of monarch butterfly, *Danaus plexippus*. Some other well-known butterflies showing return movement are: clouded yellow butterfly, *Colias croceus*; small cabbage white butterfly, *Pieris rapae*; red admiral, *Vanessa atalanta*; painted lady, *Cynthia cardui*; and long-tailed blue, *Lampides boeticus*. The hummingbird hawk moth, *Macroglossum stellatarum*; bogon moth, *Agrotis infusa*; and convergent lady beetle, *Hippodamia convergens* also show similar movements.

Weather factors and migration. Weather appears to be an important factor influencing migration. Factors like temperature, humidity, light intensity, and wind velocity have pronounced influence on migratory behaviour. These factors may also have a cumulative effect on migration. Insect migrations are usually completely confined to the lowest 2 km of the atmosphere, the planetary boundary layer (PBL). Migratory insects within the PBL, are subject to various weather effects. The migratory black bean aphid, *Aphis fabae* disperses with local winds. Winged adult forms take-off on strong vertical flights, attracted to the blue light of the sky (Johnson, 1963). After several hours of flight and often many kilometres from their take-off site, the aphids begin to descend, now attracted to yellow or green colours. Some insects can rise above lower air turbulence by wind currents or powered flight. Migratory flights at these altitudes can allow insects to disperse against wind directions found at lower altitudes.

Locusts are normally diurnal and the swarm flights have been recorded up to dusk in winter. They rest during night hours on any plant and become so inactive in desert area due to low temperature that they can be collected in huge numbers. After the sunrise, temperature increases gradually and they become active and fly *en masse* in a direction depending upon wind current and solar radiation.

8.6 Migration in Lepidoptera

The monarch or milkweed butterfly, *Danaus plexippus* (Fam. Nymphalidae) cannot survive extended periods of freezing temperatures, so from late August to early September, millions of adult monarch butterflies of the eastern North American population cease reproducing, become highly gregarious and begin migrating southwards. In the spring, these overwintering monarchs fly north toward their breeding range. It is one of the most remarkable migrant and the details of its migration have been discussed by Urquhart (1960, 1976) and Brower (1977). Monarchs of east of the Rocky Mountains generally fly to overwintering sites in the mountains of Central Mexico, while monarchs of west of the Rocky Mountains typically overwinter along the California coast. Thus, the wintering areas are distinctly located in two different areas of North America for eastern and western populations of butterflies (Fig. 8.1). First the butterflies migrate southward in small numbers and gradually joined by large groups. Monarchs make frequent stops during migration, forming roosts at night and during inclement weather. It has been suggested that levels of energy reserves may influence monarch migration

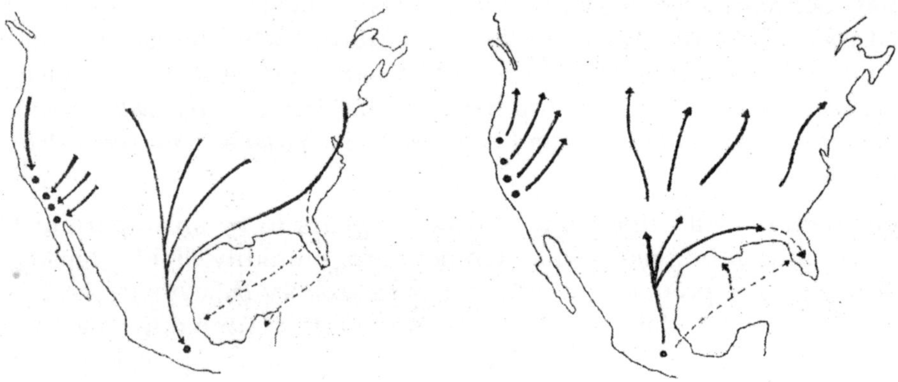

a b

Fig. 8.1. Migratory routes of monarch butterfly, *Danaus plexippus* in north America. The butterflies mainly occur in North America and southern Canada. However, more concentrated areas are in the eastern and western areas. They travel in autumn and the wintering areas for eastern and western populations are different. In these areas they congregate and pass winter in semi-hibernation. Return movement starts in spring and females lay eggs on milkweed hosts in the way. (a)- The fall Migration (broken lines indicate possible migration routes and dots indicate winter roosting sites), (b)- Spring migration.

and stopover decisions, with monarchs staying longer at stopover sites when their lipid reserves are small. Finally they reach some suitable area for passing winter in a state of semi-hibernation. In March they start return journey and arrive home about beginning of June. Upon returning, female lays eggs on milkweed hosts coming in the way. The butterflies have also invaded in European countries, Australia and other areas.

The painted lady, *Cynthia cardui* (Plate 8.1) is another widely distributed butterfly, which breed during winter along North African desert belt and move northward in spring reaching up to northern Britain or even may go far to the north. It also makes return flight to the south in the autumn. Being widely distributed it has also been reported migrating in other areas of the world.

The clouded yellow butterfly, *Colias croceus* has permanent home in Mediterranean areas. It migrates northward in spring, which continues up to July. They reach in abundance in south Britain and may extend up to southern part of Scotland. However, it is not a regular migrant and has been observed in abundance during 1877, 1947 and 1983) (Pollard *et al.*, 1984).

Red Admiral, *Vanessa atalanta* (Plate 8.2) migrate to central Europe from end of March. They form new generations here so their numbers increase until the end of September or early October when they start their way back south. In Belgium and the UK the first Admirals typically are sighted towards the end of March or early April. In the Netherlands and in Germany they first occur in April or early May, and from end of May in Estonia, Sweden, Finland, and Denmark. The immigrants form new generations, so their numbers in central and north Europe increase until they start migrating back south.

In Florida, *Ascia monuste* makes a few feeding flights and then a long migratory flight along the coast in the shelter of the dunes. Thereafter it settles down, again makes short local feeding and oviposition flights.

The hummingbird hawk moth, *Macroglossum stellatarum* is distributed in areas of Oriental, Palaearctic and Afrotropical regions. A large number of moths migrate toward north in spring and then return back in the autumn.

The silver-Y moth, *Autographa gamma* is commonly available in the Mediterranean area and flies northward in spring and early summer to central and northern Europe. In Australia the bogon moth, *Agrotis infusa* aggregate in Brindabella hill range for aestivation. They began to arrive in October and population reach to its peak by mid-December and then all move back to the plains in March.

The adults of diamondback moth, *Plutella xylostella* are known to be transoceanic migrants in Europe since ancient times. The moth, originated in Asia, spread to other parts of the world with the spread of the cultivation of its crucifer hosts, and by using its own migrational abilities over long distances. The moths are known to be able to migrate a distance of over 3000 km in continuous flight for several days. French (1967) reported a migration of 3,680 km, with the moths in continuous flight for several days. Finland and northwest Russia are the major sites of origin of the migratory moths. The insect also migrates from the USA to Canada and also in the Oriental region. Miyata (1983) assigned diamondback moth to type F category-the super wide-distributing type-which has an omniphagous habit and strong dispersal capacity. The present cosmopolitan distribution of the moth is attributed both to the extended cultivation of its host plant and to its own migration.

8.7 Migration in locusts

Migratory locusts illustrate the most complex migratory behavioural patterns. The best studied desert locust, *Schistocerca gregaria* has been causing greatest havoc and extend from northwest Africa to all countries of Middle East, Arabia, India, southern Europe and other adjoining areas. The major seasonal breeding areas during plagues and recessions are depicted in figure 8.2. It exists in two biologically and morphologically distinct phases, namely 'solitary phase' during the period of recession, and the swarm forming 'migratory or gregarious phase'. In solitary phase it breeds like ordinary grasshoppers in certain desert and semi-desert areas. Under certain environmental conditions there occur changes in the morphology, physiology and behaviour, and the hoppers in crowded conditions become very active, strongly attracted by each other and transform into migratory individuals. Even the nymphs have been observed moving; however, in winged locusts migratory behaviour reaches at its peak. Uvarov (1966) proposed comprehensive details of locust phases. He in 1977 discussed behaviour, population dynamics, distribution and ecology of locusts. Pradhan (1965) discussed the role of biotic factors in locust cycle and proposed biotic theory of periodicity of locust cycles.

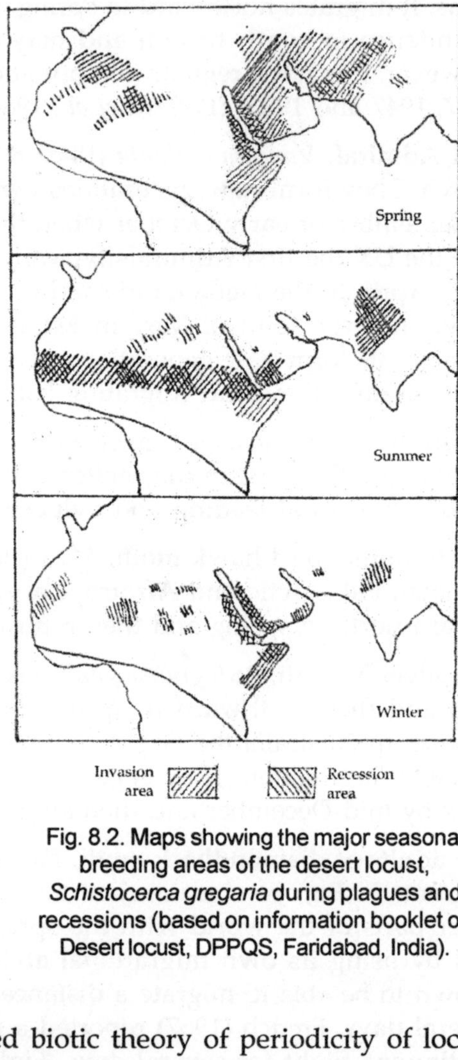

Fig. 8.2. Maps showing the major seasonal breeding areas of the desert locust, *Schistocerca gregaria* during plagues and recessions (based on information booklet on Desert locust, DPPQS, Faridabad, India).

The Australian plague locusts, *Chortoicetes terminifera* undertake long distance migrations. Migrations of several hundred kilometres occur at night and often take place on strong warm winds associated with rain-bearing fronts or low pressure systems. The mass take off and long-distance migration occurs, when the surface temperature is above 25°C, to arrive in areas (where there is rainfall) suitable for successful breeding. Migration only occurs with immature individuals, which have completed their body growth. If fat body reserves are low due to dry conditions, flight will be fuelled by the locust's limited carbohydrate reserves and will last only about half an hour. This can lead to short-range dispersal. If green vegetation is available the young adult will accumulate fat and then be capable of long distance migration. Under optimum conditions adults will be ready for migration about a week after fledging, and migration will occur on the first suitable night thereafter. Observations of migrations using radar have shown that the locusts usually fly at a height of 300 m to 1000 m.

8.8 Migration in other insects

The large milkweed bug, *Oncopeltus fasciatus* ranges over North American continent, extending from south Canada to Central America and from the Atlantic to the Pacific. It is a long distance migrant and in spring and summer migrates northward in the northern USA and Canada, where produces 2-4 generations. When the photoperiods are short it undergoes reproductive diapause and has sufficient time for long flights. Long photoperiods are favourable for breeding and in combination with high temperature flight is inhibited. In autumn it again returns southward (Dingle, 1972).

Johnson (1963) discussed migration in black bean aphids, *Aphis fabae*. After the last moulting the winged or wingless adults formed, of which only with wings wait for a few hours until the cuticle hardens and then take vertical flight. They are carried up by wind quite above and after 1-2 hours when the locomotory drive is reduced they come down and land on green plants. If the plant is not a host, then the aphids again take-off and may make several flights till finds a suitable host.

Migratory behaviour of white grub cockchafer, *Melolontha melolontha* was discussed by Stengel (1974). The adult females have 2-3 ovarian cycles; each characterized by migration towards feeding areas and then return to oviposition sites. It makes a pre-feeding flight usually over a forest and after settling down, feed for about 13 days and then mate. During feeding period the ovaries are developed. The reverse flight is called oviposition flight and on completion she lays eggs. The female migrates to feeding site in each ovarian cycle. The duration of subsequent cycles is rather short. The males do not participate in return journeys. They have been observed showing photohorotaxis and attracted by silhouettes of those, in their environment. The sun also plays a part in their orientation and they do not migrate in cloudy days.

Milne (1960) observed that in garden chafer, *Phyllopertha horticola* the females after oviposition of 70-100 per cent eggs, feed for the first time and then fly quickly and deposit remaining eggs, if any at a short distance.

Among coccinellid beetles: two-spotted lady beetle, *Adalia bipunctata*; the seven-spotted, *Coccinella septempunctata*; and the eleven-spotted, *Coccinella undecimpunctata* appeared suddenly in many areas of the world, thereby indicating them to be migratory. Among Diptera many hover flies, *Syrphus* species, and among the dragonflies, *Libellula*, *Sympetrum* and *Pantala* species are perhaps the most interesting insects known to migrate.

Hagvar (2001) reported that scorpionflies, *Boreus westwoodi* and *B. hyemalis* can be found on the snow surface in coniferous forests throughout winter, mainly in calm, cloudy weather around or slightly above 0°C. The insects may show active migration behaviour, especially on warmer, more or less sunny days. During migration, each insect moves in a fixed direction by continuous jumping, covering up to 1.2 m per minute. Different individuals, however, migrate in different directions, both in relation to the cardinal (compass) direction and in relation to the position of the sun. It was also suggested that continuous jumping as migration behaviour may be a further evolution of jumping as an escape mechanism away from predators.

8.9 Insect migration in India

Studies on migrational behaviour of insects in India are scanty. The information are widely scattered and the details are not known for most species, except locusts and a few others. Migrations of butterflies have been observed at Kodaikanal in southern India. From September-November a large number of butterflies migrate from the plains to the north, cross the mountain range and fly away down again to the south. Again the phenomenon is repeated from February-June. Other scattered records of mass flight of butterflies in other parts of India are available and a number of butterflies move upward the hills of Himalayas at the beginning of hot weather during April-May and return back again in September-November. Important species participating in these movements are: *Pieris brassicae*, *Lampides boeticus*, and *Colias fieldi*. Some species go up to a remarkable height as small tortoiseshell butterfly, *Aglais urticae* had been reported on the Zamu Glacier in north Sikkim at a height of 19,000 feet, and the painted lady, *Cynthia cardui* at 17,000 feet. The Indian tortoiseshell, *Aglais cashmirensis* (Plate 8.3) was reported at 18,000 feet in Himalayas (Williams, 1958).

The black cutworm, *Agrotis ipsilon* migrate from the plains to the hills during summer for aestivation (Woodhouse and Dutt, 1913). Singh (1949) considered it as non-migratory and postulated the possibility of thriving it on alternate hosts during summer in the plains of Bihar state. Venkatraman (1954) proposed that the outbreaks of cutworms in *chaur* lands of Bihar at the end of autumn suggest the possibility of these moths migrating during summer to adjoining low mountain ranges, where they presumably breed on some alternate hosts and return to the plains at the end of autumn to attack winter crops.

The locusts form huge congregations and then fly about in swarms of immense magnitude and devastating vast areas in distant regions up to many miles away from the place of their origin. Their swarming occurs in cycles of variable duration and intervals and a number of such cycles are recorded in India. They are most active in northwest areas of India; however, also extend up to extreme southern and eastern regions. Three species, viz. Bombay locust, *Patanga succincta;* migratory locust, *Locusta migratoria;* and desert locust, *Schistocerca gregaria* have been commonly reported in India.

8.10 Causes of migration

Southwood (1962) pointed that migration is an adaptive phenomenon for species occupying temporary habitats that become periodically unsuitable. Occupants of such habitats must be adapted to migrate in order to survive and avoid the unsuitable variability of the environment. Factors like photoperiod, temperature, high population density, changes in food quantity and climatic conditions seem to involve in migratory behaviour of insects. The migration strategies in insects are greatly influenced by environmental and physiological factors. In all cases reproduction and population growth are the eventual results.

Johnson (1966) proposed two views to explain causes of migration: (i) the insects respond to the onset of adverse environmental conditions by flying away, and (ii) the hormonal changes in accordance with environmental effects, like crowding, food shortage and short days are responsible for migration. Evidences are being produced in favour of second view; however, it has been pointed out earlier in the discussion that many migrants migrate even in uncrowded conditions and when ample food is available.

It is well-established that environmental cues influence the physiology and development. They also have a significant influence on the hormonal milieu of insects. Pener (1985) discussed the role of hormones in migratory behaviour (section 17.7). It has been suggested that in some insects having complex life histories and rapid response to environmental factors juvenile hormone (JH) also controls migration. Wigglesworth (1970) proposed that JH is concerned with maintaining juvenile stage and inhibit development of gonads and also affect the process, oogenesis. In large milkweed bug, *Oncopeltus fasciatus* intermediate titres of JH promote migration, while high titres stimulate reproduction (Rankin, 1974). JH also regulates yolk deposition in the eggs and less concentration of JH inhibits the ovarian development, which may be correlated with reduction in body weight and thus assisting in flight.

In monarch butterflies, decreasing day length and low temperature in autumn are responsible for reduction in activity of corpora allata and neurosecretory cells of brain (Herman, 1973). The monarchs of autumn and summer are physiologically and behaviourally differ. In autumn their gonads are reduced; courtship, mating and ovipositional responses are decreased; show a tendency to form clusters and move southward by September. It has been speculated that the migratory and reproductive cycles of monarchs are correlated with the activity of corpora allata.

In locusts prothoracic glands also have a decided influence on migration as in crowded locusts these glands are rather small and diminish within a few days after final moulting. In isolated individuals these glands persist until sexual maturation. Crowding of young locusts results in highly active gregarious hoppers, which are readily responded by migratory stimuli. Injection of prothoracic gland homogenates in gregarious hoppers reduces their movement and it indicates that both solitary and gregarious phases have different level of activity of prothoracic gland (Carlisle and Ellis, 1963). The fat bodies are utilized as reserve fuel during migratory flight and an adipokinetic hormone (AKH) of corpora cardiaca has also been shown to be necessary to mobilize fat body lipid reserves for longer flight (Mayer and Candy, 1969). Neurophysiological investigations of the flight systems in locusts also indicate a period of maturation in the nervous programme controlling flight (Kutsch, 1974).

Chapter 9
Feeding

Feeding is one of most basic behaviours for survival of an insect. Insects need food to fuel their daily survival and to acquire resources required for growth, development, maintenance of biochemical processes and reproduction. Smith (1985) defined feeding mechanisms as "the processes used by insects, after appropriate food is located and selected, to secure the food, gain access to or select the ingestible parts, modify them if necessary, and transfer them into the storage or processing part of the gut." They use almost all types of organic material, whether living or dead. The feeding generalists (euryphagic) use a wide variety of food; however, most feeding specialists (stenophagic) explore only a particular type of food. Some insects are even specialized on a particular host, while some utilize a wide array of foods, e.g. adults of dacine fruit flies feed on glandular secretions of plants, nectar, secretions of plants, plant sap exuding from injured plant parts, honeydew, rotting fruits, bird dung, decaying material, etc. (Agarwal and Kapoor, 1986). Exploitation of a wide range of food resources is a major factor for their dominance over the globe. The physiological need for body and various morphological, anatomical and behavioural adaptations varies constantly with a change in nutritional state, developmental stage, and experience. The adults of apterygotes and exopterygotes feed throughout their life, but in endopterygotes considerable variations are observed in their feeding patterns. The pupal stage is non-feeding and the structural functions of mouthparts are changed in many adults. Therefore, the dietary requirements of such insects differ in their larval and adult stages, e.g. larva of fruit fly has cephalopharyngeal skeleton (Plate 9.1) and feeds inside the host while the adult has proboscis and feeds on honeydew, nectar, etc. Such adaptation avoids competition within a species and promotes its abundance and success. The diverse patterns of feeding evolved by endopterygotes enable them to more efficiently utilize a broader range of organic matter as food than can other groups. Their success is reflected by the fact that nearly 85 per cent of all insects are Endopterygota (Hsiao, 1985).

The physiological adaptations of insects are also important for feeding, as fluid feeders inject enzymes into the food and blood-sucking insects inject anticoagulant in the host. Predators restrain the prey by force or by injecting venom or by stinging. The mouthparts of insects are variously modified and adapted to feed on various foods and used for biting, cutting, chewing, lapping, piercing, sucking, scraping, rasping, siphoning, etc. The feeding adaptations of different insect species to specialized diets have often been accompanied by structural modifications of their mouthparts and profound changes in their food finding behaviours. The plant feeders consume plant material while the animal feeders may be parasites or predators. A number of insects use dead and decaying organic matter and are scavengers.

The subject of feeding in insects has been discussed in the manuscripts of Brues (1946), Barton Browne (1975), Dadd (1963), Dethier (1966), Hsiao (1985), Bernays (1985), Chapman and Boer (1995), and Schoonhoven *et al.* (1998).

9.1 Feeding habits of insects groups

The feeding habits in different insect orders differ depending on the types of mouthparts, habitats, adaptations, dietary requirements, and foods; and provide insight into evolutionary relationships and links between different groups. Southwood (1973) provided a detailed analysis of the same. Two basic types of mouthparts are found in insects, which are adapted for biting and chewing, and sucking. Biting and chewing mouthparts are primitive; however, most food habits of present day insects evolved during Mesozoic and after that they have only perfected themselves. Insects may be categorized into phytophagous (feeding on plants), zoophagous or carnivorous (feeding on animals), and saprophagous (feeding on dead and decaying material). Feeding habits of different insect orders are depicted in table 9.1. Pantophagy (feeding on a broad-spectrum of diets) is not observed in insects. In many insects fungi form an essential part of diet and they have been termed as 'mycetophagous'.

Brues (1946) recognized four categories based on feeding, viz. plant feeders, predators, scavengers and parasites. Within these groups different types of feeding can be observed. Nearly 50 per cent insects are phytophagous and categorized into mono-, oligo- and polyphagous depending upon their food range. These terms are generally used to indicate the degree of diet breath of any particular insect species. Monophagous are specialist and feed only on one kind of plant taxon, e.g. boxelder bug, *Boisea trivittata;* oligophagous feed on many plant species within a family, e.g. monarch butterfly, *Danaus plexippus;* and polyphagous feed on plants belonging to different families, e.g. migratory locusts, Mediterranean fruit fly. Though the separation of oligophagy and polyphagy is arbitrary and in some cases it is difficult to limit their boundaries. The phytophagous insects may be grouped depending upon the type of plants they feed.

- **Algophagous.** Algae feeders.
- **Bryophagous.** Moss feeders.
- **Lichenophagous.** Lichen feeders.
- **Herbiphagous.** Herbaceous plant feeders.
- **Dendrophagous.** Woody tree and shrub feeders.

Phytophagous insects can also be grouped on the basis of plant parts on which they feed.

- **Phyllophagous.** Feeding on leaves.
- **Anthophagous.** Feeding on flowers.
- **Clethrophagous.** Feeding on seeds.
- **Xylophagous.** Feeding on woody tissues.
- **Pollenophagous.** Feeding on pollens.
- **Nectariphagous.** Feeding on nectar.
- **Conophagous.** Feeding on seeds and cones of conifers.
- **Heteroconophagous**. Occasionally feeds on seeds, cones of conifers; usually live, feed on stems and needles.
- **Carpophagous (Fructivorous/ Frugivorous).** Feeding on fruits and fruit seeds.
- **Rhizophagous.** Feeding on roots.
- **Pharmacophagous.** Feeding on plant-derived lures.

Phytophagous insects depending upon the nature of their attack can also be grouped as:

- **Miner.** Feed inside the blade of a leaf, between the epidermal layers or beneath the bark of plants, excavating a mine into these tissues (Plate 9.2).
- **Gall former.** That induces the formation of gall and feeds on their tissues (Plate 9.3a,b).
- **Borer.** That bores into and feed on woody and non-woody portions of plants (Plate 9.4).
- **Sap-feeder.** Suck fluids from plants, cause deformities and kill affected plant parts. Also termed as phytosuccivorous.
- **Leaf roller.** Hides and feed inside a leaf or the tip of a leaf that it has rolled up into cigar-shaped tube (Plate 9.5).
- **Leaf folder.** That folds a leaf to make a shelter for feeding or hiding.
- **Free-living defoliator**. Feeds on and moves about freely on foliage.
- **Case bearer.** Moves and feeds on foliage while hidden in a case or sheath made from leaf epidermis and silk.
- **Leaf skeletonizer.** Feeds on leaf tissues leaving the ribs (Plate 9.6).
- **Webworm.** Spin a silk shelter in which to hide and feed.
- **Gall-inhabiting**. Lives in a gall made by a different insect.
- **Free-feeding.** Moves about freely to feed.

Feeding on dead and decaying materials is common in insects, particularly in primitive orders. Microorganisms usually contaminate the decaying food and often they have symbiotic microorganisms, which assist in digestion. The food is of low nutritive value and insects require long developmental period. Saprophagy is a habit of primitive insect orders, though it has been retained in many advanced orders. In some cases contradictions arise with regard to determine the types of feeding. Feeders on wood that is dead but is still retaining its structure are usually classified with phytophagous. However, if insects consume decomposed wood, they are regarded as saprophagans. The Saprophagous may be of following types.

- **Detritophagous (Schisophagous).** Feeds on ground remains of plants and animals.
- **Coprophagous.** Feeds on excrement of animals.
- **Necrophagous.** Feeds on dead animals and their remains.
- **Saproxylophagous.** Feeds on partially decomposed woody debris.

The zoophagous insects are predators and parasites. Insect predators are free-living and not only consume other insects but many also feed on other animal groups. The parasites may be ectoparasites and acquire food from host's body. The association of a parasite with host may be permanent, e.g. in lice; or temporary only for feeding purpose, as in case of mosquito and bed bug. The endoparasites are often referred as 'parasitoids' because their immature stages completely consume the host. They are usually parasitic in immature stages only and their adults are free living. Askew (1971) termed the parasitism by immature stages as 'protelean parasitism'. Insects of the orders: Anoplura, Mallophaga and Siphonaptera feed exclusively on vertebrate hosts. The zoophagous insects may be grouped as follows:

- **Predator.** Kills and consumes the prey immediately.
- **Parasite.** Living on or in host's body and feeds in such a manner that the host remains alive.
- **Parasitoid.** Living on or in host's body and eating it gradually, but eventually resulting in its death.

The zoophagous insects are termed differently depending on the type of host or its stage consumed by them, e.g. nematophagous (feeding on nematodes), malacophagous (feeding on molluscs), acariphagous (feeding on mites), entomophagous (feeding on insects), acridophagous (feeding on grasshoppers), aphidophagous (feeding on aphids), coccidophagous (feeding on coccids), myrmecophagous (feeding on ants), oophagous (feeding on eggs), haematophagous/ haemopahgous (feeding on blood), etc.

Based on the host species utilization the parasites/parasitoids may also be grouped as:

- **Monoxenous.** A parasite or parasitoid restricted to one species of host.
- **Oligoxenous.** A parasite or parasitoid restricted to a few hosts.
- **Pleioxenous.** A parasite or parasitoid restricted to hosts of one family. Also termed as heteroxenous.

- **Polyxenous.** A parasite or parasitoid with a wide range of hosts.

The fungi are now placed in a separate kingdom and the fungus feeders are considered as a distinct group termed as mycetophagous or mycophagous. The fungi feeders may be:

- **Myxomycetophagous.** Feeding on myxomycetes.
- **Sporophagous.** Feeding on fungal spores.
- **Xylomycetophagous.** Feeding on fungus growing in wood.
- **Copromycetophagous.** Feeding on fungus growing on excrement of animals.
- **Sapromycetophagous.** Feeding on fungus growing on dead and decaying materials.

Table 9.1. Feeding habits of important insect orders.

Mode of feeding		Insect orders
Phytophagous (Herbivorous)		Orthoptera, most Lepidoptera, most Hemiptera, Thysanoptera, Phasmida, some Isoptera, some Coleoptera, some Hymenoptera, some Diptera, some larval Trichoptera, some Collembola
Zoophagous (Carnivorous)		
	Parasites	Anoplura, Mallophaga, Siphonaptera, some Diptera, some Coleoptera, some Hymenoptera, some Hemiptera, some Mecoptera
	Parasitoids	Some aculeate Hymenoptera, some Diptera, some Coleoptera, some Neuroptera
	Predators	Odonata, Dictyoptera, larval Neuroptera, some Diptera, a few Coleoptera and Lepidoptera, some Hymenoptera, some Hemiptera, some Dermaptera
Mycetophagous		Some Diptera, some Coleoptera, some Hymenoptera, some Isoptera
Saprophagous		Collembola, Protura, Thysanura, Diplura, some Orthoptera, Dictyoptera, some Isoptera, some Dermaptera, some Embioptera, some Psocoptera, immature Plecoptera and Ephemeroptera, some Hemiptera, some Mecoptera, some Trichoptera, larval Siphonaptera, some Coleoptera, a few Lepidoptera and Hymenoptera, some larval Diptera

9.2 Feeding in chewing insects

The chewing insects that feed on plants tissues cut the food into small fragments and ingest the same. Their mouthparts are primitive biting and chewing type and mandibles are used to cut the food. Other appendages are used for the purpose of ingestion. Smith (1985); Chapman (1995) discussed the mechanism of food handling in chewing insects. The feeding process may involve: (i)) to hold the food or secure relative to the mouthparts, (ii) detachment of food pieces from the source, (iii) further reduction of the size of food pieces, (iv) to bring the pieces to the mouth, (v) ingestion from the mouth, and (vi) lubrication of food and mouthparts. The mandibles are heavily sclerotized with a distal incisor region (used for cutting) and proximal molar region (used for grinding). The cutting edges contain metals like Zn, Mn or Fe in some insects (Fontaine, *et al.*, 1991), which provide further hardening. The mandibles are innervated with mechanoreceptors and strong muscles. The motor pattern controlling mandibular movements is generated in suboesophageal ganglion. Structures of mandibles vary in different insect species depending on the type of food they use. Movements of food through the mouth appear to be mechanical process involving other mouthparts. Maxillae functions as constraining and manipulating devices and are mainly involved in passing the food to the mouth. In insects like grasshoppers fore legs are also used to guide food into the mouth. Labrum and labium act as lips and help in ingestion.

9.3 Feeding in sucking insects

A number of insects fulfil their nutritional requirements by taking liquid diet. They feed on a wide range of liquids, e.g. nectar, cell sap, phloem and xylem fluids, blood, etc. The viscosity of these fluids determines the rate of fluid flow and nutrient uptake. Smith (1985) divided the process into eight steps.

(i) *Purchase*. Making a mechanical contact with host surface.

(ii) *Initial puncture*. Penetration of superficial layer of host's surface.

(iii) *Deep penetration*. To reach up to the liquid food.

(iv) *Recognition of a target*. Mechanical and chemical sensilla are used for determining the suitability of food.

(v) *Uptake of food*. Flow of food through the food canal.

(vi) *Pumping*. By creating a pressure gradient between food source and insects gut. Energy is also required for pumping.

(vii) *Saliva*. Salivary secretion may be needed to assist penetration, breaking of host tissues, dilution of food, etc.

(viii) *Ensheathing of mouthparts after feeding*.

Kingsolver and Daniel (1995) discussed the mechanics of food handling by fluid feeders. Most adult Lepidoptera feed on nectar, plant saps and other similar resources. They have highly modified mouthparts in which maxillary galeae form an elongated coiled-tube. During feeding the tube is extended into nectar source and nectar is drawn through the food canal by the action of muscular suction pump.

The proboscis of bees is a complex structure with a long and slender tongue or glossa. The food canal through which nectar moves is the region bounded by outer surface of the glossa and inner surfaces of the galeae and labial palpi. Repeated extension and retraction of the tongue form the 'licking cycle'. During extension nectar moves onto the glossa, and then the nectar-laden glossa is retracted into the galeae and labial palpi. Nectar is finally removed from the glossa into the food canal.

Many phytophagous insects take nutrients directly from liquids present in xylem, phloem and mesophyll cell cytoplasm. The viscosity of such liquids is low because of less available nutrients. A number of such feeders are important crop pests and vectors of plant diseases. The hemipterans have a long, slender beak consisting of labium and the mandibles and maxillary palpi, which form a set of long slender stylets. Maxillary stylets form both food canal and salivary canal while the mandibular stylets form piercing structures. Most phytophagous Hemiptera discharge saliva while feeding and use plant liquids from xylem or phloem tissues, or directly from leaf mesophyll cells.

The feeding apparatus of thrips is unique. Only one mandible is present and another mouth structures form a stylet or tube through which food is drawn. The single mandible punches a hole in the plant surface through which the stylet is inserted.

The larvae of fruit flies (Fam. Tephritidae) have fluid feeding mouthparts and inhabit the bacterial soup in the decaying areas of infested fruit. They bore their way into interior of the host; macerate the tissues with the help of cephalopharyngeal skeleton (Plate 9.1), and ingest the broken-down tissues and associated bacteria. Different hosts also have pronounced influence on larval development rates (Agarwal and Yazdani, 1991) in the melon fly, *Bcatrocera cucurbitae*.

The haematophagous insects exhibit great diversity and the blood feeding involves penetration of host skin, location of blood and ingestion of blood. Mosquitoes are well-known for feeding on blood, and a mosquito after landing on the host applies tip of labium on host's skin that has many sensilla. Maxillae and mandibles move with alternating thrusts to pierce the skin, and labrum (containing the food channel) and hypopharynx (containing the salivary channel) also enter in the skin. The labium remains pressed against the point of entry and the stylets probe for a capillary. The blood is sucked up as it oozes into the tissues (Fig. 9.1). The saliva is pumped into the wound

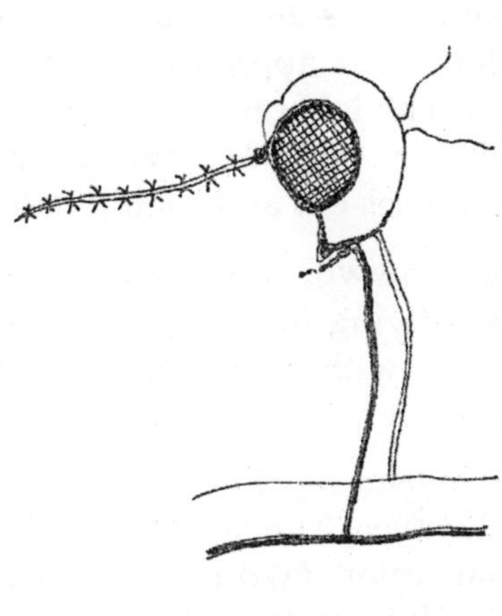

Fig. 9.1. Feeding by a blood sucking mosquito.

to prevent the coagulation of the blood. In mosquitoes, black flies, biting midges, tabanids, and snipe flies only female suck blood. In contrast, both sexes of horn flies, stable flies, and sheep keds suck blood. The blood passes directly to the midgut and the abdominal walls of a biting insect are exceptionally distensible; a mosquito may commonly double its own weight with a single blood meal.

Sutcliffe and McIver (1984) discussed the structure and interrelationships of the mouthparts and of the food canal and its accessory cephalic structures in females of black fly, *Simulium venustum* and recognized four phases of biting (i) initial penetration of the skin effected by the mandibles, (ii) consolidation of mouthpart position involving anchoring the syntrophium (composed of the mandibles, maxillary laciniae, hypopharynx and labrum) into the wound by means of the barbed laciniae, (iii) diet sampling and active blood-feeding, and (iv) mouthpart disengagement involving removal of the laciniae.

9.4 Other mechanisms of feeding

Dragonfly nymphs are predaceous and found in aquatic habitats. Their food catching behaviour is unique among insects. The labium has evolved into a hinged extended device called mask and is equipped with terminal pincers (Fig. 9.2). The modified labium is capable of being thrust forward by haemolymph pressure coupled with release of muscular tension to grasp prey with a pair of toothed lateral lobes (Snodgrass, 1954). Feeding involves three phases: (i) detecting prey and adjusting position in relation to it, (ii) ejecting the labium to grasp the prey, and (iii) employing the mandibles to devour the prey. The labium, when extended is approximately as long as the fore legs. The labium is designed for grasping and holding the prey while the maxillae and mandibles cut and manipulate it for ingestion.

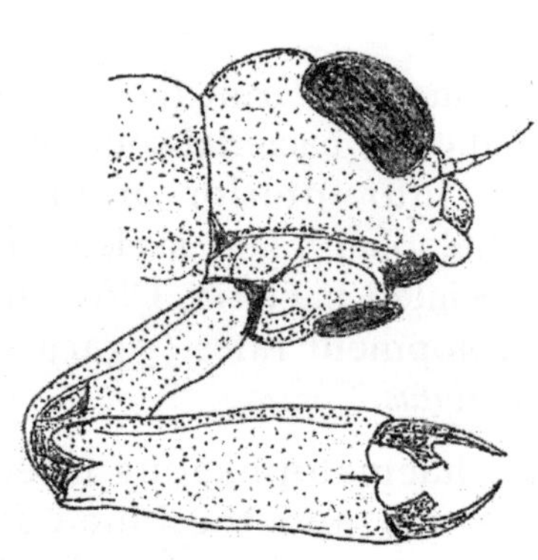

Fig. 9.2. Head of a dragonfly nymph showing partly extended mask.

The glowworm lets down long sticky lines to catch flying insects (Fig.16.2a). These lines are strong, have elasticity, and are beaded with very sticky glue. Once the insects get caught the glowworm senses the movement and pulls up the line and the struggling insect.

Filter feeding if commonly observed in larvae of mosquitoes, black flies and caddisflies. The suspended particulate matter is filtered, collected and ingested from water. Two types of structures are involved for this purpose: (i) modified cuticular structures to form 'fans', 'comb', 'brush' or 'sieves' and found in Diptera and some Trichoptera, and (ii) silknets (found in some other Trichoptera).

The feeding strategies of fungus feeders are mainly of two types, (i) microphages-feeding on small particles, and (ii) macrophages- feed on fruiting bodies. Many insects use fruiting bodies of fungi for oviposition and their larvae develop in these bodies. Such insects are obligate fungivores, while some others use fungi as supplementary diets. Many insects also consume yeasts, e.g. fruit flies, drosophilids, some nitidulid beetles, etc. The fungus gnats (Fam. Mycetophilidae) are diverse, delicate flies and inhabit damp and dark places with abundant fungal growth on decaying vegetation. Their larvae feed on fungi and some are pests of edible mushrooms. Insects like ants and termites cultivate their own fungi for feeding.

Trophallaxis is the regurgitation of food by one animal for another. It is highly developed in social insects such as the ants, in which individual colony members store food in their crops and regularly exchange it with other colony members and larvae to form a sort of 'communal stomach' for the nest. In termites traphallaxis has different functions. Lower termites share both 'stomodeal food' from the salivary glands and crop and 'proctodeal food' from the hindgut. The former is a source of nutrients, while the later contain symbiotic flagellates. In each moult the symbionts are lost and the nymphs acquire new ones by trophallaxis.

The excreta of many plant feeding Hemiptera, particularly aphids and coccids are specially interesing not only because of the relatively enormous amount produced, but also because of the high carbohydrate content. These insects feed on plant sap that is rich in sugars but poor in proteins essential for their growth. To acquire sufficient proteins they have to ingest large volume of plant juice, and the residual solution is ejected to the exterior in the form of a droplet of honeydew. Honeydew form important part of diet of many ant species and the association between honeydew secreting hemipterans and ants is mutual; the ants protecting the honeydew secreting insects. Ants collect honeydew by touching the hind end of aphid repeatedly with their antennae until a droplet appears. The ant suck it up at once and goes to another aphid for more (Fig. 9.3).

Fig. 9.3. Worker of ant *Formica* sp. attending aphid and imbibing the honeydew droplet.

Many insects require microbial symbionts for adequate nutrition, which are transferred on eggshells and eaten by neonates. Wood wasps (Fam. Siricidae) sequestered cellulases from fungi. Lower termites (wood feeders) obtain cellulase from enteric Protozoa and bacteria that thrive in hindgut.

A number of insects form galls on various plant parts. The most pertinent gall-formers are: gall midges (Fam. Cecidomyiidae) and gall wasps (Fam. Cynipidae).

Gall begins as small swelling at the point of oviposition by the female and grows with the feeding of larva. It has often been proposed that plant growth hormones are also involved in gall formation.

9.5 Host location by phytophagous insects

Females of many insects lay eggs on appropriate hosts, and the hatched young never presented with the challenge of finding a suitable host plant. In other cases newly hatched larvae must find and recognize a suitable host plant. Even when the female choice has been appropriate, insects are often required to move to new plants as their food supply is diminished (Bernays and Chapman, 1994). Selection of a suitable host plant poses two main challenges for insects: (i) detecting and locating the host from a distance, and (ii) confirming the appropriateness of the plant in terms of its species and quality (Bernays and Chapman, 1994; Schoonhoven *et al.*, 1998). The sequence of host selection by a phytophagous insect is depicted in figure 9.4.

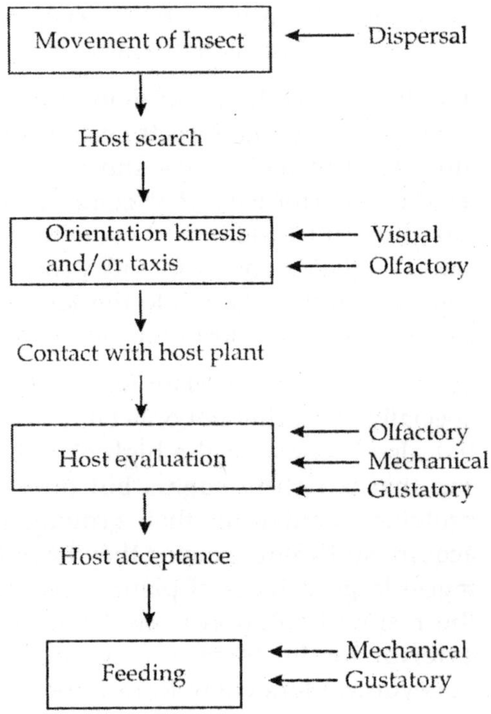

Fig. 9.4. Sequence of host selection behaviour in a phytophagous insect.

When the herbivorous insect arrives at a plant and touches it, the insect enters what is commonly called the 'contact phase' of host-plant finding (Schoonhoven *et al.*, 1998). This phase consists of a series of behavioural elements that serve to evaluate physical and chemical traits of a plant. Upon encountering the plant, locomotion is stopped and the insect restricts movement to a small area on the plant surface; this is referred to as 'arrestment phase'. In the arrestment phase, plant structures, *e.g.* leaf edges, veins, or stems, guide walking movements. The insect will then thoroughly explore the plant by repetitive contacting of sensilla located on various body parts. The exploration allows the insect to perceive physical and chemical cues offered by the plant. The surface morphology, texture, and wax structure are all characteristics evaluated by the insect during this phase. The next step is test biting (or probing) of the plant tissue. A test bite is typically smaller than a regular bite and the plant material may be held in the pre-oral cavity for a longer period of time than during regular food intake (Schoonhoven *et al.*, 1998). The plant cell contains large numbers of different chemicals and some may serve as phagostimulants while others may be deterrents to feeding. The key phagostimulants are nutrients and sugars for most insects (Bernays and Chapman, 1994). Plant secondary compounds in low concentration also may act as phagostimulants. Besides, many secondary compounds are present in one or a small number of plant taxa. Such secondary compounds provide indicators to monophagous or oligophagous insects that they have indeed located the correct host.

Herbivorous insects employ one of two search patterns, referred to as random and directed searching (Prokopy, 1986). Random searching is a function of central motor programmes, and in this behaviour no evident orientation can be demonstrated when the insect is at some distance from a host plant and the insect must literally bump into the host. When searching randomly; frequency, rate, and direction of movement appear unrelated to the suitability of plants within the insect's perceptual range. Random searching has been described for insects like locusts, caterpillars, etc. This searching type may be the best possible option when environmental cues provide no directionality or when the sensory capacity of the insect is insufficient to obtain the required stimuli (Schoonhoven *et al.*, 1998). Directed searching is possible when the host plant emits signals that, either alone or in combination with a second cue, allow direction to be perceived by the sensory system of the searching insect (Schoonhoven *et al.*, 1998).

Herbivorous insects use visual, olfactory and tactile cues for recognition of host plants. Some insects are host specialists while most feed on more than one host. No insect feed on all plants occurring in its range. Host range of some species may be very wide, e.g. desert locust, *Schistocerca gregaria* (> 400 hosts) (Uvarov, 1977); gypsy moth, *Lymantria dispar* (> 500 hosts) (Lance, 1983); tobacco whitefly, *Bemisia tabaci* (506 hosts) (Cook, 1986); Mediterranean fruit fly, *Ceratitis capitata* (253 hosts), oriental fruit fly, *Bactrocera dorsalis* (173 hosts), melon fly, *B. cucurbitae* (125 hosts), and Queensland fruit fly, *B. tryoni* (106 hosts) (Metcalf and Metcalf, 1992). Olfaction stimuli work over only short distances in host selection (Schoonhoven, 1968); however, in some cases insects fly considerable distances from downwind to host odour, e.g. in Douglas fir beetle (McMullen and Atkins, 1962). Both contact and gustatory chemoreception are important in host recognition. Many insects are guided by the smell of their hosts. Contact chemoreceptors are usually located on mouthparts and tarsi, and assist in recognition of suitable host. The final selection of the host for feeding is made after alighting on the host.

9.6 Host location by blood feeders

Haematophagy or the utilization of blood as food occurs in six insect orders. The order Diptera (true flies) has the largest number of haematophagous species and includes endoparasitic and ectoparasitic forms. Other insects that feed on blood are Hemiptera (Fam. Cimicidae and Reduviidae), blood sucking lice, a few wasps, fleas, a few Coleoptera, and a single species of noctuid moth, *Calpe eustrigata.* The blood feeders exhibit great diversity. Some like *Glossina* and *Rhodnius* feed on blood for their reproductive requirements. Some mosquitoes can feed on blood as well as plant resources. In some mosquitoes only females suck the blood while in fleas and tsetse flies both sexes feed on blood (Lehane, 2005). The tsetse fly, *Glossina palpalis* (Fam. Glossinidae) probably has no match among haematophagous insects in its vertebrate host range as the fly can feed on any vertebrate it contacts. Blood feeding insects extract blood from both vertebrates and invertebrates. Many live permanently on/in the hosts' body while others come in association with host only for feeding. Some blood feeders exhibit host specificity. Among insects most researches on feeding behaviour of blood sucking insects have been conducted on mosquitoes, black flies, tsetse flies and a few bugs. Hocking (1971), Friend and Smith (1977) reviewed feeding in blood sucking insects and factors affecting the same. Blood feeding insects use their sensory system to locate a host. Olfactory and temperature cues are more important in host finding than visual cues. Brown (1966) found moisture to be a powerful attractant for mosquitoes. Host selection is the primary crucial element in insect blood feeding. An improper host may not have sufficient nutrients needed in the blood stream, or may carry parasites. The feeding patterns vary according to the availability of hosts and to ecological factors. Host location is a complex series of behavioural patterns that gains momentum as the host is tracked down. The patterns are not a strict sequence of events, but rather a mixture of all possible avenues of action. These behavioural patterns have been classified into three phases (Sutcliffe, 1987). However, different species may exhibit variations along these lines.

(i) *Appetitive searching.* It is being driven by hunger. The insect indulges in non-oriented behaviour likely to bring it into contact with stimuli derived from a potential host.

(ii) *Activation and orientation.* Upon receipt of host stimuli (activation) the insect switches from behaviour patterns driven from within (appetitive searching) to oriented host location behaviour.

(iii) *Attraction.* The host stimuli are used to bring the insect into the host's immediate vicinity, and the decision of whether or not to contact the potential host is made.

9.7 Host location by zoophagous insects

Both predators and parasites exhibit highly variable patterns of food habits. Predators usually feed on a variety of insects; however, a few have restricted food habits. Many predators consume the prey both in adult and larval stages, but in some cases their feeding habits differ. The host location behaviour of a predatory insect comprises a stereotyped sequence of components, which lead the insect towards the prey, and on contact, to recognize and use it. Sit-and-wait predators find a suitable patch and wait for prey to come within striking range.

The praying mantis is able to judge the distance and position of prey relative to the long axis of the body. Thus prey can be caught in many different positions relative to the mantis. The mechanism is based on an interaction between compound eyes, sensory spines in neck region that monitor the head position, and the muscles of raptorial fore legs (Romoser, 1981). Fast-flying insects like dragonflies and robber flies also spend much time perched prominently on vegetation. As a prey comes in the visual range it is seized by making rapid and controlled flight. Predators tend to capture and feed the prey available in its visual range and their visual cues play a prominent role in prey capture. Water striders catch the prey, trapped in surface film by using mechanoreceptors. Ant lion constructs pitfall trap for catching the prey (Fig. 9.5). Many insects locate prey or host by active foraging and their locomotory patterns may be non-directional (random) or directional (non-random). The foraging of coccinellid beetles and larvae of syrphid flies is random and if they encounter the prey, feeding ensues. First-instar larvae of *Episyrphus balteatus* exhibit directed search over short distances, guided by olfactory cues from aphids, but not from honeydew. Aphid extracts, honeydew and sucrose were found to be feeding stimulants for the second and third instars larvae (Bargen *et al.*, 1998).

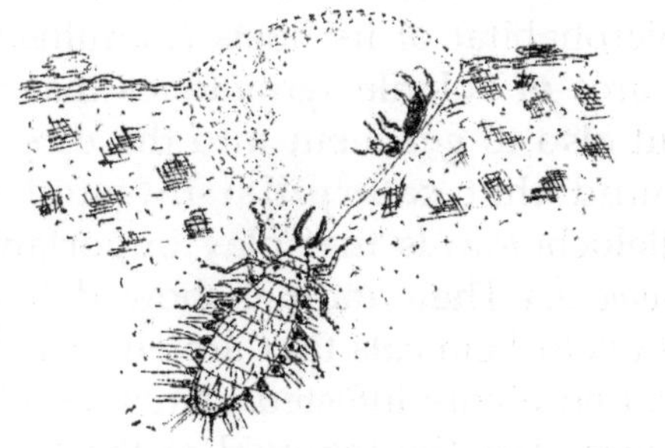

Fig. 9.5. A pit trap constructed by ant lion larva. The larva lying in wait for the prey.

Many insects use the cues like sound, chemicals and visual for prey location and their search is directional. The chemicals produced by prey or hosts (kairomones) allow a predator or parasitoid to locate these resources. Parasites and predators of bark beetles are most strongly attracted to volatiles produced by their host bark beetle, although some of the attractive compounds may be derived from monoterpene precursors in the tree. *Temnochila virescens* feeds on western pine beetle, *Dendroctonus brevicomis* and the predator is almost as sensitive to *exo*-brevicomin as is the bark beetle to this pheromone component (Byers, 1988). Some plants release compounds (synomones) after being fed upon that are attractive to predators or parasites of their herbivores, e.g. certain pteromalid parasitoids locate their hosts, the scolytid timber beetle in this way. All biting flies are extremely sensitive to CO_2 and can detect and follow 'odour plume' containing it, for great distances. Other odours from a potential host may also play a role in long-distance detection. The fly is also sensitive to heat and moisture as it approaches its host, and when quite close vision apparently comes into play. Black flies respond to human sweat. Sound signals produced by prey or hosts are used by parasites. Tachinid fly, *Euphasiopteryx ochracea* locate cricket hosts by the songs (Cade, 1975). In case of parasitic hymenopterans ovipositor has a number of sensilla, which assist in host location. Females of many parasitic hymenoptera insert their long ovipositor in the plant tissues to locate developing hosts' larvae.

Many parasitoids are known to discriminate, and respond to volatile chemicals produced by uninfested plants and plants infested with a particular herbivore species. Once the herbivore feeds on a plant, plant starts producing large amount of volatile chemicals, which are used as cues leading female parasitoids to the microhabitat of its hosts (Tumlinson *et al.*, 1993a). The plants are the principal source of volatile synomones that not only direct parasitoids to the host patch, but also to get them into the direct location of the host. Steinberg *et al.* (1993) found that caterpillar-infested cabbages initiate the release of volatile allelochemicals that play important role in long-range host location by *Cotesia glomerata*. Their findings showed that cabbage plants are involved in the production of allelochemicals that are emitted by *Pieris* damaged plants. They also reported that herbivore-infested plants remain attractive to *C. glomerata* for at least several hours after the removal of the larvae. Accessory gland materials from the fall armyworm, *Spodoptera frugiperda* and corn earworm, *Helicoverpa zea* contained kairomones that influenced the host recognition or acceptance behaviour of the parasitoids, *Telenomus remus* and *Trichogramma pretiosum*, respectively (Nordlund *et al.*, 1987). Honeydew, the faeces of feeding aphids, induces and/or prolongs searching behaviour of aphid parasitoids. These results indicate that hyperparasitoids may use aphid honeydew, a conspicuous cue from the second trophic level, as an infochemical to locate their hosts (Buitenhuis *et al.*, 2004).

Vinson (1976, 1984a) reviewed host selection by parasitoids, and discussed their relations with hosts. Successful parasitoidism depends on five factors, viz. (i) host habitat location, (ii) host location, (iii) host acceptance, (iv) host suitability, and (v) host regulation. He subdivided the factors i-iii, which make up the host selection process, into (a) habitat preference, (b) potential host community location, (c) host location, (d) host examination, (e) ovipositor probing, (f) ovipositor drilling, and (g) oviposition (Vinson, 1984b).

9.8 Receptors and host finding

The sensory receptors are basic units of insects used for interacting with the environment. Visual and chemical receptors are mainly associated with feeding. Besides, tactile and thermoreceptors also play an important role in some cases. In addition to compound eyes, the receptors associated with feeding are present on mouthparts and tarsi. Many insects appear to employ the antennae in monitoring their food.

9.8.1 Visual receptors

Visual attraction to a host plant results from responding to the colour or the form of the plant. Due to the large amount of variation that can exist within a given plant species, visual responses often occur only in conjunction with an appropriate olfactory stimulus. The development of visual receptors varies in different groups of insects. The compound eyes are capable of form perception, motor detection, and colour perception. Three optical characteristics of plants, viz. reflectance (colour), dimension (size), and pattern (shape) influence the host plant selection behaviour (Prokopy and Owens, 1983). Prokopy and Owens (1978) speculated that many mono- and oligophagous insects may tend to be visual specialists in comparison with polyphagous insects, especially those polyphagous species whose preferred feeding, mating, or oviposition sites within an individual plant are of diverse physical characteristics. They reported that European apple sawfly, *Hoplocampa testudinea* (apparently monophagous) and apple maggot fly, *Rhagoletis pomonella* (oligophagous) are more specific in orientation to hue and/or form for feeding on apple than the tarnished plant bug, *Lygus lineolaris* (polyphagous) which also feeds on apple. Compound eyes of pollinators and predators are prominent and in many groups can discriminate between different wavelengths. The spectral sensitivity of insects ranges from 350-650 nm. Colour vision is reported in many herbivores and other insects, and plays an important role in host location, but not in host discrimination. Prokopy and Owens (1983) concluded that when an insect reaches nearer to a host, many additional stimuli, e.g. hue and intensity are also perceived. Spatial patterns of a plant like dimension and background may provide additional cues. Thorsteinson (1960) suggested that narrow spectrum of light reflected from green plants is not helpful in final recognition of host. Tephritid flies locate vegetation primarily by vision, e.g. green colour of foliage and/or darkness of tree silhouette are the principal stimuli eliciting response of *R. pomonella* on trees (Prokopy, 1977). Green leaves reflect maximum energy in the infrared part of the spectrum. This energy is beyond the presumed 650 nm limit of detection by the insect eye, but Callahan (1975) postulated that insect antennae might detect it. Many insects like aphids, leafhoppers, etc. are attracted to yellow and yellow-green colour and these insects may discriminate their hosts by using colour contrast. Colours of flowers may also provide important visual cues for some insects; however, final detection of hosts is done with the help of chemical cues. Prokopy and Boller (1971) evaluated response of European cherry fruit fly, *Rhagoletis cerasi* to 27 colours or shades under orchard conditions. More flies were captured on sticky coated rectangles painted with yellow enamel. Daylight fluorescent yellow captured even more flies than yellow enamel. Prokopy and Owens (1983) demonstrated the importance of leaf reflectance properties, in which the uses of visual discrimination in the cabbage root fly, *Delia radicum* was observed. They found that visually based landing responses occurred when the fly was offered artificial leaves painted with colours mimicking host plant leaf reflectance. The yellow and green colours initially attract the mustard beetle, *Phaedon cochleariae* to the cruciferous plant and then the odour of mustard oil provokes feeding (Bernays and Chapman, 1994). The importance of colour is also apparent in many aphid species, e.g. the black bean aphid, *Aphis fabae* lands preferentially on yellow leaves and it is common for this insect to be caught in yellow traps (Jolivet, 1998). Size and shape of plant leaves plays a critical role in host plant finding for some insects. Rojas and Wyatt (1999) investigated the landing responses of female cabbage moths, *Mamestra brassicae* to visual cues from artificial plant leaves of different shapes and presence/absence of cabbage plant odour. Females showed oriented flight toward plant odours but landed significantly more often when the odour was presented with an artificial leaf. The shape of the leaf targets (circle, square or triangle) did not influence the female response; however, the size of the target did influence the insect response.

Blood feeders, particularly daytime feeders also employ their visual receptors for host location, but colour vision is less important for these insects. In predators vision also plays an important role in prey capture. Preying mantis responds and attack prey only when it is in the visual range.

9.8.2 Chemoreceptors

Smell (olfaction) is the most important stimuli during host plant finding. Air coming from an odour source contains pockets of odour-carrying air, carried in a meandering plume (Bernays and Chapman, 1994; Schoonhoven *et al.*, 1998). These pockets are carried downwind from the plant and the insect perceives a series of bursts of odour separated by periods of no odour. The duration of these bursts and the intervals between them generally increases at greater distances from source. Bernays and Chapman (1994) suggested that the arousal of an insect by the odour plume puts an insect in a state of readiness to respond and then insect is activated for flight, and finally orientation takes place after take-off. Walking insects also use odour signals to find the hosts. The Colorado potato beetle, *Leptinotarsa decemlineata* has a strong preference for the cultivated potato. During first seven days of adult life it locates host plant by walking, as the flight muscles are not fully developed. Visser (1986) found that when clean air was blown over the beetle, it displayed a menotactic response to the wind, maintaining a constant angle to the wind direction.

The chemoreceptors associated with feeding are mainly located on antennae, mouthparts and tarsi. The distribution and number of chemoreceptors on these organs vary among different insects. Chapman (1982, 2003) made detailed analysis of chemoreceptors with respect to feeding of different insects. Maximum chemoreceptors are found in Orthopteroid orders and some apterygotes, while in endopterygotes the number is comparatively less. More number of chemoreceptors in orthopteroids is probably associated with their polyphagous habit. The modification of chemoreceptor system is a major step in the evolution of diverse feeding habits (Chapman, 1982). Some chemoreceptors located on the mouthparts also act as gustatory receptors. Two basic encoding mechanisms have been proposed for sensory input of chemoreception (Dethier, 1976; Boeckh, 1980).

(i) *Labelled-line patterns*. Some chemoreceptors exhibit high degree of specificity and respond to one or a few related chemicals. These are called odour specialists. A good example is responses of large white cabbage butterfly, *Pieris brassicae* to gucosinolates (Schoonhoven, 1967).

(ii) *Across-fibre patterns.* The receptors are generalists showing varying degree of responsiveness to chemicals. It envisages that chemical discrimination depends on an analysis by the insect of the overall pattern of sensory input from an array of receptors with different qualities.

Schoonhoven and Loon (2002) suggested that all caterpillars, even closely related ones, have different taste systems. Taste cells responding to general phagostimulants (e.g., carbohydrates) have been found in all species studied. In some species, highly specialized taste cells have been found that respond to plant taxon-specific secondary plant substances that act as 'token' stimuli for plant recognition. Though the response profile of these taste cells is best described as generalized, they nevertheless show species-specific stimulus spectra. These generalist deterrent cells often play a crucial role in feeding behaviour. Both neural coding mechanisms, 'labelled lines' and 'across-fibre patterning' have been inferred to operate in caterpillars. The first type is a likely coding mode in oligophagous species employing token stimulus receptors, whereas across-fibre patterning most probably operates in all species confronted with choices between plant foods of varying quality. Variability in taste cell responsiveness is dependent on developmental stage, time of day, and feeding history. This indicates that caterpillar taste cells are not rigid systems, and even possess a 'peripheral memory'.

Barrozo and Lazzari (2004) investigated behavioural responses of nocturnal bug, *Triatoma infestans* larvae to CO_2 and other odours of vertebrate origin. *T. infestans* oriented towards airstreams enriched with CO_2. The accuracy of the oriented response to CO_2 improved with stimulus intensity. Many female mosquitoes actively seek potential hosts in the semi-darkness or the darkness, and depend upon olfactory, rather than visual, information to locate potential hosts. CO_2 is a key component of host odours and stimulates two aspects of host seeking behaviour. CO_2 activates mosquitoes, that is, it induces take-off and helps to maintain flight behaviour. Second, CO_2 also helps female mosquitoes to orient themselves with respect to a potential host. The CO_2 receptors are located on the labial palps. Lactic acid (LA) is probably the primary host odour to which female mosquitoes orient themselves during flight. LA evaporates from the skin of humans and other mammals. Mosquitoes have LA receptors on the sensilla basiconica on their antennae. Mosquitoes detect CO_2 via sensilla located on maxillary palpi (Kellogg, 1970). The mosquito, *Culex quinquefasciatus* is attracted to 3-octen-1-ol from cattle breath (Kline *et al.*, 1990).

9.8.3 Tactile and thermoreceptors

Tactile or mechanoreceptors are sensitive to vibrations and contact. These receptors assist in anemotactic orientation of insects. Tactile receptors located on tarsi identify the host's surface structure. Receptors on mouthparts assist in coordinating biting and chewing movements.

Thermal stimuli are important for blood-sucking insects in their host finding behaviour. These are usually located in antennae. Warm-moist convention currents act as stimuli and induce landing and probing.

9.9 Insect development and feeding behaviour

In most insects the feeding behaviour changes in different developmental stages with respect to their nutritional and quantitative requirements. As discussed earlier larvae and adults of most endopterygotes have different dietary requirements and mode of feeding. The mosquito larva, for example, has mandibulate mouthparts and feeds on planktons and aquatic algae; the adult has piercing-sucking mouthparts and is a terrestrial blood feeder. Larvae of many parasitic Hymenoptera are carnivorous on other insects, while the adults feed on nectar and honeydew. During the larval development substantial changes occur within and between instars. Simpson and Raubenheimer (1993a) discussed the concept of growth, nutrition, and intake targets. Barton Browne (1995) suggested two ways for developmental stage related changes in feeding behaviour. First, with respect to developmental stage there occur changes in the 'rate of intake'. Second, insects at different developmental stages may consume different relative amount of foods, which differ in nutrient and/or water content. He also suggested that the nutritional requirements of insects may be 'demand-mediated' or 'non demand-mediated'. In demand-mediated mechanisms the amount and kind of food ingested by an insect can be matched to the nutritional requirements of its body tissues, while changes in rate of nutritional uptake do not occur in non demand-mediated category. Neural or hormonal signals also involve in such elements.

The larval period of insects mainly supports growth and development and food intake of a larva increases progressively with development. Usually the total amount of food intake increases with each successive instar. Last larval instars of Lepidoptera are usually voracious feeder and consume maximum food. In many other groups feeding activity decreases in larval instar. In some phytophagous insects different instars feed on different plant parts, and thus consume different nutrients. In predators the size of prey likely to be consumed increases over successive instars.

Hochuli (2001) discussed the ontogenic changes and feeding behaviour in herbivorous insects. The larvae of herbivorous insects usually undergo significant size changes over several orders of magnitude during development, often exhibiting dramatic changes in feeding behaviour and appearance. Reavey (1993) reviewed the role of body size in caterpillar foraging and concluded that it was a key character influencing foraging. Among the general trends described, modes of nutrient acquisition (i.e. the way leaf material is ingested) and levels of gregariousness were identified as traits that often changed, as larvae grew larger. Early-instar caterpillars show a greater tendency to feed gregariously, with many species feeding solitarily in later instars.

The major factor influencing the mode of nutrient acquisition is head and mandible morphology, as smaller heads and mouthparts will facilitate selective feeding on leaves in a manner that is impossible in later instars (Reavey 1993). Caterpillars replace mandibles at each moult, offering the potential for several morphological strategies to be adopted throughout their feeding life. Similarly, the larger heads and mandibles of late-instar larvae facilitate the efficient ingestion of greater quantities of leaf material that is not possible for early instars (Bernays 1986). The consequence of this general shift is reduced assimilation efficiency in later instars, probably as a result of ingesting a higher proportion of low quality food (Slansky 1993). Many early-instar lepidopteran larvae selectively consume portions of the leaf that are low in fibre and easily digestible, but as they develop, become less discriminating and consume all.

Adult insects need nutrients for somatic and reproductive development. The nutritional requirements of females are more than males, because of ovarian development. Mature oocytes have high protein content and the females require protein rich foods. Females of mosquitoes require at least one blood meal in every ovarian cycle. The food requirements of adults may be one or more kinds of foods. Though, adults of some insects do not feed at all and use internal reserves acquired during larval feeding. This condition is known as 'aphagia', which is common in Ephemeroptera, some Lepidoptera, some Diptera and some Coleoptera. In these insects adults are primarily concerned with reproduction, mating and oviposition. They lay eggs on the hosts, which subsequently supply food for the hatched young.

9.10 Coevolution of plants and herbivorous insects

About 400 million years ago, plants and insects have been engaged in a continuing struggle for ascendancy in terrestrial environment. The plants are synthesizers of nutritional wealth, and herbivores are users of this wealth. In the course of evolution, each herbivore has become associated with a particular plant species or a group of plants. The maintenance of such associations is achieved by recognition of the host plant through successive herbivore generations. Any plant possesses a set of qualities, e.g., form, colour, texture, water content, and chemical compounds that characterize it and assist in recognition by the herbivore. The herbivore's sensory system (its detecting and coding abilities) is the underlying mechanism that allows for the recognition of given plant characteristics (Dethier, 1982). Ehrlich and raven (1964) described plant herbivore interface as the major zone of interaction responsible for generating the diversity of terrestrial life forms. Price (1984) pointed that all species must acquire sufficient energy to facilitate reproduction and against competitors. An attack on a plant by an insect selects for an adaptation for new defences by the plant, and this cycle of attack and counter attack between insects and plants is the essence of coevolutionary process. Plant secondary compounds acting as allomones and kairomones are key regulating factors of the coevolutionary processes and these relationships between plants, chemicals and insects are seminal in understanding insect speciation, distribution, ecology and behaviour (Metcalf and Metcalf, 1992).

Terrestrial insects have evolved remarkably efficient organs for feeding on plants, and a great variety of mouthparts with which they can pierce, chew or suck plant materials. They eat the leaves; bore stems and roots; make mines; form galls; devour flowers, fruits and seeds. The plants on the other hand, have obvious defences against their enemies including insects. The main lines of defence in plants are chemical and morphological. Spines or pubescence on stems and leaves, silica content are morphological attributes that deter some herbivores. Among chemicals secondary plant compounds, noxious phytochemicals or allelochemicals include: nitrogenous compounds (alkaloids, amines, amino acids, cyanogenic glucosides, glucosinolates); terpenoids (monoterpenes, sesquiterpenes, diterpenes, saponins, limnoids, cucurbitacins, cardenolides, carotenoids); phenolics (phenols, flavonoids, quinines); polyacetates (polyacetylenes), etc. (Harborne, 1993). These compounds may repel the insects or inhibit feeding. Insects on the other hand are adapted to the chemicals either by tolerating the biochemical stresses, detoxifying or even sequestering them.

The concept of warfare between butterflies and plants enlightens the details of evolutionary development. The plants must deploy their limited resources to protect themselves, while the insects reply with specializations to cope with plant defences. The success of angiosperms, which presently dominate the world, is probably due to presence of protective secondary plant metabolites (Ehrlich and Raven, 1967). As the flowering plants diversified, the insects also underwent a tremendous diversification with them. The intimate relationship between plants and butterflies leaves no doubt that the two groups evolved together each influencing the development of other.

Ehrlich and Raven (1967) discussed coevolutionary process in butterflies and plants. The host selection in butterflies is mainly chemical based. Small cabbage white butterfly, *Pieris rapae* larvae are attracted to mustard oil and feed mainly on cruciferous plants. The larvae consume other foods also, if mixed with mustard oil. In contrast, the larvae do not feed on plants of coffee family, as they are rich in alkaloids. Monarch butterflies have adapted themselves to feed on plants containing chemicals (cardiac glycosides) and such plants are unpalatable to others. Ehrlich and Raven (1964) used the term coevolution to confer the relationships between families of butterflies and their host plants. Coevolution is defined as 'evolution in two or more evolutionary entities brought about by reciprocal selective effects between the entities'. Janzen (1980) defined coevolution as 'an evolutionary change in a trait of the individuals in one population in response to a trait of the individuals of a second population, followed by an evolutionary response by the second population to the change in the first'. Thompson (1994) used the term coevolution to mean 'reciprocal evolutionary change in interacting species'. The extreme variations in interactions between plants and insects provide a unique wealth of data for theoretical considerations on the evolutionary processes that may have led to insect-plant relationships.

Besides the example of coevolution in butterflies and plants as given by Ehrlich and Raven, many other examples have been cited to discuss coevolution. Mutualistic relationships between insects and plants appear to have developed by coevolution. A really interesting case of mutualism is that of the relationship between ants of the genus *Pseudomyrmex* and plants of the genus *Acacia*. The ant protects the plant against herbivory and the plant provides the ants with protein, carbohydrate and a place to live (Janzen, 1966).

Another fascinating example of coevolution is that of figs, *Ficus* and fig wasps (Fam. Agaonidae) (Baker, 1961; Galil and Eisikowitch, 1969). The fig flowers are on the inside of a structure called a syconium, which develop into the ripe fruit. The female fig wasp enters the syconium through an opening called the ostiole. It pollinates the flowers inside the syconium and lays eggs into flowers with short styles. The eggs hatch in the ovaries of the short-styled flowers and the larvae consume the ovary tissue as they grow. The long-styled flowers do not have eggs laid in them, apparently because their ovaries are too deep in the fig wall for the wasp ovipositor to reach. The larvae develop into adult wasps. The male wasp is a wingless; it fertilizes the female, chomp a hole in the wall of the now mature fig, and dies inside the fig. The female leaves the fig through this hole, taking pollen with her.

The well-known obligate pollination mutualism between yuccas and yucca moths of the genera *Tegeticula* and *Parategeticula* is a major model system for studies of coevolution, and it relies on the key innovation in the moths of complex tentacles used for pollen collecting and active pollination. In this association yuccas are pollinated exclusively by yucca moths, whose larvae in turn consume some of the developing yucca seeds. This has been an evolutionarily and ecologically highly successful association. The evolution of tentacular mouthparts in the yucca moths is central to understanding the coevolutionary history of diversification and changing interactions between yuccas and yucca moths (Pellmyr and Krenn, 2002).

The parsnip webworms, *Depressaria pastinacella* (Lepidoptera: Oecophoridae) feed exclusively on a handful of furanocoumarin-containing hosts, including wild parsnip *Pastinaca sativa* and several species of *Heracleum*. These compounds are found throughout the parsnip plant, including its reproductive parts, which are the favourite food of the parsnip webworm. Few herbivores are capable of coping with the high concentrations of furanocoumarins found in wild parsnips. The webworms have a highly efficient detoxification system involving cytochrome

P450s and are well-adapted to deal with these compounds. Furanocoumarins are considered as potent defence chemical and survival of parsnip webworms on these chemicals is considered a case of coevolution between the herbivore and its host plant (Carroll and Berenbaum, 2002.) For coevolution to progress in this pair of interacting species, several conditions must be satisfied: (i) there must be genetic variation for characters in both plant and insect that influence the interaction between the species, (ii) each species must be a selective force on the other species (i.e., affect the other's fitness), and (iii) there must a response to selection in each species. The black swallowtail, *Papilio polyxenes* also feeds exclusively on plants that contain furanocoumarins, including wild parsnips. It copes with furanocoumarins via cytochrome P450-mediated detoxification.

Following theories have been proposed for coevolution of plants and insects.

I. Coevolution

Ehrlich and Raven (1964) proposed the theory of coevolution and quoted that:

(i) Herbivory is detrimental to plants and plants through mutations, recombinations evolve new chemical defence not directly related to their metabolic pathway.

(ii) Evolution of a new chemical defence could reduce herbivore attack and the compound allows plant to enter 'new adaptive zone'.

(iii) Subsequently, insect lineage might evolve a counter adaptation to plant defence; promotes species radiation of insect lineage into new adaptive zone created by the newly diverse plant lineage.

(iv) Ehrlich and Raven finally concluded: importance of 'reciprocal selective responses' between ecologically linked organisms has been greatly underrated as factor in origin of organic diversity

Cornell and Hawkins (2003) collected literature data on the floristic distribution and toxicity of phytochemicals to herbivores and on herbivore specialization and supported coevolution theory.

Thompson (1994) also used the term coevolution to mean 'reciprocal evolutionary change in interacting species'. He challenged the conventional co-speciation model in his book '*The Coevolutionary Process*' and mentioned that- interactions between plants and insects may not result in fixed traits in species that can then be mapped onto a phylogenetic tree, but instead there may be a shifting geographic mosaic, combining differences in the physical environment and the local genetic and demographic structure of populations.

II. Diffuse coevolution

Fox (1988) proposed that the coevolution must be considered in a community context and not simply as an isolated two species interaction. The community affecting one plant species includes several types of herbivores, alternative host plants, potentially competing plants and species in higher trophic levels that may respond directly to the herbivores or to the plants, changes induced by one member of the community may affect many species. For example, a particular family of insects feeding on a group of plant species may change hosts in evolutionary course.

III. Community coevolution

Valen (1973) proposed that interactions between the members of a community are so tight that the evolution of one member affects the evolution of all other members. Thus, in tightly coevolved interactions, evolutionary change by one species (e.g. a prey or host) could lead to extinction of other species (e.g. a predator or parasite), and that the probability of such changes might be reasonably independent of species age. Valen named the idea 'the Red Queen hypothesis' because, under this view, species had to 'run' (evolve) in order to stay in the same place (extant).

IV. Geographic mosaic coevolution

The geographic mosaic theory of coevolution suggests that much of the dynamics of coevolution between pairs or groups of species often occurs at a geographic scale above the level of local populations and below the level of the fixed traits of interacting species (Thompson, 1994). The theory predicts differences in the advance or trajectory of the coevolutionary process between local communities due to their composition and the strength of ecological selection pressures through competition and resource availability. The theory, therefore, suggests that the coevolutionary process is much more dynamic than is apparent from the study of individual populations or the distribution of characters found in phylogenetic trees (Thompson, 1994).

The geographic mosaic theory of coevolution provides a framework for asking how coevolution continually reshapes interactions across different spatial and temporal scales. It produces specific hypotheses on how geographically structured coevolution differs from coevolution at the local scale. It also provides a framework for understanding how local maladaptation can result from coevolution and why coevolved interactions may rarely produce long lists of coevolved traits that become fixed within species. Long-term field studies of the same interaction across multiple communities and spatially structured mathematical models are together beginning to show that coevolution may be a more important ongoing process than had been indicated by earlier empirical and theoretical studies lacking a geographic perspective (Thompson, 1999).

V. Sequential evolution

Jermy (1976) proposed that 'the evolution of phytophagous insects *follows* the evolution of plants, the latter being one of the most important selection factors in the evolution of insects'. He pointed that this theory seems to be suitable for the evolutionary interpretation of all existing insect-host plant relationships. Mitter and Brooks (1983) termed this as 'sequential colonization'.

Schoonhoven *et al.* (1998) discussed these theories and concluded: the evolutionary interactions between plants and herbivorous insects are asymmetrical. Plant evolution provided a profusion of niches: the chemically and structurally diverse plant species that enabled the radiation of herbivorous insects. Conversely, the insects exercised little if any influence on plant evolution.

9.11 Plant chemicals and feeding

Plant chemicals that affect behaviour have also been classified into two main groups, first the insect can utilize as nutrients as well as behavioural cues, and second which have no apparent nutrient value but which serve only as sign stimuli (secondary plant chemicals), enabling the insect to select the appropriate food or host. A number of plant chemicals have a pronounced influence on feeding behaviour of insects. These chemicals have been named with respect to the responses shown by insects. Dethier *et al.* (1960) proposed five terms for such chemicals.

Attractant. A chemical that causes an insect to make orientational responses towards its source.

Repellent. A chemical that causes an insect to make orientational responses away from its source.

Arrestant. A chemical that causes a decrease in the rate of movement and eventually lead insects to aggregate at the food source.

Stimulant. A chemical that causes an increase in the rate of movement and eventually lead insects to disperse away from the food source.

Deterrent. A chemical possessing the property of inhibiting the feeding of an insect.

Beck (1965) added two more terms 'incitant' and 'suppressant' to describe the chemicals affecting feeding behaviour in phytophagous insects. The chemicals that evoke or prevent biting or piercing actions are termed incitant and suppressant, respectively. The termed feeding stimulant (phagostimulant) and feeding deterrent are commonly used to refer the compounds that initiate or inhibit feeding. The feeding co-factors may enhance feeding by producing a synergistic effect when combined with feeding stimulants (Hsiao, 1985). The above designation of chemicals is not as rigid as many chemicals may have multiple effects. The relative importance of chemicals in feeding behaviour of insects varies significantly. Some insects do not require any chemical stimuli for feeding, while some depend on visual cues. Among phytophagous insects, mono- and oligophagous rely more on chemical signals.

Host ranges of phytophagous insects are determined to a large degree by plant chemistry. Specialist insects are often closely associated with plants that produce characteristic chemicals, which may act as attractants or stimulants to aid in finding or recognizing a host. Generalist insects are commonly believed to rely on the presence of repellents or deterrents to ensure avoidance of unsuitable plants. However, the chemistry of any plant can be highly variable, as a result of growth characteristics, genetic variation, or environmental factors. Such variable chemistry may provide windows of opportunity for non-adapted insects to utilize a plant or for a plant to become resistant to a normally adapted herbivore. Differences in insect responses to plant constituents may also result from genetic variation or environmental factors. Thus, the dynamics of plant biochemistry along with plasticity in the sensory system of insects might play a major role in the evolution of new plant-insect relationships (Renwick, 2001).

9.11.1 Attractants

Many secondary plant chemicals, belonging to different groups assist insects in their food finding behaviour. Some of these chemicals are widely distributed in a number of plant species, while others are restricted in a few species only. Green leaf volatile is a mixture of many organic compounds and acts as olfactory attractant for a variety of phytophagous insects. The feeding specialists are narrowly tuned to a few specific chemicals that are characteristic of their hosts. These chemicals play a decisive role in host locating behaviour and are regarded as token stimuli. Several cruciferous feeders are attracted to isothiocynates. Although the Australian tomato weevil, *Listroderes costirostris* is considered to be polyphagous, it inflicts particularly severe damage on plants in the families Cruciferae and Umbelliferae. The principal odours of these plants are characterized by volatile mustard oils and of these, allyl isothiocyanate and phenyl isothiocyanate are strong attractants for *Listroderes* larvae. These compounds also attract walking adult weevils and are found in mustard, cauliflower, cress, turnips, and nasturtium (Matsumoto, 1970). Similarly organic sulphurs of onion and leek attract onion fly, *Delia antiqua* and leek moth, *Acrolepiopsis assectella*, respectively. The Colorado potato beetle, *Leptinotarsa decemlineata* has a strong preference for cultivated potato. In the attraction of young codling moth the active attractant proved to be an obscure sesquiterpene called α-farnesene present in the cuticular coating of apples. Young silkworm larvae, which feed exclusively on mulberry leaves, are also attracted by the odour of their host plant.

Adults of bark beetles bore into the host plant for the purpose of egg laying and feeding. In temperate climates they generally feed on phloem/cambium of conifers while majority of them feed on xylem and phloem of broad-leaved dicot trees and shrubs. Monocot plants are fed by only a few tropical species. Species that feed on phloem have restricted host range. Some xylomyceptophagous ambrosia beetles carry symbiotic fungi to break down the xylem (Wood, 1982). Semiochemicals from host trees and bark beetles affect their behavioural actions. After emergence the adult bark beetles attempt to locate a host tree for colonization and reproduction. They find suitable host trees by orienting to host odours, especially ethanol, verbenone, and monoterpenes, as well as aggregation pheromones (Byers, 1995). Monoterpenes are most effective volatiles for attracting beetles, though in high concentration they are toxic. Quantities of volatile compounds such as ethanol and verbenone that are released from killed trees later during colonization by beetles and microorganisms are probably not regulated by the tree's genetics. Thus, trees are not under selection pressure from bark beetles because the tree has no control over how the insects use these compounds. The sensory organs of bark beetles play important role in host finding, host acceptance, and feeding. It is expected that the beetles have evolved their behavioural responses with respect to evolution of volatile chemicals in host trees. Two theories have been proposed on how beetles find suitable host trees (McMullen and Atkins, 1962). First, they locate such trees by orienting over several metres to volatile chemicals usually released by damaged or diseased trees (primary attraction). Second, the beetles fly about and encounter suitable host trees at random, whereupon they land and test them by short-range olfaction and gustation.

9.11.2 Repellents

Use of repellents is common for the protection of humans and domesticated animals. Though most known insect repellents are compounds selected artificially for pest management purposes, the negative orientation responses due to presence of chemicals in plants are not much known. Fewer examples are known of plant-produced volatile repellents. The catnip plant, *Nepata cataria* produces a cyclopentanoid monoterpene called nepetalactone that repels at least 17 species of phytophagous Hemiptera and Coleoptera (Eisner, 1964). Among polyphagous insects the desert locust, *Schistocerca gregaria* does not feed on lavender because the odour of the plant contains a volatile repellent, which causes even the hungry locusts to move away.

9.11.3 Arrestants

In the feeding sequence, once the insect has located the food plant an arrestant may prevent further searching behaviour. They may be perceived as odours or tastes and only a few such compounds have been isolated and identified. The coumarin produced by sweetclover arrests flight by the sweetclover weevil, *Sitona cylindricollis* (Heidweg and Thorsteinson, 1961). The ultimate effect of such a chemical is aggregation of the insects on the host plant.

9.11.4 Stimulants

Phagostimulants or gustatory stimulants fall into two broad categories: nutrient chemicals and secondary chemicals. In general it is true to say that phagostimulants for polyphagous insects tend to be ubiquitous nutrient chemicals (primary chemicals) and those for oligo- and monophagous insects both nutrient and secondary chemicals. Some secondary metabolites are products of plants metabolism and are characteristics of certain taxa. Plants produce them for defence, but specialized phytophagous insects that adapt to feed on these plants are able to overcome the defensive action and utilize the secondary metabolites as stimulants. Nutrient chemicals, because of their general occurrence, provide the basic ingredients for some key phagostimulants while secondary chemicals furnish specific cues in host recognition. Oligophagous insects respond only to a few nutrients and secondary chemicals, and are more rigid in their requirements of phagostimulants.

Numerous primary nutrients like sugars, amino acids and lipids act as feeding stimulants. Sucrose is a common phagostimulant for most phytophagous insects. Sugars are also important in regulating feeding of immature stages and polyphagous insects tend to respond to more sugars than oligophagous. In grasshopper as sucrose level increases up to 3-4 per cent (dry weight), consumption rate also increases. Aliphatic amino acids are also important phagostimulants and reactions of insects to amino acids are more specific than sugars. Other nutrients like vitamins, sterols, phospholipids, and nucleotides also serve as phagostimulants. Insects need ten essential amino acids (arginine, histidine, isoleucine, leucine, lysine, methionine, phenylalanine, threonine, tryptophan, and valine); among these, phenylalanine appears the most important for larval growth. Cuticle formation requires large levels of phenylalanine and it can be limiting to growth, survival, and reproduction. They require: cyclitols such as *myo*-inositol for membrane synthesis (especially plant feeders); vitamins like ascorbic acid, B complex, vitamin A; free fatty acids like oleic and linoleic acids; and sterols. Insects cannot synthesize sterols. Plant feeding insects modify phytosterols to cholesterol that is essential for synthesis of ecdysteroids (moulting hormones), surface waxes of the epicuticle and for membrane production. Synergism of phagostimulants with other chemicals is also recognized in some instances. Methyl linolenate, glucose and fructose act as feeding stimulants for phytophagous large twentyeitht-spotted lady beetle, *Epilachna vigintioctomaculata*. Although methyl linolenate alone was inactive, it acted synergistically with sugars, and plays an important role in the host selection of the beetle (Endo *et al.*, 2004).

Grass grubs, *Costelytra zealandica* respond strongly to sucrose, glucose, aspartic acid, and glutamic acid, all of which are common to most plants. The African black beetle, *Heteronychus arator* shows an unusual and strong response to maltose and sucrose. The phagostimulants for *Bombyx* larvae are principally sucrose, inositol, and cellulose that together with silica and phosphate promote continuous feeding (Sutherland, 1977). The oligophagous large white cabbage butterfly, *Pieris brassicae* relies largely on secondary plant chemicals for food plant selection. Sucrose certainly plays a role, but the presence of mustard oil glucosides is essential for continued feeding (Schoonhoven, 1967).

9.11.5 Deterrents

Feeing deterrents are vital in determining the host range of insects as these chemicals prevent insects from feeding. Factors like concentration in a plant and degree of tolerance in an insect determine that whether an insect would feed or not on a particular plant. Some phagostimulants also act as deterrents, if present in high concentration; or a compound may be deterrent for one species and phagostimulant for another (Chapman, 1974). Most insects detect deterrents via chemoreceptors and their reaction is dose dependent. In some insects an increase in tolerance may be acquired by behavioural manipulation (Schoonhoven, 1982). Deterrents may produce deleterious effects like abnormal growth and development, or even death. Mono- and oligophagous insects are highly sensitive to deterrents. Terms like antifeedants and feeding inhibitors are commonly used for feeding deterrents. A variety of chemicals exhibit such attributes and many of them occur naturally. Plant secondary chemicals like alkaloids and terpenoids are potential deterrents. Triterpenoids like azadirachtin and warbugnal are well-known for their deterrents actions.

Feeding specialists have greater sensitivity to deterrents than generalists. Bernays *et al.* (2000) supported this hypothesis by examining sensitivity of caterpillars of *Heliothis virescens*, a feeding generalist, and *H. subflexa*, a specialist, to eight different plant secondary compounds. All compounds deterred feeding, at least at the higher concentrations, but the generalist was less affected than the specialist.

The Indian neem tree, *Azadirachta indica* is a promising source of feeding deterrents. The leaves, twigs, and seeds of this tree contain at least 25 biologically active compounds that act as insect repellents, feeding deterrents, or growth regulators. Azadirachtin is the most abundant of these active ingredients. If the chemical is applied to the soil in which young bean plants are growing, it is absorbed by the bean roots, translocated to the growing points and protects treated beans from attack of desert locusts, *Schistocerca gregaria* (Gill and Lewis, 1971). The Colorado potato beetle, *Leptinotarsa decemlineata* has a strong preference for some *Solanum* species, but rejects other plants probably because of feeding deterrents. *Leptinotarsa juncta* and *L. texana* do not feed on potato as solanine act as deterrent for them (Hsiao, 1986). Similarly, the monophagous silkworm, *Bombyx mori* caterpillars do not feed on other plants because deterrents are present in virtually every other plant. Larvae of the tobacco cutworm moth, *Spodoptera litura* are polyphagous but there are some plants that they do not eat, and if extracts from the leaves of these plants are painted on to otherwise acceptable leaves, these are then rejected by the *Spodoptera* larvae (Wada and Munakata, 1968). Cucurbitacins, the bitter triterpenes common to all Cucurbitaceae, are thought to be potent feeding deterrents for all insects not adapted to exploiting cucurbits. Tallamy *et al.* (1997) suggested that cucurbitacins are generalized antifeedants for non-adapted mandibulate insects but may actually be phagostimulants for phytophagous insects with haustellate mouthparts.

9.12 Regulation of feeding

Feeding in insects is governed by a central excitatory state, which is affected by sensory input. External stimuli like olfaction, gustation; and internal stimuli like crop volume, nutrient concentrations in haemolymph regulate the feeding. Haemolymph plays a central role and provides a constantly updated summary of nutritional state (Simpson and Raubenheimer, 1993b). Responsiveness of sensilla can vary with other factors, viz. state of hunger (time since last meal, nutritional state, neurohormone levels, and associative learning). Centres for control of feeding reside in the suboesophageal ganglion, which is formed by the fusion of mandibular, maxillary and labial ganglia and regulate the activities of these mouthparts.

Insects exhibit variety of ingestion mechanisms, and therefore, a wide variety of behaviours are associated with feeding. In most cases insects depend on some direct stimulus from the food, i.e. a phagostimulant as perceived by the sensilla. Sensilla located on mouthparts initiate the feeding process. In black blow fly, *Phormia regina* tarsal sensilla stimulated by sugars leads to extension of the proboscis while labellar sensilla determine further consistency of food. In *Schistocerca* tarsal, maxillary and labial palp sensilla all need to be stimulated to initiate biting; after the first bite, quality of the food is monitored by cibarial receptors. Specific chemicals in the food may act as phagostimulants or deterrents, and their perception is species-specific. As a food may contain both stimulants and deterrents, its acceptance and feeding is determined from the ratio of both. Ablation and denervation cause a lot of redundancy in gustatory reception, e.g. surgical removal of maxillae in caterpillars can lead to hyperphagia or acceptance of normally rejected hosts. Stimulants/deterrents did not change time to initiate feeding, but did affect total consumption. Starvation had the strongest influence, as it decreases time to initiate feeding and increases food consumption.

Duration of feeding depends upon quantity and nature of previous meal, amount of energy spend in the interim, and food quality. In most insects feeding occurs at short intervals, insects like filter feeders feed continuously. In blow fly an increase in sugar in haemolymph slows the rate of crop emptying, which in turn increases meal interval. Termination of feeding is based on an end of the excitatory state, through chemosensory adaptation or habituation; negative feedback from gut stretch receptors; and negative feedback from nutrients or hormonal factors in the haemolymph. Feeding behaviour of any insect is regulated not only by its own endogenous rhythms and developmental physiology, but also by the life cycle and activity rhythm of its host species (Atkins, 1980). Environmental factors are also important in the establishment and maintenance of feeding periodicity.

In grasshopper regulation of food consumption depends on meal size, meal duration, and ingestion rate (Simpson, 1990). Palatability of food, duration of prior food deprivation, developmental stage, elapsed time within a developmental stage, and presence of other individuals nearby all affect meal size or duration. In addition, internal controls such as fluxes in amino acid concentration in the haemolymph can regulate feeding based on nitrogen needs through a series of physiological feedbacks (Simpson and Simpson, 1990).

The regulation of feeding has been thoroughly studied in black blow fly, *Phromia regina* (Dethier, 1976). The larva of blow fly derives all essential nutrients during feeding while adult need only water and sugars to fulfil their energy requirements. At the initiation of feeding taste receptors on tarsi are stimulated, and on finding suitable food proboscis is extended. Further consistency of food is evaluated by, mechano- and chemoreceptors on labellum. Food quality, and central and peripheral adaptive processes regulate the sucking activity. The flies maintain constant daily intake at a definite sugar concentration; and with a change in concentration the intake is affected. When the gut is filled; negative feedback from gut stretch receptors results into cessation of feeding.

Chapter 10

Reproduction

Insects are most successful and highly dominant animals on the earth. They are most abundant in terms of their number and species composition. The features of insects that facilitate their great ecological impact are: short life cycle, ability to move over impressive distances and amazing reproductive potential. Most adults are short-lived and much of their life is concerned with feeding and reproduction; however, some do not feed in adult stage. Reproductive behaviour of insects involves complex responses to environment, monitoring of internal stimuli and role of neuroendocrine system. These complex regulatory mechanisms and integration of various external and internal factors play important role in reproduction of insects. The reproductive behaviour of insects entails many intricate steps like mate finding, mating and oviposition, and insects synchronize these steps in accordance with cues from their environment.

Various aspects of insect reproduction have been dealt in the texts published by: Richards (1927), Alexander (1964), Davey (1965), Manning (1966), Wilde and Loof (1973a,b), Blum and Blum (1979), Thornhill and Alcock (1983), Cade (1985), Bailey and Ridsdill-Smith (1991), Choe and Crespi (1997a,b).

10.1 Mate finding

Mating involes the events occurring during insemination of the female by the male. A number of closely interlinked sequences of events occur during mating. The prerequisite of mating is that indivuduals of both sexes must aggregate or should be in the same vicinity. Insects' receptors play important role in aggregation. Mating behvaiour is characterized by signalling between individuals of opposite sex and by using these signals, recipient individuals exhibit their response. Mate finding by acoustical, mechancal, chemical and visual signals is common in insects. Bailey (1991b) discussed the role of sensory cues in mate finding in insects. Flashing of fireflies, sound production by crickets and cicadas, and use of chemical signals are spectacular examples which assist in mate finding. Their signals are species-specific, serving to attract individuals of opposite sex of the same species.

10.1.1 Aggregation

The simplest way to find a mate involves either active search by an insect or to wait for a mate. Some insects also respond to moving objects of appropriate size. In some cases recations of insects to environment bring them together. Movements of insects are useful for understanding of optimal searching for mates or oviposition sites. Males of some insects form substrate-based aggregations where they may defend a territory against conspecific males. The fruit flies (Fam. Tephritidae) have a lek mating system and mate on the host trees. The monarch butterflies, *Danaus plexippus* aggregate in large numbers and the aggregation is beneficial for mating as the females mate with as many males as possible. Each time the female mates with a male, the male passes to the female an energy rich spermatophore. It is believed that the female tries to get as many of these spermataphores as possible to replenish her energy supply for the remigration (Wells *et al.*, 1990).

10.1.2 Swarming

Swarming behaviour is commonly observed in insects like mayflies, caddisflies, stoneflies, many Diptera, some Lepidoptera, and some Hymenoptera. This behaviour is commonly regarded as an instance of insects reaction to environment. Swarms are usually species-specific, but mixed swarms have also been reported. Swarms are evident when a large number of indivduals are involved. Though swarming is not always concerned with mating and often involves male individuals only, but it acts as a visual marker. Usually the marker is some feature which contrasts with its background, e.g. a dark or light patch; silhouette of a branch against the sky; or tall objects. The insects maintain their positions by balancing themselves against the wind and adjusting their speed so as to remain at one place. Time of day and light intensity also affect swarming behaviour. Swarming is considered to increase sexual excitement and strengthens the signals received by femlaes in the vicinity. Females of some insects flying into or near the swarm may be captured by males and mating occur, while in insects like *Culicoides nubeculosus* and *Mansonia fuscopennata* females also fly in the swarm and mate (Downes, 1969). Svensson and Petersson (1987) showed that dance flies, *Empis borealis* form competitive mating swarm above bushes or trees. In this case swarm members are females and a sex-role reversal is observed as the 'lekking swarm' consists of females. Sexually active males carrying nuptial gifts (prey items) approach the swarm and mating followed after the male handed the prey to the female.

10.1.3 Acoustic signals

The night is a time when sound is the channel for communication. We can hear some of the sounds that insects use for communication. Noises in the night usually have something to do with sex. Insects advertise their presence in the dark by scent, light or sound. As a means of communication sound has one big advantage compared to visual display. Sound travels around corners so the caller can stay hidden while communicating with its own species and remaining concealed from predators.

Acoustic signals are used by number of insects, and in most cases males produce such signals to attract conspecific females. Males of insects including crickets, katydids, long-horned grasshoppers, cicadas, some flies are well-known for their sound production behaviour, while in mosquitoes females produce sound. Besides, some Coleoptera, Hymenoptera and Lepidoptera also produce sound. Some Hemiptera and Orthoptera emit acoustic signals as well as vibratory signals, which are usually distinct in terms of frequency and temporal pattern, and are produced by different mechanisms. Both signals may be important for localization of a conspecific mate. In tettigoniids, substrate vibrations are accessory signals in finding the precise place of the sound emitting conspecific, particularly within the complex three-dimensional environment of a woody plant (Stiedl and Kalmring, 1989). Other insects communicate primarily with either airborne sound, like many grasshoppers or with vibratory signals, like ants.

The most common method of sound production is stridulation or rubbing together of various body parts. Such frictional mechanisms are common in Orthoptera and Coleoptera. Many insects produce sound by vibrating their wings. Cicadas are well-known for their acoustic communication, but not for vibratory communication with the exception of some primitive species. They have specialized membrane that vibrate by muscular action. Male cicadas produce loud airborne signals, which attract conspecific females, or males in a few cases (Moore, 1993). In cicada, singing male usually sits on woody plant parts such as branches, or the trunk of trees. The receptive females of cicadas take short flights in response to calling songs of males and land on trees with calling conspecific males. Sueur and Aubin (2004) showed that courtship signals could act as distinctive species mating recognition systems in seven cicada species belonging to the Palaearctic genus *Tibicina*.

Among sound producing insects, crickets are the most thoroughly studied. Their sound production is concerned with calling, courtship, copulation, aggression and alarm. Male crickets often maintain and defend a territory that serves as a courtship and mating arena. They produce their songs by stridulation of their wings, and singing itself is an energy-intensive process. Before singing, males of some crickets make a small burrow. The calling behaviour is under circadian rhythm control. Crickets defend mating territories where they attack other males and prevent them from calling to females. Fighting behaviour results in a dominance hierarchy among males, and the dominant males call more and probably mate more than other males (Burk, 1983). Males express their calling behaviour on a daily basis in a predictable rhythm. The females are attracted to the calling song and males must be prepared physiologically and behaviourally to court and mate. After receiving the sound stimuli the female crickets also exhibit a number of sexual behaviours. The call song can stimulate virgin female into locomotor activity to find a mate. Female cricket phonotaxis for the conspecifics has been demonstrated in many species. Cade (1978) reported that phonotaxis response of female crickets increases several-fold when they are kept isolated from males.

After mating, the female is no longer responsive to calling songs, and she no longer expresses her nightly locomotor activity.

Haskell (1974) reviewed sound production in insects. Many species of sound producing insects form chorusing aggregations where males attract females and compete for them. The female of a grasshopper respond to a sound signal only when she is sexually mature. The sound signals are received by special sensilla located on tympanum or Johnston's organ. These songs are species-specific and usually used to attract the conspecific females. In *Amphiacusta maya* (Fam. Gryllidae) the calling song involve male-male aggression. The dominant male produce sound and the song maintain dominance within groups of males (Boake and Capranica, 1982). Temporal pattern of sound production and interval between each chirp are also important. The calling songs of male katydids are more complex in terms of their frequencies. Besides acoustical signals, grasshoppers also use visual cues in mating.

10.1.4 Mechanical signals

The vibrations produced by calling of males strike with the tympanum or other sensory structure and thus causing a mechanical change in the receptor organ. Some other mechanical signals are also important in mate finding in insects. The males of water striders (Fam. Gerridae) create surface waves by oscillations of legs, and these waves attract the females. Wilcox (1979) showed that these vibrations induce mating response in water striders and vision was of limited significance. Many true bugs also use vibrational signals for mate finding. Male green shield bug, *Nezara viridula* emit a volatile sex pheromone that acts as a long-range attractant to females. Both sexes communicate through vibrations once they are on the same plant. Males respond to the female calling song by emitting a male courtship song, and they orient to the female calling song on a plant. Simultaneity between vibratory and chemical communication during the last steps of mate finding suggests that pheromone emission might be modulated by signals from conspecifics (Miklas *et al.*, 2003). The sex-specific vibrational signals produced in *N. viridula* during the courtship are well suited for propagation through plants and to transmit the relevant information about the species and sex of the sender as well as provide the directional cue for locating the mate (Virant-Doberlet and Cokl, 2004). Cokl and Virant-Doberlet (2003) suggested that vibratory signals of plant-dwelling land bugs of the families Cydnidae and Pentatomidae are produced mainly by stridulation and/or vibration of some body part. Signals emitted by the vibratory mechanisms have low-frequency characteristics and are well tuned to the transmission properties of plants.

10.1.5 Visual signals

Many insects use visual signals for close-range mate finding and courtship. Short-range signals are more important as compound eyes of most insects are not well-suited for lonag range perception. Besides, the signals appropriate for noctural activity may not be suitable for day. The nocturnal species with the exception of bioluminescent displays, use other signals like acoustic or chemical. Visual signals like body movements, colours, patterns, and light production are used in sexual context. Magnus (1958) demostrated that males of silver-washed fritillary, *Argynnis paphia* respond to orange colour objects of any shape. In close range mate finding and courtship visual signals are very important. Otte (1970) reported that male of twostriped grasshopper, *Melanoplus bivittatus* jumps on the back of nearby female by using vision and if lands successfully on the female it holds the female tightly and make specific jerking movements. Males of North American grasshoppers, *Syrbula* species move their body appendages if the receptive females are nearer to them. The appendage nearest to the female is moved at a higher amplitude (Otte, 1972). Movement of body parts in sexual behaviour of *Drosophila* species is also well-known.

The fireflies (Fam. Lampyridae) are well-known for use of visual signals in mating. They have a light producing organ (section 16.1) and at night emit flash of light. The light producing patterns vary in colour, rate, length, and intensity. Receptive females respond to flashing males. In some species individuals of both sexes produce light and have a complicated signalling system. The interval between a male's flash and a female's answer is often species-specific. Females of fireflies of the genus *Photuris* mimic the signal pattern of *Photinus* fireflies. If an unfortunate male of *Photinus* species attempts to mate, the female eats him (Lloyd, 1965). Lloyd (1971, 1979) reviewed mating behaviour in fireflies and other luminescnt beetles.

10.1.6 Chemical signals

Behavioural responses to pheromones include searching for mates by one sex, aggregation of both sexes at a host plant, and dispersal of both sexes away from a specific area. Chemical signals are the most persistent and remain in the environment for extended periods, but their transmission is rather slow. A number of insects find their mates by using chemical signals. Chemicals emitted with the intension of altering sexual behaviour of recipient individual are termed as sex pheromones. These pheromones are very potent chemicals that are used by insects in very low quantity. The production and role of sex pheromones in mate finding behaviour of insects has been described in many insect orders (Jacobson, 1965, 1972). Usually female to pronounce her presence produces such chemical. Recipient males detect the odour plume, become aroused and orientate from downwind towards the source. In some cases males also produce pheromones. Insects of many species orientate towards volatiles emitted by their hosts to find mates in the vicinity of hosts. Of all the communication modalities, chemical communication is the slowest. Once chemicals are emitted into the environment they linger, and are no longer under the control of emitter (Baker, 1985). Sometimes these chemicals are detrimental if predators or parasites orient to a mating signal.

Males of bumble bee, *Bombus* employ their mandibular gland secretions to mark territory for mating. The marks produced by one species do not attract individuals of other species, but individuals of both sexes of a species come together and resulting into an increase in mating probability. The scent marks probably are also important for sexual stimulation.

Byers (1991) reported that a number of pine shoot beetles, *Tomicus piniperda* of both sexes aggregates *en masse* on fallen Scot spine that have been damaged during winter storms. Females soon locate suitable sites where they bore into the bark. Also at this time males wander in search of females that are boring into the bark. Upon encountering a female, male attempts to monopolize her by jousting with and displacing any resident male, during that time both males stridulate repeatedly. He presented a mate-finding model, which is simple in that it does not rely on spatial memory or on long-range orientation of the animal. The searching sex (male) is 'captured' when an individual enters or intercepts the circular area of the female. In cockchafers, *Melolontha* mate finding is chemically mediated. Mate location in cockchafers occurs during a spectacular swarming flight performed mainly by males at sunset. The males orientate towards volatiles emitted by damaged green leaves in order to locate females that stay during this time within the host trees and continue to feed (Ruther *et al.*, 2002). In order to distinguish between unspecific leaf damage and damage caused by feeding females, male cockchafers orientate by a sex attractant, 1,4-benzoquinone, a compound that has been known as a highly effective defence compound used by numerous arthropod species to repel enemies. Thus, the sex pheromone of cockchafer, *Melolontha hippocastani* might have evolved from a defence compound (Ruther *et al.*, 2001).

Eggert (1992) observed that male burying beetles, *Nicrophorus vespilloides* (Fam. Silphidae) use two alternative mate-finding tactics: (i) search for carcasses that serve as oviposition sites, or (ii) attract mates via pheromone emission. But there were significant differences among males in the time they spent employing either tactic. Time of day also affected the tactic used. The search tactic results in higher reproductive success than the attraction tactic. Gupta and Sehgal (1997) observed that in red cotton bug, *Dysdercus koenigii* olfaction appeared to be the basis of mate finding.

In many cases natural enemies detect their hosts or prey by using the mate finding signals. Thus, reproductive and foraging behaviour may conflict with defence behaviour. Many insects take risk to predators in relation to either mate finding or foraging behaviour. However, the level of risk-taking might depend on the behavioural context in which the insect is involved and the sex or the mating status of the animal. Males of silver-Y moth, *Autographa gamma* orient towards a pheromone-releasing female conspecific. While engaged in these olfactory-induced activities moths defend themselves against their most common predator, echolocating bats (Skals *et al.*, 2003). Most moths are equipped with a pair of ultrasound sensitive ears by which they can detect echolocating cries of insectivorous bats (Roeder, 1965). Acharya and McNeil (1998) found that pheromone-based mate searching behaviour in moths depends on the level of predation risk. Female moths also take risks when exposing themselves to bat predation when foraging for nectar or searching for oviposition sites. Svensson *et al.* (2004) demonstrated that males of *Agrotis segetum* and *Plodia interpunctella* orienting towards a sex pheromone accepted the predation risk when attracted to pheromone sources of high quality. The risk taking behaviours of these moth species seem to be similar in spite of the fact that their ultrasonic hearing systems have evolved independently.

Tooker and Hanks (2004) observed that males of the gall wasp, *Antistrophus rufus* use olfaction rather than visual or tactile cues in searching for mates. The plant stems that contain larvae soon to emerge as female wasps produce chemicals that differ from the mix produced by uninfested plant stems. The male wasps can exploit the chemical composition of plants for the purpose of locating mates. Female wasps lay their eggs inside the stems of several species of perennial prairie plants. When the eggs hatch, larvae feeding inside the stems cause the plant to alter the balance of (+)- and (-)-enantiomers of several terpenes, including α- and α-pinene. Months later, the male wasps emerge from the stems before the females do and use these altered ratios to locate stems containing females.

10.2 Territoriality

An organism's ability to hold a territory is referred to as resource holding power. The term territoriality is defined as 'the pattern of behaviour associated with the defence of a territory'. The resources likely to be defended may be food, females and oviposition sites. Markl and Lindauer (1965) defined a territory as "the area in which a male, ready to mate, awaits a female, or an area to be used for oviposition". In many insect species territoriality is a strategy used by a male to obtain increased mating. Insects may simultaneously defend multipurpose territories with mates, shelter sites, and food. Some insects only defend a food resource while some social species stake out and protect foraging territories. Males of bees, wasps, butterflies, true flies, dragonflies, crickets, fireflies, some bugs and some social insects establish territories. Insect territoriality has not been extensively studied because physical aggression is less obvious in insects. Besides, insects such as butterflies, dipterans and many others are poorly equipped for physical aggression (Baker, 1983). The territorial activities are incurring an expense of energy, and selection for such behaviours tends to operate only under a limited set of conditions (Otte and Joern, 1975). Many kinds of insects including flies, beetles, wasps, bees, some dragonflies, and butterflies are known to congregate at locations termed 'leks', where there is no food, water or beneficial resource, and only the individuals of the same species are present. These leks are simply a place to locate mates. The lek mating system of tephritid fruit flies has represented a fascinating area. Tyschen (1977) described lek formation in Queensland fruit fly, *Bactrocera tryoni*. Lekking studies in these flies include aspects like interspecific lek formation, lek size, and interactions between two species of flies on the same host tree, male-male aggressiveness, territory defence and acoustic signalling.

The resource defence occurs when the resource is limited or the benefits derived from restricting the access of competitors to these resources must be greater than the cost of resource defence. An insect can defend a resource without showing obvious physical aggression, but many insects defend resources by physical aggression or other forms of advertisements like physical conspicuousness, acoustic signalling and scent marking. The wing colouration of many male butterflies, especially the ultraviolet flashing that result from the flapping wings advertises the presence of a territorial male to other males (Baker, 1983). Males establish two types of territories, the first and most common are territories that contain a resource, e.g. food or oviposition site while the other contains no resource. The type of territory and the resource being defended greatly affects the behaviour of the territory holder. Establishment and defence of territories have been extensively studied in dragonflies and crickets. Physical combat occurs in insects like predatory flies, some beetles, crickets, and some Hymenoptera. The physical combat may results in minor injuries but usually does not harm seriously to either side.

Dragon- and damselflies establish aerial territories often on a plant part that overhangs the water surface whereas some remain in flight while looking for both prey and mates. Odonata species that remain at one site often patrol the area for mating opportunities. This territorial behaviour increases reproductive fitness. When a male has established a territory, several other males make attempt to enter the site hoping to find a receptive female. Once these males enter the territory, several types of territorial behaviours may be displayed. The territorial male will leave his perch to chase away conspecific males and flutters its wings to warn intruder. Brief clashes lasting only a few seconds may also occur resulting into one male leaving the territory. The winner returns to the territory and perches, waiting for the arrival of a receptive female. Of the many displays of territorial behaviour, variation exists to a large degree possibly due to several factors such as larval ecology, residency, time of day, weather conditions, and age and size of the dragon- or damselfly. It is suggested that larger males exhibit greater success of defending larger territories that are highly desirable to female conspecifics. Despite this fact, variation persists in the insect orders.

Alexander (1961) discussed aggressive and territorial behaviour of male crickets. Encounter of two males may result into a number of interactions but finally one has to withdraw. The most intense encounters occur between evenly matched males. Fighting may be in the form of antennal lashing, sparring with fore legs, kicking, wrestling with mandibles; however, in spite of apparent violence of the battle, defacement rarely occurs. The occupation of a territory is indicated by the occupation of a burrow and the same also results in the reduction of male-male encounter. Stridulation is known to influence the territory defence of many Orthoptera species (Baker, 1983). Besides, in many beetles stridulation has been shown to act as an antipredator defence device. Another possible effect of acoustical signalling is to advertise that this area is occupied.

Females of some insects also establish feeding territories as food resources that not only influence their survival but also directly determine the number of eggs to be produced. A few females also establish mating territories, but this behaviour is more pronounced in males. The territorial behaviour difference between males and females of the water strider, *Metrocoris histrio* suggests that competition for space be more severe in males than in females. Since only, territorial males excluded intruders of the same sex from holding territory in the same pool therefore, males' territories were much larger than those of females (Koga and Hayashi, 1993). For mating, males of fruit flies, *Bactrocera* species take up individual territories on leaves and sexually active males produce high-pitch buzzing sound by a stridulatory mechanism, which probably acts as specific releaser (Bess and Haramoto, 1961). They also release pheromones from rectal glands, which attract females and evoke copulatory responses and their role in courtship and mating of *Bactrocera* species has been well documented (Koyama, 1989).

Davies (1978) described the territorial behaviour of male speckled wood butterfly, *Pararge aegeria* that defend sunlight spot on forest floor, and if an intruder enters the occupied territory, the butterfly challenges it. The two butterflies make spiral flight upward and then one return to the light spot. He concluded that spiral flight communicates that the space is occupied and is readily accepted by the intruder. The behaviour assures occupation of mating territory with minimum energy expenses. Males of butterfly, *Vanessa kershawi* occupied perch territories in the late afternoon in sunspots on hill-tops and other locations. The perching males responded to intruders with chases and ascending flights, with some individuals defending the same perch site for several consecutive afternoons. The females visited the territories strictly to mate and not to use any resource in these areas (Alcock and Gwynne, 1988).

Territory size can vary greatly among insects and is determined by how much an individual can afford in its maintenance. A territory with average resources and small foraging area is easier to defend than a larger territory with scarce resources. Body size also plays an important role in territoriality. In water strider, *Aquarius remigis* (Fam. Gerridae) the winged females are considerably larger than wingless females and the territorial behaviour in winged females is more distinct than wingless females (Kaitala and Dingle, 1992).

The males of cicada killer wasp, *Sphecius speciosus* emerge before females and each male establishes a territory and drive off any intruder, and sometimes grapples with conspecific male. They exhibit some tendency to mark the boundaries between adjacent territories so that the grappling is reduced. If a receptive unfertilized female enters the territory the male follows her and mate (Lin, 1963). Territories held by male bees are either holding territories or the bees patrol a home range. The solitary anthophorid bee, *Anthophora pauperata* uses physical conspicuousness as advertisement of territoriality. Males establish territories in areas where females are likely to occur. Some males will patrol their territory while others wait and perch on a site watching over their territory. In either case these territorial males will actively confront any intruders. They also exhibit aggressive behaviour if any other insect enter in their area. When a male encounters another male of the same species within his territory, a chase results which continues until the intruder leaves the territory. Sometimes the confrontation escalates to a point when the two bees are grappling and fighting with one or both males falling to the ground (Willmer *et al.*, 1994).

10.3 Aggression

Aggression is overt behaviour intended to inflict damage on another individual. Konrad Lorenz argued that aggression in animals is an inherited, spontaneous tendency much the same as the motivation to eat, drink and make love. Males of many insects exhibit aggressiveness if the females are scarce. Males usually compete for access to females while females also show some form of mate choice. Thus, the reproductive success of males is comparatively less as compared with females. The aggressive responses of males may be intraspecific or sometimes to males of other species. Aggression behaviour has been studied in insects like honey bees, ants, termites, bees, wasps, beetles, crickets, grasshoppers, dragonflies, and vinegar flies. Males defending their territories, searching for mates or engaged in courtship enter into clashes with other males. Some Orthoptera respond to other calling males with a rival song while rivalry clashes also occur among insects. In crickets fighting occurs between rivals and becomes fiercer with time. The status of a cricket during a clash depends on age, mating status, burrow occupancy, and his dominance in recent fighting (Alexander, 1961).

The aggressive encounters between males of Madagascar hissing cockroach, *Gromphadorhina portentosa* are quite impressive. Males ram into each other with their horns and/or they push each other with their abdomens. Larger males usually win. Hissing plays an important role during male-male encounters and winners hiss more than losers. Bhandari and Pathak (1997) observed violent movements between two males of German cockroach, *Blattella germanica* whenever they were kept with a female (>4 days old). When two mature males were kept with one mature female, the males always displayed aggressive behaviour followed by submissive behaviour in one. Aggressive behaviour included violent movements of the antennae and head pushing. Fighting is often composed of discrete agonistic displays. Tachon *et al.* (1999) evaluated the relationships between the fighting behaviour of male field cricket, *Gryllus bimaculatus* and its potential benefits in mating success. They observed that: (i) the presence of females elicited an increase in agonistic displays characteristic of intermediate levels of escalation, (ii) female scents did not produce a similar increase in the frequency of agonistic displays, and (iii) in the presence of females, the frequency of agonistic displays was positively correlated with mating success. Aggressive stridulation, an energetically low-cost display, was more strongly associated with mating success, as compare to other more costly displays.

Hercules beetles are sexually dimorphic in respect to the head and pronotum, the males bearing horns and/or deep excavations, the females either without, or with less strongly developed armature. The spectacular clash between two male beetles for dominance and access to a nearby female is well-known (Plate 10.1). The loser is slammed down or carried some distance and banged to earth. The female remain inert during the clash, even though the victor may pick up her with glee and carry her aimlessly about (Beebe, 1947).

Spieth (1974) discussed aggression and territoriality in certain Hawaiian *Drosophila* species. The components of fighting behaviour in adult male fruit flies are: (i) wing threat, in which both wings are elevated for a sustained period of time; (ii) fencing, in which flies push off with one of their legs in a sideward or forward direction; (iii) boxing, in which both flies rise on their hind legs and thrash at each other with their fore legs; and (iv) chase, seen when a hierarchical relationship has been established, with the winner pursuing the loser who will either fly or retreat.

In bottle brush weevil, *Rhinostomus barbirostris* males use their elongate rostrum to flip male rivals from the log on which receptive females are mating and ovipositing. The males are of variable sizes and small males are inferior fighter (Eberhard, 1983). Males of Dawson's burrowing bee, *Amegilla dawsoni* vary greatly in size and the occurrence of unusually large males appears to be the product of sexual selection for fighting ability. Males regularly competed aggressively to mount sexually receptive, virgin females as they emerged from the ground. The larger males usually won fights for potential mates and the fighting advantage of large males translated into greater mating success (Alcock, 1996).

10.4 Female stimulation

In some insect species, there are no signs of rituals preceding mating; however, in many cases the female is not immediately receptive and need to be stimulated by the male prior to mating. Numerous courtship rituals in insects have developed through evolution, and these must be precise routines that are unique characteristics of a species. Many non-receptive females repulse males by exhibiting various behavioural displays. The males often stroke females, by use of legs or antennae. Dance patterns may be performed, wings may be fluttered or moved in circles, or short flights may occur. Female of *Drosophila* requires intense stimulation from a given male. She permits a male to mount her when she spreads her wings and genitalia while non-receptive female flicks her wings and twists the abdomen sideways (Petit and Ehrman, 1969). In many *Drosophila* species the mating process begins with dancing (rhythmic movement). They perform elaborate courtship dance that ends with tapping and stroking of the female's body by the male with his legs. The male also touches the female with the mouthparts or even stimulate the female's genitalia with his mouthparts. In predatory insects before mating, appeasement of female by the male, i.e. inhibition of predatory instinct of the female is necessary. Courtship feeding or nuptial feeding where a male provides a food gift to the female is another mechanism of female appeasement. The male of bug, *Stilbocoris natalensis* offers a seed to the female as a nuptial gift. The seed is partially predigested with saliva by the male and he stimulates the female by offering the seed (Carayon, 1964). Scorpionflies, some *Drosophila* and many other insects are nuptial-gift giving insects (Alcock and Gwynne, 1991).

10.5 Sexual selection and courtship

Darwin (1871) explained the concept of sexual selection, which explains the reason for sexual divergence in terms of differential ability of individuals for different genetic types to acquire mates. He argued that various anatomical, physiological and behavioural mechanisms are involved in mate selection. Richards (1927) discussed the subject of sexual selection in insects. In some insects strong selection pressure exists for a potential mate to be convinced. The females are particular in discriminating potential mates, and the mates should satisfy them by performing elaborate courtship displays or should satisfy the females in terms of their superiority. Sexual selection involves choices between sexes which is mainly based on interactions between males or sometimes also between females. The theoretical basis of sexual selection is that individuals attempts to do best which yields the greatest number of surviving progeny. This expectation has been responsible for a major reinterpretation of several elements of insect reproductive behaviour (Alcock and Gwynne, 1991). Earlier such patterns were explained by taking acoustic or visual signalling in terms of benefits gained by insects; however, modern studies suggest that such patterns can be outcome of competition among males to avoid masking of their individual signals or may arise to prevent other males from monopolizing the channel of communication to females (Thornhill and Alcock, 1983).

In many species, females prefer mates with extreme characteristics that are apparently useless or deleterious for survival, such as bright colours, elaborate ornaments and conspicuous songs (Andersson, 1994). Such mate choice by the females can be understood as a means for females to increase their immediate fecundity whenever females receive direct benefits, such as improved territory quality or paternal care, by choosing certain males over others as mates (Price *et al.*, 1993). Females of most species; however, do not seem to receive any appreciable direct benefits from exercising mate choice, and the evolutionary mechanisms responsible for the origin and maintenance of such female mate choice remain an evolutionary enigma (Andersson, 1994). It has recently been suggested that sexual conflict over mating may result in the evolution of mate choice. In this scenario, males evolve to attract females into mating and females evolve resistance, rather than attraction, to males in order to reduce direct costs associated with mating. Bonduriansky (2001) synthesize the empirical evidence and theories pertaining to the evolution of male mate choice and sex role reversal in insects, and examine the potential for male mating preferences to generate sexual selection on female phenotypes.

Mating systems are viewed as evolutionarily dynamic resolutions of sexual conflicts or interests (Alexander *et al.*, 1997). The individuals display behavioural options and interactions, in a system that each individual's selection of tactics for profiting more from a given interaction, while paying less, thereby directly affects each participant's fitness. Sexual conflict is thus at the heart of evolutionary ecology. Parker (1979) stressed that a clear demonstration of sexual conflict requires a quantitative assessment of the potential costs involved for both sexes; however, it is difficult to perform such assessments. Costs of various components of mating, such as male display behaviour or female coyness, are key considerations in most models of sexual selection, since some form of evolutionary balance between benefits and costs is assumed (Andersson, 1994). Again, despite their critical importance for testing sexual selection models, few studies have made quantitative estimates of the costs of sexual behaviours or traits. In models of sexual selection by female choice, costs of mating and mate assessment alter optimal mate choice behaviour (Andersson, 1994). The potential costs of mating to females include time devoted to mating, energetic costs of reproductive behaviours, increased risk of predation while mating, risk of disease or parasite transmission, and risk of death or injury inflicted by the male.

Many studies in insects have shown that males provide material resources to their mates. These resources that directly benefit the female include the nutrients that female consumes, care for her offspring, or protection from predators. Females that mate with multiple males can, therefore, acquire more resources and increase their reproductive success. Vahed (1998) reviewed the function of nuptial gift in insects. Nuptial feeding encompasses any form of nutrient transfer from the male to female during or directly after courtship. These gifts may be in the form of food captured or collected by the male; parts, or even the whole of the male's body; glandular products of the male; spermatophore; and substances in the ejaculate. There has been considerable debate over the current function of nuptial feeding in insects, centred on the issue of whether nuptial gifts function as paternal investment (i.e. function to increase the fitness and/or number of the gift-giving male's own offspring) or as mating effort (i.e. function to attract females, facilitate coupling, and/or to maximize ejaculate transfer). Experimental evidences suggest that many diverse forms of nuptial feeding in different insect taxa function, at least in part, as mating effort. However, in some mating systems, males provide only sperm (genes) and no direct benefits. In these systems, females might benefit indirectly by obtaining good genes, and when it is difficult to assess male quality prior to copulation, the female may mate with multiple males and rely on post-copulatory cues, such as sperm competitiveness, to select the ideal father.

When the costs of mating are high relative to the benefits, a multiple mating system may be selected. The benefits of multiple mating may be:

(i) Prevention of sperm depletion.

(ii) Encouraging oviposition.

(iii) Finding a genetically superior male.

(iv) Genetic diversity among offspring.

(v) Encouraging sperm competition.

(vi) Regularly obtaining nutrients from males.

(vii) Avoiding mating with mates having low sperms or nutrient reserves.

(viii)Increasing effectively population size.

(ix) Higher reproductive success.

Bateman (1998) reported that female of the cricket, *Gryllus bimaculatus* prefer to mate with a novel partner, rather than one with which she had mater earlier. Arnqvist and Nilsson (2000) reviewed the evolution of multiple matings and female fitness in insects.

Body size is a signal used by many animals to assess the quality of a mate. Measurements of mate quality, including overall energy intake, genetic quality, and other components of fitness, are associated with larger body size in both sexes (Andersson, 1994). Larger males often achieve higher mating success by winning competitions for mates or through pre- or post-copulatory mate guarding. Females may benefit by choosing larger males because these males have higher lifetime mating success, higher genetic quality or more vigorous copulatory courtship (Eberhard, 1996). Males may benefit by choosing larger females when larger females have higher quality offspring or greater fecundity (Andersson, 1994). Females of chocolate moth, *Ephestia elutella* reject small males and mate choice is adaptive because females mating with larger males produce more offspring and also progeny that are significantly larger (Phelan and Baker, 1986). Thus, females select genetically superior males, whose offspring will be more robust.

Adult male *Onthophagus* beetles are dimorphic and can be divided into two distinct types: 'major' males with head horns (Fig. 10.1), and 'minor' males with rudimentary horns or are hornless. Larvae growing larger than a genetically determined threshold size metamorphose into adults with long horns, and larvae not reaching this size metamorphose into adults lacking or with very short horns. Beetle horns have been shown to operate as a tool in intrasexual combat. The smaller males may engage in an alternate, less aggressive behaviour. Large, horned beetles are found guarding entrances of tunnels that contain females. Small, hornless males adopt a sneaking behaviour by digging new tunnels that intercept the guarded tunnels, remaining undetected. Cook (1988) showed that female of dung beetle, *Onthophagus binodis* mated to horned males produce more progeny than those mated to males without horns. Males are highly variable in this species and may assist females in parental duties.

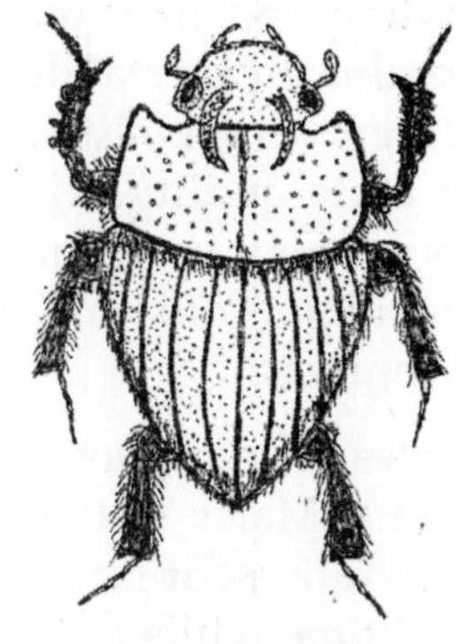

Fig. 10.1. Male *Onthophagus* beetle with head horns.

The males of winter moth, *Operophtera brumata* had a higher probability of attempting to copulate with large, more fecund females, while females did not show any mate discrimination. Mate choice was possibly based on the variation in female pheromone quality and/or quantity (Dongen *et al.*, 1998). Female of bella moth, *Utetheisa ornatrix* (Fam. Arctiidae) mates preferentially with larger males, thereby potentially deriving both direct phenotypic benefits and indirect genetic benefits. Its larva feeds on plants of the genus *Crotalaria*, containing poisonous pyrrolizidine alkaloids. The larva stores the chemicals systemically, retaining them through metamorphosis into the adult stage. At mating, the male transfers a substantial fraction of his alkaloidal load to the female with the spermatophore. The gift is transmitted by the female in part to the eggs, together with a supplement of her own alkaloidal supply. Having a larger father results in the eggs being more richly endowed with defensive pyrrolizidine alkaloid; and having a larger father also results in the sons and daughters themselves being larger (body mass is heritable in *Utetheisa*). Iyengar and Eisner (2002) provided evidence that these consequences enhance the fitness of the offspring. Eggs sired by larger males are less vulnerable to predation (presumably because of their higher alkaloid content), whereas sons and daughters, by virtue of being larger, are more successful in courtship and more fecund.

Thornhill (1976) demonstrated adaptive mate choice by female of scorpionfly, *Hylobittacus apicalis*. The female goes from one male to another and the male provides a prey as food gift to female. The female favoured the male that provided larger nuptial gift and rarely mate with male that offers small prey. The mating duration is also determined by the size of prey and if the prey is small, mating terminates earlier. Thus, male presenting small prey might not mate at all or if mate, he would lose sperm competition with other male able to provide larger prey.

Brown (1999) reviewed the evidence for female mate choice in tree crickets. The evidence shows that female choice may occur during several different stages of mating and may target several different criteria. Song preferences are perhaps dominated by stabilizing preferences for the cues of species recognition, but there is a growing body of evidence for directional preferences based on sensory biases or mate quality. Mate rejection during courtship and forms of postcopulatory choice may favour males, based both on phenotypic quality and on the amount of nutritious courtship gifts they provide, and may differ with the value of mating incentives.

Males of many insect species choose their mates according to their reproductive status. Males of the mustard white butterfly, *Pieris napi* transfer large ejaculates that represent on average 15 per cent of their body mass when mating for the first time while in second mating transfer only a small ejaculate. The females use male-derived nutrients for egg production and somatic maintenance. Mating with a recently mated male increased female mating cost by increasing time in copula and mating frequency. Virgin females that mated with virgin males remated after an average of 6 days, whereas virgin females that mated with recently mated males remated after an average of 2 days. Moreover, copulations involving virgin males lasted on average 2 hours. Recently mated males were eager to remate, in spite of the fact that the size of the ejaculate was small, and they remained in copulation for a long time (Kaitala and Wiklund, 1995). Wedell and Karlsson (2003) demonstrated that in *Pieris napi* females increased their reproductive effort after receiving large male donations. By contrast, in the speckled wood butterfly, *Pararge aegeria* where males do not provide nutrients, female reproductive effort was independent of male ejaculate. Increased reproductive effort in *Pieris napi* resulted from the production of more eggs, rather than from investing more resources per egg. In this species donating ability is heritable, hence females laying more eggs after mating with high-donating males benefit both through higher fecundity and through the production of high-donating sons.

Carazo *et al.* (2004) tested male mate choice in yellow mealworm beetle, *Tenebrio molitor* and observed that males prefer to court and mate with mature and virgin females. The males were able to assess female reproductive status on the basis of chemical cues. Bateman and Ferguson (2004) reported that males of the tettigoniid, *Acanthoplus discoidalis* produce a large sperm ampulla and associated spermatophylax. Female mates before laying eggs and subsequent matings are random. The males, therefore, only have a high confidence of paternity when mating with virgins and reject non-virgins more often. Males seem to prefer females of a lower mass, as the higher mass may reflect increasing female age and her likelihood of previous matings. Pureswaran and Borden (2003) suggested that reproductive success in the mountain pine beetle, *Dendroctonus ponderosae* is determined by pheromone production and body size. The males paired with females produce more antiaggregation pheromones, thereby suggesting that after pairing, males take over the role of pheromone production. The males could also select large females by evaluating gallery size, and females could select large males on the basis of their strength in stridulation, but the mate choice occurs primarily by olfaction.

Males of Mediterranean fruit fly, *Ceratitis capitata* visually display courtship behaviour while perching on the underside of a leaf by curling the abdomen upward and releasing a pheromone attractive to females. Upon seeing a female, the male orients toward her, tucks his abdomen under his body, and begins to vibrate his wings. When the female approaches nearer to male, he begins a second type of wing movement termed 'fanning'. In addition, the male begins a series of side-to side movements with its head. The male then moves closer to the female and if the female remains motionless, leaps onto the female back and attempts to mate. Courtship behaviours by wild males usually follow a predictable sequence of male and female behaviours:

Male calling ⇒ female approaches ⇒ male wing vibration ⇒ female standing ⇒ male wing fanning ⇒ copulation (based on Liimatainen *et al.*, 1997).

Males of scorpionfly hunt for prey and food items that are offered to the female, and eaten by her during copulation. The female is first attracted by a pheromone emitted by one or more vesicles at the end of the male's abdomen. The female examines the offering while the male searches for her genitalia. If the gift is rejected, the female flies away. If the gift is accepted, the genitalia of the male couples with that of the female, who lowers herself until she is hanging upside down. She consumes the offering during copulation. The male supports the female by holding her legs or the prey. Both sexes mate several times per day and mating duration depends on the size of the gift. It has been observed that prey 3 to 14 mm long will provide 1 to 17 minutes of copulation in *Harpobittacus australis* (Alcock, 1979).

Eggert and Sakaluk (1994) reported that female of sagebrush cricket, *Cyphoderris strepitans* feed on the fleshy hind wings of the male during copulation, and consume haemolymph flowing from the wounds they inflict. Males without hind wings obtained fewer matings and more vulnerable to injuries inflicted by females on other body parts. The focus of providing nuptial gift remains gender-specific, with a male donor and female recipient.

The female praying mantis attracts the male by secreting a pheromone and shows her receptiveness to mate. The male slowly approaches her with watchfulness. Usually the male slowly approaches the female from front or back for mating. The male jumps on her back, they mate, and he flies away quickly. The mating lasts for about half an hour but sometimes the mating duration is long. The female praying mantis is well-known for its cannibalistic mating process but it does not happen in all cases. Sometimes the both sexes perform a ritual dance before mating, stroking each other with their antennae. In European mantid, *Mantis religiosa* usually larger female eats the male before, during or after mating (Fig. 12.1). Decapitated male is even known to copulate more vigorously and ejaculate faster because of loss of suboesophageal ganglion. Sexual cannibalism occurs most often if the female is hungry. Further, suggestions of gains in reproductive fitness of the male via indirect nutritional benefits to his offspring are negated by the obvious unwillingness of the male to participate in the ultimate nuptial sacrifice- his own life (Gullan and Cranston, 2000).

Arnqvist *et al.* (2003) showed that females of the Zeus water bugs, *Phoreticovelia disparata* provide glandular secretions for males during mating. This previously undescribed reversal of gender roles indicates that nuptial feeding might not be related to paternal investment.

Sometimes intrasexual mounting has also been observed in insects. Harari *et al.* (2000) reported that the citrus weevil, *Diaprepes abbreviatus* shows three kinds of same-sex mountings: males mount other unpaired males, males mount males already engaged in copulation and females mount other females. They evaluated four hypotheses in order to explain same-sex matings by males: (i) female mimicry by inferior males, (ii) dominance of larger males which affects the behaviour of small males, (iii) sperm transfer in which smaller males gain some reproductive success by 'hitchhiking' their sperm with the sperm of larger males, and (iv) poor sex recognition. Data from mate choice and sperm competition experiments supported the poor sex recognition hypothesis.

Otte (1974) suggested following conditions for the choice of a male mate by the female.

(i) One ? and one ?- no choice, ? selects available ?

(ii) One ? and more ?- ? selects nearby ?

(iii) One ? and more ?- ? selects victor of competition among ??

(iv) One ? and more ?- ? selects the ? based on comparison against some absolute standard

(v) One ? and more ?- ? selects an extreme phenotype, e.g. largest ?

(vi) One ? and more ?- ? selects a ? from a territory or position in a lek

10.6 Copulation

Copulation involves close range courtship and transfer of sperm. Insects exhibit a wide range of copulation patterns and behaviours by which sperm transfer occur. Insects show variations in positions of male and female during copulation, copulation duration, number of times females mate, and the behaviour during and after copulation.

10.6.1 Pairing

In copulation one sex commonly mount on the back of the other; however there are exceptions in positions of male and female individuals during copulation (Fig. 10.2a-e). Most positions are derived from a situation in which the female

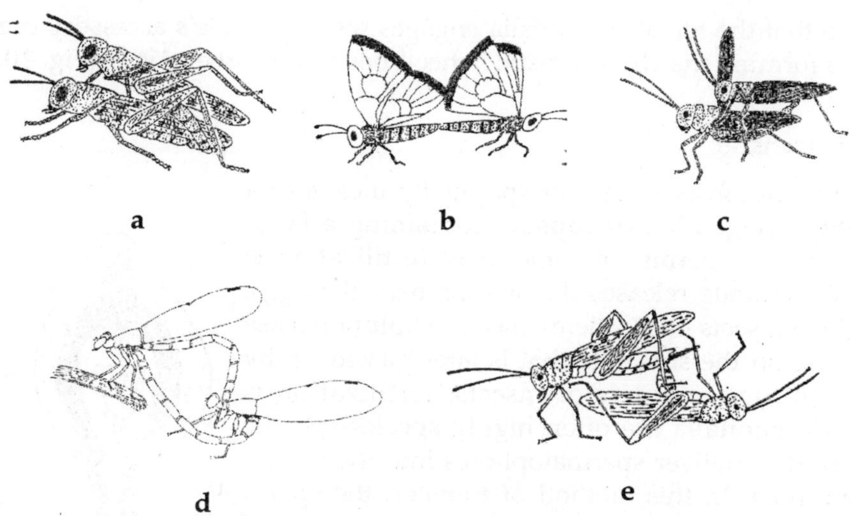

Fig. 10.2. Copulatory positions shown by some insects. (a)- Female above the male, (b)- End to end mating, (c)- Female above the male in a cricket in which male raises his wing to facilitate the female to feed on secretions from the metanotal gland, (d)- Mating pair of dragonflies. When a male encounters a receptive female he uses the appendages at the extreme tip of his abdomen to clasp the female and both partners curve their abdomens so that the female's genitalia engages with the male's accessory organ, the pair thus forming the characteristic 'wheel' position, (e)- Ventre to ventre mating position in grasshopper (redrawn from various sources).

rests on the back of the male and Alexander (1964) considered it as the primitive condition since it occurs in lower insect orders. Chapman (1972) cited under noted copulatory positions in insects.

(i) ? above the ?

(ii) ? above the ?

(iii) False ? above- ? above the ?, but twist his abdomen below the ?

(iv) Ventre to ventre

(v) End to end

(vi) End to end- male inverted

(vii) End to end- male abdomen twisted

In an insect order considerable variations in mating positions can be observed. Once the genitalia of both sexes make contact the male uses his claspers to hold the female. Some insects also use their legs and other appendages for grasping the females. In many cases the insects change their position after linking of the genitalia. The mating of Odonata is virtually unique among insects. Prior to mating the male transfer his sperm on an accessory organ present underside of his second abdominal segment. When a male encounters a receptive female he will use the appendages at the extreme tip of his abdomen to clasp her securely behind her head (dragonflies) or neck region (damselflies) and both partners will curve their abdomens so that the female's genitalia engages with the male's accessory organ, the pair thus forming the characteristic 'wheel' position (Corbet, 1962) (Fig. 10.2d, Plate 10.2).

10.6.2 Sperm transfer

Males of terrestrial insects transfer sperms by means of a spermatophore (Fig. 10.3) or capsule containing a large number of mature sperms. In some cases fertilization is external and the male releases the sperms near the eggs. Some primitive insects like Collembola and Diplura release spermatophore on the substrate that is later picked up by the female. In higher orders of insects, fertilization is internal. Their genitalia are often highly species-specific and specialized to deliver spermatophores into the female reproductive tract. In this method of transfer, the sperm are not diluted by the surrounding medium in case of aquatic forms, nor do they suffer from rapid desiccation on land. Either the female is attracted to the spermatophore chemically or the deposition of the spermatophore occurs during the course of a nuptial dance, and the male afterward manoeuvres the female into a position in which she can take up the spermatophore within her genital opening. The sperm are stored (and maintain viability) in the female's spermatheca- a pouch connected to the median oviduct. They can be stored for months (in moths) or years (in ant queens). The spermatophore apparently evolved in insects with respect to their terrestrial mode of living and to facilitate the fertilization. In most cases for transfer of spermatophore, the male directly insert it in the bursa copulatrix of the female. The structure and size of spermatophore varies greatly in different insect groups. Male spermatophore size also decreases with number of achieved matings. Usually the spermatophores are provided with hooks, which keep them in position while sperm migrate into female's reproductive tract. In some insects spermatophore form a plug that prevents loss of sperm from the female while they move to spermatheca. In most bush crickets males transfer a large spermatophore during copulation that is later consumed by the female. In many insects the male guards the female to avoid its consumption by the female while many females obtain nutrients by way of spermatophore. Male bush crickets, *Requena verticalis* feed their mates with an elaborate spermatophore consisting of a spermatophylax, which is eaten by the female after mating, and a sperm ampulla, eaten after the ejaculate has emptied. The spermatophore nutrients are important to females, and females prefer to mate with males that are able to supply larger spermatophores. The females also aggressively compete for males capable of producing large spermatophores. Feeding on the spermatophore enhances the

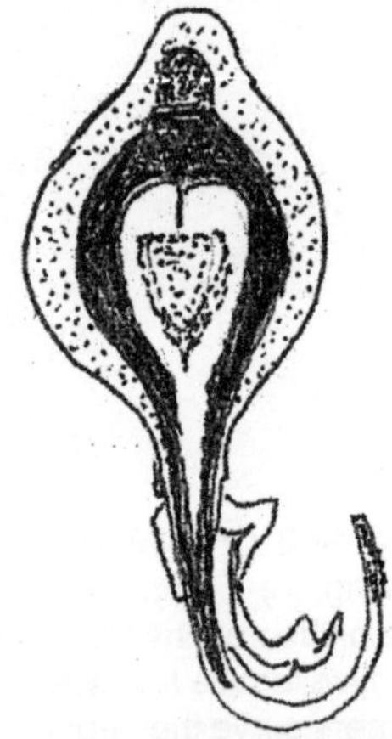

Fig. 10.3. Spermatophore of *Gryllus campestris*.

reproductive success of female by increasing the number and size of eggs produced (Gwynne, 1984). The male bush crickets, *Ephippiger ephippiger* invest heavily in matings by producing large, nutritious spermatophores. On high-quality diets such species show conventional sex roles, but under low-quality diets males become the choosier sex. *E. ephippiger* has one of the largest described spermatophores (up to 40 per cent of the male body weight). In the reversed condition of poor-quality diet, rejections of mating attempts were observed and males sang less frequently and at a much-reduced intensity (Ritchie *et al.*, 1998).

Males of many katydids transfer larger spermatophores, which remain attached externally to the female genitalia. Each spermatophore has a large proteinaceous sperm free portion called 'spermatophylax'. The female consume spermatophylax and sperm ampulla after the sperm are being transferred to the female (Gwynne, 1990). In decorated crickets, *Gryllodes sigillatus* cerci play a role in coordinating movements of males and females during copulation. The cercectomized males were unable to induce females to mount and even when succeeded in securing an initial mount, they took significantly longer time to transfer spermatophore (Ritz and Sakaluk, 2002).

In whirligig beetles, *Dineutus* sp. head of sperm are planted on a rod-shaped structure produced by male accessory glands and the same is transferred to the female during copulation (Breland and Simmons, 1970). The secretion of male accessory glands (MAG) of insects serves a variety of functions like spermatophore and mating plug formation, sperm activation, provision of nutrients to females, production of fecundity-enhancing and/or receptivity-inhibiting substances, and modification of female reproductive behaviour (Gillott, 1996). Lamunyon (1997) reported that female of bella moth, *Utetheisa ornatrix* may receive up to thirteen spermatophores, which provide the female not only with sperm but also with a nuptial gift of pyrrolizidine alkaloid that she transmits to the eggs, protecting them against predation.

Male fireflies of the genus *Photinus* transfer a nuptial gift in the form of a spermatophore to females at mating, and nutrients from the spermatophore are incorporated into the female's eggs. *Photinus* fireflies do not feed as adults and spermatophores represent the only nutrient input for adult females. This male investment may be an important component of mate quality. *Photinus* females may choose mates through differential responsiveness to the bioluminescent flashes produced by males and may prefer particular flash characteristics because they are indicators of male spermatophore size. The ability of male *Photinus ignitus* to produce a spermatophore declines over his lifespan and may constrain male mating success (Cratsley *et al.*, 2003). Thus, seasonal changes in firefly reproductive investment might be associated with changes in courtship behaviour and mate choice. Cratsley and Lewis (2005) examined seasonal variation in operational sex ratio, courtship behaviour, spermatophore mass, egg count, and the relationship between morphological traits and mating success in firefly, *Photinus ignitus*.

Successfully mating males had larger lanterns than unsuccessful males when the operational sex ratio was male-biased. In addition, female responsiveness to male signals increased as the number of courting males decreased, and male spermatophore mass decreased with body size across the mating season. Successfully mating females had larger body mass than unsuccessful females. Female body mass predicted egg count and female rejection by males increased as the season progressed and female size decreased.

Some Hemiptera, Diptera, Coleoptera and Hymenoptera transfer spermatozoa directly to the female tract or spermatheca. They usually have a long flagelliform aedeagus to facilitate the transfer. The copulation involves physical apposition of male and female genitalia, followed by insemination. In many species first the aedeagus is extended, then the intromittent organ is everted or extended to produce an elongate structure called 'endophallus' or 'flagellum' or 'vesica' which is used for depositing semen deep within the female's reproductive tract. Many insects use their modified appendages to hold the female during sperm transfer. The female receives sperms in bursa copulatrix or directly into spermatheca.

Bed bugs use traumatic insemination, which means instead of being sperm deposition in the female's reproductive tract they use their hypodermic needle like genitalic structures to pierce through the female's abdomen to inject her with sperm. There occurs variation between the species practicing this method of haemocoelic insemination and blood phagocytes digest some sperm. Hinton (1964) suggested that these sperm are of nutritional value and the insemination facilitate prolonged survival of the female in the absence of food.

10.6.3 Copulation frequency

Monogamy, in which each adult is mated once with the member of opposite sex, is rare in insects. Females of a few species such as screwworm, *Cochliomyia hominivorax* are known to mate only once. In polygamy, one individual mates with two or more individuals of opposite sex. Polygamy is of two types, one in which a single male mates with many females (polygyny) and the reverse is 'polyandry'. Polygyny is most common in insects while polyandry is also prevalent in some insects although the sperm received by the female in a single mating are sufficient to fertilize all female's eggs. Male insects may copulate many times in rapid succession. A male *Aedes aegypti* mated 30 times in 30 minutes under experimental conditions (Chapman, 1972). The queen honey bee mates with several males in succession during nuptial flight (Gary and Martson, 1971). Multiple mating by a female incur some cost like energy and time expended to find a mate, risk of physical injury from incorrect or repeated coupling of genitalia, higher probability of parasitism or predatism, physical aggression from males, etc. Many insects copulate for a long duration and in such cases temporal cost of mating are very high.

Parker (1970, 1979) suggested that the female must be benefitted in multiple mating.

The multiple mating may provide: additional sperm, avoid depletion of sperm, physiological or morphological expense of sperm storage, induction of genetic diversity in offspring, and acquisition of resources provided by conspecific males. Sometimes sufficient numbers of sperm are not transferred in a single mating and if so the female requires additional matings. Sakaluk and Cade (1980) reported that in house cricket, *Acheta domestica* doubly mated females produced significantly higher offspring than the females mated only once. Some females live for exceptionally longer duration and their sperm storage run short to fertilize all eggs, and they require additional sperms. Multiple mating in *Callosobruchus subinnotatus* by both males and females affected the number of eggs laid, duration of mating, and uncoupling time at the end of mating. Females that mated two or three times laid more eggs than females that mated once or more than three times (Mbata *et al.*, 1997). Fjerdingstad and Boomsma (1998) reported that in leafcutter ant, *Atta colombica* queen sperm stores increased on average by 30 million sperm for each additional mate. In *Dinarmus basalis*, an ectoparasitoid of cowpea seed beetle, *Callosobruchus maculatus* once-mated females suffer sperm depletion about 21 days after mating and thereafter are constrained to produce only haploid males. On the other hand, three-times-mated females store more sperms in their spermathecae and produce daughters during the major part of their reproductive life (Chevrier and Bressac, 2002). Males in some insects engage in two distinct behaviours, multiple mating (with different females) and repeated copulations (with the same female). The males of the red flour beetle, *Tribolium castaneum* were found to be capable of mating with as many as seven different virgin females within 15 minutes. Across sequential copulations with virgin females, there was no decline in either male insemination success or average female progeny production over 48 hours. However, when males copulated with previously mated females, there was a significant decline in male paternity success across sequential copulations, possibly due to male sperm depletion (Lewis, 2004).

Repeated mating lessens the likelihood of inbreeding in the population and assists in maintaining genetic variability. In species where males feed the female or provide nutrients during courtship or copulation, multiple mating obviously provides additional nutritional benefits to the female (Thornhill, 1976). The monarch butterflies, *Danaus plexippus* survive on stored lipids during the winter and it has been argued that the multiple matings, coupled with the nutrient absorption of the spermatophore by the females increases their fecundity (Wells *et al.*, 1990). Females also acquire resources other than food by multiple mating. The female of dung fly, *Scatophaga stercoraria* lays eggs in fresh cattle dung. Male patrol fresh dung, locate and copulate with the female approaching the dung.

Gupta and Sehgal,(1997) observed that sexually mature adults of *Dysdercus koenigii* mated throughout their life span. Males exhibited brief courtship behaviour to accomplish end-to-end mating. A single female mated up to seven times. The first mating was longer and subsequent matings were of shorter durations. In between two matings there were bouts of oviposition. Female mating status governed the duration of matings.

10.6.4 Sperm storage and fertilization

Female usually stores sperms received from mate(s) in the spermatheca. Usually a female has one spermatheca while some flies have more, often two or three. The sperms remain viable for considerably long duration and secretions from spermathecal gland maintain their viability. In honey bees sperms remain viable for up to three years. Eggs are fertilized as they pass down from the ovary to the oviduct, and sperm enters the egg through the micropyle. Many females are able to control fertilization and sperm release activities. Sex determination is predominantly genetic and most insects are diploid. Usually an insect with XX chromosomes is female and XO or XY is male, but this allocation varies within and between insect groups. In insects like mosquitoes and midges a single gene determines sex, and it is difficult to recognize sex chromosomes in these insects. Haplodiploid sex determination is found in honey bees in which workers and queens develop from fertilized eggs and drones (males) from unfertilized eggs.

10.6.5 Sperm competition

Factors like multiple matings by females, sperm storage and internal fertilization lead to competition amongst sperms for access to unfertilized eggs. This competition mainly occurs within the female's reproductive tract when sperms from two or more males attempt to fertilize the eggs. Parker (1970) listed four conditions for sperm competition: (i) multiple mating by the female prior to fertilization of all eggs, (ii) storage of sperms, (iii) sperms remain viable for over the length of female's life, and (iv) efficient sperm utilization. Sperm competition is common in insects and usually the sperms of last male mate fertilize majority of female's eggs, and this is termed as 'last male precedence'. Results of sperm competition are determined by both physiological and behavioural mechanisms. Attributes of mating and events inside the female reproductive tract determine which sperm will fertilize an egg. The numbers of eggs fertilized by the sperms determine the male reproductive fitness. Various mechanisms are selected by males to ensue sperm precedence. Male of the damselfly, *Calopteryx maculata* not only transfer the sperms with its aedeagus but also remove the pre-existing sperms, if any (Waage, 1979). Following sperm removal, male injects his own sperms into female's bursa copulatrix. Other strategy to reduce subsequent insemination is use of mating plugs that switch off female's receptivity. Mating plug of male bumble bee, *Bombus terrestris* prevents female (queen) from further mating. The active substance is linoleic acid, a ubiquitous and rather unspecific fatty acid. The plug is highly efficient and allows the males to determine queen-mating frequencies (Baer *et al.*, 2001). Prolonged copulation, guarding female and improved structures for gripping the female during copulation to prevent take over by other males are other strategies to lessen sperm competition. In case a male achieves both sperm precedence and able to prevent other males to inseminate, he would ensue significant selective advantage. Female choice also has a significant role in male precedence as many females select potential mates.

Lewis and Jutkiewicz (1998) examined last-male sperm precedence patterns in doubly and triply mated red flour beetle, *Tribolium castaneum* females. Both doubly mated and triply mated females exhibited high last-male sperm precedence for progeny produced during the first 48 hours following the last mating, with females in both groups exhibiting parallel declines in sperm precedence one and two weeks later. Seth *et al.* (2002) studied sperm utilization by tobacco cutworm moth, *Spodoptera litura* and observed that after second mating, the sperm utilized for subsequent fertilization were almost exclusively from the last mating with little mixing.

10.7 Post-copulatory behaviour

Mate guarding, which may increase a male's paternity assurance, is a common post-copulatory behaviour in many insects. The phenomenon of 'last sperm precedence', which occurs when only sperm from the last mating is used to fertilize the eggs, may lead to the evolution of such behaviour. However, other aspects of mate guarding, e.g. guarding duration may change according to more variable conditions such as sex ratios, densities of conspecifics, and the number and quality of available females. Theoretical studies have investigated how post-copulatory mate guarding evolved and how different social environments should affect optimal mate guarding behaviour (Parker, 1974; Yamamura, 1986). Parker (1974) used time spent as a cost of mate guarding and found that post-copulatory mate guarding would be favoured by a male-biased sex ratio. Yamamura (1986) predicted that a male should guard his mate when the sex ratio is male-biased, when density is high, and when the female lays eggs soon after the copulation. Alcock (1994) discussed diversity of postinsemination associations in insects and reviewed mate-guarding hypothesis. He suggested conditions for the evolution of mate guarding; some of these conditions include last male advantage, high potential of taking over previously paired females, and a male-biased sex ratio. These theoretical predictions have been tested in many studies. Stoks *et al.* (1997) demonstrated that the mate guarding behaviour of the emerald damselfly, *Lestes sponsa* supported the predictions of Alcock (1994) for the evolution of mate guarding.

In many Odonata males guard their mates after copulation and during oviposition. Two main types of mate guarding behaviours are observed. First 'contact guarding', or tandem oviposition, in which the male stays physically attached to the female after mating, and the pair moves about and oviposits as a unit (Fig. 10.4). Second 'non-contact guarding', in which the male detaches from the female after mating and hovers near her as she oviposits, chasing other males attempting to seize and inseminate her. Mate guarding facilitates uninterrupted oviposition and has obvious advantages for the female. Sperms received from the last male fertilize most eggs, and the 'last-male phenomenon' makes it advantageous for a male to protect his sperm investment in a female by guarding her after mating and during oviposition. McMillan (1991) discussed variable mate guarding behaviour in the dragonfly, *Plathemis lydia*. When male densities are high, a guarder hovers close to his female and repels rivals, but if he is the only male, frequently he may go away from his mate or abandon the female altogether. The males make both subtle and major adjustments in guarding intensity, fine-tuning their behaviour in response to moment-to-moment changes in their immediate surroundings.

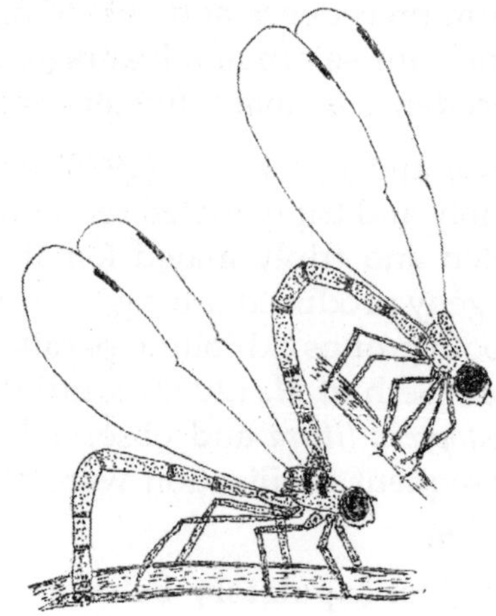

Fig. 10.4. Tandem oviposition in dragonfly. The male continuing to hold his mate while she deposits her eggs.

Crickets offer an ideal model to examine the evolution of post-copulatory female choice. In many crickets a male guards his mate by standing motionless next to her, usually draping his antennae over her dorsum, and producing violent shaking motions upon any attempt by the female to move away or upon disturbance of the pair by an intruder male. In crickets the ejaculate of a male typically remains attached outside the female's genital opening after mating in the form of an externally attached spermatophore, and females are thus well positioned to determine the fate of their mates' gametes through the judicious removal of spermatophores. Sakaluk (1991) discussed post-copulatory mate guarding in decorated cricket, *Gryllodes sigillatus*. In this cricket, the spermatophore includes a large gelatinous mass, the spermatophylax, which the female detaches and feeds on after mating. While the female consumes this nuptial food gift, sperm are evacuated into her reproductive tract from the remaining portion of the spermatophore, the sperm ampulla. Smaller spermatophylaxes require less time to consume, and males providing such gifts are penalized in the form of premature ampulla removal and reduced sperm transfer. Wynn and Vahed (2004) reported that in the field cricket, *Gryllus bimaculatus* presence of a guarding male also delayed the female from mating with the rival male. Female mating status had a significant effect on the duration of spermatophore attachment. Females mating for the first time retained the spermatophore for a significantly longer period of time than females that had mated previously. Mature male desert locusts, *Schistocerca gregaria* in the gregarious phase release phenylacetonitrile (PAN). PAN serves as a kind of olfactory concealment during mating and to prevent competing males from homosexual encounters. Seidelmann and Ferenz (2002) concluded that PAN is a courtship-inhibiting pheromone exclusively used under crowded conditions when high sperm competition occurs among males. By chemically enhancing their mate guarding, gregarious males improve the protection of their mate from rivals and ensure their reproductive success.

10.8 Oviposition

Insects are mostly oviparous and fertilized eggs pass from the external genital opening and laid in/on a suitable site. Females usually deposit their eggs on or near the food source likely to be consumed by hatched offspring. Insects not investing in parental care can influence the success of their progeny by actively selecting sites that are most favourable for their growth and survival. Primarily through the study of phytophagous insects, it has become clear that oviposition behaviour, and choice of oviposition sites in particular, can increase the performance and survival of progeny. Such oviposition behaviour is largely driven by variation in the environment. Jones (1991) discussed host location and oviposition behaviour in insects on plants.

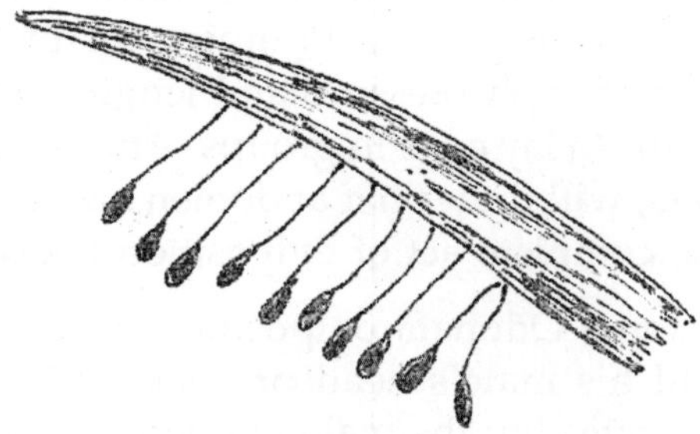

Fig. 10.5. The stalked eggs laid on a leaf by common lacewing, *Chrysopa*.

Many females have special appendages which form ovipositor or egg-laying organ. The morphology of the ovipositor is extremely varied among insects. In some cases last abdominal segments are modified to form a tube-like substitutional ovipositor (Plate 10.3). Hymenopterans have most diverse ovipositor morphology. Some females of ichneumon wasps have a very long, thin ovipositor, which may be up to 7 cm in length, allowing the female to inject her eggs into hidden hosts. Some parasitoid wasps have large teeth on the ovipositor in order to 'drill' through wood to reach their host. Insects like termites, lice, stoneflies lack an ovipositor and directly lay eggs on hosts or ground. In workers of social Hymenoptera ovipositor has lost its original function and modified as sting. Ovipositors of many insects are either pointed or have serrations that assist in piercing. A number of sensilla are present on the ovipositors of many insects, which help out in evaluation of egg deposition site. Some insects also produce sticky secretion or cement that fastens the eggs to a substrate or host. The eggs of common lacewing, *Chrysopa* are laid on leaves, and raised on stalks (Fig. 10.5). Ootheca of praying mantis (Fig. 10.6a), and other

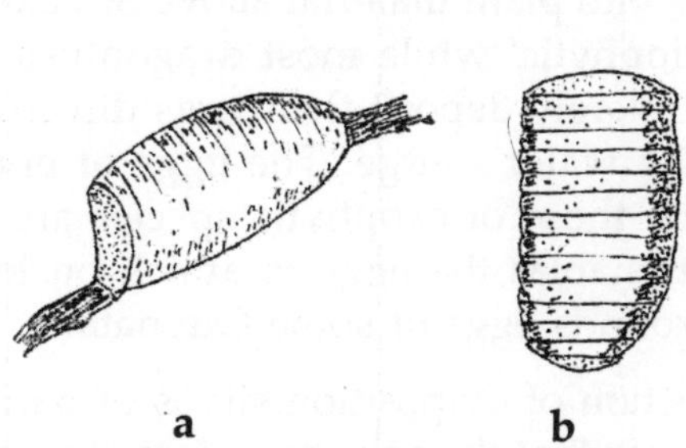

a b

Fig. 10.6. (a)- Ootheca of praying mantis, (b)- Ootheca of German cockroach.

orthopterans is foamy and protects developing embryos from desiccation. Eggs of cockroaches are enclosed in a purse-like ootheca (Fig. 10.6b) formed by colleterial glands.

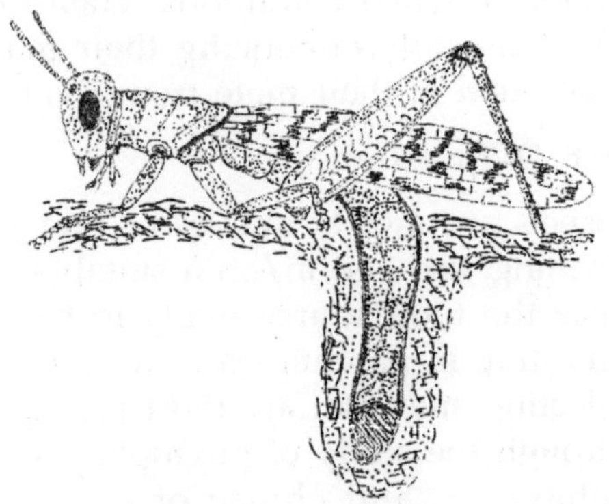

Fig. 10.7. Oviposition by a desert locust. The eggs are laid deep in the soil by stretching and extending its abdomen to its maximum.

Oviposition behaviour varies enormously in different insects and has been the subject of many studies on ecology and evolution of insect-plant interactions (Thompson and Pellmyr, 1991). Gravid females of locust and grasshopper select suitable soft and sandy soil for oviposition. The female arches tip of her abdomen downward and scrape the soil with ovipositor valves. She forms a deep hole in which the long abdomen is inserted (Fig. 10.7). Pumping air in the tracheal system assists in maintaining haemolymph pressure and lengthening of the abdomen. If the soil is moist, she begins to lay eggs in groups. After laying all eggs she plugs the hole with a frothy mass, withdraws her abdomen and scrapes soil over the hole with her hind tibiae. The complete act of oviposition takes about two hours (Uvarov, 1966).

In some Odonata oviposition is carried out in tandem, the male continuing to hold his mate's head or neck while she deposits her eggs. In others, the pair separates but the male will guards her and discourages other males from mating with her. In a third group, the female oviposits totally unattended but, in these cases, she will do so in isolated places. Some females lay their eggs underwater for safety. Even as she goes underwater to lay, the male will remain sexually attached to her. Some Odonata have a well-developed ovipositor and insert their eggs into plant material above or below the surface of the water. They are termed 'endophytic' while most dragonflies are 'exophytic', lack functional ovipositors and merely deposit their eggs directly onto the surface of the water or into mud at the water's edge. The eggs of endophytic species are long and cylindrical, while those of exophytic species are broad and elliptical. A jelly-like substance that enables the eggs to attach on leaves or stones and rocks under the water surrounds eggs of some Odonata.

Selection of oviposition site is of paramount significance by a female insect as it ensures that the eggs are adequately protected and suitable food will be available to larvae/nymphs when they emerge. Many Lepidoptera and Hemiptera lay eggs on larval hosts. It involves two steps, first is an initial selection based on a reaction to the environment, and then final selection mediated by specific responses. For final selection receptors located on tarsi, ovipositor and antennae are used by females. Host plant factors, viz. colour, texture, surface configuration, odour, etc. play an important role in egg laying of many phytophagous insects. Female of codling moth, *Cydia pomonella* lays some eggs in the absence of any olfactory stimulant at all, but the presence of the apple volatile α-farnesene stimulates a much heavier egg laying (Wearing and Hutchins, 1973). Females of spruce budworm, *Choristoneura fumiferana* are very selective during oviposition. They prefer white spruce to balsam fir or red or black spruce. Several factors including needle spacing, light, and gravity have been implicated in this selectivity but the odour of the needles is especially important. A crude mixture of natural pinenes induces a strong ovipositional response, and of these D-α-pinene and L-α-pinene are each highly active (Städler, 1974). Wallace *et al.* (2004) documented behavioural patterns associated with oviposition in *C. fumiferana*. Apart from resting, the most pronounced behaviours were probing, drumming + probing, and egg laying. Probing was a bending of the abdomen and extension of the ovipositor to touch the substrate in a rocking, back and forth motion. Drumming + probing involved tapping of the substrate with the fore and mid legs with concurrent bending of the abdomen as above. Egg laying was the actual process of egg deposition onto the needle surface of a balsam fir twig. Both the frequency and the duration of these behavioural elements varied depending on the mating status of the female. In tephritid fruit flies visual stimuli have been studied from the perspective of colour and shape. Foliage colours, tree shape, tree size, all have been shown to play a role in influencing arrival of flies to host trees for oviposition. Host colour has an important effect on fly searching behaviour, and background characteristics have a substantial influence on the ability of flies to distinguish colours (Prokopy and Owens, 1983). The Mediterranean fruit fly, *Ceratitis capitata* lays eggs underneath the host epidermis. Females alighting on a fruit thoroughly examine the potential host before oviposition. After oviposition, females always mark the oviposition site. The effects of food odour on fly attraction and oviposition would be greater for relatively visually inconspicuous fruits than for relatively visually conspicuous fruits (i.e. that proximity to food odour would enhance fly discovery of inconspicuous fruit to a greater extent than conspicuous fruit) (Prokopy *et. al.*, 1996). Many females mark the host with oviposition-deterring pheromone (ODP) to check the further oviposition by other females, thereby assuring proper development of their larvae. This behaviour has been observed in many tephritid fruit flies. Straw (1989) observed that females of *Tephritis bardanae* avoid attacked flowerbuds of common burdock, *Arctium minus* for oviposition, as following egg-laying the buds were marked with ODP by the female. Raghu *et al.* (2004) found that abundance and oviposition behaviour of fruit fly, *Bactrocera cacuminata* is significantly influenced by host plant (*Solanum mauritianum*) structure (number of leaves, branches and height of plant) and microclimate (temperature, relative humidity and light intensity). The number of ovipositing females was positively influenced by temperature and negatively by relative humidity.

Female golden egg bugs, *Phyllomorpha laciniata* show unique flexibility in their oviposition behaviour. They can lay eggs on plants, where they are left unattended, or on the back of conspecifics, where they remain until hatching. The survival rate of eggs is higher when carried by an adult than when laid on plants, as they avoid natural enemies. The main causes of egg mortality are predators and a parasitoid wasp. The females prefer to lay eggs on the backs of conspecifics than on plants (Reguera and Gomendio, 2002). Most adult butterflies are nectar feeders and require frequent refuelling of the carbohydrates that the nectar provides. Their females could potentially use the more conspicuous inflorescence cues for both adult feeding and oviposition. Female of common blue butterfly, *Polyommatus icarus* prefers to oviposit on birdsfoot trefoil, *Lotus corniculatus* plants with flowers. Observations of behavioural sequences also revealed that oviposition often follows immediately after nectaring (Janz *et al.*, 2005).

Wellings (1991) discussed host location and oviposition behaviour in insects on animals. In several parasitoid species females show behavioural plasticity towards host acceptance and oviposit preferentially in the most profitable host. Recent theories on host-parasitoid relationships have placed more emphasis on determinants related to the physiological and informational state of the foraging female such as her egg load, previous experience, perception of the environment and life expectancy that might influence the dynamic expression of host selection for oviposition by parasitoids. The great majority of parasitoid species can only parasitize a single host stage and have evolved specific adaptations to exploit that particular stage. However, some parasitoids are able to attack both immature and adult hosts from the same species. For example, the females of aphid hyperparasitoid, *Syrphophagus aphidivorus* display a dual oviposition behaviour as they have the ability to attack either the primary-parasitoid larva when the aphid is alive or the primary-parasitoid prepupa or pupa after the parasitoid has killed and mummified the aphid (Kanuck and Sullivan, 1992). Many parasitic insects locate and oviposit on their hosts by using various physical and biochemical factors. In the host location and selection process, cues associated with the hosts have a hierarchical array, with volatile compounds (synomones) emitted by plants as a consequence of herbivore activities playing a role in attracting parasitoids to patches infested with hosts (Vinson, 1985). Oviposition can also induce a change in the profile of the plant volatiles, and these volatiles have been shown to attract egg parasitoids. Plants that are able to attract egg parasitoids soon after herbivore eggs are laid get benefit by increasing the parasitoid's effectiveness. Feeding and oviposition activity of the green shield bug, *Nezara viridula* on broad bean and French bean leaves resulted into emission of synomones which attracted the egg parasitoid, *Trissolcus basalis* (Colazza *et al.*, 2004). Gravid female mosquitoes are influenced by the presence of vegetation and amount of light reflected from the water surface. They determine salt concentration and pH of water through their tarsal sensilla and then finalize whether or not to lay eggs (Matthews and Matthews, 1978).

Some insects are viviparous, and ovoviviparity and three types of viviparity are observed in insects.

(a) *Ovoviviparity*. The fertilized eggs are incubated in female's reproductive tract and fully developed eggs hatch just after being laid or just prior to ejection, e.g. in some cockroaches, some hemipterans, some flies, thrips, some lepidopterans, and a few beetles.

(b) *Pseudoplacental viviparity*. Yolk-deficient eggs obtain nourishment from the female tract through placenta-like tissue, e.g. in many aphids, some earwigs, and a few psocids and hemipterans.

(c) *Haemocoelous viviparity*. Embryos develop in haemolymph of female and obtain nourishment through osmosis, e.g. in Strepsiptera and some gall midges.

(d) *Adenotrophic viviparity*. The larvae hatch and feed on accessory gland secretions (milk) in the female's uterus. Full grown larvae are deposited, e.g. in dipterans like tsetse flies, keds, and bat flies.

10.9 Hormonal regulation of reproduction

Regulation of reproductive functions in any insect is a complex phenomenon and involves environmental factors, nutritional status, sensory receptors, transmission of impulses, and chemical messengers secreted by neurosecretory cells and endocrine glands. Neurosecretory cells of brain produce neurohormones. Barth and Laster (1973) discussed neurohormonal control of sexual behaviour in insects. Neurohormones are also produced, outside the brain, in the ventral nerve cord ganglia. The best-known release sites of neurohormones are the cephalic and metameric neurohaemal organs. Moreover, the neurohormones are also released in close contact with or within the target organs (Raabe, 1982). Neurohormones act according to two distinct modes, either directly upon the target organs or by controlling the activity of the endocrine glands that in turn, influence the different reproductive steps. Thus, these hormones also regulate activities of prothoracic gland and corpus allatum, which secrete ecdysteroids and juvenile hormone (JH), respectively. All these hormones play a vital role in regulation of reproduction in insects (Table 10.1). Neurohormones also influence reproductive functions like ovulation and oviposition. JH mainly regulates the functioning of organs like gonads, male accessory glands, and vitellogenesis in female, egg production, and fat bodies production while ecdysteroids regulate morphogenesis and functions of gonads. Wyatt and Davey (1996) reviewed the specific actions of JH in adult insects and explored the evidence relating to its cellular and molecular modes of action. They reviewed effects of JH in relation to reproduction, with reference to the fat body production, vitellogenesis functions of, gonads, accessory glands functions, and behaviour of various insect groups. In insects the male accessory glands (MAG) exhibit a wide range of form, and their secretion serves a variety of functions, including spermatophore and mating plug formation, sperm activation, provision of nutrients to females, and through production of fecundity-enhancing and/or receptivity-inhibiting substances, modification of female reproductive behaviour (Gillott, 1996).

Table 10.1. Insect hormones and their role in reproductive events in insects.

Endocrine gland	Reproductive events
Neurosecretory cells in brain (secrete neurohormones)- neurohormones also regulate activities of corpus allatum, prothoracic gland and ovary	• Ovulation • Oviposition • Pheromone production • Sexual behaviour • Previtellogenesis • Vitellogenesis
Corpus allatum (secrete juvenile hormones)	• Pheromone production • Sexual behaviour • Male accessory gland function • Previtellogenesis • Vitellogenesis
Prothoracic gland (secrete ecdysteroids)	• Germ cell division • Ovariole differentiation • Oogenesis • Accessory gland differentiation • Previtellogenesis • Vitellogenesis • Spermatogenesis
Ovary (secrete ecdysteroids)	• Vitellogenesis

10.10 Unusual modes of reproduction

Parthenogenesis is a common phenomenon found in many insect groups. Three types of parthenogenesis can be distinguished:

(i) *Arrhenotoky*. Unfertilized eggs develop into males, fertilized eggs develop into females.

(ii) *Thelytoky*. Unfertilized eggs develop into females.

(iii) *Deuterotoky or Amphitoky*. Unfertilized eggs develop into both sexes.

Thrips show all three types of parthenogenesis. Other groups like many Hymenoptera; some Coleoptera; some Diptera; and some aphids, whiteflies, scale insects reproduce parthenogenetically.

Larval paedogenesis, i.e. production of young by larva is observed in a few gall midges and coleopterans. In mushroom cecid, *Heteropeza pygmaea* eggs develop into female larvae, which may develop into adult females or paedogenetically produce more larvae. These larvae may become males, females or individuals of both sexes. Pupal paedogenesis is also observed in some gall midges, in which embryos form in the haemocoel of mother pupa. In neoteny, a larval instar develops adult's characters, i.e. mate finding, copulation and oviposition in a predictable manner. This condition is observed in some scale insects.

Many self-fertilizing cottony cushion scale, *Icerya* species are gynomonoecious hermaphrodites. They are female-like and their gonads have both ovarian and testicular tissues. Polyembryony, a form of reproduction involves production of two or more embryos from one fertilized egg is observed in some insects. In some parasitic hymenopterans like *Platygaster, Aphelopus, Litomastix* and others it is a regular phenomenon.

10.11 Parental investment

The amount of energy used for the purpose of reproduction is termed as reproductive effort. It includes the amount of energy used to acquire mates and parental effort (Thornhill and Alcock, 1983). Individuals of each species make efforts to increase their fitness by increasing the number of offspring produced. Trivers (1972) defined parental investment as "any investment by the parent in an individual offspring that increases the offspring's chance of surviving (and hence reproductive success) at the cost of the parent's ability to invest in other offspring." Varying degree of parental investment has evolved in many insect species.

Zeh and Smith (1985) suggested that paternal care consists of three components: prezygotic investment including indirect contributions to offspring, biparental care, and exclusive care provided by the male. Thornhill and Alcock (1983) pointed that prezygotic investments consist of nuptial gifts and spermatophores and should be considered simply as mating efforts. They believed that males evolved these traits to secure mating, or to increase the probability that the female will use their sperms.

In the giant water bugs (Fam. Belostomatidae) male carry the eggs on their backs and help the young to hatch (Fig. 10.8). These eggs need to be covered by water, have water movement and frequently need to be exposed to atmospheric oxygen, and the eggs on the back of a healthy male had a very high per cent hatch. Males also brush their hind legs over the egg mass to supply the eggs with source of dissolved oxygen. In adverse conditions when the egg mass is disturbed the male remove and cannibalize the egg mass. The cannibalism of the eggs is thought to be a method of gauging the energetic cost of care when the probability of a successful hatch is low (Smith 1976b). Smith (1979) studied mating behaviour of giant water bug, *Abedus herberti* and found that the female first and repeatedly approaches the initially passive male. Males usually showed sign of rejecting females, but while mating he controlled the mating sequence.

Fig. 10.8. A male giant water bug, *Belostoma* sp. carrying eggs on his back.

After copulation, he directed and assisted the female during oviposition and the female lays eggs on the back of male (Plate 10.4). Males incur high costs when caring for eggs including, risk of predation, a reduction in ability to hunt, a decrease in ability to disperse, and a probable reduction in the number of eggs that can be fertilized. Thomas and Manica (2003) investigated filial cannibalism in the assassin bug, *Rhinocoris tristis* in which males guard eggs of a number of females, cannibalizing some of their offspring within the brood. Males typically ate eggs around the periphery of the brood, which were those most likely to have been parasitized by wasps. However, cannibalism also persisted in the laboratory in the absence of parasites, and the number of cannibalized eggs was related to the length of care and overall brood size, suggesting that males use eggs as an alternative source of food.

Maternal care is the most prevalent form of parental investment in insects. Though in most insects the egg contains half of the female's genetic material but the female contributes all the nutrition components. In contrast, sperm production has a relatively low energetic cost to the male. The female investment must also increase the probability of survival of offspring. An extreme example of this is the case of the viviparous tsetse fly that internally nourishes one offspring until it reaches the pupal stage. Thus, the individual offspring has a high probability of reaching reproductive stage.

Parental protection of eggs represents one of the most basic forms of parental care. It has been suggested that even such basic parental investment represents a trade-off between current offspring survival and future reproductive success. Females of treehoppers, *Publilia concava* exhibited a range of guarding durations for their first clutch and longer guarding duration resulted in greater hatching success. Besides, the females that remained to guard eggs had a lower number and size of future broods, suggesting that parental care may reduce lifetime fecundity (Zink, 2003).

Many insects guard their offspring from natural enemies and adverse environmental conditions. The female of *Erixestus winnemana* (a parasitic wasp of the Colorado potato beetle) after injecting her eggs into the egg of the beetle guards the parasitized egg until her offspring emerge so hyperparasitoids do not kill the larvae (Schroder *et al.*, 1996). Maternal guarding and provisioning through out the larval development is also found in some species of dung beetles. In some insects female lays as many eggs as she can defend. Females of mothering bug, *Elasmucha grisea* defend eggs (Fig. 10.9) and small nymphs, and small

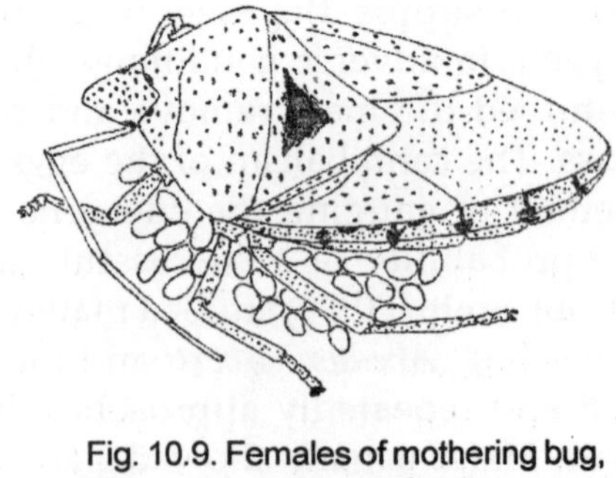

Fig. 10.9. Females of mothering bug, *Elasmucha grisea* defending eggs.

females guarding large clutches lost significantly more eggs while large females could defend, on average, 16 per cent more eggs against invertebrate predators (Mappes and Kaitala, 1994).

Postovipositional parental care in insects has evolved in response to a variety of environmental stresses and ranges from briefly guarding eggs after oviposition to elaborate nidification and provisioning behaviours. Investment in parental care bears various costs, and should not continue beyond the point at which the costs to future reproductive success exceed the benefits to current reproductive effort. Filippi *et al.* (2001) reported that females of the subsocial shield bug, *Parastrachia japonensis* progressively provision their nymph-containing nests with drupes of the single resource, *Schoepfia jasminodora,* and this provisioning drastically enhances offspring survival. A female rears only one brood throughout her lifetime and continues provisioning the brood until about the third larval instar, when the female dies. Thus, the female's entire reproductive effort is expressed in the success of that one brood, which suggests a reproductive strategy with enormous costs and risks.

Biparental care occurs in a number of species of both dung beetles and burying beetles. Parental care in dung beetles, *Onthophagus* species is usually restricted to brood mass construction. In the majority of the species, the female alone is capable of successfully provisioning larvae; however, wide levels on cooperation occur between males and females. Males have been observed helping move dung from the dung pad to the site of burial and oviposition. Offspring size upon emergence reflects the size of the brood ball, but not their parental sizes. Size is largely determined by the quantity of dung provided in the brood mass. Sowig (1996) examined the biparental brood care behaviour of the dung beetle, *Onthophagus vacca* and reported that when a male does not assist the female, the survival of a single offspring is unchanged. However, if a male provides assistance in brooding the young, the number of offspring is increased, which in turn increases of both parents reproductive successes.

In social insects, a special caste exhibits guarding, and or provisioning depending on the species involved. However, this investment may still be considered parental investment because the amount of energy used in rearing the offspring is spent for the fitness of the colony that can be translated into the fitness of the individual. Evolution of parental or perhaps more exclusively, sibling care in social colony may be due to nepotism.

Chapter 11

Communication

Communication may be defined as "an action or condition on the part of one organism that alters the behaviour of another organism in an adaptive way." Insects use visual, acoustic, chemical and tactile means of communication. However, their 'language' is almost entirely innate. Each individual is born with a distinctive 'vocabulary' that is shared only with other members of its own species. Thus, intraspecific communication is a prerequisite for any behaviour that involves the participation or cooperation of two or more individuals. Intentionally or not, insects may also communicate with members of other species (interspecific communication). The adaptive significance of communication may include: (i) recognition of kin or nestmates, (ii) finding or identifying a member of the opposite sex, (iii) to facilitate courtship and mating, (iv) in finding food or other resources, (v) regulating spatial distribution of individuals- aggregation or dispersal, (vi) territorial display, (vii) alarming behaviour, (viii) in advertising one's presence or location, (ix) express of threat, and (x) deception or mimicry. The book edited by Lewis (1984) includes valuable discussions on insect communication.

Most insects communicate by using chemical signals. The chemical aspect of insect communication is mainly based on pheromones. Pheromones are used for sexual attraction, defence, alarming, aggregation, and spacing. Visual communication is vital to insects and visual signals are used for aggregation, recognition, defence, territorilaty, and sexual selection. Well-illustrated example of visual communication is light production by lightning bugs, or fireflies. Mating behaviour of fireflies is governed by a combination of sound patterns and flashes of light. Many insects produce sound and communicate acoustically for their survival. Sound signals specify the behaviours like aggregation, recongnition, isolation, alarm, defence, aggressiveness, social communication, and sexual attraction and copulation. Besides these, some insects also communicate by making contact with body or its parts of the conspecifics.

11.1 Tactile communication

Many insects communicate by making physical contact with each other. Tactile communication can occur obviously only if the insects are together. 'Keep in contact' is a strategy adopted by some insects as a channel of communication. Theoretically, tactile communication could form the basis for a complex and important communication system, transmitting a great variety of messages by varying frequencies, pressure, and time of contact (Matthews and Matthews, 1978). However, this system of communication appears to be less important when compared with other communication means. Besides this, it is also difficult to investigate the sense of contact in a communicatory perspective.

The compound eyes are reduced or lost in many insects and many are unable to perceive the auditory signals, physical contact in such insects provides an important avenue of communication. The tactile communication provides immediate feedback, and is an effective means of communication in dark areas like nests, caves, wood galleries, etc. The prerequisite for tactile communication are that the insects must be in direct contact, the means is not effective over distances, and the message must be repeated to each recipient. Insects like cockroaches are highly sensitive to tactile stimuli. Nerves from leg sensilla are connected to giant axons for rapid transmission of nerve impulses. They show extremely rapid movement when handled.

Insects are engaged in elaborate forms of contact behaviour for the purpose of courtship and sexual stimulation. Most contacts involve the antennae and mouthparts, but not much information is passed by this form of communication. In *Drosophila* the male is attracted to the female by using visual cues but final species recognition is facilitated by tapping the female with fore legs of male. In many species female is not immediately receptive to the male's copulatory advances and will allow the male to mount after some appropriate foreplay. The female is appeased by offering some food or nuptial gift by the male and then the male copulate when she is busy in eating the offering. In blister beetles (Fam. Meloidae) courtship begins with a series of antennal taps by the male on each side of the female's body. She signals her receptivity by lifting her elytra and allowing him to climb on her back. But to complete his quest, the male must continue tapping, alternating from side to side at just the right frequency until the female is stimulated to extend her genitalia and begin mating.

Tactile cues generated by ripples on the water surface allow whirligig beetles (Fam. Gyrinidae) to constantly monitor the location of dozens of other nearby beetles. By communication, the whirligigs can swim rapidly in circles, avoid bumping into other members of their own species, and still detect the presence of nearby predators or prey.

In the complex nests of social insects, workers perform all duties for maintenance of social organization. In insects' societies all colony members are always acquainted with social situations and communicate with each other as per social requirements to maintain a balance. Touching can be used for recognizing nestmates, especially those that live in darkness. A female mole cricket recognizes her nymphs in a dark burrow by touching them with her antennae and even picking them up with her mouthparts. Social insects, such as ants, often stroke and groom each other with their antennae and mouthparts. However, both tactile and chemical signals may be involved in these behaviours. Antennal tapping is an essential component of communication in both ants and termites. It's not clear exactly what information may be exchanged, but it certainly involves nestmate recognition and leads to exchange of food through trophallaxis. Antennal tapping on the hind legs is used during 'tandem running' in both ants and termites. This is a 'follow-the-leader' behaviour in which the tapping informs the leader that she has not lost her disciple. If tapping stops, the leader instinctively turns around and searches in ever-widening circles until she re-establishes contact with the follower. Honey bees perform 'round' and 'waggle' dances to intimate the foragers regarding the location of food sources. The sound signals produced during dance assist in recognition of dancing pattern. Besides, the tactile communication system is also involved for recognition of dancing patterns performed in total darkness on the vertical surface of the honeycomb. At the hive entrance certain honey bees perform the duty of guarding and confirm the identity incoming bees. If the bees are of same hive they are allowed to enter and the bees of other hives are not acceptable. For recognition of incoming forager bees both tactile and chemical cues are used by the guard bees.

11.2 Acoustic communication

Comprehensive publications on sound production and its role in insect behaviour are: Pierce (1948), Haskell, (1961, 1974), DuMortier (1963), Busnel (1963), Alexander (1967), Michelsen and Nocke (1974), Otte (1977), Ewing (1984, 1989), Kalmring and Elsner (1985), Bennet-Clark (1984, 1989), Bailey (1991a), and Drosopoulos and Claridge (2006).

Whether it's the buzz of bees, the chirp of the crickets or the hum of the mosquito, all sounds are produced by insects to send a message to conspecifics. A sound might be used to attract or alarm the conspecifics or frighten individuals of different species. Sound is produced when an object vibrates rapidly, disturbs surrounding molecules, and generates compression waves that travel in the medium in all directions away from the source. Some sounds become so familiar to us that we can name them instantly from the first tones. The ability to recognize a sound and discern where it comes from is fundamental to communication by sound. Many insects produce sound which are recognized by conspecifics and serves as an effective means of communication. Insects may produce sound as a by-product of some activity like flight, by rubbing two body parts, by striking a substrate, by activating a vibrating membrane or by pulsation of an air stream (Chapman, 1972). Most insect sounds are produced by males to attract the females. Singing males of cicada and katydid just try to impress the females. The male cicada has a tymbal membrane that vibrates over a hollow cavity. Stridulation is another approach used by some insects like grasshoppers, katydids and crickets, in which they rub one surface over another to produce the sound. Most stridulators rub wing surfaces and legs on each other. Sound is also produced by insects like bees and mosquitoes, by vibrating the wings or as a consequence of flight. Each flying insect produces a different pitch sound because of differences in wing-beating frequencies. The sound produced by female mosquito is very attractive to males and in this instance sound produced by the female insect is used to attract the male.

The human ear can detect sound frequencies only within the range of about 20-20,000 hertz. However, some insects produce and detect ultrasonic sounds as high as 80,000 hertz. Insects like grasshoppers and moths detect sound with a tympanic membrane in the abdomen while in crickets and katydids the membrane is present in the tibiae of the fore legs. Mosquitoes have antennal hairs that resonate to certain frequencies of sound. Insects like ants, bees, termites, and treehoppers can detect substrate vibrations with mechanoreceptors, and as the sound signals are 'felt' rather than 'heard', they are usually regarded as a form of tactile communication.

Bailey (1991a) divided sound production in insects in three categories: (i) near-field sound, (2) substrate-borne sound, and (iii) airborne sound. In some cases two or more forms of sound production may act synergistically. Any sound generated in air or water and produce displacement waves close to its source is near-field sound. Striking the ground with body parts is a direct method of substrate-borne sound. The airborne sound is produced by one insect and the signal is detected by another at a distance from the caller. Vibration of tymbals of cicada and arctiid moths is commonly used mechanism for production of airborne sound signals.

11.2.1 Mechanisms of sound production

Sound production in insects includes not only vibrations carried through air or water, but also through substratum. Insects use different mechanisms for the production of communicative sound at some stage of their life cycle. Among these, frictional methods are most common and the hardened cuticle of insects is used for the purpose of sound production. Many insects produce sound during locomotion, feeding, cleaning and other activities. In some cases, almost every part of the body has become modified to produce sound.

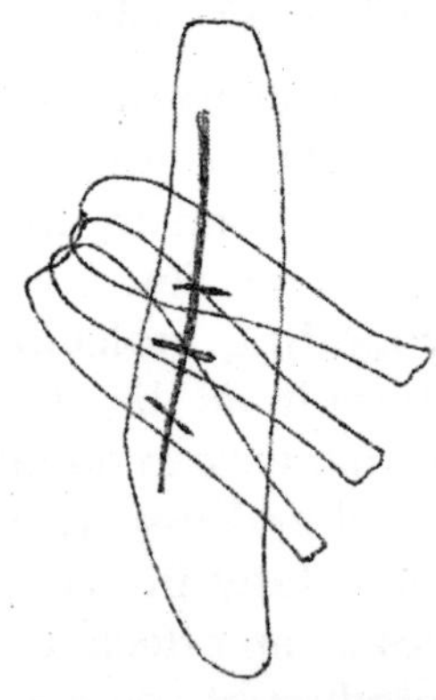

Fig. 11.1. Courtship song in grasshopper (Subfam. Gomphocerinae) is produced by stroking a stridulatory file situated on the hind femur against the radial vein located on the fore wing. The vibrations produced differ with position of the femur.

11.2.1.1 Stridulation

Stridulation is rubbing of one body part against another. One body part, the 'file' is usually a series of 'pegs' or 'teeth'; while the other part is 'plectrum' or 'scraper' (Fig. 11.1). This system of sound production is common in crickets, grasshoppers, katydids, true bugs, beetles, some ants, moths and caddisflies larvae.

Males of the suborder Ensifera (Order Orthoptera) often produce acoustic signals in the context of mate attraction. The acoustic signals or calling songs produced by the males are highly species-specific, with each cricket species producing a unique song. Crickets provide excellent systems to study the perception, function, diversity and evolution of animal communication signals because they use a repertoire of acoustic and non-acoustic signals for both long- and close-range communication in the context of mate finding. Bennet-Clark (1989) discussed the mechanism of sound production by crickets. Males of field crickets, *Gryllus* species produce distinctive calling songs. To produce a song, the male raises its fore wings above its abdomen and opens and closes them in a species-specific rhythm. On each closing stroke, the 'scraper' or 'plectrum' (a sharp edge near the wing base) of one wing engages the 'file' (a vein with many evenly spaced small teeth beneath) of the other wing, causing the wing membranes to vibrate and produce a pulse of sound. The sound is then amplified by special resonating areas of the fore wing, such as the harp and mirror (Fig. 11.2). Thus, field cricket calling songs consist of some pattern of sound pulses that correspond to fore wing closures. Some crickets produce long runs of pulses, termed 'trills', whereas the others produce short groupings, termed 'chirps'. Besides the calling songs, males also produce loud fight songs when they encounter other males and softer-sounding courtship songs when they encounter females. In katydids (bush crickets) left fore wing overlaps right, and in most fully winged forms, only left file and right scraper are present. In species lacking hind wings, short fore wings are retained to produce sound. They have a file and scraper on each fore wing, although only the left file is functional. These species produce very high frequency sound and the wide spectrum of sound frequencies produced by them indicate that a number of resonating systems operate in these insects. Females of some species also stridulate but the system is less developed and the file is on upper surface of right tegmen. The mole crickets also produce sound and use tunnel as resonating chamber.

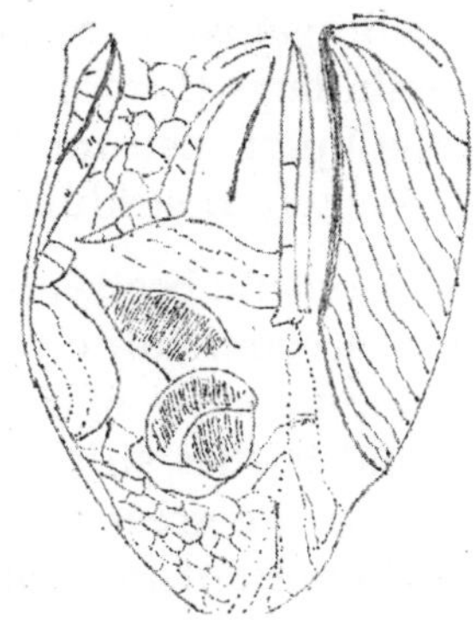

Fig. 11.2. Stridulatory apparatus of cricket, *Gryllus campestris*.

In grasshopper (Fam. Acrididae) the sound is produced by stroking a file situated on the inside of each hind leg femur, against the edge of the wings which form the plectrum. Some grasshoppers have a ridge on the inside of the hind femur that rasps against the irregular

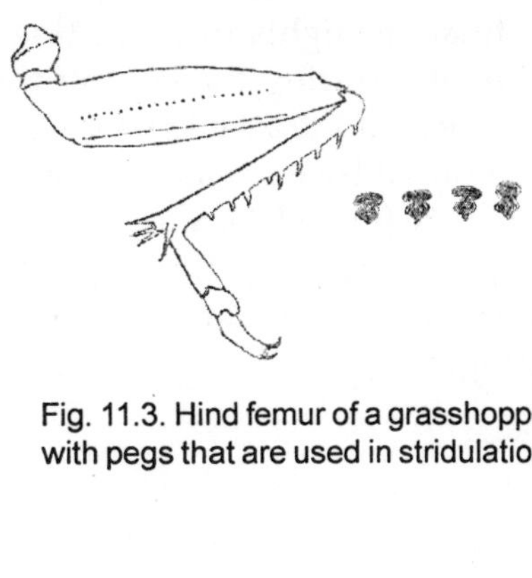

Fig. 11.3. Hind femur of a grasshopper with pegs that are used in stridulation.

intercalary vein while some others have a row of pegs on the femur (Fig. 11.3). Many grasshoppers produce ultrasonic mating calls (above the range of human hearing). In some species, the sounds may be as high as 100 kHz. Grasshoppers of the subfamily Gomphocerinae have acquired the most elaborate and complex songs due to the peculiarities of the sound production structure and mating strategy. Using both hind legs, the grasshoppers have two separate sound-producing devices that usually work with a phase shift.

In Hemiptera (Pentatomorpha) the most common mechanism of stridulation involves a file on ventral surface and scraper on leg; or a file on the wing is rubbed against a scraper on dorsum of thorax, e.g. in *Kleidocerys* species. The assassin bugs (Fam. Reduviidae) have a file between front legs, which is rasped by the tip of rostrum. Some aquatic bugs like *Ranatra, Corixa* and *Notonecta* are also known to stridulate.

In Coleoptera, beetles of several families stridulate by using different body parts, but most commonly elytra are involved. The carabid, *Adelium pustulosum* stridulates by rubbing sides of abdomen against edge of elytra. Both adults and larvae of passalid beetles stridulate by entirely different mechanisms. In larval passalids hind legs are greatly reduced and act as a scraper. The sound communication is very important to them as they live most of their lives in total darkness. The sounds used for aggression/defence are different from the courtship songs, and the courtship songs differ among species.

Adult of peacock butterfly, *Inachis io* produce sound by rubbing the wing veins. Larvae of several species stridulate with file on abdomen. Pupae have three ways of stridulating: (i) course transverse ridges on anterior edge of some abdominal segments, rubbed on tubercles on posterior edge of preceding segment, (ii) in Noctuidae rough areas are rubbed against cocoon, and (iii) pupa of hesperiid, *Gangara thyrsis* has a transverse ridge on ventral midline of fifth abdominal segment, where the elongate proboscis rubs.

Some Trichoptera larvae have a series of ridges on sides of head and a scraper on front femur. They stridulate in fights over feeding sites. If the defender stridulates during the contest then it invariably wins, regardless of the attacker's size. Many ant species have striations at base of gaster which are rubbed by scraper on petiole (Fig. 11.4). Virgin queen honey bees produce a piping sound by thoracic sclerite vibration, other queens still in their larval cells respond with their own piping.

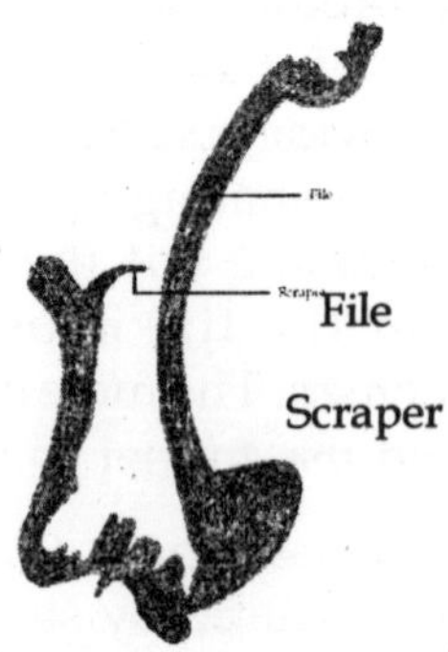

Fig. 11.4. Stridulatory apparatus of an ant.

11.2.1.2 Tymbal vibration

Vibration of a membrane by muscular action is common in certain Hemiptera and Lepidoptera. In cicadas tymbal is located on first abdominal segment. The lateral side of the tergum is modified into a striated tymbal and the tymbal muscle is attached to the sternum. An air sac is present behind the tymbal, which fills most of abdomen and opens through spiracle. Air sac is constructed to resonate with tymbal, thereby increasing intensity. Sound is produced by myogenic contraction and relaxation of muscle. The African cicada, *Brevisana brevis* produces a calling song with a mean sound pressure level of 106.7 decibels at a distance of 50 cm (Villet, 1987). *B. brevis* is likely the loudest insect species on record. Males of all Auchenorrhyncha (Hemiptera) and many females have tymbals for sound production, but are not backed by air sacs. Tymbals are also present in both sexes of some Pentatomidae, Cydnidae, and Reduviidae on dorsolateral surface of fused first and second abdominal segments.

In tiger moths (Fam. Arctiidae) tymbal is formed by a thin area of cuticle on each side of metathorax and used to produce defensive signals during flight. Blest *et al.* (1963) gave complete description of mechanism of sound production in these insects. Posteriorly tymbal is covered with scales while anteriorly has band of horizontal striations. The tymbal buckles along its length; however, the coxo-basalar muscles that control the buckling motion are not attached to the tymbal itself but act indirectly on the surrounding ring of cuticle. As it buckles, each rib yields one at a time, giving several small pulses for each muscular contraction. In wax moth, *Galleria* the tegula at the base of wing is modified into a tymbal-like structure. It works during down stroke of the wing from forces of twisting during flight (direct flight muscles). The gallerine moths use sound in social communication.

11.2.1.3 Forced air movements

The most spectacular example of sound production by forced air movements is in death's head sphinx moth, *Acherontia atropos* where swallowed air is passed backwards and forwards across a circular flash in the epipharynx. The sound is produced for defence.

In Madagascar hissing cockroach, *Gromphadorhina portentosa* a hissing sound is produced by forcibly expelling air through a pair of modified second abdominal spiracles. During normal respiration the hissing spiracle remains closed while during hissing all spiracles except the hissing spiracle are closed, and hissing spiracle opens on exhalation. This is also used in courtship or interacting aggressively.

11.2.1.4 Vibration of wings

A number of insects produce sound during flight and the signals may have communicative implications in some cases. The gnats and mosquitoes produce sound while beating their wings and locate their partners, responding to the sound thus generated. Similarly the wing sound of certain flies plays a role in attracting a mate. The hawk moth produces a sound like a modulating oscillator in flight. The social insects in swarms produce distinctly different sounds depending on their mood. Insects like hornets, wasps and bees 'roar' collectively when disturbed or deprived of their queen for any period.

11.2.1.5 Other mechanisms

Many sounds are produced as a by-product of other activities like feeding, cleaning or copulating; however, these are not of much significance. Sounds produced during flight may have some value. Striking the ground with body parts would appear a more direct method of producing substrate-borne signals. This mechanism is common in Plecoptera, Isoptera, Psocoptera, Orthoptera, Hemiptera, Coleoptera, and Hymenoptera. The death watch beetle, *Xestobium* produces sound by bending its head down and banging it against the floor of burrow in the wood. Male and female *Ephippiger ephippiger* tremble once they are within close contact on a bush, and this violent jerking of the body lures the female over remaining branches. The thornbug, *Umbonia crassicornis* communicate by shaking the surface they resting on a method scientifically called substrate vibration. The bugs send signals into the stems of trees by shaking their bodies. Various species of wood beetles produce knocking sounds as they move around inside the hollow logs they inhabit.

11.2.2 Functions of insect sound

The insect sounds may be patterned in which there is repetition of a particular basic sound or may be unpatterned sounds of high intensity. Most alarming and warning sounds are unpatterned. The sounds produced by insects may be used in aggregation, attraction, courtship, copulation, aggressiveness, alarm and defence, and social signals.

11.2.2.1 Aggregation

Aggregation provides an opportunity for individuals to assess the quality of potential mates and improves detection of predators and defence against them (Thornhill and Alcock, 1983). Aggregation may be cued by visual signals (as in fireflies), sound (in stridulating Orthoptera), or olfaction (in bark beetles). Male insects may be drawn to areas where females are concentrated, such as where they emerge, forage, or oviposit. Insects like black flies, cicadas and tettigoniids aggregate in response to conspecifics sound signals. Sound signals produced by *Magicicada* lead to assemblage of individuals of both sexes. The sound signals in different species differ and each species of cicada has a peak chorusing period. These result into formation of clearly defined group of each species.

Acoustic signals are also important for aggregation of crepuscular or night-flying insects. It has been assumed that swarms of mosquitoes and midges maintain their uniformity through reactions to flight tones.

11.2.2.2 Recognition

In acoustically active insects (e.g. crickets, grasshoppers, mole crickets, cicadas), males often advertise by means of conspicuous songs. The song may encode information about the identity of the species, which can be used by other individuals to discriminate conspecifics from heterospecific sympatric species. The ability to discriminate the conspecific signal from heterospecific songs has been studied in crickets and grasshoppers. In order to assess the relative importance of particular song parameters in song discrimination, most studies have used the phonotactic preference of reproductively receptive females or any other stereotyped response to a modified or a natural song playback. Both the time pattern (Pollack and Hoy, 1979) and the frequency spectrum (Oldfield, 1980) of the signal have been shown to be involved in species discrimination. In cicadas; however, few studies have addressed this question, and the relative importance of the song parameters in species recognition is still poorly known. Some evidence indicates that the frequency spectrum of the signal carries information about the species-identity of a calling male, whereas the temporal pattern might reveal the quality of the sender at a short distance (Doolan and Young, 1989).

Sound signals are also used by predators to recognize the prey. Haskell (1974) suggested that predators can locate prey both on land and in water by responding to the vibrations produced by their prey. Some hymenopterous parasitoids locate their hosts lying beneath the substrate by their sound signals. Tachinid parasitoids of the genus *Ormia* locate male crickets and katydids by their sound. The male songs of the hosts range in the frequency 2 to 7 kHz. In dark, female *Ormia* locate calling host, on or near which she deposits first instar larvae which burrow into the host and develop by feeding on host's tissues.

11.2.2.3 Isolation

The sound production behaviour has the effect of augmenting isolation of individuals of same or different species. Stridulations probably assist in spacing of males of many species of grasshoppers, crickets and cicadas over the largest possible area and also reduce interference during mating. Songs of two sibling species of grasshoppers, *Chorthippus brunneus* and *C. biguttulus* are of major importance in isolating these species. Both species are closely related and have same mating behaviours. The crossing readily occurs in the laboratory with production of viable offspring. However, very little crossing occurs in the field. The major difference between these species is their songs which have species-specific stimulating and orientating influence resulting into isolation of two species (Perdeck, 1958).

11.2.2.4 Sexual signals

Acoustic signals are involved in a wide range of sexual behaviours, attraction, courtship, copulation and post copulatory pair maintenance. Acoustic behaviours of a large number of Orthoptera and Hemiptera have been studied with respect to sexual signals. Chirping of cricket is a communication signal by the male for the silent female. The cricket male first produces a 'calling signal' and after arrival of the female it produces 'courtship' song. In certain diurnal insects like grasshoppers, acoustic-visual signals may operate in open grasslands. The female responds to the male only depending on her state of physiological development and sexually immature and already mated females do not respond to acoustic signals. However, nocturnal insects fully rely on acoustic signals.

The acoustic signals of crickets are used to communicate with each other and these signals take the form of stereotyped, repetitive songs produced by stridulating males. Oceanic field cricket, *Teleogryllus oceanicus* has a range that includes three structurally distinct songs termed calling, courtship and aggression song, respectively (Alexander, 1961) which serve distinct functions. Calling song serves to attract females from a far, courtship song induces the female to mount the male, and aggression song is produced during aggressive encounters with other males (Alexander, 1962a). Female crickets recognize and preferentially orient towards the conspecific calling song, which they identify based on its fundamental frequency and species-specific temporal parameters (Elsner and Popov, 1978). In *T. oceanicus* courtship is initiated when the antennae of the male contact the body of the female. The male then vibrates his antennae, adopts a typical singing posture and begins to produce courtship song, simultaneously stroking the female with his antennae. He then turns away from her and presents his abdomen, stridulating all the while. The female reacts by following the male, approaching from behind and touching his abdomen with her antennae and palpi, whereupon the male spreads his hind wings and flattens his abdomen, allowing her to mount him. Burk (1983) concluded that male mating success was correlated with the production of courtship song. Muting and playback experiments in different cricket species support this conclusion (Crankshaw 1979). Muted males by removing their tegmina perform all the acts of courtship normally, including singing, but they have little success in eliciting a mounting response from the females. The courtship song of *T. oceanicus* consists of a short, amplitude-modulated chirp, followed by a long trill of constant intensity and high syllable rate. Balakrishnan and Pollack (1996) determined that the chirp component of the song alone was sufficient to evoke normal levels of mounting, but the trill was only partially effective on its own.

In grasshoppers of the subfamily Gomphocerinae a male produces a 'calling song' when being alone while the female that is ready to mate produces a 'response song'. The response of a female initiates a duet, during which both sexes approach each other. When a male appears to be nearby a female, the male begins to court. In some species, a 'courtship song' is extremely elaborate in terms of a number of different temporal parameters, and sometimes, conspicuous visual signals are integrated into the courtship. It has been observed in the field that several competing males often sing around a female, and courtship songs are very long and frequently repeated. This may facilitate the female choice and favour competition among males.

The sound signals, known as 'courtship song' or 'love songs', are generated by wing vibration in *Drosophila melanogaster*, play prominent role in the mating behaviour of these flies. As soon as a male fly has approached and oriented towards a female, it extends and vibrates one of its wings, thereby emitting a species-specific song (Fig. 11.5). This sound enhances the female's receptivity to copulation and, thus facilitates mating (Bennet-Clark and Ewing, 1970; Ewing, 1983). Since the discovery of acoustic signalling in these flies (Shorey, 1962), the generation of these signals has been extensively studied at the behavioural, physiological, genetic and molecular levels. It has been established that the antennae of *Drosophila* serve as 'love song' detectors, and Johnston's organ, a mechanosensory chordotonal organ in the pedicel of the antenna, serves as the auditory sensory organ (Ewing, 1983).

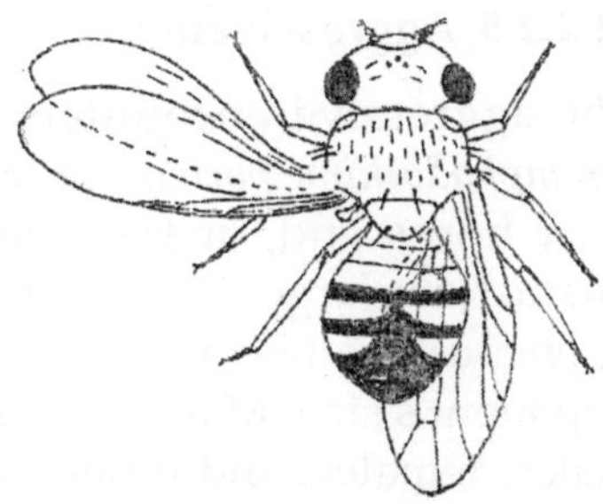

Fig. 11.5. A male *Drosophila melanogaster* during the courtship often extends one wing or the other and vibrates it.

The tephritid fruit flies also vibrate their wings during courtship. Although signals received by individuals of opposite sex are mainly visual; however, sound supplements the behaviour shown by these flies. Song parameters, calling propensity and mating success in Caribbean fruit fly, *Anastrepha suspensa* were discussed by Burk and Webb (1983). Sexually mature females of yellow fever mosquitoes, *Aedes aegypti* produce buzzing sounds during flight to attract the males (Roth, 1948), whereas immature females have a different tone to which males do not respond.

Many host-specific psyllids of both sexes produce acoustic signals during mating. The acoustic calls are specific to each species. Cryptic psyllid species can mainly be distinguished by mating signals and distinct differences in the acoustic signals produced by morphologically cryptic species have been demonstrated.

The cicadas produce communicative sound primarily by a tymbal mechanism. A singing male usually sits on woody plant parts such as branches, or the trunk of trees. Females perform phonotaxis to find the male within the complex tree habitat, and can acoustically locate the sound source during flight. The sound produced by males carries significant energy that is produced while standing on a woody plant, suggesting the possibility of simultaneous transmission or induction of signals in the plant substrate.

Late at night when the house is quiet, a tapping sound could be heard coming from the walls. This sound is made by death-watch beetles (Fam. Anobiidae) which eat the wood in old houses. The death-watch beetles strike their heads against the roof of their tunnel to make the tapping sound, and this sound attracts the opposite sex of the beetle.

11.2.2.5 Aggressiveness

The aggressive encounters between males of Madagascar hissing cockroach, *Gromphadorhina portentosa* are quite exciting. Males smash into each other with their horns and/or they push each other with their abdomens. In aggressive interactions larger males usually win and hissing plays an important role. Winners of encounters hiss more than losers. The hissing may also be used to assess the opponent's size. Males also hiss during courtship interactions with females. Adult males, females, and older nymphs hiss when disturbed or handled.

Crickets' flights have remained a subject of entertainment. Adult males fight and actual physical combat occurs when two rivals are engaged in battle. They produce aggressive rivalry songs that are quite unlike with other sounds produced for attracting the females. Male of *Oecanthus* has a territory of about 50 cm^2 in which he sings his song. If another cricket attempts to enter the territory a distinct aggressive sound is produced. The intruder also replies by producing sound, fighting may also occur until one male leaves. The dominant male stridulate more aggressively (Alexander, 1961). Phillips and Konishi (1973) observed aggressive singing in house cricket, *Acheta domestica* and fall field cricket, *Gryllus pennsylvanicus*. The conspecific male crickets were allowed fighting in cages, and their repeated encounters demonstrated clear dominance of one cricket in each species. Then the tympanal organs of losers were removed and these deafened losers were again allowed to fight with the dominant one. The deafened losers were observed to be extremely aggressive, initiated combat and after fighting shown their dominance over the former superior rivals. The results indicated that acoustic signals of dominant cricket's song inhibit the aggressive behaviour of subordinate cricket and inability to hear the sound signal frees the deaf crickets from this inhibition. The rivalry songs of grasshoppers also exhibit their aggressive behaviours, and usually two males commonly chirp alternatively in a regular manner.

11.2.2.6 Alarm and defence

Insects of nearly all orders are known to react when disturbed. Many of them produce high frequency unpatterned protest sound. These signals are elicited largely upon tactile stimulation. These protest sounds may act as predator escape mechanisms in many cases. Many protest sounds are associated with aposematic colour patterns, defensive secretions, bad taste or presence of sting and the predators learn to associate the protest sounds with earlier painful experiences. Some grasshoppers produce clicking sound during flight and it may be a flash sound. The predator following a grasshopper on the basis of this sound may get baffled. Workers of Pacific dampwood termite, *Zootermopsis angusticollis* rock back and forth their middle legs so that the mandibles tap against the tunnel floor, so they warn other members of the nest to move elsewhere (Howse, 1964).

Many insects respond to a predator by adopting a characteristic posture, which intimidate the predator. Such startle displays or frightening behaviours are of significance as they often repel a prey. Robinson (1969) discussed such displays in different species of stick insects, which often display erecting brightly coloured wings with eyespots. These displays are associated with stridulation resembling hissing of a snake or sound of a threatened scorpion. Many insects utter a shrill cry when seized. The stridulation of reduviids, long-horned beetles, mutillid wasps and certain lepidopterous pupae perhaps alarm a potential predator. It has also been suggested that in some cases sound resemblance occur, e.g. certain syrphids not only look like bees but also sound like them. Some arctiids and ctenuchid moths are distasteful, in addition they display bright colours on hind wings and abdomen, produce high frequency sound, and a few emit a repellent smelling froth from thorax when disturbed (Blest, 1964).

Bats are fast, manoeuvreable fliers equipped with sensitive sonar guidance systems and pose serious threat for night-flying moths. Moths have evolved an array of sensory and behavioural strategies that enable them to avoid the predatory bats (section 13.4.4). Kenneth Roeder and his colleagues studied the evasive behaviour of moths in response to ultrasonic pulses emitted by their predators-echolocating bats. Roeder and Treat (1957) firmly established that the tympanal organs of noctuid moths are specifically tuned to the ultrasonic signals used by bats. Many night-flying moths have pairs of ears located on both sides of their abdomens or other body parts that are tuned to exactly the sound frequencies emitted by hunting bats. These sensitive ears allow the moths to eavesdrop on the hunting cries of bats and to attempt to avoid them. The ears in most noctuid moths are most sensitive to frequencies of the echolocating calls used by the local bat population; it is thought that the auditory systems in most of these moths evolved as a direct result of bat predation (Fullard, 1987). Noctuid moths show a stereotyped two-phased behaviour in response to the sounds of an echolocating bat. Weak ultrasonic stimulation (indicating a bat at a distance) generates negative phonotaxis, or a flying away response, to take the moth out of the detection range of the bat. Strong stimulation (indicating an approaching bat) produces either erratic behaviour (e.g. unpredictable looping), or in a last ditch effort to escape, wing-folding to produce a free-fall (Fig. 11.6a-d) (Roeder, 1965).

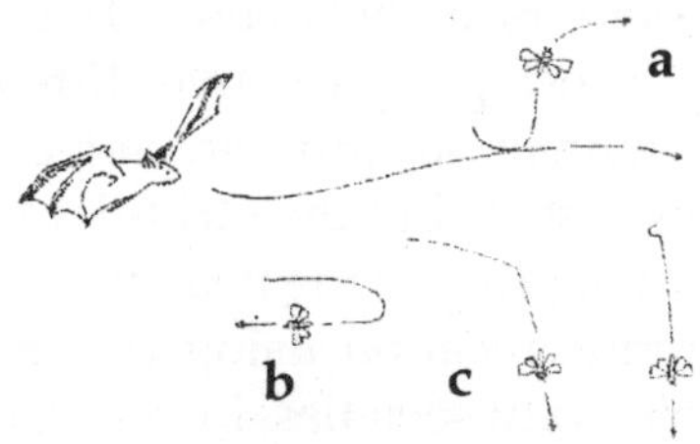

Fig. 11.6. Escape movements of moth in response to the ultrasonic sounds of a hunting bat. (a)- Erratic flight (b)- Sharp turn (c)- Abrupt dive, (d)- Free-fall.

Some moths confuse bats by emitting sounds similar to those emitted by a bat. Some species of tiger moths sequester the plant toxins that their larvae eat. Further protection is provided by their bright and conspicuous colour pattern. In addition, they also produce unique sounds that their potential bat predators can detect and associate with their poisonous nature. Besides this, when these moths detect a bat in pursuit, they issue a series of ultrasonic sounds in return which startles the bat, confuses it or interferes with its reception of the echoes. If the bat has never eaten a tiger moth before it will behave as though it had been startled by the unexpected sound and bolt away.

Intraspecific sound signals characteristic of social insects are commonly used to alarm the nestmates. Workers and soldiers of termites use substrate vibration that is produced by tapping head on roof and floor of tunnel to send alarm. They often produce these signals when encounter other termite species.

11.2.2.7 Social communication

Hickling and Brown (2000) analysed acoustic communication by ants, based on data obtained from black imported fire ant, *Solenopsis richteri*. Many ant species communicate by means of squeaking sounds from a stridulatory organ (Fig.11.4) on the ant's body, consisting of a set of ridges and a scraper. The squeaking sounds are usually very faint but they pass through ant colonies. It has been suggested that they transmit stridulation signals through the soil, or other solid substrate. However, for a number of reasons, this mode of transmission is highly unlikely. A more likely explanation is that ants communicate with each other through the air using near-field sound. The near-field is an acoustic transition zone surrounding a small source, the size of an ant, in which the characteristics of the sound change abruptly before it can propagate fully in the far-field. The acoustic receptors are believed to be concentrations of trichoid sensilla on the two antennae. By sensing the relative difference in sound displacement between the tips of the antennae, an ant can detect a stridulation signal in the near-field. It has been suggested that the stridulatory organ and acoustic receptors of ants have evolved by adapting to the near-field (Hickling and Brown, 2000). Combined with chemical communication using pheromones, acoustic communication plays an important role in ant societies. Roces and Tautz (2001) summarized vibrational communication in ants and challenged the near-field particle oscillation around a stridulating ant. They calculated the amplitude of the near-field particle oscillation around a stridulating ant, and compare it with the sensitivity threshold of the ant sensory receptors. They concluded that ants are highly responsive to the substrate-borne components of stridulatory signals, and insensitive to near-field sound. Taking into consideration the contemporary views the subject of vibrational communication in ants need further investigation.

Piping of queen bee may be important in informing the colony members about the presence of a virgin queen in the colony. Returning foragers of stingless bee,

Trigona produce a characteristic buzz and transmit the information to the hive mates.

11.2.2.8 Communication by dance

Perhaps one of the most fascinating aspects of honey bee behaviour is the dance performed by worker bees to tell others about food sources in the field. Germen Biologist Karl von Frisch (1937, 1947, 1954, 1967, 1971) published notable contributions on the communication in honey bees. He correlated the runs and turns of the dance to the distance and direction of the food source from the hive. The orientation of the dance correlates to the relative position of the sun. He was honoured with Nobel Prize in 1973. He suggested that a forager was able to communicate its hive-mates the quality, direction and distance of food source with such accuracy that the other bees after learning the message directly flew to the food. He observed that when a forager found a food source it brought a full load of food to the hive, and went out again at the same site for making confirmation. After its second return she communicated its vital information to others by exhibiting specific, repeated, stereotyped behavioural patterns, which Frisch termed as 'dance' upon the vertical face of the honeycomb. Thus, bees communicate their floral findings to other foragers. Other bees, observing her dance, will know where to find the wonderful source of food. This amazing level of learning and communication benefit both bee and plant. There are basically two types of dances.

(I) Round dance. When the food source is located at a distance of less than 50 yards from the hive the bee performs this type of dance. In this dance the bee circles in one direction, turns and then circles back in the direction from which she came (Fig. 11.7). There is no indication of direction and it indicates presence of food nearer to the hive. This simple form of dance recruits other workers to search food in the immediate vicinity of the hive.

Fig. 11.7. The round dance.

(II) Tail-wagging dance (8-fashion dance). As the distance to the foraging site increases the dance become modified by incorporation of a straight run during which worker waggle its abdomen from side to side, which is one of the most intriguing examples of complex animal behaviour. Then she runs in a half circle to the left, back to her starting point, performs another straight waggle run along the path of her first, and then circles to the right. This pattern is repeated multiple times, and is eagerly attended by unemployed bees in the hive. Shortly after such dances commence, newly recruited foragers arrive at the food source being advertised. The bee used to dance for a few seconds to 3 minutes. The long duration of dance indicates richness of food source. The bee can convey to the foragers information about :

- The odour of the food.
- Its direction from the hive.
- Its distance from the hive.

The waggle dance (Fig. 11.8) has two components:

- A straight run- the direction of which conveys information about the direction of the food.
- The speed at which the dance is performed is correlated with distance of food from the hive. If the food is far from the hive, the dancing speed is slow.

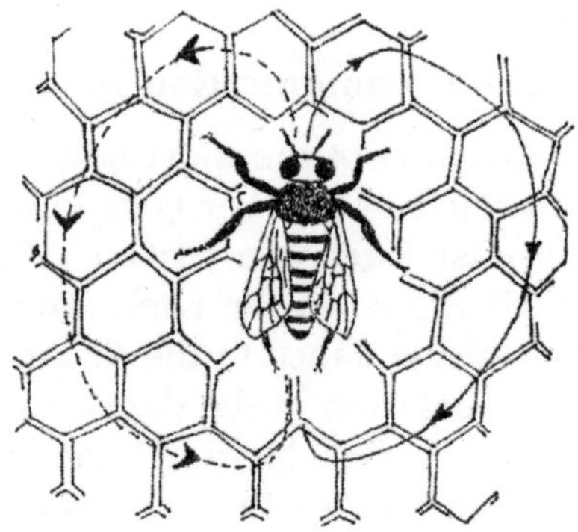

Fig. 11.8. The tail-wagging dance (modified after Frisch, 1971)

But how do bees measure the flight distance that they communicate so precisely? Two theories have been proposed about how honey bees estimate the distance to food sources. Heran (1956) believed that the bees measured the distance by assessing the amount of energy spent by them during their outward journey. However, doubts emerged when it was found that distance estimation by bees could be manipulated by altering the number of landmarks between the hive and a food source, suggesting bees were counting landmarks encountered *en route*. Esch and Burns (1995) showed that the perception of distance by the bee couldn't solely be based on energy expenditure. They conjectured that foragers process the speed with which visual contours move across the eye (optic flow), and integrate this with travel time. This theory suggests that the cue is visual, and is derived from the extent to which the image of the world has moved on the eye during the trip. Thus, the bees measure distance by the motion of images received by their eyes as they fly. The compound eyes are very sensitive to "flicker effect", and bees respond better to flowers that have complex petals and move in the breeze. In other words, bees possess a visually driven 'odometer' that is robust to variations in wind load and energy expenditure. The quality of information available about the velocity of the passing landscape will depend, of course, on the sensitivity of the eyes. The eyes of bees contain three types of colour receptors, with maximum sensitivity in the ultraviolet, blue, and green domains of the spectrum. Their excellent colour vision is optimal for flower identification (Chittka, 1996), but do they also use it to measure the image velocity of the passing landscape? Chittka and Tautz (2003) reported although bees see flowers in colour, they do not analyse the colours of the landscape image that moves across the eye as they fly. Their perception of landscape motion is colour-blind; only a single spectral receptor, the bees' green receptor, drives motion vision.

Srinivasan *et al.* (2000) tested the two theories by observing dances of bees that have flown through a short, narrow tunnel to collect a food reward. The results showed that the honey bee's 'odometer' is visually driven. They also provided a calibration of the dance and the odometer in visual terms. Esch *et al.* (2001) reported that when foragers collect food in a short, narrow tunnel, they dance as if the food source were much farther away. Dancers gauge distance by retinal image flow on the way to their destination. Their visually driven odometer misreads distance because the close tunnel walls increase optic flow. They examined that recruited bees search outside in the direction of the tunnel at exaggerated distances and not inside the tunnel where the foragers come from. Thus, dances must convey information about the direction of the food source and the total amount of image motion *en route* to the food source, but they do not convey information about absolute distances. They also found that perceived distances on various outdoor routes from the same hive could be considerably different. Navigational errors are avoided as recruits and dancers tend to fly in the same direction.

Frisch also noted that the direction of the straight portion of the waggle dance varied with the direction of the food source from the hive and the time of day. For communicating the direction of food the bee uses the position of sun. When it comes back it knows the angle between food, hive and sun. While dancing they make same angle in the hive (Fig. 11.9). Thus, the sun is used as the key reference point for the communication of direction. In case food source is in the direction of the sun then the straight run of wagging dance is straight upward on the comb. Source in opposite direction is indicated by downward straight run. The angle of straight run also changes with the direction of food source and if the forage site is located 45° to the right of the sun, the straight run of dance will be performed at an angle of 45° to the right of the vertical. The foragers do not actually have to see the sun to navigate. As long as they can see a small patch of clear blue sky, they get along fine. The light from the blue sky is polarized and the direction of polarization at each point depends on the position of the sun. When the bee cannot directly see the sun, it will infer its position from the polarization pattern of the blue sky. Bees can use this 'sense' when flying to a goal or when communicating directions to other bees. The compound eyes of bees are sensitive to ultraviolet radiation and they can detect the position of the sun and communicate the direction of foraging site, even on cloudy days.

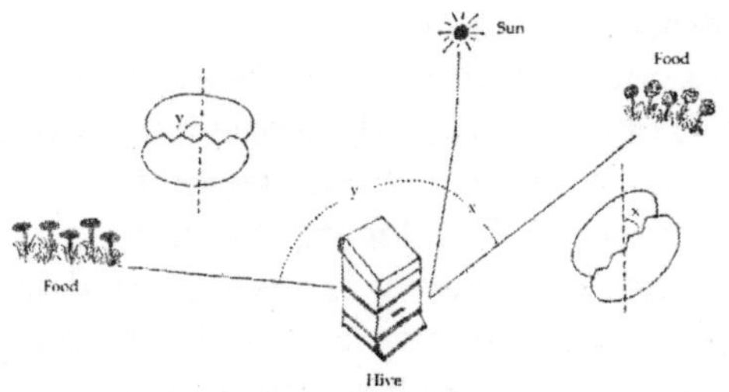

Fig. 11.9. Relationship between the angle of the dance on the vertical comb and the position of the sun with respect to the location of the food.

The dance of honey bee became an exciting example of visual communication in insect and cited in many texts. Wenner (1964) reported that sound is most important channel for distance indication. Esch, who also studied bee communication behaviour, was in dilemma that how bees were able to observe the dance in total darkness of the hive? To seek an answer he built a motor driven bee capable of performing waggle dance. After seeing its dance other bees clustered around it but none of them left the hive in search of food (Esch, 1967). Esch reasoned that this might be because the dummy bee was not producing right sound. The bees performed rightly when a microphone with record of actual dancing bee sound was placed and played simultaneously with dance of dummy bee in the hive. This helped in drawing a logical conclusion that bees might equally use sound as a cue to determine distance rather than visual or tactile cues.

The theory of Frisch (1947) claimed that the dances were correlated with the distance and direction of the food source from the hive. Frisch developed several experiments to verify the theory. Wenner (1964) and Esch (1967) suggested that as bees dance, they emit faint low-frequency sounds. This sound might play a critical role in bees' communication, as they communicate effectively in complete darkness that prevails in the nest. Wenner and Johnson (1967), Wenner (1971) reported that the bees were unable to locate the food source with dance alone. They contended that the bees could adequately achieve recruitment with the odour of nectar brought to the hive and the recruitment pheromone left by foragers in field. Wenner felt that inexperienced foragers have great difficulty to locate a food source. They locate the source by using right combination of odours from the food. His theory is also termed as 'odour theory of communication'. This scientific controversy resulted into publication of a number of articles in favour and against of Frisch's views, including Frisch's defence of his work. Gould (1975, 1976) conceded that the design of the experiments of Frisch had been inconsistent and that his earlier results could all be accounted for by Wenner's locale-odour hypothesis. Gould argued that Frisch and Wenner adopted different training techniques. Wenner's training method encouraged bees to rely on odour, while Frisch's method led them to use the dance cues. He further suggested that Frisch and Wenner investigated situations involving exploitation of an abundant food source but focussing on different stages of the process; Frisch on early stages, while Wenner on later stages. Rosin (1988) argued that the proponents of dance language now grant the locale-odour mechanism a role in the communication amongst bees.

Wenner and Wells (1990) pointed that even the weak version of the hypothesis is invalid and they argued the experiments conducted by Frisch and their analysis could be criticized on several points. The important ones are: (i) negative results were ignored, (ii) the importance of conditioning was not recognized, and (iii) there was lack of adequate controls in experimental design.

The reason for controversy is because various workers have analysed Frisch hypothesis differently. Some argued that bees communicate distance information entirely by dance, while others stated that the dance information is used only to find the vicinity of feeding station and odour cue is used for final approach to the station. It was also argued that the dance of the bees could not communicate distance information to the bees, but it might communicate approximate direction information. However, this latter possibility does not qualify as a language.

Dyer (2002) reviewed the biology of the dance language. Presently, most scientists firmly believe that the dance is a mysterious coded language that the bees use to direct their hive-mates to fly toward distant food sources. Many experiments over the decades have claimed to support this theory. The latest effort to decode the honey bee's waggle dance is being published recently by a British physicist Joseph R. Riley and his colleagues (Riley, *et al.*, 2005). They tagged bees with radar transponders to track their flight and the results of their experiments confirm that the dance is the most sophisticated example of communication. The bees clearly could read the message encoded in the waggle dance. In their first experiment, they captured workers as they left the hive after, attached a thin wire radar transponder, and then recorded their flight paths using radar. After looping flights, while bees find their bearings, they fly pretty much as the waggle dance predicts. In a second experiment, the bees were captured after they'd observed the dance and begun flying with their transponders. They were taken to three distant release points and released there; the bees continued their flight path as though they had not been moved away from their hive.

This was very strong support for the Frisch hypothesis because in this case there was no possibility the bees were following regular routes or any odours that the dancer might have left in the air, said Professor Riley. He believes his team's experiments will end debate about the function of the waggle dance.

Thus, it seems that the debates pertaining to dance language of bees have been resolved and it is confirmed that both sound and dance are needed to communicate information about the location of food. Odours too are involved but appear to lack the importance that Wenner ascribes to them.

Other forms of dances

Transition dances. Frisch (1962) reported that different races of honey bees have incorporated variations into their dance that have been termed as 'dialects'. The flattened, figure-eight pattern or 'sickle dance' or 'crescent dance' (Fig. 11.10) of Italian race of *Apis mellifera* is used when the food source is at

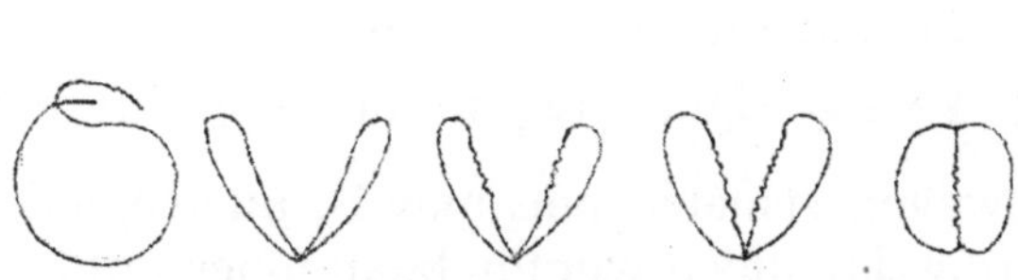

Fig. 11.10. Intermediate forms of honey bee dance that occur with an increase in distance between hive and foraging site.

an intermediate distance from the hive, the opening of the sickle facing the source of food. The dance patterns changes to some transitional forms and finally resembles two half moons placed side by side.

Joy dance or DVAV. In dorsoventral abdominal vibration (DVAV) or joy dance a bee places its fore legs on some part of the body of another bee and makes up and down shaking movements with the abdomen. Maximum dancing activity occurs during peak foraging. Schneider *et al* (1986) believed that it is performed to regulate the daily and seasonal foraging patterns according to food fluctuations and to regulate the queen activity in relation to swarming.

Alarm dance. When a forager brings nectar poisoned with insecticide, the hive bees and field bees become excited and perform alarm dance. They run in spiral or zigzag path and vigorously shake their abdomen sideways.

Cleaning dance. The forager appears to dancing while cleaning their contaminated body with dust, hairs or other foreign particles. The bee is seen stamping her legs rapidly and swinging her body side to side. The cleaner bees also assist in cleaning the body of forager.

11.3 Visual communication

Communication in visual domain involves visual system with all its peripheral functions like colour vision, pattern, movement, polarized light, and temporal pattern detection. This kind of communication is found between insects and flowers, conspecific individuals of opposite sexes, and prey and predator (Järvilehto, 1985). Insects do not appear to rely upon visual signals for intraspecific communication to as much as they do upon chemical and auditory signals. The body or colour patterns of insects that help in communication may pose problems when detected by predatcrs. Therefore, potential visual signals for communication are exploited by insects that are protected from predators. The visual communication may be of significance because of advantages such as:

(i) Effective over long distances.

(ii) Can be used when the insect is in motion.

(iii) Quick method of communication because of fast speed of light.

(iv) Effective in all directions.

(v) Passive signals require no expenditure of energy.

However, visual signals may be intercepted by predators, and are effective only during daytime (except in bioluminescent insect like fireflies). Active signals may be metabolically expensive to produce. The visual signals can be detected only at a short-range and the insect's eye has a limited aptitude to process them. The interspecific visual signs are employed in defence behaviours like crypsis, threat display, and mimicry; and also in pollination. In intraspecific perspective, most visual communication behaviours are associated with reproductive activities.

Insects perceive the light signals through three categories of sensory organs. Most adult insects have a pair of compound eyes, bulging to varying degree and occupy fronto-lateral position on the head so as to give a wide field of vision in all directions. Each compound eye is formed from a group of closely packed hexagonal units called ommatidia (see section 3.4). The number of ommatidia varies in different insects. Honey bees have 3,000 to 4,000 facets in each compound eye and have multiple photoreceptors for each facet of the compound eye. The centre facets are larger than peripheral sensors, so they can see better straight ahead. Bee eyes can also sense polarized light and UV light. They seem to see blue colours best. Dragonflies have large compound eyes with very wide-angle vision to allow them to see as they fly forward and backward. As their eyes do not move, they have to position their eyes by changing their body orientation. Each of the dragonfly's compound eyes is made up of about 30,000 eye sensors, so they have better resolution than many other insects. Butterflies need to see a wide spectrum of colours in order to find food, survive and multiply. Butterflies may have up to four different pigments in their eyes, as compared to two or three in many other insects. As a result, some butterflies have wide-spectrum colour vision allowing them to see UV light reflected from flowers while some others can also see near-infrared light.

Larvae of hemimetabolous insects and most adults have usually three simple eyes called ocelli. The ocelli are not involved in vision but are sensitive to low light intensities. Larvae of holometabolous insects do not have compound eyes, and perceive forms, to a limited extent through stemmata located on either side of the head. In larval sawflies, *Neodiprion* and tortricid caterpillars stemmata may also show responses to polarized light. The compound eyes facilitate a rather wide field of vision. They are reduced or absent in parasitic forms, many soil-dwellers, and in some species that live in very dark habitats. Light sensitivity of insect eyes varies according to the state of light and the eyes may be dark-adapted or light-adapted. Vision involves the transduction of light energy into a bioelectric signal within the nervous system. Photobiological processes occur in a narrow band of the electromagnetic spectrum between 300 and 700 nm, with two peaks of maximum response- one at 350 nm (near ultraviolet) and other at 500 nm (blue-green). Experiments on insect behaviour indicate that many insect species can distinguish colours (Burkhardt, 1964). Frisch (1967) showed that honey bees were able to differentiate most colour categories and have been widely studied with respect to their visual system. They can detect three colours UV, blue and yellow. The visual system of bees is considered to be trichromatic. However, bees are unable to visualize the red flowers reflecting strong red light. The ability to see red is rare in the most insects but appears to be quite common in butterflies (Bernard, 1979), and is now known to be an essential part in the release of courtship behaviour of some species (Struwe, 1972). A pentachromatic visual system (the eyes contain five different types of cell which react to different bands of light) has been reported in citrus swallowtail butterfly, *Papilio xuthus* (Singarajah, 1988).

Butterflies vary widely in their sensitivity of light, and are considered to have the widest visual range of any animal. Therefore, colour vision depends on the ability to discriminate between light of different wavelengths. Such discrimination is possible because compound eyes have photopigments with differing spectral properties. Many insects have two rhodopsins, one with maximum absorption in UV wavelengths and one with maximum absorption in green wavelengths. Some species have a third pigment that absorbs maximally in the blue region. A few Lepidoptera have four visual pigments.

The dorsal ocelli allow only poor, if any, perception of form. Stemmata generally do not produce clear images, but most caterpillars can discriminate some shapes and they can orientate themselves with respect to boundaries.

The colour sensitivity of insects has significant implications to understand the visual behaviour. Many flower colours reflect ultraviolet light over their petals and have patches at the bases that fail to reflect UV light. These bases are clearly differentiated by honey bees and act as 'nectar guides' (Fig. 11.11). The nectar guides, also referred to as 'pollen guides' or 'honey guides' act as guides for arriving in on the target once the flower has been chosen by an insect. In cryptic insects colour vision assist them in selecting a suitable background. The markings of wings insects like butterflies may be of communication significance during courtship. Patches of silvery hairs on insects' body reflect light and serve as communication signal. Deonier (1974) reported such hairs are present in *Ochthera mantis* (Fam. Ephydridae).

Fig. 11.11. Nectar guides which direct insects to the flower nectaries by providing both colour and odour cues.

The insect eye is better adapted to perceive the objects in motion than static form. Thus, the compound eyes are able to scan a moving object and resolve a finer pattern when insect is flying than when it is at rest. Visualization of the apparent motion of objects is termed 'flicker vision' and is a normal way of vision of compound eye. Many predators prefer moving prey and most insects prefer the shapes that flicker. Insects like honey bees are able to detect polarized light pattern. When the sky is covered with cloud, bees dances are disoriented, and if even a small patch of blue sky is visible, bees are able to orient themselves correctly by using polarized pattern of the sky. This ability has also been demonstrated for a wide range of other insects.

11.3.1 Aggregation and dispersal

In bringing together a large number of insects of a species, visual cues may play some role. However, acoustic and chemical signals are more important for aggregation in insects and their compound eyes are such that they do not permit clear long-range vision. Dispersion through visual communication is somewhat better studied than aggregation. Insects might simply avoid conspecifics of same sex at sight or they may elicit withdrawal.

11.3.2 Territoriality

Territories in dragonflies are maintained by visual communication. The male selects its breeding site and localizes his activities over a certain area and defend it against intrusion by other males. The male also courts female in the territory. In dragonflies visual sense is greatly dominant and helps in maintaining a territory. The territory is also defended by aggressive displays, and physical combat as a last resort (see section 10.2).

11.3.3 Conspecific recognition

The colour patterns and markings on the wings of Lepidoptera facilitate species recognition. The male and the female of a butterfly species are often very different, this is a necessity for quick recognition. Quick recognition during courtship is also important. Though many butterflies appear the same, but the butterflies themselves can often identify each other quite easily from UV markings. These UV markings are essential for visual communication during courtship of the two sexes of the butterfly. The males and females of the small sulphur butterfly, *Eurema lisa* differ only in the UV region, the males being strongly UV reflective and the females unreflective in UV (Rutowski, 1978). In addition to this sexual selection can occur within species, the males choosing the female on the basis of age. The younger females have less UV than the older individuals, and it is the younger ones which are preferred. The UV patches on some butterflies are directionally iridescent, as a result they appear to flicker in flight. This flickering is thought to have an important role in butterfly behaviour and communication. The patterned wings of tephritid fruit flies are often moved in complex manners during sexual encounters. However, there are few cases of sexual dimorphism in which wing movements may also occur in non-sexual contexts (Sivinski and Pereira, 2005).

11.3.4 Sexual signals

Visual sexual displays in some insects have evolved through intrasexual selection to mediate aggressive interactions between males and in others through intersexual selection to advertise qualities to the opposite sex. In the later case, displays may convey several types of information including: (i) promises of material transfer such as nuptial gifts; (ii) advertisements of genetic quality that would improve offspring fitness; and (iii) false-information that manipulate the receiver's perceptions. Because female insects may be able to decide the parentage of their offspring through the preferential use of stored sperm, males may continue visual courtships during and even after mating. Visual displays are more likely to occur in ground-based mating systems than those which occur in the air. Part of the reason may be the difficulty of keeping track of flying signal-emitters and judging the quality of their displays. Visual signals (ornaments) are often located on movable parts of the body, presumably to call greater attention to colours and forms, and are not evenly distributed among insect taxa (Sivinski and Wing, 2004).

The role of visual signals with respect to sexual behaviour has been well-documented in many insects. The females of mayflies, caddisflies, and midges are attracted visually to swarms of males engaged in dance-like up-an-down flights in compact groups. The well-decorated and coloured wings of Lepidoptera primarily act as visual communicators. In orange queen butterfly, *Danaus gilippus* courtship consists of an aerial phase in which the male pursuits the female and of a ground phase in which the female alights. In aerial phase male is attracted to female by using visual signs, and once the male has overtaken the female a chemically mediated phase begins (Brower *et al.*, 1965). Females of small cabbage white butterflies, *Pieris rapae* have UV-reflecting scales on the dorsal wing surface. When they fly, each down-stroke of the wing creates a brief 'flash' of UV that is apparently recognized by males as the flight mark of a potential mate. A 'flashing female' may attract several males who engage in aerial courtship displays. In alfalfa butterflies, *Colias eurytheme* only males have the UV-reflective scales. They flutter in front of the females to create a flickering courtship display and absence of scales reduces the reflectivity of the wings which is an indication of male's aging. Knuttel and Fiedler (2001) reported a case in which sequestered secondary plant compounds determine an insect's external appearance in the UV spectrum and thereby influence visually mediated mate choice. Larvae of the common blue butterfly, *Polyommatus icarus* specifically sequester flavonoids in different amounts and types, depending on the part or species of food plant. During late pupal development, the majority of UV-absorbing flavonoids are deposited in the wing scales. The flavonoid content of the larval diet thereby determines UV wing patterns.

The silver-washed fritillary, *Argynnis paphia* is an impressive, spotted orange butterfly that uses visual signals in sexual communication. Magnus (1958) demonstrated that male approaches objects that are orange in colour, the typical colouration of this species. He constructed paper models of various shapes and sizes and observed that the males were attracted to any shape, but attracted only to the models having approximately four times large size than the female.

Male grasshoppers use visual signals in close range courtship (Otte, 1970). Twostriped grasshopper, *Melanoplus bivittatus* male jumps onto the back of nearby female. It holds the female tightly and makes species-specific jerking movements. Male *Ageneotettix deorum* approaches the female and tap the female's body with antennae while raising and lowering hind legs. Males of *Syrbula* species move their abdomen and paired appendages in displays to female.

Male Hawaiian *Drosophila* often perch on vegetation and move abdomen up and down and simultaneously open and close the wings. Wings display demonstrates the colour bands on wings which act as visual signals (Spieth, 1974). Kimsey (1980) reported that male orchid bees of the genus *Euglossa* (Fam. Apidae) also move abdomen and wings to attract the females.

11.3.5 Defence

There are many examples in which a visual communication is passed to another species and the signals are used for defence. Most of these signals are effective only as long as they are visible in daylight. Some insects have bright red or orange colours that are exposed only when they are threatened by a predator. When the eastern lubber grasshopper, *Romalea guttata* is disturbed, it will flash its bright red hind wings. The sudden display of bright red may scare the predator and give the grasshopper time to escape. As compare to butterflies, moths are not so bright but they use body colours in a variety of ways. In diurnal species body colour is important in informing potential predators that they are bad tasting and possibly poisonous both as an adult and as a caterpillar. Many moth species use colour in order to protect themselves from predators by concealing in their habitat. Most moths are either cryptically coloured to blend in with their background or resemble parts of a plant. The caterpillars of many moths look like a bird dropping and resemble with something that a bird would never eat (Plate 11.1). Besides, some adult moths also have striking resemblance with bird dropping (Plate 11.2). Many caterpillars use other devices such as false and enlarged eyes to scare a predator. Many lepidopterans use false eyes on the hind wings as a method of startling a potential predator. Close approach to them results in sudden unveiling of the hind wings creating the sudden appearance of a large-eyed owl opening its eyes. Sudden erratic flight of a large number of individuals often warns and frustrates a predator. Adult tiger swallowtail butterflies congregate in large numbers around margins of puddles or animal faeces to obtain sodium. If disturbed by a predator they fly in masses and surround the predator with a whirling cloud of butterflies and it becomes difficult for the predator to catch a single individual (Arms *et al.*, 1973). The paper wasp, *Polistes* alert other workers in the nest by threat display in which fore legs are raised and wings are vibrated.

Some insects advertise their ability to sting, or to mimic the appearance of another unpalatable species. By blending in with the environment, the praying mantis can avoid any encounter it may have with a predator. Mantids accomplish this for the most part by blending in with the foliage they dwell in. If their first line of defence fails, they can use startling defensive displays.

Passive signals, such as eyespots and colour patterns can serve as a form of 'free advertising'. The colourful wings of a butterfly act as a 'poster' publicizing its identity. To display these messages insects incur little or no metabolic cost, as they are an integral part of the integument. It is vital to hide these signals from a potential predator, so some insects have a way to conceal their message when necessary. For example, the red admiral, *Vanessa atalanta* has bright, distinctive markings on the upper wing surface and dull, protective colouration on the underside.

11.3.6 Prey capture

For predatory insects relying on vision, it is important that they have regions in their compound eyes that are specialized to see their prey well. Many insects including mantids have a zone with the highest acuity is called the 'optic fovea'. Catching of prey by a praying mantis is a fairly complex procedure. Vision is the primary indicator of prey presence and other cues also assist in prey capture. Prey capture is accomplished by two independent behaviours that are regulated by the visual tracking of the prey item. The rapid extension and grasping motions of the prothoracic legs are the primary capturing tool used by the mantid. The second behaviour is carried out by the meso- and metathoracic legs. By moving the body forward during a strike, the legs can accomplish a leap that moves the mantid closer to the prey item. Both of these behaviours depend upon the visual input received from prey position. The proprioceptive hairs in the neck region are important in coordinating the head and body position for optimal prey capturing movements. Thus, the visual system is used to track the prey item in visual space while the proprioceptive hairs in the neck determine the placement of the impending prey capture strike.

In dragonfly larva prey capture is accomplished by shooting out the mask. The larva can capture the prey when it is at one point in space with respect to the head. The larva moves towards the prey and visualize it with increasing precision. When the prey's image falls on certain inner ommatidia and a most clear and exact image is formed within the correct striking distance, it shoots out the mask.

11.3.7 Visual communication in light producing insects

In insects the production of light is a phenomenon of chemiluminescence and highly evolved in some insects of the orders: Coleoptera (Lampyridae, Phengodidae, and Elateridae); Diptera (Mycetophilidae); Hemiptera (Fulgoridae) and Collembola. Coleoptera comprises the largest insect group in which a large number of species have highly evolved photogenic organs. Fireflies also known as 'lightning bugs' are actually beetles that can make flashes of light. The two sexes of fireflies use different codes of flashes to find and recognize each other. Each species of firefly has a different kind of flash pattern. Male fireflies flash in order to convey the message of location and to state that they are sexually mature while female fireflies flash a response to convey their location and that they are receptive. It has been hypothesized that fireflies also use their flash to locate a safe landing and oviposition sites. Different species of fireflies avoid interspecific mating by using flash signals. Flash behaviour of one species has also been observed to be mimicked by another species. This mimicking is performed by *'femmes fatales'* that eat other species of fireflies. For details of visual communication in light producing insects see section 16.10. Carlson and Copeland (1985) suggested that the firefly flash is a signal in a courtship system based upon timing patterns controlled by the brain. Both temporal and insect's state of arousal greatly affect the flashing behaviour of fireflies. Also, the firefly's visual apparatus is highly specialized for the reception of fast transient flashes. Finally, and perhaps most significantly, the flash pattern generators of fireflies can be entrained by incoming flashes to produce, in females, responses of appropriate delay; and in males, flash synchrony.

11.4 Chemical communication

For an insect, the environment contains a multitude of chemical compounds; most of them are physiologically irrelevant, whereas some may be essential because they carry information. Insects can communicate very well by sending and receiving chemical signals and rely more heavily on these signals than on any other form of communication. The chemicals involved in communication are often called 'semiochemicals' or 'infochemicals', and serve as a form of 'language' that helps to mediate interactions between conspecifics, or different species. These chemicals are required in very low concentrations and only a few molecules may be enough to elicit a response. Semiochemicals can be divided into two groups: (i) pheromones that carry intraspecific messages, and (ii) allelochemicals that regulate interspecific behaviours.

Insects represent good subjects for the study of the transfer of olfactory sense information by means of special- or group specific chemical releasers. They respond to odours of their surroundings and are especially sensitive to biologically meaningful chemical signals such as food, prey and mate. The system of communication by means of pheromones is highly developed in many insects, particularly in social insects. The colonies of bees, ants, termites, and wasps are to a large extent regulated by chemical signs. The other odours used by insects are mainly sex attracting substances produced by individuals of one or both sexes. Insects possess many specialized pheromone producing structures and olfactory receptors which assist in communication.

Pheromones include sex attractants, aggregation and alarm signals, trail or territory markers, oviposition deterrents or compounds which induce or control more complex social behaviour, and govern many other activities. Most of these species-specific messages are coded in complex multicomponent mixtures released in extremely small amounts. The importance of smell in insect communication was first recognized by the French entomologist, Jean Henri Fabre. He observed that male eggar moths, *Lasiocampa quercus* were attracted to the chemical produced by the female despite many olfactory assaults. It wasn't until the 1950's that the term 'pheromone', was coined to the chemicals 'excitement carrier' [from the Greek *pherein* (to carry) and *horman* (to excite)]. Wilson and Bossert (1963) divided pheromones into two functional types based on their mode of influence: (i) *releasers*-immediate effect on the central nervous system and behaviour, and (ii) *primers*-

trigger a chain of physiological/developmental events in the recipient individual.

Allelochemicals play important roles in interspecific communication. They constitute signals in predator-prey relations, host-parasite and herbivore-plant interactions, and may be responsible for attraction as well as chemical defence. Recognition, location, and selection of prey, hosts, or habitats as well as pollination behaviour are strongly influenced by both volatile and nonvolatile compounds. Allelochemicals can be further subdivided into three groups based on who benefits from the massage:

I. **Allomones.** A chemical or mixture of chemicals that is produced or acquired by an individual that, when it contacts an individual of another species evokes in the receiver a behavioural or physiological reaction that is adaptively favourable to the emitter but not the receiver. Most such chemicals are defensive compound.

II. **Kairomones.** A chemical or mixture of chemicals that is produced or acquired by an individual that, when it contacts an individual of another species evokes in the receiver a behavioural or physiological reaction that is adaptively favourable to the receiver but not the emitter. Such as an odour that a parasite uses to find its host.

III. **Synomones.** A chemical or mixture of chemicals that is produced or acquired by an individual that, when it contacts an individual of another species evokes in the receiver a behavioural or physiological reaction that is adaptively favourable to both receiver and emitter. Such as plant volatiles that attract insect pollinators.

The chemoreceptors in insects are located on all body parts, but are common on tarsi, antennae, palps, and ovipositor. The chemical communication is advantageous to the insects as: (i) not limited by environmental barriers, (ii) effective over long distances and around corners, (iii) effective during all the periods, (iv) long lasting as compare to other system of communication, (v) metabolically inexpensive, and (v) species-specific. These chemicals may be disadvantageous as they are not effective in upwind direction and may be detected by natural enemies to find their hosts or prey. Some male moths can pick up a female's scent at an amazing distance of 5 km or more. Just one molecule of the pheromone is enough to send him flying upwind in search of his mate. The male's antennae are furnished with special chemoreceptors that are tuned to receive the female's scent. It's not just the female insects that release such scents, in many cases males and in some others individual of both sex release pheromones. The 'active space' is the zone in which the concentration of a pheromone is at or above the level needed to evoke a behavioural response, and it may by very big for sex pheromones. Insects that live in colonies have provided model systems to study chemical communication. Social insects may exist in large groups and perform tasks as groups through cooperation among individuals. This cooperative group behaviour arises from the coordination of the actions of individuals, and this coordination is achieved by chemicals that regulate social behaviours.

The pheromones have been classified on the basis of behavioural responses elicited by insects into groups like: (i) alarm pheromones, (ii) aggregation pheromones, (iii) epideictic pheromones, (iv) trail-marking pheromones, and (v) sex pheromones. Detailed discussion on insect pheromones and their role in communication is included in chapter 18.

11.4.1 Alarm pheromones

The chemical senses of social insects are extremely well-developed and many paradigms in olfactory communication have been established using different species of social insects. The alarm signalling is an effective means of chemical communication shown by many insect groups, particularly social insects. Alarm pheromones are released by insects when they are disturbed or threatened (section 18.4.1). When an ant colony is disturbed, workers emit alarm pheromones that evoke alarm behaviour in nestmates. Many ants run about the mound to attack the invader while other will gather eggs and larvae to escape. Sometimes the workers become aggressive. Some alarm pheromones induce a 'panic' response, and individuals that detect the signal will abscond.

11.4.2 Aggregation pheromones

Many insect species aggregate for feeding, mating, hibernation, defence, overcoming the resistance of host, or in response to environmental conditions. The aggregation pheromones are released to attract conspecific individuals together, including both sexes (section 18.4.2). Bark beetles (Fam. Scolytidae) commonly release such pheromones when find a host tree suitable for infesting. The pheromone attracts large numbers of beetles to the same tree.

11.4.3 Epideictic pheromones

Some pheromones released by insects influence their population density, and bring out their dispersal or spacing (section 18.4.3). Thus, epideictic or spacing pheromones are those which stimulate dispersal from a resource once an optimal density has been reached, and limit intraspecific competition. Both immature and adult insects secrete such olfactory and tactile chemicals. Host marking pheromones secreted by many fruit flies (Fam. Tephritidae) provide excellent examples of such pheromones. Certain pheromones released by bark beetles reduce further arrivals of individuals on host trees.

11.4.4 Trail-marking pheromones

Trail pheromones (section 18.4.4) are used by many ants, caterpillars, and other insects. These chemicals are used as a road map for finding food. After the ant finds some food, it will find its way home by this chemical trail, and other ants can follow it to the food. In tropical rainforests, leafcutter scout ants search for suitable vegetation to cut and collect to rear fungus to eat. The scout ants lay chemical trails to guide their nestmates to the tree. The trail pheromone stimulates ants to leave the nest and orient to the food. The trail pheromone is discharged on the ground and has incredible biological activity.

11.4.5 Sex pheromones

Sex pheromones are attractants for members of the opposite sex of a species (section 18.4.5). The 'perfumes' of some females can be detected by males over long distances. In most cases females release pheromones to attract males from a distance, while males release pheromones only when close to the females in order to sexually stimulate her. In salt-marsh moth, *Estigmene acraea* small groups of males assemble to release a pheromone which attracts females. Sometimes a sex pheromone includes a mixture of different chemicals that only one species will recognize. If two different species use the same chemical to communicate, they may send their chemical messages at different times of the day. Because of different chemicals or different times of communicating, the male of one species is not attracted to a female of another species.

11.5 Interspecific communication

Interspecific chemical communication system involves transfer of information via chemical signals between individuals of different species. Otte (1974) classified the individuals exploiting chemical signals into illegitimate receivers and signallers. Chemical signals are commonly exploited by predators, parasites and commensals. The odours of host individuals, host's habitat, and host's products are exploited for finding food and oviposition sites. Pheromones produced by hosts are commonly used as a communication system and confer a selective advantage to natural enemies to locate the hosts. Such allelochemicals are termed kairomones and many hymenopterous parasitoids commonly exploit these chemicals for host location. Aggregation pheromones of bark beetles are commonly exploited by their predators and parasites. Prokopy and Webster (1978) reported that the parasitoid *Opius lectus* locates its host, the maggot of *Rhagoletis pomonella* by responding to host marking pheromone. The pheromone is deposited on the egg surface during egg laying, and on the host surface immediately afterward, and deters boring attempts by other *R. pomonella* females in that fruit. Other examples of interspecific pheromonal communication are cited in section 18.6. Pheromones of social insects can be exploited by insects that depend on ants and termites for their food. Ant trails are followed by a number of myrmecophiles. Illegitimate signallers like slave-making ant, release 'propaganda pheromone' which causes communication disturbance among defenders (section 18.4.1).

Some ants and their guests called myrmecophiles live in ants' nests, where they are groomed and cared. Hölldobler (1971a) discussed a well-known example of interspecific communication between ants and European staphylinid beetle, *Atemeles pubicollis* (Fig. 11.12a). As cited by Hölldobler the guests beetles have broken the ants' code, that is, attained the ability to 'speak' ants' language, which involves a diversity of contact, visual and chemical cues. The beetle's larvae are found in the nest of mound-making ant, *Formica polyctena* and are well accepted by the ants. The larvae secrete a chemical from integumentary glands (Fig.11.12b), which act

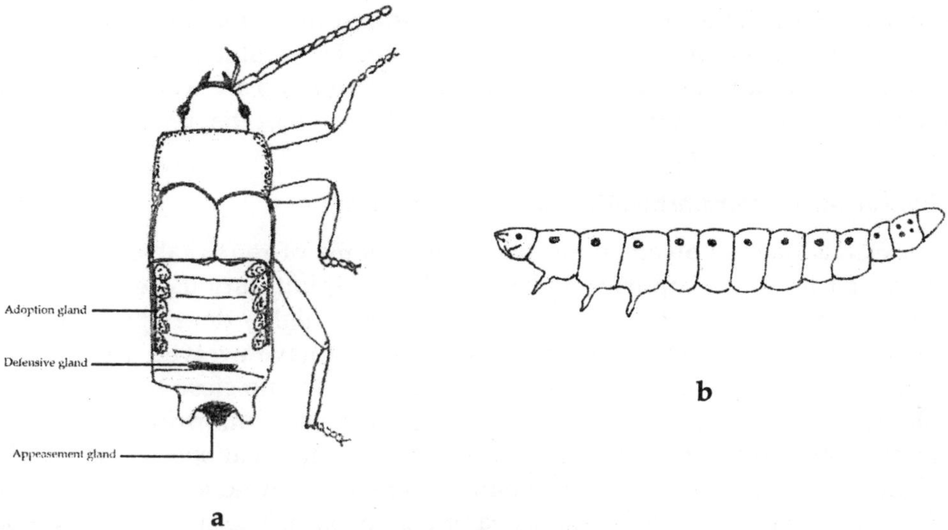

Fig. 11.12. Rove beetle, *Atemeles pubicollis*, (a)- Adult, (b)- Position of attractant secreting glands on the body of larva.

as an attractant for the ants. The ant workers groom beetle's larvae after receiving the chemical signals. Hölldobler coated a number of beetle larvae with shellac, and these larvae were ignored or deposited in the garbage dump by the ants. The beetle larvae also beg for food from ants in a similar manner as done by ant larvae and the begging behaviour of beetle larvae was more pronounced as compared with larvae of ants. It involves mechanical stimulation of ants. When the ant worker touches a beetle larva, it promptly rears up and attempts to contact mouthparts of ant; and if it makes contact with ant's head, the ant regurgitates a droplet of food. Besides this, the beetle larvae were not only predaceous but also cannibalistic, and feed on ant larvae as well as they reduce their own population. The larvae pupate in the nest and consequent upon hatching the adult beetle migrate to the nest of another ant species of the genus *Myrmica* in the winter. The *Formica* ants suspend brood rearing during winter, while the food supply and brood rearing are maintained throughout the year in *Myrmica* colonies. Before leaving the *Formica* nest the beetle takes an ample amount of food by employing a technique of tactile stimulation.

Formica nests are found in woodland, while *Myrmica* nests exist in open grassland. For finding its way to *Myrmica* nest the beetle moves in the direction of sun and using windborne odour of the host species attains final approach. The beetle obtains recognition an adoption in *Myrmica* nest with a practice involving chemical communication. It releases a chemical by raising the tip of its abdomen towards the host from the appeasement glands, which is licked by the ants and suppresses its aggressive behaviour. The ant is further attracted by the secretion of adoption glands located along the sides of beetle's abdomen. The odour of this secretion resembles with that of ant's odour, and finally the beetle is accepted and taken in the nest as a guest. The beetles are sexually matured in *Myrmica* nest, and return to *Formica* nest for mating and oviposition. The beetles exhibit a remarkable form of evolutionary adaptation and its life cycle and behaviours are well synchronized with social life of *Formica* and *Myrmica* ants in such a manner that it takes maximum advantage of host species.

11.6 Evolution of communication in social insects

Insect societies have evolved communication systems of remarkable complexity. Highly social bees use sophisticated methods to exploit various resources. The dance language of bees is clearly a very complicated, highly developed system of recruitment. To understand how such a system may have evolved attention has turned to related species. The honey bee, genus *Apis* has no close living relatives. Bumble bees and stingless bees are their nearest kin, and unfortunately, it seems that bumble bees do not recruit at all. Many related tropical species of stingless bees (Tribe Meliponini), on the other hand, do recruit but none has developed a symbolic language similar to the dance language of *Apis*. The subject of evolution of bees' language has been discussed by (Lindauer and Kerr, 1960; Esch, 1967; Lindauer, 1971; Michener, 1974, Nieh, 2004). The stingless bees use some means of alerting colony members for food, but the communication systems vary greatly. In *Trigona iridipennis* workers after finding a food source fly back to the comb and excitedly run about, thus alerting other workers. *Trigona postica* workers construct a trail and left marks at short intervals by depositing the secretion of their mandibular glands. The forager alerts other bees and leads them along the scent trail. In this way, some species of bees exploit chemical clues left by other bee species to guide their kin to food provides evidence that eavesdropping may be an evolutionary driving force behind some bees' ability to conceal communication inside the hive, using a form of language to encode food location.

Honey bees use functionally referential communication, the transformation of environmental information into coded signals understood by a conspecific receiver. Such a system may be a sophisticated form of communication because of the cognitive complexities presumably involved in transforming sensory information into coded communication signals. Stingless bees are excellent model for the study of evolution of bees' language because they possess the widest diversity of species and communication strategies, including the ability to acoustically encode the distance and height of food sources. Moreover, some species may use functionally referential communication, e.g. *Melipona panamica* foragers can communicate the three-dimensional location of food sources. To achieve this, foragers use a combination of mechanisms. Results from a series of removal experiments (segregating all feeder-experienced foragers from potential recruits as they left the nest) suggest that direction is communicated outside the nest whereas height and distance are communicated inside the nest (Nieh and Roubik, 1998).

Nieh *et al.* (2003b, 2004) proposed that chemical espionage in stingless bees might have played an important role in the evolution of communication strategies inside the nest. Some stingless bees use odour trails for communication, but there is fascinating variation in the length of these odour trails. Some species make complete odour trails, others make short odour trails, and finally some only mark the food with odour. Thus, decreasing odour trail length could be an effective counter strategy to such eavesdropping. However, this would also decrease the amount of guidance information offered to recruits. Providing encoded location information, functionally referential communication, at the nest would replace such lost odour trail information. Interestingly, all bees for which there is some evidence for functionally referential communication, including honey bees, only use point source odour marking.

All four species of honey bees use similar distance and direction codes, even though there are some differences. *Apis mellifera* and *A. cerana* build nest in dark, enclosed areas such as hollow trees or other similar cavities, and produce sound while dancing. The third sound producer, the giant honey bee, *A. dorsata* makes nest in the open and produce sounds that are particularly low in pitch (Kirchner, 1993). The single species that dances silently, the dwarf bee, *A. fiorea* dances in the open only during the day and probably use visual signals. It has been suggested that *A. florea* bees are the most primitive; and the complicated acoustical communication system of the other three species most likely evolved from a visual display when these bees developed a habit to live in dark.

11.7 Integration of communication methods

Communication in insects is a complex phenomenon and involves methods like tactile, visual, acoustic, and chemical. These methods of communication do not act independently and in many instances more than one methods are used by insects. Communication by means of chemical signals is a commonly occurring phenomenon and the odours are carried to long distances on the wind. After detecting the odour, the receiving individual is oriented to the signal and consequent upon its arrival nearer to the releasing individual, visual or acoustic signals may assist in providing appropriate information. Chemical signals, in this way are more important in long-range communication while insects rely more on other methods for short-range communication. Insects tend to show complex behavioural patterns composed of a sequence of responses when they receive a variety of communication signals.

For mating insects cover long distances and remain exposed to the attack of natural enemies. This danger of attack is considerably lessened when they are in concealed habitats and able to communicate with each other by means of chemical or acoustic signals. Thus, these signals jointly are helpful for them to serve this strategy. For recognition of conspecifics they may jointly use visual, acoustic or tactile signals. In social insects existence and survival of the colony mainly depend upon integration of various methods of communication. All individuals of a colony respond in a coordinated fashion of mass behavioural phenomena. Honey bees use elaborate system of communication by using all means, and among these systems chemical communication predominates. At the nest entrance some 'guard bees' check the intruders and recognize the bees of their own colony by using tactile and chemical cues. Besides this, bees are also able to exploit the odour cues of substances present in their food. How do honey bees recognize their nestmates? Guard bees at the colony entrance encounter incoming bees and exclude, by grappling, biting or stinging, bees that don't belong in the colony. Guards recognize colony identity of an incoming bee by odours carried by the bee. These odours come primarily from the wax comb in the bee's colony. A newly emerged bee carries no recognition odour and is acceptable in any honey bee colony. After it has walked around on the wax a while (no more than a few minutes), its surface chemicals are modified. The wax absorbs some chemicals, removing them from the bee, and gives up other chemicals, adding to the bee's burden of surface chemicals. Because all the bees in the colony walk around on the same combs, all bees in a colony carry essentially the same recognition odours. This simplifies life for the guard, as it needs to learn only one identity, rather than several thousand unique identities (Breed *et al.*, 1995; Smith and Breed, 1995).

Chapter 12

Insect Interactions

Interacting species have a tremendous influence on the size of each other's populations. The various mechanisms for these biotic influences are quite different from the way in which abiotic factors effect the size of populations. Biotic factors also regulate the size of populations more intensely, and the influence of biotic interactions can occur at two different levels. Intraspecific interactions occur between individuals of a single species while interspecific interactions occur between individuals of groups of species. Insect species are intimately associated with other insects, vertebrates, other invertebrates, plants, fungi, and microorganisms. Many of them transmit animal and plant diseases. In this section their intra- and interspecific interactions with other insects, interactions with plants and fungi, and tritrophic interactions between plants, insects and their parasitoids are discussed.

12.1 Intraspecific interactions

These interactions occur between individuals or groups within the same species. These interactions can be positive for both parties, negative for each or positive for one and negative for the other. Intraspecific interactions exhibit association of individuals of same species for aggregation, cooperation, competition, mating, parental care, protection, sociality, cannibalism, trophallaxis, etc., the various methods by means of which the members of same species interact are related in part to population density.

12.1.1 Aggregation

Many insects aggregate in response to environmental cues and also for the purpose of mating, feeding, hibernation and defence. The aggregation behaviour of many insects is a fascinating occurrence. These aggregations may include from a few individuals to many thousands. Usually, except in social insects, these are temporary and open groupings. The phenomenon is of common occurrence in springtails, cockroaches, locusts, flies, bugs, butterflies, beetles, and social groups. The notable example of assembly is exhibited by fertilized queen of social insect, which remains surrounded by a retinue of workers.

Many insect species use aggregation pheromones to communicate which elicits aggregation (section 18.4.2). Certain cockroaches may live together naturally. Nymphs of German cockroach, *Blattella germanica* aggregate in response to the pheromone located in the faeces.

Every fall the monarch butterflies, *Danaus plexippus* migrate to one of about fifteen sites in inland Mexico. During migration they travel a distance of up to 4500 km and migration occurs because the monarch is unable to withstand the harsh winters at northern latitudes. These conditions apparently provide a climate for them to overwinter in suitable sites. It is thought that the monarch has been using the same overwintering sites since possibly the Pleistocene (Masters *et al.* 1988). The monarchs reach the overwintering sites around November and each site can contain from 50000-200000 individuals. Members of these aggregations are fairly inactive, survive on stored lipids, and mating is also very rare. At the beginning of spring as the monarchs began remigrating there is a mating frenzy. The females mate with as many males as possible and each time she gets an energy rich spermatophore, which is important to replenish her energy supply for the remigration. It has been argued that these multiple matings coupled with the nutrient absorption of the spermatophore increases the monarchs fecundity (Wells *et al.*, 1990). Multiple matings in such a short span of time would not be possible if the monarchs were not living in aggregations.

The mountain pine beetles, *Dendroctonus ponderosae* feed on the inner bark of their preferred host lodgepole pine, *Pinus contorta*. Their adults emerge in spring and attack trees that are physiologically weakened, diseased, or dead. The first beetle after discovering a suitable host releases an aggregation pheromone, which in combination with host's odours is very attractive to other beetles. They use this odour to aggregate (Borden, 1982). Other coleopterans like convergent lady beetle, *Hippodamia convergens*; lucerne weevil, *Hypera postica* and seven-spotted ladybird, *Coccinella septempuctata* have been found aggregating in many areas. The aggregation of *H. convergens* appears to be temperature related.

Many aposematically coloured insects aggregate for protective defence against predation. A solitary insect has a greater risk of being eaten by a predator, than if it is a member of conspicuous group. Predators learn to avoid these insects after experiencing their unpleasant defence chemicals or their warning colouration. The aggregation functions directly to enhance the effectiveness of warning signals, or the chemical defences on which they depend.

The caterpillars of jack pine sawfly, *Neodiprion pratti* feed in groups upon jack pine trees. The isolated larvae suffered 80 per cent mortality while those allowed remaining in groups only 53 per cent died. In group the larvae successfully cut through the internal tissues of pine needles for feeding (Ghent, 1960). Besides, in later instars they synchronize their vigorous defensive reactions into a group display which in highly effective in discouraging the predators.

12.1.2 Competition

When two or more organisms in the same community seek the same resource (e.g., food, nutrients, light, mates, water, nesting space, ground space, etc.), which is in limiting supply to the individuals seeking it, they compete with one another.

The requirements may vary depending on age and/or sex. The competition among members of the same species is 'intraspecific' while among individuals of different species it is referred to as 'interspecific' competition.

Intraspecific competition is a major selective force in the evolution of behavioural strategies (Krebs and Davies, 1997). Competition with conspecifics may be active in which competitors limit access of other individuals to the needful resources, or passive in which the limited resource is shared among all individuals. The competition may results into reduction of fitness, as survivors may be small sized and have lower fecundity and/or mating success. Intraspecific larval competition may be reduced by avoiding hosts where conspecific eggs or larvae are already present. In apple maggot fly, *Rhagoletis pomonella* the occipied sites are avoided by releasing an oviposition deterring pheromone (Prokopy, 1972a). Besides, epideitic pheromones are also used by many insects for avoiding intraspecific competition. Individuals of a species often search other food patches in order to avoid intraspecific competition; however, this activity is energetically costly, may increase the risk of predation, and also results in a decrease in fitness. Averill and Prokopy (1987) reported intraspecific competition among larvae of *R. pomonella* in the native host hawthorn, *Crataegus mollis*. Survivorship dropped sharply when >1 larva developed per fruit. Multiple infestations also caused decreased size of pupae, which resulted in prolonged maturation of females and lower daily fecundity. They suggested that interference competition occurs among rivals *R. pomonella* larvae within the same fruit and that older larvae may be competitively dominant. The female marks oviposition site with a contact pheromone which deters most females from laying more eggs.

Intraspecific competition has been shown to affect development of insects, usually because variation in density affects the amount of food available to larvae. Long larval development periods are generally considered to be disadvantageous since they may limit opportunities for reproduction, and lengthen the larva's exposure time to predators and parasitoids. Under some conditions; however, there may be advantages of longer development. Longer development can result in an increase in body mass because of the ability to consume more food over a longer period. However, in most cases larval competition resulted in the production of small adults over long development periods. Generally, small adult size is considered disadvantageous because large size is often positively correlated with fecundity and/or competitive capacity. Thus, intraspecific competition during the larval period has the potential to reduce overall fitness.

Gibbs *et al.* (2004) manipulated larval densities of a Madeiran population of the speckled wood butterfly, *Pararge aegeria* and confirmed that intraspecific competition affected growth. As densities increased, *P. aegeria* adults were smaller and larval development periods were longer. Sexes responded differently to rearing density. Females were more adversely affected by high density than males, resulting in females having smaller masses at pupation. Survivorship was significantly higher for larvae reared at low densities. Intraspecific competition during the larval stage would appear to carry a higher cost for females than males.

12.1.3 Sociality

Sociality in insects includes interaction and cooperation among members of a social group. The eusocial (true social) insects exhibit characteristic features like: division of labour, with a well-defined caste system, (ii) cooperation among the colony members, (iii) tending of young, and (iv) overlapping of generations. Subsocial insects on the other hand, have less strongly developed social habits, falling short of extensive cooperation and reproductive partitioning. Eusociality is found in ants and termites and some bees and wasps while subsociality is widespread in many insect orders. For details of sociality in insects see chapter 14.

12.1.4 Cooperation

Cooperation in nature persists in spite of individuals being also under continuous selection to pursue selfish interests. Cooperation in insect societies (ants, termites, social bees and social wasps) has evolved to such levels that most individuals have given up their personal reproduction to become 'helper workers' in colonies and only queen lays all the eggs. All aspects of natural cooperation by evolutionary models were derived from Hamilton's inclusive fitness principle (Hamilton, 1964a,b). His evolutionary theory of cooperation is also commonly referred to as 'kinship theory' or 'kin-selection' theory, because it is through kinship with breeders that helpers realize copies of their own genes in the next generation even if they do not reproduce themselves. Thus, queen and workers in the same colony are often each others close relatives, but express greatly different phenotypes. In many tropical ants, the workers have different forms and it is believed that these differences were exclusively determined by the way in which larvae were reared by their sister workers.

12.1.5 Cannibalism

Many termite species are known to exhibit a high degree of cannibalistic behaviour. The cannibalism probably acts as a protein-conserving device, as they usually feed on low protein diet. *Coptotermes* reproductives which are unable to leave on normal nuptial flight are killed and eaten by workers. Some termites consume alien conspecific workers. Egg cannibalism in cotton bollworm, *Helicoverpa armigera* larvae may make a significant contribution to its population suppression. Kakimoto *et al.* (2003) suggested that cannibalism is a behaviour of larval *H. armigera* that has evolved to remove potential intraspecific competitors and thus to avoid unsuitable conditions. Some insects display more than one feeding habits as both grubs and adults of skirt and blouse beetle, *Paederus fuscipes* are important predators of many insect pests throughout the world. After locating the prey, the predator grabs it by inserting its mandibles and devours it within seconds. Besides this, cannibalism, necrophagy and fluid feeding habits are also displayed by them (Devi *et al.*, 2003).

Brood cannibalism is common in social Hymenoptera. Ants commonly consume injured immature stages, and under starved conditions workers quickly eat healthy brood. Oophagy or egg cannibalism is also widely observed in social groups. The trophic eggs laid by ant workers are usually eaten by the queen.

The praying mantis is known for its unique look and very interesting aspects of behaviour. A number of reports indicate that in mantids cannibalistic mating occurs and male is eaten by the female during or after mating (Fig. 12.1). Females usually have a heavier abdomen than males. Its front legs are modified as graspers with strong spikes used for grabbing and holding prey. The mantid has an enormous appetite and is carnivorous and cannibalistic. Decapitate male copulate more vigorously because of loss of suboesophageal ganglion that normally inhibits copulatory movements. The frequency of sexual cannibalism is higher in nutritionally deprived mantid females. It has been suggested that the ultimate nuptial sacrifice by the male may results into reproductive fitness of the male via the indirect nutritional benefits to his offspring. Another explanation for eating a male during or after copulation is assurance of paternity. Although the praying mantis is known for its cannibalistic mating process but in actuality it does not occur in all cases. However, in European mantid, *Mantis religiosa* it is necessary that the head of male be removed for the mating to take effect properly. Sexual cannibalism occurs most often if the female is hungry and eating the head does cause the body to ejaculate faster.

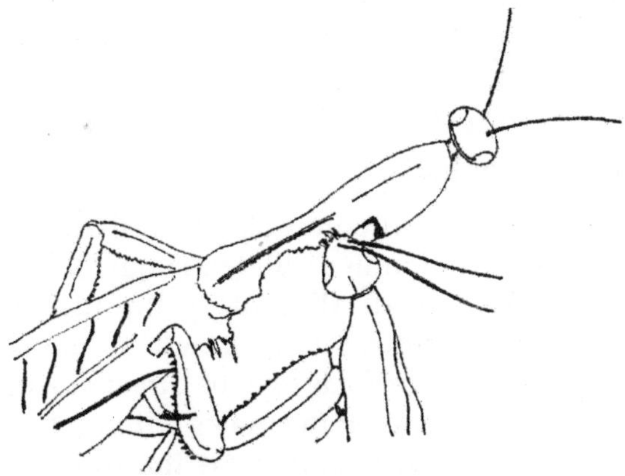

Fig. 12.1. Anterior body portions of praying mantids showing cannibalistic mating in which the male is eaten by the female during.

The male webspinners (Order Embioptera) are also sometimes cannibalized by the females during copulation. Earwigs also cannibalize her 2nd and 3rd instar offspring. In many biting midges, the female eats the male during copulation, treating her mate the way she does with any other insect prey. She pierces his cuticle and dissolves and sucks out his body contents. When the female disengages from her lifeless mate, a portion of his body remains attached to her. This 'plug' may keep away other males from mating with her, thus assuring not only paternity of the now-deceased male but also fertilization of all the female's eggs. By dying in this way, the male increases his reproductive success.

Thomas and Manica (2003) reported 'filial cannibalism' in the assassin bug, *Rhinocoris tristis* in which males guard eggs of a number of females, cannibalizing some of their offspring within the brood. The cannibalism has typically been explained as a response to high energetic costs of brood defence and decreased feeding opportunities during the period of care. They monitored guarding males in both the field and the laboratory. Males typically ate eggs around the periphery of the brood, which were those most likely to have been parasitized by wasps. However, cannibalism persisted in the laboratory in the absence of parasites, and the number of cannibalized eggs was related to the length of care and overall brood size, suggesting that males use eggs as an alternative source of food.

12.1.6 Trophallaxis

North American entomologist William Morton Wheeler in 1918 used the term trophallaxis for uni- or bilateral liquid food exchange. Trophallaxis usually includes direct transfer of alimentary liquids, including suspended particulates and derivatives, from one nestmate to another via regurgitation or anal feeding. However, trophallaxis may be direct in which the trophallactic substance is directly transferred between the mates without being placed on an intermediate substartae while in indirect trophallaxis the trophallactic substance is transferred from the donor to an intermediate substartae before being picked up by the recipient.

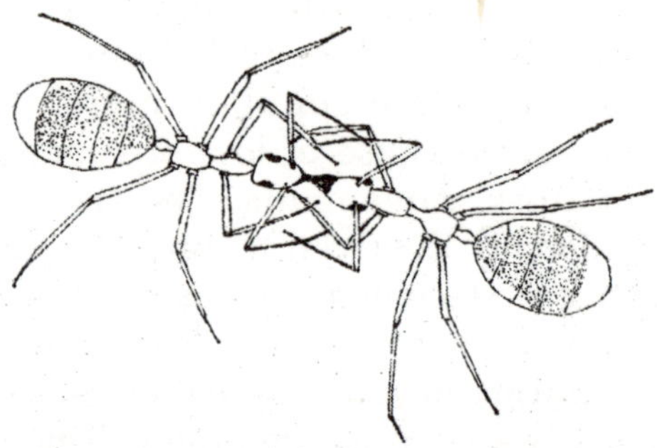

Fig. 12.2. Trophallaxis between two members of an ant colony. This behaviour is important in the nutritional dynamics and communication of many social insects.

This behaviour is important in the nutritional dynamics and communication of many social insects. Trophallaxis is a mechanism for the transfer of nutrients, symbionts, pheromones, and information within social insect colonies. The transfer of materials may occur both among adults and between adults and larvae. Trophallaxis (Fig. 12.2, Plate 12.1) enables feeding of dependent castes and instars in termites, ants, wasps, and bees; transfer of symbionts among nestmates, and a more efficient division of labour. Trophallaxis also is the mode of transfer of caste regulatory pheromones and permits the signalling of food resource profitability from foragers to recruits in honey bees. In lower termites both stomodeal and proctodeal trophallaxis occur by which they share stomodeal food and a milky liquid containing symbiotic flagellates, respectively. The gut symbionts in termites are lost with each moult and the nymphs acquire new ones through trophallaxis.

Mating trophallaxis constitutes an array of behaviours in which males provide females with nuptial gifts which are then consumed. It may occur before, during or after copulation. Mating trophallaxis includes the exchange of both solid and liquid substances which may have originated from a variety of organs or even from outside the body of donor. Mating trophallaxis has been reported in *Rivellia melliginis* (Fam. Platystomatidae) during mounting. Males fan their wings, sway their bodies, and rotate. After the arrival of the female they form the droplet and continue fanning. This is followed by foretarsal tapping, circling, and eventual mounting (McMichael *et al.*, 1990).

12.2 Interspecific interactions

These interactions occur among individuals or groups of different species. Typically, these interactions are classified based on whether they are beneficial to one or both of the species involved or whether they are detrimental to one of the species involved. Different species, even closely-related ones, will have different food preferences, seasonality, daily feeding rhythms, and location within the habitat. Important interspecific interactions can be described by the following chart:

Interaction	Species A	Species B	Consequence of interaction
Neutralism	0	0	Both species are independent and do not interact
Mutualism	+	+	Both species get benefit
Competition	-	-	Both species compete for same resource and have a negative effect on each other
Predation	+	-	The predator (A) benefits while the prey (B) is killed
Parasitism	+	-	The parasite (A) benefits while the host (B) is negatively affected
Commensalism	+	0	One species benefits from the relationship but the other is neither harmed nor benefitted

12.2.1 Neutralism

Neutralism is the most common type of interspecific interaction in which neither population directly affects the other. The interactions are simple, slight and indirect. The simple presence of the two species should not directly affect the population level of either. The term is used in ecology to describe biological interactions when the population density of one species has absolutely no effect whatsoever on the other. However, when dealing with the complex networks of interactions presented by ecosystems, one cannot assert positively that there is absolutely no interaction between two species. If two species do not interact directly, links can be found through other species and shared resources. Since true neutralism is rare or nonexistent, its usage is often extended to situations where interactions are merely insignificant or negligible.

12.2.2 Competition

Interspecific competition takes place when the requirements of two or more different species for a given resource are same or when their niches overlap. Competition may be the result of two different processes, namely exploitation and interference. Competition by exploitation occurs between individuals when the indirect effects of two or more species or individuals reduce the supply of the limiting resource or resources needed for survival. The exclusion of one organism by another can only occur when the dominant organism requires less of the limiting resource to survive. Further, the dominant species must be able to reduce the quantity of the resource to some critical level with respect to the other organism. Resource exploitation; however, does not always cause the exclusion of a species from a community. It may just cause the species involved in this interaction to experience a reduction in their potential growth. Competition by interference occurs when an individual directly prevents the physical establishment of another individual in a portion of a habitat.

Park (1948) studied the competition between red flour beetle, *Tribolium castaneum* and confused flour beetle, *T. confusum*. When both the species were reread together, both never survived. Out of 74 trials *T. confusum* became extinct 8 times and *T. castaneum* became extinct 66 times. This victory is achieved because of interaction of a pathogenic coccidian parasite, *Adelina tribolii*. The extinction curve of *T. castaneum* typically shows a dramatic drop in numbers around 300th day. In single culture *T. castaneum* recovers, but in presence of *T. confusum* is unable to recover and goes extinct soon after 300th day. He found that at high humidity and temperature *T. castaneum* invariably wins out while low humidity and temperature favour *T. confusum*. The example of competition between Mediterranean fruit fly, *Ceratitis capitata* and oriental fruit fly, *Bactrocera dorsalis* in Hawaii is well-known. The larvae of *B. dorsalis* completely suppress the development of larvae of *C. capitata* in hosts, thus resulting in the suppression of *C. capitata* by *B. dorsalis*. This may possibly be because both species have same host requirements and *C. capitata* is unable to compete with *B. dorsalis* except in situations which are unfavourbale to *B. dorsalis* (Keiser *et al*., 1974). Another interesting case of competition between *B. zonata* and *Carpomya vesuviana*, a monophagous species infesting only *ber* (*Ziziphus* species) fruits, is reported. *C. vesuviana* is on the verge of replacement by *B. zontata* as its larvae suppress the development of larvae of *C. vesuviana* in *ber* fruits (Agarwal and Kapoor, 1986).

Inouye (1978) discussed the exploitation of shared resources by two bumble bee species, *Bombus appositus* and *B. flavifrons*, in the Rocky Mountains of Colorado. The bees live in the same area, but *B. appositus* forages primarily from the flowers of larkspur, *Delphinium barbeyi*, and *B. flavifrons* forages primarily from monkshood, *Aconitum columbianum*. However, each species appears to specialize only in the presence of the other species. When one or other was removed, the remaining species quickly modified its behaviour to include much more feeding from the previously less-preferred flower and it stayed longer at each visit. This indicated that each bee's fundamental niche includes both species of flower; it is the presence of the competitor that restricts each bee's range to a realized niche of one flower type. Once again, therefore, the species coexist in the habitat as a whole, but exclude one another from particular parts of it. This reciprocal competitive effect is almost certainly related to differences in the length of proboscis of each species. *B. appositus* has a longer proboscis, making it better able to forage from larkspur whose flower has a longer corolla. *B. appositus* therefore diminishes the nectar in larkspur to a level that makes it unattractive for *B. flavifrons*; *B. flavifrons* consequently specializes on monkshood, making it less attractive for *B. appositus*.

Bark beetles compete for food and space in thin phloem layer of their host tree. It has been observed that the competition results in a decrease in the number of offspring produced per parent. The six-spined spruce bark beetle, *Pityogenes chalcographus* and the eight-spined spruce bark beetle, *Ips typographus* are the most serious pests of Norway spruce trees, *Picea abies*. Male *P. chalcographus* bore into the host tree and produce two pheromone components, chalcogran (CH) and methyl (*E,Z*)-2,4-decadienoate (MD), that act synergistically to cause an aggregation of both sexes. *I. typographus* also aggregates on same host in response to male-produced pheromone components, 2-methyl-3-buten-2-ol (MB) and (4S)-*cis*-verbenol (cV). The two species can be found alone or together in colonized trees, but when they occur together, *P. chalcographus* usually segregates in the top, smaller-diameter areas. The intra- and interspecific competition by these bark beetles for the phloem of their host tree can be severe. The beetles preferred to colonize uninfested logs rather than logs that are already occupied. The aggregation pheromone components of *P. chalcographus* inhibit the attraction response of *I. typographus* to its pheromone components, while the converse was not true (Byers, 1993).

12.2.3 Mutualism

Mutualism is a positive reciprocal relationship between two species. Through this relationship both species enhance their survival, growth or fitness. To a certain extent the relationship is more a reciprocal exploitation rather than a cooperative effort on the part of the individuals involved. The individuals in the populations of each mutualist species grow and/or survive and/or reproduce at a higher rate when in the presence of individuals of the other species. This type of interaction is an extremely widespread phenomenon. Mutualism can take on two main forms:

(i) **Symbiosis.** In symbiosis both organisms live together in close proximity, and both generally derive benefit. If both species are obligate symbionts, it means that neither can survive alone. True symbiosis is illustrated by the relationship of insects like cockroaches and termites with the cellulose-digesting protozoans that live in their intestines; neither organism could survive without the other.

(ii) **Non-symbiotic mutualism.** The species do not live together, nor are completely dependent on each other; the relationship is facultative or opportunistic but does profit the organisms when together.

The well-known mutualistic relationship occurs between ants and sap-sucking hemipterans which excrete a sugary substance called 'honeydew' (Plate 12.2). Certain gregarious aphids feed by inserting a pointed, straw-like mouth structure called a stylus into the vascular tissues of the plant and sucking the plant juices out. Plant sap, a combination of water and sugars, is low in other nutrients; however, the aphid must process a great deal of plant sap in order to get the amino acids and other nutrients it needs. This semiprocessed phloem sap is excreted as waste through a pair of structures called cornicles and is exploited by ants. In return, the aphids receive protection from predation. In some cases ants tend colonies almost like ranchers with their cattle, not only protecting the aphids, but moving them around from plant to plant (Plate 12.3).

12.2.4 Parasitism

Parasites obtain food at the expense of their hosts. Parasites generally reduce the fitness of their hosts by feeding in or on the host, resulting in death or reduced fecundity. The condition of being parasitized, either by a parasite or parasitoid is termed parasitization. A parasite that lives on or in another parasite is 'hyperparasite' and the phenomenon in which a parasitic species grows parasitically on other parasitic species is termed 'hyperparasitism'. When an individual host is attacked more than once is termed 'superparasitism' while 'multiparasitism' is parasitization of a host by two or more parasites or parasitoids.

Parasitoid is an organism that lives at the expense of another (host) and eventually killing the host and becoming free-living as adult. Insect parasitoid has an immature life-stage that develops on or within a single insect host, ultimately killing the host, thus the parasitoid acts as natural enemy. Adult parasitoids are free-living or may be predaceous. Most beneficial insect parasitoids are wasps or flies, although some rove beetles and other insects may have life-stages that are parasitoids. Most insect parasitoids only attack a particular life-stage of one or several related species. The immature parasitoid develops on or within a host, feeding on body fluids and organs, eventually killing the host. The parasitoids may attack egg, larval or pupal stages of the host and are termed as egg, larval and pupal parasitoids, respectively. In larval-pupal parasitoid, the parasitoid which deposit egg(s) in or on host's larva and whose progeny emerge from the host's pupa. Hyperparasitoid is a secondary parasitoid that lives/develops upon another parasitoid and the condition is termed 'hyperparasitoidism'. Hyperparasitoids belong to two families of Diptera (some Bombyliidae and Conopidae), Coleoptera (a few Rhipiphoridae and Cleridae) and noticeably the Hymenoptera. 'Idiobiont' is a parasitoid that prevents further development of its host, by paralysis or death while 'koinobiont' lays egg(s) in the young host, which grows continuously thereby providing an increasing food resource to the developing parasitoid(s). Hosts attacked by parasitoids die more slowly. Some hosts are paralysed, while others may continue to feed or even lay eggs before succumbing to the attack. Parasitoids can be the dominant and most effective natural enemies of some pest insects, but their presence may not be obvious. Important characteristics of insect parasitoids are: (i) specialized in their choice of host, (ii) smaller than host, (iii) only the female searches for host, (iv) different parasitoid species can attack different life-stages of host, (v) eggs or larvae are usually laid in, on, or near host, (vi) immatures remain on or in host; adults are free-living, mobile, and may be predaceous, and (vii) immatures almost always kill host.

Insect eating (entomophagous) parasitoids show a range of strategies for development on their selected insect hosts. They deposit eggs on or in the hosts and the larvae develop externally or within the hosts. The hymenopteran endoparasitoids lay eggs within the hosts by using a piercing ovipositor while parasitoid dipterans use substitutional ovipositor. Most parasitoids are able to discriminate the hosts that are already parasitized.

12.2.5 Predatism

Predator is a natural enemy that preys and eats more than one organism during its lifetime. Hunting live prey is so widespread among insects that they are always busy killing each other. The worst enemies of insects are insect themselves. Some predators locate prey by chasing it by scent and others, especially the winged ones by sight. Most orders of insects include predators except Isoptera, Phasmida, Phthiraptera, Strepsiptera and Siphonaptera (Hagen, 1987). Some predators are feeding specialists while most are generalists. Some are extremely useful being natural enemies of insect pests. Most predatory insects feed on insects, but some may consume other animals. The predator insects consume their prey immediately after attack except for predatory Hymenoptera that provision their nests with prey. Major characteristics of insect predators are: (i) adults and immatures are often generalists rather than specialists, (ii) predator is generally larger than the prey, (iii) a predator kill or consume many prey in its life, (iv) both immatures and adults may be predatory, and (v) predator attacks both immature and adult prey.

Active chase of the prey by a predator may be on the ground, in water or in the air. The ground beetles, fireflies and digger wasps grasp their victims on the ground while dragonflies are perhaps most powerful winged hunters. The robber flies (Fam. Asilidae) capture the flying prey and suck them. Many predators paralyze the prey by stinging and then feed on it or carry off it to their nests. Insects like praying mantis while waiting for the prey, remain motionless, assume innocent-looking, and innocuous posture in a place often frequented by the prey. Its fore legs are long, powerful and armed with stout and sharp spines for grasping the prey (Fig. 13.3). In aquatic larva of dragonfly labium is modified into a fang-mask (Fig. 9.2) which suddenly spurts out when the prey is in its striking range.

The ant lion larva constructs a pit and sucks the body fluid of ants that fall in the pit (Fig. 9.5).

Predator insects exhibit great variation in their dietary range. Some like the vedalia beetle, *Rodolia cardinalis* and the green lacewing, *Chrysopa slossonae* are highly specific and feed on only one species of prey. Others like aphid-feeding convergent lady beetle, *Hippodamia convergens* and two-spotted lady beetle, *Adalia bipunctata* are stenophagous or oligophagous and restrict their feeding to a range of related taxa. Finally, other predators, e.g. the spined soldier bug, *Podisus maculiventris* and the spotted ladybird beetle, *Coleomegilla maculata* are polyphagous and consume a wide variety of prey and other materials like plant fluid and pollen.

Adult predators may use a variety of cues, including volatile chemicals from both plants and insects, to find prey. The role of chemical cues in locating prey is relatively well-known for the green lacewing, *Chrysoperla carnea*. Several species of beetles also use chemical cues to find their prey. Larval predators have evolved various behavioural tactics of locating and recognizing their prey. A predator may reject the prey with low nutrition and thereby influencing its development, reproduction or survival. Many predators exhibit slow development and reduced fecundity when they eat suboptimal prey. In some cases, feeding on a prey having toxins in its body may result into death of predator. After the predator discovers prey, it must pursue, subdue, and consume it. Prey defences (e.g., behavioural, morphological, chemical) may determine whether a predator accepts or rejects prey.

It is easy to believe that the predator-prey interaction is somehow detrimental to the prey population. However, functional relationships between predator-prey within natural ecosystems have coevolved over long periods, creating a dynamic balance between their interacting populations. Thus, the population sizes of predator and prey species are interregulated by delicate feedback mechanisms that control the densities of both species. Predation is complex phenomenon and includes coevolution, specialization, food preference, effects on fitness, distribution in time and space of predators and prey, optimal foraging, etc. Each also interacts with its abiotic environment and the other organisms in the community which leads to complex multiple interactions. Prey have evolved over time to avoid predators, while predators have evolved to better capture their prey. This coevolution has often led to specialized characteristics of both prey and predator. Dynamic models of the interactions between predators and prey have provided testable predictions about the role of coevolution with regard to the stability or instability of ecological interactions in general (Abrams, 2000). Most models predict that coevolution can produce cycles in population densities and trait values, and that cycling may be evolutionarily stable under some conditions.

12.2.6 Phoresy

Phoresy is an example of 'commensalism' in which individual of one species gets benefit and the other is neither helped nor harmed. In phoresy an individual of one species is transported by the individual of another species. This interaction benefits the carried while the carrier remains unaffected. It provides a means of finding a new host or food. An often-cited example of phoresy is ischnoceran lice transported by the winged adults of hippoboscid fly, *Ornithomyia* (Fig. 12.3). The lice have secondarily lost their wings in favour of a more passive form of transmission from one host to another. The lice use a number of alternative possible mechanisms for transmission, and perhaps the most famous of these involves the louse-fly (Fam. Hippoboscidae). These flies are frequently found around birds and the lice attached to the abdomen of the hippoboscid fly by their mandibles, particularly after the death of the host. This behaviour by the lice is known as phoresy and is in fact a process of redistribution in which the lice get to travel freely on the flies without feeding on them. There are many records of lice found attached to hippoboscid flies, although in almost all instances the lice concerned are members of the suborder Ischnocera. This type of transmission allows the louse to escape a dead or dying host in an attempt to find a more favourable host environment.

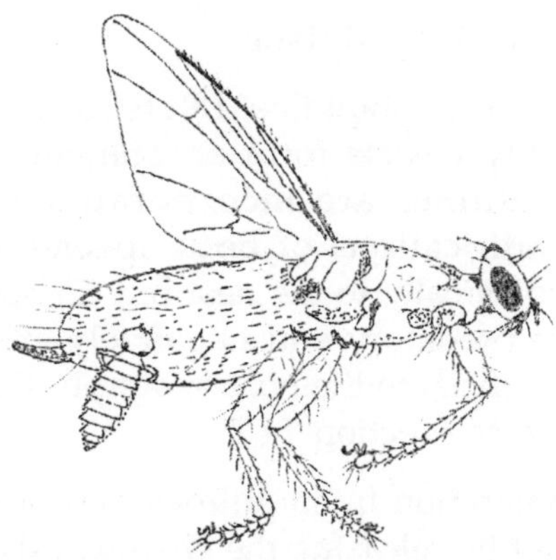

Fig. 12.3. The bird lice attach themselves to the abdomen of hippoboscid parasitizing bird, thereby obtaining phoretic transport to new bird hosts.

In some cases the male acts as dispersing agent for small wingless female (Fig. 12.4) e.g. in parasitic wasp, *Dimorphothynnus haemorrhoidalis* the male carries small wingless female and inseminate her while flying. It has been termed as 'phoretic copulation' and provides effective dispersal of female into new areas (Evans, 1969).

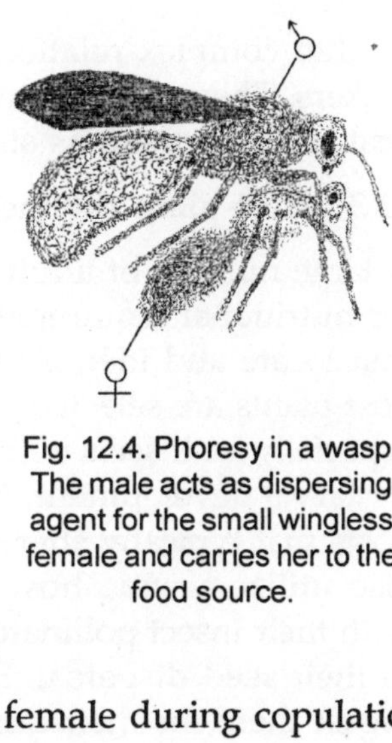

Fig. 12.4. Phoresy in a wasp. The male acts as dispersing agent for the small wingless female and carries her to the food source.

The female human bot fly, *Dermatobia hominis* uses services of blood sucking mosquitoes and muscoid flies. It attaches her eggs to the legs and body of mosquitoes, and in this way transporting them to human host for hatching and larval development. Many egg parasitic insects attached themselves to the host species, thereby gaining immediate access to the eggs while the female is ovipositing. *Mantibaria manticida* (Fam. Scelionidae), an egg parasitoid of European mantid, *Mantis religiosa* is predominantly phoretic on female hosts. Individuals of *M. manticida* that are phoretic on male mantids may transfer to the female during copulation.

12.2.7 Inquilinism

The term 'inquiline' refers to an organism that shares home and food of another. Many insects form an obligate association with social insects. The integrated inquilines are incorporated into their hosts' social lives by behavioural modifications of both species while non-integrated inquilines are adapted ecologically to the nest but do not interact socially with the host. Some inquilines are predatory and negatively affect the host's population whereas other may be beneficial to the host or simply share the host's nest. For discussion on inquilinism also see section 14.7.

Integration by inquilines may be achieved by breaking the chemical code of the host by releasing the chemical signal similar to that of host or by tactile signalling that releases social behavioural responses. The term 'Wasmannian mimicry' includes some or all of the chemical or tactile mimetic features that allow the inquilines to be accepted by the host.

Inquilinism is also displayed by some social insects like ants that live as permanent social parasites in the nest of other species. Social parasites exploit the socially managed resources of their host's society. Inquiline social parasites are dependent on their host throughout their life cycle, and so many of the traits inherited from their free-living ancestor are removed by natural selection. Obligate social parasites cannot initiate their own colonies, but invade established nests of other species. Some wasps are facultative social parasites and their females can start their own nests or they can be parasitic. Some social ants live as temporary or permanent social parasites in nests of other species. Several ant genera have the habit of entering nests of other species of ants or termites to prey upon the inhabitants and/or their food. This is termed 'trophic parasitism' and is also exhibited by many bees and other insects.

Further complex relationship is showed by some species of ants that are slave-makers. They capture immature stages of other species, bring them to their own nest and rear them as slaves. The phenomenon is termed as 'dulosis'.

12.3 Insect-plant interactions

A large number of adult and immature insects are closely associated with plants for nutritional requirements. Insects searching for an acceptable host plant must first locate and identify the appropriate plant species. The accuracy with which host plants are selected, and individual plant quality assessed are also important, especially for insects with narrow host ranges and specific nutritional requirements for larval development. Majority of insect species use a very restricted number of hosts that typically share characteristic phytochemicals while in many cases they also utilize a wide host range. Flowering plants exhibit mutualistic interaction with their insect pollinators in which both are benefitted. Insects also assist plants in their seed dispersal. Some plants are enemies of insects as they capture and digest them for their nutritional requirements.

12.3.1 Herbivory

Majority of plant species are consumed by insects for their dietary requirements. The plant feeding insects may be: monophagous- specialist, feeding on only one kind of plant; oligophagous- feeding on a few plants or a group of plants; or polyphagous- generalist, feeding on many groups of plants. Many complex interactions occur between insect herbivores and plants. Some of these factors limit and inhibit herbivory by insects while many host plant traits influence the success of insect herbivores. Herbivores play an important role in the ecology of any area, influencing plant communities and their growth. The great diversity of insect herbivores that exist, reflects the diversity of plants. Each plant provides a wide range of feeding opportunities and insects show many adaptations, like modified mouthparts and behaviours that allow them to feed on plants.

Herbivory occurs above and below ground. Insect herbivores are the most numerous and approximately half of all living insects are herbivores. Some large insect groups are almost exclusively plant-feeders. Most insects have evolved specialized features and feeding strategies to utilize plants as food source. Some insects have developed alliances with bacteria or fungi that assist in breaking down complex chemicals. The plants on the other hand, have obvious defences against their enemies including insects. The main lines of defence in plants are chemical and morphological. The 'battle' between insects and plants is likened to an arms race and has led to strong relationships between many insect herbivores and plants. It has been proposed that the herbivorous insects and plants evolved through a coevolutionary process. The concept of warfare between insects and their host plants is discussed in section 9.10.

12.3.2 Mutualism

The most obvious example of mutualistic interaction is the relationship between flowering plants and their insect pollinators (Plate 12.4a,b.). Sexual reproduction in plants involves transfer of pollen (male germ cells with a cover) from the anthers of a flower to the stigma. Transfer of pollen usually occurs either by an animal or by the wind. The role played by insects in pollination of plants is well-known and the success of angiosperms is directly related with the development of interactions between plants and their pollinators. The benefits of insect pollination over wind pollination include: increase in pollination efficiency including reduction of pollen wastage, successful pollination under conditions unfavourable for wind pollination and maximization of the number of plant species in a given area. It has been estimated that more than 65 per cent of the flowering plants are pollinated by insects. Insects like beetles, thrips, wasps, bees, flies, butterflies and moths, ants, etc. are specialized visitors of flowers and flower-frequenting taxa are termed 'anthophilous' (means flower-loving) while the insect pollination is termed as 'entomophily'. The terms related with entomophily are: cantharophily (beetle pollination), myophily (fly pollination), melittophily (bee pollination), sphecophily (wasp pollination), myrmecophily (ant pollination), phalaenophily (moth pollination) and psychophily (butterfly pollination). Melittophily plays the dominant role in pollination in majority of plant species. Flowers and their pollinators are a common and ubiquitous form of mutualism. To achieve successful pollination, a plant species needs to fulfil two requirements.

(i) The plant must attract a pollinator. The flower is a structure designed to attract a pollinator by its shape, colour, and smell. The flower and its aroma provide a series of visual clues to the pollinator associates with a food reward.

(ii) A mutualistic relationship requires a positive interaction between both species. A flower gets its pollen passed from one individual to another. In turn the flower provides a food reward in the form of nectar (a sugar rich solution), or pollen as a solid food source. If the food reward is large enough, the pollinator is likely to go looking for flowers of the same plant species after visiting the flowers of the first plant.

Though nectar and pollen are the most common rewards, floral tissues may also be offered as brood sites for pollinators in some plant groups. Plants such as *Ficus, Yucca, Trollius, and Heuchera* sacrifice a portion of their ovules for brood sites of their pollinators. On the other hand, other plants have developed brood sites and nutrients for pollinators, independent from ovules. Androecial parts of plants function as brood sites for pollinators in *Peltandra, Zamia,* and *Eupomatia.* In most of the cases; however, plants and pollinators directly exchange service and rewards.

A fascinating example of pollination is that of figs (genus *Ficus*) and fig wasps (Fam. Agaonidae). Figs have a very unusual flowering system and the flowers are on the inside of a structure called 'syconium', which develop into the ripe fruit. The female fig wasp enters the syconium through a small opening termed 'ostiole'. The wasp carries pollen and pollinates the flowers on the inside of the hollow syconium. At the same time, it lays eggs into flowers with short styles. The eggs hatch in the ovaries of the short-styled flowers and the larvae consume the ovary tissue as they grow. These flowers develop galls, which are regions of tissue which are stimulated to undergo mitosis by chemicals released by the female wasp, thus providing the larvae with an ample food source. The long-styled flowers do not have eggs laid in them, apparently because their ovaries are too deep in the fig wall for the wasp ovipositor to reach. But, the long-styled flowers do get pollinated by the fig pollen carried by the female wasp.

The larvae develop into adult wasps. The male and female wasps are strikingly different as the male is a wingless creature. The males fertilize the females, chomp a hole in the wall of the now mature fig, and die inside the fig. The female wasps emerge from their seeds, gather pollen and depart the fig through this hole, taking pollen with her to locate another conspecific fig tree in the phase of fig development suitable for oviposition.

It has been assumed that the fig and wasp have closely coevolved. The reciprocal selection pressure presumably results in matching of fig and fig wasp traits. For example only particular kinds of wasps can enter the syconium and the sensory receptors of the wasp respond only to volatile chemicals of the host fig. The size and morphology of ostiole allow entry of a wasp of correct shape and size. The timing of flowering is such that the female flowers, in which wasp eggs are laid, mature earlier than the male flowers. So, the male flower is not mature until close to the time at which the larvae have produced the next generation of wasps and the females are ready to leave the fig with the pollen. The female wasps have evolved just the right shape of ovipositor to place eggs in the ovary of the short-styled flowers. In both figs and wasps these adaptations work together to ensure reproduction and development of both species.

The 'bottle gentian' or 'closed gentian' (*Gentiana andrewsii*) is an unusual woodland flower and its mature flowers look just like large buds, as the petals never open. Although these flowers produce a rich source of pollen and nectar, most insect pollinators are not able to get inside. The flowers are pollinated almost exclusively by large bumble bees that are strong enough to force the petals open and crawl inside. The mutualistic association between the bees and flowers is evidenced by the benefit received by bees as they are able to access a bountiful nectar supply, and the plants benefit by attracting 'loyal' pollinators that improve the chances for cross pollination.

Yucca flowers are of certain shape so that only tiny moth can pollinate them. The pollinating yucca moths of the genera *Tegeticula* and *Parategeticula* lay their eggs in the yucca flowers and the caterpillars live in the developing ovary and eat yucca seeds. The obligate mutualistic association between moths and yucca plants provide one of the most apparent examples of coevolution which was established at least 40 million years ago. The association relies on the key innovation in the moths of complex tentacles used for pollen collecting and active pollination. Female yucca moths possess unique tentacles on their mouthparts that are used to actively pollinate host flowers where they oviposit. The female gathers the glutinous pollen of yucca flowers by scraping it off the anthers with her tentacles. The pollen is immediately compacted by using tentacles and sometimes the fore legs as well, and placed as a solid batch on the concave posterior-ventral surface of the head. The pollen mass may approach 10,000 grains and weigh up to 10 per cent of the moth body mass. Prolific pollen coating maintains batch cohesion, and the tentacles are not involved in its retention. After pollen gathering, the moth seeks out flowering yucca plants where she oviposits into or near pistils. As oviposition is completed, the female flexes her tentacles and uses the apical portion to remove a small pollen load from her batch. She walks to the floral stigma and very deliberately places the pollen on it.

Hawk moths and sphinx moths (Fam. Sphingidae) constitute an important class of nocturnal pollinating agents around the world. Sphinx moth pollination is termed as 'sphingophily'. The adult hawk moth is a heavy insect with an aerodynamically formed body. It is a strong, vigorous flyer that moves rapidly across long distances, darting skillfully and with clear purpose from target to target. During flight the long, narrow fore wings are extended from the thorax to generate great speed. Hovering at the face of a flower, it uncoils a long tubular proboscis to insert into the flower centre and sucks copious nectar (Fig. 12.5, Plate 12.5). With a few daytime exceptions, hawk moths typically perform these foraging acrobatics at night. Hovering hawk moths need no landing platform but require a high-energy supply to maintain active flight. Sweet scents dispersed as odour plumes attract the fast-flying insect to the plant, often in total darkness.

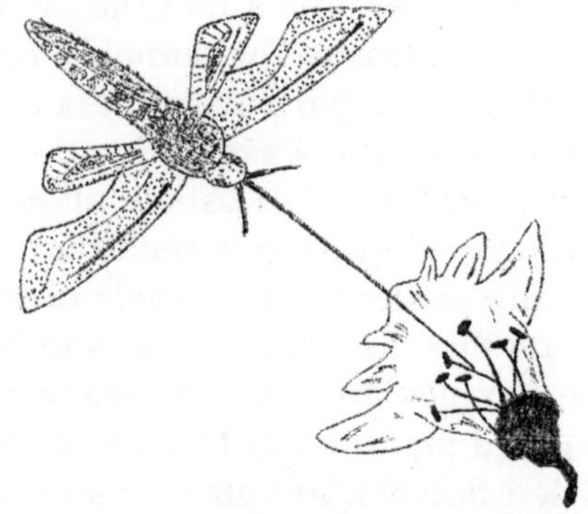

Fig. 12.5. A hawk moth with extended proboscis. The long proboscis help to extract nectar and the insect does not land on flower.

Sakai *et al.* (2000) reported the pollination system of jack tree, *Artocarpus integer* (Fam. Moraceae). The male inflorescences of *A. integer* on a tree are infected by the fungus, *Choanephora* sp., and gall midges, *Contarinia* species (Subfam. Cecidomyiinae). Gall midges feed on the fungal mycelia and oviposit on the inflorescence. Their larvae also feed on the mycelia and pupate in the inflorescence. The gall midges are also attracted by female inflorescences lacking mycelia, probably due to a floral fragrance similar to that of male inflorescences. *A. integer* is thought to be pollinated by the gall midges. In this pollination mutualism a fungus plays an indispensable role.

An obligatory mutualistic relationship is found between ants and myrmecophyte plants (Fig. 12.6). Some plants have hollow stems, tubers, swollen petioles or thorns which are termed domatia (little houses) and used by ants for feeding, nesting or both. Besides, ants also use natural tunnels or hollows made by borers. Myrmecophytes have true domatia, provide nesting sites and food resources to ants. Food is provided either from extra floral nectarines (EFNs) or fruit bodies or via honeydew excreted by hemipterans living within the domatia. Myrmecophytes may have both EFNs and

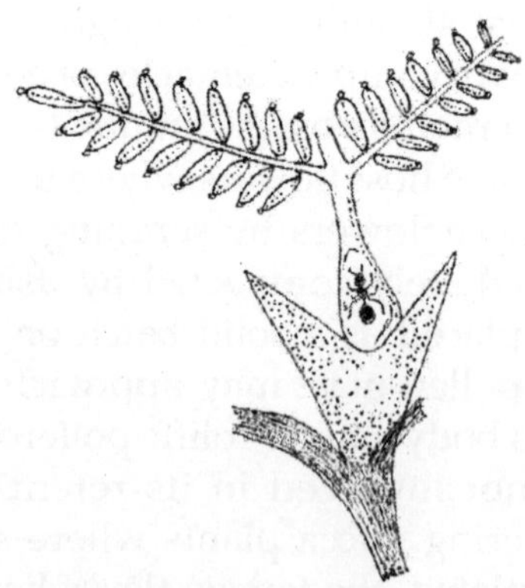

Fig. 12.6. *Pseudomyrmex* ant occupying the hole in the thorn of the bull horn acacia, *Acacia cornigera*. An obligatory mutualistic relationship is found between ants and acacia plant.

fruit bodies while some may lack both. The ants protect these plants from herbivores and plant competitors. The ant, *Pseudomyrmex* protect *Acacia* plants in tropical America. Some myrmecophytes absorb nutrients and nitrogen from ants' excreta. The bull horn acacia, *Acacia cornigera*, a common shrubby tree in Central America and Mexico has swollen horn-like thorns at leaves bases. *Pseudomyrmex* ants occupy the holes in the thorns. The ends of each leaf segment are modified into small, oval 'Beltian bodies' (discovered by Naturalist Thomas Belt) which are eaten by ants or fed to young. Janzen (1966, 1967) suggested that the *Acacia* plants are protected by ants, and herbivores learn to recognize the alarm odour of ants and avoid contact with the plant.

12.3.3 Myrmecochory

Myrmecochory is seed dispersal by ants and is an important factor in the survival and reproduction of some plants. Ants are common seed dispersers, and plants that depend on ants for seed dispersal are termed myrmecochores. Seed dispersal by ants may also occur accidentally when seeds are lost in transport or seed stores are abandoned. The seeds have an attached food-bodies called elaiosomes, that the ants feed to their larvae in the nest. Besides, some seeds release special chemical attractants that stimulate ants to collect them. The elaiosomes contain nutritive lipids, proteins and carbohydrates in varying proprotions and ants carry the seeds away from the maternal plant. After removing the elaiosome, the undamaged seed is usually deposited in ant colony's trash pile. That trash pile is typically nutrient-rich, and those nutrients can provide a second benefit to the plant. The association is of obvious benefit to the ants and may reduce intra- or interspecific competition amongst plants by dispersing seeeds to new sites. An example of myrmecochory is seed dispersal of black wood, *Acacia melanoxylon* plant. The seed has a long red stalk which is very nutritious and remains attached to the seed after falling. Ants gather the seed, take it to their nest, remove the nutritious portion and then discard the unwanted seed either in a disused part of the nest, or on the rubbish heap.

12.3.4 Insectivorous plants

Some plants can capture and digest live insects to obtain nitrogenous compounds that are lacking in its usual marshy habitat. Insectivorous plants have adapted to grow in poor soil conditions where the numbers of microorganisms recycling nitrogen compounds are very much reduced. Besides, some near-carnivorous plants are unable to digest insects, but still trap them on their sticky coated leaves. The insects die and decay naturally, with the nutrients eventually becoming washed into the soil where they finally benefit the plant.

Venus flytrap. The Venus flytrap, *Dionaea muscipula* is endemic to the Atlantic coastal plain of North America. Many minute red glands cover its inner surface and perhaps discharge sweet liquor which tempts the poor insect to taste them. When the tender plant parts are irritated by insect's legs, the two lobes rise up, grasp it fast, lock the rows of spines together, and squeeze it to death. The sweet nectar and bright colouration of the trap interior are important to allure the prey.

Pitcher plants. The pitcher plants, *Sarracenia* species occur in the Atlantic coastal regions of North America. The hollow trap leaves arise directly from the rhizome above the ground and are carefully constructed pitfalls designed to attract and capture small insect prey with amazing efficacy. The leaves exhibit all alluring elements of real flowers and their visual lures are: striking colours and patterns, copious nectar secretions, and a convenient landing site for flying insects. Some pitcher plants are known to release sweet attracting odour. The insect resting on landing site is deceived, gets exhausted in attempting to fly and finally falls into the pitcher bottom.

Pitcher plants being insect-pollinated must rely on the visiting insects for pollination, and at the same time must consume the insects as prey to supplement their nutritional need. In many pitcher plant species flowers and traps are produced in different times of the season. Besides, some species deploy spatial separation to resolve, at least in part, the dilemma of pollinator/prey differentiation. In *S. psittacina* and *S. purpurea* the positions of the flower is well above the trapping space. A few specialist insects live and survive in inhospitable environment and can move freely in and out of the pitchers. Many of these inquilines live in a mutualistic association with the plant, digesting trapped prey and microorganisms and assist in availability of nutrients to the plant.

Cobra plant. In cobra lily plant, *Darlingtonia californica* tubular pitcher leaves arise directly from the rhizome, forming a rosette. The hood of the pitcher is well-developed to form a dome, with the pitcher opening facing downward. The pitcher lid is modified into a two-lobed, fishtail-like appendage projecting downward from the front edge of the opening which has numerous nectar glands. Its surface is covered with short, stiff hairs, pointing towards the pitcher opening. The fishtail appendage provides a convenient landing and feeding site for a flying insect. The insect enjoying nectars on the fishtail appendage enter into the dome interior and finally tumble down into the spiral depth of the cobra leaf.

Sundews. The sundews, *Drosera* species are normally perennial with several leaves forming a rosette. The upper surface of the leaf blade is covered with numerous glands raised on a stalk. These hair-like stalked glands often referred to as 'tentacles', hold a large droplet of sticky mucilage at the tip which form the basic trapping mechanism of the adhesive trap. The viscous secretion attracts and traps many insects. As the prey struggles to free itself, tentacles begin to bend over the prey and entrap it.

Butterworts. The butterworts, *Pinguicula* species are rosette of thin leaves and the upper surface of the leaf is covered with short stalked glands that secrete clear sticky mucilage to trap small insects like ants and mosquitoes. When an insect lands or crawls on the upper surface of the leaf, the insect is caught up by the sticky substance secreted by the stalked glands.

Bladderworts. The bladderworts, *Utricularia* species are commonly found floating freely in ponds and ditches. Many tropical terrestrial species are epiphytic. The trapping bladder is a modified leaf. Each trap has some antenna-like hairs on one side that attract and trap tiny insects.

12.4 Insect-fungal interactions

Fungi serve as food for many groups of insects particularly, Collembola, Isoptera, Coleoptera, Hymenoptera, and Diptera. The fungi eating insects are termed fungivorous or mycophages. The insects may be microphages which feed on small particles like spores, hyphal fragments or liquid media while macrophages use fruiting bodies as food. Insects that develop as larvae in fruiting bodies are obligate fungivorous, whereas those feeding on decomposed fungi are saprophagous. Insects like drosophilid fruit flies, larvae of nitidulid beetles are known to feed on yeasts. Among Diptera the fungus gnats (Fam. Mycetophilidae) are diverse and have oligophagous relationships with fungi. Their larvae feed on fungi and some are serious pests of mushrooms. Other dipteran larvae like certain phorids and cecidomyiids also feed on commercial mushrooms.

12.4.1 Fungus farming

Among insects three groups, termites, ants, and beetles have independently evolved the ability to grow fungi for food. Like humans, the insect farmers became dependent on cultivated fungi for food and developed task-partitioned societies cooperating in gigantic agricultural enterprises. Agricultural life ultimately enabled all of these insect farmers to rise to major ecological component. Indeed, the fungus-growing termites of the Old World, the fungus-growing ants of the New World, and the cosmopolitan, fungus-growing beetles are not only dominant players in natural ecosystems, but they are also major insect pests (Mueller and Gerardo, 2002). For details of fungus farming in termites and ants also see sections 14.4.2.3 and 14.5.3.1.

These insects appear to have made the transition to fungiculture via different evolutionary avenues. In the termites, fungi probably were an important food source before true cultivation, and fungiculture arose when the termites secondarily developed an ability to manipulate fungal growth in their nests. Many nonfarming termite species are attracted to feed on fungus-infested wood, and termite fungiculture therefore may represent an elaboration of such ancestral feeding habits. It is unclear whether fungiculture in attine ants arose from ancestral mycophagy or from a system of fungal vectoring by ants (Mueller *et al.*, 2001). In ambrosia beetles multiple origins of fungiculture has been suggested that resulted in diversity of beetle species and their feeding specializations (Farrell, 1998).

Termite fungiculture. A number of termite species are obligately dependent on the cultivation of a specialized fungus, *Termitomyces*. The fungus is grown on termite faeces in subterranean combs that the termites construct within the nest mounds. Combs are supplied with faeces of workers that forage on wood, grass, or leaves (Batra and Batra, 1979). Spores of consumed fungus are mixed with the plant forage in the termite gut and survive the intestinal passage. The addition of a faecal pellet to the comb therefore is functionally equivalent to the sowing of a new fungal crop.

Ant fungiculture. Among the attines, ants in the two genera of leaf cutting ants (*Acromyrmex* and *Atta*) exhibit the most complex fungicultural systems. Fungiculture among *Atta* (Plate 12.6) supports colony populations in the millions of workers and explains why leaf cutters are the dominant herbivore in the Neotropics. Colonies of *Atta* species can survive for extended periods, often living for 8-10 years or more after initial nest founding. These ants have developed a mutualistic relationship in which the vegetative mycelium of a particular fungus is passed from one generation of ants to another. The colony of a fungus growing attine ant is typically founded by a mated female who takes a fungal inoculum from her mother's colony to start her new garden. These ants grow fungi in subterranean chambers, manuring the gardens with dead vegetable debris, or in the case of the leafcutter ants, with leaf fragments. Attine ants are obligately dependent on their fungi; their brood is raised on an exclusively fungal diet. The ants are unable to digest cellulose, but the fungus can and is thereby able to convert it into nutrients that the ants can utilize.

Beetle fungiculture. Many fungi are known to be associated with ambrosia beetles and bark beetles. Most of them are conidial (asexual) states of Ascomycetes. Ambrosia beetles produce piles of shredded wood and debris outside their galleries. There are several fungi that grow within this debris. The ambria fungi have an obligate ambrosia fungi have an obligate mutualistic relationship with the beetle farmers. The fungi serve as the beetle's primary food source and are essential for the completion of the beetle life cycle. The fungi influence the hatching of eggs, provide food for the developing larvae, and prevent the encroachment of pathogenic fungi. The beetles in turn provide plant debris and insect excrement as a food source for the fungi, and eventually are instrumental in disseminating the fungus.

12.4.2 Symbiosis

A case of symbiotic interaction is known between Septobasidial fungi (Basidiomycota) and scale insects. The fungus provides protection to insects from natural enemies and it depends on the insect for nutrients dispersal. The fungus grows over the colony of scale insects, providing a tough barrier. It also penetrates the insect and enters in the haemocoel. Insects continuously provide a flow of nutrients to the fungus directly from the plant. The parasitized insects do not reproduce while the non-parasitized individuals reproduce within the protection of the fungal shield. The fungus releases a chemical that attract insects. In addition, the fungus is dispersed by scale insects carrying basidiospores from infested trees to uninfested trees.

The wood wasp, *Sirex noctilio* (Fam. Siricidae) is known to cause shoot deformation in *Pinus* species. Females bore into shoots and deposit eggs and the fungus *Amylostereum*. The fungus causes a white rot of the timber. The fungus digests both cellulose and lignin. Newly hatched larvae feed on the fungus, and also utilize fungal cellulases to digest components of timber. The fungal inoculum is carried by the adult female on the outer cuticle, and is deposited into timber during oviposition.

12.5 Tritrophic interactions

Organisms in an ecosystem interact with each other and at least three trophic levels are involved in a simple food web which have feeding link (Ahmad *et al.*, 2004). Tritrophic interactions, as these are known, essentially focus on the abilities of natural enemies to use plant traits to locate insects, besides emphasizing on the ecological and evolutionary implications of the food chain (Price *et al.*, 1980). The lower trophic levels are forced to evolve, to reduce feeding by higher trophic levels; whereas the members of higher trophic levels evolve to increase consumption. The natural enemies of herbivores favour plants by reducing the herbivore populations and plants favour the natural enemies by making herbivores more vulnerable to them. Semiochemicals play an important role in mediating these tritrophic interactions. These chemicals elicit behavioural and physiological responses in the receiver, which results in the interaction between them (Nordlund, 1981).

Plant odorants dominate the atmospheric chemical environment forming 'aerial bouquets' and insects including parasites select a few critical signals that stimulate their behavioural patterns. Plants produce semiochemicals act as intrinsic defence against herbivores; however, these chemicals are utilized by third trophic level resulting in the plant-natural enemy interaction. The floral scent and volatile organic compounds (VOCs) also assist the natural enemies in finding their potential host (herbivore) feeding on the same plant, e.g. many parasites and predatory ants are attracted towards plants due to nectar (Bentely, 1977). Kesten (1969) reported that the eye-spotted lady beetle, *Anatis ocellata* feeding on pine aphid is attracted towards its prey by the odour of pine needles. Numerous ecological and physiological factors tend to influence this process of host location directed by signals originating from the microhabitat, plant, host insect or microorganisms. In this exercise they respond and use specific cues from host plants to locate them, including those from damaged plant leaves and odours from host insect faeces (Vinson, 1985), so that natural enemies base their foraging strategies on chemical information from the first and second trophic levels (Vet and Dicke, 1992). Volatiles released by damaged plants after insect feeding tend to increase the efficiency of natural enemies and plants producing such alarm responses attract more parasitoids and predators (Karban and Baldwin, 1997). Studies show that plant chemicals induced by the host's feeding and by-products of the host derived from the plant are vital to the parasitoid's host-finding and assessment of the profitability of the habitat (Lewis and Martin, 1990). Substance(s) in the oral secretions of lepidopterous larvae and other insects were found to act on wounded plant tissues to induce the plants to release the volatile signals that attract the parasitoids (Dicke *et al.*, 1993).

Immature stages of many parasites and predators are carnivorous while the adults feed on nectar. These adults are attracted to plant nectaries and lay eggs, the hatched larvae feed on the herbivores providing benefit to the plants. Plants also increase the fitness, lifespan and fecundity of natural enemies by providing the best quality food in the form of pollens and nectar. Extrafloral nectaries in cotton appear to be an important source for hymenopterous parasitoids involved in the control of budworms and bollworms of cotton. These parasites are twice as abundant in cotton cultivars with extrafloral nectaries which act as an energy source, with sugars and varying types of amino acids, increasing the longevity three to four times of parasitoids like *Trichogramma* (Pemberton and Lee, 1996). Some plant volatiles act as allomones to herbivores and mediate tritrophic interaction, e.g. interaction between the mustard aphid, *Brevicoryne brassicae* and its parasitoid, *Diaeretiella rapae*. *B. brassicae* is attracted to mustard plant due to sinigrin, but *D. rapae* uses the odour of sinigrin to locate the mustard and ultimately the aphid host (*B. brassicae*) (Read *et al.*, 1970). Many predators are attracted towards the sugar-rich honeydew and ultimately consume the prey. Moreover, some plant chemicals when incorporated in the body of herbivores act as kairomones between insect and natural enemies, especially the parasitoids (Vinson, 1984a). Corn earworms, *Helicoverpa zea* derive trichosane from corn, which is passed unchanged into its eggs. Its parasitoid, *Trichogramma evanescens* uses this chemical as a kairomone and finds eggs of the host (Lewis *et al.*, 1972). Tritrophic interaction is also affected by the quality of host diet. Growth and survival of the host is adversely affected by its diet, which ultimately affects the third trophic organisms. Adverse effect on herbivore survival results in greater mortality of the parasitoid population.

In certain cases semiochemicals produced by plants may act against the third trophic organisms. For example, the body odour of larch, *Larix* species and larch sawfly, *Pristiphora erichsonii* are attractant of the parasitoids leading to 86 per cent parasitization in pure stand of larch, but if other trees are grown in association, the parasitization is decreased to 12 per cent due to masking effect of the volatiles produced by associated trees (Monteith, 1960).

Herbivores form genuine link in tritrophic chain and interact with both plants and natural enemies. Thus, they have to cope with the plant's intrinsic defence and the natural enemies attack. They also produce semiochemicals either to cope with the plant defence or to deter natural enemies. However, many kairomones produced by herbivores are attractive to parasitoids and assist them in host location. For example bee wolf, *Philanthus triangulum*, a predatory wasp detect the bee only by using the body odour of the bee (Tinbergen, 1972). Noldus and Lenteren (1985) have shown that the sex pheromone of cabbage moth, *Mamestra brassicae* females were absorbed on to the leaf surface of Brussel sprouts, *Brassica oleracea* var. *gemmifera* to such an extent that it was capable of eliciting behavioural responses from the parasitoid, *Trichogramma evanescens*.

Herbivore produced allomones are also important in tritrophic interaction and these may act against either plants or natural enemies, e.g. gall insects produce chemicals, which result in gall formation, providing favourable habitat for the herbivore. Certain plant produced chemicals may also assist herbivores like tomatine, an alkaloid produced by tomato plant is acquired by corn earworm, *Helicoverpa zea*. This alkaloid produces adverse effects on parasitoids, *Hyposoter exiguae* which results in reduced survival, longevity and delayed larval growth of parasitoid larvae inside the host body (Campbell and Duffey, 1981). Toxic plant metabolites are often sequestered by herbivores and thereby become co-opted by the herbivore for defence against its enemies (Cortesero *et al.*, 2000). Tobacco hornworm, *Manduca sexta* feeding on tobacco, *Nicotiana tabacum* plants retains nicotine in their haemolymph, which detrimentally affects the performance and survival of the endoparasitic wasp, *Cotesia congregata* (Kester and Barbosa, 1991). Some predators release synomones which are acquired from plants. In contrast, some kairomones produced by natural enemies are detected by herbivores and assist them in avoiding or escaping.

The fitness of natural enemies is related to the quality of hosts, which in turn is affected by plant nutrition and allelochemicals so that the diversity and abundance of the hosts influence the diversity and abundance of natural enemies (Dicke *et al.*, 1990). Host plant nutritional characters indirectly alter the fitness of parasitoids through change of herbivore quality (Ananthakrishnan, 1992). Besides the influence of host physiology on the fitness of parasitoids and the effects of plant physiology on the biology of natural enemies, the effects of plant allelochemicals on parasitoid success and fitness are essential aspects, since some of them may prevent normal nutrient use or cause inhibition of enzyme systems (Schoonhoven *et al.*, 1998). In some cases parasitism of herbivores may not be significantly useful for the host plants. Parasitism of caterpillars of large white cabbage butterfly, *Pieris brassicae* by the endoparasitoid *Cotesia glomerata* does not benefit the *Brassica oleracea* host plant. The parasitized *Pieris* caterpillars consume leaf material at a higher rate than do unparasitized caterpillars and may even extend the length of their larval period. In such cases parasitoids may have adapted to utilize a signal that is a by-product of herbivore activity (Coleman *et al.*, 1999).

Chapter **13**

Defence

The concept and mechanism of defence in insects have been discussed variously in literature. The subject has been dealt by considering structural, behavioural, ecological, genetical and physiological aspects. Whatsoever defences adopted or exhibited by insects have evolved because of various selective advantageous pressures in the course of evolution. However, in most cases they use more than one method, either in sequence or in combination. Insects are inveterately persecuted by a number of omnivorous or carnivorous animals like birds, amphibians, reptiles, mammals, spiders and other predatory insects. The prey insects, on the other hand, respond in various ways to avoid or escape the ultimate capture by predators. The diverse and complex mechanisms adopted by insects are incredible and responsible for their dominance on the earth. Such mechanisms are of great survival value and in many cases their protective devices are astonishingly dizzy. In many cases not only predators but also even we have hard time to detect them. The subject has been discussed by Cott (1940), Imms (1947), Wigglesworth (1964a), Edmunds (1974), Matthews and Matthews (1978), Atkins (1978, 1980), Evans (1984), Evans and Schmidt (1990), and Ruxton *et al.* (2005).

Two major types of defence mechanisms have been recognized in insects, viz. primary and secondary. Primary defence operates prior to attack of the prey, thereby decreasing chances of being attacked while secondary mechanism operates when a prey encounters a predator. Insect defences against predators often include both a primary defence that reduces the probability of being attacked and a secondary defence; typically escape behaviour, employed if the primary defence fails. These two mechanisms have also been termed as passive and active defences. Robinson (1969) considered primary defence mechanism as those operate before a predator initiates any prey-catching behaviour. Edmunds (1974) defined primary defence mechanism as 'mechanism, which operates regardless of whether or not there is a predator in the vicinity'.

The defence mechanisms adopted by insects are mainly categorized into mechanical, visual, chemical and behavioural strategies. However, these patterns may overlap with each other as in many cases more than one tactics are adopted by insects for their protection.

13.1 Mechanical defence

The heavily sclerotized integument provide protection against all forms of natural enemies in many Coleoptera. The mouthparts, cerci, setae, spines, spurs are other sclerotized and hard body parts which not only provide protection to insects but also used for conflict. Some insects deceive their enemies by detaching their body parts or coverings. Besides, many construct protective structures and get physical protection.

13.1.1 Structural defence

The highly sclerotized integument in some insects particularly in coleopterans provide effective defence against almost any form of natural attack. Besides, in many of them modified head and its appendages are effective organs of defence (Fig. 13.1a-b). Some ants block the access to the nest entrance with their modified heads and the phenomenon is termed 'phragmosis'. Soldiers of some formicines and many myrmecines made entrance of such a size that the head of single worker could seal it (Fig. 13.2). In some others the entrance is larger and the guarding blockers act as 'gatekeepers'. Many have sharply pointed mouthparts, and other structures, e.g. modified setae, cerci, spines, and forceps. The forceps present at the abdominal tip of earwig picks up adversaries such as ants and throw them to one side with a flick of the abdomen. The adults of tiger moths have poison glands associated with their leg spines that greatly increase their effectiveness. Many caterpillars have a cloak of hair-like setae that make them unpalatable. Short fragments of these setae break off and cause irritation. Cuticular spines or horns may deter a predator or have use in fighting. Laterally compressed body in fleas and dorsoventrally flat

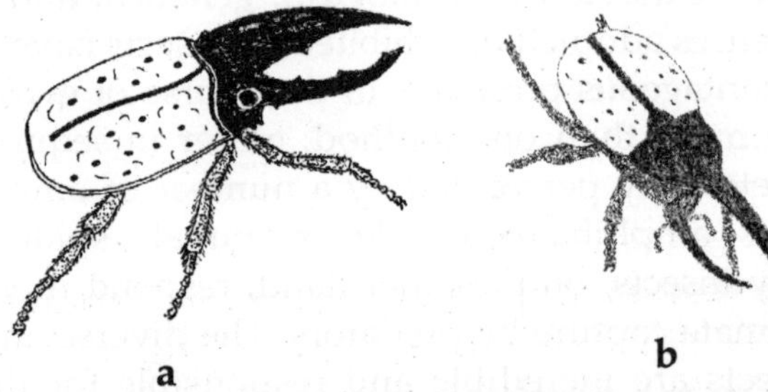

Fig. 13.1. (a)- The Hercules beetle, highly sclerotized integument and modified head are effective organs for defence, (b)- A horn beetle, the highly modified head is used for defence.

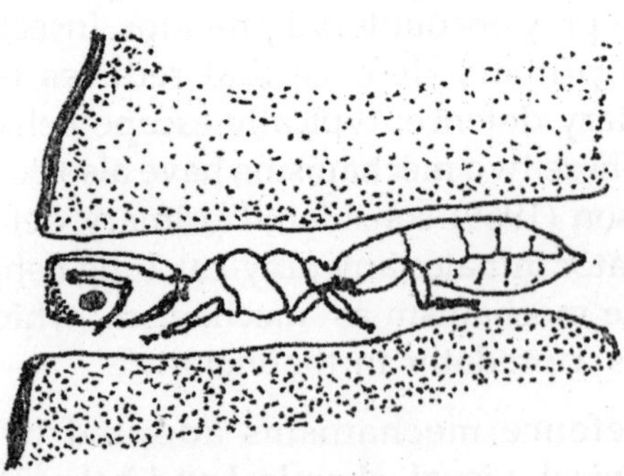

Fig. 13.2. Blocking of the nest entrance with the head by worker of European ant, *Camponotus truncatus*.

shapes in biting lice make these insects difficult to dislodge from hairs or feathers. The raptorial fore legs of mantids (Fig. 13.3, Plate 13.1) and powerful mandibles of biting insects may be employed in defence. Caste skin and faecal material of larvae of chrysomelid beetle, *Cassida rubiginosa* on a fork of posterior abdominal appendages that dries and form an effective faecal shield (Eisner *et al.*, 1967). The larvae of the tortoise beetle, *Eurypedus nigrosignata* also construct faecal shields using cast skins and faecal strands. Survival of larvae with intact shields was higher in the field than for larvae with shields removed. The chemical composition of the larval faecal shield is influenced primarily by the host plants' secondary chemistry.

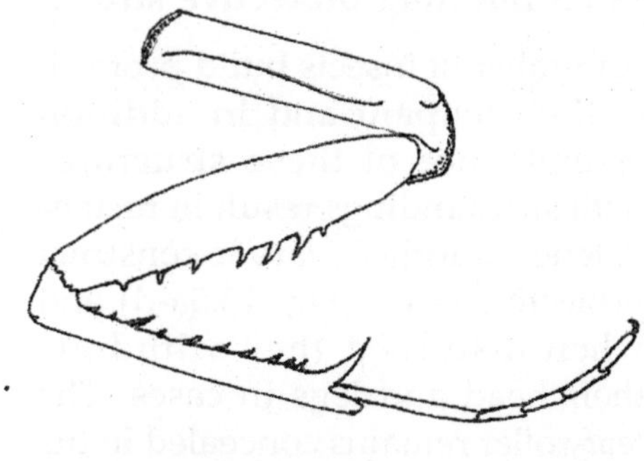

Fig. 13.3. The raptorial fore leg of a praying mantis.

13.1.2 Detachable body parts

Integumentary outgrowths of certain insects detach from their body and may be of protective value. Waxy and powdery coverings of some aphids and whiteflies; hairs of caddisflies; and scales of bristletails, thrips, moths and butterflies assist in their escape. Probably scales first evolved as escape structures and later assumed other functions (Evans, 1984). The caddisflies and lepidopterans are not usually entangled in the spider webs as their scales stick to the sticky strands of the web, thereby permitting these insects to escape. Certain moths shed scales when pecked by birds.

In daddy-long-legs (Fam. Tipulidae) (Fig. 13.4) and stick insects legs break off readily. Immature stick insects and mantids can regenerate lost limbs at moulting, and even certain autotomized adults can induce an adult moult at which limbs can regenerate.

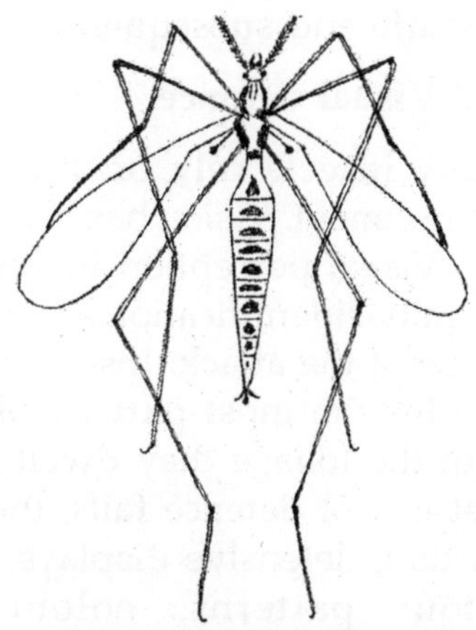

Fig. 13.4. Daddy-long-legs (Fam. Tipulidae). The legs are detached readily from the body and thus divert a predator.

Social hymenopterans, e.g. honey bee can autotomize the sting. When a honey bee stings, the barbed sting apparatus remain embedded in the victim and is ripped away from the bee's body, the bee subsequently dies. The behaviour is rather advantageous to colony defence.

13.1.3 Building protective structures

A number of insects build protective structures that give some physical protection to the occupant and in addition resemblance of these structures with surroundings result in further defence. Caddisfly larvae construct protective cases (Fig. 13.5a-d) and when disturbed they withdraw their head and legs in cases. The leaf-roller remains concealed in the tubular sheaths of leaves. The bagworms (Fam. Psychidae) are prominent case-makers and use bits of leaf or bark, small twigs held together with silken fabric for case making (Fig. 13.6a-b). They also pupate within the case. It has been assumed that the construction of such cases probably evolved in response to environmental pressure and subsequently provided defence against enemies.

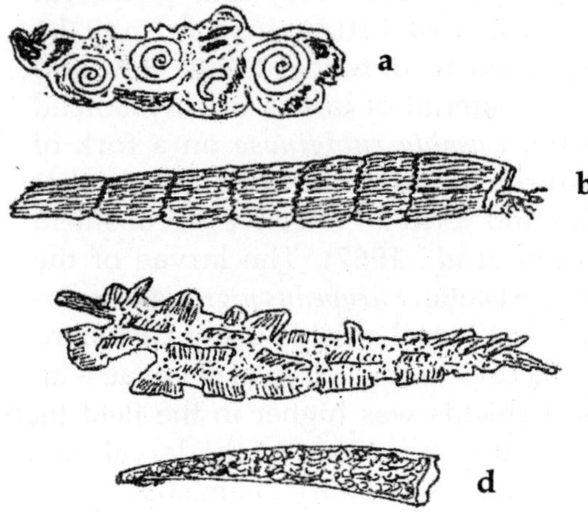

Fig. 13.5a-d. Protective cases of Trichoptera larvae.

13.2 Visual defence

Many insects rely on their visual system to function successfully in their environment. Vision has become incorporated in almost every aspect of the insect's life. Visual perception involves the most effective predatory defence that deters the initial identification and recognition phase of the attack. Insects accomplish this for the most part by blending in with the foliage they dwell in. If their first line of defence fails, they can use startling defensive displays. Protective colour patterns, colour change; resemblance with background, some inanimate object, frightful species; display of unpleasant attributes; flash patterns are amazing behaviours that discourage the attacker.

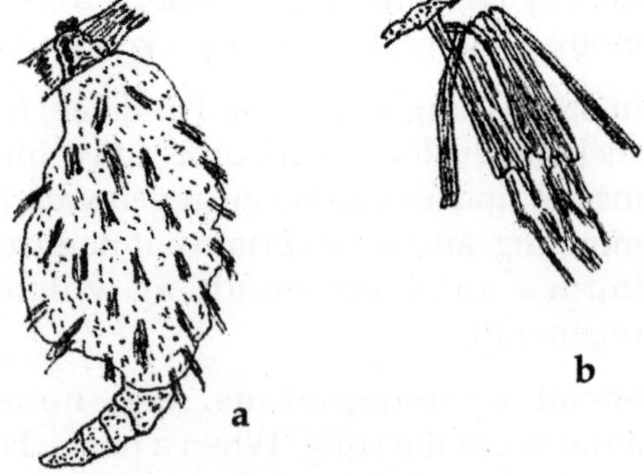

Fig. 13.6a-b. Protective cases of bagworms.

13.2.1 Crypsis

The phenomenon has also been termed as 'camouflage', and the animals that resemble part of the environment are called 'cryptic'. Many insects resemble to inanimate objects in such a way that the predators fail to recognize them or their body pattern, colour, and form, which are similar to the background. The cryptic insect requires selecting an appropriate and best-suited background as the success of crypsis depends on close resemblance between colour and pattern of a species with its background. It also includes appropriate behavioural fashions. Though gut content analysis of some predators indicates presence of cryptic insects; however, further increase in crypsis results in increasing success in escaping when detected by the predators.

Hugh B. Cott (1940) published his monumental work on adaptive colouration in animals, in which he described numerous ways by which colour patterns can serve as defence mechanisms. A number of evidences suggest that protective colouration; form or markings cannot be merely accidental but are results of evolutionary processes. The process of natural selection produces such highly specialized and intricate patterns. The crypsis may involve 'protective colouration' or 'protective resemblance' to inanimate objects or both. Atkins (1978) referred to these as 'homomorphism' (pertaining to shape), 'homochromism' (pertaining to colour), and 'homotypism' (pertaining to both form and colour).

For success of crypsis, habitat selection is an important consideration, e.g. a green insect may be cryptic in green vegetation but becomes conspicuous in other sites. Sargent (1968) conducted experiments for the background colour choice by moths and observed that dark moths prefer to rest on black surface while pale moths rest often on white surface. He concluded that background selection in experimental moths is genetically determined. In some species different forms are found which rest on different coloured background. Kettlewell (1955a) showed that melanic peppered moths, *Biston betularia* rest on black whilst the typicals rest often on white backgrounds. Many insects also select a proper background by comparing its own body colour with that of background and if the posterior half of eye is painted they cannot see their own colour and the background preference ability is lost.

For an effective crypsis some insects exhibit 'disruptive colouration' and 'countershading'. Disruptive colouration is a visual breaking of the insect's outline so that parts of it appear to fade separately into the background, whilst countershading is compensatory deepening or lightening of body colour to counteract for apparent colour changes due to light intensity (Matthews and Matthews, 1978).

13.2.1.1 Protective colouration

Protective colouration is 'the colouration or colour pattern of an insect that affords it protection from the observance of a predator' (Plate 13.2). The most common form of protective colouration is the similarity of colour between the insect and its surroundings, which allows the insect to remain undetected when in the background of its habitat. A form of colouration that is similar in purpose is 'disruptive colouration', in which irregular patches of contrasting colours serve to distract the observer's eye from the outline of the insect. Another type of protective colouration occurs in insects whose colouration or markings purposely contrast with their habitat in order to serve as a warning of the toxic qualities of species to the predators. These forms of colouration increase the statistical chance for survival of the species. Protective colouration is usually based on qualities of the natal environments, if the species is placed into a foreign environment that bears little or no resemblance to the natal environment the protective colouration will not aid in the survival of the species; the insect may no longer be concealed.

Many green coloured insects derive protection from resemblance to the foliage of their surroundings (Plate 13.3). The green grasshoppers, bugs, aphids gain protection by virtue of their colour. Crypsis is most evident in the banded-winged grasshoppers. These ground-dwelling grasshoppers superbly match the colour of the soil they live on. The pallid-winged grasshoppers, living on red soil have a predominantly reddish colour, those living on white soil are white, and those on dark or black soil are dark brown or black. Species of the genera *Achurum* and *Mermiria* (Subfam. Acridinae) live and feed on grasses. They are long and slender and typically have bands running the length of the body, mimicking grass stalks. It is hard to observe them by a casual observer when one of these insects clinging to a stalk of grass. The grasshopper, *Philippiacris* from rocky desert areas of Chile has mottled colour perfectly matching with background colour (Atkins, 1980). The creosote bush grasshopper spends all its time among leaves of the creosote bush. It is olive green, with shiny, pearly spots resembling the green leaves of creosote bush with their shiny, oily secretions. The creosote bush katydid, *Insara covilleae* also lives solely on creosote bushes; it too is olive green with pearly, shiny patches. The Indomalayan orchid mantid, *Hymenopus coronatus* blends beautifully with the pink flower spike of an orchid, where it sits waiting for prey (Plate 13.4). The mantid's femora resemble with flower's petals. Crypsis enables the mantid to avoid detection by its prey as well as conceal itself from predators.

13.2.1.2 Protective resemblance

Protective resemblance is common in many insect groups. This is another form of crypsis and also termed as 'masquerade' or 'mimesis' in which the organism deludes a predator by resembling an object that is a particular specific feature of its environment, but is of no inherent interest to a predator. Katydids are nocturnal insects, which use their cryptic colouration to remain unnoticed during the day when they are inactive (Fig. 13.7). They remain perfectly still, often in a position that

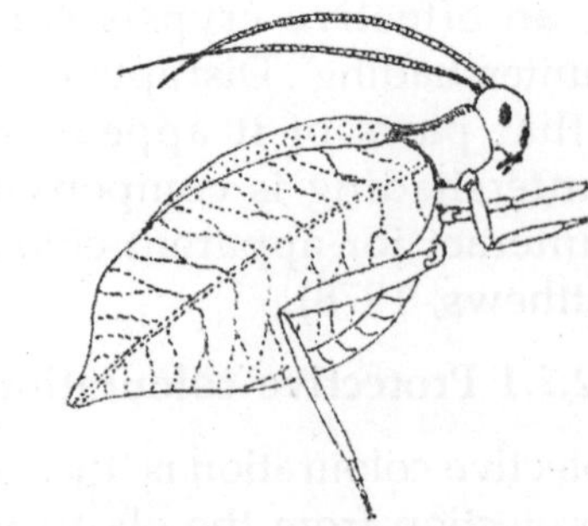

Fig. 13.7. Katydids with fore wings resembling a leaf.

makes them blend in even better. Katydids have evolved to the point where their body colouring and shape matches leaves, including half eaten leaves, dying leaves, and leaves with bird droppings, sticks, twigs, and tree bark. The stick insects (Fig. 13.8) resemble colour of foliage and twigs on which they rest (Plate 13.5a,b). In spring they are green and become brown in autumn. The leaf insects (Order Phasmida) have striking resemblance with leaves (Fig. 13.9, Plate 13.6). Some long-horned grasshoppers (Subfam. Pseudophyllinae) simulate leaves, giving an exact appearance of the leaf on which they commonly rest (Fig. 13.10). Praying mantids are usually green and blend with the foliage; however, these may have close resemblance with flowers, twigs, debris or dead litter (Plate 13. 7). Many moths resemble with bark, leaf (Plate 13.8) or feather (Plate 13.9) during rest. Geometrid caterpillars, stick insects, mantids resemble sticks and grasses and avoided by predators. Geometrid caterpillars often resemble twigs both in colour and form (Plate 13.10). They are so mixed in surroundings that sometimes it is rather hard to find them. The buff-tip moth, *Phalera bucephala* when resting gives an appearance of a broken pale-coloured piece of a dead twig. The moth is difficult to detect because of a disruptive pattern at the wing-tips (Imms, 1947). Young caterpillars of citrus butterfly, *Papilio aegeus* resemble with bird-dropping. Scale insects resemble gall-like or other structures on stems of woody plants. Among flies many resemble wasps or bees. The syrphid flies particularly have black and yellow bands similar to many hymenopterans. The drone fly, *Eristalis tenax* is usually mistaken as a honey bee. The deadleaf butterflies of the genus *Kallima* have extraordinary perfect resemblance to leaves, which is exhibited by undersurface of the closed wings (Plate 13.11). Many other butterflies have close resemblance to dry or green leaves (Plate 13.12 and 13.13). Many bugs have a close resemblance with ants. Some moths have transparent wings, e.g. in bee hawk moth scales fall off during first flight and they come to resemble bumble bees.

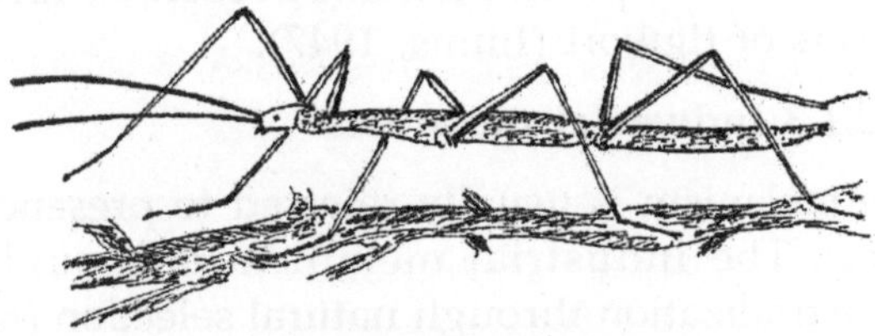

Fig. 13.8. A stick insect.

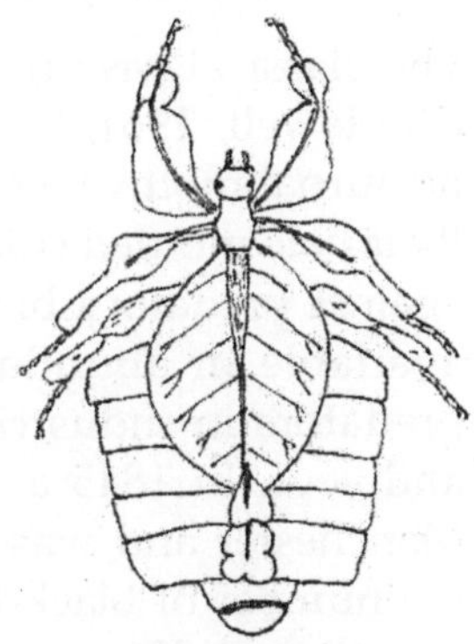

Fig. 13.9. A leaf insect, *Phyllium* sp.

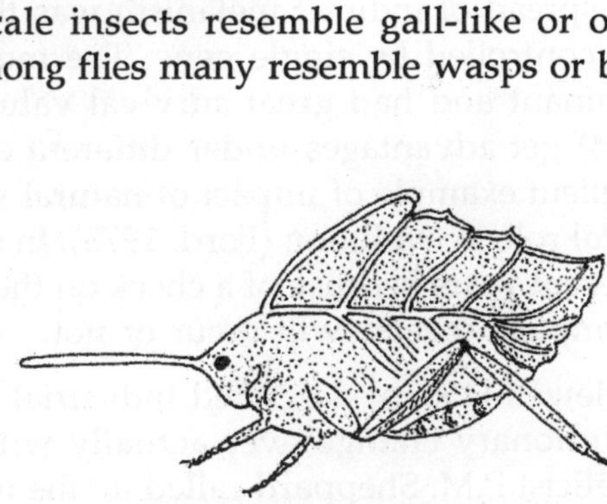

Fig. 13.10. Long-horned grasshopper (Subfam. Pseudophyllinae), the wings resembling with leaves.

The larvae of *Cionus* weevils resemble green tissues of figwort leaves except mid ribs and veins. The full-grown larvae ascent among the developing seed capsules and form neat parchment-like cocoons, which resemble in colour and size to seed vessels of figwort (Imms, 1947).

13.2.1.3 Industrial melanism

The melanism is usually referred to presence of dark pigment in different body parts. The industrial melanism is brought about by the indirect effects of industrialization through natural selection (Kettlewell, 1955b, 1961). Air pollution is underlying cause of this evolutionary change, and so the phenomenon is called 'industrial melanism'. Both the genetic and environmental factors have been considered responsible for the appearance of melanics. The melanics depend on crypsis for their survival and occur in those species, which normally rest in exposed situations. Ford (1955) referred to it as 'transient polymorphism'.

The classical case of peppered moth, *Biston betularia* has been studied in detail (Kettlewell, 1961, 1973), Bishop (1972) and Bishop and Cook (1975). Earlier the nocturnal moths rest quietly during day on tree trunks covered with lichens and their pale mottled colour blended with the background, thereby getting protection against predatory birds. By the middle of 19th century such normal forms were available in abundance in Great Britain. These forms were conspicuous to predators in industrial areas when resting on tree trunks covered with smoke and soot. In 1845 a black or melanic form was observed in industrial area of Manchester and was termed as 'carbonaria'. Subsequently there was increase in the number of black form mainly in the vicinity of industrial area and by 1895 it constituted 95 per cent of the moth population of this region. By 1950, their population became nearly 99 per cent around Manchester. While in non-industrial areas of the countryside the light-coloured form 'typica' was clearly distinct.

The spread of indirect melanism was the direct consequence of pollution fallout and controlled by single gene. The melanic character inherited as a Mendelian dominant and had great survival value. The melanic and normal forms (Plate 13.14) get advantages under different environmental conditions and provide an excellent example of impact of natural selection on biological adaptation, playing a vital role in evolution (Ford, 1975). In some areas reverse has also been reported in recent years because of a check on the pollution and the time will state whether a complete reverse will occur or not.

Kettlewell (1955b) described industrial melanism in moths as "the most striking evolutionary change ever actually witnessed in any organism." Later, British geneticist P.M. Sheppard called it "the most spectacular evolutionary change ever witnessed and recorded by man, with the possible exception of some examples of pesticide resistance" (Sheppard, 1975), and famed population geneticist Sewall Wright called it "the clearest case in which a conspicuous evolutionary process has actually been observed" (Wright, 1978).

It was reported that the moth colour is inherited and it is controlled by a polyallelic gene where the alleles for black colour are progressively more dominant to the allele for peppered colour. Following points were considered in the evolution of melanic forms of peppered moth.

(i) The selective pressure came from predators (insect eating birds). The moth's colour provides camouflage.

(ii) Moths rest during the daytime on trees.

(iii) In rural areas where the air is clean the trees are covered in lichen. The peppered variety is better camouflaged when it rests on these trees. The melanic form is easily seen.

(iv) In industrial areas air is polluted and the lichens die.

(v) The tree trunks may also be covered in soot giving a black background.

(vi) This favours the melanic form giving them a selective advantage.

Majerus (1998) suggested that the peppered moth story is more complex than original thoughts and evidences gathered in the 40 years since Kettlewell's work shows that "the basic peppered moth story is wrong, inaccurate, or incomplete, with respect to most of the story's component parts." What follows is a review of some of the flaws that have been discovered in the classical story. Kettlewell's successors, found unexpected under mentioned problems with his conclusion that industrial melanism is due to cryptic colouration and selective predation.

1. The geographic distribution of melanic peppered moths did not fit the theory: the frequency of melanics was not as high as it should have been in some places, and higher than it should have been in others.
2. Melanism is not correlated with lichen cover, in the U.K., it declined before lichens returned to the trees, while in the U.S., it occurred despite the presence of lichens and declined without any perceptible changes in lichen cover.
3. Peppered moths do not normally rest on tree trunks: instead, they normally rest under horizontal branches high in the canopy, not where Kettlewell and his successors had carried out their experiments on selective predation.

Cook (2000) proposed that in peppered moth story (i) nonvisual components of selection have not been investigated directly in this species, and (ii) the experimental and observational evidence cannot on its own carry the burden of a particular view of evolution. However, the peppered moth story continues to provide an exceptional opportunity to analyse a pattern of selection. Majerus *et al.* (2000) suggested that early work evaluating the relative crypsis of the forms of *B. betularia* on tree trunks with different lichen flora was the reliance on human observers. Humans; however, do not have the same visual capabilities as birds. Birds have well-developed UV vision, an important component of their colour processing system that affects many aspects of behaviour, including prey detection. They examined the UV characteristics of the two forms of *B. betularia* and a number of foliose and crustose lichens. In human visible light the speckled form typica appeared cyptic when seen against a background of foliose lichen, whereas the dark form carbonaria was conspicuous. Under UV light the situation was reversed. The foliose lichens absorbed UV and appeared dark, as did carbonaria. Typica; however, reflected UV and was conspicuous. Against crustose lichens, typica was less visible than carbonaria in both visible and UV light. These findings were considered in relation to the distribution and recolonization of trees by lichens and the resting behaviour of *B. betularia.*

Judith Hooper (2002) strongly opposed Kettlewell's results. Besides, some experts on moth behaviour remain convinced that the story of the peppered moth is sound. Kettlewell was a distinguished naturalist whose studies on predation in peppered moths were a landmark in demonstrating natural selection in the wild. His classic studies are widely accepted and quoted in most texts of natural selection, ecology, evolution, population genetics, and defence mechanisms. More pertinently, Kettlewell's data are completely consistent with normal experimental variation, and Hooper's insinuations are groundless. Hooper's claims are moonshine; they are based on a lack of understanding of Kettlewell's experiments in particular and experimental science in general (Young and Musgrave, 2005).

The melanic mutation of the peppered moth is not the only example of industrial melanism, nor is the phenomenon confined to urban England. More than 70 species of night-flying moths in Britain have exchanged light markings for dark ones, and the same is true of moths in industrial areas of Europe, Canada and Unites States (Bishop and Cook, 1975). Other well-known case of pale brindled beauty moth, *Apocheima pilosaria* occurring in Britain and nearby countries was discussed by Less (1971). Both sexes of the moth are polymorphic for two melanic forms. The females are with 2-4 mm long vestigial wings and creeps into crevices of the bark. One form 'monacharia' is intensely black, while other intermediate forms with less black colour have also been reported. In industrial areas black colouration becomes an advantage to this cryptic species, and factors like mean bark reflectance, smoke and rainfall are considered to be responsible for development of melanics.

The two-spotted lady beetle, *Adalia bipunctata* exhibits widespread polymorphism in Britain with three main forms namely, 'quadrimaculata' (black with four red spots on each elytron), 'sexpustulata' (black with three red spots on each elytron), and 'typical red form' (with one black spot on each elytron). Presently a number of colour morphs of *A. bipunctata* (Fig. 13.11) are known and over 100

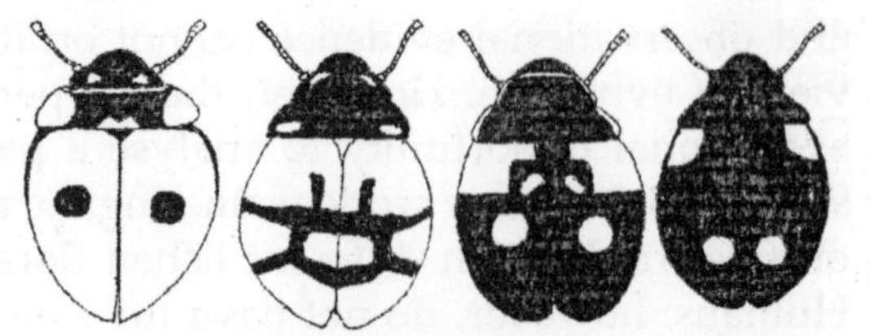

Fig. 13.11. Some colour morphs of *Adalia bipunctata*.

have been named. Creed (1966, 1971) suggested that phases of beetle are associated with atmospheric pollution produced by different types of coal. The beetles get further protection by their warning colouration. In this case the industrial melanism has nothing to do with cryptic colouration and selective predation. Ladybird beetles are extremely distasteful to most birds, and exhibit warning colouration rather than cryptic colouration; there is no evidence that they are significantly affected by selective predation (Ford, 1975). Industrial melanism in ladybird beetles has been attributed to non-visual selection on the ability of melanic and non-melanic forms to absorb solar radiation- a phenomenon known as 'thermal melanism' (Brakefield, 1985).

13.2.2 Mimicry

British naturalist Henry Walter Bates (1862) made some remarkable observations in the Amazon area and proposed the concept of mimicry that has been termed as 'Batesian mimicry'. Later other forms of mimicry were enunciated by various workers and valuable contributions on the subject have been made by Carpenter and Ford (1933), Carpenter, (1946), Ford (1965, 1975), Clarke and Sheppard (1960), Rothschild (1961, 1967, 1971), Brower (1958), Brower *et al.* (1963, 1967, 1971), Rettenmeyer (1970), and Alcock (1971). Pasteur (1982) published a classificatory review of mimicry systems.

Wolfganag Wickler's (1968) book '*Mimicry in Plants and Animals*' includes nearly all types of defence mechanisms. However, the term has restricted applications as that originally proposed by Bates in which a palatable species has evolved a colour pattern and/or behaviour similar to that of a distasteful species. Mimicry is the resemblance of one palatable organism (the mimic) usually in colour, pattern or behaviour to another unpalatable or dangerous (the model).

13.2.2.1 Batesian mimicry

In Batesian mimicry natural selection favour the evolution of distinctive colouration in species unpalatable to predators, as the distinct colour pattern provides a clue to the predator regarding distastefulness of the prey. Because of this prey is less prone to the predator's attack and predator tend to search other useful prey. Genes producing changes in colour pattern may need a fairly large selective advantage before mimicry can evolve. The next step in the evolution of mimicry is mimicking by the palatable species, which is brought forth by natural selection. Other factors like numerical balance between the model and mimic, degree of distastefulness of model, predation pressure, their sympatric occurrence and genetic and ecological factors are also important for Batesian mimicry to remain affective (Rettenmeyer, 1970).

Viceroy butterflies, *Basilarchia archippus* (Plate 13.15) provide the best-known example of a mimic. They have a colour pattern similar to orange wing pattern of monarch butterflies, *Danaus plexippus* (Plate 13.16) belonging to a separate family

(Fig.13.12a-b). The monarch caterpillars (Fig. 13.13) feed on milkweed (Fam. Asclepiadaceae), many of which contain cardenolides (cardiac glycosides). These chemicals are incorporated in monarch body until the adult stage and if eaten by birds, the birds vomit and get ill for a short period (Brower, 1969). Grazing animals usually avoid the milkweed plants; however, certain species of milkweed are without or very little concentration of toxins and monarchs feeding on them are palatable to birds. Thus, the concentration of cardenolides also varies in monarch butterflies and they show variation in palatability depending upon the milkweed species; however, they have perfect resemblance and birds avoid all. Brower (1969) termed this intraspecific mimicry as 'automimicry' in which one species includes both models and mimics. Brower further tested the phenomenon by providing viceroy to inexperienced Florida jays. Birds readily ate these butterflies and when monarchs were provided suddenly, the jays tested and rejected them. After a few such repetitions jays rejected both. Finks and Brower (1981) demonstrated that maximum concentration of toxin is present in wings of monarch and in some cases birds avoid toxin by simply eating the thorax and abdomen muscles, which contain the minimum toxin.

Fig. 13.12. Mimicry in butterflies. (a)-Monarch butterfly, *Danaus plexippus*, (b)-Viceroy butterfly, *Basilarchia archippus*. The monarch butterfly (model) is unpalatable to birds and gets protection. The viceroy butterfly is palatable, but birds avoid it because of its resemblance with monarch butterfly.

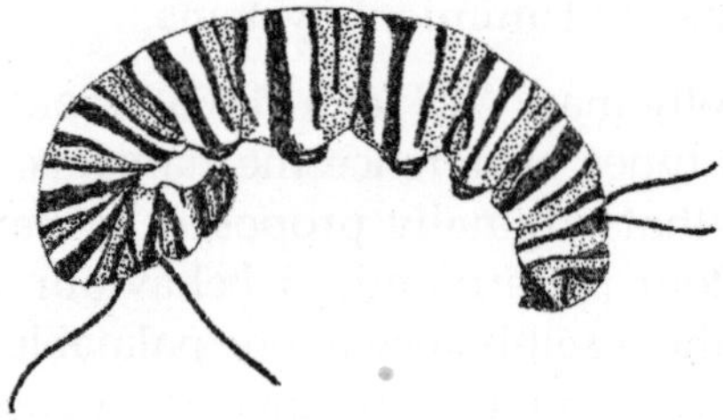

Fig. 13.13. Caterpillar of monarch butterfly, *Danaus plexippus*.

The resemblance between monarch and viceroy is not the result of a close genetic relationship. They belong to different groups and their habits are also very different. The monarch migrates over three to five generations between summer sites in the northern United States and Canada, and overwintering sites in Mexico. The viceroy is territorial, overwinters in its home environment at the first and second growth stages (instars) of the caterpillar, in a form of hibernation (diapause). The caterpillars are not at all similar. The viceroy caterpillar resembles a bird dropping while the monarch caterpillar is striped orange, yellow, and black, with horns on both the head and tail.

Another extensively studied instance of mimicry in Lepidoptera is provided by swallowtail butterfly, *Papilio dardanus* in Afrotropical region. The males are invariably tailed, monomorphic and non-mimetic. In its distributional range eight races of butterflies are recognizable based on wing markings and male genitalic armature. Some of these races interbreed and natural selection has also produced a number of morphs in these races, e.g. race 'polytrophus' has 13 morphs alone. A number of factors need consideration to answer the probable existence of so many forms in *Papilio dardanus*. The adults are palatable to birds and take advantage of several sympatric models to avoid predation. For each model a different mimic morph has evolved and highly remarkable similarity between model and mimic has been observed. Studies on genetics of these morphs and mimetic patterns of *Papilio dardanus* have been discussed by Clarke and Sheppard (1960) and Sheppard (1962).

Brower *et al.* (1967) investigated mimicry by using artificial mimics and wild predators in their habitat. The diurnal palatable North American silk moth, *Callosamia promethea* were taken to Trinidad, where they did not occur earlier. Their wings were painted artificially to produce some colour pattern as that of local unpalatable butterflies. The results indicated that the experimental mimics survived better in natural environment than the non-mimetic controls. However, this protective advantage is lost if the relative frequency of the model becomes too low.

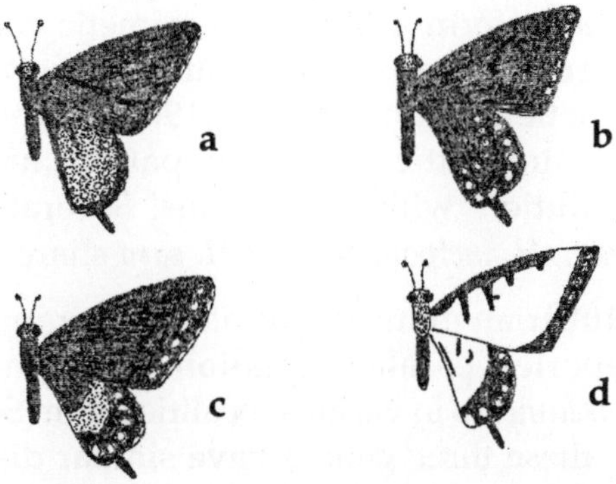

Fig. 13.14. *Battus philenor* and some of its mimics. (a)- Male *Battus philenor* (the model), (b)- Female *Papilio glaucus*, (c)- Male *Papilio troilus*, (d)- Male *Papilio glaucus*. *Papilio troilus* is the mimic and only the female of *P. glaucus* is mimetic.

Excellent examples of mimicry are found in United States. The model pipevine swallowtail, *Battus philenor* and mimetic black form of tiger swallowtail, *Papilio glaucus* coincides with areas of abundant *B. philenor*. Non-mimetic *P. glaucus* are abundant in areas where model is rare (Fig. 13.14). Jays experienced with the model also avoid mimetic forms of *P. glaucus*. Besides, *Limenitis astyanax* and *Papilio troilus* are also close mimics of *B. philenor*.

13.2.2.2 Müllerian mimicry

German Zoologist Fritz Müller (1879) first noticed its occurrence in Brazil. It involves a resemblance between a numbers of relatively protected species, which by their uniformity of appearance reduce the depredations made by predators. In the New World Tropics, Müllerian mimicry rings can involve literally hundreds of species. The basis of Müllerian mimicry lies in the response of the receiver; if the predator has the ability to learn to avoid a species due to previously failed attempts, then the mimic can gain protection against the predator by using the same warning signals as the species that the predator previously encountered. Therefore, by copying a similar pattern of warning colouration and markings, various emetic species can increase their protection against would-be predators by reducing the variation in warning patterns, which must be learned by these predators. A single lesson of distastefulness by the predator is shared in varying proportion by the whole group (Ford, 1953).

The tiger stripe butterflies complex in Neotropical region includes sympatric members of different families and all these unrelated butterflies are distasteful. Such a pattern is advantageous as predators because of common warning signal avoid all. The butterflies belonging to this complex are characterized by yellow, brown, black and orange stripes on wings and mostly belong to different families. In addition several species of *Heliconius* are also included in this complex. These species form a Müllerian mimetic complex in which all butterflies have varying degree of unpalatabiliy and are with specious similarities. Morphological and genetic work of Emsley (1965) supports the hypothesis that in genus *Heliconius* the similarities of colour pattern are due to close relationship and 'explosive evolution' within the genus. Several species like *Heliconius aoede, H. elevatus, H. erato, H. melpomene* and *H. sara* share common colour pattern (Brower *et al.*, 1963).

Müllerian mimicry has also been reported in insects of other groups. Vecht (1961) reported parallel variation in colour pattern in wasps *Eumenes, Pseumenes* and *Pareumenes* in various localities from South India to New Guinea. Sympatric species of these three genera have similar distinct black and yellow patterns that tend to decrease or become monocoloured in marginal or isolated populations. Other examples are found in lycid beetle complex (Linsley *et al.*, 1961; Emmel, 1965). Lycid beetles are often brightly coloured and discarded as food by predators due to nauseous smelling fluid secreted from hind femora. In North America and some other areas two or more species of beetles occur together forming a Müllerian complex. Besides, some cerambycid beetles, lepidopterous and other insects may also have similar colour patterns. Of these, the palatable are Batesian mimics, while unpalatable are Müllerian mimics.

In these two types of mimicry selection is operating quite differently. In Batesian mimicry resemblance of mimic to model is disadvantageous to model. Consequently there will be divergent selection away from the mimic. Whereas, in Müllerian mimicry there will be convergent selection of both model and mimic toward a common pattern (Ruiter, 1958). For predator Batesian mimicry is disadvantageous as they avoid palatable prey and Müllerian mimicry is advantageous because they do not waste time in pursuing unpalatable prey.

13.2.2.3 Wasmannian mimicry

The resemblance that facilitates co-inhabitation of mimic and model with its host (the model) has been termed Wasmannian mimicry (Rettenmeyer, 1970). It is a true deceptive mimicry in which host species is both model and selective agent, while the inquilines or commensals are mimics and resemble with hosts. In ant colonies a number of beetles and other guests live. Seevers (1965) listed 33 species in 11 genera of staphylinids found with Old world doryline ants, and 47 species in 27 genera with New world Dorylinae. Similarly, in termite nests some guests have a close resemblance with termites.

Aggressive mimicry wherein predator mimics their prey (the model) to facilitate prey capture is advantageous to predator and disadvantageous to prey. However, the reverse in which prey mimic their predators to avoid predation is also reported in some cases. Such behaviour in which the prey is a proverbial 'sheep in wolf's clothing' has been reported in two tephritid fruit flies *Rhagoletis zephyria* and *Zonosemata vittigera* (Mather and Roitberg, 1987; Greene *et al.*, 1987). In these flies both the wing pattern and wing-waving displays may be important for effective defence and they resemble with predatory jumping spiders. This striped wing patterns are similar to markings on spiders' legs. This behaviour is quite different from the classical Batesian and Müllerian mimicry.

13.2.3 Aposematism

E. B. Poulton (1890) coined the term 'aposematism'. Certain insects exhibit dangerous or unpleasant attributes by means of characteristic structures, colour or signals and thereby avoided by predators. The aposematics that display their dangerous attributes by peculiar colour and such features have been often termed as 'warning colouration' (Plate 13.17). Black and yellow are prominent colours in insects having warning significance.

Why should aposematic signals usually be conspicuous, and use simple colour patterns of red, yellow or black? Several hypotheses have been put forward to explain the correlation between bright colours and unprofitability in insects. Bright contrasted colours are thought to be: (i) easier to learn, (ii) more difficult to forget, and (iii) as different as possible from edible prey, thereby avoiding recognition errors. Because both predators and aposematic prey benefit from correct identification, aposematic colours are believed to take advantage of any bias in the predator's cognition system. However, some distasteful preys are not particularly conspicuous (Joron, 2003).

Aposematics may also have additional secondary mechanism like sting, distasteful body, etc. Many hymenopterans have painful stings and conspicuous coloured bands on abdomen, which provide warning signals to predators. Birds and mammals often reject soldier beetles and ladybird beetles with brightly coloured and spotted elytra. Predators easily learn the colour patterns of butterflies and beetles. Certain insects have characteristic smell; others produce warning sounds and these characters associated with bright colours possibly act as warning signals. Brightly coloured bees and wasps while crawling produce an effective warning buzz. Many aposematics are gregarious so as to give a more conspicuous effect.

Colour patterns of unpalatable butterflies along with presence of toxins in body provide them effective defence. Caterpillars with long stinging hairs vibrate these hairs noticeably while crawling and are unacceptable to predators. Some caterpillars break off the hairs, which enter the skin of predator and cause irritation. The Bihar hairy caterpillars of *Spilarctia obliqua* moths have hairs (Plate 13.18) on body and live in aggregated condition and remain protected up to 2nd instar, and when become somewhat larger they disperse and feed apart (Plate 13.19). Some apterous females of velvet ants walk over the ground and have cryptic resemblance with sand; while others have bands of brilliant velvety-yellow, golden or red pubescence. Additional protection is provided by sting and tough integument. The harlequin or rainbow grasshopper, *Dactylotum variegatum* is small, short-winged, and black with bright blue, red, yellow, and white markings. They sequester toxins from the plants they eat, making them unpalatable for most predators. Their aposematic colouration informs potential predators that they are poisonous. Whatever route to distastefulness is taken, a general correlation between clades of distasteful insects and toxicity in host plant families is observed. In butterflies, the distasteful Troidinae (Papilionidae) tend to feed on Aristolochiaceae, monarchs (Nymphalidae: Danainae) usually feed on milkweeds (Asclepiadaceae), longwing butterflies (Nymphalidae: Heliconiinae) feed on Passifloraceae, while clearwings (Nymphalidae: Ithomiinae) mainly on Solanaceae and Apocynaceae. In contrast, butterfly clades feeding on chemical-free plants did not evolve distastefulness. In many cases bioluminescence can only have evolved through a defensive function. Light signals could be used in many ways to reduce predation, but for spontaneous glowing species in particular, aposematism seems the main functional defence strategy.

For effective aposematism either the predator must learn to avoid the prey or it must have an innate pattern to avoid aposematics. In the former case a few prey have to sacrifice themselves and the predator tend to avoid them after a few unpleasant experiences. Variations in unpalatability among prey species, along what is called the 'unpalatability spectrum', affect the rate at which predators modify their behaviour with experience.

13.2.4 Colour change

The phenomenon of adaptive colour change is not much pronounced in insects. However, in some cases it occurs by movements of pigments in integument due to metabolism of pigments or changes in environmental conditions. By changing its colour an insect can harmonize with two or more different backgrounds but the change appear gradually.

In stick insect, *Carausius* individuals are brown in daytime and become black at night due to movement of pigments in epidermal cells (Dupont-Raabe, 1957). Similar colour change is also observed in Australian grasshopper, *Kosciuscola tristis* in response to temperature. Above 25°C males are blue and below 15°C they are dull black (Key and Day, 1954). Colour patterns of pupae of large white cabbage butterfly, *Pieris brassicae* are determined in advance by the background colour perceived in the larval stage (Fig. 13.15). Basically, the pupa is green but assumes the whitish-grey colour with black pigmentation if reared on black background, and destruction of larval eyes always resulted in the formation of green pupa (Wigglesworth, 1964a).

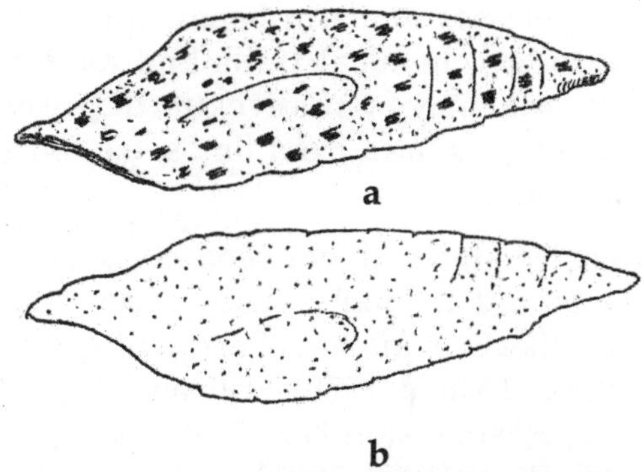

Fig. 13.15. Pupae of large white cabbage butterfly, *Pieris brassicae*, (a)- Whitish-grey with enormous black pigmentation, reared on black background, (b)- Green with less pigmentation, reared on green background.

Many orthopterans can adopt whitish-grey, yellow, brown, greenish-brown, brownish-black or black colour if their adults are kept on corresponding coloured backgrounds. Food, temperature, humidity, crowding, and aging also influence their body colour. The *Acrida* change form straw-yellow to green and *Aulacobothrus* from grey to coal black within three days or more if placed on appropriate background (Uvarov, 1966). The riparian earwig, *Labidura riparia* appears paler in arid zones of Rann of Kutch while it has blackish-brown colour in other areas (Ramamurthy, 1966). African armyworm moth, *Spodoptera* species exhibit phase variation of dark and velvety green to pinkish-grey colour analogues to that seen in locusts. Similar colour change patterns have also been reported in pupae of pipevine swallowtail, *Battus philenor* and common mormon, *Papilio polytes* (Clarke and Sheppard, 1972).

In American grasshopper, *Syrbula admirabilis* green nymphs are formed if fed on a diet with high moisture content and changed to brown with low moisture diet. In nature grasshoppers in green vegetations thereby indicate adaptive colouration. Some grasshoppers and mantids produce black forms on burnt ground and termed as 'fire melanics'. The acridid, *Aulacobothrus wernerianus* turned black within 2-4 days after being kept on burnt ground (Burtt, 1951).

13.3 Chemical defence

Many insect species have evolved mechanisms to discharge odourous or irritating fluid when handled, disturbed or attacked. The chemical defence systems adopted by insects' exhibit great diversity accompanied by morphological, metabolic and behavioural specializations. These chemicals are synthesized either within the insect (endogenous) or may be obtained from some outside source (exogenous). The endogenous compounds are probably produced metabolically as in many insects food does not contain such compounds. They are usually expelled out of the body in various forms to avert the predator. The exocrine constituents that confer protection to insects have also been termed as 'defensive allomones'. Insects use certain exocrine compounds both as pheromones and deterrent allomones. Formic acid, the earliest known secretion of ants is used both as effective deterrent and a releaser of alarm behaviour. Similarly, benzaldehyde is used by stingless bees to lay chemical trails and as defensive chemical. The chemical defence mechanism may by passive- the defensive chemical is released from haemolymph; secretive- exudates from limbs, e.g. blister beetles; or active- ejected or exploded from body.

Extensive studies have been conducted on effectiveness of chemical defence in insects and in many cases these chemicals have been identified and their biosynthesis is known. Pertinent contributions to this effect are: Pavan (1959), Kaiser and Michl (1958), Roth and Eisner (1962), Beard (1963), Eisner and Meinwald (1966), Weatherston (1967), Weatherston and Percy (1970), Eisner (1970), Duffey (1977), Blum (1978, 1981), and Schmidt (1982).

13.3.1 Defensive structures

The structures that produce defensive secretions vary extraordinarily in insects. Many such secretions are stored in glands and in other cases haemolymph or enteric discharges are utilized as defensive secretions. The glands are multicellular, located in different body regions and each consists of a reservoir and a duct opening outside. In many cases highly toxic substances are produced in these glands. The reflex bleeding has also been reported in many Orthoptera, Coleoptera, Lepidoptera and other groups, in which small drops of fluid or blood are released from various points of their body, due to an increase in hydrostatic pressure. In *Dictyophorus* (Order Orthoptera) the blood is mixed with air to form brownish foam (Fig. 13.16). Most meloids and aphids also exhibit similar reflexes. Many caterpillars have 'urticating hairs', containing irritating substances secreted by glands located at their bases. The secretion is expelled out by muscular compression of reservoir, due to increase in fluid pressure in haemocoel, muscular compression of body parts or regions, combination of muscular and hydrostatic pressure, eversion of gland or reservoir, action of tracheal system spray mechanism, by special structure, etc. Smedley *et al.* (2002) reported that the pupal surface of herbivorous coccinellid beetle, *Subcoccinella vigintiquattuorpunctata* bears glandular hairs that produce a secretion known to consist principally of three polyazamacrolide alkaloids and serves as a potent antipredator defence.

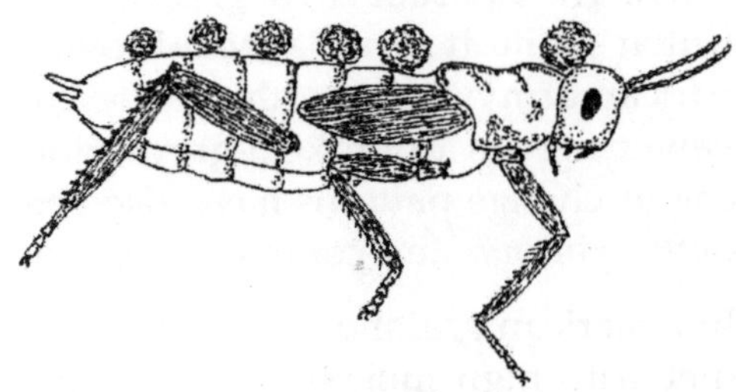

Fig. 13.16. Short horned grasshopper, *Dictyophorus* frothing out an offensive brown secretion from glands.

The use of poisonous apparatus for defence is well seen among bees and wasps, where poison glands associated with ovipositor secrete venoms, and the ovipositor is transformed into sting. The caterpillars of some papilionid butterflies have a forked organ called 'osmeterium' (Fig. 13.17) which is withdrawn into a pouch on the prothorax. On disturbance the osmeterium is everted and a repellent odour from a volatile substance is emitted. One well-known example of caterpillar with osmeterium is the third instar caterpillar of zebra swallowtail, *Eurytides marcellus* (Damman, 1986).

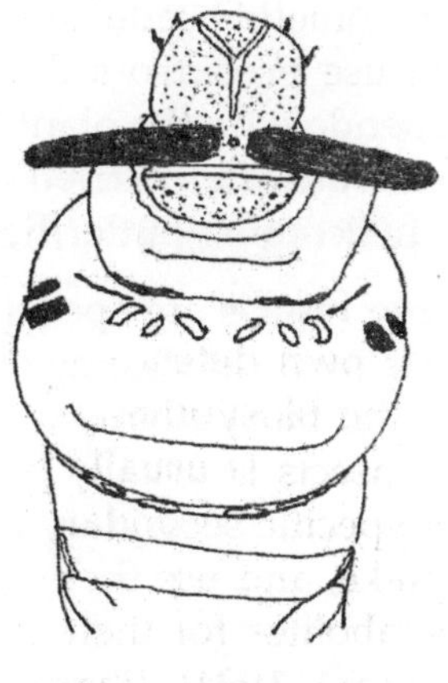

Fig. 13.17. Anterior end of caterpillar of *Papilio polytes* showing osmeterium.

13.3.2 Sources and release

The defensive compounds may be obtained by exogenous or endogenous sources. Many insects of orders: Lepidoptera, Coleoptera, Hemiptera and Orthoptera sequester toxic plant compounds in their tissues. In soft insects, e.g. larvae, toxins are usually stored near the teguments or into special glands ready to release their contents upon handling. The toxicity of insects that extract and sequester plant chemical compounds is dependent on the concentration of these compounds in the host plant. Sawfly larvae, for instance, reflex-bleed drops of haemolymph when touched; unpalatability of such larvae is shown to be directly dependent on the glucosinolate concentration of their crucifer host plant. Insects feeding on milkweeds sequester cardenolides and store in their body tissues. In monarch butterflies these compounds are even retained by the adults and provide effective defence. Caterpillar of the spurge hawk moth, *Hyles euphorbiae* feeds on highly toxic spurge plants and when disturbed, it vomits and spreads the bitter, irritating terpene containing spurge juice over the predator to defend itself (Fig. 13.18). Similarly larvae of pine sawfly sequester terpenes from pine hosts. The painted grasshopper, *Poekilocerus pictus* ejects a fluid that produces irregular R-waves and bachycardia in predatory garden lizard, *Calotes nemoricola*. The fluid contains cardenolides which the grasshopper obtains from food plants (*Calotropis* sp.) (Sreenivasulu *et al.*, 1996). Insects feeding on oleander sequester cardiac glycosides from oleander leaves. The larvae of spotted tortoise beetle, *Aspidomorpha miliaris* remain in cluster during rest and evade predator by simulating the form of a giant spider. During feeding, they disperse and arrange themselves in a linear fashion. They are unpalatable to probable predators because of their feeding on toxic plants *Ipomoea fistulosa* (Chattopadhyay and Sukul, 1994). This ability of insects' is highly selective and all toxic substances are not absorbed and stored in body. Duffey (1980) gave detailed account of sequestration of plant products by insects. Insects like many

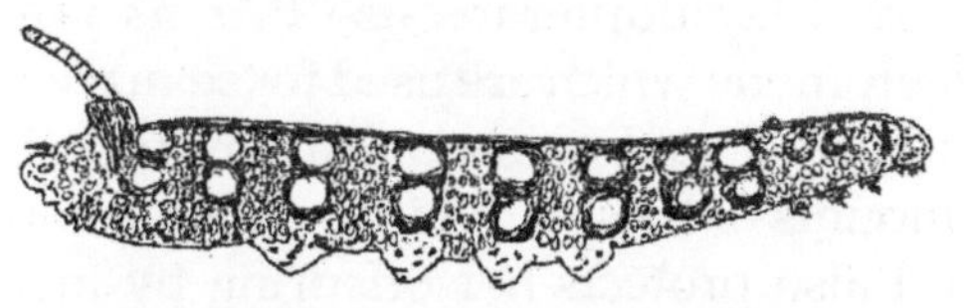

Fig. 13.18. Caterpillar of the spurge hawk moth, *Hyles euphorbiae*. It feeds on highly toxic spurge plants and when disturbed, it vomits and spreads the bitter, irritating terpene containing spurge juice over the predator to defend itself.

chrysomelid beetles synthesize their toxins *de novo*. Although many of these species still use precursors derived from their food plant, these insects are usually less dependent on the plant's toxicity to develop their own noxious compounds. Larval toxins are often carried over to the adult stage, although some groups like ithomiine or heliconiine butterflies also get toxin precursors in their adult diet.

Some insects are specialized to the extent of storing unconverted plant toxins for their own defence against predators, and/or even requiring them as precursors for the biosynthesis of pheromones. Sequestration of secondary plant chemicals by insects is usually linked with feeding. The 'pharmacophagous' insects search for specific secondary compounds directly, consume them independently of food uptake, and use them to increase their fitness, again either by storing these plant metabolites for their own defence and/or using them as pheromone precursors (Boppré, 1984). Pharmacophagy is non-nutritional and exclusively mediated by and directed towards plant secondary compounds. Pharmacophagy in adult Lepidoptera usually involves unrelated plants and it reflects an insect-plant relationship separate from those plants which provide nutrients for larvae or adults. Adults of numerous Danainae, Ithomiinae, Arctiidae, and Ctenuchidae are attracted to plants parts which contain pyrrolizidine alkaloids (PAs). They apply a fluid on the dry plant material with their proboscides to extract PAs. PAs are usually repellent to non-adapted insects and serve as protective devices for the plants that produce them. By sequestration and storage of PAs many, if not all, PA-pharmacophagous taxa gain protection from predators. Besides, males of several lepidopterans use PAs as precursors for the biosynthesis of major sex pheromone which are used for sexual communication. Thus, pharmacophagy assists in both sexual communication and chemical defence. The male provides large amounts of PAs with his spermatophore to the female. The female gains protection and also protects her offspring by incorporating PAs into her eggs.

In some cases protection is achieved because of lack of a particular chemical in the insect's body. Moraes and Mescher (2004) reported that the larvae of the specialized lepidopteran herbivore, *Heliothis subflexa* reduce their vulnerability to natural enemies through adaptation to a remarkable feature of their host plant, *Physalis angulata*. The fruits of this plant lack linolenic acid (LA), which is required for the development of most insects. By overcoming this nutritional deficiency, *H. subflexa* larvae achieve numerous advantages. First, they gain near-exclusive access to a food resource on which other larvae cannot develop. Second, they reduce their vulnerability to enemies and are far less attractive to female *Cardiochiles nigriceps* parasitoids. Finally, they render themselves nutritionally unsuitable as hosts for enemies that require LA for their own development. Thus, *H. subflexa* larvae not only overcome a serious dietary deficiency but also reduce their vulnerability to natural enemies through a form of 'biochemical crypsis'.

Some insects discharge intestinal products that may be constituent of food or formed from digestive metabolism to repel the predator. Secretion of salivary glands in some cases may contain toxic substances or may become viscous to entangle the predator. Some termites discharge secretion of mandibular glands on predator that becomes viscous and hard on exposure to air. The water bug, *Velia* and beetle, *Stenus* expel secretion from salivary and anal glands, respectively which lowers the surface tension of water and the insect shoots forward. Many fireflies synthesize deterrent bufadienolides, but *Photinus* females cannot do so. They acquire this deterrent by eating males *Photinus,* which are attracted by females by flashing sexual signals.

Many beetles discharge blood from joints on stimulation by 'reflex bleeding' or 'autohaemorrhage'. The adult Mexican bean beetle, *Epilachna varivestis* releases blood from tibio-femoral joints and its larva also releases blood, which is strongly repellent to ants. The blood after coagulation may entangle the assailant; act as deterrent or sometimes discharged along with secretion of defensive gland. Many hymenopterans possess poison glands and their secretions are stored in reservoir associated with sting or seta. Insects like cockroaches, bugs, caterpillars, beetles and earwigs have characteristic odours, which emit from stink glands or repugnantorial glands. Cantharidin (also known as Spanish fly) in blister beetles and pederin in staphylinid of the genus *Paederus* are some commonly known defensive compounds found in haemolymph of these beetles.

In many insects defensive secretion ooze out as a liquid droplet, while in others it is expelled as a spray. The stimuli responsible for release may be from direct contact or other received from a distance. The discharge may be random or at a particular aim. Some insects withdraw the secretion back into gland or reservoir. The range of spray varies in different insects. The cockroach, *Eurycotis decipiens* spray its secretion up to a distance of 3 ft (Roth and Willis, 1957). The predaceous reduviid bug, *Platymeris* sprays highly toxic saliva by spitting up to a distance of 30 cm (Edwards, 1961). Red imported fire ants, *Solenopsis invicta* disperse venom through the air by raising and vibrating gaster, i.e. 'gaster flagging'. Eisner (1965) demonstrated that Florida walking stick, *Anisomorpha buprestoides* shows its spraying response in various ways, and elicits both bilateral and unilateral discharge response directly on the attacker.

13.3.3 Chemistry of defensive secretions

The first secretion investigated chemically was probably formic acid from the formicine ants. It occurs as aqueous solution along with free amino acids and small peptides and is applied as a spray or into wounds caused by ants with mandibles. Some formicine ants release citronellol from mandibular glands. Formic acid has also been reported in exudates of notodontid caterpillars and adult carabids. Coreid bugs and carabid beetles secrete acetic acid, while benzoic acid and p-hydroxybenzoic acid and methyl p-hydroxybenzoate are synthesized in pygidial glands of many dytiscid beetles. Secretions from notonectids are odourless and brownish in colour and contains two major components, viz. p-hydroxybenzaldehyde and methyl p-hydroxybenzoate. The gyrinid beetles produce norsesquiterpene that are strongly repellent to fishes, birds, and small mammals. Defence mechanisms in sphingid moths include high-speed flight, crypsis, and aggressive warning signals (associated with toxic body fluids, bitter taste and unpleasant or evocative odours). A pyrazine (which is highly aromatic) emitted by some sphingid moths can provide a warning smell that boosts flavours and can potentiate prey-avoiding learning in certain vertebrate predators (Rothschild, 1989).

The adult cerambycid beetle of the genus *Phoracantha* secrete an alcohol namely phoracanthol. Many bugs secrete α–β unsaturated aldehydes. Among lactones, γ-dodecalactone and marginalin are major constituents in defence secretion of staphylinid and dytiscid beetles, respectively. Phenolic compounds have been identified in many beetles. From earwigs, cockroaches and termites, 1-4 quinones have been isolated. In Dufour's gland of ants many hydrocarbons are formed. Hydrogen cyanide and benzeldehyde are formed in secretion of some larval chrysomelids. Terpenoids have been isolated from Argentine ant, *Iridomyrmex*. Poison gland secretion of ants of the genus *Solenopsis* contains mainly alkaloids. The larvae of certain fungus gnats spin silken webs whose threads bear droplets of oxalic acid, which quickly kill small insects that contact with them.

The venoms formed in sting glands of hymenopterans mainly constituted of proteins and/or peptides. The *Apis* venom contains amines (histamine, dopamine, and noradrenaline), polypeptides (apamin, melittin, mast cell-degranulating peptide) and enzymes. Serotonin has also been identified in venom of the genera *Vespa* and *Vespula*. In *Vespa* venom acetylcholine is also present. Among enzymes hyaluronidase and phospholipase A widely occur in venoms of hymenopterans (Blum, 1978).

Bombardier beetles are well-known for their chemical defence mechanism. Bombardier beetles earned their common name from their ability to defend themselves against predators by firing a mixture of boiling-hot toxic chemicals from special glands. The bombardier beetle, *Metrius contractus* discharges its defensive secretion as a froth that clings to its body. When attacked from the rear, it allows the froth to build up over the gland openings near the abdominal tip (Fig. 13.19); when attacked from the front, it conveys the secretion forwards along special elytral tracks. *M. contractus* has two-chambered defensive glands, and the defensive secretion is quinonoid and hot. Its frothing mechanism; however, is

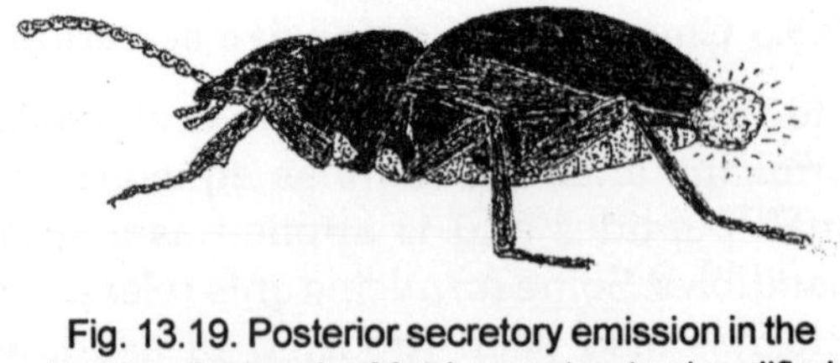

Fig. 13.19. Posterior secretory emission in the bombardier beetle, *Metrius contractus* (modified after Eisner *et al.*, 2000).

unique for bombardiers and possibly illustrative of the ancestral glandular discharge mechanism of these beetles (Eisner *et al.*, 2000). In bombardier beetles of the genus *Brachinus* glands are located in the posterior body region. The defensive system of these beetles consists of a gland, a reservoir and a chitinous capsule (Fig. 13.20). Secretory cells produce hydroquinones and hydrogen peroxide (and perhaps other chemicals, depending on the species), which collect in a reservoir. The reservoir opens through a muscle-controlled valve onto a thick-walled reaction chamber. This chamber is lined with cells that secrete catalases and peroxidases. In the reservoir hydroquinones and hydrogen peroxide are mixed and in dangerous situation, the mixture pass into chitinous chamber containing catalytic enzymes, which oxidize hydrogen peroxide to give water and oxygen and hydroquinones to give benzoquinones (Fig. 13.21). Much heat is liberated and the temperature of mixture may reach up to 100°C (Aneshansley *et al.*, 1969; Eisner and Aneshansley, 1999). Under pressure of the released gases, the valve is forced closed, and the chemicals are expelled explosively through openings at the tip of the abdomen.

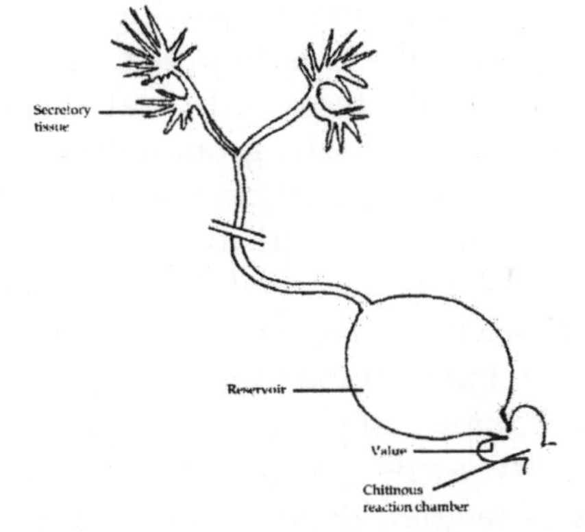

Fig. 13.20. Organs in bombardier beetle. Defence organs of bombardier beetle.

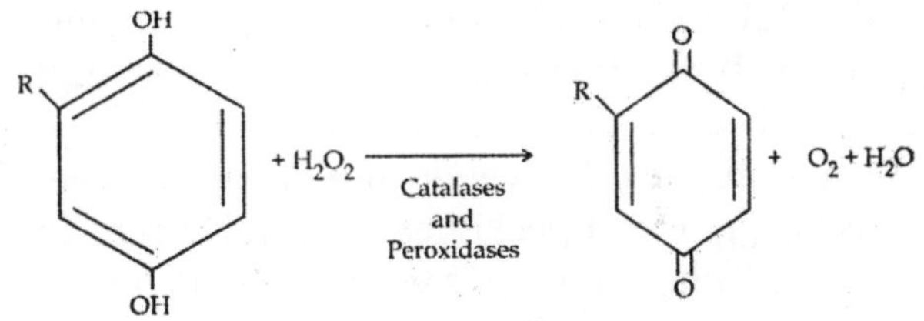

Fig. 13.21. Chemistry of defensive explosion in bombardier beetle.

13.4 Behavioural defence

Behaviour is an important component of defence and some insects rely completely on behavioural patterns for protection. The simplest and most effective approach of defence is to flee, and a number of small-sized and quick accelerating insects get protection by escape. Jumping, death-feigning, construction of protective structures, living in protected places, adoption of threatening posture, startle displays, and flash patterns are main behavioural defences adopted by insects.

13.4.1 Death-feigning (Thanotosis)

A characteristic death-feigning is exhibited by a number of insects which inhibit the killing response of the predator. Some scientists believe that the widespread animal behaviour called death-feigning is a form of sleep. Death-feigning is known, in popular language, as 'playing possum' because possums are so good at it. Many beetles, weevils, grasshoppers, stick insects, mantids, caterpillars become inert when they are disturbed or attacked. Some drop from a twig or leaf and lie completely stiff on the ground. In many cases, insects draw their appendages and resemble to stationary inanimate objects, which may be of survival value as many predators are attracted to moving prey only. Many caterpillars assume strange postures when disturbed. For instance, a cutworm curls in a ball and plays dead. Rapid evasive movement usually follows a death posture. Most feints are of short duration but in some insects the feint lasts for hours. A giant water bug may feint up to 8 hours. In sweet potato weevil, *Cylas formicarius* the proportion of death-feigned weevils was significantly lower at night than in the daytime. Resting weevils feigned death more frequently and took a longer duration to recover compared with walking ones. A significantly lower proportion of death-feigned adults was reported in starved weevils compared with unstarved.

13.4.2 Anachoresis

Many insects live in protected places like crevices, holes, in soil, beneath stones, in burrows, plant tissues, etc. thereby avoiding effective predators. A number of soil-dwelling insects remain away from the reach of predators in their larval, adult or both stages. Some come out at night for feeding, e.g. black cutworm, ground beetle. Most borers, miners, gall-formers, etc. remain well protected within the host tissues. However, the enemies have also devices to find such protected insects, e.g. wood pecker can extract borers and many hymenopterous parasitoids are able to lay eggs on such protected hosts by using long ovipositors.

13.4.3 Deflection of attack

Some insects have markings on body, which draw attraction of a predator towards non-essential part of the body. Butterflies having small spots along with edge, tail-like structure on wing induce the predator to attack on these body parts. The Kaiser-I-Hind butterfly, *Teinopalpus imperialis* found in eastern Himalayas has thin antenna-like projections at the tip of hind wing and spot resembling the eye at base of false antenna (Fig. 13.22). The silverline butterfly, *Spindasis* species has antennae, legs and eye-like structures on the posterior end of the body (Plate 13.20). Thus, it seems to have a head on the posterior end. If attacked by a bird only the predator seizes a part of wing.

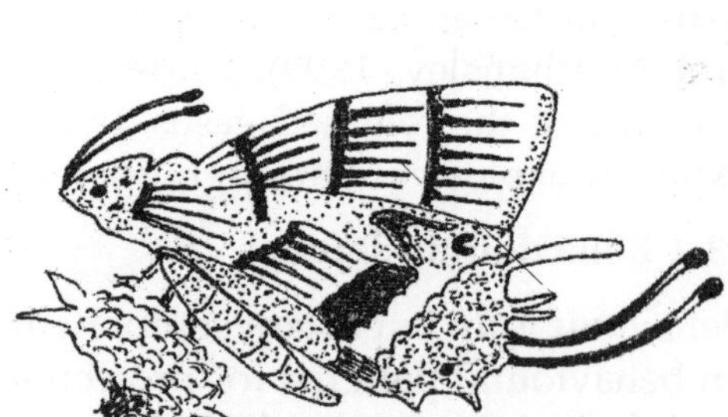

Fig. 13.22. Kaiser-I-Hind butterfly, *Teinopalpus imperialis* on flower with false antennae-like extensions of the hind wings.

Eyespots on wings (Plate 13.21) and body of many insects may deflect attack of predators. Beak-marks on the wings of many butterflies near eyespots help in escape from birds attack.

13.4.4 Escape

Rapid movement of an insect away from the predator is a common response. It may be in the form of flying, running, jumping, swimming or passive dropping. The presence of wings is advantageous to insects as a means of escape from predators. Many swift fliers like butterflies and moths escape from their pursuers by flight. They also avoid predation by abrupt colour change on settling, as both sides of wings are variously coloured.

Hawk moths are predated by bats, which use sonar to echolocate their insect prey at night. While flying usually moths provide a quick snack for these bats; however, some moths, such as belonging to subfamilies Acherontiinae and Choerocampinae, have the ability to 'hear' the high frequency sonar sounds emitted by the bats and respond. These moths have developed anti-bat tactics, which include: (i) moth stealth technology in which fuzzy wings of moth do not reflect the bat's pulses, (ii) many moths have 'tympanic organs' or 'ears' to detect bat echolocation, which allow flying moths to avoid predatory bats (Roeder, 1965). These ears are membranes stretched over sensors. They are found on different body parts of moth species. Moths are able to compute their flight path to avoid bats. Moths hear the bats first, and before the bat can pinpoint a moth's location it hide or get away (Fig. 11.6). If the bat is too near, the moth flies in loops, make surprising noises or simply folds up its wings and power dives to the ground, and (iii) tiger moths return a short burst of pulses, mainly to advertise their distasteful nature. Rubbing a wing against a leg, or vibrating castanet-like structures on the wings makes the sound. However, higher frequency sonar, which indicates a bat is close, triggers a 'drop dead' response in which the moth simply stops flying and does a nosedive.

Over 14 families of moths have ears that are adapted to detect the ultrasonic echolocation calls of bats. On hearing a bat, these moths respond with an escape response that reduces their chances of being caught. As an evolutionary response, bats may then have evolved behavioural strategies or changes in call design to overcome the moth's hearing. Waters (2003) discussed the role of the echolocation calls of bats in the shaping of the structure, neurophysiology and behavioural responses of moths. The interaction between bats and moths has much to interest general biologists, and may provide a useful model in understanding the neurophysiological basis of behaviour, including protean escape behaviours.

Many quick runners, e.g. silverfish, cockroach, ants, some beetles are able to avoid their enemies. Grasshoppers, crickets, fleas, some beetles, springtails have modified legs or other structures supplied with powerful muscles and are able to leap.

Another effective means of escape is reflex dropping. Many beetles, weevils, caterpillars, hemipterans drop freely from host plants. Some caterpillars that live on trees use this very smart way to escape from predators. When disturbed, the caterpillar drops to the ground and disappear form predator's eyesight. It jumps with a silk thread, attached to its body and hangs itself in the mid-air (Fig. 13.23, Plate 13.22). After a minute or so, it climbed via the silk back to the same place easily. This behaviour is advantageous as if the caterpillar just drops to the ground, the ground is surely not a safe place for the caterpillar. Also it is very hard for the caterpillar to climb back up to the tree. The worst is, if it drops to the ground, the predator may follow and find it. The click beetle when disturbed drop to the ground and lies for sometime as in death. Then it gives a stretch and goes off suddenly with a click. This behaviour may be useful as a means of escape.

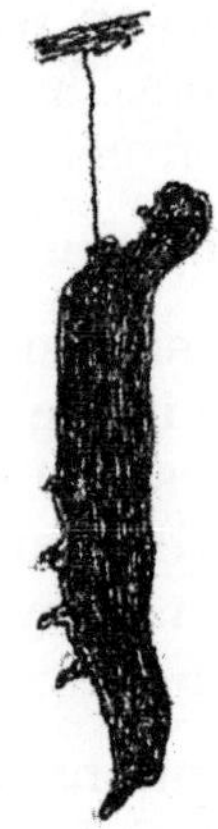

Fig. 13.23. A caterpillar when disturbed jumps with a silk thread, attached to its body and hangs itself in the mid-air. The strategy by which it disappears form predator's eyesight.

13.4.5 Flash pattern

Certain insects have adopted an additional course of defence in addition to colour patterns. In dangerous situation they display a rapid change to bluff a predator. They fly for a short distance and then come at rest and exhibit cryptic pattern so that the predator is unable to detect them. Thus, a rapid and sudden change occurs in search image of the predator. The hind wings of banded-winged grasshoppers are often brightly coloured with red, orange, yellow, or white bands, in sharp contrast to the often drab brown of the fore wings. When startled by a predator, they take to the air. The predator focuses attention on the brightly coloured and flashy hind wing, but the grasshopper disappears from sight when it folds its wings, lands, and again cryptically blends into the background. Many moths, butterflies, grasshoppers, and cicadas are cryptic while at rest, but during flight exhibit conspicuous wing patterns. Some grasshoppers produce clicking sound during flight and it may be a flash sound. The predator following a grasshopper on the basis of this sound may get baffled when the grasshopper comes at rest.

Eyespots present on wings (Plate 13.21) or other body parts are apparently effective defence structures (Blest, 1957). They are not only an effective means of altering a predator's search image but also sudden display of eyespots is effective in frightening a predator. Hawk moths particularly are cryptic while at rest; however, when disturbed they exhibit prominent eyespots on hind wings by spreading the fore wings. The lycaenid butterflies have eyespots on ventral surface of hind wings and exhibit them while at rest. In some cases eyespots are displayed in such a fashion that they have close resemblance with eyes of snakes or owls. The peacock butterfly, *Nymphula io* at rest keeps the wings folded, which resemble a dead leaf. When disturbed wings are opened and closed repeatedly and the eyespots are exposed in open wings.

13.4.6 Startle displays

Many insects respond to a predator by adopting a characteristic posture, which intimidate the predator. Such displays have been termed as startle displays, dynamic displays, or frightening behaviour (Edmunds, 1972). These patterns are of significance as they often repel a prey. Such patterns may be real warning or a bluff and may occur in both cryptics and aposematics.

Robinson (1969) discussed such displays in different species of stick insects, which often display erecting brightly coloured wings with eyespots. These displays are associated with stridulation resembling hissing of a snake or sound of a threatened scorpion. In addition they also produce repellent secretions. Many insects utter a shrill cry when seized. The stridulation of reduviids, long-horned beetles, mutillid wasps and certain lepidopterous pupae perhaps alarm a potential predator. In some mantids, grasshoppers, butterflies and moths such sound production is associated with display of warning colouration. It has also been suggested that in some cases sound resemblance occur, e.g. certain syrphids not only look like bees but also sound like them. Some arctiids and ctenuchid moths are distasteful in addition they display bright colours on hind wings and abdomen, produce high frequency sound, and a few emit a repellent smelling froth from thorax when disturbed (Blest, 1964).

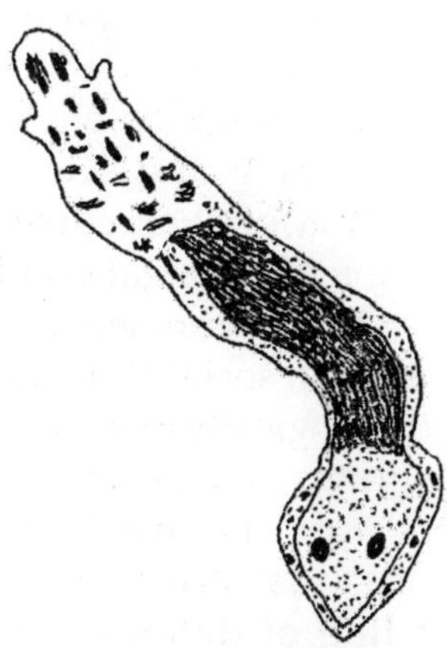

Fig. 13.24. Caterpillar of *Hemeroplanes ornatus* when disturbed give appearance of a small snake.

13.4.7 Threatening posture

Such poses can be quite effective in frightening the predator. Bizarre structural features that make the insect appear grotesque often enhance threatening postures. Some threatening postures are similar to defensive behaviour of other species. Caterpillar of a Neotropical sphinx moth, *Hemeroplanes ornatus* that resemble with its lichen-covered background while at rest exhibits a remarkable pattern. Upon disturbance it swells the anterior portion of body revealing the hidden eyespots, sways back and forth and assumes the threatening attitude of a small snake (Fig.

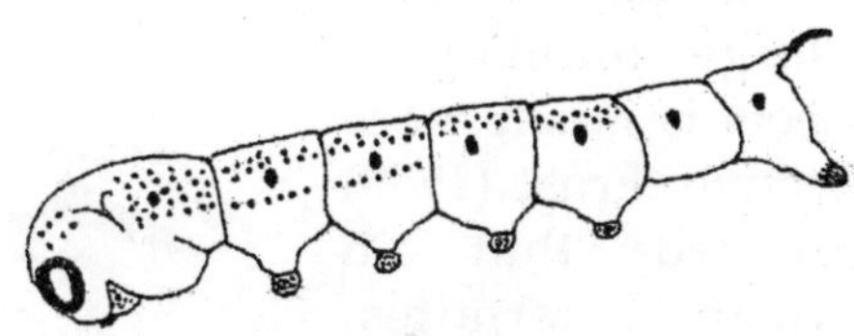

Fig. 13.25. Caterpillar of oleander hawk moth, *Daphnis nerii* showing threatening posture.

13.24). The greater arid land cricket, *Neobarrettia spinosa* advertises a threatening posture and seems to be aggressive in response to attack of a predator. Caterpillar of oleander hawk moth, *Daphnis nerii* shows threatening posture to terrify the nearby predator (Fig. 13.25, Plate 13.23)

13.5 Group defence

Many insects exhibit group defence patterns; however, a group may be consisting of one or more species. Single species group or collective defence can be commonly observed in gregarious and eusocial insects. Orderly clusters occur in phytophagous larvae of sawflies and some chrysomelid beetles which form defended circles 'cycloalexy'. Some lie within the circle while others form an outer ring and secrete toxic chemicals. Formation of such clusters is facilitated by secretion of aggregation pheromones in some cases. The phenomenon of group defence is common in social Hymenoptera and termites. Morphologically modified individuals termed as workers or soldiers with the task to protect the nest and its inhabitants assume the defence actions. Nest structure and location is often a first line of defence. In termites soldiers, enlarged heads have well-developed mandibles (Fig. 13.26), or are nasute with small jaws but an elongate 'nasus' or 'rostrum' (Fig. 13.27). They use strongly developed mandibles to injure an attacker. Some termites snap elongated mandibles against one another. In *Capritermes* and *Homallotermes*, the mandibles are asymmetrical. These soldiers can only strike to their left and can deliver a powerful blow in a confined tunnel. It has also been observed that in alarming situation all members come out of their nest and thrust away the intruder. Thus, the joint action of a large number of insects is much more effective. Ernst (1959) observed that in African termites, *Nasutitermes* 70-90 per cent of nasuti soldiers aggregate outside the nest and protect the

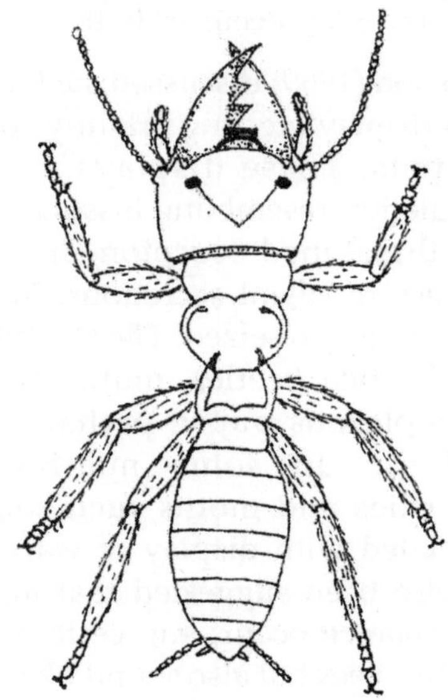

Fig. 13.26. Madibulate soldier of a termite.

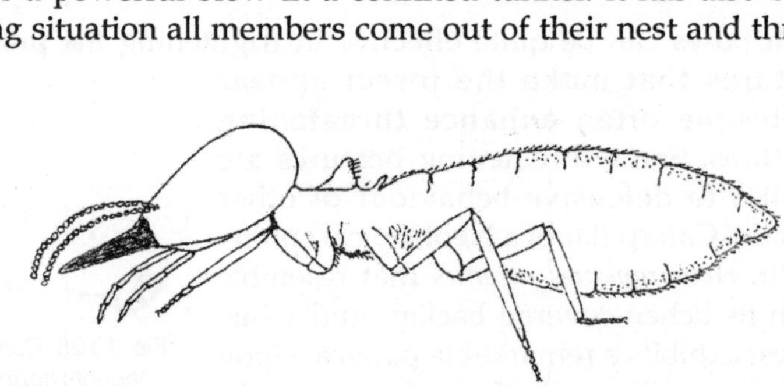

Fig. 13.27. Nasute soldier of a higher termite.

workers on foraging trail by aligning themselves on each side of foraging trail. Intruding insect is encircled by nasuti, all of which point their snouts toward the intruder and discharge their secretion. The secretion is ejected as a single jet that forms quick-drying fine threads, which impairs the movement of the intruder. The appendages of intruder stuck together, spiracles are occluded and intruder eventually dies.

Termite soldiers also use sticky fluids for defence, e.g. *Globitermes sulphureus* soldiers release sticky yellow fluid to entangle the predator. Some *Camponotus* ants use hydrostatic pressure in the gaster to burst the abdomen and release sticky fluid from salivary glands. In European ant, *Camponotus truncatus* well-sclerotized head capsule of soldier is flat and soldiers in-group may block the entrance hole in the nest with their head.

A number of aposematics exhibit more conspicuous effect in gregarious condition. Larvae of sawflies of the genera *Neodiprion* and *Diprion* discourage enemies by synchronous jerking of anterior and posterior body ends and by secreting sticky material from mouthparts (Prop, 1962). In aggregation they are avoided by the predatory bug, and when dispersed attacked by the bug. Hinton (1955) discussed group defence by ceratopogonid larvae of *Forcipomyia nigra* that lives in very humid environment. Larvae have paired setae on head, thorax and abdomen through which a hygroscopic substance is secreted. The secretion extracts water from atmosphere and causes a drop solution at tip of each seta. When the ant tries to seize the larva it touches the liquid, drops the prey and clean itself. The pupa also retains shed larval skin over posterior abdominal segments containing hygroscopic substance and similar drops form over the pupa. Larvae pupate in a circle with their heads pointed toward the centre and protection is confronted with a ring of larval cuticle armed with defensive fluid. Locust species exhibit phenotypic plasticity in numerous morphological, physiological and behavioural traits as their population density increases. These density-dependent changes in colour in nymphs of the desert locust, *Schistocerca gregaria* results in warning colouration and a decrease in predation on locusts.

Social hymenopterans frequently use venom, which is injected through an ovipositor modified as sting. Many social and sub-social hymenopterans cooperate to sting an intruder *en masse*, thereby escalating the effect of an individual attack. They also release alarm pheromone, often with communication role that stimulate one or more defensive actions; e.g. by marking an attacker so that the attack is targeted, or as last resort they may encourage the colony to flee from the danger. The stingless bees and formicine ants use their mandibles and defensive chemicals if attacked.

Insects of different species living in groups may get defensive advantages. Mutualistic association of ants with many hemipterans is well-known in which the benefit for hemipterans is the protection they receive. Aphids that live in association with ants are less prone to attack of predators. Ribeiro (1989) reported that nymphs of the bug, *Jadera haematoloma* remain aggregated during the premoulting period. These were found to be distasteful to toads and blue jays and their red colouration functions as a warning signal to predators. Adults that are black and dispersed on the host trees also appeared to be distasteful. Many lycaenid caterpillars have protective association with ants and in other cases ants carry eggs and larvae of insects to their nests where they get protection. The association of rove beetle, *Atemeles pubicollis* with *Formica* and *Myrmica* ants is rather complex in which beetles gain both protection and food in the nests of ants (Hölldobler, 1971a). In these cases insects, which get protection either, secrete some sugary fluid from their body (honeydew by aphids) or other chemicals from certain glands, which suppress aggressive behaviour of ants. The larvae of spotted tortoise beetle, *Aspidomorpha miliaris* remain in clusters, each in the form of a disc during resting periods. Most members of the cluster react synchronously by moving the tip of their exuviae attached to their tail upward to any object approaching them, thereby simulating a giant spider. During midday feeding, the larvae disperse and arrange themselves in a linear fashion. Predators like birds, lizards and mantids avoid the larvae probably because of their feeding on a toxic plant, *Ipomoea fistulosa* while the spiders were found to prey an isolated larvae but not a cluster (Chattopadhyay and Sukul, 1994). Clark and Faeth (1997) reported that females of the nymphalid, *Chlosyne lacinia* clump eggs in a few large clusters on their host plant, sunflower. Resulting larvae form sibling aggregations up to at least the 3rd instar. Larvae developed faster and survived better in larger groups. Group defences and aposematism of *C. lacinia* larvae are probably ineffective against predatory ants that attack *en masse* and recruit other colony members; however, protection from solitary predators was achieved due to aggregation. The effects of various predator guilds, viz. ground-dwelling arthropods, aerial arthropods and avian predators on survival of larvae was tested and it was observed that the southern fire ant, *Solenopsis xyloni*, reduced larval survival greatly but other solitary invertebrate and avian predators did not alter survival.

13.6 Integrated defence systems

Some insects achieve defence by adopting a number of defence strategies. They may exhibit different response at different growth stages or by adopting more than one strategy in a sequence against a predator. Some species preyed by several predators exhibit different strategies against different predators. The young caterpillars of citrus swallowtail, *Papilio demodocus* resemble bird-dropping and rest on upper surfaces of leaves, while the mature larva are green and black and rest on stems. Similarly larva of a viceroy butterfly resembles bird-dropping, pupa to a dried leaf and adult to a distasteful monarch butterfly. Some tiger moths have cryptic fore wings, brightly coloured hind wings, which provide flash pattern and also discharge toxic fluid in dangerous situations. Social wasps display warning colouration of their abdomen and have biting or stinging apparatus. Termites are able to build very hard termitarium; some of their soldiers have strong biting mandibles while others secrete noxious chemicals. Praying mantids exhibit cryptic colouration, escape mechanism, startle displays and in adverse situations at last may attack the predator with raptorial fore legs. Crane (1952) studied defence mechanisms adopted by 15 species of praying mantids and observed all of them to be cryptic, with some types of escape mechanism and many have startle displays and also attack predators.

The scarlet tiger moth, *Panaxia dominula* is cryptic and well mistaken for a dark patch of leaves at rest, while flying scarlet on hind .wings may be a flash colour and disappears on landing. If the resting moth is disturbed it exposed brilliant hind wings and may also exude a drop of extremely repellent fluid from thaoracic glands (Fraser and Rothschild, 1960).

Chapter 14

Insect Society

The social organization of insects is inferior to human society because of intellect and absence of so-called culture; however, they are far superior with respect to cooperation, polymorphism, division of labour and altruism. The insect's society is well organized and is also termed as superorganism, because it displays so many social phenomenons that are analogous to the physiological properties of various organs and tissues in an organism (Wilson, 1971). Social insects, particularly ants are ecologically most successful insects and dominate the land in tropical areas. They are considered to be chief earthmovers in the tropics. The numerical dominance of social insects can be astonishing and there may be many millions of individuals in a supercolony. Nearly 75 per cent of all insect biomass consists of social insects (termites, ants, bees, and social wasps). Their biomass and energy consumption exceeds that of vertebrates in most terrestrial habitats. Various types of social behaviour occur in at least 13 insect orders. The true social insects, also termed 'eusocial insects' are mainly hymenopterans, and isopterans. Besides these, some aphids and thrips are also considered as eusocial. These are characterized by:

(i) Mutual cooperation for brood care.

(ii) Polymorphism with a well-defined caste system and division of labour.

(iii) Overlapping of generations.

The other groups possessing any level of sociality, but not possess all the above three characters are termed 'presocial'. Michener (1969) provided a classification that includes various levels of social stages.

Solitary. Showing none of the above three traits.

Subsocial. The adults care for their own nymphs or larvae for some period of time.

Communal. Members of the same generation use the same composite nest without cooperating in brood care.

Quasisocial. Members of the same generation use the same composite nest and cooperate in brood care.

Semisocial. In addition to characters of quasisocial, there is also reproductive division of labour, i.e. a worker caste cares for the young of reproductive caste.

Eusocial. In addition to characters of semisocial, there is also an overlap in generations so that offspring assist parents.

Other than eusocial, all five categories have been considered as 'presocial', while communal, quasisocial and semisocial categories are together termed as 'parasocial'- in which members of same generation interact (Wilson, 1971). Batra (1966) used the term 'hypersocial' for the social arrangement of honey bees and stingless bees in which the queen is structurally and behaviourally specialized only for egg laying; she cannot begin a new nest without the help of swarm of workers (considered as eusocial by many ethologists). Levels of social organization can also be grouped as:

14.1 Levels of social organization

I. **Solitary.** Living alone or as mated pair; no extended contact with brood.

II. **Aggregation.** More than one mated pair living together; no cooperation for other activities; no overlap of generations.

III. **Presocial**

- (a) **Subsocial.** Some adults care of immature stages, but not up to reproductive life.
- (b) **Communal.** Same generation uses same nest without cooperating in brood care.
- (c) **Quasisocial.** Individuals of a generation use same nest and cooperate in brood care.
- (d) **Semisocial.** Same as quasisocial with reproductive division of labour (worker and reproductive castes).

IV. **Eusocial.** Same as semisocial with overlap of generations; offspring assist parents in rearing young and do not reproduce themselves.

Many definitions have been proposed for social insects. Wheeler (1928) defined as 'social progeny are not only protected and fed by the mother, but eventually cooperate with her in rearing additional broods of young, so that parents and offspring live together in an annual or perennial society'. Richards and Richards (1951) proposed for social wasps that 'real social life appears when the mother and her offspring (or, less probably, a number of sisters) cooperate in making a nest and feeding their young'. Richards (1965) stated that 'in social insects the term social is reserved for communities in which there is more than one female, one or more usually being sterile, unfertilized and nursing the young derived from one or a few of them'. Emerson (1959) suggested that 'true social insects are those that live in populations exhibiting division of labour for various functions among mature individuals separated into reproductive and sterile castes'. Wilson (1971) defined eusociality as 'individual of the same species cooperate in caring for the young; there is a reproductive division of labour, with more or less sterile individuals working on behalf of fecund individuals; and there is an overlap of at least two generations in life-stages capable of contributing to colony labour, so that offspring assist parents during some period of their life'.

Significant contributions published on social insects are: Wheeler (1910, 1923, 1928), Michener and Michener (1951), Richards (1953), Michener (1969, 1974), Wilson (1971, 1975b), Hermann (1979-82), Breed *et al.* (1982), Andersson (1984), Hölldobler and Wilson, (1990), Carpenter (1991), Billen (1992), Crozier and Pamilo (1996), Choe and Crespi (1997a).

14.2 Subsocial insects

Many authors have used the term 'subsociality' and they consider that subsociality has arisen independently in many insect orders. In advance subsocial groups behavioural patterns like nest building, nest guarding, nest cleaning, parental care, transport of food to brood cells, transport of young individuals, alarm signalling can be observed. Subsocial behaviour is of interest as in some respects it resembles the behaviour of true social insects, and analysis of simple parental responses may suggest ways in which certain aspects of social behaviour could have been derived from less complex habits. The term 'presociality' is used for the society that does not fulfil all the criteria of eusociality. Since, presociality is not always a precursor to eusociality, many authors avoid using the term presociality. Some important behavioural patterns like aggregation parent care and nestting are well demonstrated by subsocial insect groups.

14.2.1 Aggregation

In aggregation more than one mated pair living together but there is no cooperation for other activities and no overlap of generations. Aggregations of non-reproductive adults and immature stages are observed in many groups and represent primitive social interactions. These aggregations may be for feeding, mating, overwintering or defending themselves (also see section 12.1.1). The aggregates face less risk of being encountered by predators as they share common warning signals. Many butterfly species form roosting aggregations. Aposematic insects often form feeding aggregations and in some cases pheromones are used to lure conspecifics to a favourable feeding site. The aggregation of monarch butterfly, *Danaus plexippus* in their overwintering sites is well-known to all. The mountain pine beetle, *Dendroctunus ponderosae* feed on the inner bark of pine trees. After diapause, the adults emerge in spring and seek out new hosts. The first beetle that discovers a suitable host is called 'pioneer beetle' and after deciding that the host is suitable he releases an aggregation pheromone. This pheromone combined with natural odours of the host is very attractive to other beetles and they aggregate for mating and feeding. Ladybird Beetles like *Hippodamia convergens* and *Coccinella septempuctata* have also been found in large aggregations.

14.2.2 Parental care in subsocial insects

Insects, like all other animals face a number of calamities. The adults can defend themselves up to some extent, but the immature are mostly helpless. In many insects the parents generally guard the young from danger, shield them from exposure, feed them and care them till they are able to fend themselves. In some insects there is relatively high degree of parental solicitude. Though many insects do not care their offspring, but the female while depositing the eggs select an appropriate site, affording protection to the eggs and ensuring availability of food after hatching. Some insects depositing eggs on leaves also fix the leaf to the twig with a hard secretion, so that the leaf may not fall down. Some insects bury their eggs into the ground. The female tiger beetles excavate tubular tunnels in the ground, lay eggs at bottom, and refill the tunnel with mud and even take trouble to make the surface smooth. Many insects lay eggs inside the hosts while others cover their eggs with special secretions often mixed with foreign material; others make egg case, egg cocoons, egg rafts, etc. The females of praying mantids prepare a great variety of oothecae for their eggs from a special secretion. The ootheca is essentially air-filled solidified foam that serves as thermo-insulator for the eggs, and it is fixed to a twig. Hydrophilid water beetles prepare egg rafts of silken webs and even provide a ventilator tube for the eggs inside.

The males of many species of Belostomatidae carry the eggs (Fig. 10.8). Smith (1976a,b) discussed paternal care in the giant water bugs, *Abedus herberti* and *Belostoma flumineum*. Water bug eggs need to be covered by water and frequently need to be exposed to atmospheric oxygen. He found the eggs on the back of a healthy male had a very high per cent hatch. Eggs left under water developed a fungus and did not hatch while eggs left in the air, desiccated and could not hatch. The eggs exposed to water and air started hatching, but not successfully. Males also exhibited behaviour, brushing their hind legs over the egg mass for the supply of dissolved oxygen to eggs. In adverse conditions, such as a loose egg mass or when disturbed, the male remove and cannibalize the egg mass. The cannibalism of the eggs is thought to be a method of gauging the energetic cost of care when the probability of a successful hatch is low (Smith 1976a). Some pentatomid bugs deposit eggs on leaves and guard over them (Fig. 14.1).

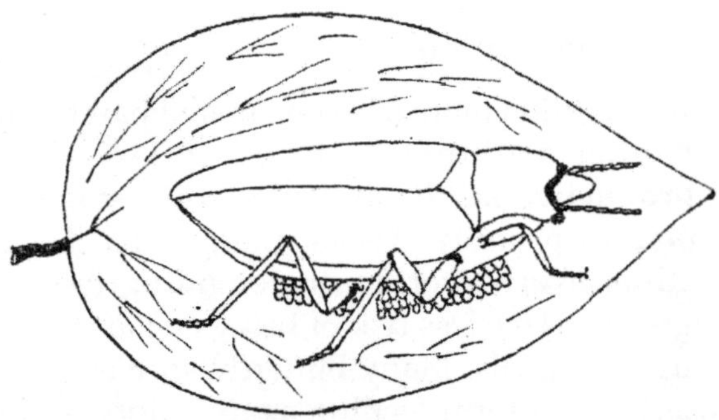

Fig. 14.1. A pentatomid bug brooding her eggs laid on a leaf.

Insects of many orders take care of their immature stages until the mature larvae or nymphs are formed. The parental care may be exclusive by an individual of one sex. Though parental care is primarily is a function of females, yet in some cases males also attend developing young stages. Maternal investment ranged from nutrition provided by the mother, to guarding, to provisioning, to communication between mother and offspring, to joint guarding by females, and combinations of these behaviours. More complex behaviours are also found in subsocial insects and insects showing biparental investment.

Maternal care is the most prevalent form of parental investment in insects. In case of the viviparous tsetse fly the female internally nourishes one offspring until it reaches the pupal stage. In this strategy, the individual offspring has a high probability of reaching reproductive age.

Guarding of immature stages is observed in many orders and families of insects. In the ovoviviparous leaf beetle, *Gonioctena sibirica* the female remains with the offspring until they reach their final larval instar (Kudo *et al.*, 1995). Females protect their offspring from predators by positioning themselves at the base of the leaf and exhibiting defensive behaviour that include rushing the intruder, swinging the body from side to side, and stamping the legs. They tested the percentage offspring surviving in three conditions, with the female present, with female absent, and with a sticky barrier at the base of the leaf. Their results found a significantly higher percentage offspring surviving when the female or the sticky barrier was present. The stink bug, *Antiteuchus tripterus* lays eggs in compact mass on the undersurface of leaves. The female lives with the clutch without feeding till the eggs hatch and nymphs develop up to second instar.

Parental guarding was shown in *Erixestus winnemana*, a parasitic wasp of the Colorado potato beetle. After injecting her eggs into the egg of the beetle, the female stand guards until her offspring emerge so hyperparistoids and other parasites do not kill the larvae (Athanas *et al.*, 1996).

The shield bug, *Parastrachia japonensis* exhibits both maternal guarding and progressive provisioning. It carries the egg mass away when confronted by a disturbance and may also flee to draw the predator away (Filippi *et al.*, 1995). Females also provision young with drupes, which is a type of fruit. The female treehopper, *Umbonia crassicornis* provides protection, provision, and communicates with her offspring. The mothering bug, *Elasmucha grisea* is a subsocial shield bug in which the females frequently jointly guard their offspring.

Fig. 14.2. Adult wood roach, *Cryptocercus penctulatus*.

Two species of wood roaches, *Cryptocercus clevelandi* and *C. penctulatus* are found in USA, and feed on nutrient-poor wood. Wood roaches (Fig. 14.2) are monogamous and exhibit considerable parental care. After a sexually mature wood roach finds a mate, the pair establishes a nesting site in a dead log on the forest floor. They will probably stay in one log for the rest of their lives and raise a single set of offspring. To digest they have cellulose-digesting symbiotic protists in their gut. They live in colonies consisting of a mated pair and 15-25 nymphs of same age group. Nymphs obtain the protists from its mother or father by 'proctodeal trophallaxis' (feeding on fluids from the adult's anus). Initially the nymphs are completely dependent on their parents for their nutrients. As they mature, they acquire their own gut flora and begin feeding on dead wood directly (Nalepa, 1984). Parental care involves maintenance of nest tunnels, defence and feeding of young on hindgut fluid. The grown nymphs also assist in nest maintenance, but because only one brood is present they cannot assist in feeding and brood care. Their cellulose-digesting ability is similar to that of termites and social behaviour involves parental care. It has also been suggested that it displays early step in the evolution of termite like eusociality. Like termites, they feed on dead wood and live in galleries they construct within fallen logs. However, recent morphological and molecular sequencing studies have indicated that *Cryptocercus* is evolutionary distant from termites.

Biparental care does occur in a number of species of both dung beetles and burying beetles. Burying beetles, *Nicrophorus* species conceal small vertebrate carcasses underground and prepare them for consumption by their young. Scott (1998) reviewed complex social behaviour of burying beetles in an ecological context. Eggert and Sakaluk (1995) examined that in burying beetle, *Nicrophorus defodiens* both sexes independently search for carrion, to feed their expected larvae. Once a suitable carcass is found, a sex pheromone to attract the opposite sex is emitted. Immediately the pair will bury the carcass under ground and then mates. Both parents remain to feed regurgitated tissue to the young and protect the larvae from other predators or fungus that may grow on the carcass. The female will then construct a short chamber above the carrion in which she lays her eggs. When the larvae hatch, they will then crawl into the brood ball. Here the larvae concentrate and are fed regurgitated carrion. If the carcass is larger than that likely to be used by the offspring, the mating male attempts to emit sex pheromone to attract another female. In such a situation the female often attempted to interfere with male pheromone emission. The interference behaviour observed most frequently was mounting of the male by the female, pushing the male, undercutting him or pinching his abdomen with her mandibles (Trumbo and Eggert, 1994). A monogynous association best serves the female's interests and if there will be an overall increase in the number of offspring produced in a polygynous association, the number of offspring per female will be reduced. Males do not seem to be any more helpful in protecting the brood against insect predators than a female alone. In fact, a female parent alone was sufficient to protect the brood against a staphylinid predator (Scott, 1989). Males can also assist in defence against parasites and fungi, which will attack the brood ball. Oral and anal secretions are used to keep these parasites at bay and thereby keep the brood ball intact for the larvae.

The dung-rolling species construct food balls (for their own consumption), and brood balls for burial and oviposition, within each of which a single larva develops (Fig. 14.3). Many larvae have colonies of microorganisms in an expanded region of their digestive tract; these break down cellulose and help in digestion.

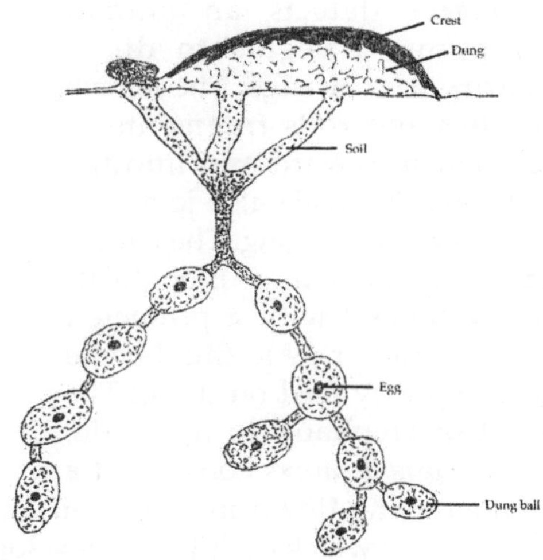

Fig. 14.3. Nest of a dung beetle, *Onthophagus* sp. with brood balls.

Many wood-mining beetles (Fam. Passalidae) live in decaying logs and form communities of adults and immature stages. Adults defend their young and provide them with macerated wood pulp. They do not have gut protozoa to digest cellulose, but fungi and other microbes decompose the macerated pulp. Some ambrosia beetles prepare larval galleries in which larvae are fed upon cultivated fungus and are protected by male that guards tunnel entrance.

14.2.3 Nesting in subsocial insects

Nest building or 'nidification' is a common behaviour in certain earwigs, crickets, beetles, solitary and social hymenopterans and termites. In a nest the female lays eggs and brings food supplies for the developing young. Nest building behaviour may involve only a female, individual of both sexes or many individuals. Earwigs of both sexes overwinter in a nest. Before oviposition, the male is expelled and mother tends the eggs (Fig. 14.4). In some earwig species the female forages and brings food for the developing young. Certain crickets also demonstrate similar behavioural patterns. The common mole cricket, *Gryllotalpa* prepares an underground nursery chamber in which she deposits her eggs (Fig. 14.5). Some insects do not prepare nests themselves but use nests of other insects or animals. Many parasitic insects enter the nests of their hosts, paralyze them and lay eggs on them.

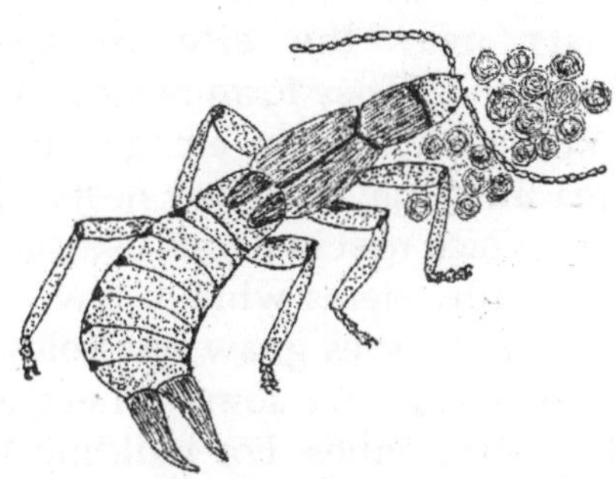

Fig. 14.4. Nursing of eggs by a female earwig.

Nest building behaviour is predominantly displayed by dung beetles (Fam. Scarabaeidae) and carrion beetles (Fam. Silphidae). When a big green dung beetle,

Phanaeus detects an animal dropping it begins to dig a burrow near its edge. The female pushes and rolls fragments of excrement toward and into the burrow. The male also joins her and assists in digging. The female prepares one or more brood balls of excrement with a protected egg chamber. As the larvae develop they feed on dung. The nest of elephant dung beetle, *Heliocopris dominus* consists of a tunnel going down into the soil and opening into a horizontal, ovoid brood chamber in which each brood develops inside individual brood balls, constructed by the female out of fresh elephant dung. The mother continues to live inside the brood chamber, taking care of developing larvae until pupation, contributing to the reproductive success of subsocial insect (Joseph, 1998).

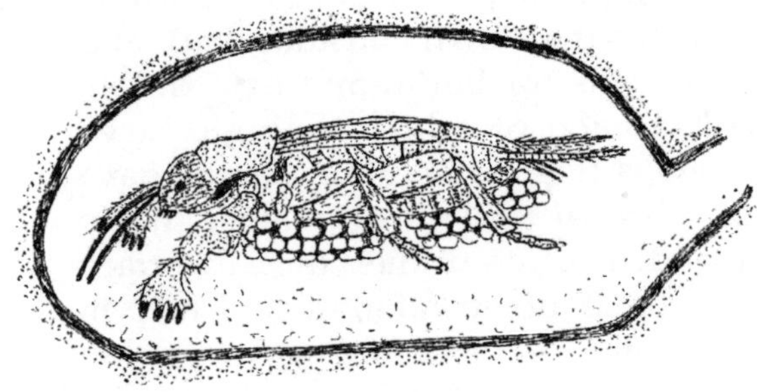

Fig. 14.5. Common mole cricket, *Gryllotalpa* female brooding over her eggs in the underground nest.

Some solitary bees and wasps are prolific nest builders. Nest complexity may vary from a simple burrow (Fig. 14.6) to multicellular nests. Many solitary bees build nests near the site of their emergence. They form nests in soil, wood, plants, walls, etc. and in an area there may be many nests. The bees which nest in plants use plants with pithy stems while a few, like *Xylocopa* species gnaw out holes in solid wood, and some others use different cavities. For building the nests they may use various materials like, mud, leaves, plant fibres, pebbles, rotten wood, plant pulp, wax, pine needles, faecal pellets, hairs, cotton, resin, wooden pieces, etc. They transport the building materials to the building site that has already been surveyed and selected. In most of the solitary bees the cells are lined with wax or varnish-like waterproofing material secreted by salivary or abdominal glands. Many solitary wasps construct vase-like nests of mud which are attched to twigs and other objects (Plate 14.1). The nest may be divided into cells by mud wall (Plate 14.2). The potter wasp, *Eumenes* is a well-known brood nester. The female builds a complex of many pot-shaped cells of clay. She makes beautiful pots with circular mouth and hangs an egg by a fine thread from the roof of the pot (Fig. 14.7). The pot is provided with green caterpillars as food for

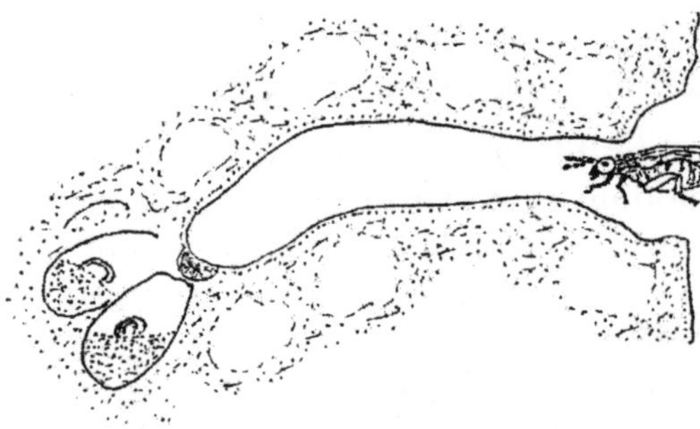

Fig. 14.6. Nest of a solitary bee in loose soil.

future larvae, and then closed. Evans and Eberhard (1970) suggested that nests of solitary bees and wasps are broadly of three types: (i) nests made in a substrate e.g. soil, rotten wood, plant pith, (ii) nests made in pre-existing hollow cavities, and (iii) nests made from foreign materials. In some subsocial hymenopterans many individuals participate in the construction of nest. The females also perform duties like cleaning, guarding and provisioning.

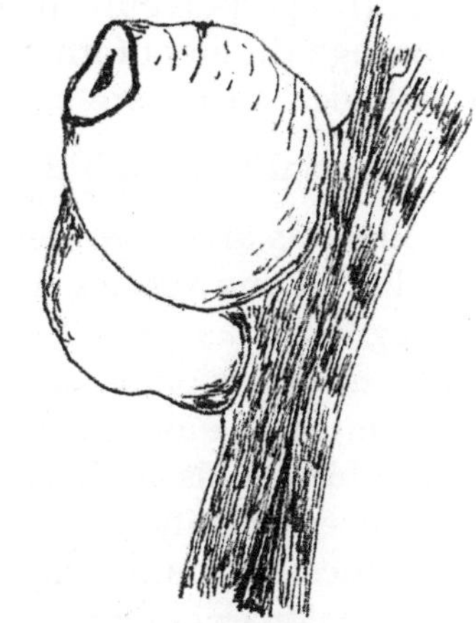

Fig. 14.7. Pot-shaped nest of the potter wasp, *Eumenes* sp.

14.3 Quasisocial and semisocial insects

In quasisocial insects, a communal nest is inhabited by members of the same generation all of which assist in brood rearing, and all females can lay eggs. In semisocial groups, the communal nest contains members of the same generation, which cooperate in brood care, but there is reproductive division of labour. Some females can lay eggs while others act as workers and remain sterile. Thus, the workers are sisters of egg-laying females while in eusocial groups the workers are daughters produced by the queen. The subsociality in insects may be evolutionary precursors for eusociality.

14.4 Eusociality in insects

Eusocial insects are characterized by a well-defined division of labour involving a caste system, live in colonies, make nests, and overlapping of generations. The caste system comprise of reproductive groups of one or more queens and males which are assisted by workers- the sterile individuals. Besides, in termites and many ants there is an additional defensive group termed as 'soldiers'. There may be further subdivision of castes into subcastes, which undertake specific jobs. Workers perform most duties in the colony or nest. In social Hymenoptera primary differentiation between male and female individuals is determined by haplodiploidy. Males are haploid and produced from unfertilized eggs, while from fertilized eggs diploid females are formed. The queen is able to determine the sex of offspring. The queen- reproductive female or 'gyne' is long-lived and perform the main duty of deposition of eggs while males are short-lived and usually die after mating. In termites individuals of both sexes may be equally represented and contribute to worker caste. A single male- the king may permanently attend the queen. Members of a colony are genetically and morphologically similar if derived from a single pair. They may differ because of environmental or other factors (polymorphism). Individuals within a caste or subcaste often differ behaviourally, in what is termed 'polyethism'. The polyethism may be: (i) age polyethism- temporal division of labour, in which individuals pass through different forms of specialization with age, and (ii) caste polyethism- morphological castes are specialized to perform different functions.

14.4.1 Eusociality in aphids and thrips

Eusociality has also been demonstrated in aphids (Aoki, 1977). The discovery of eusociality in any species of animal outside the Isoptera and Hymenoptera has been regarded as sensational. Some aphid species clearly have a distinct reproductive division of labour. Earlier it was considered that they do not exhibit all attributes of eusocial insects. However, many authors considered these polymorphic aphids as eusocial and claim that the Hemiptera is third insect order displaying eusociality. These aphids are diploid by nature, but also have the ability to reproduce asexually, thus producing genetically identical individuals. There are three major areas presently studied in the behavioural ecology of social aphids: (i) nesting behaviour- the social aphids either produce galls or live inside hollow plants such as bamboo, (ii) defence- many species of social aphids produce a specialized soldier caste which are sterile and defend the nest from predation, and (iii) life histories- the colonies are able to adjust the production of the various castes throughout the colony cycle.

Aoki (1977) for the first time demonstrated that the aphid, *Colophina clematis* produced instars that defend the colony from intruders. Since then many species of social aphids have been described in the two families, viz. Pemphigidae and Hormaphididae (Stern and Foster, 1996). Andersson (1984) proposed that all social species should be predisposed in having a nest. This centralizes the reproductive investment of a colony. Many times the nest serves as the first line of defence of a colony. Social aphids produce galls, which are tough pockets, induced in a plant by the aphids. All of the alates and reproductive instars are normally found inside the galls. The individuals on the outside are the soldiers who defend the gall from any predator.

Stern and Foster (1996) described several types of soldiers based on instar, physical characteristics, and behaviour. In some subsocial aphids first and second instars exhibit aggressive behaviour and never develop into adults while in some others all individuals become temporary soldiers, which later grow into normal non-aggressive, reproductive form. Mainly the soldiers are monomorphic or dimorphic in cases where defending first instars are morphologically different from reproductive destined instars. The dimorphic soldiers can be further subdivided into:

(i) *Dimorphic biting instars*. These soldiers have hairs and with longer hind or front legs. These soldiers not only pierce their victim with stylets, but also rip into a victim with their legs.

(ii) *Pseudoscorpion instars*. These instars have elongated front legs which allow them to grasp their victim.

The soldiers lack reproductive system and have a large number of fat bodies. Besides, they also lack the endosymbiont which help the aphids to produce certain amino acids that the aphid could not normally produce on its own. The lack of amino acids in soldier affects their physiology and behaviour. The soldiers not only defend the nest but assist in its maintenance. They clear excess honeydew in the gall, which can be detrimental to the colony. Soldiers are capable of defending against different types of predators and their bites are very irritating. The stylet using soldiers are not only effective against predator insects' larvae and adults but also some vertebrates.

Thrips comprise the only order besides Hymenoptera where females are diploid and males are haploid. This makes them useful insects for studying the roles of kin selection and ecology in social evolution. A few species of gall thrips also exhibit eusocial behaviour. Crespi and Mound (1997), Crespi *et al.* (2004) published their significant contributions on ecology and evolution of social behaviour among Australian gall thrips that make nests in *Acacia.*

Bernard J. Crespi (1992) demonstrated eusociality in two species of Australian gall thrips, *Oncothrips tepperi* and *O. habrus.* Gall (the alike of a nest in eusocial Hymenoptera) is initiated by single fertilized macropterous female in spring in growing phyllode tissue of *Acacia oswaldii* and *A. melvillei.* The foundress oviposits inside the gall in which the offspring hatch, feed, develop and eclose. She produces four kinds of offspring: macropterous females, macropterous males, micropterous (short-winged) females, and micropterous males. The important feature of micropterous adults is their enlarged and armed fore legs specialized for fighting (Fig. 14.8). Micropterous adults attack and attempt to kill *Koptothrips* species (inquiline thrips), lepidopteran larvae and Argentine ants, *Iridomyrmex humilis.* The micropterous adults are therefore termed 'soldiers'. Besides, the ovarian development in micropterous adults was clearly poorer to that of the foundresses. There is overlap of generations, the morphological specialization and defensive behaviour of the micropterous female and its consequence for the survival of the foundresses certainly implies cooperative brood care and there is some level of reproductive caste differentiation. Thus, both *Oncothrips* species appear to satisfy the criteria required to consider them as eusocial. Crespi pointed "Australian gall thrips provide remarkable new opportunities for analysing the causes of the evolution of eusociality."

Fig. 14.8. Anterior portion of micropterous soldier daughter of a typical eusocial Australian gall thrips on *Acacia*. The enlarged and armed fore legs are specialized for fighting.

In a bark-dwelling species of the genus *Anactinothrips*, thrips live in a community, cooperate in brood care and forage with their young. Non-reproductive females are not found in this group. In certain gall-forming thrips defender females are with reduced wings and enlarged fore legs. Kranz *et al.* (1999) reported that in an Australian gall thrips, *Kladothrips hamiltoni* a single foundress, who rears her brood and remains within the enclosed gall for life, initiates galls. The adults of both sexes of her first brood cohort are morphologically distinct from the second generation, comprising a nondispersing soldier caste. The foundress and sib-mated soldiers jointly produce a second, dispersing generation, approximately 60-80 per cent of which are produced by the soldiers.

14.4.2 Termites

Howse (1970); Krishna and Weesner (1969, 1970) published their significant contributions on termites. Termites, commonly termed as 'white ants', (Order Isoptera) are highly social and predominant insects in tropical areas. They are considered to be closely related to cockroaches. The primitive termite genus, *Mastotermes*, like cockroaches, lays its eggs in an ootheca-like mass and has a lobe in the hind wing. They are well-known for construction of huge mounds or 'termitaria' and their ability to digest cellulose, which is done by symbiotic intestinal protozoans. Termites loose their intestinal protozoans after every moult and have to reinoculate themselves from the anal secretion of another member of the nest. The caste system of termites (Fig. 14.9) is somewhat similar to that of ants, though phylogenetically they are far distinct groups. In both the groups, number of workers is very large and they have a specialized caste- soldier. Termites are hemimetabolous and their immature stages are similar in form and behaviour with that of adults. Termites form two major groups, viz. lower and higher termites. Higher termites (Fam. Termitidae) lack symbiotic flagellates in

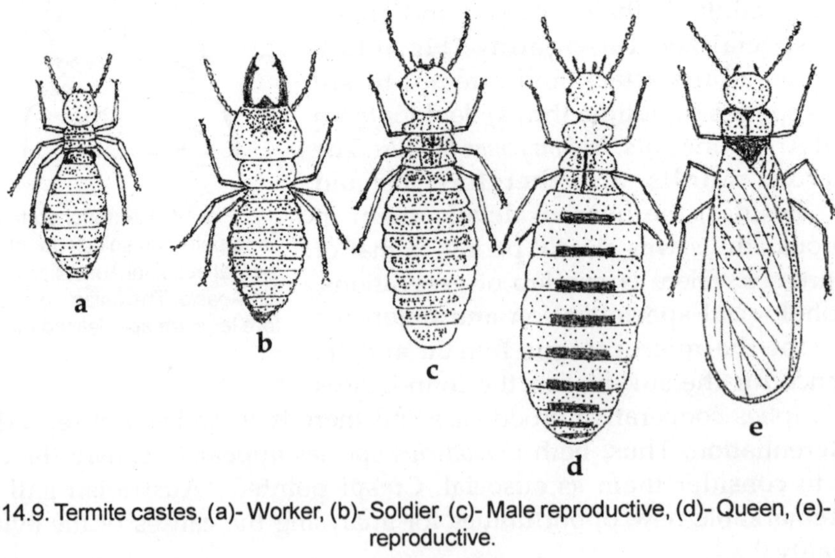

Fig. 14.9. Termite castes, (a)- Worker, (b)- Soldier, (c)- Male reproductive, (d)- Queen, (e)- Winged reproductive.

their gut and grow fungi inside their nests on piles of faecal pellets; they use this cultivated fungus to predigest food. Their queens undergo physogastry (with distended abdomen) and live in royal chambers (Fig. 14.10). These termites form the largest groups with millions of inhabitants in a colony.

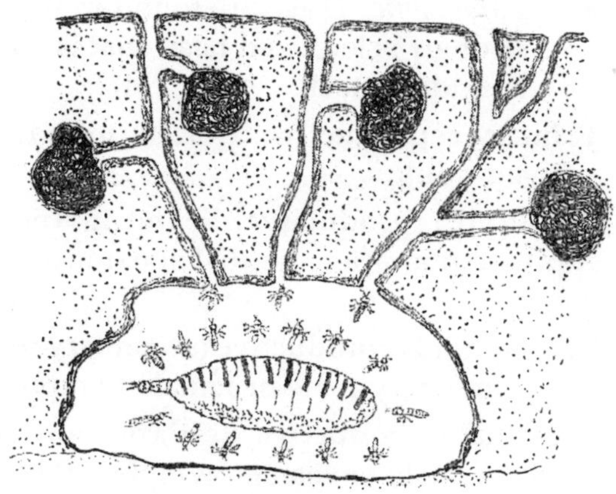

Fig. 14.10. A termitarium showing royal and fungus chambers.

The young stages in primitive termites are active, able to fend themselves and contribute significantly in the work of colony. Thus, they are employed as 'child labour'. However, in higher termites young are dependent on well-defined workers. In termites, individuals of both sexes participate in colony labour, which is their distinct character with that of hymenopteran social system. In a termite colony under noted castes/stages are present.

(i) **Larva.** Immature nymphal stage lacking any external trace of wing buds.

(ii) **Nymph.** Derived from larva, with external wing buds, capable of developing into functional reproductive.

(iii) **Worker.** Apterous individuals in which compound eyes and ocelli are reduced or absent. The workers build and maintain the nest; conduct all the foraging; care for the eggs, young and the queen.

(iv) **Pseudergate (False worker).** Present in lower termites and are either regressed nymphs or derived from larvae that undergo stationary moultings.

(v) **Soldiers.** Morphologically specialized with large modified mandibles, stopper-like head or glands capable of discharging defensive secretions. The soldiers may be of following types.

- *Mandibulate.* They are most common form with large mandibles and powerful head muscles, the mandibles are of different shapes.
- *Nasute.* They have a long frontal rostrum, through which they can spit a poisonous or sticky solution produced by their 'frontal gland', they occur only in the Nasutitermitinae.
- *Nasutoid.* These are found among the Rhinotermtidae and have mandibles of different sizes as well as an elongated, grooved, snout like structure through which they can spit saliva and a sticky solution secreted by the frontal gland.

- *Phragmotic.* These are found in the primitive Kalotermitidae and have a strongly sclerotized head that is used to block the nest entrance when it is under attack.

(vi) **Primary reproductive.** Colony founding queen or male derived from the winged adult.

(vii) **Adultoid reproductive.** Found in Termitidae and assume the function of reproduction after the disappearance of primary reproductive of the same sex.

(viii) **Nymphoid reproductive (secondary reproductive/ brachypterous neotenic).** Supplementary male or female nymphs with wing buds.

(ix) **Ergatoid reproductive (tertiary reproductive/ apterous neotenic).** Supplementary male or female nymphs without wing buds.

14.4.2.1 Caste differentiation in termites

In lower termites caste differentiation is rather flexible. They employ pseudergates, which are able to differentiate into other castes by moulting. The moulting may result into: (i) morphological changes, (ii) no morphological change (stationary moulting), (iii) formation of new caste, e.g. reproductive, (iv) development to a new morphology, missing a normal intermediate instar, (v) addition of an instar, and (vi) reversion of morphology.

It becomes rather difficult to determine the instar because of this flexibility in moulting. In some lower termites soldiers are absent while in some others more number of queens may exist in the same colony. Genetic differences among individuals of different castes must be either sex-related or due to differential expression of genes. Caste differentiation is greatly influenced by hormones, environmental factors, interaction between colony members, and social requirements. Termite colonies remain in a state of homeostasis and able to restore number of individuals of any particular caste after any disturbance. Homeostasis is also maintained by pheromones. In *Kalotermes* primary reproductives inhibit differentiation of supplementary reproductives and alate nymphs. Pheromone produced by reproductives is disseminated to pseudergates by anal trophallaxis and they transfer it to other colony members by oral trophallaxis. The juvenile hormone is considered to be responsible for differentiation of pseudergates into soldiers.

In higher termites routes for caste differentiation are rigid. Watson and Abbey (1985) discussed the routes of caste differentiation in *Nasutitermes exitiosus*. The differentiation occurs at first larval stage and three different routes exist, one leads to formation of reproductives and the others to sterile castes (Fig. 14.11). The differentiation may occur at first larval stage; however, some castes may not be morphologically recognizable until later moults.

I. Egg → Larva 1 → Nymphal instars 1 to 5 → Winged reproductives
II. Egg → Larva 1 → Small larva 2 → Small worker 1
Small worker 1 - - → Small worker 2
Small worker 1 → Small presoldier → Small soldier
III. Egg → Larva 1 → Large larva 2 → Large worker 1
Large worker 1 → Large presoldier - - → Large soldier
Large worker 1 → 4 moultings → Large worker

Fig. 14.11. Developmental routes of *Nasutitermes exitiosus*. Dotted arrows indicate minor routes of development (based on Watson and Abbey, 1985).

14.4.2.2 Division of labour

The primary reproductive couple constructs a 'nest cell' or 'royal cell' after nuptial flight and start brood rearing. Copulation takes place in nest cell and the queen lays the eggs. First offspring are workers that are fed on regurgitated wood and plant matter. They are also provided with gut symbionts. In lower termites the royal couple feed themselves and is capable of surviving for long periods; however, in many higher termites couple takes no food. Besides, higher termite pairs also provide food to their first brood from nutritional reserves. In early stages of colony development workers are produced while soldiers are produced in later stages. The workers and nymphs of first brood take over all activities, viz. nest construction, nursing, foraging, etc. They also fed royal pair with their salivary secretion. As the colony grows, the abdomen of the queen becomes large due to enlargement of ovarioles and fat bodies. The Queen cannot move so wherever she is placed that will be her permanent resting place. The reproductives are produced only when the colony is fully grown. These reproductives form new colonies in favourable season.

In different species of termites workers are morphologically similar but behaviourally diverse; however, soldiers are morphologically diverse but behaviourally uniform. The soldiers perform specialized function of defence. In some species such as *Macrotermes estherae* and *Nasutitermes costalis* the workers are dimorphic having large and small forms, while in the Macrotermitinae the larger workers are the males and the smaller workers the females. Most soldiers are 'mandibulate' and characterized by large, heavily sclerotized heads; powerful head muscles; sharp, elongated mandibles which are especially designed for biting and cutting. In different termite species shapes and sizes of mandibles differ. In some termites 'snapping soldiers' have asymmetrical mandibles and in *Pericapritermes* only left mandible strikes out, so that the target can be hit only if it is at the right side of soldier's head.

Soldiers are also specialized to use chemical defence. Many termites use salivary secretions for defence. The cephalic or frontal glands of certain termites are modified and open through a small frontal pore near the centre of head. In *Nasutitermes* chemical defence has evolved to its extreme and head of soldiers form a conical structure 'nasus' through which defensive chemicals are discharges (Fig. 13.27). Certain termite genera like *Rhynchotermes* and *Armitermes* have soldiers that possess both large mandibles and nasute head. These individuals are 'double threats' in their defensive role.

14.4.2.3 Termite nests

Most primitive termites live in dead wood, their homes are just the tunnels created while they are acquiring food, and their nests have no real structure. Species of *Rhinotermes, Reticulitermes* and *Coptotermes* are known as subterranean termites and have their nest below ground but also attack above ground wooden structures (Plate 14.3). They are soft-bodied insects and need protection from the outside environment. The workers build the shelter tubes from tiny pieces of soil, wood, and debris that are glued together using secretions and faecal material. Termites tend to have an extensive tunnelling system underground that allows them to carry food resources back into the colony. These shelter tubes also protect termites from predators. The more advanced Termitidae build the largest and most complex termitaria (Fig. 14.12). These termitaria are built of either soil excavated in course of digging underground tunnels or from soil and sand collected on the surface mixed with saliva and faeces (Plate 14.4). The shape of the termitaria depends upon the species. Some termitaria may be as high as 17 feet, while other termite species can construct mound that look like giant mushrooms. The whole mound is engineered, in such a way as to control the internal temperature for the colony and the growing fungus. The inside of the structure is built in such a way that not only the warm air can expel but also to allow air intake near the base of the structure. Lüscher (1961) demonstrated that in African fungus-farming termite, *Macrotermes natalensis* air circulation in a series of passageways give microclimatic control of temperature and carbon dioxide for fungus culture and brood rearing. The termite, *Amitermes meridionalis* is known as compass termite because it builds 3 metre high mound with one long axis and one short axis, the long axis always runs north/south and the short axis east/west. Thus, the termitarium has a large surface area facing the sun in the morning and the afternoon and very small surface area receiving the sun in the middle of the day. This helps in producing a steady temperature for the longest possible time (Hadlington, 1987).

Fig. 14.12. A termitarium.

Ball termites build their nest from soil and make covered walkway tunnels to travel through. This is to stop predators from picking them off while foraging. The termite ball is found in what looks to be in a precarious position balanced upon a branch or on the side of the tree (Fig. 14.13). The termites attach the ball to the wood using the soil and saliva.

Fig. 14.13. A termite nest at the base of a tree trunk.

Some social insects are always found in association with fungus, and some fungi can survive only in association of insects. Both have mutualistic association. Termites (Fam. Termitidae; Subfam. Macrotermitinae) are most conspicuous fungus-growing social insects. A fungus, *Termitomyces* develop as monoculture in their nests. Each nest of fungus-growing termite contains one or more fungal gardens, the shape and number of which depend on termite species and colony size. Each fungal garden (Plate 14.5a,b) is enclosed in a close fitting cavity lined with a mixture of saliva and dirt. The fungal chambers of certain termites are well ventilated. The workers deposit partly digested faecal matter on fungus for its growth. As the fungus-growing termites have no Protozoa in their gut it is presumed that the fungus assists in conversion of plant nutrients and digestion of cellulose (Batra and Batra, 1966). These termites feed continually on fungus. Batra (1975) observed nymphs and workers of *Microtermes obesi* eating spherules of its mutualistic fungus. These viable spherules were also transferred among individual termites.

14.4.2.4 Evolution of eusociality in termites

Evolution of sociality in termites is thought to be different from that in social Hymenoptera, possibly through feeding communities for exchanging cellulose digesting symbionts and intense sibling competition (group selection). Besides this, in termites both males and females are diploid and thus there is no genetic predisposition toward kinship-based eusociality.

Another possible explanation for the evolution of sociality in termites may be need-based, as they require symbiotic microorganisms for the digestion of cellulose. The need to transfer these symbionts might have encouraged a high level of interaction through trophallaxis. The transfer of symbionts in the individuals of next generation requires overlapping of generations. These factors might have encouraged them to lead a colonial life. But the concept does not explain evolution of polymorphism and altruistic behaviour in termites.

Thorne (1997) discussed the origin of eusociality in termites. Eusociality in termites converges along many lines with colony organization and highly social behaviour. Termite families lack asymmetric degrees of genetic relatedness generated by meiosis and fertilization. The evolution of eusociality in termites likely occurred in small families in which most helpers retained developmental flexibility and reproductive options. A suite of ecological and life history traits of termites and their ancestors may have predisposed them toward eusocial evolution. These characteristics include familial associations in confined, food-rich habitats; slow development; overlap of generations; monogamy; high-risk dispersal for individuals; opportunities for nest inheritance by offspring remaining in their natal nest; and advantages of group defence. The evolution of eusociality in termites probably evolved in response to a variety of contributing elements and the selective pressures that they generated. Thorne *et al.* (2003) demonstrated that in *Zootermopsis* inevitable meetings between termite families living within the same piece of wood result in assassination of some reproductives, merging of remaining populations of the two colonies, and differentiation of some helpers into new reproductives. This competitive circumstance provides a missing link in explaining how offspring helpers in ancestral termite families might have had opportunities to inherit their parents' nest resources. The same context of battles between neighbouring colonies frequently produces reproductive soldiers, a highly unusual termite caste that can be aggressive during intercolony fights and therefore appears to be a precursor to modern, sterile soldier termites.

Lo *et al.* (2000) determined the multiple sequences of genes in a diverse range of dictyopterans to know the evolutionary origin of eusociality in termites. Maximum parsimony and likelihood analyses of these sequences revealed strong support for a clade consisting of termites and subsocial, wood-feeding cockroaches of the genus *Cryptocercus*. This clade is nested within a larger cockroach clade, implicating wood-feeding cockroaches as an evolutionary intermediate between primitive non-social taxa and eusocial termites.

14.5 Social Hymenoptera

Social insects of the order Hymenoptera constitute a highly diverged group, with cosmopolitan distributions. Because of the existence of different castes, division of labour, altruism, and a haplodiploid genetic system, these social insects represent a unique system to study kin selection. The social Hymenoptera show primitive to high degree of eusociality. Paper wasps, hornets, yellowjackets, certain species of sweat bees, bumble bees are primitive eusocial insects while in ants and honey bees social organization has reached at its zenith.

14.5.1 Social wasps

Notable contributions on social wasps were published by Spradberg (1973); Ross and Matthews (1991). Eusocial behaviour among wasps is found only in certain members of the family Vespidae. Social wasps, which are commonly known as hornets (*Vespa* species), yellowjackets (*Vespula* and *Dolichovespula* species), and paper wasps (*Polistes* species), live together in communities called colonies and construct nest (Plate 14.6a-e). The yellowjackets usually built nests in underground cavities, such as old rodent burrows, tree cavities or other places. In hornets nests are always located above ground. Some species colonize hollow trees while others hang brood comb from a tree branch and surround it with paper walls for protection against the weather and natural enemies. Nests of common paper wasps are typically found under sheltered overhangs where they are protected from wind and rain. They use paper (wood pulp) to construct their nests. A wasp collects wood fibre by using its mandibles, but sometimes it also collects fibre from paper products. The insect then chews the wood and mixes it with saliva to make it moist and soft. They make combs, which are primarily used for brood rearing; however, they do not store food. The brood comb is always constructed like an inverted umbrella with open ends of the hexagonal cells facing downward. Workers usually cling to the underside of the comb as they guard the nest, feed the larvae, and perform other housekeeping chores. Hornets and yellowjackets built a paper envelope around the combs, which serves as good insulation material (calyptodomous nest). The nest with exposed or uncovered combs is termed 'gymnodomous'. All social wasps are carnivores, and stinging is an excellent defence in wasp colonies and a wasp is quite capable of stinging repeatedly.

The caste structure of the wasp's social system appears to be maintained primarily through aggressive interactions among the colony members. If a queen fails to sustain her dominant role, she will be replaced by another fertile female who assumes primary responsibility for egg production while males develop only from unfertilized eggs. In tropical areas, social wasps are active throughout the year. New colonies often form by swarming in which a fertile queen and a group of workers leave the parent nest and form new colony. In temperate areas, wasp colonies are founded in early spring by one or more queens who mated the previous summer and hibernated throughout the winter. The foundress constructs a small nest containing just a few brood cells. After laying a small complement of eggs, she feeds and cares for her offspring until they emerge as adults. These individuals (all females) become her workers and assume all brood care, foraging, and housekeeping duties. The queen continues to lay eggs and the colony grows larger throughout the summer. In early fall, the colony structure begins to break down. Unfertilized eggs give rise to males who mate with newly emerging females that will overwinter and found new colonies the following year.

14.5.1.1 Paper wasps

The subfamily Polistinae (paper wasps) is entirely composed of social wasps. This subfamily is characterized by two major behavioural groups, based on their mode of colony founding and mechanism of reproductive dominance. Independent founding is characterized by small, simply constructed nests without a protective paper envelope, founded by a single queen or few queens, without the assistance of workers. By contrast, swarm founding species have large colonies initiated by swarms consisting of multiple queens and a large number of workers.

Primitive eusociality is exhibited by polistine or paper wasps, *Polistes* species (Fam. Vespidae) and some others. They are social insects; since individuals of the same species cooperate in caring for the young, labour is divided between reproductives and workers, and generations overlap in such a way that offspring assist their parents in contributing to colony labour (Evans and Eberhard, 1970). Colonies are composed primarily of females while males are produced only for mating purposes. Mated, overwintered females that emerge from their hibernation sites on warm days start colonies in the spring. The colony is founded either by a single female or more gynes. If involving more females, they fight amongst themselves to establish a dominance hierarchy. The foundress (alpha female) and subordinate reproductives differ little, but show behavioural differences. The alpha female dominates all other individuals of a colony and devotes much of her time to social interaction. These behavioural divisions are not permanent, and if an alpha female is removed from a nest then another female (usually the second-most dominant, beta female) assumes the role and behavioural profile of the removed dominant. Finally the colony becomes monogynous, i.e. dominated by one queen with other reproductives either departing the nest or reverting to a worker-like status, and take on the tasks of foraging for water, plant fibres, nectar and caterpillars. The queen establishes a dominance and gain monopoly rights to lay eggs and construct cells. The other gynes may also lay some eggs, which are eaten by the queen or allowed to develop as workers. Sexual retardation in subordinates is reversible and in the absence of queen either a subordinate foundress takes over, or a worker can mate. Some species of these wasps are 'polygynous' as several queens are retained in the colony while some others are 'serial polygynous', having succession of functional queens. Drastic depredation to the colony may result from the destruction of the nest by predators. Ants frequently raid nests and take all the brood for food.

Reeve (1991) discussed the general life cycle of a *Polistes* and divided into four phases:

(i) *Founding (or pre-emergence) phase.* Involves a solitary female initiating a nest.

(ii) *Worker phase.* It starts with the emergence of the first workers, which on emergence take up most of the colony's foraging, brood care and nest maintenance. At this stage, the auxiliary (subordinate) foundresses are driven from the nest.

(iii) *Reproductive phase.* Emergence of the first female reproductives (gynes) is the start of the reproductive phase of the colony.

(iv) *Intermediate phase.* The period in which the gynes and males mate and then disperse from the natal colony, before overwintering and the start of the next colony cycle.

Ropalidia marginata is a primitively eusocial polistine wasp widely distributed in the Peninsular India. Gadagkar (2001) discussed division of labour and work organization in this wasp. A single egg-laying queen heads the colony, while the remaining females function as sterile workers. There is lack of morphological caste differentiation between queen and workers, which is compensated by a system of behavioural caste differentiation. The wasps in a colony may be classified into three behavioural castes, called 'sitters', 'fighters' and 'foragers', and the queens are almost always in the sitters caste. The queens are relatively inactive, behaviourally subordinate individuals. The workers continue to remain active, bring food and feed the larvae, even if the queen is removed. Their activity appears to be regulated by the workers themselves. In undisturbed colonies, intranidal workers who also unload food and pulp bearing foragers, appear to regulate foraging rates. In spite of primitive eusocial status, *R. marginata* has a well-developed age polyethism. Workers show strong preferences to feed larvae, build the nest, bring pulp and food, in that order, as they age. Sumana and Gadagkar (2003) reported that *R. marginata* queens are strikingly docile individuals who show little or no physical dominance. When such a behaviourally docile queen is removed from a colony, one of the remaining individuals becomes extremely aggressive, to go on to become the next queen if the original queen is not returned. When the original queen is returned after a day's absence, she re-establishes herself as the queen and she usually manages to do so with little or no aggression.

The paper wasp nest is open-faced, single-layered, and inverted umbrella-shaped (Fig. 14.14). These nests are constructed above the ground and can be found hanging with a petiole from parts of trees, various objects at home, and other protected areas. The nest material is a water-resistant gray paper made up of a mixture of masticated wood and the salivary secretion of the female wasps. The nest construction started with the emergence of overwintering queen. She has underdeveloped ovaries and first feed on nectar and plant sap. She finds a suitable nesting site and brings masticated wood pulp to the site. The pulp is applied to the under site of a masticated support, on which the nest is formed. At first two cells are formed and further, new hexagonal cells, wider at open end are added. At first the queen mainly forages for the collection of construction material, but with the development of larvae she also brings food to nourish them. The workers, after emergences undertake various duties, while the queen will then devote all her time to egg-laying. At the end of season males and gynes are produced, and with the failure of queen, the colony dwindled and males mate with gynes. The mated females seek suitable overwintering site.

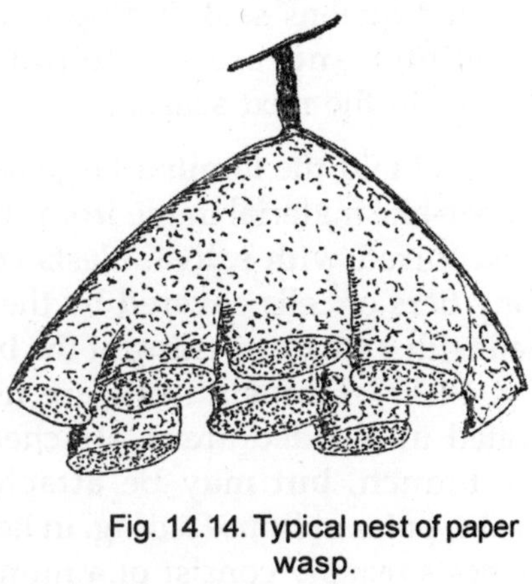

Fig. 14.14. Typical nest of paper wasp.

14.5.1.2 Yellowjackets and hornets

The term yellowjacket refers to a number of different species of wasps in the genera *Vespula* and *Dolichovespula* (Fam. Vespidae) with distinct yellow and black markings (Fig. 14.15). In these insects a young fertilized queen builds a small nest from wood fibres, below soil surface (Fig. 14.16), in tree cavities or other places. Nests may be aerial or terrestrial, depending in part upon the species of the wasp. Some species may construct both types of nest. Regardless of location, each nest is a series of horizontal combs completely surrounded by a paper envelope. She lays a single egg into each nest's cell. The queen feeds the developing larvae with nectar and insect prey. After emergence the first workers take over the work of gathering food and enlarging the nest and the queen just lays eggs. For most of the season the colony consists of sterile worker females, which are noticeably smaller than the queen. Each worker tends to perform a given task, for a given day, but may change tasks if the need arises. In autumn, larger cells are constructed for new queens, and at the same time, the queen begins to lay unfertilized or male eggs in either large or small cells. Then the social structure of the colony declines, as new queens and males are produced. Mating occurs during nuptial flights. The newly fertilized queens seek hiding places to overwinter and are the founders of colonies in the next season.

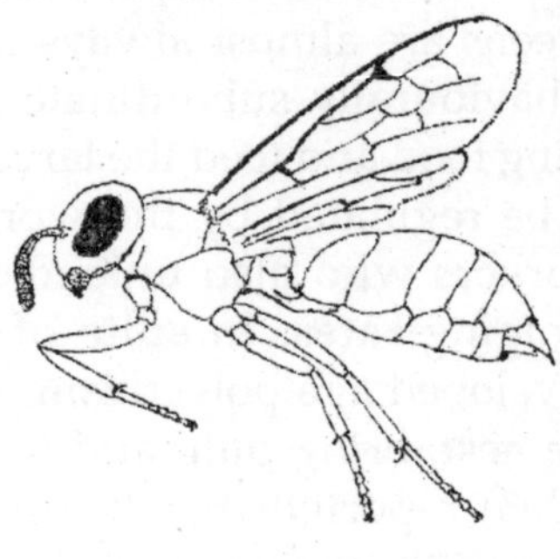

Fig. 14.15. A yellow jacket, *Vespula* sp.

Hornets build the familiar large nests of a paper-like material made from chewed wood mixed with saliva. Nests contain many tiers of cells covered by the outer shell with a single opening at the bottom (Fig. 14.17). Hornets nests are usually located in wooded areas, attached to a tree branch, but may be attached to shrubs, poles or house siding. In hornets, the nests usually consist of a number of stories or 'tiers', one below the other and completely enclosed by spherical walls. Each cell may be used for two or three successive batches of brood. Hornets can be aroused to sting in great numbers, but only when the nest is disturbed or threatened.

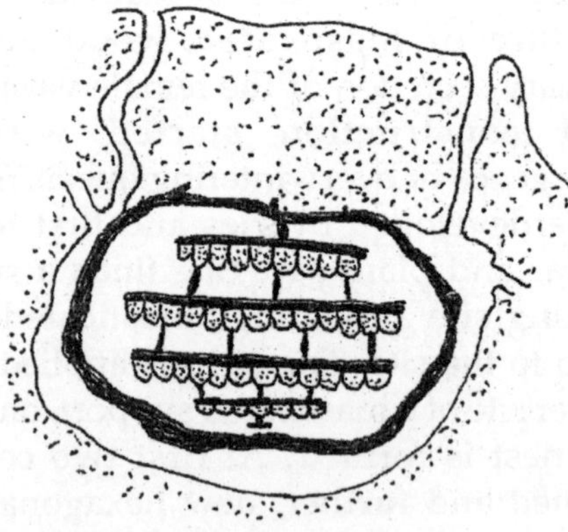

Fig. 14.16. Underground nest of a wasp.

14.5.2 Social bees

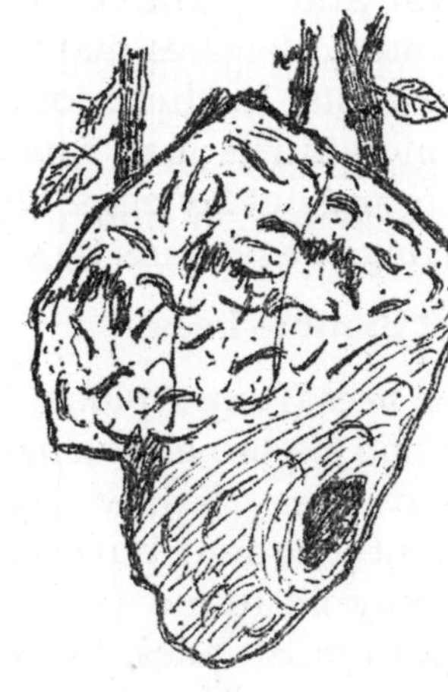

Fig. 14.17. A hornet nest.

Bees are diverse in habits and forms, and exhibit different levels of social organization. The nesting habits of bees are interesting and they use materials like wax, dung, mud, resin, sand, pebbles, plants parts, glandular secretions, fibres, etc. for nest construction. The nests may be built on trees, rocks, and walls; in cavities, hollow trees, self-made holes, soil, and other places (Batra, 1977). Males ordinarily do not help with nest building, defence or provisioning. All bees feed on nectar, honeydew and pollens. For feeding they may be dependent on one species or genus of flowers (monolectic), or a few related genera (oligolectic); or may visit flowers of many unrelated genera (polylectic) (Linsley, 1958). The evolution of social behaviour in bees has attracted much attention. However, about 80-85 per cent bees are solitary, each female making and provisioning her own nest without any assistance. Social bees exhibit primitive to highly organized social behaviour and in honey bees it evolved to its zenith.

14.5.2.1 Sweat bees

Halictine sweat bees (Fam. Halictidae; Subfam. Halictinae) exhibit an astonishing range of social behaviour and are significant model organisms for studying the evolution of altruism and eusociality. Their social behaviour is not only extremely variable, but also ecologically and evolutionarily labile. Many species are solitary, with females constructing nests and raising brood, but they also exhibit many types of social organization, ranging from more or less egalitarian communal societies, to semisocial and eusocial types (Batra, 1966). The communal societies do not have queen and worker castes, being composed of multiple females that may or may not be related. Eusocial societies are characterized by queen and worker reproductive castes, and queens lay all or most of the eggs. The communal societies are composed of adults of the same generation, whereas eusocial societies are formed when a queen's offspring become workers. A striking aspect of halictine social variation is the mutual exclusivity of communal and eusocial types of colony social organization within the same species. Richards *et al.* (2003) reported an exception to this rule in a population of *Halictus sexcinctus* that contained both communal and eusocial colonies. Moreover, communal and eusocial females exhibit morphological differences that imply a preimaginal developmental switch, which could also underlie the two types of social behaviour.

Eusocial sweat bees make nest in ground (Fig. 14.18a,b) or in rotten wood. Their soil nests rarely have more than a few occupants. The fertilized female constructs the nest and is joined by other overwintered females that have not started nests. In these joint nests some individuals may function as workers and not lay eggs. Only one egg-laying bee remains by summer.

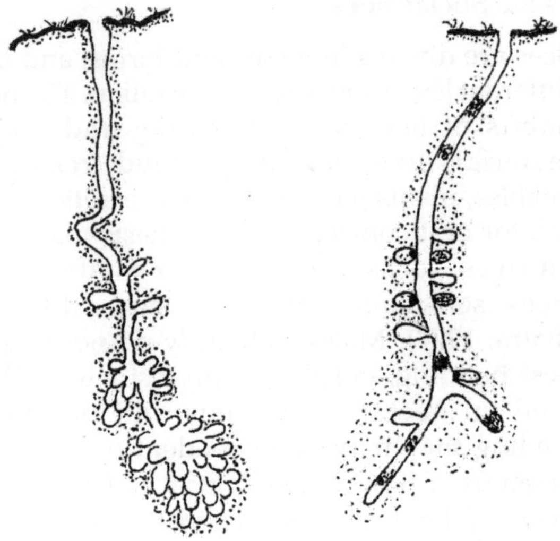

Fig. 14.18a,b. Nests of soil dwelling halictine sweat bees.

14.5.2.2 Bumble bees

Bumble bees (Fam. Apidae) feed on pollen and nectar, and usually nest in cavities below ground (frequently rodent burrows), but nests may be found on or above ground in brush piles, trash heaps, bird houses, etc. They form colonies like honey bees; however, their hives are usually much less extensive than that of honey bees, because of the small size (Fig. 14.19). Single fertilized female starts nest in spring. The female builds a large irregular cell of wax and pollen and stocks it with pollen and nectar. Several eggs are laid in the cell; the female then enlarges it and supplies the young with additional food. They become workers when mature and take over the tasks of pollen and nectar collection, feeding young, and other nest duties. Foraging bumble bees tend to avoid flowers already visited by other bumble bees of the same species. This is because bumble bees can scent mark the flowers they have visited. The queen maintains dominance purely by aggressive behaviour.

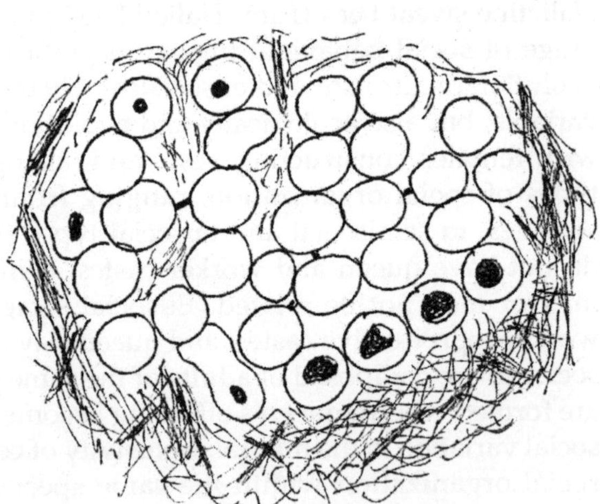

Fig. 14.19. Nest of bumble bee, *Bombus* sp.

Honey and pollen may be stored in vacated cells. These bees only store a few days' worth of food and so are much more vulnerable to food shortages. Nests rarely have more than a hundred workers at a time. Bumble bees mostly do not preserve their nests through the winter in temperate areas, though some tropical species live in their nests for several years. The last generation of summer includes a number of queens who overwinter separately in protected spots. Only future queens survive the winter by hibernating in the ground. Bumble bees can sting and aggressively defend their colonies.

14.5.2.3 Honey bees

Honey bees are known to be the fascinating eusocial creatures. Extensive studies have been conducted on social behaviour of honey bees because of their economic significance and ease to observe them. Man on one hand, is a friend of bees, as he provides better living conditions to them, facilitate the bees to increase their broods, manage them in a proper manner and grows crops from which bees gather their food; while on the other hand the greedy man is also pertinent enemy of bees as he destroys the bees and their progeny by adopting crude method of honey extraction (Agarwal and Singh, 2005). People admire honey bees for their diligence, unity, sacrifice, tolerance, cooperation, coordination, division of labour, social service and usefulness. They live in large colonies, each consisting of a queen (mother of the colony), a few hundred of drones (males) and several thousand sterile female workers (Fig. 14.20a-c). Workers, as their name implies, do most of the 'work' around the hive. Four species of honey bees, viz. giant honey bee, *Apis dorsata* (Plate 14.7); Indian honey bee *A. cerana* (Plate 14.8); small honey bee, *A. florea*; and European honey bee, *A. mellifera* (Plate 14.9) are well-known in India.

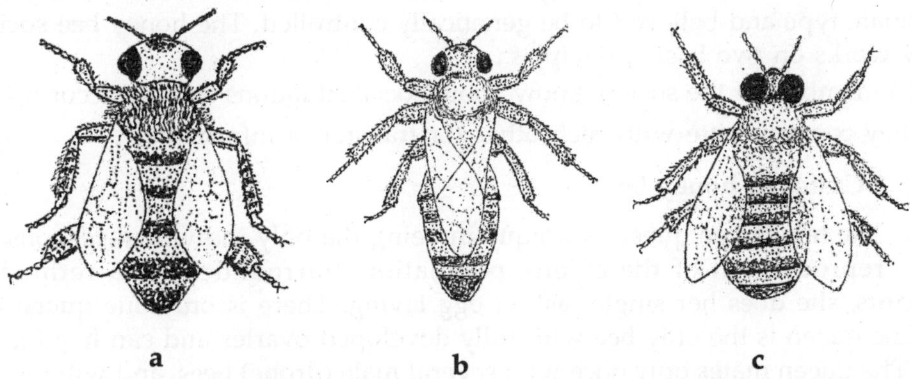

Fig. 14.20. Honey bee, *Apis mellifera* castes, (a)- Worker, (b)- Queen, (c)- Drone.

During its development honey bee undergoes complete metamorphosis and the development is indirect. The queen is capable of laying fertilized eggs that produce worker bees or queens, and non-fertilized eggs that result in males 'drones'. The eggs look alike in appearance; however, they are laid in cells made for rearing of different castes. During the first three days, the embryo inside the egg will develop rapidly. Then, just prior to hatching, the egg is provided a minute drop of royal jelly. This occurs to all brood regardless of caste. The hatched larvae are nourished by workers and grow in size. In worker brood, the larva is liberally fed glandular food or 'royal jelly' for two or three days, followed by a honey and pollen diet. Thus, caste differentiation is largely 'trophogenic' and quality and quantity of food determines the worker caste. When the diet is not changed and remains royal jelly throughout the larval period, she becomes a queen. However, the trophogenic effects cannot always be separated from hormonal effects. The role of juvenile hormone (JH) in polymorphic caste differentiation in social insects is well-known, as the nutritional status is linked with the activity of corpora allata, which secrete JH. The growing larvae are similar in early stages of development, but the queen and drone larvae become larger during later half of their development. When the larva is fully grown and fed, and at this point workers seal the cell with a porous capping. This is called the 'capped brood' stage. The pupae of three castes differ from each other and can easily be distinguished. The adult bee hatched from the pupa. Workers and drones are reared in hexagonal shaped cells that comprise the comb, but the queen is reared in an acorn-shaped cell, normally protruding vertically from the comb surface, with the opening at the bottom.

In the honey bee society all the three castes, viz. workers, drones and queen perform specific functions. Queen has the sole responsibility of laying eggs, drones mate and fertilize the queen and the workers perform all other duties in the hive. The honey bee colony acts as a single unit and the social behaviour of honey bees is of innate type and believed to be genetically controlled. The honey bee society mainly works on two basic principles.

- All members of the society know about social situations and act accordingly.
- They communicate with each other for transfer of information.

14.5.2.3.1 Castes of honey bee

Queen. The honey bee queen is unique in being the only individual responsible for the reproduction of the colony population. Surrounded by a retinue of attendants, she does her single task of egg laying. There is only one queen per hive. The queen is the only bee with fully developed ovaries and can live for 3-5 years. The queen mates only once with several male (drone) bees, and will remain fertile for life. Her sole function is to serve as the reproducer; she acts as an 'egg laying machine' and lays 1000-1500 eggs per day but does not nurse her brood.

A multicomponent pheromone termed as 'queen substance' is secreted from the madibular glands of queen bee, coats her body and is transferred to workers by trophallaxis. The main compounds of queen substance are: 9-oxo-(*E*)-2-decenoic acid (9-ODA) and 9-hydroxy-(*E*)-2-decenoic acid (9-HDA). It has also been reported that both enantiomers, the (+) and (-) of 9-HDA are present along with two other substances namely, methyl *p*-hydroxybenzoate and 4-hydroxy-3-methoxyphenylethanol.

Worker. All worker honey bees are female, but they are not able to lay eggs. Worker bees may live for 3 months during the summer season, but only 6 weeks during the busy winter months (they literally work themselves to death). Her six-week life span in winter is devoted to carrying out the many tasks necessary for colony development and survival. Many duties carried out by the worker bee are the result of physiological changes that take place during the worker's life. The worker bees sequentially take on a series of specific chores during their lifetime: housekeeper, nursemaids, construction worker, storage duty, undertake guarding; and finally, after 21 days they become forager. The worker bee differs morphologically from other castes in having pollen collection apparatus, barbed sting, presence of certain glands and other features. The stinging results in her death; therefore, she can only sting once. During stinging, the worker releases alarm pheromone which attracts other bees to the location in their defensive behaviour.

Three distinct phases may be observed in the life of the worker bee:

1. The 'nurse bee' phase during which she helps in incubating the brood.
2. The 'domestic' phase during which the worker bee takes on various hive duties including the storing of honey, building and repairing comb, temperature regulation, cleaning the hive and guard duties at the hive entrance.
3. The 'field' bee or 'forager' that collects pollen, nectar, propolis or bee glue, and water. At this final stage in life, most worker bees die in the field while performing their duties.

Drone. The drone is haploid and develops from an unfertilized egg. All his genetic characteristics originate from his mother, the queen. They do not perform any duty within the hive and their feeding requires considerable resources from the colony. There are only 300 or more drones in a hive. Because drones are of no use during dearth period, they are expelled from the hive.

14.5.2.3.2 Age polyethism in workers

Honey bee workers are monomorphic but exhibit age polyethism. The age polyethism has reached its greatest development in honey bees (Table 14.1).

Table 14.1. Age polyethism among worker honey bees.

Age of the worker	Duties
1st-3rd day	The workers are quite young and unable to take their food, but perform the duties of cleaning and temperature regulation
4th day	Start taking food and also provide food to growing larvae and young bees
5th-6th day	Royal jelly glands develop in the head region, which secretes a milky liquid known as 'royal jelly' or 'bee milk'.

12th day	Wax glands become functional and the wax is used for making cells and repair
13th day	Worker starts coming out of the hive, takes short orientation flight, note the landmarks and comes back in the hive
18th-20th day	Perform the duties of cleaning, temperature regulation and guarding
21st day	Bee becomes forager and starts collecting nectar, pollen, water and propolis

14.5.2.3.3 Honeycomb

The workers construct the honeycomb (Plate 14.10) using the wax secreted from the wax glands situated ventrally on 4th-7th abdominal segments. A honeycomb is a mass of hexagonal wax cells built by honey bees in their nests to contain their larvae and stores of honey and pollen. Wax is mixed with saliva and used for making hexagonal cells in a parallel series or comb. The axes of honeycomb cells are always quasi-horizontal, and the non-angled rows of honeycomb cells are always horizontally (not vertically) aligned. Thus, each cell has two vertical walls, with 'floors' and 'ceilings' composed of two angled walls. The cells slope slightly upwards towards the open ends. The architectural design of the comb (Plate 14.11) of all honey bee species is essentially similar. It is likely that the honey bees construct the honeycomb based on instinct, and the appearance of such efficient shapes in nature is a result of natural selection. Worker bees are reared in small cells while drones' cells are large. Queens are reared in acorn-shaped cells. Bees store pollen in lower and central cells whereas honey is stored in upper and peripheral cells.

14.5.2.3.4 Formation of new colony

With an increase in the number of workers swarming occurs. With the onset of swarming season, secretion of queen substance decreases and workers construct new queen cells. The queen lays eggs in these royal cells and hatched larvae are fed special diet. The queen leaves the hive along with a number of workers soon after the first royal cell is capped. The swarming bees fly *en masse* for a short distance and settle on a suitable site. Scout bees then fly to find a new permanent nest site. On return, they signal the distance and direction of the site by dance. Consequently the entire swarm shift to the permanent site.

The workers at the parental nest remain without queen for some time. The first virgin queen after hatching finds her rival sisters in the hive and if other queens are present fighting ensues, which continues until only one queen is left. Around third day of her life the virgin queen takes nuptial flight and release small quantity of pheromone to attract drones. The mating takes place during flight and the drone, which leaves his sex organs in the queen, dies shortly afterwards. The queen makes a number of flights and every time she mates with a new male. Thus, she receives enough sperms for her lifetime. Then she takes over the nest and starts egg laying. The queen is continuously surrounded by young worker attendants, who also lick her body for the pheromones that are needed for well being of the colony.

14.5.2.3.5 Working behaviour of forager bees

Forager bees show a fidelity to a source that is they restrict themselves to the blossoms of plants of on species at a time. They shift to the other source when nectar and pollen availability ceases. Sometimes nectar and pollen loads are mixed to a small extent. They restrict their foraging to a small area and make it possible with the help of landmarks. The foraging range of *Apis cerana* is up to nearly one km while the *A. mellifera* may go up to a distance of 3 km or more. Number of flowers visited by a bee depends upon many factors like amount of nectar and pollen, stage of flowering, climatic conditions and also the competition with other foragers. The workers used to go for foraging only when the temperature is above 15°C. The honey bee forages around the hive in search of resources. Several landmark studies were performed in the middle of the 20th century to analyse how these social organisms communicated exact distances and direction to one another in order to effectively locate these resources. Honey bees perform distinct dances to communicate the approximate distance and direction from the hive to the food resource (see section 11.2.2.8).

Mechanism of nectar collection. Foragers survey the vicinity for the presence of nectar plants or may be guided by dancing bees. Odour is also an important cue for the bees in finding the food source. Mouthparts in workers are modified for sucking and lapping. These are highly specialized organs assisting in collection of pollen and nectar. With the lapping motion of the tongue the nectar is drawn into the food canal. If the bee finds the nectar is to be more viscous then mixing with saliva dilutes the same. The ingested nectar is stored in 'honey chamber'. After returning to the hive the nectar load is transferred to hive-bees and they store it in proper cells. *Apis mellifera* can carry a maximum nectar load up to 80 mg while the average load is 40 mg. In case of *A. cerana* the average load is 20 mg; however, it may carry a maximum load of 40 mg. A load is collected from a number of flowers and the total number of flowers likely to be visited by a bee depends on the type of flower and availability of nectar in the flowers. The unripe honey is deposited on the sidewall of the comb cells by hive bees. Further evaporation of water takes place in the hive by fanning and manipulating by the bees.

Mechanism of pollen collection. Forager bee scribbles the anthers with her tongue and mandibles. Pollen collection is quick in case of flowers with loose pollen but sticky pollens are collected by biting with mandibles. Sometimes saliva and some nectar are also added to the pollens so that they become moist. The pollen grains also stick to the body parts and hairs on bee body. The legs also assist in pollen collection. Fore and mid legs have pollen brush which help in collection of pollens and cleaning of body parts. The tibiae of hind legs are modified into pollen baskets, which are used for storage of pollens (Fig. 14.21). The weight of a pollen load may be 11-30 mg. When the bee goes out for the collection of pollen it also carries required amount of honey for fulfilling its energy requirements. The pollens are stored in cells near the brood area. The hive bees pack the load into cells after mixing it with honey or nectar. These pollens form essential protein component of diet of larvae and adults.

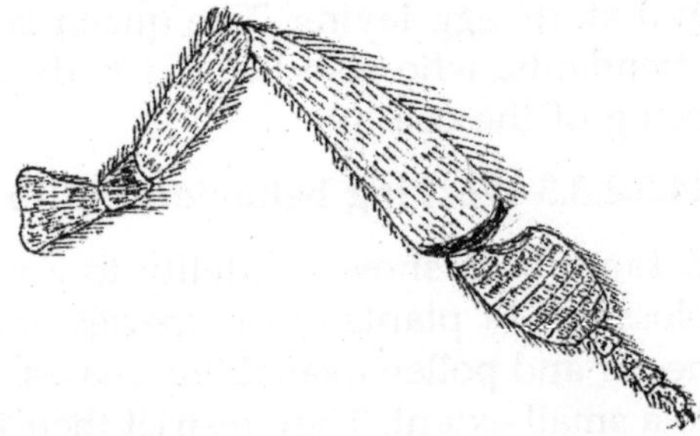

Fig. 14.21. Modified hind leg of honey bee, *Apis mellifera* for collection of pollens.

Economics of foraging. Foraging requires energy and the honey bees are able to calculate the economics of energy consumption and the net gain of food to the colony. Generally bees fly only as far as necessary to secure an acceptable food source from which there is a net-gain. Factors that influence foraging behaviour and determine profitability are: (i) weather (wind, temperature, and sunlight), (ii) distance of the food source from the hive (including differences in elevation), and (iii) food quality (concentrations of sugars, protein content of the pollen).

Foraging range. Bees may fly as far as 12 km, but usually foraging is limited to food sources within 3 km. Approximately 75 per cent of the bees from a colony forage within one kilometre while young field bees only fly within few hundred metres.

Constancy to species. Foraging bees tend to limit their visits to a single species of plant during each trip. This behavioural adaptation is critically important for plants and foraging constancy is essential in optimizing seed set and fruit development.

Speed of work. Bees visit up to about 40 flowers per minute depending on floral type, nectar availability and weather conditions. If more nectar is available the time spent on a flower will be more. It takes twice as much time to collect a load of nectar compared to a load of pollen.

Temperature conditions. Below 8°C - no foraging, 8°C - 16°C - some activity, 16°C - 32°C - maximum foraging, and above 32°C - reduction in foraging, increase in water collection.

Speed of flight. A bee while leaving the hive flies at a speed of 20 km/h while loaded bee flies at a speed of 25 km/h. Foraging will stops at a wind speed of 40 km/h.

Number of trips per day. The number of trips will depend on conditions, viz. weather, forage availability, colony strength, etc. In general, 5-15 trips are made while a water collector may make as many as 100 trips per day.

Nectar flow. It is the period when there is much abundance in nectar production that the bees gather a surplus beyond the immediate needs of the colony, which is converted to honey and stored in the combs.

14.5.3 Ants

Ants (Fam. Formicidae) are the dominant group of social insects and flourish on all land areas of the earth. An ant colony may contain from a few to 20 million individuals. Most ant species live in the soil. Some, like the carpenter ants, also live in wood (they excavate, but do not actually eat the wood). Some ants live beneath stones, in cavities made inside plants such as acorns, twigs, and galls. They feed on a variety of small living and dead insects, nectar, plant juices, grains, seeds, sugars, meat, honeydew, etc. They can take even large prey, cut them into small pieces and transport them to the nest. Many ant species live in association with honeydew producing hemipterous insects and protect them from other insects. Ants need a balance diet of carbohydrates and protein. Proteins are especially needed for the queen to produce eggs and for the larvae to grow. Some ant species are able to maintain steady state conditions in their colonies or nests, e.g. in temperature and humidity. This is called homeostasis and is essential for colony health.

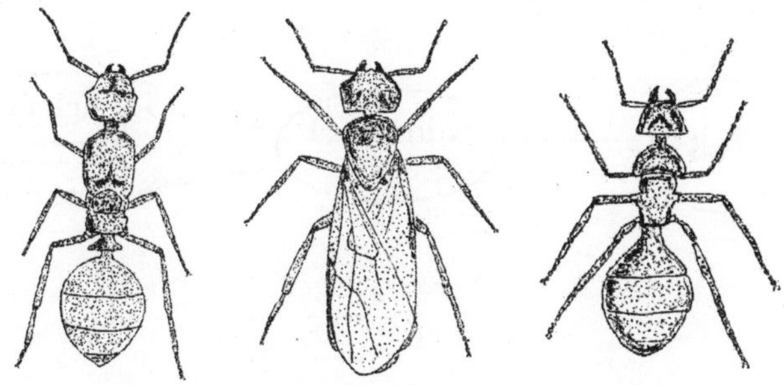

a b c

Fig. 14.22. Castes in *Formica polyctena* (a)- Wingless female (queen), (b)- Winged male, (c)- Infertile worker.

Ants are polymorphic social insects (Fig. 14.22) with reproductive queen and workers which are females and differ structurally. The males are haploid and produced from unfertilized eggs. Many ants have monomorphic workers, but others have different subcastes based on their size- major, media and minor (Fig. 14.23). Similar to other social insects queen's job is only to lay eggs while all other duties are undertaken by the workers. In ant species with monomorphic workers, there may be age polyethism and young workers perform tasks in the nest while older workers act as foragers. If workers are polymorphic large workers act as soldiers.

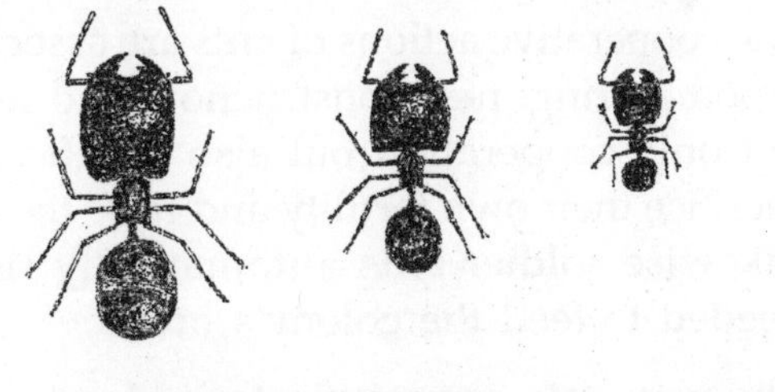

Fig. 14.23. Forms of workers of ant (a)- Major, (b)- Media, (c)- Minor (modified after Wheeler, 1910).

Caste differentiation is trophogenic and a high protein diet promote differentiation of queen and less nutritious diet results in formation of workers. The JH also plays an important role in caste differentiation (Fig. 14.24). In workers of some ant species ovaries are functional and they can lay haploid eggs, from which males are formed. Under favourable conditions the winged reproductives make nuptial flights to found a new colony. One or more males inseminate the queen, and after mating she lost her wings and found a protected place or excavated a chamber. In this protected place 'the nest', she starts to lay eggs, from which after a certain period, emerge the larvae. Initially the queen is responsible for all colony tasks, not only feeding the larvae and herself, but also for the maintenance of the nest and for colony defence. The hatched larvae are fed by stomodeal or oral trophallaxis, i.e. regurgitation of liquid food from her internal food reserves.

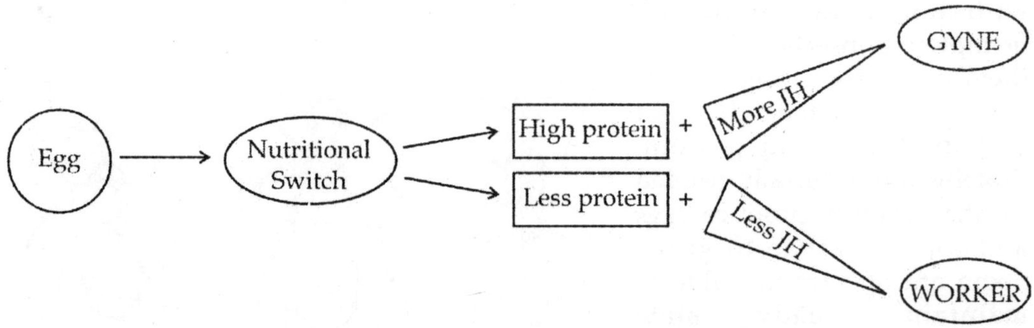

Fig. 14.24. A model of caste differentiation in ants. High protein diet leads to increase in juvenile hormone levels which promote differentiation of queen and less nutritious diet results in formation of workers.

After the eclosion of the first workers, the queen does not perform most behaviour necessary to nest maintenance and colony development any longer, and restricts herself only to egg laying, while the workers do all other tasks, including the feeding of the queen. The colony passes through a growing phase which results an increase in population, nest size and foraging area. In some ants more than one queen found the colony and this is termed as 'pleometrosis'. Multi-queen nests may persist as polygynous colonies.

The cooperative actions of ants are associated with food gathering, colony defence, brood rearing, nest construction, and rearing the next generation. This demands not only cooperation but also sacrifice. The female worker ants (Fig. 14.25a-h) sacrifice their own fertility and raise the eggs of a single, continually laying queen. Likewise soldier ants automatically risk their own lives to defend a territory needed to feed the colony's brood.

The way ants communicate is also very interesting. They produce an array of pheromones that serves as the ant's language. Through these pheromones ants can convey messages ranging from the location of food to the presence of danger.

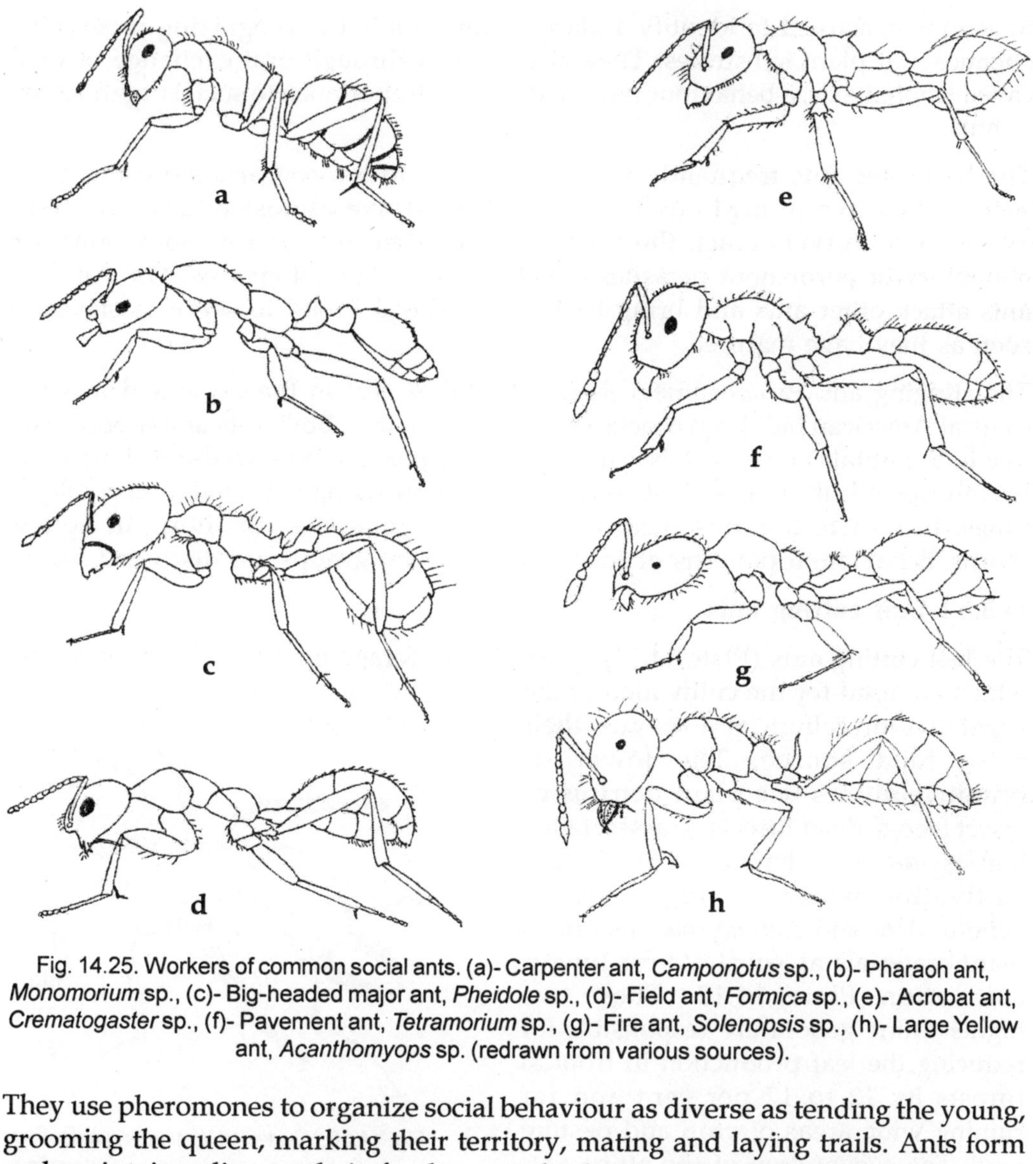

Fig. 14.25. Workers of common social ants. (a)- Carpenter ant, *Camponotus* sp., (b)- Pharaoh ant, *Monomorium* sp., (c)- Big-headed major ant, *Pheidole* sp., (d)- Field ant, *Formica* sp., (e)- Acrobat ant, *Crematogaster* sp., (f)- Pavement ant, *Tetramorium* sp., (g)- Fire ant, *Solenopsis* sp., (h)- Large Yellow ant, *Acanthomyops* sp. (redrawn from various sources).

They use pheromones to organize social behaviour as diverse as tending the young, grooming the queen, marking their territory, mating and laying trails. Ants form and maintain a line to their food source by laying a trail of pheromone, i.e. a chemical to which other members of the same species are very sensitive. They deposit a certain amount of pheromone (from the Dufour's gland) while travelling back to the nest via a direct route, and each ant prefers to follow a direction rich in pheromone. This enables the ant colony to quickly find the shortest route (see section 18.4.4). It has also been suggested that ants also use extra social chemical learning. The 'topochemical sense' hypothesis proposes that foraging ants also use local 'chemical emanations' for navigation. Helmy and Jander (2003) observed that black carpenter ants, *Camponotus pennsylvanicus* on their way to and from a food source, learned to identify a place on their route by recognizing preexisting chemicals of plant substrates. They also interact through oral exchange of food called trophallaxis, a behaviour pattern of critical importance to social organization in ants.

The harvester ants frequently visit grass fields to harvest and store the grass seeds. They live in arid environments and survive almost totally on seeds. Specialized workers crack the seeds for the other ants to eat. Some ants are temporary or permanent parasites in colonies of other ant species. The Amazon ants attack other ants and bring back unconsumed brood to serve as slaves as soon as they have matured.

The stinging ant, *Pseudomyrmex ferruginea* makes nest in the expanded thorn of Central American bull horn acacia, *Acacia cornigera* tree. Both ants and acacia plant live in a mutualistic symbiosis. The colony occupies the hollowed-out thorns and fiercely defends the tree against ravaging insects, browsing mammals and epiphytic vines. In return, the ants feed on nectar and get space for living in hollow thorns. Behavioural patterns of some important ant groups are discussed below.

14.5.3.1 Leaf cutting ants

The leaf cutting ants (Plate 14.12) (Tribe Attini) forage on a variety of substrates which are used for the cultivation of the vegetative mycelium of a fungus, their main food source. The lower or primitive attines use plant detritus or insect faeces, dead insects, grasses, fallen leaves as a substrate for fungal cultivation while the higher attine genera, *Atta* and *Acromyrmex* use fresh plant materials as substrate for fungus cultivation (Plate. 14.13). These two higher attine genera are responsible for reducing the leaf production in tropical forests by 12 to 17 per cent and for ruining wide areas of crop and pasture land. The coevolution of the attine ants and their fungi dates from 50 million years back, reaching varying degrees of co-dependency in each case. It wasn't a single acquisition event, but rather something that happened over a period of time as the ants were driven into the role of associating with fungi (Mueller *et al.*, 1998). They harvest plant material and take to the underground chambers in which the fungus is cultivated. The workers (Fig. 14.26) cut out pieces of leaves with their strong mandibles and head clumsily back to the nest (Plate 14.14). All around them smaller ants weave to and fro, constantly scanning the surrounding

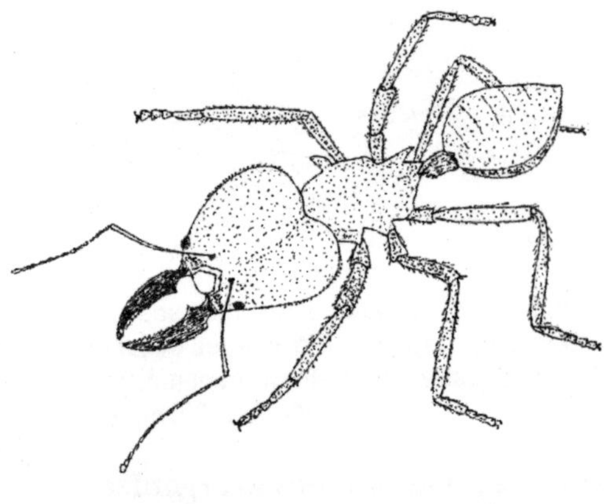

Fig. 14.26. Worker of leaf cutting ant.

terrain for danger. Some of these smaller ants may also hitchhike on the leaf fragments carried by their bigger sisters and act as defender (Fig. 14.27). The plant material is licked, cleaned and then chewed into a pulp before using as a substrate for the fungus. The exploitation of fresh plant material has contributed significantly to the success of the leaf cutting ants and has made them one of the most dominant herbivores in Central and South America; where they are highly abundant and have become severe agricultural pests (Hölldobler and Wilson, 1990). The ants provide the fungus with a suitable and stable environment, in that it is cultured in nearly pure gardens. The fungus is continuously manured with fresh plant material, protected against abiotic conditions, parasites, competitors and predators. Little *et al.* (2003) reported that leaf cutting ants have evolved complex disease management strategies to reduce their garden's exposure to potential parasitic microbes. The infrabuccal pocket, a filtering device located in the oral cavity of all ants, is an integral part of the mechanisms that they use to prevent the invasion and spread of general microbial parasites and the specific fungal-garden parasite, *Escovopsis*. Fungus-growing ants carefully groom their garden, collecting general debris and pathogenic spores of *Escovopsis* in their infrabuccal pocket, the contents of which are later expelled in dump chambers inside the nest or externally.

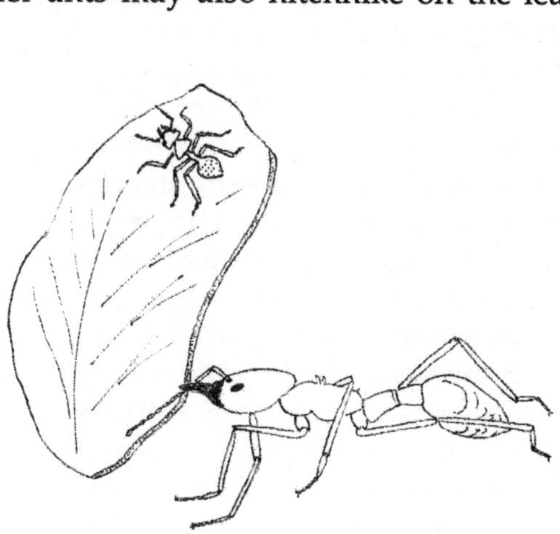

Fig. 14.27. A worker of leaf cutting ant, *Atta cephalotes* carrying a cut leaf to her nest. A guarding minor ant hitchhikes on the leaf fragment.

This fungus serves as the exclusive food source for the ant broods, which are fed with the fungal hyphae and gongylidia. In the ant and fungus mutualistic relationship, both species have a higher 'fitness' and cannot survive alone. The fungus cannot break down fresh leaves into exploitable nutrients. The ants are incapable of digesting the leaves, they allow the fungus to metabolize the material and then gorge themselves on the nutritious upwelling of the fungus. They also remove undesirable fungi if grown and maintain a monoculture. Though an ant colony grows only one kind of fungus, but amongst different nests of a single species different species of fungus are tended. The fungus garden is an efficient factory for turning indigestible material into food that can be used by the ant colony. The fungus grows at a constant temperature of about 25 ± 2°C. The relative humidity is also important, as high humidity has a detrimental effect on the fungus, and in dry conditions the fungus dries out and is then of no use to the colony.

The lower attine genera exhibit monomorphism among workers while higher attine genera exhibit division of labour and task specialization. The worker caste system in *Acromyrmex* was previously believed to be trimodal (a minor, a media and a major caste) (Weber, 1972). Wetterer (1999) suggested that workers of intermediate sizes (media) do not belong to a distinct caste, rather is a group of intermediate size workers. In *Atta* worker polymorphism is most pronounced and four distinct castes are observed, with the largest one (soldier caste) being specialized in mechanically defending the colony from predators and small workers involved in fungus garden maintenance and brood care. Most leaf cutting ants have only one queen in a nest. Polygyny is found in a few species of higher ants *Atta* and *Acromyrmex* and the lower attines such as *Mycocepurus*. The queen spends her entire life in the fungus garden. The larvae develop and pupate in the fungus garden and finally emerge as adult ants, either sterile female workers or winged reproductives. Before leaving the nest, the new queens collect a small amount of the fungus from their colony in their infrabuccal pockets, a pocket underneath the mouth area. The new queen after mating tries to found new colony (Fig. 14.28). She sheds the wings, selects and excavates a suitable nest site, in which she expels the contents of the mouth and makes use of the fungus from the colony she has left as an inoculum to start a new fungus garden (Mueller *et al.*, 2001). At colony founding stage queens of *Atta cephalotes* and *Acromyrmex octospinosis* use their own faeces as a substrate for growing the fungus. In the genus *Atta* wing muscles are used as energy resources she needs to produce the first brood of worker ants. In the other fungus-growing ants, the queen must leave the nest from time to time to collect food. The newly emerged workers care totally for the queen, fungus, leaf substrate collection and brood.

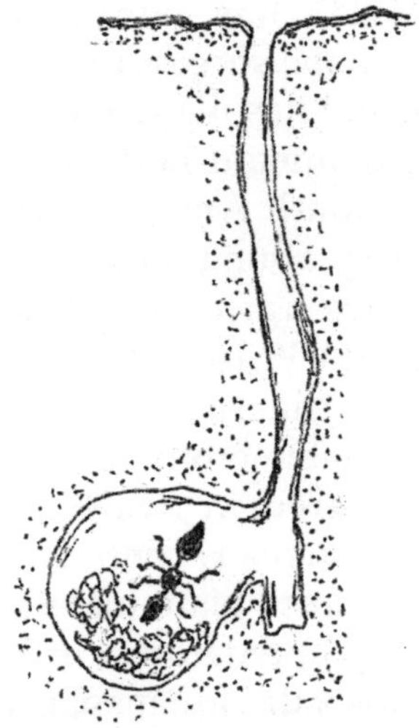

Fig. 14.28. Colony founding in fungus-growing ant, *Atta* sp. The colony is typically founded by a mated female who takes a fungal inoculum from her mother's colony to start her new garden.

The nests of the higher attines are complex structures, often with several chambers, each with a fungus garden, connected by tunnels. They also have specialized dump chambers that are excavated below the nest. There may be many entrances to a single nest. In a large nest of *Atta* there can be millions of workers, which may excavate a large area of earth. Much of this is deposited as nest mounds above the nest, the shape of which gives the colony extra protection against rain and floods as well as providing a structural defence against attack by ant eaters and other predators.

The leaf cutters also act as seed dispersers. Seeds carried away by leaf cutters may be deposited on a foraging trail or taken a far distance away from the parent tree. The seeds that do survive have a better chance of dispersal than those dropped to the base of the parent tree (section 12.3.3).

14.5.3.2 Honeypot ants

Myrmecologists have long been enthralled by 'honeypot ants' or 'honey ants' as these amazing ants develop specialized workers, called 'repletes' or 'honeypots'. The repletes gorge on food that is collected outside the nest and brought to them by worker ants. Deep in underground chambers, the repletes hang quietly in clusters. Their nests almost always have only a single surface opening surrounded by a small accumulation of excavated substrate. The genus *Myrmecocystus* occurs exclusively in western North America and contains many closely related species.

These ants are highly social creatures and live in well-organized colonies dominated by females. A typical honeypot ant colony contains a single fertile queen and thousands of sterile female workers. Workers are polymorphic- the smallest workers, called minims, remain in the nest as nursemaids caring for the queen, the developing brood and the repletes. Medium and larger workers are involved with nest excavation, food acquisition and colony defence. The largest workers, called majors, are the individuals most likely to develop into 'repletes' or 'honeypots' (Plate 14.15) with tremendously swollen abdomens (Fig. 14.29).

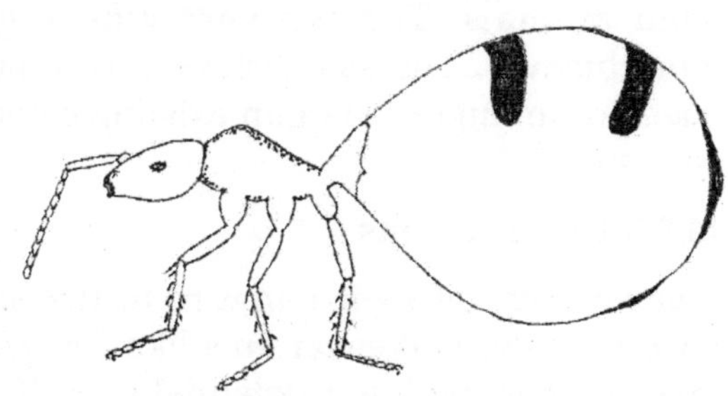

Fig. 14.29. A honeypot ant.

Honeypot ants are generalized scavengers and predators; however, they need sugars for their energy requirements. They visit flowers, extrafloral nectarines, and honeydew producing hemipterans for the sweet rewards. The workers return to the nest with crops filled with honeydew, which is fed to repletes by oral trophallaxis. The abdomens of repletes become distended and they act as food reservoirs. A large number of repletes can be found in a mature honey ant colony, each patiently waiting to serve up dinner when called upon by its nestmates. They are unique in using their own bodies as living storage vessels and their crops serve as 'communal stomach'. Sometimes other creatures burrow into and plunder honey ant nests. Even some honeypot ant colonies raid one another, stealing not only repletes but also brood, which when mature, are 'enslaved' and used as colony's worker force. Protein sources also are necessary, particularly for the growing larvae. Foraging honey ants search for live or recently dead invertebrates and small soft-bodied insects.

14.5.3.3 Carpenter ants

Carpenter ants, *Camponotus* species excavate galleries in both moist and dry wood, but they prefer moist wood. Under natural conditions, carpenter ants nest in live and dead trees and in rotting logs and stumps. Besides this, they also construct their nests in houses, telephone poles, and other man-made wooden structures. Workers constantly make irregular galleries in the wood to accommodate the enlarging colony. Access to the outside may be through natural cracks in the wood or the ants cut special openings, or 'windows'. Occupied galleries are kept clean and tidy. Shredded wood fragments from the excavations are carried from the nest and deposited outside. They construct two different kinds of nests, (i) parent colonies, which contain a queen, brood and 2000 or more worker ants, and (ii) satellite colonies, which may have large numbers of worker, but no queen, eggs or young larvae. These ants utilize a wide variety of food sources, viz. insects, other arthropods, household food items, honeydew, plant fluids, and dead animals. The workers who regurgitate the food material feed the reproductives. The establishment of a new colony is done either by a fertilized queen or immigration of an existing colony. A colony if disturbed may move to a new site.

14.5.3.4 Weaver ants

Weaver ants get their name from the habit of binding fresh leaves with silk to form their nests. Weaver ants like *Oecophylla longinoda* and *O. smaragdina* display excellent nest building potential even if within a confined space. They use fresh leaves for building the nest and make most complex nest among ants ((Plate 14.16). The ants pull the leaves into just the right position, holding them until workers bring larvae, which then proceed to secrete a silky substance from their well-developed silk glands, which acts as a 'glue', binding the leaves together (Plate 14.17). Only the larvae have silk glands. The ants then take up residence within the newly established leaf nest. In the nest construction a remarkable collaborative effort follows, in which leaves are manipulated into a tent-shape by linear ranks of workers, often involving 'living chains' of ants that bridge wide gaps between the leaf edges. A colony may be dispersed over several nests, which may be placed in various locations in a tree, or even span several trees. The queen is located in one nest and her eggs are distributed to the other nests.

The colony stays mostly within the nest. But on days when it is raining the ants come down from the tree in great numbers and collect water droplets, dampened soil, and dislodged insects. Weaver ants are also well-known for their territorial behaviour. To guard their area the workers mark their territory and also stand at various points waving their antennae. The queen mates with one or a few males and finds a suitable nest site. She seals herself in a small chamber and lays eggs. Fertilized eggs develop into females and unfertilized into males. The larger workers act as soldiers. The larvae also feed on unfertilized eggs. Ants eat any small creatures that they can find, but they are particularly attracted to nectar. They do not have a sting, but inflict a painful bite, which is aggravated by irritating chemicals secreted from their abdomen.

14.5.3.5 Army ants

The army ants form huge colonies with millions of individuals. Foraging efficiency of ants can be very high and they collect large amount of food. They are well-known for their predatory activities and attack any animals that get in their path (Gotwald, 1995). These ferocious creatures have been known to kill lizards, birds, snakes, pigs, and sometimes even large animals. They feed mainly on insects, which they capture by highly organized group raids. All the time huge soldier ants with massive jaws watch over the workers and protect them (Plate 14.18). The jaws of the soldiers are so huge that they are unable to feed themselves and must rely on workers to give them all the nourishment they need. These ants show two types of raiding patterns (Rettenmeyer, 1963). *Eciton hamatum* is a column raider. The ants produce branching trails by dividing into small foraging groups, forming a tree like pattern (Fig. 14.30a). Each trail has a two-way stream of ants

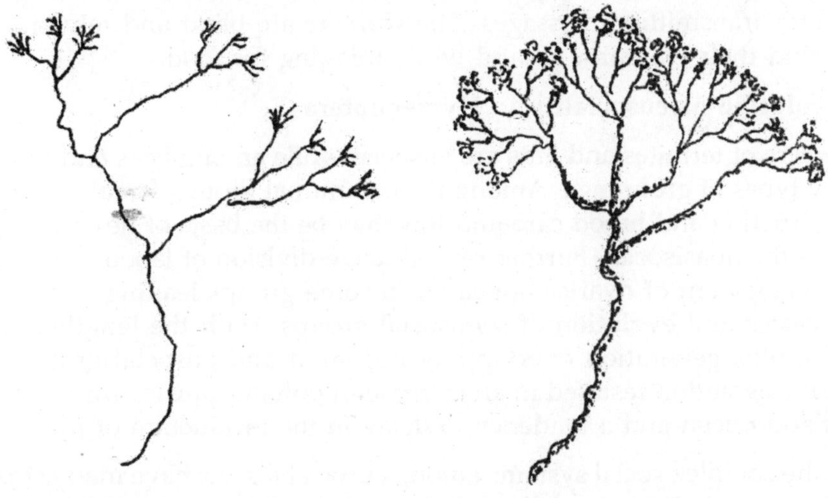

a b

Fig. 14.30. (a)- Raiding patterns of army ants. *Eciton hamatum* is a column raider. The ants produce branching trails by dividing into small foraging groups, forming a tree like pattern, (b)- *E. burchellii* is swarm raider, foraging in dense fan-shaped swarms, attached to the temporary nest (bivouac) by a single column (based on Rettenmeyer, 1963).

running along it, some ants are advancing whilst some are returning with prey. *E. burchellii* are swarm raiders, foraging in a dense fan-shaped swarms that can span several metres across, attached to the temporary nest (bivouac) by a single column (Fig. 14.30b).

The 'nests' of army ants are not like those of other ants; they are made out of the ants themselves. They are therefore referred to as 'bivouacs' describing the resting state of the colony rather than a physical nest structure. *Eciton* colonies have been observed to contain beetles, mites, and other arthropods. *Eciton burchellii* is specifically known to have relationships with beetles that live in the nest, travel with the bivouac, groom adult workers, and feed the ant brood.

They exhibit alternate cycles of 'nesting' or 'sedentary' (statary), and 'migratory' (nomadic) phases (Schneirla *et al.*, 1954). In nomadic phase, a nightly bivouac is formed and entire colony forms a cluster. In the morning all members of the colony, carrying the larvae move. While moving, the queen and the broods are protected by the huge soldiers who will kill everything that come into their way. The massive group also raids on a wide range of hosts and group predation results into even preying on large prey. After a period of nomadism, a statary period commences during which the queen lays large number of eggs. During statary phase, the larvae of previous oviposition emerge as new workers, and their addition results into start of next migratory phase. Army ants appear to have the highest fecundities among ants, though the longevity of the queens is not known. The army ant, *Eciton burchellii* can lay a new brood of 120,000 eggs every 36 days (Schneirla, 1957b). Ants communicate with each other using different methods for transmitting messages. The workers are blind and rely on chemical trails to find their way around and while foraging for food.

14.5.4 Evolution of eusociality in Hymenoptera

All members of termites and ants are eusocial while among bees and wasps there are many types of groupings. Among the communal groups females cooperate in nest construction and brood care and this may be the basis of next level of social evolution- the quasisocial. Further reproductive division of labour and variations in the development of ovaries appeared in some groups leading to formation of definite castes and evolution of semisocial groups. With the lengthening of life span of adults, generation overlapping appeared and eusocial groups formed. Progressive evolution resulted in an increase in polymorphism, with well-defined workers and queen and a tendency to delay in the production of males.

Though the complex social systems among eusocial insects have many similarities, but phylogenetic indices clearly revealed that the social groups are distantly related and routes of origin of eusociality are separate in different groups.

Lin and Michener (1972) summarized three possible routes for the evolution of eusociality in hymenopterans.

(i) Social insects evolved directly from solitary.

(ii) Wheeler (1923) and subsequent workers considered 'subsocial route' as cited by Lin and Michener (1972). Wilson (1975b) termed this as 'subsocial sequence'. In many cases eusocial hymenopteran society starts with a single founding female or a royal pair. The parents remain alive after the birth of young and then division of labour appear.

(iii) Michener and Lange (1958) suggested 'semisocial route' or 'parasocial sequence' as cited by Wilson (1975b). In such a society females of similar age form the society and while hierarchical relationships between these individuals are forming, a new generation and division of labour appear.

The resultant of evolution of social behaviour was the appearance of altruistic or self-sacrificial behaviour. In altruistic acts an individual is spending time which could be better used in courting or gathering resources; at worst, it could incur injury (or even death) without any tangible reward in exchange. Many wasps and hornets sacrifice themselves for the protection of nest and brood. The honey bee colony consists of thousands of female individuals and only one of them is queen. All other females are workers which have morphological and anatomical adaptations for performing all tasks of the colony. The queen is simply an egg-laying machine and drones have no work in the colony. The workers have no reproductive options and spend whole lives working and caring for queen and her brood- an act of supreme sacrifice or altruism. Besides this, honey bee worker has barbed sting which remains in the body of victim and cannot be withdrawn. When the bee attempts to fly away after stinging, the sting along with poison gland and part of digestive system are torn away which leads to death of worker bee. For the bee stinging is an act of suicide to protect the colony. The worker's altruism is not just remarkable from a human point of view, but also from the point of view of Darwin's theory of natural selection. Darwin (1859) suggested that in case of social insects natural selection had operated on the level of a colony or group, rather than an individual organism. This type of selection is also termed as 'group selection'.

William Donald Hamilton (1964a,b), a British evolutionary biologist, proposed a 'kin-selection theory' for genetic evolution of social behaviour. At the time, it was so innovative that it almost failed to be published and was largely ignored for a decade. When finally noticed, its influence spread exponentially until it became one of the most cited papers in all of biology. It is the key for understanding the evolution of altruistic cooperation among related organisms, such as the social insects. Hamilton was considered one of the greatest evolutionary theorists of the 20th century and he became famous for his theoretical work expounding a rigorous genetic basis for the existence of kin selection. His findings introduced a new term- 'inclusive fitness', which maintains the logic of genetics over evolutionary time. Inclusive fitness is defined as an individual's relative genetic representation in the gene pool in the next generation. Thus, fitness denotes the capability of an allele to spread in a population: if the fitness value for a given allele is larger than one it will increase in frequency and if it is smaller it will decrease in frequency.

The prime-mover in the evolution of social behaviour among the Hymenoptera is considered to be 'haplodiploidy' (Trivers and Hare, 1976). Many eusocial insects, including ants, bees, and wasps, are haplodiploid. Therefore, each female has two alleles at a locus, while each male has only one. This leads to a different kin relatedness in haplodiploid species and can be cited as given in table 14.1.

Table 14.1. Genetic relatedness in haplodiploid species.

Relatives	Mother	Father	Sister	Brother	Son	Daughter
Female	0.5	0.5	0.75	0.25	0.5	0.5
Male	1	0	0.5	0.5	0	1

The sisters share more genes with each other and therefore, the inclusive fitness of a sterile female (worker) is greater than its classical fitness. As the selection is operating an individual should maximize its inclusive fitness, and a worker should invest in the survival of his sisters, rather than in the production of her own offspring. This explains why there are sterile castes in eusocial insects; these workers may give up potential direct benefits associated with raising their own children, because indirect benefits are much beneficial. The haplodiploid females get benefit from helping in raising their sisters, which they are related to 3/4. These females, on the other hand, are related to their brothers only 1/4. Therefore, they will prefer for their mothers to have daughters.

Other factors involving in promoting eusociality are recognized in "Hamilton's Rule", which emphasize the ratio of costs and benefits of altruistic behaviour as well as relatedness. The condition under which selection will favour altruism can be expressed as follows:

$$rB - C > 0$$

or

$$C < r \times B$$

Where:

r = coefficient of genetical relatedness.

B = the additional reproductive benefit gained by the recipient of the 'altruistic act'.

C = the reproductive cost to the individual performing the altruistic act.

Kinship calculations assume that all offspring of a mother in colony share an identical father. However, in higher eusocial insects, queen may mate several times with different males. In such cases, intercolony relatedness will be less than predicted by the model. However, this difficulty is unrelated to the origin of eusociality, being more concerned with the maintenance of an already existing eusocial system. Thus, current evidences favours kin selection but roles for other factors cannot be excluded.

The other idea concerning the origin of eusociality involves maternal manipulation of offspring (both behaviourally and genetically) as the sterile worker castes result from maternal manipulation of offspring. The queen is in a position to control the development and behaviour of offspring by releasing pheromones and by laying fertilized or unfertilized eggs. The food supply also determines the caste as well-known in case of honey bee. Thus, the control by queen over caste determination in many hymenopterans provides support for the importance of parental manipulation in eusociality.

A further well-supported situation emphasizes the role of competition and mutualism. Mutualism in a society arose because of adaptive advantages of societies. Each individual in the colony is benefitted through communal defence. Thus, mutualism (including the benefit of defence and nest construction) and kinship encourage the establishment of group living. However, it fails to account for sterility or altruism and therefore cannot be considered as a major factor in the origin of eusociality.

Gadagkar (1990) proposed a hypothesis for the route to eusociality, which focuses on the factors selecting for group living at different stages in social evolution. He argued that group living originates owing to the benefit of mutualism (gambling stage) but parental manipulation and subfertility soon become important (manipulation stage) and finally the highly eusocial state is maintained because genetic asymmetries created by haplodiploidy are exploited by kin recognition (recognition stage).

Korb and Heinze (2004) suggested that hymenopteran societies are highly suitable model systems for investigating the details of social evolution and to resolve within-group conflicts. In insect societies conflicts among nestmates arise, e.g. about the partitioning of reproduction and the allocation of resources. Variation in relatedness among group members, therefore, appears to have a profound influence on the social structure of groups. They suggested that constraints, such as the lack of power or information, prevent individuals from pursuing their own selfish interests. Applying a multilevel selection approach shows that these constraints are in fact group-level adaptation preventing or resolving intracolonial conflict. Insect societies can thus be regarded as a level of selection with novelties that provide benefits beyond the scope of a solitary life. Therefore, relatedness is less important for the maintenance of insect societies, although it played a fundamental role in their evolution.

Eberhard (1978) proposed polygynous family hypothesis for the evolution of social behaviour in wasps. He suggested that marking of reproductive division of labour are present in primitive social wasps, in the form of differences in sizes, reproductive success and social dominance. In polygynous wasps a gradual development of reproductive division of labour appeared. This hypothesis is well suited for wasps, for which polygyny is universal with the exception of some temperate *Vespa* and *Vespula* species. Some authors believe that polygynous family hypothesis could well include other groups of social insects, in addition to wasps.

14.6 Dominance of social insects

Social insects have attained numerical and ecological dominance in certain areas. They are most dominant in tropical and subtropical areas of the world; however, in summer their activities are also conspicuous in temperate and hilly areas. Evolution of caste system, high fecundity, social communication, and mutual cooperation has contributed significantly in their dominance. Attributes like brood care, foraging, defence strategies, and nest construction are other major factors for their success. As they perform various activities in a parallel way, one activity does not jeopardize other.

Their ability to involve all workers in emergency assists them to overcome adverse situations. The homeostatic regulation of the colony, storage of food reserves, and manipulation of microclimatic conditions are also specialized features of their joint activities. They have a well-defined communication system and use pheromones, which permit a high level of control over huge number of individuals. Some species also acquire parasitic inquilines to assist them.

14.7 Social alliances

Social insects have alliances with individuals of other species. The impact of such alliances may vary as these may be beneficial for both, detrimental to one and beneficial to another, or having no impact on each other. Social parasitism is a common and fascinating phenomenon in social insects. It occurs when a host species raises young of a parasitic species. Matthew and Matthews (1978) defined social parasite as a species that uses its host as a work force rather than a direct source of food. It may involve individuals of one (intraspecific) or more species (interspecific). Social parasitism may be obligatory or facultative. Social parasites exploit the socially managed resources of their host's society. Inquiline social parasites are dependent on their host throughout their life cycle, and so many of the traits inherited from their free-living ancestor are removed by natural selection. A few inquiline parasites that have retained a worker caste are thought to be at a transitional stage in the evolution of social parasitism, and their worker castes are considered vestigial and non-adaptive. Permanent parasitic species are phylogenetically very close to the host species (Emery's Rule). There are several forms of social parasitism. In some cases workers raid the nests of their own or other species to take food resources (e.g. robbing behaviour of honey bees). A parasite that feeds on the food reserve of other species is termed 'Cleptoparasite' (also as Kleptoparasite). Some species show only temporary social parasitism in the nest founding phase, when mated queens usurp the nests of host species instead of establishing nests by themselves (e.g. wood ants of the genus *Formica*). Other species take slaves, by stealing brood from host's nests. The host brood is raised in the slave maker nest and host workers perform all tasks necessary for the maintenance of the parasite colony. The phenomenon is known as 'dulosis' and is found in many parasitic species.

A number of social hymenopterans have lost their nest construction and provisioning behaviours and rely entirely upon the nests and labours of others. Many wasps exhibit social parasitism. Obligate social parasites cannot initiate their own colonies, but invade established nests of other species. The parasitic queen kills the queen of the host colony and uses the host colony's workers to produce her own offspring. Some wasps are facultative social parasites and their females can start their own nests or they can be parasitic.

Some social ants live as temporary or permanent social parasites in nests of other species. The queen invades to a host nest and kills the resident queen. The queen produces workers, which in due course take over the command of the colony. In others, queen produces only males and reproductives and workers of host perform various duties; however, the colony continues to exist until the workers of host survive. In red Amazon ant, *Polyergus* all ants are soldiers and have sharp, sickle-shaped mandibles (Fig. 14.31). They are completely dependent on their slaves, because their sharp jaws, so well suited for fighting, are useless in taking care of their own larvae or carrying out much of the work around the nest.

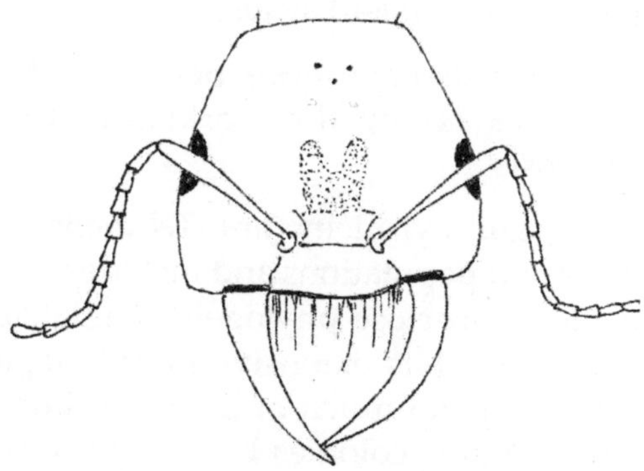

Fig. 14.31. Head of the soldier of red Amazon ant, *Polyergus* showing sharp, sickle-shaped mandibles.

When *Polyergus* ants are in their nest their only activities are begging food from their slaves and cleaning themselves. Their behaviour changes completely when they raid, and destroy the resisting defenders and carry off their cocoons containing pupae of workers. The hatched workers accept their captors as sisters. As the workers are sterile, the ants must periodically conduct additional raids to maintain an adequate labour force (Wilson, 1975a). Some ants also rob the nectar-filled honeypots ants. Several ant genera have the habit of entering nests of other species of ants or termites to prey upon the inhabitants and/or their food. This is termed 'trophic parasitism' and is also exhibited by many bees and other insects. Different ant species intrude beehive and prefer to feed on honey, pollen, brood and wax. The attack results in weakening of the colony.

In some species of slave-making ants, the workers are strictly bred for the purpose of going out and conquering other nests. The colony cannot survive without slaves as they lack the abilities to tend the queen, raise young and hunt for food. The Amazon slave-makers, if separated from their slave, they will starve to death even though food is available to them. Different species of slave-making ants may adopt different tactics, but their goal is the same. Slave ants gather food, feed their hosts, groom and feed the larvae and queen, and defend the colony against attacks by other insects. If the colony moves to a new location the slaves carry their masters to their new home. Wilson (1975a) found that workers of ant, *Leptothorax duloticus* are highly efficient for raiding and fighting. In home, workers are almost inactive and leave their all duties to their captives. When the slaves were temporarily removed from the nest, the workers undertook most of the tasks formerly carried by slaves. Thus, they retain a capacity to work, but not in a well-defined manner. These workers were unable to forage, and the colony displayed signs of starvation and deterioration. When the slaves were placed back in the nest, they put the nest in good order, and the slave-makers quickly resumed their indolent ways.

Some small ants have nests near or in the nests of larger species, and feed on refuse or waylaying the workers and stealing their food. This is termed as 'Cleptobiosis'.

The European myrmicine ant, *Teleutomyrmex schneideri* (means 'final ant') live in small isolated populations and include no worker. The ants always live as parasites in the nest of its close phylogenetic relative pavement ant, *Tetramorium caespitum.* The queen exhibits extensive morphological degeneration, except reproductive system and spend much of her time riding upon back of queen of host's colony. The infested host colonies become weaker and produce no sexual forms of their own. *Teleutomyrmex* female produce an attractant, which is licked and distributed by host's workers, and check the development of gonads in host individuals (Kutter, 1950).

In many cases social parasites have evolved communication systems, which are very similar to their host species (Stowe, 1988). Lenoir *et al.* (2001) suggested that the social parasites exploit semiochemicals for tolerance and association with their hosts. The obligate parasites are odourless (chemical insignificance) at the time of usurpation, like all other immature ants, and this 'invisibility' enables their entry into the host colony. By chemical mimicry, they later integrate with odour of host's colony (chemical integration). Thus, both host and parasite are likely to be related chemically, thereby facilitating the necessary mimicry to permit bypassing the colony odour barrier. Some parasitic ants release a pheromone at the time of attack, which is similar to a warning substance that exists in that colony and when it is secreted in large quantity by parasite ants, it results in running away of the ants instead of protecting their colonies. In their haste to leave the colony, pupae will be left behind. These developing ants are then taken back to the slave-maker nest.

Cuckoo wasps (Fam. Chrysididae) are parasites and lay their eggs in the nests of bees, wasps, and certain other insects. They often wait near the burrows of their hosts, waiting for an opportunity to sneak in and lay eggs. Their hard cuticle provides protection from stings and strong biting mandibles of host insects, which may attack the cuckoo wasp. In addition, the underside of the abdomen is concave and allows the wasp to roll up into a protective ball, which is another defensive mechanism (Fig. 14.32). Cuckoo wasps may feed on the larva of the host or steal the host's food. The Pacific cuckoo wasp, *Chrysis pacifica* parasitizes mason bees. The wasp's larva feeds on the bee's larva after the bee has spun its cocoon. The large blue cuckoo wasp, *Chrysis coerulans* (a cleptoparasite) lays its eggs in the nests of solitary wasps. The cuckoo wasp larva kills the host larva and eats the caterpillars that have been captured by the host for food.

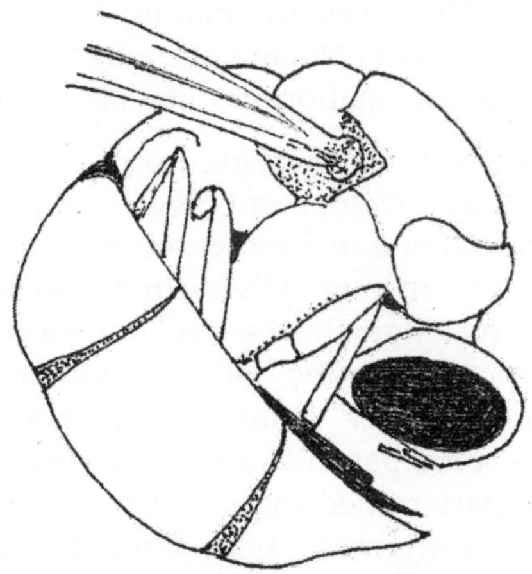

Fig. 14.32. Cuckoo wasp, a parasite that lays her eggs in the nests of bees, wasps, and certain other insects. Its hard cuticle provides protection from stings and strong biting mandibles of hosts. The wasp can roll up into a protective ball, which is another defensive mechanism.

The acacia thrips induce galls on the young phyllodes of *Acacia* species. The cleptoparasitic thrips belonging to the genus *Koptothrips* invade these galls and usurp the domiciles. The larvae of a tachinid fly, *Strongygaster globula* develop as endoparasites inside the gaster of queen of common black ant, *Lasius niger*. The behaviour of the infected queen is not noticeably affected except that they cease to lay eggs. When the last instar larva of the parasite leaves the body of the host, it quickly pupates and is groomed and tended by the host ants. After hatching of the fly this friendly attitude is abandoned and the fly is forced to leave the nest. The larvae of syrphid fly, *Microdon piperi* make a hole in the silken cocoons of *Camponotus* ant and feed on the ant pupae. The emerged adult is recognized as an intruder and removed from the colony.

The greater wax moth, *Galleria mellonella* is a serious threat to raised combs in honey bees' hives. The caterpillars make tunnels through near the midrib of a comb. In later stage of attack the comb looks like a tangled mass of silken galleries in which excreta of caterpillars is enmeshed. Some social insects also support parasitic mites and their effects have been devastating. The endoparasitic honey bee mite, *Acarapis woodi* attacks honey bees' tracheae by entering through the spiracles. The attacked bee has 'K' winged condition, unable to fly, crawl in front of the hive and her abdomen is swollen. The attack may cause dwindling and ultimate death of the colony. The ectoparasitic *Varroa* mites live on drones and workers. They feed by making holes in intersegmental membrane. *Tropilaelaps clareae* is another ectoparasitic mite that attacks bees. Infected colonies has irregular brood pattern, malformed life-stages and the adults crawl around the hive.

Social insects also provide hospitality to a number of other organisms for their growth and development. The association is obligatory in many cases. Organisms that share home of another are termed 'inquilines'. The term inquilines has been defined as, "an animal living in the home of another and sharing its food, especially a guest or lodger in the nests of social insects"; and also as "an organism that lives on or within the body of another, or in its nest or abode, without benefit or damage to either". Matthews and Matthews (1978) also termed inquilines as extreme social parasites and these appear to have arisen convergently along various evolutionary routes. Inquilines in wasp nest are termed vespophiles, and similarly in ant nest myrmecophiles. Gullan and Cranston (2000) classified inquilines into two groups, based on the degree of integration. The first, 'integrated inquilines', which form close association with their hosts' social lives, because of modification in behavioural attributes of both parties; whereas the second 'non-integrated inquilines' are ecologically adapted to the nest, but do not interact with the hosts. Some predatory inquilines result into deprivation of society, while others being scavengers may be beneficial.

The integrated inquilines using chemical signal similar to that of host exhibit Wasmannian mimicry. These inquilines produce appeasement and/or adoption chemicals. The larvae of European staphylinid beetles, *Atemeles pubicollis* live in the nest of wood ant, *Formica rufa* where they pupate and adults emerge. The adults migrate to the nests of *Myrmica* ants and take shelter until become mature. Consequent upon maturity the adults shift to *Formica* nests for oviposition. The glandular secretion of larvae and adults beetles assist in interspecific communication and survival (Hölldobler, 1971a) (see section 11.5).

The milkman ants, place aphids which suck phloem sap, on the leaves. The aphids convert plant sap in the honeydew. A hungry ant approaches the aphid and starts brushing its abdomen with its feelers and antennae. The aphid likes this very much and emits a droplet of honeydew and gives it to the ant. In return, the ants look after their aphids very well.

Some inquilines of termite nests show physogastry (dilation of abdomen), as in queen termite. The phorid fly, *Termitoxenia* is well-adapted to survive in termite nest. The female shed its wing before entering the nest and develop physogastry in termite nest. Termites treat its eggs as their own eggs.

Some syrphids of tribes Volucellini and Merodontini are remarkably specialized inhabitants of social insect nests. Members of the syrphid subfamily Microdontinae, are all inquilines in the nests of social insects (mostly ants). The *Microdon* species has a myrmecophilous larva that lives unharmed amongst nest debris, but the emerged adult is recognized as an intruder. Adults of the hover fly, *Volucella* are similar to *Polistes* wasps or *Bombus* bees. The females fly in and out of hymenopteran nests and lay eggs. The hatched larvae feed and develop in the nests. The thrips, *Advenathrips inquilinus* has evolved as a true inquiline, in that it apparently breeds within the colonies of a particular domicile creating species without unduly disturbing the original inhabitants.

14.8 Thermoregulation

Social homeostasis is a phenomenon of social insect colonies, in which the collective activities of the colony's inhabitants regulate the colony environment. It arises through the coordination of the numerous and nearly autonomous individuals that together comprise the colony. Social insects have astonishing capacity to maintain their numbers and static microclimatic conditions in their nests. The regulation of nest temperature is an ability of social insects which assist in their successful survival and dominance. Regulation of nest temperature is also important to the fitness of eusocial insect colonies. To maintain appropriate conditions for the developing brood, workers must exhibit thermoregulatory responses to ambient temperature. Because nestmate workers differ in task performance, thermoregulatory behaviour provides an opportunity to test threshold of response models for the regulation of division of labour. Most social insects which form large groups are able to maintain the nest temperature within a constant range. The temperature regulation is achieved through structural design of nest and/or behavioural manipulations, or both.

Honey bees are able to regulate their hive temperature in an efficient manner. The selection of nest site, like a hollow tree trunk, shady areas is the first step in their thermoregulatory activity. Workers of *Apis mellifera* seal all cracks and crevices of their hive with propolis, which not only keeps the enemies out but also maintain temperature and humidity inside the hive. In winter, they are able to maintain hive temperature between 20-30°C, while in summer the temperature is efficiently lowered. When the outside temperature is low they raise hive temperature by their metabolic activities. They form clusters and further increase in temperature is achieved by cluster tightness. Temperature regulation in summer is more sophisticated. The workers cool the hive by fanning with their wings to circulate the air over the brood combs and if the temperature rises above 35°C fanning is no longer sufficient, then they bring water and sprinkle it over the brood cells. The excess of moisture is also removed by fanning. Fanning is also employed by halictine bees, bumble bees, and some stingless bees. Among wasps, *Polistes, Vespula, Vespa* species use both fanning and water to lower the hive temperature. O'Donnell and Foster (2001) found that worker bumble bees, *Bombus bifarius nearcticus* responded to changes in ambient temperature by altering their rates of performing two tasks- wing fanning and brood cell incubation. With an increase in temperature the rate of incubating decreased, and the rate of fanning increased.

Ants lack wings and unable to ventilate by fanning, and also cannot employ droplet evaporation. They mainly rely on location and nest designing to achieve temperature regulation. Many ant species shift their brood in best suited chambers with a change in temperature, and as a last resort the whole colony is shifted to a new nest site with favourable climate. This ability is not found in bees and wasps. Soil-inhabitating ants get the advantage as the temperature and humididty below soil surface are subject to less fluctuations and the soil microclimate is best-suited for brood rearing. The leaf litter or humus covering the soil surface provide additional temperature regulation. For making their nests many ants use rotting wood on the ground, some make nests in rotting arboreal logs, and a few others underneath the rocks. These ants are less susceptible to high temperatures than those that nest in soil. A few ants make mounds, which are piles of excavated soil, rich in organic matter, perforated with dense system of interconnected galleries and chambers that serve as living quarters for the ants (Wilson, 1971). These mounds are mainly constructed for microclimatic regulation. The shape of mound make it exposed to more sunlight, which enhances warming.

Thermoregulation is not so pronounced in lower termites that live in wood. Soil-dwelling termites too mainly rely on natural insulation. The higher termites build mounds by transporting soil from deep within the ground to the surface. The architecture of mounds varies depending on their their thermoregulatory ability. The African fungus-farming termite, *Macrotermes natalensis* makes temitarium which allows circulation of air to give microclimatic control of temperature and carbon dioxide (Lüscher, 1961). Lüscher's elegant scheme provided a mechanistically sound model for social homeostasis in termite. The mound architecture is dynamic, changing through time as termites transport soil up into the mound and erosion removes soil and spreads it onto the surface. As mound architecture varies, so too does its ability to capture wind for ventilation. Sen-Sarma (1988) reviewed different aspects of homeostasis in termite society. Korb and Linsenmair (2000) analysed the thermoregulation in mound of fungus-cultivating termite, *Macrotermes bellicosus*. Mound structure resulted in a relatively constant nest temperature. Abiotic heat production via solar radiation alone yielded nest temperatures that corresponded to mean ambient temperatures. However, only the production of metabolic heat by the termites and fungi increased these temperatures to the actual nest temperature. Only large colonies attained constant nest temperatures of 30°C, which were largely independent from ambient temperatures and optimal for the growth and development of termites and fungi.

Chapter 15

Biological Rhythms

Most living beings on the earth are subjected to environmental changes resulting from daily cycles of light and dark, seasonal cycles, lunar periodicities and other climatic changes. Animals including insects have evolved strategies to counteract or exploit such periodicity. In majority of cases daily or annual cycles of responses have been observed, as their particular activity can only be envisioned at particular time of the day or during a period of the year. Insects may be active during day, night, dusk or dawn or may perform various life activities during a fixed season. Some of their activities may be direct response of environmental factors; however, most such responses are expressions of an endogenous periodicity. Thus, such responses are innate and determined by genes. Their life processes are temporarily organized in interwoven biological rhythms, i.e. repetitive incidences with characteristic periods. The rhythmic pattern may be spontaneous or may be exhibited only once in a period of several years. One of the most important biological rhythms are those endogenous oscillations which follows the periodicity close to 24 hours, with its regular changes of illumination and temperature. These daily rhythms (circadian rhythms, Latin *circa*, meaning 'around' and *dies*, 'day', meaning literally, 'around a day') affect most of behavioural and physiological processes in such an apparently self-evident way, that they allow an insect to perform a particular activity at the right time of the day or to attain synchrony with conspecifics. It has long been assumed that daily rhythms such as wake/sleep cycles are only passively driven by the changes of day and night. The formal study of biological temporal rhythms (such as daily, weakly, seasonal, etc.) is called 'chronobiology'. Besides the circadian rhythms, which have evolved to match 24 h periodicity of earth's rotation, other endogenous oscillations with tidal (~12.4 h), semilunar (~14.7 days), lunar (~29.4 days) cycles or annual (~a year) are also found in animals (Saunders, 1982). These cycles can also be divided into three categories: ultradian rhythms with a period shorter than 24 h, circadian rhythms with a period close to 24 h and infradian cycles with a period longer than 24 h.

The cycles of light and temperature serve to regulate these endogenous oscillators, but even in absence of such temporal cues, i.e. in complete darkness and at constant temperature the rhythms 'free-run' and reveal their own natural period (τ) which is close to, but significantly different from that of the solar day (Saunders, 1982). In an individual insect, τ is subject to variation. Page and Block (1980) reported spontaneous and induced changes in ô in Madeira cockroach, *Rhyparobia maderae*. They showed that exposure of adults to 22 h (11 h light and 11 h dark, i.e. LD 11:11) and 26 h (LD 13:13) caused shortening or lengthening of τ. Pittendrigh (1960) referred these τ changes as 'after effects'. The circadian rhythmicity is at most only slightly affected by changes in constant level of ambient temperature. Pittendrigh (1954) demonstrated the near independence of τ from temperature for eclosion rhythms of *Drosophila pseudoobscura*. Temperature compensation has also been reported in other insects like American cockroach, *Periplaneta americana* and yellow mealworm beetle, *Tenebrio molitor*.

In many insects free-running rhythms also persist in LL, but the period of rhythms in LL differs from the period in DD. Thus, for diurnal species τ decreases with an increase in light intensity ($\tau_{LL} < \tau_{DD}$) and is converse in case of nocturnal insects ($\tau_{LL} > \tau_{DD}$) 'Aschoff's rule' (Pittendrigh, 1960). However, there are exceptions to this rule, e.g. diurnal flight activity rhythm of *Aedes aegypti* (Taylor and Jones, 1969). The patterns of free-running activity rhythms can also be influenced by factors, viz. light intensity, temperature, reproductive or developmental state of an insect. Animals also exhibit the endogenous rhythms even if the environmental cues (*Zeitgebers*) are eliminated. Other endogenous oscillations are associated with semilunar, lunar, tidal and annual cycles. It is assumed that the organisms measure the time by using the 'biological clock'. Circadian clocks are omnipresent and considered to be of adaptive significance. Sharma (2003) gave preliminary circumstantial evidence to support adaptive significance of circadian clocks. It is believed that an organism possessing circadian clocks gains fitness advantage in two ways:

(i) *Extrinsic adaptive value*. By synchronizing its behavioural and physiological processes to cyclic environmental factors.

(ii) *Intrinsic adaptive value*. By coordinating its internal metabolic processes.

Each circadian timing system consists of three parts: the input pathway, the oscillator (pacemaker), and the output pathway (Fig. 15.1). The input pathway

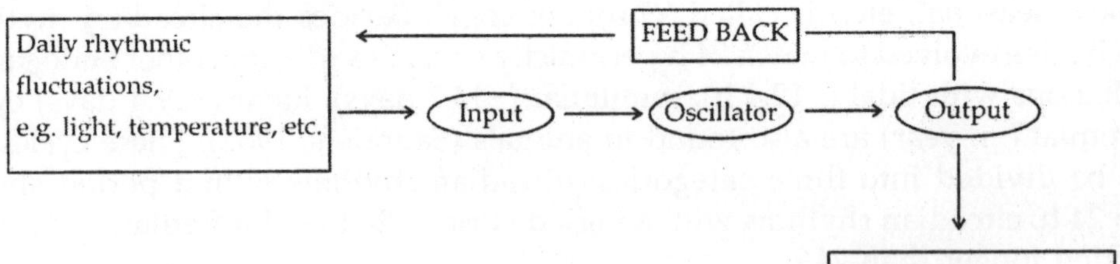

Fig. 15.1. The circadian system model. The input pathway links the internal cycle to external light-dark, temperature cycles. The second is an autonomous, self-sustained and endogenous pacemaker or oscillator, which generates the circadian oscillations. The final one is an output pathway, which relays the circadian rhythmicity to overt physiological and behavioural rhythms.

provides the connection of the clock with the environment and is necessary for entrainment. The oscillator represents the time-keeping device, the endogenous clock, which can run freely without any external stimuli. The output pathway ensures the transfer of the timing information from the core oscillator to the rest of the organism, and partly feedback to clock entrainment.

In insects a number of behavioural phenomenon, viz. locomotion, feeding, mating, oviposition, pupal eclosion, etc. are known to be controlled endogenously by a clock and only exhibited at a particular time of the day, e.g. adults of fruit flies (Fam. Tephritidae) emerge early in the morning. The time memory (*Zeitgedächtnis*) of bees is also a classical example and they can also be trained to visit a food source at a particular time. Similarly, mating in a number of insects also takes place at a fixed time and enhances the chances of successful encounters between conspecific individuals of opposite sex.

The first recorded circadian rhythm was for the sleep movements of the leaves of the tamarind tree by the Greek philosopher Androsthenes. Carolus Linnaeus constructed a 'floral clock' based on his knowledge of the opening of flowers at different hours of the day. The French Naturalist Jean-Jacques d'Ortous de Mairan (1729) observed daily leaves movements of touch-me-not plant, *Mimosa pudica* for several days even under constant darkness and the sleep movements were endogenous. Nevertheless, it took two more centuries until the existence of internal clocks became a generally accepted opinion among life scientists. Brown (1965) proposed a theory for biological rhythms, stating that 'subtle geographical forces' associated with solar cycle of the day exogenously control all such periodicities. However, the endogenous regulation of circadian periodicity was well established by Hamner *et al.* (1962) as they observed pupal eclosion periodicity of *Drosophila pseudoobscura* at the South Pole, where diurnal variables were eliminated. The pioneer works of Ernst Bünning (1973), Jürgen Aschoff (1984) and Colin S. Pittendrigh (1993) revealed that living organisms are in fact able to self-generate rhythms with periods very close to, but not exactly matching the 24 h of a day (circadian rhythms). These rhythms can be observed under constant environmental conditions, where no external time information is available for an organism under surveillance (free-running condition).

Significant contributions on insects' clocks have been published by Pittendrigh and Minis (1964), Pittendrigh (1958, 1960, 1981), Aschoff (1965, 1984), Brown *et al.* (1970), Bünning (1973), Palmer (1976), Saunders (1982), Page (1985), Winfree (1986), Denlinger *et al.* (2001) and Dunlap *et al.* (2003).

15.1 Types of rhythms

Pittendrigh (1958) differentiated rhythms in three categories:

(i) *Pure rhythms*. Such as colour change in animals.

(ii) *Interval timers*. In which a particular event occurs at a particular time of the day, e.g. pupal eclosion.

(iii) *Continuously consulted clocks.* Such as time memory (*Zeitgedächtnis*) of bees and time-compensated sun orientation.

Truman (1971a) proposed two types of animal clocks.

Type I. Clocks associated with developmental rhythms, e.g. moulting, hatching etc.

Type II. Rhythms that include locomotor activity, e.g. time memory and time-compensated sun orientation.

Matthews and Matthews (1978) cited two main types of biological rhythms having relevance to the programming of behaviour.

Reiterative behaviour. These behavioural rhythms have a periodicity of about 24 h (circadian rhythms), e.g. daily activity rhythm of cockroaches, which begins shortly after dusk.

Gated rhythms. Behavioural events occurring only once in the life of an individual, but appear as an obvious rhythm when one views the synchronous population, e.g. emergence of cicada.

The biological rhythms have also been classified as:

(i) Circadian rhythms, with a periodicity of approximately 24 h.

(ii) Tidal and lunar rhythm.

(iii) Circannual rhythms, with a periodicity of about a calendar year.

15.2 Circadian rhythms

Insects perform their activity at certain time of diel cycle. They may be nocturnal (night-active), diurnal (day-active) or twilight-active (crepuscular). Such rhythmic patterns may be exo- or endogenous, or may be because of both components. The endogenous aspects of rhythms provide temporal organization for a wide range of behavioural, physiological and metabolic processes. Saunders (1982) provided an exhaustive list of insects exhibiting various rhythms of activities.

15.2.1 Locomotor activity rhythms

Locomotor activity of an individual insect is restricted to a particular period of the day, e.g. cockroaches are nocturnal and commence their activity soon after dusk in cycles of 12 h light and 12 h dark (LD 12:12) while honey bees forage only during day. In tsetse fly, *Glossina morsitans* the activity is bimodal with peaks during dawn and dusk (Brady, 1972). The endogenous nature of rhythm is revealed when the insect is shifted from a light/dark cycle (LD) to continuous light (LL) or continuous dark (DD) under constant conditions. Under these conditions an endogenous oscillation controlling a rhythmic activity will free-run and reveal its natural periodicity (τ) (Saunders, 1982). Usually the periods of free-running rhythms are close to, but not exactly equal to 24 hours. Roberts

(1960) placed three cockroaches, *Rhyparobia maderae*, *Byrsotria fumigata* and *Periplaneta americana*, in DD at 25°C and found that rhythms persisted for at least 3 months with periodicity between about 23-25 hours. Roberts (1962) demonstrated that locomotor rhythm in *R. maderae* was entrained to a 24 h period by locomotor LD cycles with submultiples of 24 h. Some individuals became entrained to a cycle of LD 2:2 (T = 4) or LD 4:4 (T = 8), but entrainment failed with LD 1:1 (T = 2). The phenomenon is termed as 'frequency demultiplication'. Sokolove (1975) found that rhythms of locomotion and stridulation in male common field cricket, *Teleogryllus commodus* were circadian with a periodicity of 23.4 h in DD and 25.1 h in LL.

In *Glossina morsitans* high (35°C) and low (19°C) temperatures reduce evening peak activity by about 80 per cent (Brady and Crump, 1978). In stick insect, *Carausius morosus* rhythms of locomotion and oviposition are lost when individuals are shifted to DD (Godden, 1973). Alpatov *et al.* (1999) discussed behaviour and motor activity rhythms of beetle, *Trigonoscelis gigas*. In order to avoid extreme temperatures, this beetle developed a peculiar behavioural strategy: narrow time windows for activity (in the morning and in the evening). The rest of the time it spends deep in the sand, at constant comfortable temperature and in continuous darkness, in temporal isolation. They showed that there are no available *zeitgebers* for the beetle during its rest interval. Hence, a reliable circadian clock is the only way for *T. gigas* to implement the proper timing of its activity. Thus, the circadian timing system of this beetle is not just a source of relative advantage, but is a crucial prerequisite of survival.

15.2.2 Rhythms and reproductive behaviour

A number of reproductive processes, viz. sex pheromone release, calling behaviour, mating, oviposition, hormonal secretion, spermatophore production, etc. are under the control of circadian system. Rhythmicity in sex pheromone release under constant conditions has been reported in noctuid moths, viz., alfalfa looper, *Autographa californica*; cabbage looper, *Trichoplusia ni*; beet armyworm, *Spodoptera exigua* and tobacco budworm, *Heliothis virescence* (Shorey and Gaston, 1965). In many other insects of the orders: Coleoptera, Lepidoptera, Diptera and Hemiptera circadian rhythms of pheromone release have been described. The Indian meal moth, *Plodia interpunctella* and Mediterranean flour moth, *Ephestia kuehniella* are closely related and often sympatric species, but their pheromone release patterns differ, and thereby pheromone communication shows drastic differences between the two species. *P. interpunctella* calls around lights-off (dusk) and *E. kuehniella* around lights-on (dawn). *E. kuehniella* females also show a free-running rhythm in calling behaviour when maintained in DD, while *P. interpunctella* females turn practically arrhtyhmic (Kuwahara and Casida, 1973). Rosén *et al.*, (2003) observed the pheromone-mediated upwind flight of male turnip moths, *Agrotis segetum* in a flight tunnel at different times of day under conditions of LD cycle, DD, and a shifted photoperiod. Under both LD and DD conditions, a significantly larger number of males flew to the pheromone during both the scotophase and the subjective scotophase than during the photophase and the subjective photophase for 2 consecutive days. This showed that male behavioural responses to the pheromone are under the control of an endogenous oscillator.

In common field cricket, *Teleogryllus commodus* production of spermatophore is in juxtaposition with stridulation (Loher, 1974). Spermatophore production is cyclic in LD, LL and DD and is therefore fully innate. Stridulation in *T. commodus* also occurs at night in crickets maintained in LD, LL and DD (Sokolove, 1975). Oviposition at particular time of the day has been reported in yellow fever mosquito, *Aedes aegypti*, pink bollworm, *Pectinophora gossypiella* and *Drosophila* species. In *A. aegypti* oviposition occurred in well-defined peak near dusk (Haddow and Gillett, 1957) and in *P. gossypiella* exclusively during dark (Minis, 1965).

Sharma *et al.* (2004b) monitored the locomotor activity rhythm of different castes of the ant, *Camponotus compressus* under LD and DD cycles. The virgin queens and males display rhythmic activity patterns, but the mated queens were arrhythmic while laying eggs, with the rhythmicity resuming soon after egg-laying. Under LD major workers showed nocturnal patterns, while about 75 per cent of the media workers displayed nocturnal patterns and about 25 per cent showed diurnal patterns. Under DD regime, most major workers exhibited circadian rhythm of activity with a single steady state, whereas media workers displayed two types of activity patterns, with activity patterns changing after 6-9 days in DD (turn-around). Sharma *et al.* (2004a) studied locomotor activity rhythm of virgin and mated queens of *C. compressus* under constant laboratory conditions. The locomotor activity rhythm of virgin queens entrained to a 24 h (12:12) LD cycle and free-ran under constant dim red light (RR) with a free-running period of approximately 24 h. The locomotor activity of the mated queens on the other hand was arrhythmic during the period when they were laying eggs, and robust rhythmicity appeared soon after the egg-laying phase was over. The locomotor activity rhythm of mated queens was significantly greater than that of virgin queens.

15.2.3 Developmental rhythms

Developmental event, viz. hatching from the egg, larval ecdysis during each moulting, pupation and eclosion occurs only once in the life of an insect and generally timings of such events are under the control of circadian rhythmicity. Pittendrigh (1967) showed that in 12 replications of asynchronous populations of *Drosophila pseudoobscura* (in LD 12:12) with pupae of mixed age, the flies emerged shortly after dark to light transition. When transferred to DD the periodicity of eclosion persisted and emergence occurred at a sharply defined peak each day. Such rhythmicity is referred to as 'gated rhythm' in which an event is controlled by a circadian oscillation. The gate is opened for a specific period once each day, during which the flies emerge. The flies that are not fully matured must wait for the next gate. In developmentally asynchronous populations if the circadian oscillators of individuals are synchronous, gates will open synchronously at ô intervals in constant conditions (Page, 1985). Gated rhythms of pupal eclosion have been described in many Diptera, Lepidoptera and Coleoptera. The gating of certain developmental stages by such a mechanism is probably omnipresent in insect population rhythms (Saunders, 1982) and is evident from the examples: pupal eclosion in *Antheraea pernyi* (Truman, 1971b), egg hatch and oviposition in *Pectinophora gossypiella* (Minis and Pittendrigh, 1968; Pittendrigh and Minis, 1964), oviposition in large milkweed bug, *Oncopeltus fasciatus* (Rankin *et al.*, 1972), pupation in *Aedes vittatus* (McClelland and Green, 1970) and pupation in *Aedes taeniorhynchus* (Nayar, 1967). Insects living in temperate regions are exposed to drastic changes in photoperiod and temperature. Studies on strains of *Drosophila littoralis* originating from a wide range of geographic locations at different latitudes revealed a mild latitudinal trend in the phase and period of eclosion rhythm. Similar latitudinal clines for phase and amplitude of eclosion rhythm were also reported in *Drosophila auraria* (Pittendrigh and Takamura, 1989). Since amplitude of circadian rhythm responds to changes in photoperiod as well as temperature, it was concluded that these insects use circadian clocks to sense seasonal changes in their environment. Truman (1972a) described fortuitous gating in tobacco hornworm, in which each larval instar ecdysed at a particular time of the day. The timing of each ecdysis occurred later in the day than the previous one, which was correlated with the release of prothoracicotropic hormone (PTTH). Sheeba *et al.* (1999a) assayed egg to eclosion development time and survivorship on 4 laboratory populations of *Drosophila melanogaster* that had been reared for over 600 generations in LL and constant temperature. The assays were performed in LL, LD (12:12) and DD. Development time in LL was significantly less than that in LD, which in turn, was significantly less than that in DD, whereas survivorship did not differ significantly among the 3 treatments. Individuals from *Drosophila* populations routinely maintained in LL did not suffer any deleterious effect pertaining to pre-adult fitness. Vafopoulou and Steel (1999) reported that the synthesis of ecdysteroids by the prothoracic glands (PGs) of *Rhodnius prolixus* occurs with a circadian rhythm throughout most of larval-adult development. A circadian oscillator within the PGs generates this rhythm. The principle regulator of the PGs is the cerebral neuropeptide prothoracicotropic hormone (PTTH), which is also released with a circadian rhythm.

15.2.4 Metabolic and physiological rhythms

Rhythms of various metabolic and physiological activities, viz. oxygen consumption, luminescence, colour change, cuticular growth, etc. exhibit circadian periodicity. Rhythmicity in O_2 consumption in LD cycles has been observed in cockroaches, *Blattella germanica, Periplaneta americana*; European corn borer, *Ostrinia nubilalis*; yellow mealworm beetle, *Tenebrio molitor*; *Drosophila melanogaster*; codling moth, *Cydia pomonella* and confused flour beetle, *Tribolium confusum*. The stick insect, *Carausius morosus* is light in colour during day and dark at night, and this rhythm persists in DD for several days (Schliep, 1910). Deposition of cuticle in desert locust, *Schistocerca gregaria* is under the control of circadian system and a lamellate layer of cuticle is formed at night while non-lamellate layer is laid down during day. The periodicity of about 23 h persists for at least 2 weeks in DD (Neville, 1965). Rhythmic cuticular growth has been described in many species of nine insect orders (Neville, 1970). Dreisig (1975) suggested that the daily onset of activity in fireflies and glowworms is a complicated process involving both circadian rhythm and triggering mechanism controlled by ambient illumination. In firefly, *Photinus pyralis* endogenous rhythm of luminescence occurred in constant dim light (LL) for a short duration once every 24 h (Buck, 1937). The glowworm, *Lampyris noctiluca* also exhibit similar rhythm, which is inhibited by light above about 10 lux (Dreisig, 1978). Groups of honey bee workers, *Apis mellifera* show endogenous circadian rhythms in metabolic activity. Workers entrained to two different photoperiods, when put together in a group, coordinate their individual metabolic activity cycles into a synchronized group oscillation. Either physical interaction among workers, or a low volatility contact pheromone, is implicated in the control of this oscillating system (Southwick and Moritz, 1987).

15.2.5 Rhythmic sensitivity to insecticides

Daily rhythmic patterns of insect sensitivity to insecticides were first reported in the German cockroach, *Blattella germanica* maintained under LD (12:12) for 7 days and then injected with DDT, dimetilan, dichlorvos and KCN at different times of the day, and then held in LL. Sensitivity to KCN was highest after dusk and dawn and dimetilan was most toxic in mid night. Sensitivity to dichlorvos showed a bimodal rhythm with peaks before dawn and dusk, and for DDT it was unimodal but very low (Beck, 1963, 1980). The cotton boll weevil, *Anthonomus grandis* maintained under three different photoperiods were treated with a standard dose of methyl parathion at different times of the day. Regardless of the photoperiodic regime itself, a distinct peak of resistance occurred at the beginning of the scotophase and recurred at 6-hour intervals (Cole and Adkisson, 1964). The house fly, *Musca domestica* was found to display a daily rhythm of sensitivity when treated with trichlorfon and the rhythm was maintained when the insects were kept under different LD regimes (Fernandez and Randolph, 1967). Ware and McComb (1970) described circadian susceptibility of *Pectinophora gossypiella* to azinphos-methyl. *Musca domestica* and *Rhyparobia maderae* reared under alternating cycles of LD, exhibited daily sensitivity rhythm to pyrethrum. Both species were most susceptible to insecticide during the last quarter of the daily light span, about mid-afternoon (Sullivan *et al.*, 1970). The sensitivity rhythm of house flies was tested using malathion, DDT and dieldrin. The flies were most susceptible to the three insecticides at the end of the scotophase, one hour before light-on (Shipp and Otton, 1976a). Adults of the yellow mealworm beetle, *Tenebrio molitor* reared under LD (12:12) showed bimodal susceptibility rhythm when treated with methyl parathion. Peaks of maximum sensitivity occurred at the onset of both light and dark phases (Fondacaro and Butz, 1970).

It has been suggested that circadian rhythms pertaining to insect sensitivity to toxicants might be expected to be reflections or rhythmic biochemical processes involved in the mode of action and metabolism of the toxicants. Maximum house fly sensitivity to DDT coincides with respiratory metabolism conversion of DDT to DDE increases with rate of oxygen uptake. (Shipp and Otton, 1976b). The sensitivity of house crickets, *Acheta domestica* to the anesthetics: ether, chloroform and carbon tetrachloride was also found to be at its daily maximum during the early scotophase, which corresponds with the time of greatest locomotor activity (Nowosielski *et al.*, 1964). Metabolic relationships studied in the confused flour beetle, *Tribolium confusum* showed a significant relationship between oxygen consumption and sensitivity to the insecticide dichlorvos. In LD (12:12) regime, the highest sensitivity to dichlorvos occurred shortly after midnight, while maximum oxygen consumption occurred about 2 hours later (Vea *et al.*, 1977). Onyeocha and Fuzeau-Braesch (1991) reported circadian changes in toxicity of the dieldrin in the fifth instar of the migratory locust, *Locusta migratoria* housed under light LD (12:12). The test insects were more susceptible to dieldrin when dosed during the night rather than during the day, and females were less susceptible than male. Pszczolkowski and Dobrowolski (1999) tested rhythmic locomotor activity and daily susceptibility to deltamethrin in the large pine weevil, *Hylobius abietis*. In constant darkness, beetles revealed a free-running circadian pattern of locomotor activity (average period 22 h: 20 min). Under long photoperiod, LD 18:6, entrainment of motor activity was observed. In the entrained population of the weevils, a peak of locomotor activity occurred at about the beginning of the dark phase, and the minimum activity occurred in the middle of the light phase. Fluctuations of susceptibility to a standard dose of deltamethrin (0.5μg/g body weight) varied across the day, and were inversely correlated to the changes in locomotor activity.

15.2.6 Time memory of honey bees

Buttel-Reepen (1900) suggested that honey bees, *Apis mellifera* had a time sense (*Zeitsinn*) and visit blossoms secreting nectar in the morning hours. Forel (1910) proposed that bees had a time memory (*Zeitgedächtnis*). Beling (1929) successfully trained bees for an artificial feeing site at a fixed time. She trained bees to feed sugar solution at a fixed location for a brief period each day at the same time. These bees were marked and then an empty dish was set out. Bees were observed to approach the empty food dish only during the period for which they were trained. She also showed that bees could also be trained to feed at two or more separate time during a day. Besides, the bees were also trained successfully by eliminating external daily cycles of temperature, light, etc. which indicated that an endogenous clock is involved in rhythmic behaviour. Wahl (1932) also obtained similar results by training bees in a salt mine 180 m below the earth surface. These results suggested that bees were relying on an internal clock with a periodicity of about 24 h and whose motion continued in constant conditions.

Renner (1957) conducted notable experiments on bees; he trained bees to a food source in Paris in a closed chamber under constant light and temperature conditions. The trained bees were transported to New York and tested under similar conditions. Though the bees were transported to a different longitude and with a time difference of 5 h, but they came to the feeding site exactly after 24 h of their feeding in Paris. Renner (1959) repeated related experiments in open conditions and obtained similar results for two days. Bennet and Renner (1963), Beier (1968), Beier and Lindauer (1970) also demonstrated endogenous nature of time memory in bees.

Attempts were also made to increase or decrease the metabolic rate of bees; however, such changes did not effected time sense of bees (Renner, 1957), but chilling of bees to 4 or 5°C for about 5 h caused a 3-5 h delay in their arrival at the feeding site (Medugorac and Lindauer, 1967). Martin *et al.* (1978) successfully transplanted corpora pedunculata from the trained bee to inexperienced bee and observed that after 1-2 days recovery, she exhibited a preference to approach the food source at the same time for which the donor bee was trained.

15.2.7 Time-compensated sun orientation

Many insects are known to orient themselves by maintaining a fixed angle to the sun, artificial light or polarized pattern of sky. Santschi (1911, 1913) demonstrated that ants maintain a straight course by using position of the sun (light compass reaction). He showed that if the ants were exposed to reflected light with the help of a mirror, they altered their route in a predictable manner. He suggested that the insects must be able to compensate such changes in sun's azimuth and light compass reaction is useful for longer-term orientation. Brun (1914) detained workers of common black ant, *Lasius niger* in a dark box and then released, they continued at the same angle to the sun. The works of Brun (1914) on ants and Wolf (1927) on bees suggested that such time-compensation was occurring in insects. Frisch (1950) demonstrated that honey bees possess an innate biological clock for such time-compensation. Frisch and Lindauer (1954) trained honey bees to visit a sugar source 180 m north-west of the hive and then they shifted the colony to new territory. The bees were provided with four feeding sites in different directions, but the bees visited the feeding site in the direction for which they were trained (Fig. 7.2). Thus, the bees were able to compensate the movement of the sun and maintained a constant compass direction. Meder (1958) captured the bees and maintained them in darkness for one or more hours and when they were released, the flight to the hive was in the appropriate direction, suggesting that the bees have compensated for the movement of the sun occurred during their captivity. The use of a biological clock to compensate the changing azimuth of the sun has also been demonstrated for ants (Jander, 1957); beetle, *Geotrupes stercorosus* (Birukow, 1953); white grub cockchafer, *Melolontha melolontha* (Cuoturier and Robert, 1955); beetle, *Phaleria provincialis* (Pardi, 1955); pond-skater, *Velia currens* (Birukow, 1956); bumble bee, *Bombus terrestris* (Jacobs-Jessen, 1959) and wasp,

Polistes gallicus (Ugolini, 1985, 1986). The results of experiments related with time memory and time-compensated sun orientation suggested that a circadian oscillation can function as a 'clock' that can be continuously consulted for time-of-day information (Page, 1985).

15.3 Lunar, semilunar and tidal rhythms

The position of moon with respect to earth and sun varies each day and leads to periodicities, viz. moonlight intensity, tidal cycle and semi-lunar cycles. The brown clouds rising out of Lake Victoria represent a mass emergence of 'lake flies' and billions of flies emerge synchronously. These swarms of synchronously emerging flies have also been observed at other lakes in East Africa, such as Lake Malawi and Lake Nyasa. Usually, *Chaoborus edulis* (Fam. Chaoboridae) and other chaoborids are the predominant species in such swarms. One of the common names for these flies is 'phantom midges' comes from the menacing 'clouds' that are visible on the horizon from miles away. The timing of the mass emergence of chaoborid flies is controlled by circadian rhythms. *C. edulis* exhibit lunar emergence rhythms. These flies swarm monthly near the time of the new moon (Corbet, 1958, 1964).

Palmer (1995) discussed the biological rhythms and clocks of intertidal animals. The most extensively studied are semi-lunar rhythms of emergence in intertidal chironomid midges of the genus *Clunio*. The life cycle of a marine chironomid midge, *Clunio marinus* is positively baroque. The midge inhabits tidal waters along western Europe and environmental cues can play critical roles in timing developmental events that enable the perpetuation of the species. Adults of both sexes are short-lived (~2 h) and emerge only at a low tide. Females must mate and lay their eggs within this short time. The males emerge before the females, assist the wingless females in emergence, copulate, and then carry the females to the larval habitat. The egg laying is confined to sandy masses of red algae that are exposed only during the lowest ebbing of the spring tide. Here the eggs are deposited and quickly covered by the sea. Therefore, mature females must emerge from their pupal cases during the few days of the spring tide and at the correct hour for the ebb tide.

This complicated life cycle is possible only by coordinating the insect's lunar (i.e. tidal) and circadian (photoperiodic) adaptations. Neumann (1963, 1966) showed that rhythm of adult emergence in *C. marinus* is governed by the superposition of a circadian rhythm controlling pupal eclosion and a semilunar rhythm determining the beginning of pupation. Neumann (1976) summarized adaptations of midges to intertidal environments. He showed that *Clunio* has integrated a 15-day lunar-regulated rhythm of adult emergence to a sun-regulated circadian rhythm. In this species a number of genetically distinct geographical races exist, each with its own emergence rhythm that is accurately adapted to local tidal conditions. Moreover, each population along the European coastline has made adjustments to local tidal conditions. In the southern portion of its range, pupation is controlled by the 15-day 'semilunar' rhythm. However, the moon is a poor cue in the northern portion of its range (since it does not rise much above the horizon during summer months and the sun stays out much longer). Here, the turbulence of the tides appears to control pupations. Thus, the emergence of adults from pupae appears to be regulated by a combination of photoperiodic, lunar, and tidal rhythms. Neumann (1989) analysed semilunar and lunar reproductive rhythms in different geographic populations of *Clunio*. In stocks of three populations differing in period and phase relationship with the lunar month, these long-term rhythms were synchronized in the laboratory by using artificial moonlight cycles of 30 days in otherwise 24 h light-dark (LD) cycles (0.4 lux during 4 successive nights every 30 days in LD 12:12). In LD cycles of various periods, a strong synchronization was only possible in LD 12:12 and LD 11:11, whereas in LD 10:10 and LD 15:15 the synchronization by the 30-'days' moonlight cycle was weak or even absent. The study demonstrates a limited range of circadian periods for entrainment of the long-term rhythms. He concluded that an LD cycle with a period near 24 h is an essential *zeitgeber* condition for semilunar and lunar timing in this marine insect. Further, the underlying physiological timing mechanism of *Clunio* consists of a circadian function for the perception of the monthly moonlight *zeitgeber* cycles that entrain the endogenous, temperature-compensated oscillator of the circasemilunar (or circalunar) period. The long-term oscillator triggers the metamorphosis of the insect, and thereby determines the time of its eclosion and reproduction on the shorelines, in correlation with days of spring tides recurring about every 14-15 days. Similarly, the Japanese midge, *C. takahashii* is exposed twice daily by the tidal cycle throughout the season, and its emergence occurs on every ebb tide (Hashimoto, 1966).

Evans (1976) reported that intertidal carabid beetles, *Thalassotrechus barbarae* forage and mate at night only, during periods of low tide. Hartland-Rowe (1955) found that mayflies, *Povilla adusta* emerges regularly on second night after full moon and the emergence time was maintained even if the larvae were kept in complete darkness. Youthed and Moran (1969) reported that in ant lion, *Myrmeleon obscurus* peak pit-building activity occurs at full moon day.

Danthanarayana (1976) observed diel and lunar periodicity in flight of light brown apple leafroller, *Epiphyas postvittana*. The moths showed two peaks, one large, 2-3 h after sunset, and one small, 3-4 h after sunrise. Males were more active than females at both flight periods. Peak activity occurred shortly after new moon, around full moon, and shortly before new moon. As flight activity is controlled by light intensity, he argued that the flight 2-3 h after sunrise and the peak in nocturnal flight activity at full moon are behavioural adaptations towards wind-assisted dispersal.

15.4 Circannual rhythms

Some long-lived insects exhibit seasonal cycles, which were reviewed by Farner

(1985). An endogenous circannual rhythm controls diapause and development rate in varied carpet beetle, *Anthrenus verbasci* (Blake, 1958). The beetles larvae are found in house sparrow nest and feed on animal matters. Young larvae spent first winter in diapause, and again in next winter they enter in diapause as mature larvae. The adults emerge during following spring. Blake (1958) reared the larvae in laboratory at constant temperature and humidity and in complete darkness (DD), and observed that the rhythm diapause persisted, thereby indicating its endogenous control. The larvae were also kept at different temperatures in DD and at high temperature conditions (22.5° and 25°C) all larvae pupated in the first year, whereas at 15°C all larvae underwent two cycles of diapause and pupated in the second year, about 41 weeks after the first diapause. Blake recognized that this endogenous 41 weeks interval as rhythms 'basic periodicity', as in the absence of natural *Zeitgeber*, the rhythmic period was between 41-44 weeks. These experiments obviously established that the circannual rhythm of pupation in the beetles is a gated phenomenon. Blake (1960, 1963) further confirmed that natural seasonal changes in temperature acted as important *Zeitgeber*. He maintained four groups of larvae at different combinations of outdoor temperature, constant temperature (20°C), and changing daylengths and complete darkness. In complete darkness and constant temperature the endogenous rhythm free-ran and second peak occurred after 40-44 weeks. In constant temperature and natural daylength the second peak was delayed by about 13 weeks. Thus, second cycle was of about 52 weeks. In outdoor conditions of temperature with natural daylength or complete darkness the larvae pupated at the same time, suggesting that light played no role in synchronization. Nisimura and Numata (2001) described circannual pupation rhythm in *A. verbasci* determined by laboratory experiments under various photoperiods and temperatures. The frequency distribution of larval duration showed a periodic pattern over 2-3 years and the period was 37-40 weeks under a constant short-day photoperiod (LD 12:12) at 20°C. This rhythm showed temperature compensation to some extent under a short-day photoperiod between 17.5°C and 27.5°C. Under alternations of a long-day (LD 16:8) and a short-day photoperiod, pupation occurred 21-24.5 weeks after transfer from a long-day to a short-day photoperiod.

Lees (1960) reported that in vetch aphid, *Megoura viciae* overwintering diapause egg develops into a virginopara (the fundatrix) that gives birth to viviparous and parthenogenetic offspring (Fig. 15.2). If it is exposed continuously to a short photoperiod at 15°C, no sexuals (males and oviparae) were produced for at least 90 days. The aphids may possess a non-oscillatory and transgeneration timer that serves to prevent them from producing premature oviparae during the long nights of spring. The monarch butterflies, *Danaus plexippus* undertake migratory flights every fall from northeastern America to their over-wintering grounds in central Mexico. Brower (1996) demonstrated that circadian clocks play a key role in time-compensated navigation of migratory flight in these butterflies.

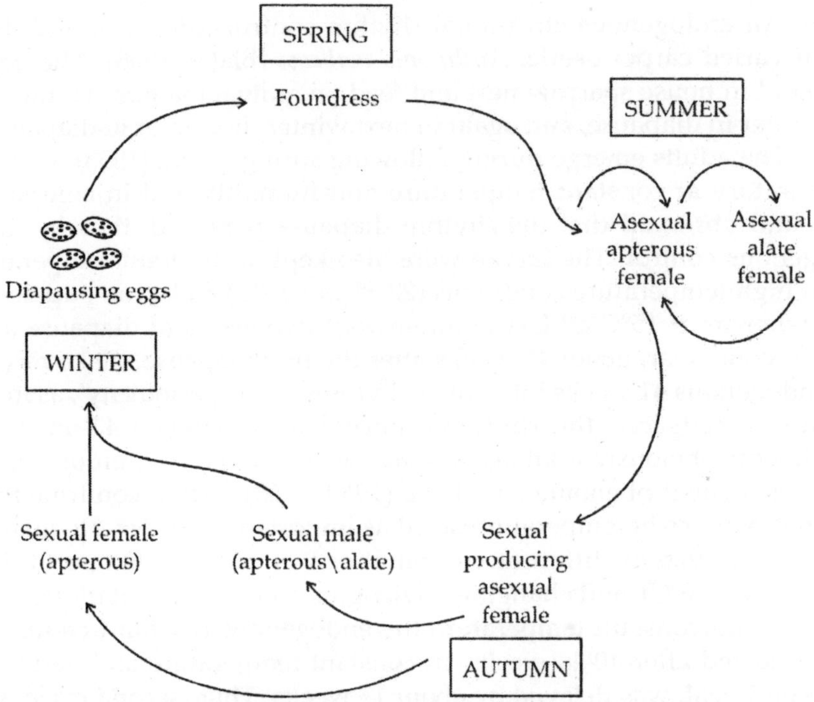

Fig. 15.2. Life cycle that in vetch aphid, *Megoura viciae*. Diapausing egg develops into a foundress that gives birth to viviparous and parthenogenetic offspring.

15.5 Entrainment

Circadian oscillations provide an organism with an innate temporal programme, which allows the organism to predict environmental fluctuations. The oscillation must be correctly phased with respect to geographical cycles of immediate surroundings. Usually the oscillations are synchronized with external environmental factors and circadian oscillations assist to assess the environmental cues (*zeitgebers*), and these results in their synchronization (entrainment) to local conditions. Thus, circadian clocks enable living organisms to keep track of time in their local environment. These clocks are ubiquitous and must provide adaptive advantage to their owners. Circadian clocks enhance the innate ability of organisms to survive under ever-changing environments by enabling them to efficiently anticipate periodic events such as availability of light, temperature, sound, food and mates. It is believed that circadian clocks have evolved under selection pressures comprising of periodic cycles of the environment, which act on these clocks under the entrained state. As a result, precisely timed rhythmic activities confer greater adaptive advantage to organisms as compare to randomly occurring activities (Sharma, 2003; Roenneberg *et al.*, 2003). Hence, the free-running phenotypes of circadian clocks are considered to be an evolutionary outcome of natural selection on entrained clocks (Pittendrigh, 1993). Although circadian clocks are believed to have arisen as a result of adaptive evolution under periodic environments, there has been hardly any rigorous and conclusive empirical study to support this (Sharma, 2003).

Daily periodicity of light and dark is most important entraining stimulus. Besides, for insects temperature is also effective in entrainment of circadian system. In some insects an increase in light intensity results in loss of rhythmicity. In *Drosophila pseudoobscura*, the eclosion becomes aperiodic at light intensity above 0.05 lux (Pittendrigh, 1981) while in *Rhyparobia maderae* rhythmicity persists at high light intensities (Roberts, 1960). Peak activity of different parasitoids of *Drosophila* occurs at different times of the day, which significantly reduces intrinsic competitive disadvantage for the inferior competitor (Fleury *et al.*, 2000). In the colony of the *Apis mellifera*, young workers perform 'nursing' task around the clock without taking any rest, while foragers visit flowers in a rhythmic manner timed by well-developed circadian clocks (Moore *et al.*, 1998). It appears that honey bee workers use circadian rhythmicity to match age-dependent behavioural development. In the ant, *Camponotus compressus* virgin queens and males use circadian clocks to time their mating flights by maintaining appropriate phase relationship with the cyclic environments, perhaps to facilitate cross-breeding between colonies and to avoid inter-species mating (Sharma *et al.*, 2004b). The monarch butterflies, *Danaus plexippus* undertake migratory flights every fall from northeastern America to their over-wintering grounds in central Mexico. Froy *et al.* (2003) demonstrated that circadian clocks play a key role in time-compensated navigation of migratory flight in monarch butterflies.

15.6 Photoperiodism and seasonal cycles

Photoperiodism comprises a miscellany of clock phenomena in which organisms distinguish the long days (or short nights) of summer from the short days (or long nights) of winter, and thereby obtain information on calendar time from the environment (Saunders, 1982). Beck (1980) discussed photoperiodic regulation of behavioural and rhythmic processes in insects. The seasonal changes exert a strong selective pressure on the physiology and behaviour of animals. It has been observed that many insects have evolved systems that can measure photoperiodic duration in a precise manner. Responses to photoperiod allow insects to survive periods of unfavourable environmental conditions. Since the early 1950s, light-sensitive and temperature-compensated circadian clocks have also been known to regulate the timing of many events (behavioural, physiological, biochemical) in insects. Considerable efforts have been directed at elucidation of the nature of the clock mechanism that underlies photoperiodism (Takeda and Skopik, 1997). Diapause is a common photoperiodic response and is known to occur in many insects. Other photoperiod determined phenomena are: growth, sexual behaviour, migration, colouration, fecundity, formation of seasonal morphs, etc.

Many insects in a state of cold inertia become quiescent, and quiescence is a sudden, short-term, non-cyclic, interruption in the growth and development of an individual that is directly imposed by adverse conditions. The recovery occurs soon after these restrictions are removed. In quiescence, slowing down of metabolism occurs in response to adverse environmental conditions. The chironomid, *Polypedilum vanderplanki* breeds in small pools that alternately become dry and flooded. During the dry periods the larvae dry out on the mud under 4 to 8 mm of plant debris in depressions. Larvae are exposed to temperatures as high as 70°C. Hinton (1951) brought the larvae into the laboratory to determine how the larvae survive. The larvae were dried to <3% moisture and were heated at several temperatures for varied amounts of time. Some of the larvae metamorphosed after exposure to 102-104°C for 1 minute, and some recovered temporarily after exposure to 106°C for 3 h or 200°C for 5 min.

Many insect species have evolved a strategy called diapause. Diapause is a neurohormonally mediated dynamic state of low metabolic activity. It occurs in genetically determined stages and is species-specific in response to 'token stimuli'. Token stimuli must be perceived by an insect, and then translated into a biological response. Once diapause is initiated it does not terminate with immediate return of favourable conditions. Diapause is a suspension of development that can occur at the embryonic, larval, pupal, or adult stage, depending on the species. In some species, diapause is facultative and occurs only when induced by environmental conditions, while in others it has become an obligatory part of the life cycle. The latter is often seen in temperate-zone insects, where diapause is induced by changes in the photoperiod. The day length when 50 per cent of the population has entered diapause is called the 'critical day length'. Insects entering diapause when the day length falls below this threshold are called 'long-day insects'. Those insects that develop normally when there are only a few hours of sunlight and that enter diapause when exposed to longer days are called 'short-day insects'. The critical day length is a genetically determined property (Tauber *et al.*, 1986). The diapause has recognizable stages. The 'prediapause period' is regarded as a time of feeding and metabolic reserve accumulation, as well as growth to the species-specific diapause stage. The 'induction period' is usually characterized by reduction in feeding, metabolism and growth. The 'maintenance period' sets the length of diapause, and is genetically determined.

It is commonly agreed that diapausing adult insects have reduced corpora allata activity, and hence, reduced haemolymph concentration of juvenile hormone (JH). However, JH action in regulating diapause in juvenile insects is just one aspect. In adults JH is involved in regulating aspects of sexual maturation, egg development in some species and male accessory gland development. In some cases treating diapausing adults with JH can terminate adult diapause. In Colorado potato beetle, *Leptinotarsa decemlineata* short day lengths induce adult diapause and corpora allata activity is regulated by stimulatory neurohormones from the brain. Thus, complete regulation of diapause involves both neurohormones and JH. In the bug, *Pyrrhocoris* activity of JH is directly inhibited by inhibitory neurons from the brain, and thereby maintains low JH titre and diapause. In the rice stem borer, *Chilo suppressalis* JH titres remain high throughout diapause, and corpora allata activity decreases as diapause draws to an end.

In the silkworm, *Bombyx mori* embryonic diapause appears to be regulated by diapause hormone, a 24-amino acid peptide that is produced in the suboesophageal ganglion (Fukuda, 1952). In cecropia moth, *Hyalophora cecropia* pupal diapause follows cessation of brain neurosecretory activity, which stops prothoracicotropic hormone (PTTH) release (Williams, 1952). In many butterflies, this inhibition of PTTH is due to a continued elevated titre of juvenile hormone. Similarly, the lack of PTTH and ecdysone secretion, once pupation has occurred will cause diapause during this part of development. Adding back 20-hydroxyecdysone can reactivate diapausing pupae. However, under normal conditions, the brain of diapausing pupae (as in *Hyalophora cecropia*) is activated by the exposure to cold weather for a particular duration. Moth pupae kept in warm conditions will remain in diapause until they die (Nijhout, 1994). The pupal diapause of corn earworm, *Helicoverpa zea* is independent of PTTH. Low temperatures somehow act directly upon prothoracic glands, which reduces titres of moulting hormone.

Photoperiod is also a decisive environmental factor involved in the formation of seasonal morphs mainly found in the insects of the orders Lepidoptera, Hemiptera, Orthoptera and Thysanoptera. The nymphalid butterfly, *Araschnia levana* occurs in two distinct morphs *levana* and *prorsa*. The *levana* form is orange-brown with black spots and emerges in autumn from a diapausing pupa, while *prorsa* is black and white and emerges in summer from a nondiapausing pupa (Müller, 1955). The cabbage white butterfly, *Ascia monuste* is melanic in LD 16:8, but white in LD 8:16. Morphs have also been reported in silver-lined moth, *Bena prasinana;* some *Pieris* species; alfalfa butterfly, *Colias eurytheme;* vapourer moth, *Orgyia thyellina* and many other lepidopterans.

A number of hemipterans like leafhoppers, whiteflies, psyllids and aphids exhibit seasonal polymorphism. In vetch aphid, *Megoura viciae,* apterous virginopara is formed under long day while alate virginopara is formed in long day and crowded conditions. In short day sexual forms (oviparae and males) are formed (Lees, 1959). Other important hemipterans in which such morphs have been reported are: leafhoppers, *Euscelis plebeja, E. lineolatus, Nephotettix cincticeps;* aphids, *Aphis forbesi, Aphis fabae, Brevicoryne brassicae* and *Acyrthosiphon pisum;* pear psylla, *Cacopsylla pyri;* smaller brown planthopper, *Laodelphax striatellus;* whitefly, *Aleurochiton aceris;* and dusky stink bug, *Euschistus tristigmus.*

Many insects display seasonal movements and migrate over long distances. In many cases migrations are closely associated with the induction of diapause, and probably controlled by photoperiod. The photoperiods had a marked influence on the migration strategies of large milkweed bug, *Oncopeltus fasciatus* (section 8.8).

In some insects photoperiod had a marked influence on development and sexual behaviour. Females of diamondback moth, *Plutella xylostella* raised in LD 15:9 laid more eggs than those reared in LD 9:15. The caterpillars completed their development in 15.5 days at LD 15:9 while took 18 days at LD 9:15 (Atwal, 1955). Males of migratory locust, *Locusta migratoria* raised in LD 12:12 spent about 70 per cent of their time in sexual behaviour for first 5-6 weeks while those raised in LD 16:8 showed only insignificant sexual behaviour or none for first 2-3 weeks. Females maintained in LD 12:12 laid more eggs than those raised in long days (Perez *et al.*, 1971).

15.7 Location of rhythms

In insects much work has been done on the clock controlling locomotor activity in cockroaches. Harker (1954) reported presence of an endocrine gland in suboesophageal ganglion (SEG) of American cockroach, *Periplaneta americana* and concluded that its secretion, carried either in blood or tissues is involved in production of diurnal activity in cockroach. She found that headless cockroach did not exhibit any rhythmic activity; however, the activity was restored for a few days when the SEG was implanted in headless cockroach from a donor insect (Harker, 1956). Further attempts to repeat such transplantation experiments were failed, and other evidences argued persuasively against location of a hormonal clock in SEG. Roberts (1966) failed to restore rhythmicity in headless cockroaches after transplantation of SEG. Nishiitsutsuji-Uwo and Pittendrigh (1968) also showed that removal of neurosecretory cells from SEG was not affecting the rhythm. It in fact was an insect, the Madeira cockroach *Rhyparobia maderae*, where for the first time a circadian clock could be localized. Lesion studies found the pacemaker controlling circadian rhythmic behaviour in the brain's optic lobes of the cockroach (Nishiitsutsuji-Uwo and Pittendrigh, 1968). Roberts (1974) attempted to find more precise site to locate driving oscillation in *R. maderae* and *P. americana* and found that the synaptic areas, lobula and medulla were important in the control of rhythmic behaviour. Page (1978) concluded that cells in ventral half of the lobe near the lobula are crucial in maintaining the rhythmicity.

Truman and Riddiford (1970) demonstrated that in moths, *Hyalophora cecropia* and *Antheraea pernyi* circadian pacemaker that control eclosion time is located in the brain. The emergence in these moths is gated, but becomes irregular with the removal of brain. Normal rhythmicity was restored when the brain was removed and reimplanted in the abdomen (Truman, 1972b). The results suggested the brain as a site of the clock, which control emergence time via release of eclosion hormone. Loher (1972, 1974; Sokolove and Loher, 1975) reported that ablation of optic lobes in *Teleogryllus commodus* completely disrupted stridulation, spermatophore formation, and locomotion.

Circadian rhythms are generated in nerve cells termed 'clock neurons'. The presence and location of these neurons have been identified in many insects by testing the persistence of rhythms in locomotor activity after lesions of different brain structures, and by recording spontaneous circadian oscillations in neuronal electrical activity. Clock neurons have been found in brain of *Gryllus bimaculatus* (Tomioka and Chiba, 1992), and these structures are termed circadian pacemakers. In flies, the circadian clock that regulates the temporal aspects of behaviour has best been described in *Drosophila melanogaster* which is composed of two groups of neurons, the dorsal and ventral-lateral neurons located in the protocerebrum. In house fly, *Musca domestica* the circadian clock that regulates locomotory activity is located outside the optic lobe, since lesion of the optic lobes does not disturb the circadian rhythm in locomotor activity (Helfrich *et al.*, 1985). Circadian rhythms have; however, been found beneath the compound eye of the fly, in the first optic neuropile, the lamina, of the optic lobe (Pyza and Meinertzhagen, 1995). In the lamina, two classes of monopolar cell interneuron, L1 and L2, have lamina axons that change their calibre during the day and night. L1 and L2, exhibit cyclic size changes. In the house fly, *Musca domestica* these cells swell during the day and shrink at night and, because these cyclical changes persist in constant darkness (DD) as well as in continuous light (LL) (Pyza and Meinertzhagen, 1995). Circadian oscillators located inside and/or outside the optic lobe may generate these rhythmic size changes. Bays and Pyza (2001) demonstrated that circadian oscillators located inside and/or outside the optic lobe might generate these rhythmic size changes in house fly.

A number of evidences suggest that the organized retinal photoreceptors are not involved in response to light. In many insects cauterisation or covering of retinal photoreceptors (compound eyes, ocelli) had no interference with normal photo period response, and it has been presumed that extraretinal photoreceptors are located in the brain of many insect species. Saunders (1997) suggested that photoperiodic induction of larval diapause in *Calliphora vicina* and ovarian diapause in *Drosophila melanogaster* are based on the circadian system but seems to involve a separate mechanism at both the molecular and neuronal levels. The compound eyes and ocelli are neither essential nor necessary for photic entrainment, and the circadian clock mechanism is not within the optic lobes. In *D. melanogaster*, a group of lateral brain neurons act as important circadian pacemaker cells, which are possibly also photo-sensitive. Similar lateral brain neurons were also identified in *C. vicina*.

In vetch aphid, *Megoura viciae* development of female offspring into sexual or asexual morphs is determined by the photoperiod to which mother is exposed. Lees (1964) reported that cauterisation of the compound eyes had no effect on the photoperiodic response of the aphid. Williams and Adkisson (1964) demonstrated location of photoreceptors in the brain of *Antheraea pernyi*. However, the carabid beetles, *Pterostichus nigrita* with ablated compound eyes responded as if they were in constant darkness (Ferenz, 1975).

15.8 Genetics of circadian rhythms

The real impetus to molecular genetic analysis of the circadian clock has come from *Drosophila* species and *Drosophila melanogaster* was the first insect model system in classical physiological studies of circadian rhythms. By analysing the circadian patterns of activity and eclosion behaviour of mutant flies, several genes have been identified that encode essential elements of the clock (Hall, 2003). The genes that control circadian rhythms are called 'clock genes'. These genes are responsible for the generation of endogenous circadian oscillations by controlling circadian expression of other 'clock-controlled genes' (Young, 1998). The first circadian gene was discovered in 1971 in *Drosophila* (Konopka and Benzer, 1971). Three mutant lines in circadian rhythmicity were isolated which mapped to the same locus on the X chromosome and the gene was named the *period* (*per*). The first mutation, called per^0, exhibited an arrythmic behaviour. The second mutation had behavioural rhythms shorter than wild type flies, about 17.5 h, and was called per^S. The last mutant line, per^L, exhibited long behavioural rhythms, more than 26.5 h. Myers *et al.* (1995) cloned the second *Drosophila* clock gene and named *timeless* (*tim*). Recent studies have revealed several of the key genes involved in central clock of *Drosophila*. Wilsbacher and Takahashi (1998) compile these findings into a model of the *Drosophila* clock. The *period* (*per*) and *timeless* (*tim*) genes are two of the key players in this cycle. Clock genes involved in the *Drosophila* circadian timing system are: *period* (*per*), *timeless* (*tim*), *double-time* (*dbt*), *clock* (*clk*), *cycle* (*cyc*), *vrille* (*vri*), *shaggy* (*sgg*), and *cryptochrome* (*cry*). Mutations of these genes can either speed up or slow down the clock, giving flies with days of 20 or 28 hours. Alternatively, mutations can destroy altogether the ability of a fly to be rhythmic.

The proteins encoded by these genes are components of a self-sustaining negative feedback loop, which is now thought to form the driving oscillation of the timing system (Fig. 15.3). The fly's clock consists of a core system of four regulatory proteins that interact to give the clock periodicity. Four factors cooperate to generate sustained molecular oscillations with a period of 24 h. The cycle begins when two of these proteins, CLOCK (CLK) [encoded by the gene *clock* (*clk*)]; and CYCLE (CYC) [encoded by the gene *cycle* (*cyc*)], bind together and increase the production of two other proteins: Period (PER) and Timeless (TIM). The PER and TIM proteins (synthesized on ribosomes in the cytoplasm) form dimers and the levels of these proteins slowly accumulate over time. The level of protein products Period (PER) and Timeless (TIM) reach their maximum in the late night. Period (PER) and Timeless (TIM) proteins move around the cell, and their abundance and location defining circadian time. When the concentration of these gets high enough (early evening), they are transported into the nucleus. In the nucleus PER binds to the CLK/CYC transcription factors, thus shutting off transcription. Because these genes include *per* and *tim*, the result is a negative feedback look. In this way the products of the *per* and *tim* genes inhibit their own synthesis. So a rising level of PER/TIM dimers turns off further synthesis of them. Consequently, PER and

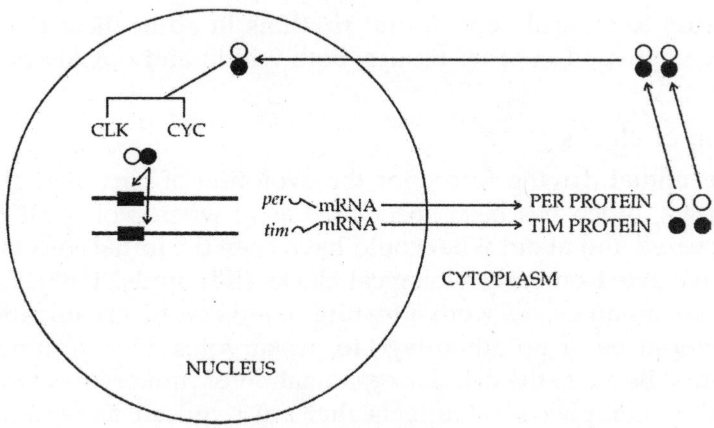

Fig. 15.3. *Drosophila melanogaster* circadian clock model.

TIM levels decline due to degradation of both proteins, ultimately resulting in a release of CLK and CYC inhibition. As the TIM and PER proteins turn over and their level falls, this inhibition is lifted and transcription begins again in the morning. Under natural conditions, the clocks are precise (~24 h) as they are 'set' (synchronized) by environmental cues, of which light is one of the most important. The cycles persist even without any external cues (alternating light and dark) although they tend to drift away from environmental time. Edery *et al.* (1994) reported that daily oscillations in apparent molecular mass as well as abundance of per protein (PER) for *Drosophila melanogaster* are largely or exclusively due to multiple phosphorylation events. As change in abundance and phosphorylation persists under constant environmental conditions, they reflect or contribute to a free-running rhythm. They suggested that phosphorylation status of PER is an important determinant in the *Drosophila* clock's time-keeping mechanism.

An endogenous clock controls the daily fluctuation in the mating activity of some insects. Sakai and Ishida (2001) demonstrated that wild-type *Drosophila melanogaster* display a robust circadian rhythm in the mating activity, and that these rhythms are abolished in *period*- or *timeless*-null mutant flies (per^{01} and tim^{01}). Circadian rhythms were lost when rhythm mutant females were paired with wild-type males, demonstrating that clock genes govern female mating activity. They also found that the mating activities of *disconnected* (*disco*) mutants, which have a severe defect in the optic lobe and are missing lateral neurons, are arrhythmic. Besides, they also detected an antiphasic relationship in the circadian rhythms of mating activity between *D. melanogaster* and its sibling species *D. simulans*.

Giebultowicz (2000) reviewed molecular mechanism and cellular distribution of insect circadian clocks and suggested that in *D. melanogaster* several genes are involved in the mechanism of the circadian clock. Periodic transcription and translation of these clock genes form the basis of a molecular feedback loop that has a 'circa' 24-hour period. Rhythmic expression of clock genes in specific brain neurons appears to control behavioural rhythms in adult flies. However, clock genes are also expressed in other tissues, both within and outside of the nervous system.

15.9 Evolution of clocks

Though the eventual driving forces for the evolution of circadian clocks remain largely unknown, much has been speculated as to when biological clocks might have first appeared and about what could have been the initial selection pressures that might have acted on early biological clocks (Edmunds, 1988). Earlier it was believed that circadian clocks were a feature of eukaryotic organization, and that 24 h clocks would be of no advantage to prokaryotes, whose numbers double every few hours. Besides, the cellular organization of prokaryotes was too simple to accommodate complex mechanisms that are required to regulate circadian systems. However, it has been well demonstrated that plants, fungi, protista, invertebrates and vertebrates have circadian clocks. In many cases even primitive unicellular organisms such as cyanobacteria have functional circadian clock machinery (Johnson and Golden, 1999). It is now believed that circadian clocks may have appeared on earth along with primitive life forms. In experiments designed to test the genetic basis of biological rhythmicity many generations of laboratory organisms were raised and spent all their lives in aperiodic environments, yet the rhythms persist (Sheeba *et al.* 1999b). Period of the rhythm was species-specific, highly heritable, and very amenable to artificial selection. So, if it is in the genes, the clock must have evolved due to some kind of selective pressure. It was often assumed that same selective pressures which keep the clocks ticking all over biosphere today and are responsible for the initial discovery of timing mechanisms by early forms of life. Pittendrigh (1993) believed that circadian rhythms had evolved under selection pressure presented by environmental LD cycles, wherein photophobic processes were confined to darkness and photophilic processes to light. According to Pittendrigh's 'escape-from-light' hypothesis, circadian rhythms evolved to protect organisms from harmful photo-oxidative effects of the environment by helping them in reorganizing light-sensitive reactions during the night.

Chapter 16

Bioluminescence

Bioluminescence has been observed throughout the ages with awe and intrigue. Christopher Columbus, during his historic voyage across the Atlantic, noticed mysterious patches of luminescent light around the waters of his ships. Bioluminescence refers to the visible light emission in living organisms that is the result of chemical reactions and accompanies the oxidation of organic compounds, luciferins (*Luciferin* refers to any light-emitting compound, *Lucifer* means 'light-bringing' in Latin) mediated by an enzyme catalyst, luciferase. The light production does not involve production of heat. Luminescent organisms include bacteria, fungi, fishes, insects, algae and squids that are found in marine, freshwater, and terrestrial habitats. It has been observed that the light-emitting reactions are quite distinct for different organisms, with the only common component being oxygen. Therefore, significant differences have been found between the structures of the luciferases and the corresponding genes from one luminescent organism to another. Light emission by insects particularly fireflies has fascinated the man from centuries. In insects the production of light is always a phenomenon of chemiluminescence and highly evolved in orders: Coleoptera (Lampyridae- lightning bugs or fireflies; Phengodidae- railroad worms; Elateridae- wireworms, click beetles, firebeetles), Diptera (Mycetophilidae- fungus gnats, glowworms), Hemiptera (Fulgoridae- lanternfly) and Collembola (springtails). Chemiluminescence, as the name implies, generates light through chemical reactions (Biron, 2003). Coleoptera comprises the largest insect group in which a large number of species have highly evolved photogenic organs and commonly termed as fireflies, glowworms, railroad worms, automobile bugs, firebeetles, lightning bugs, blinkers, etc.

Detailed accounts of various aspects of insect bioluminescence are discussed in the publications of Lund (1911), McElroy (1951), Harvey (1920, 1952), McElroy and Hastings (1955), McElroy and Seliger (1962), McElroy *et al.* (1974), Buck (1948, 1978), Lloyd (1966, 1971, 1977, 1978, 1979, 1981, 1983, 1997), McElroy and DeLuca (1973, 1985), DeLuca and McElroy (1981), Viviani and Bechara (1997), Ghiradella (1998), Sivinski (1998), Viviani (2001, 2002), and Trimmer *et al.* (2001).

16.1 Fireflies

Firefly is the common name for the nocturnal luminous insect belonging to the beetle family Lampyridae (which means 'shining fire') and about 1900 species of fireflies inhabit the tropical and temperate regions. The greatest numbers of firefly species are found in warm, humid areas of the tropical Asia and Central and South America. Both male and female have special light organs on the underneath of the abdomen. The rhythmic flashes produced vary with sex and species. Fireflies emit green-yellow flashes, characterized by their duration, interval, and frequency, from ventral lanterns, mainly for sexual attraction purposes. Some fireflies also flash their lights at only certain times during the evening in order to attract a mate. Some believe that the flashing is a warning system to predators. During summer fireflies rest on plants or in trees during the day and fly around between dusk and midnight. The firefly light is called a 'cold light' because it produces no heat. The light is produced in special cells called 'photocytes'. Females lay eggs in the soil and the hatched larvae are carnivorous. The larvae sometimes also glow and are commonly known as 'glowworms', as are wingless adult females of certain Lampyridae. Their eggs are also reported to glow. Most firefly larvae are found in rotting wood or other forest litter or on the edges of streams and ponds at night. Some Asian species are fully aquatic (due to the presence of tracheal gills) and live underwater, feeding on aquatic snails. In males of aquatic firefly, *Luciola substriata* luminescent organs are present on abdominal segments V and VI while females have only one on segment V. Larvae have a pair of spot-like light organs on abdominal segment VII. The courtship flashing activity reached peak one hour after sunset. The light spectrum emitted by *L. substriata* was measured as 425-603 nm with the peak 504 nm (Fu *et al.*, 2004). The larvae of several tropical firefly species in the genus *Pyractomena* (Plate 16.1) are strictly arboreal, feed on arboreal snails and pupate while hanging under living leaves (Lloyd, 1991). In general, *Pyractomena* species are rarer and their larvae are occasionally seen glowing in damp woods as they walk up the trunks and along the branches of shrubs and trees.

A firefly's luminescence first appears in the embryo stage when the beetle is still within the egg 'shell'. The larva of fireflies glows and even the larvae of species in which the adults are not producing light the larvae have ability to glow. The function of larval light could be related to the predatory behaviour of the larvae. They hunt snails, earthworms, larvae of other insects, and probably other soft-bodied animals on and in the soil, depending on what kind of firefly they are. Once it was suggested that the glowworms (firefly larvae) attract prey with their light. It has also been suggested that larvae sometimes use their glows to coordinate life history events, such as pupation. Most adult fireflies do not eat. A few adult fireflies are believed to feed upon pollen or nectar. Females of fireflies of the genus *Photuris* (the so-called firefly '*femmes fatales*') mimic the signal pattern of *Photinus* fireflies. Lying in wait, the female continues to signal to a male. When the *Photinus* male arrives to mate, the female pounces upon and devours him (Plate 16.2) (Lloyd, 1965). Female *Photuris* fireflies contain defensive chemicals of two types. They sequester steroidal pyrones (lucibufagins) from male fireflies of the genus *Photinus* that they eat, and themselves produce the defensive betaine

N-methylquinolinium 2-carboxylate. Chemical analyses of *Photuris* eggs showed that females that fed on *Photinus* males endow their eggs with both lucibufagin and betaine, while those that did not feed on *Photinus* lay eggs that contain only betaine. Both types of defensive compound decreased egg predation in the field. By endowing their eggs with both exogenous and endogenous chemicals, *Photuris* females are essentially maximizing their defence options (Gonzalez *et al.*, 1999).

The flashing fireflies try to attract mates, and males fly about flashing while females perch on vegetation, usually near the ground. If the female sees a flasher and ready to mate she responds by flashing right after the male's last flash. A short flash dialogue takes place as the male flies closer and closer, and then mating takes place. Each species has its own flash pattern and the flashing is species-specific. Lall and Worthy (2000) determined that behavioural action spectra of the female's response in big dipper firefly, *Photinus pyralis* were narrow and peaked in the yellow region of the spectrum. The females responded only to stimuli of wavelengths longer than 480 nm and not to stimuli in the blue (420-460 nm) part of the spectrum. This suggests that the yellow receptors of the female mediate the detection and processing of the optical signal of the conspecific male and the signal detection is achromatic. Cratsley and Lewis (2003) observed variation in female flash responsiveness as a function of male flash duration, female condition, lantern size, and lantern distance, as well as the relationship between male characteristics and spermatophore mass in fireflies, *Photinus ignitus*. The results suggested that female *P. ignitus* may prefer long flashes in order to obtain the direct benefit of larger spermatophores and may adjust their overall flash responsiveness as the relative importance of this benefit varies with changing female condition.

In European glowworm, *Lampyris noctiluca* (Plate 16.3) the larva is most active at night and devotes much of its life to feeding and building up it food reserves. The adult does not feed and concentrate only on the task of finding a mate and reproducing. The larva has quite a specialized diet and feeds almost exclusively on snails and slugs, and its body and behaviour are well-adapted to hunting, killing and eating them. The food is processed into fat and then stored in fat-bodies. The glowworm larva is able to produce a yellow-green light from a pair of organs on the underside of its body, near the tip of its abdomen. The larva appears to use its light in at least three different ways: (i) if disturbed, it switches on its lights for a few seconds and then turns them off again. It seems that this is a defensive reaction, intended to scare off a predator, (ii) some fully grown larvae glow continuously for hours, and (iii) sometimes it produces definite pulses of light. A number of views have been proposed for use of light production by the larva; however, the most reasonable is the idea that by producing light the larva deters potential predators. Female glowworm (Plate 16.4) has prominent resemblance to the larva and is wingless. The light producing organs consist of large luminous bands on both the sixth and seventh segments and a luminous spot on either side of the eighth segment. The female usually begins to glow soon after dusk and usually keeps up her display for two or three hours. If the female has not been successful in attracting a mate she will stop glowing, move back into the grass and return for another performance the following night. The male (Plate 16.5) is active partner with well-developed wings and actively seaches for displaying females. He has two small luminous spots at the tip of his abdomen. After reaching nearer the female he uses his antennae to get his way onto her back and begins to mate with her. The mating may carry on for several hours, but it is quite common to find a female with more than one male. The mated female seeks out a suitable place in which to lay her eggs and she dies soon after laying her eggs.

Some fireflies show variations of this type of communication system, as well as courtship behaviours that include pheromones as well as photic signals (Ohba 1983). Branham and Greenfield (1996) considered that male flashing patterns are affected by sexual selection. Female fireflies have been shown to prefer certain characteristics of a male's photic signal (such as increased flash rate) and respond preferentially to males that possess these 'sexy' signal components. Among some species individuals of both sexes flash, but among others only the members of one sex do it. Some adult fireflies do not flash at all but the larvae of all species have photic organs and produce light. The behavioural function of the larval light has received considerable assumption and it has been suggested that it helps in luring prey, defence against predators, illumination, territorial advertisement, adult-juvenile communications and cooperative warning. These functions of larval light remain untested speculation (Underwood *et al.*, 1997; Viviani, 2001). However, it is generally accepted that the larvae use their luminescence as a warning signal. The intensity and frequency of their glow increase when disturbed (Sivinski, 1981).

The flashing signals can take the form of anything from a continuous glow, to discrete single flashes, to 'flash-trains' composed of multi-pulsed flashes (Fig. 16.1). Most studies on fireflies have focussed on the importance of light in adults where it is known to be used in courtship signals (Carlson and Copeland, 1985). They not only recognize flash codes, but also timing and place of activity whether combined or not with the colour of the emitted light (Buck and Buck, 1978). In some cases, the colour of the adult light shows sexual dimorphism, or the colour of emission appears to be evolutionarily tuned for maximum discrimination of conspecific signals from spectrally broader backgrounds, depending on the type of habitat and time-dependent ambient light conditions. The larvae of all firefly species produce light and use the same luciferase-luciferin system to produce light, their luminescence differs from adults because of the use of isozymes, differences in location, morphology and physiology of the light organs, often resulting in a different colour of bioluminescence and behavioural displays.

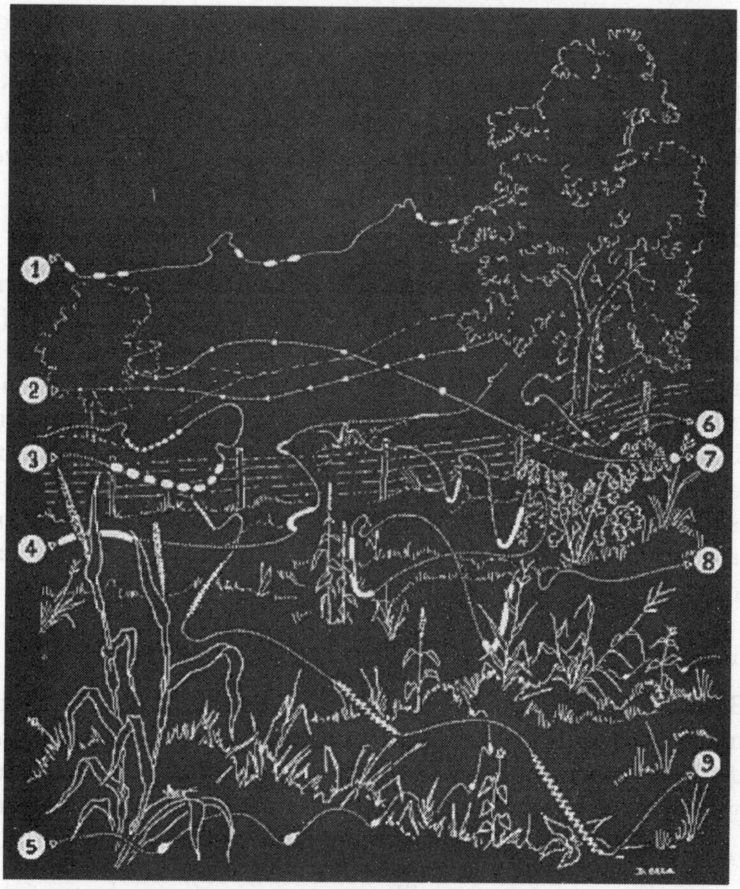

Fig. 16.1. The flashes and flight paths of different *Photinus* species as they would appear in a time-lapse. (From: Lloyd, J.E. 1966. *Univ. Michigan Mus. Zool. Misc. Pub. No.* 130: 95 pp.). (Figure courtesy and with permission from J.E. Lloyd).

One clear advantage of the use of light allows fireflies, and other bioluminescent beetles, to mate under both hazy and dark environmental conditions, outweighing the costs of predation which can follow (Buck, 1978). Lloyd (1978) discussed two main bioluminescent communication systems in fireflies. In signal system I, females have more developed lanterns, stay on the grass, and emit a primary intense luminous signal to flying males, who are attracted to it. In system II, males have more developed light organs and emit a primary signal, to which stationary females reply, and the male then approach the female. Because system II allows the female more discretion in choosing her mate, system I appears to be the evolutionary precursor to system II (Lloyd, 1978).

One of the most fascinating questions about bioluminescence is how such a process originated in insects during evolution. Early theories proposed that bioluminescence arose as an accidental by-product of oxygen detoxification when photosynthetic oxygen started to rise in the atmosphere. According to such a hypothesis, luciferases arose from luminescent oxygenases, which helped to remove toxic oxygen (Buck, 1978). Recent evidence suggests that luciferases did not originate from oxygenases, but rather, the luminescent phenotype caused the evolution of the new oxygenase functions (Viviani, 2002).

16.2 Phengodids

Larvae and the larviform females of 'railroad worms' or 'giant glowworm beetles' belonging to the family Phengodidae are amazingly bioluminescent (Teimann, 1970). Males with their elaborate antennae fly at night in search of females (Lloyd 1979), and males occasionally turn up in light traps. Females and larvae apparently spend most of their lives underground, and are only rarely encountered. Their embryos can be seen glowing insight the eggs and the larvae and larviform females have multiple light organs, the pattern varying somewhat among species. Their pupae are also luminous and have larva-like light pattern, but its intensity is greater. Female phengodids have the same compliment of light organs as the larvae, but are often more brightly luminous. Members of the genus *Phengodes* and *Zarhipis* are important and common phengodids. These glowworms are specialized hunters of millipedes.

Viviani and Bechara (1997) reported that the life history of Brazilian phengodids resembles in many aspects that of the North American species, *Zarhipis integripennis*. In larvae and in larviform females of the species studied, the bioluminescence ranges from yellow-green to red (λ_{max} = 562-638 nm) for the head and from green to orange (λ_{max} = 535-592 nm) for the lateral lanterns. In adult males, the emission colour varies from green to yellow (λ_{max} = 549-580 nm), except *Euryopa* species, who emit orange light (λ_{max} = 592-598 nm). Luminescence was observed among all life-stages of the species studied, in contrast to the North American Phengodini, whose adult males do not glow. Bioluminescence from the larval and female head lanterns may serve for illumination of the nearby environment, whereas the lateral lanterns of larvae and adults seem to be important for self-defence.

16.3 Click beetles

Beetles of the family Elateridae are known as click beetles. The click beetle's name derives from a jumping mechanism composed of a prothoracic projection that fits into a socket in the middle part of their thorax, which makes a clicking sound when the beetle rapidly flexes its head and thorax in an attempt to right itself or escape from a predator. The well-known luminous click beetles belong to the genus *Pyrophorus* that contains over 100 luminous species. They have two pairs of dorsal bioluminescent light organs and one ventral light organ. The light organs appear to be used during mating in a similar manner as in fireflies, although click beetles do not flash. Both male and female click beetles have functional dorsal and ventral organs, and different species differ primarily in the colour of light organs. Maximum production of light is reported by a click beetle, *Pyrophorus noctilucus* with a brightness of 45 millilamberts (Harvey and Stevens, 1928). This insect is also known as the Jamaican click beetle and the 'Cucujo' beetle of the West Indies. Nicol (1978) reported that *Pyrophorus* as having the greatest number of photons emitted per unit area per unit of time.

Another species of click beetle, *Pyrophorus plagiophthalamus* is unique among all bioluminescent organisms in displaying a striking light colour polymorphism. Beetles vary in the colour of their ventral light organs from yellow-green to orange and their dorsal organs from green to yellow-green. The genetic basis for the colour variation involves specific amino acid substitutions in the enzyme luciferase. Stolz *et al.* (2003) showed that dorsal and ventral light colours in *P. plagiophthalamus* are under separate genetic control and resolved the allelic basis for colour variation. Each colour phenotype is coded by the specific nucleotide differences between different alleles. These nucleotide differences are solely responsible for the colour variations. Sexual selection is considered to be the reason for the causation of the apparent shift in frequency among colours.

16.4 Dipteran glowworms

The term 'glowworm' has generally been used loosely. Glowworms are usually defined as crawling, luminous insects that emit light either continuously or in prolonged glow rather than in brief flashes like that of the firefly. Some authors have given the wingless females and larvae of terrestrial luminescent beetles (fireflies) the name glowworm. Lloyd (1971) defined glowworm as a luminescent larva or larviform female, usually of the family Lampyridae but also of other beetle families. Besides, luminescent larvae of Mycetophilidae are also termed glowworm while their adults are termd glowworm flies.

Glowworms (Diptera: Mycetophilidae) inhabit a variety of terrain including shady protected places with high humidity like in crevices and cuttings, under rocks, and the root masses of large fallen trees. These are carnivorous larvae of certain luminescent mycetophilid fungus gnats and have exploited the light production to obtain prey. They produce mucous webs (Fig. 16.2a) of various forms, sometimes decorated with adhesive or poisonous droplets (oxalic acid found in *Platyura* and *Orfelia* species) to snare a variety of small arthropods.

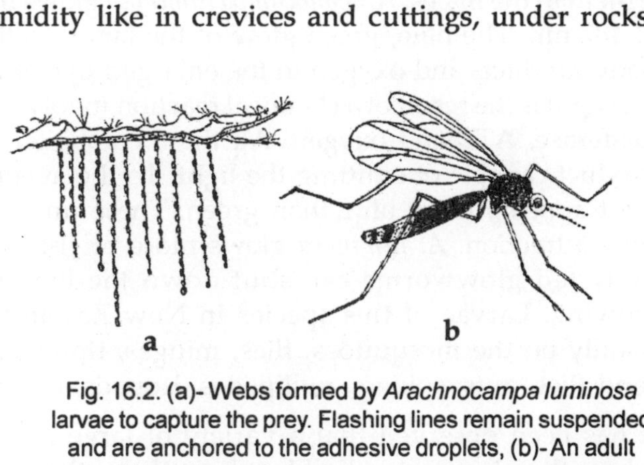

Fig. 16.2. (a)- Webs formed by *Arachnocampa luminosa* larvae to capture the prey. Flashing lines remain suspended and are anchored to the adhesive droplets, (b)- An adult male of New Zealand glowworm, *Arachnocampa luminosa*.

The entire life of the glowworm is spent inside a suspended mucous tube with it's head facing the escape route into a crevice or safety haven in the rock or earth wall into which they move at remarkable speed when disturbed. The mucous tube insulates and prevents the glowworm from desiccation. At night inside the tube it moves back and forth breaking through the tube to repair the web or to feed on trapped insects. During the day the glowworm hides inside its safety haven to avoid daylight predators. The colours of mycetophilid luminescence differ from the usual greens and yellows of other insect lights (Sivinski 1981). *Keroplatus sesiodes* and *K. japonicus* larvae emit a bluish-white light while *Orfelia fultoni* larvae emit a bright blue light. In many fungus gnats the intensity of luminescence changes following disturbance. *K. testaceus* brighten their lights when stimulated while *K. nipponicus* remains constant despite pressing, puncturing, and cutting. Their lights may also protect them from their own negatively phototropic predators and/or be used as aposematic signals (Sivinski, 1998). On the other hand, lights may aid hymenopterous parasitoids to locate fungus gnat hosts. This concentration of bioluminescence may be due to the unusual, sedentary nature of prey capture (i.e. use of webs) that allows the 'mimicry' of a stationary abiotic light cue, or the strangely potent defence webs and associated chemicals, which might provide an aposematic display of unpalatability.

The New Zealand glowworm, *Arachnocampa luminosa* (Fig. 16.2b) lives in dark, humid, sheltered places such as caves or riverbanks and emits light from the Malpighian tubules. In larvae the light is thought to attract small nocturnal prey (Plate 16.6). In pupae (Plate 16.7), especially female ones which shine brighter, light may serve a double function: it could be a sexual attractant for males which are known to copulate immediately with emerging females, and it could ward off or alarm small predators (Plate 16.8). Adult of both sexes glow apparently in the same colour as the larvae, but the female adults produce a considerably brighter light than the males. *Arachnocampa luminosa* glows blue-green, with a peak emission at 487 nm. The blue/green glow of the larvae is the result of a reaction between body products and oxygen in the enlarged tips of the Malpighian tubes of larvae. The light is the result of a chemical reaction involving several components: luciferin, luciferase, ATP and oxygen. All these combined make an electronically excited product capable of emitting the light. To the average person's sight, up close the light appears more blue than green. These unusual colours might contribute to prey attraction. *A. luminosa* glows more brightly when hungry (Richards 1960). Fully fed glowworms can shut down the bioluminescent reaction and cease glowing. Larvae of this species in New Zealand's famous Waitomo cave feed mainly on the mosquitoes, flies, midges, tipulids, moths, stoneflies, caddisflies, sand flies, ants, spiders, millipedes, isopods, and small snails.

It has been observed that a variety of large and small spiders share the areas where the glowworms build webs and use the light produced by the glowworm to catch insects. However, the larger spiders appear to protect the glowworms from intrusion by larger insects like moths and beetles that are attracted by the glowworms bioluminescence. The deprivation of their food source may be a contributing factor to migration for finding more favourable locations. Overpopulation of glowworms may also result into migration or cannibalism. Extremes of temperature particularly very hot winds in the summer or cold dry winds in the winter have a striking influence on their survival and may cause heavy mortality of them in exposed locations.

16.5 Springtails

Self-luminescence has been reported in some springtails (Order Collembola). Some species glow continuously while others emit flashes when stimulated (Lloyd, 1971). *Anurida granaria* exhibits unique bioluminescent display. It produces pure green bright light and flashes more impressively when disturbed. *Supraphorura furcifera, Sinella curviseta* are other light producing springtails.

16.6 Lanternflies

A few fulgorids (Hemiptera: Fulgoridae) are self-luminescent. The well-known luminous lanternfly, *Fulgora lanternaria* (Plate 16.9) is reported from tropical South America and appears to give light only when male and female are placed together. It has been indicated that lanternfly emit light only during mating season when individuals of both sexes are together. The lanternfly is a rare insect found in the South American rain forest of Peru, and also called as the 'peanut-head bug' or 'alligator bug'. The most identifiable structure of this insect is the large snout-like feature on the head that is brightly coloured and has been reported to glow at times.

16.7 Organs of light production

The location of light organs is very variable in insects but they are situated very close to the body surface behind a window of translucent cuticle (Fig. 16.3). They may be present anywhere from the head to the tip of abdomen and are usually different in male, female and larva. In size, the organs may also vary from minute pinheads to masses. In adult insects, these organs are mostly on the ventral side of the thorax or abdomen. In *Photuris* (Lampyridae), males have two pairs of the light organ while females possess only one pair. In *Pyrophorus* (Fam. Elateridae) two kinds of luminous organs are present, one pair on the posterior-lateral margin of prothorax and the other on the posterior abdominal segments. Larvae and adult females of railroad worm have 11 pairs of light organs on the lateral sides of the thorax and abdomen and another pair on the head. In *Fulgora* (Fulgoridae) the light organ is situated only on the head. The light organs generally originate from fat bodies, except in *Arachnocampa* (Mycetophilidae) where these stem from the enlarged distal ends of Malpighian tubules. Photogenic organs consist of a large number of specialized cells called photocytes, arranged cylindrically at right angles to the translucent cuticle, which permit light to pass through it. Behind the

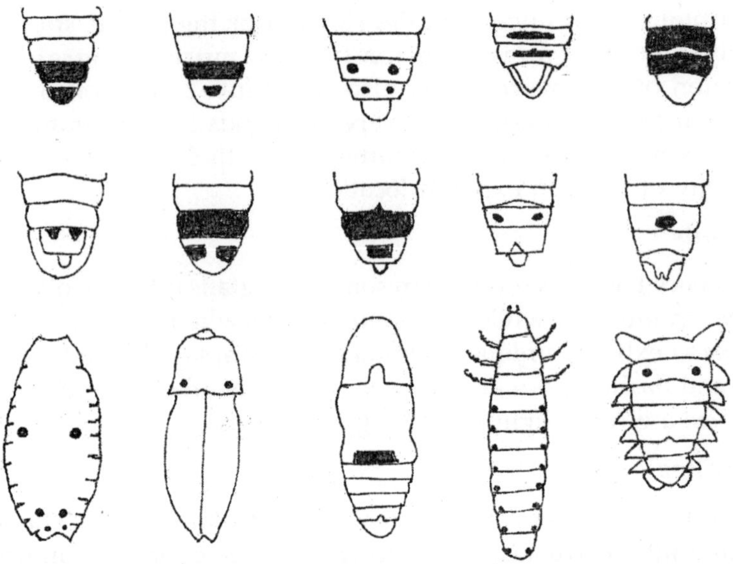

Fig. 16.3. Positions of light producing organs (indicated by dark areas) in luminescent beetles (redrawn from Lloyd, 1983 and other sources).

photocytes there is a reflecting surface chiefly consisting of urate granules. The photocytes receive oxygen through air tubes or tracheoles. They also contain a heavy aggregation of mitochondria that supplies ATP required for the chemical reaction.

16.8 Patterns of light emission

The light emitting patterns in insects may be as cited below.

(i) *Continuous glow.* Light is produced without interruption and continuous glow is of uniform intensity.

(ii) *Intermittent glow.* Light appears and disappears at regular intervals.

(iii) *Pulsation.* Light is emitted in pulses of different ranges. Often the pulses are fused, giving effect of a continuous light.

(iv) *Flash.* Flash is a short burst of light that remains for a short duration. This is most common type of light production in lampyrid beetles. The light intensity in a flash increases abruptly and the duration of flash varies in different species.

16.9 Mechanism of firefly luminescence

Most studies on luminescence control mechanisms have involved the American fireflies *Photuris* and *Photinus*, and biochemical studies have principally involved big dipper firefly, *Photinus pyralis* (Plate 16.10) (Lloyd, 1978). Raphael Dubois, a French Physiologist, in 1885, demonstrated that the luminescence of *Pyrophorus* was due to the action of a heat-labile and a heat-stable substance. If the luminous organ was ground up with water, the homogenate would glow for several minutes but the light finally disappeared. Addition of an extract of a fresh lantern to this cold-water homogenate, resulted into restoration of light (McElroy and DeLuca, 1985). Dubois was able to isolate two different substances that are required to generate bioluminescence and he characterized the luciferase-luciferin system (Harvey, 1920). Luminescence involves a biochemical reaction in which chemical energy is directly converted into light energy. The Firefly lantern is located on the ventral abdominal surface beneath translucent cuticle. It is innervated by octopaminergic neurons in the posterior ganglia of the nerve cord. Bursts of action potentials in these neurons control the normal flash pattern. The process begins with bursts of neural activity stimulating the release of the primary neurotransmitter, octapamine. The light organ contains thousands of photocytes (light cells) subdivided into an inner region full of organelles housing luciferin and luciferase (Viviani, 2002). These react to generate light when oxygen is available, which is provided by thousands of mitochondria packed along the photocyte edges. Timmins *et al.* (2001) concluded that gating of oxygen to the photocytes controls the flash of the adult firefly, and this control mechanism is likely to act by modulating the levels of fluid in the tracheoles supplying photocytes, providing a variable barrier to oxygen diffusion.

The photocytes contain two chemicals that are essential to making light, luciferin and luciferase. In the biochemical process, luciferases are the key enzymes that catalyse the light-emitting reactions (Fig. 16.4). They catalyse the oxygenation of compounds generically known as luciferins using ATP and magnesium ions (Mg^{++}), generating the energy-rich adenylluciferin. In presence of oxygen and luciferase, the adenylluciferin is rapidly converted to adenyloxyluciferin product that decays to oxyluciferin with the emission of a photon of visible light. The chemical reaction is controlled by the amount of oxygen the firefly lets into its abdomen. The firefly luciferase-luciferin reaction is so efficient that 98 per cent of the energy released is in the form of light, with little energy is released as heat (Biron, 2003).

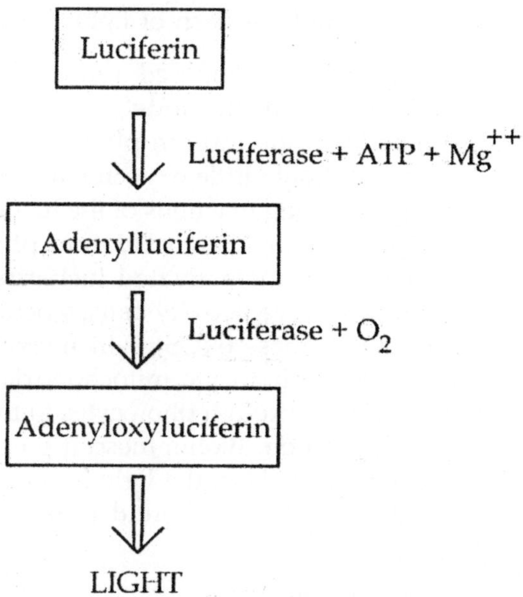

Fig. 16.4. Sequence of reactions involved in light production in an insect.

Until recently, it was unclear how this process was switched on, particularly because the firefly abdominal nerve endings are not in direct contact with the photocytes; instead, they end on tracheolar cells that lie between the tracheolar airway and the photocytes. There are several theories on how fireflies control the 'on' and 'off' of their photic organs.

(i) *Oxygen control theory*. The firefly turning on and off its light by controlling the oxygen supply to the photic organ for use in the chemical reaction.

(ii) *Neural activation theory*. The fireflies have neural control of structures called 'tracheal end cells' which, upon stimulation, may release a messenger molecule in the photic organ that initiates the activation of the chemical reaction.

(iii) *Nitric oxide control theory*. Trimmer *et al.* (2001) suggested that the simple molecule nitric oxide (NO) acts as the on-off 'button'. Whenever the *Photuris* fireflies were exposed to NO they glowed or flashed almost continuously, and they stopped once the NO was turned off. *Photuris* fireflies can be stimulated by NO gas in the presence of oxygen, and that NO scavengers block bioluminescence induced by the neurotransmitter octopamine. NO synthase is robustly expressed in the firefly lantern in cells interposed between nerve endings and the light-producing photocytes. NO is generated to inhibit oxygen extraction by mitochondria, allowing oxygen to pass through to oxidize the luciferin-adenylate intermediate (Trimmer *et al.*, 2001; Aprille *et al.*, 2004). Essentially, NO is responsible for the neural action by which firefly lantern yields a sudden flash of light.

Aprille *et al.* (2004) proposed a model for NO control of the firefly lantern flash. In 'quiescent or dark mode', oxygen delivered through lantern tracheae is consumed by respiration in photocyte mitochondria clustered in the peripheral cytoplasm and only little oxygen reaches peroxisomes (the organelles that contain the light-producing reactions of the luciferin-luciferase pathway). In this quiescent state, ATP produced by oxidative phosphorylation promotes formation and accumulation of the activated luciferin-adenylyl intermediate by luciferase. In 'flash mode', nerve activity causes octopamine release that transiently activates lantern NO synthase (NOS), which creates NO. NO diffuses rapidly and inhibits oxygen use by photocyte mitochondria. Now oxygen delivered by the tracheal system diffuses through photocytes to the peroxisomes where it triggers the light-producing reaction. Bioluminescence ends once the neural signal cease and NO production stops. Also, the light acts as a form of feedback inhibition, promoting degradation of the NO bound to the light-sensitive cytochrome (Aprille *et al.*, 2004).

16.10 Significance

The evolution of bioluminescence in nature has resulted in several behavioural adaptations. Although at least 20 separate functions of bioluminescence have been proposed (Harvey, 1952), the most common associated with predation, defence, and communication.

Communication. Use of bioluminescence as a tool for communication can be observed in fireflies and is possibly the earliest recognized function of bioluminescence (Buck, 1978). Communication between the sexes of fireflies can be used for mating. The flashes of the fireflies have been duplicated experimentally and the rhythms and frequencies of the flashes are key factors to firefly attraction (Buck, 1978). Light is used as a mating signal in fireflies. Luminescence assists in aggregation of individuals of same species, thereby indirectly improving the chances of mating. Some tropical species congregate in large numbers and flash in harmony. In some fireflies, the females are wingless and sedentary; light production is therefore essential for them to attract the winged males. Flash patterns in bioluminescent insects are species-specific and vary between species and between sexes. Male and female fireflies, *Photinus pyralis* emerge at dusk, emitting a single short flash at regular intervals. The flashes are usually from male fireflies seeking mates. Females climb a blade of grass, flashing when males flash and exchange of signals is repeated 5 to 10 times until they start mating.

Predation. The most distinctive example of light acting as a lure for prey is found in the New Zealand glowworm, *Arachnocampa luminosa*. The female fly deposits eggs on the ceiling of dark caves. Upon hatching, the larvae hang down by a sticky thread and produce light (Plate 16.6). During night, the entire cave may glow with this light, attracting other insects. These attracted insects get entangled in the sticky threads and are preyed upon by the larvae. The caves inhabited by flies are popularly known as 'luminous caves' and are tourist attraction spots in New Zealand.

Broadley and Stringer (2001) tested the hypothesis that bioluminescence produced by larvae of *A. luminosa* attracts prey, in Reserve Cave, Waitomo, New Zealand, and in its bush-clad entrance over a total of 200 days during winter, spring, and summer. They placed transparent adhesive traps either over glowworms or over areas from which glowworms had been removed. Adhesive traps over glowworms caught significantly more invertebrates per trap per day than did control traps. Glowworms in bush attracted greater numbers and types of invertebrates than did glowworms in the cave. Diptera predominated in both bush (86 per cent of the total catch) and cave (89 per cent). Besides, small numbers of acarines, spiders and other insects were also trapped. No adults of *A. luminosa* were caught. Glowworms under adhesive traps survived with little or no food for up to 78 days.

Defence. In railroad worms the continuous glow of the head region when the larvae are walking, suggests a possible illumination function, whereas the circumstances under which the lateral light organs are switched on suggest a defence function. Sudden flashes can repel potential predators. The railroad worm larvae live at high densities, confined to small areas, and may use simultaneous emission to frighten potential enemies, or they may also use the light to intimate the mated females about to lay eggs about over crowding and competition for food sources.

Cock and Matthysen (2003) demonstrated that lampyrid larvae use luminescence to signal unpalatability to nocturnal, visually guided predators. Wild-caught toads, *Bufo bufo* were more reluctant to attack luminescent artificial prey. After being exposed to larvae of *Lampyris noctiluca,* which the toads experienced as disagreeable, attack latencies to luminescent prey increased.

Aggressive mimicry. The female of the species of some fireflies can falsely communicate flashes of attraction in order to lure and eat the unsuspecting male of other species (Lloyd, 1978). Aggressive mimicry is a phenomenon where one organism (a mimic) tricks another organism (the dupe) into thinking it is another (the model), with the result being a negative outcome for the dupe, as well as the model. A few species in the genus *Photuris* mimic the female responses of other fireflies in the same area in order to attract males of the mimicked species. When these males are tricked (or duped) into landing near these mimics to mate, they are pounced upon and eaten (Plate 16.2) (Lloyd, 1965). Eisner *et al.* (1997) suggested that these female mimics are not only acquiring food but also defensive steroidal pyrones called lucibufagins from their prey, which they themselves do not produce in large quantities.

16.11 Uses for humans

Besides their beauty and scientific interest, bioluminescent species are very important to the human society because their luciferases and luciferins are currently used as very sensitive tools in biotechnology and biomedicine. During World War II, the Japanese used fireflies' powder to read maps and messages. Scientists hope that introduction of luciferin and luciferase genes can be used to identify and control plant diseases. Both luciferin and luciferase have been synthesized and these chemicals can be used to detect diseases in people. When the chemicals are introduced into a blood or urine sample of someone with a bacterial infection, it glows. Firefly luciferin and luciferase are used to determine the freshness of blood. Scientists are studying ways to use bioluminescence in other medical ailments. The technique is now used to study heart ailments, muscular dystrophy, urological problems, AIDS, cancer, etc., and also for testing the effectiveness of drug treatments.

These chemicals can also be used to check for bacteria in food. The enzyme apyrase along with luciferin and luciferase is added to the food. If bacteria are present, the food glows.

The larvae and adults of some species eat so many snails, slugs, and cutworms and thus assist in reducing pest populations. Bioluminescence is also used as a tool for mapping organism distribution patterns. Genetically modified strains of pink bollworm, *Pectinophora gossypiella* with green fluorescent protein (GFP) derived from the jellyfish, *Aequora victoria* have been produced. The GFP transgenic pink bollworm strain fluoresces strongly green when viewed in its larval stage. This strain of pink bollworm can be used for field performance studies and to map the distribution of the pest. The future objective is to ultimately add a temperature-sensitive lethal gene along with the GFP gene into the pink bollworm that could be used for its management.

These light-producing chemicals may be useful for finding extraterrestrial life. The chemicals could be mixed with Martian soil, for example, and used to test the presence of ATP, necessary for life.

Chapter **17**

Hormones

Insect hormones are chemical messengers secreted by endocrine glands or other cells and transmitted by the haemolymph to the sites of action where they influence a remarkable variety of physiological processes. Although typically produced in very small quantities, hormones may cause profound changes in their target cells. Their effect may be stimulatory or inhibitory. In some cases, a single hormone may have multiple targets and cause different effects in each target. Gade *et al.* (1997) suggested two main classes of hormones in insects: (i) the true hormones produced by epithelial glands and belonging to the ecdysteroids or juvenile hormones, and (ii) the neuropeptide hormones produced by neurosecretory cells. These hormones regulate physiological, developmental, and behavioural events in insects, viz. metabolism, muscle activity, reproduction, growth and development, diapause, eclosion, pheromone biosynthesis, and cuticle tanning and colour change. The hormone-secreting structures form an endocrine system that helps in maintaining homeostasis, coordinate behaviour, regulate growth and development, and other physiological activities. There are at least four categories of hormone-producing cells/glands in an insect's body:

(i) *Endocrine glands.* Secretory structures adapted exclusively for producing hormones and releasing them into the circulatory system.

(ii) *Neurohaemal organs.* Similar to glands, but they store their secretory product in a special chamber until stimulated to release it by a signal from the nervous system or another hormone.

(iii) *Neurosecretory cells.* Specialized nerve cells found throughout the nervous system, but they occur in major groups in brain. These cells respond to stimulation by producing and secreting specific chemical messengers and functionally serve as a link between the nervous system and the endocrine system.

(iv) *Internal organs.* Hormone-producing cells are associated with body organs, including ovaries and testes, fat bodies, and parts of the digestive system.

In insects, the largest and most obvious endocrine glands (Fig. 17.1) are found in the prothorax, just behind the head. These glands secrete ecdysteroids (a group of closely-related steroid hormones including ecdysone) that stimulate synthesis of chitin and protein in epidermal cells and trigger a flow of physiological events that culminates in moulting. The ecdysteroids are often called 'moulting hormones'.

A pair of neurohaemal organs or neuroglandular bodies located on the walls of the aorta just behind the brain and termed as corpora cardiaca (sing. corpus cardiacum), store and release prothoracicotropic hormone (PTTH). PTTH stimulates the secretory activity of prothoracic glands. The corpora cardiaca (CC) release their store of PTTH only after they receive a signal from neurosecretory cells in the brain.

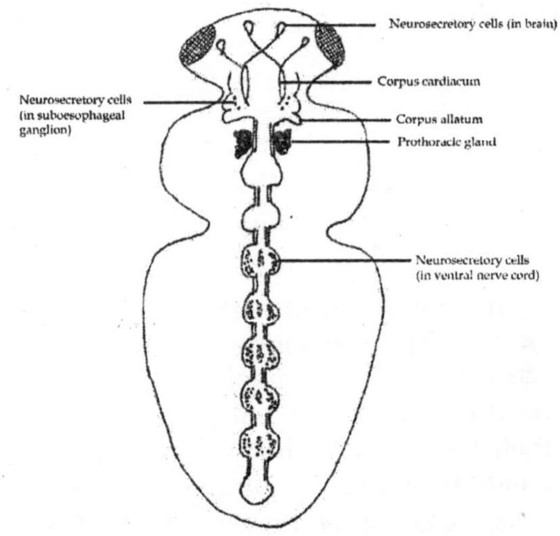

Fig. 17.1. Hormone secreting glands and neurosecretory cells in a generalized insect.

The corpora allata (sing. corpus allatum), another pair of glandular bodies, lie just behind the corpora cardiaca and secrete juvenile hormones (JH) that has regulatory roles in both metamorphosis and reproduction. Neurosecretory cells in the brain regulate activity of the corpora allata (CA).

The neurosecretory cells (NSC) are found in clusters, both medially and laterally in the insect's brain and secrete neurohormones. Axons from these cells can be traced along tiny nerves that run to the corpora cardiaca and corpora allata. Large numbers of NSC also occur in the ventral ganglia of the nerve cord. A number of other neuropeptides secreted by these cells mediate important physiological processes.

17.1 Hormones

Ecdysone is released from the prothoracic glands and usually is converted into active 20-hydroxyecdysone, which is most common and physiologically significant ecdysteroids. Ecdysteroids are derived from sterols such as cholesterol, and cannot be synthesized *de novo* and are obtained with diet. These hormones play a crucial role in moulting in all insects. Juvenile hormones form a group of related sesquiterpenoid compounds and regulate the metamorphosis and reproductive functions. Larval characteristics are maintained and metamorphosis is inhibited by JH. JH also stimulates deposition of yolk in eggs, affects accessory gland activity, and pheromone production. Neurohormones are usually peptides (small proteins) and also termed as neuropeptides. These act as messengers and regulate many aspects of development, homeostasis, metabolism, reproduction and physiological functions. Besides, they also regulate secretion of JH and ecdysteroids.

Insect hormones are regulated by external as well as internal cues via the brain by means of two- or three-tiered cascades with feedback loops that begin with brain neurohormonal peptides and end at peripheral target organs (Nijhout 1994). In one axis, neuroglandular bodies produce PTTH, which is stored in and released from the corpora cardiaca. PTTH in turn stimulates the prothoracic gland, which secretes ecdysone, which is converted to ecdysteroids when it reaches its targets. In another axis, brain neurosecretory cells produce the peptides namely allatotropin and allatostatin, which regulate the production of JH by the corpora allata. JH in turn acts on multiple targets including the nervous system.

Behaviourally, hormones act as 'primers' setting the 'mood' or internal motivation of an insect to perform a particular behavioural act. Evidence for this role comes from three types of observations: (i) correlation of onset of hormone secretion and appearance of a specific behaviour, (ii) precocious induction of a specific behaviour by hormone application or implantation of active glands, and (iii) disappearance of a particular behaviour after removal of the hormonal source (Matthews and Matthews, 1978). Effects of hormones have been studied in insects with respect to sexual maturation, reproduction, social behaviour, migration, rhythms, colour change, diapause, polymorphism, pheromone production, etc.

Major contributions pertaining to hormonal regulation of insect behaviour are summarized by Gilbert and King (1973), Wigglesworth (1964b, 1970), Wilde and Loof (1973b), Truman and Riddiford (1974), Novák (1975), Riddiford and Truman (1978), Nijhout (1994), and Awasthi (2005).

17.2 Diapause

Diapause represents a syndrome of behavioural and physiological characteristics, all of which enhance the insect's survival during the period of adversity. It is characterized by the cessation of reproductive development, reduced metabolic activity, movement to an appropriate hibernaculum, and enhanced fat body content in response to seasonal changes. Once the diapause programme is initiated, metabolic activity is suppressed even if conditions conducive to development prevail, thus ensuring that the insect life cycle is kept in phase with changing seasons. Diapause is a periodic event and is also controlled by hormones. The neuroendocrine system serves the crucial role of conveying the environmental signals into the many aspects of the diapause programme. Populations of univoltine species enter diapause in each generation regardless of environmental factors (obligatory diapause) while the majority of species have the capacity to decide whether or not they will enter diapause (facultative diapause). Diapause may occur at any developmental stage; however, most insects enter diapause in the pupal stage. For each species the developmental stage in which diapause occurs is quite specific, but some enter diapause in two or more stages. Temperature, photoperiod, and quality and quantity of food exert significant effect on diapause in many cases.

Denlinger (1985) discussed hormonal control of diapause. Many studies specify environmental and hormonal control of diapause, but substantial variation in its mechanism reflects the multiple evolution of diapause. Generally in larva and pupa, and probably also in late embryonic stages, the production of ecdysteriods from the prothoracic glands ceases so that growth and moulting do not occur. JH plays an important role in termination of larval diapause. Increased levels ecdysteroids appear essential for the termination of pupal diapause. JH is important in diapause regulation in adults, but may not be the only regulator. Neurosecretory cells of the brain control the activity of both corpora allata and prothoracic glands, and these cells cease to secrete the hormones during diapause.

In the silkworm, *Bombyx mori* embryonic diapause appears to be regulated by 'diapause hormone' (DH), a 24-amino acid peptide that is produced in the suboesophageal ganglion (Fukuda 1952). The developmental fate of the egg depends on the photoperiod to which the mother was exposed during larval development. Females reared under short-day conditions lay non-diapausing eggs, whereas those exposed to long-day conditions lay diapausing eggs. Under long-day conditions neurosecretory cells in suboesophageal ganglion secrete diapause hormone, which acts on ovary. In *B. mori* complex antagonistic interactions occur between diapause hormone and JH. The adult female produces non-diapause eggs in the absence of diapause hormone and presence of JH whereas the diapause eggs are produced under the influence of diapause hormone.

In the Colorado potato beetle, *Leptinotarsa decemlineata* long days induce reproduction and short days induce onset of diapause 10-12 days after adult emergence. Adults feed actively during prediapause; however, no reproduction occurs and the onset of diapause is characterized by burrowing behaviour. Extensive studies have been conducted to examine neuroendocrine regulatory factors involved in diapause control. Early work suggested that the absence of JH was responsible for the diapause syndrome, in that when long-day beetles were allatectomized, their reproductive activity was arrested and they subsequently entered diapause (Wilde and Boer, 1961).

Taub-Montemayor *et al.* (1997) discussed the role of JH and juvenile hormone esterase (JHE) in the control of the different life history strategies of the cotton boll weevil, *Anthonomus grandis*. Elevated levels of haemolymph JHE were found to be positively correlated with survival throughout the winter in South Texas population of weevils. Winter weevils were examined for haemolymph vitellogenin (Vg) and their subsequent survival was monitored. The majority of weevils surviving beyond ten weeks had no haemolymph Vg. They concluded that elevated JHE and the absence of Vg are good predictors of survivors in the South Texas population.

Readio *et al.* (1999) reported that the synthesis of JH in diapausing and nondiapausing *Culex pipiens* differ greatly. Paired corpora allata from diapausing females synthesized much less JH per hour during the first 3 weeks after emergence as compared to nondiapausing females. Subsequently JH synthesis decreased in both groups of females, but corpora allata from nondiapausing mosquitoes still were 3 times more active than those from diapausing mosquitoes. Allatectomy (removal of corpora allata) of young diapausing females prevented follicle growth and blood feeding when diapause was terminated prematurely, demonstrating that the physiological events associated with diapause termination were associated with JH biosynthesis.

Munyiri and Ishikawa (2004) observed that fifth instar larvae of yellow-spotted longicorn beetle, *Psacothea hilaris* pupate at the next moult at 25°C and under a long-day photoperiod. Under a short-day photoperiod, in contrast, they undergo one or two additional larval moults and enter diapause; the 7th instar larvae enter diapause without further moult. JH titres of the 5th instar pupation-destined larvae decreased continuously and became virtually undetectable on day 13. Ecdysteroids exhibited a small peak on day 8, 1 day before gut purge, and a large peak on day 11, 2 days before the larvae became pre-pupae. The two ecdysteroid peaks are suggested to be associated with pupal commitment and pupation, respectively. Diapause induction in this species was suggested to be a consequence of high JH and low ecdysteroid titres.

17.3 Moulting

In insects, growth is punctuated by moults, which provide a new cuticle to accommodate larger or different body forms. Moulting is a complex process involving hormonal, behavioural, epidermal and cuticular changes that lead up to the shedding of the old cuticle. The complex events occurring during moulting are affected by hormones on the epidermal cells to control the cuticular changes and also on the nervous system to coordinate the behaviours associated with ecdysis (Fig. 17.2). The motor patterns and behaviours associated with the shedding of the old cuticle are collectively called the ecdysis sequence and depend on the action of a number of neuromodulators and neurohormones.

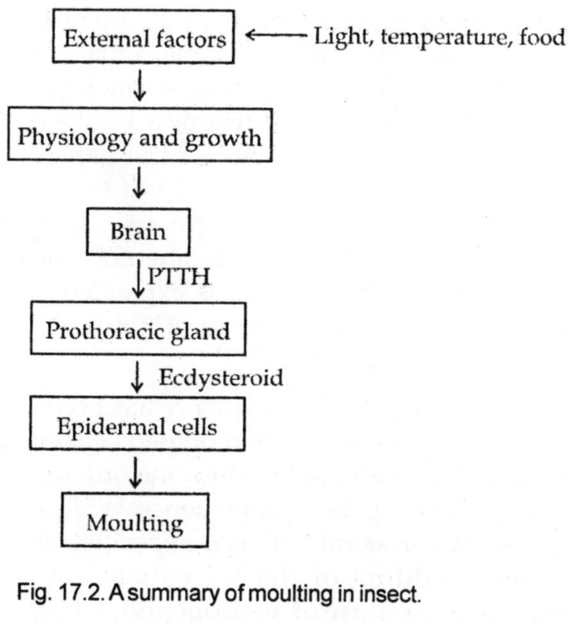

Fig. 17.2. A summary of moulting in insect.

The hormones involving in moulting are: prothoracicotropic hormone (PTTH), ecdysis triggering hormone (ETH), eclosion hormone (EH), ecdysteroids, and JH. The PTTH stimulates ecdysteroid synthesis and secretion, which initiates changes in the epidermal cells that lead to production of new cuticle. JH inhibits the expression of adult features and its high titre is associated with larval-larval moult. Ecdysis is mediated by ETH and EH. EH acts on steroid-primed central nervous system (CNS) to evoke the coordinated motor activities associated with the escape from old cuticle. Four key players in the regulation of the ecdysis sequence have been identified. These include pre-ecdysis-triggering hormone (PETH) (Zitnan *et al.*, 1999), ecdysis-triggering hormone (ETH) (Zitnan *et al.*, 1996), eclosion hormone (EH) (Truman and Riddiford, 1970) and crustacean cardioactive peptide (CCAP) (Gammie and Truman, 1997). These peptides probably play a role in the ecdysis behaviours of many, if not all, insects, but their roles have been especially well described in the tobacco hornworm, *Manduca sexta* (Ewer *et al.*, 1997).

Bollenbacher *et al.* (1987) suggested that a larval moult in *M. sexta* involves an endocrine cascade that begins with the release of PTTH. The release of PTTH is gated, occurs during the scotophase and appears to be developmentally cued. In fourth instar hornworm larvae, PTTH release into the haemolymph occurs as a single burst over a few hours during the head critical period, i.e. the time during which the head (brain) is needed for the initiation of the moult to the fifth (last) instar. Released PTTH activates the prothoracic glands (PGs), and within a few hours the cumulative effect of this event results in a dramatic increase in the haemolymph ecdysteroid titre, which then elicits the moult.

Riddiford (1991) discussed relationship between hormonal environment and epidermal activities that control moulting and cuticular deposition in *M. Sexta* and suggested that both ETH and EH are important for ecdysis and stimulate each other's release. Besides this, there is correlation between ecdysteroid and JH titres and the cuticular changes that occur in the last two larval instars and in prepupal development. These hormones regulate moulting both at cellular and molecular levels. In addition, a rising concentration of EH stimulates other neurosecretory cells in the ventral ganglia to secrete bursicon, a hormone that causes hardening and darkening of the integument. As moulting progresses, the inner layers of the old cuticle are digested away. Then a soft new cuticle is deposited. When these developmental events are complete, the shedding processes begin, and the new cuticle is expanded and hardened. The shedding processes are behavioural, and the role of EH is to release a pre-programmed sequence of behaviours.

In *M. sexta,* the ecdysis sequence has become a model for studying the neural and endocrine regulation of stereotyped behavioural programmes. A neuromodulatory network has been proposed for the ordering and timing of the various behavioural phases that comprise this sequence. While the network itself appears to be activated by a decline in steroid titres, the specific behaviours and motor programmes leading to the shedding of the old cuticle are coordinated and maintained by the subsequent release of neuropeptides from the central nervous system and the periphery. Zitnan *et al.*, (1996) discovered the existence of a second ecdysial hormone, called Mas-ETH (Mas comes from *M. sexta*). ETH is produced by a system of epitracheal glands. In *M. sexta* there are 8 pairs of epitracheal glands, each located on the outer surface of the large tracheal tube immediately adjacent to each spiracle. The hormone is produced and released from Inka cells. Ecdysis is now thought to follow from the sequential release of two hormones. EH is released from the brain, and it primes the Inka cells to release ETH. ETH acts on the abdominal ganglia to release the behaviours required for ecdysis. Successful ecdysis requires that all behaviours are coordinated. Failures in coordination result in deformities and death and each ecdysis episode is a crucial event in an insect's life.

In *M. sexta,* there is a working model for the organization of the pre-ecdysis and ecdysis behaviours. Declining steroid titres and possibly an early release of EH are thought to initiate the release of PETH and ETH. PETH triggers pre-ecdysis behaviour I, while ETH triggers pre-ecdysis behaviour II, both of which help the insect to loosen the old digested cuticle (Zitnan *et al.*, 1999). Once ETH has been released, there is a positive feedback cascade between ETH and EH in which ETH stimulates release of EH and EH stimulates further release of ETH (Ewer *et al.*, 1997). This feedback results in the massive release of both hormones from their respective cells prior to the onset of ecdysis. EH then stimulates the release of CCAP from a group of cells in the ventral nerve cord (Davis *et al.*, 1993). Two major functions of CCAP are to release the ecdysis motor pattern itself, enabling the insect to shed its old cuticle, and to inhibit the ETH-induced pre-ecdysis behaviours (Gammie and Truman, 1997).

17.4 Metamorphosis

Wigglesworth (1934) first demonstrated that blood-borne hormones are responsible for the induction of metamorphosis in the blood-sucking bug, *Rhodnius prolixus.* He observed that the corpora allata produces a hormone that counteracts the tendency to undergo metamorphosis and if the corpora allata was removed from a third-instar larva, the next moult turned the larva into a precocious adult. Conversely, implantation of corpora allata into fifth-instar larvae produced a large sixth-instar larvae rather than into adult. The metamorphosis appears to be regulated by effector hormones, controlled by neurosecretory peptide hormones in the brain. The neurosecretory cells release PTTH that stimulates the production of ecdysone by the prothoracic gland. Ecdysone is first changed to the active hormone 20-hydroxyecdysone and one or more pulses of 20-hydroxyecdysone results into each moult.

JH is responsible for averting metamorphosis and as long as JH is present a new larval instar is formed. In the last larval instar, the medial nerve from the brain to the corpora allata inhibits the corpora allata from producing JH, and there is a simultaneous increase in the body's ability to degrade existing JH (Safranek and

Williams, 1984). This triggers the release of PTTH from the brain and PTTH, in turn, stimulates the prothoracic glands to secrete ecdysone. The resulting hydroxyecdysone, in the absence of JH, commits the cells to pupal development. A complex interplay between JH and ecdysteroids thus allows the larva to progress to a pupa, and finally ecdysteroids alone drive the transformation to adult. In lepidopteran larvae absence of JH is a prerequisite for the onset of cocoon spinning behaviour. The removal of corpora allata from young larvae results in the cessation of feeding and the precocious construction of a tiny cocoon. Likewise, application of JH late in larval life serves to delay cocoon spinning or to force spinning of a moulting pad, characteristic of early larval instars. Although a low JH titre is required to display premetamorphic behaviours, nevertheless such behaviours are apparently triggered through the action of PTTH (Truman and Riddiford, 1974). During metamorphosis larval-specific mRNAs are not replaced, and new mRNAs are synthesized whose protein products inhibit the transcription of the larval messages. After the second ecdysone pulse, new pupal-specific gene products are synthesized (Riddiford, 1982), and the subsequent moult shifts the insect from larva to pupa. It appears then that the first ecdysone pulse during the last larval instar triggers the processes, which inactivate the larva-specific genes and prepare the pupa-specific genes to be transcribed. The second ecdysone pulse transcribes the pupa-specific genes and initiates the moult (Nijhout, 1994).

It has been suggested that JH inhibit expression of adult characters so that high levels of JH induced larvae, intermediate levels of JH produced pupae, while low levels of JH produced adults (Fig. 17.3). However, the hormonal regulation of metamorphosis appears to be more complex as JH levels fluctuated during the final instar period. In the tobacco hornworm, *Manduca sexta* there are specific times when different cells are sensitive to JH and if JH is present during a JH-sensitive period, the current developmental state is maintained, whereas if JH is absent during that period, this

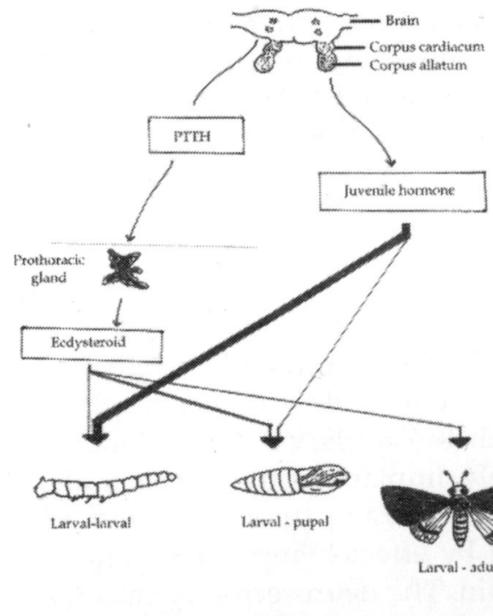

Fig. 17.3. Schematic representation of endocrine control of moulting and metamorphosis in an endopterygote insect. Juvenile hormone (JH) works in conjunction with ecdysteroid to determine the fate of each moult. With high concentration of JH, juvenile characteristics are maintained. Less concentration of JH results into formation of pupa. For the metamorphosis of pupa to adult only ecdysteroid is required (modified after Richards, 1981).

tissue will progress to a more mature developmental state. The onset and duration of the JH-sensitive period appears to be an autonomous state of the cell and is not controlled by hormones (Nijhout, 1994). Two groups of peptides, allatotropins (AT) and allatostatins (AS), which either stimulate or inhibit JH synthesis, respectively, have been found in insects from different orders. Allatotropic activity has been described in biological extracts from a number of species, and the allotropin (Mas-AT) has been chemically identified in *M. sexta* (Kataoka *et al.*, 1989). It also stimulates JH biosynthesis in other lepidopteran and some non-lepidopteran species.

17.5 Reproduction

Reproduction in insects is often accompanied by the expression of specific behaviour that is, in many cases, directly or indirectly regulated by JH. Generally the effects of JH are diverse, but they predominantly involve in courtship behaviour, female sexual receptivity or egg-laying behaviour (Wyatt and Davey, 1996). Wigglesworth (1936) first showed the hormonal control of reproduction in *Rhodnius prolixus*. He reported that the newly fed female without brain but with intact corpora allata laid eggs normally. In the females without corpora allata, no egg development took place; however, implantation of corpora allata in such a female caused ovarian maturation. These experiments clearly indicated that corpora allata that are inactivated in the last larval instar to permit metamorphosis, are reactivated in the adult to stimulate reproduction and JH is required for reproduction. In adult females that do not feed, no hormonal control for egg maturation seems necessary but in most feeding adult females JH is required for oocyte growth, vitellogenin (yolk protein) synthesis and vitellogenin uptake into the oocyte (Riddiford and Truman, 1978). The making of vitellogenesis requires the controlled massive synthesis of the yolk protein precursor, vitellogenin, as well as its controlled incorporation into the oocytes in a timely fashion. In most species of insects the predominant yolk protein, vitellin is derived primarily from the extraovarially synthesized vitellogenin that is usually only identified in the adult female. In majority of the investigated insect species JH stimulates the transcription of the vitellogenin genes and the consequent control of vitellogenin production (Engelmann, 1983). However, it has also been shown that ecdysteroids (ecdysones) can be involved directly or indirectly in vitellogenin production in a limited number of insect species, particularly certain Diptera (Dhadialla and Raikhel, 1994). Both JH and ecdysone occur in all immature insects simultaneously, but peak levels often do not coincide. It is also known that vitellogenesis is not observed in most immature insects, even in the presence of high titres of JH and ecdysone. This indicates that additional factors must play crucial roles in regulating the timing of vitellogenin synthesis in the adult female. For example, competency of the fat body to produce vitellogenin appears to be attained only during the pre-adult period as part of the metamorphic development of the insect. Also, high concentrations of free ecdysones do not occur in the haemolymph of the non-dipteran adult insects in a timely pattern, and therefore, JH appears to be the major controlling agent associated with vitellogenin synthesis in the mature female of most insect species. The role of JH and ecdysteroid in insect reproduction was discussed by Koeppe *et al.* (1985) and Hagedorn (1985).

In the female mosquito, *Aedes aegypti* the JH is secreted after emergence and causes oocyte to grow. Telfer (1954) showed that the blood of cecropia moth females, *Hyalophora cecropia* contained a protein, which was not found in males. These female-specific proteins are found in mature eggs of most insects. Engelmann (1979) demonstrated hormonal stimulation of vitellogenin synthesis in Madeira cockroach, *Rhyparobia maderae.* Injection of JH I, II or III into immature females caused the synthesis of a haemolymph protein. The JH appears to act primarily by stimulating the production of a specific mRNA for the production of yolk protein.

In *A. aegypti* after the female takes a blood meal, the neurosecretory cells release 'egg development neurosecretory hormone' (EDNH) which in turns stimulate the ovary to secrete α-ecdysone. The α-ecdysone acts on the fat body (presumably after being converted to β-ecdysone) to stimulate vitellogenin synthesis and release (Hagedorn, 1974). However, JH also has a role in priming the fat body for vitellogenin synthesis. Ecdysones have also been reported from the ovaries of insects like *Locusta, Drosophila, Bombyx, Galleria, Rhyparobia, Macrotermes* queen, etc. Engelmann (2002) provided experimental information regarding JH and ecdysone interactions in immature and adult females of the *R. maderae* and showed that ecdysone inhibits vitellogenesis by two independent mechanisms: (i) inhibition of the corpora allata resulting in the inhibition of JH production, and (ii) inhibition of vitellogenin synthesis by the fat body. Both of these mechanisms appear to be operative in immature and mature insects.

Sorge *et al.* (2000) studied vitellogenesis and its regulation in female adults of the fall armyworm, *Spodoptera frugiperda.* A single female-specific protein, likely to be the *S. frugiperda* vitellogenin (Vg), appeared approximately 5 h after adult eclosion in the haemolymph of virgin females and its concentration increased with age. Vg could not be detected in the haemolymph of females decapitated directly after eclosion. Injections of JH-mimic methoprene in decapitated virgin females were resulted into formation of Vg, indicating that vitellogenesis in *S. frugiperda* depends on JH. Injection of 20-hydroxyecdysone (20E) promoted Vg production in decapitated female moths, but never resulted in the production of vitellogenic oocytes. They suggested a crucial role for 20E in the induction of vitellogenesis in the *S. frugiperda,* while JH seems to be essential for the continued uptake of Vg by developing oocytes.

JH also promote the uptake of vitellogenin into the eggs. The proteins reach the oocyte surface and penetrate the follicular epithelium, which surrounds the oocyte. The follicle cells change their shape, so large gaps are formed between them, through which the proteins migrate to the egg surface and there be taken into oocyte by pinocytosis. Bell and Barth (1971) transplanted ovaries and injected vitellogenin in male cockroach, *Eublaberus* and observed that no vitellogenin uptake occurred into the ovaries until JH was administered. After JH treatment the oocytes accumulated yolk, but could not develop fully. *In vitro* incubation of follicular epithelium of *Rhodnius prolixus* showed that JH I stimulated an increase in intercellular spaces (Davey and Huebner, 1974). JH may also be needed for the maturation of female accessory glands, such as colleterial glands in the cockroach, *Periplaneta*. It is also required for the production of sex pheromones in females of many insects. Kozhanova (2000) discussed the role of JH, ecdysone and brain in the regulation of oogenesis and spermatogenesis in insects. The early period of gametogenesis in both sexes is controlled mainly by ecdysone and neurosecretory cells of the brain. In periods of cytoplasmic growth of oocytes and vitellogenesis, the main role in the regulation belongs to JH.

In male insects JH has no role in spermatogenesis, and meiosis and sperm maturation occur in response to ecdysone in the absence of JH. Schmidt and Williams (1953) demonstrated that a material in the blood of silkworm was essential for the maturation of sperm. The appearance of this material 'macro-molecular factor' (MF) in blood was age-specific as it was absent in fifth instar larvae and diapausing pupae but present in pupating insects and developing adults. Kambysellis and Williams (1971) confirmed that MF and ecdysone together promote sperm maturation and incubation of testes with both MF and ecdysone led to complete spermatogenesis. Although JH is not required for sperm maturation in males; however, it is required for proper development of accessory glands. The secretions of these glands make up the spermatophore and help in sperm transport.

JH also affects the neuronal elements responsible for the phonotactic response in female *Acheta domestica*. Cayre *et al.* (1994) showed that neurogenesis in the mushroom bodies of adult crickets is stimulated by JH, and this might be the basis for hormonal control of oviposition behaviour. Males are less often dependent on the action of JH, although a few studies provide clear evidence for effects of JH in males.

Female locust uses her abdomen as an ovipositor; she extends abdomen segments in a telescopic manner up to six times of its normal length. During this highly coordinated behaviour, telescopic extension mainly involves abdominal segments 4-7 in which the longitudinal muscles are required to follow and tolerate lengthening (superextension). Rose *et al.* (2001) found that, during reproductive development, the properties of the longitudinal muscle of females are subject to change, possibly to adapt them for oviposition behaviour. These properties are segment- and gender-specific and are controlled by JH.

Yin *et al.* (1999) reported that both dietary protein and the CA were required for normal mating behaviour in both male and female black blow fly, *Phormia regina*. Protein diet activated the CA in both sexes. A midgut peptide hormone activates the endocrine cascade leading to oogenesis. Allatectomy suppressed mating behaviour more than 2-fold in both sexes and topical application of S-methoprene (a juvenile hormone analogue) at 12 h after the onset of liver feeding restored sexual activity of both allatectomized males and females. Incidence of successful insemination increased as the oocyte development progressed. Ovariectomy suppressed sexual receptivity more than 3-fold in liver-fed females. Thus, in addition to nutrition and the CA, ovaries and their developmental status can also affect the sexual receptivity in female *P. regina*.

17.6 Polymorphism

Ford (1940) defined polymorphism as 'the occurrence of two or more well-marked forms, capable of appearing among the offspring of a single female, occur(ring) with frequencies high enough to exclude the maintenance of the rarest of them by recurrent mutation'. This definition includes genetic basis of polymorphism. In many insects two or more forms may arise by different developmental routes, which are mainly determined by environmental signals. Such insects may have different developmental potentialities and only one is expressed while the other remains dormant. Therefore the occurrence of different forms in any insect may be either genetically determined (polymorphism) or environmentally controlled (polyphenism). In many cases, polymorphism is under strict genetic control, but often environmental factors play a major role in the determination of the actual morph. Though obviously the potential to exhibit polymorphism is genetic but in many instances the genotype may modify phenotype response to environmental stimuli. Polymorphism is discontinuous when normally no intermediate forms are found between the different morphs, e.g. in case of apterous/alate polymorphism. However, in continuous polymorphism intermediate forms can occur between extreme forms, e.g. in locusts (Pener, 1985). Hardie and Lees (1985) defined polyphenism as 'the occurrence of two or more phenotypes which can be induced in individuals of the same genotype by extrinsic factors'. In social insects many behavioural patterns are not associated with morphological differences and individuals of one caste exhibit behavioural differences (polyethism). Hormones are known to play an important role in determination of polymorphism in different insects.

17.6.1 Wing dimorphism

In variegated grasshopper, *Zonocerus variegatus* two distinct short-winged and long-winged forms are found. McCaffery and Page (1978) showed that wing dimorphism is controlled by JH and removal of CA in the sixth instar results in the production of macropterous whereas injection of JH analogue ZR 515 entirely prevents their appearance. In the cricket, *Gryllus rubens* application of JH III to larvae intended to mature into long-winged insects resulted in the formation of brachypterous adults, corresponding to the higher JH titre in the last larval instars of the brachypterous as compared to the macropterous morph. Thus, high JH titres in the last larval instar are thought to inhibit metamorphosis of the wings and development of flight muscles (Zera and Denno, 1997).

Zhao and Zera (2004) reported that in flight-capable and flightless, adult females of black cricket, *Gryllus firmus* the haemolymph JH titre varied greatly. The flight-capable morph exhibited a large-amplitude daily cycle in the haemolymph JH titre, while the flightless morph exhibited a barely perceptible cycle. The JH titre cycle was observed on all days in the flight-capable morph, but the large amplitude cycle (>15-20 fold increase in mean titre, >100-fold increase in some individuals), began on day 5. The ecdysteroid titre did not exhibit corresponding large amplitude daily cycle. As JH regulates flight behaviours, morph-specific daily cycles in the JH titre may be especially common in dispersal-polymorphic insects, in which flight is restricted to one morph during a limited period of the day or night.

17.6.2 Locust phases

The migratory locusts, *Locusta migratoria* and *Schistocerca gregaria* appear in two phases (green solitary and brown gregarious) depending on population density and habitat quality, with a continuous range of intermediates between the extreme phases. These phases differ in behaviour, colouration, morphometric traits, metabolism related to mobilization of energy resources for flight activity, and fertility (Pener, 1991). The first evidence that some phase characters could be influenced by the endocrine system was given by Joly and Joly (1953) who showed that implantation of CA into crowded hoppers of *Locusta* promoted the green colouration of the integument. Surgical removal of CA in the fifth instar solitaria had the reverse effect leading to the formation of brown adult gregaria. JH is directly and indirectly involved in the determination of many phase characteristics of locusts. JH modulates cuticular melanization and the rate of reproductive maturation, specifically regulating vitellogenesis. The fact that the green colouration typical for the solitary form could be induced by CA implantation into nymphs kept under crowded conditions, and the demonstration that the JH titre and CA activity during critical phases of development were indeed higher in isolated than in crowded locusts (Injeyan and Tobe, 1981). Behavioural phase characteristics could also be inverted by JH treatment. The marching activity of crowded hoppers (a gregarious phase characteristic of immatures) was intensified by JH treatment. JH increased fecundity in crowded adult females and also accelerated oocyte maturation. Thus, JH may be stimulating oogenesis in general and vitellogenesis in particular in adult females.

Tanaka (2001) discussed the hormonal mechanisms for the control of body-colour polymorphism in locusts. Implantation of corpora cardiaca taken from normal nymphs of *Locusta migratoria* induced dark colour in nymphs of an albino strain. A neuropeptide, [His7]-corazonin, was identified as a dark colour-inducing factor for *L. migratoria* and *Schistocerca gregaria*. In *L. migratoria* this peptide and JH developed various body colours. In *S. gregaria*, on the other hand, the peptide induced black patterns, but the orange or yellow background colour observed in gregarious forms was not induced when the peptide was injected into solitary individuals. [His7]-corazonin also induced darkening in other grasshoppers and locusts, but not in katydids. Albino *L. migratoria* developed dark colour when implanted with brains or CC taken from other insects belonging to 10 major insect orders, but not with those from Coleoptera. This indicates that [His7]-corazonin or a similar compound is widespread among insects and plays a pivotal role in controlling body colour in some species and presumably other physiological roles in other species. Tanaka (2004) suggested that [His7]-corazonin and JH are involved in the control of black pattern expression and green-brown polyphenism, respectively, in American grasshopper, *Schistocerca americana*.

17.6.3 Aphid forms

Distinct morphological forms exist in the life cycle of an aphid which enables them to adapt and survive in their habitats. Aphids are well-known for parthenogenetic reproduction and alteration of sexual and parthenogenetic generations (Fig. 15.2). The apterous/alate wing dimorphism can occur in combination with switches between parthenogenetic and sexual modes of reproduction, and also with seasonal oscillation between host plant species. These forms also differ in behavioural characters. The sexual forms are: males and oviparae- sexual egg-laying females while other forms are: virginopara- parthenogenetic form which may be winged or apterous and give birth to parthenogenetic progeny; vivipara- parthenogenetic form producing parthenogenetic or sexual females; gynopara- a winged form which produces oviparae; and sexupara- winged form producing both oviparae and males. Thus, the problem of how environmental cues are translated into endocrine signals to switch between the different modes of reproduction becomes very complex (Hardie and Lees, 1985). Although JH plays a part in the determination of the characters of various forms, its role may differ radically in different species. The most convincing evidence for a role of JH in morph determination has been obtained for photoperiod-mediated wing polymorphism in black bean aphid, *Aphis fabae* where JH application during a critical phase sensitive to photoperiod strongly redirected development from the winged to the apterous form.

17.6.4 Polyethism

Temporal polyethism is widespread in the social Hymenoptera (ants, bees, and wasps) and typically there is a correlation between a worker's age and the tasks it performs. In honey bees, an adult worker spends roughly its first 20 days performing tasks inside the hive and then switches to outside tasks, mainly foraging, for the rest of its life. The caste development and social regulation of division of labour has been best studied in the honey bee, *Apis mellifera* (see section 14.5.2.3). Detailed analyses of JH titres during embryonic, larval and pupal development of queen and worker bees have revealed dramatic differences in the third to early fifth larval instars, along with a switch in the feeding programme for queen and worker larvae. The queen larva is continuously fed with copious amount of royal jelly while the worker larvae are reared on a mixed diet. This probably results in high rates of JH synthesis by the CA of queen larvae and low rates of JH synthesis in workers (Hartfelder and Engels, 1998). JH plays an important role in the regulation of behavioural development; haemolymph titres of JH are usually correlated with behavioural state, and treatments involving either the addition or removal of JH cause acceleration or retardation of behavioural development. JH has been described as a pacemaker since it affects the rate and timing of developmental changes but is not required for maturation into a forager (Sullivan *et al.*, 2000). Octopamine is also involved in the regulation of temporal polyethism and its levels are higher in foragers than nurses. Treatment of colonies with octopamine induces precocious foraging (Schulz and Robinson, 2001).

The age-correlated changes in worker behaviour (age polyethism) have been reported in the Neotropical eusocial wasp, *Polistes canadensis.* Application of JH analogue methoprene accelerates the onset of guarding behaviour, an age-correlated task, and increases the number of foraging females. JH titres correlate with both ovarian development of queens and task differentiation in workers of this wasp (Giray *et al.* 2005).

The range of morphs and mechanism of caste determination differ considerably in termites. Lüscher (1960) transplanted active CA from reproductives into pseudergates and this resulted into induction of soldiers. This indicates that the production of soldiers requires a high titre of JH.

Brian (1974) suggested a role of JH in queen differentiation and a possible role for ecdysteroids in worker differentiation in red ant, *Myrmica rubra.* In a number of ant species JH appears to influence caste determination both in egg and larval instars. In these instances JH induces queen development. Treatment of workers with JH analogue can also induce the development of majors. Some evidences also indicate that ecdysteroids favour worker development (Hardie and Lees, 1985).

The role of JH in colony organization of the more primitively social bumble bees exhibits striking differences from the role it plays in the highly social honey bees: (i) JH is associated with dominance and reproduction (vitellogenesis) in the former but not the latter, (ii) age polyethism and division of labour among workers is not affected by JH in bumble bees, (iii) foraging workers in bumble bee colonies have lower JH titres than workers that perform tasks within the nest, and (iv) mating flight activity of virgin queens is not affected by JH in bumble bees (Hartfelder, 2000).

17.7 Flight and migration

Kennedy (1951, 1961) considered persistent locomotion as the major characteristic of migration and migrants are highly sensitive to stimuli leading to 'straightened-out' movements, and responsiveness to vegetative stimuli like feeding and reproduction, etc. is suppressed during migration. Many insect species migrate before the onset of reproduction and hormones of these insects affect their migratory ability and/or performance. It is difficult to conduct experiments on insects that are actually migrating and most studies pertaining to hormonal regulation of flight and migratory behaviours are restricted to laboratory studies. The results so obtained are extrapolated with the assumption that these reflect hormonal effects upon natural migration (Pener, 1985). Although adults often undertake migration, dispersal and flight; however, insects like army ants migrate by walking. Johnson (1969) suggested a hypothetical relation between flight, ovarian development, and activity of corpora allata and prothoracic glands. He also discussed the role of hormones in alary polymorphism, and development and degeneration of flight muscles. Truman and Riddiford (1974) reviewed hormonal effects on flight, migration and flight orientation.

Pener (1985) suggested following modes of action of hormones on dispersal, flight and migration:

(i) Most insects migrate at well-defined stage of development, and their development is under the influence of hormone.

(ii) Effect on polymorphism and the flight abilities of different morphs.

(iii) Effect on morphology of organs responsible for flight, e.g. hormonal induction of degeneration of flight muscles.

(iv) Effects on behaviour, e.g. locomotor activity, rhythms, flight direction, which are component of migratory behaviour.

(v) Effect on metabolism, e.g. intense fuel supply for intense locomotion and migration.

Locusts show phase polymorphism and gregarious locusts migrate and exhibit more locomotor activity. Locust flight involves mobilization, transport and utilization of endogenous energy reserves at extremely high rates. The peptide, adipokinetic hormone (AKH) and synthesized and stored in neuroendocrine cells, integrate flight energy metabolism. The AKH is needed in fuel mobilization and transport, and is also necessary for long migratory flights of locusts. AKH promotes triglyceride breakdown in the fat body and formation of specific diglycerides which are released in haemolymph (Jutsum and Goldsworthy, 1976). Min *et al.* (2004) found that in the lesser migratory grasshopper, *Melanoplus sanguinipes* JH levels increased after long-duration flight performance. The increased JH titre induced by performance of long-duration flight is thus at least one component of flight-enhanced reproduction. They also reported that though

AKH is released during flight and controls lipid mobilization, it is not the factor responsible for increased JH titres after long-duration flight.

Effects of hormones on migratory flights and other behaviours has been studied in the large milkweed bug, *Oncopeltus fasciatus* (Dingle, 1972). Three major factors, viz. photoperiod, temperature and food affect the timing of flight and oviposition. Long photophase (LD 16:8 or 17:7) and moderate temperature 27°C enhance the onset of reproduction and reduce prereproductive flight while short photoperiod LD (12:12 or 11:13) and low temperature 23-24°C delayed oviposition and bugs exhibit long flights (Caldwell and Rankin, 1972). Rankin (1974) showed that administration of JH increased the percentage of males exhibiting long flights. Implantation of CA in early period of adult females also gave similar results. This suggests that JH induces long flights in *O. fasciatus*.

The observation that migratory flights in search for more suitable sites for reproduction are undertaken by sexually immature adults in many species has led to the proposal of an 'oogenesis-flight syndrome', meaning that migration and reproduction should be mutually exclusive and could not occur at the same time. Strong evidence for a role of JH as a master regulator for the oogenesis-migration syndrome comes from studies on the rice armyworm, *Mythimna unipuncta* which exhibits migratory behaviour in Northern populations (North American) and is non-migratory in the South (Azorean). Comparison of onset of female calling behaviour, progress in oogenesis, and rates of JH biosynthesis by CA of females during the first days after emergence revealed a time shift for these two populations, with the non-migratory Southern population being more advanced (McNeil *et al.*, 1996).

Relative to summer adults, migrant monarch butterflies, *Danaus plexippus* are characterized by a large wingspan and body mass, high lipid content, and arrested accessory gland (in both sexes) and ovarian development (in females) (Herman, 1985). The migrant adults feed intensively for several weeks following eclosion before their southern fight to winter roosts (Urquhart, 1960). Monarch butterflies' reproductive diapause is centrally regulated by suppressed production of JH. Both neck ligation and allatectomy elicit traits of the diapause syndrome in adults raised under summer conditions, like reduced accessory gland volume, arrested oogenesis and increased fat body lipid content (Herman, 1975). These traits are rescued in part by therapy with JH I or JH II (Lessman *et al.*, 1982). The JH titres of summer adults based on direct and indirect measures are 100 times greater than the measures from migrant adults that are wild caught in November to January (Lessman and Herman, 1983). Herman and Tatar (2001) reported that in *D. plexippus* migrant adults persist during a state of reproductive diapause in which both male and female reproductive development is arrested as a consequence of suppressed synthesis of JH. They showed that migrant adults are longer lived than summer adults when maintained under standard laboratory conditions, and the longevity of migrant adults is curtailed by treatment with JH and that the longevity of summer adults is increased by 100 per cent when JH synthesis is prevented by surgical removal of CA. Thus, monarch butterfly persistence through a long winter season is ensured in part by reduced aging that is under endocrine regulation, as well as by the unique environmental properties of their winter roost sites.

17.8 Circadian rhythms

Behavioural rhythmicity was reviewed by Brady (1974) and Saunders (1974). Biological Clocks enable insects to live in harmony with the rhythms of nature, such as the cycles of day and night and of the seasons. Such biological 'timers' exist for almost every kind of periodicity among insects, but most of what is known about them comes from the study of circadian rhythms. Circadian rhythms or daily behaviour patterns depend on internal timers for their periodicity. Light is believed to be the most powerful synchronizer of circadian rhythms. Many insects also make use of rhythmic variations in temperature or other sensory inputs to readjust their internal timers. A master clock appears to exist in the brain of most insects that communicates its timing signals chemically. For example, a brain removed from a moth pupa and exposed to an artificial sunrise of one time zone, then implanted into the abdomen of a headless pupa on a different time zone schedule, will cause the second pupa to emerge at the time of day appropriate to the disconnected brain floating in its abdomen. The clock in the brain triggers the release of a hormone that switches on all the complex behaviour involved in pupal emergence.

In the circadian system, this clock comprises a circadian pacemaker, which generates an endogenous rhythm of about 24 h. The circadian system also includes an entrainment pathway for the synchronization with '*zeitgebers*' and output pathways to effector organs. The hormone synthesis is modulated during the course of development and by a circadian rhythm. The insect endocrine glands contain a circadian clock and the specific hormones released from the glands vary with the time of day. The appearance of the receptor for ecdysteroids in target tissues also has a daily rhythm, enabling the rhythm in the blood to elicit time-of-day specific responses from target cells. PTTH release also occurs with a circadian rhythm which is controlled by a clock located within the brain. Melatonin, a central hormonal mediator of internal time signals in higher animals has also been identified in insects. The melatonin has a circadian rhythm in insect blood.

Bowen *et al.* (1984) reported that pupal diapause in the tobacco hornworm, *Manduca sexta* is a photoperiodically induced event expressed by the absence of release of the cerebral peptide PTTH. A short-day photoperiod (light-dark, 12:12) during larval development programmes induces the brain to forego the release of PTTH in the pupal stage. PTTH release can occur in the pupae and diapause can be averted if developing larvae experience an intervening long-day photoperiod (light-dark, 16:8) for 3 days early in the last larval instar. Implanted brains from larvae that received such long-day stimulation can reverse the diapause commitment of short-day-reared diapause-destined larvae, thus overriding the diapause programme of the recipient. These findings demonstrate that the photoperiodic clock of an insect can be reprogrammed by a light-dark cycle *in vitro*.

Vafopoulou and Steel (1991) studied rhythmic synthesis of ecdysteroids by prothoracic glands in *Rhodnius prolixus* during the last larval instar and observed a massive daily rhythm, with synthesis at night being three- to fivefold higher than during the day. This rhythm of ecdysteroid synthesis by prothoracic glands (PGs) free-runs in continuous darkness with a temperature-compensated period length close to 24 h and is therefore controlled by a circadian system. The synthesis rhythm also free-runs in continuous light, but with an inverted phase and shorter period length. They concluded that the circadian system controlling synthesis of ecdysteroids constitutes a pacemaker which drives various rhythms in the target cells of ecdysteroids via the rhythm in the haemolymph titre. Ecdysteroids are viewed as 'hormonal *zeitgebers*', imposing temporal order on development.

Pener (1992) suggested that environmental cues, mostly photoperiod and temperature, mediated by effects on the neuroendocrine system, control reproductive diapause in female insects. Arrest of oocyte development characterizes female reproductive diapause, which has two major adaptive functions. First, it improves chances of survival during unfavourable season(s), and/or it confines oviposition to that period of the year that is optimal for survival of the eggs and progeny. Second, the selection pressure directs the male's reproductive strategy toward maximum chances of fertilization of the female's eggs with minimum waste of energy. Therefore, in species with female reproductive diapause, the males may or may not exhibit diapause, but if they do, their diapause must be adapted to that existing in conspecific females. If there is a reproductive diapause in the male, it is controlled by similar environmental cues (photoperiod, temperature) and by the same intrinsic mechanism (neuroendocrine system, especially CA).

PTTH is the central cerebral neurohormone in insect development and released to 1 (or 2) critical moments early in each developmental stage, at which time it triggers prolonged activation of the prothoracic glands to synthesize and release the ecdysteroids, which elicit developmental responses in target tissues. Vafopoulou and Steel (1996) used an *in vitro* assay for PTTH released from excised brains of the bug *Rhodnius prolixus* and reported that release of PTTH does occur at the expected time on day 6, but that this release is merely the first in a daily rhythm of release that continues throughout most of the 21 days of larval-adult development. Vafopoulou and Steel (1999) reported that the synthesis of ecdysteroids by the PGs of *R. prolixus* occurs with a circadian rhythm throughout most of larval-adult development. This rhythm is generated by a circadian oscillator within the PGs. The principle regulator of the PGs, i.e. PTTH is also known to be released with a circadian rhythm. There is a daily rhythm in the responsiveness of PGs to PTTH; high responsiveness occurs around dusk and PGs are largely unresponsive to PTTH during the daytime. They concluded that the daily releases of PTTH are detected by the PGs and participate in the regulation of daily ecdysteroid synthesis. Therefore, PTTH plays a continuing role in the regulation of PGs throughout development.

17.9 Social behaviour

Division of labour in honey bees. Division of labour in honey bee is based on a stereotyped pattern behavioural development by workers (Winston, 1987). Young bees perform the task of nursing and hive maintenance for the first 2-3 weeks. They also take a series of orientation flights prior to onset of foraging to learn location of hive and landmarks. Workers older than 3 weeks forage for nectar and pollen. The regulation of division of labour in honey bees involves several physiological changes, including changes in level of JH (Robinson, 2002). Division of labour in the European honey bee, *Apis mellifera* has been extensively studied and workers typically change their jobs as they age in a colony. This progression of tasks, a form of behavioural development, is correlated with a secretion of JH. In adult honey bees its function is 'maturing' rather than 'juvenilizing'. JH titres in blood typically increase with age; they are low in bees that perform in-hive tasks such as nursing, comb building and other activities, and high in foragers (Rutz *et al.*, 1976). When corpora allata, the organs that produce JH, are removed from workers and incubated *in vitro* and measured for rates of JH biosynthesis, they also show the same upward change with age, i.e. low in young bees and high in foragers (Huang *et al.*, 1991). Applying JH and JH analogue to bees causes workers to forage earlier (Jaycox, 1976). These results suggest that not only is JH correlated with behavioural development in *A. mellifera*, but it also plays a key role in modulating the pace of behavioural transition in workers. Huang *et al.* (2001) studied the mechanisms underlying social behaviour and the role JH plays in behavioural development in Indian honey bee, *Apis cerana*. They observed that adult workers in *A. cerana* have similar JH profiles as *A. mellifera* workers performing different tasks. *A. cerana* and *A. mellifera* seem to show very similar physiological attributes: workers performing different tasks have different levels of JH and older bees (foragers) tend to have the highest JH titres. JH treatment also accelerates behavioural development in workers.

Sullivan *et al.* (2000) reported that JH does not activate foragers, but rather is involved in controlling the pace at which bees develop into foragers. They also reported that allatectomy results in a deficit in either the physical aspects of flight behaviour or the ability to learn/remember the location of the hive. Maleszka and Helliwell (2001) also reported that JH had an influence on learning of honey bees. Sullivan *et al.* (2003) observed that removal of the corpora allata causes honey bees to fail to return to their hive upon initiating flight because:

(i) Allatectomized bees had a significantly slower ground speed than sham and untreated bees during orientation flights, but otherwise attributes such as flight range and area were normal.

(ii) Allatectomized group had a significantly smaller percentage of flightworthy bees than did the sham and untreated groups.

(iii) Allatectomy causes a decrease in metabolic rate and allatectomized bees had significantly lower metabolic rates than untreated and sham bees.

Honey bee aggression. Apis mellifera bees vary substantially in the degree of their aggressive response. This can be attributed to geographic variation among populations, differences among colonies, and differential expression of defensive behaviour among bees within the same colony. Breed (1983) showed that older bees have higher JH levels and are generally more aggressive than younger bees with lower JH levels. Bees treated with a JH analogue exhibited an earlier response to alarm pheromone, and the proportion of bees acting as guards also increased. Workers reared in isolation have elevated level of JH and are more aggressive. Besides JH titres of guard bees are higher than all other middle age bees.

Pearce *et al.* (2001) suggested that JH is an endocrinological correlate of aggression in honey bees. They determined seasonal differences in aggressiveness in honey bees and their association with JH levels and correlated JH titres with aggressive behaviour at two levels. First, winter bees and summer nurses, known to have lower JH titres, both showed less aggression toward foreign bees than did summer guards. Second, aggressive individuals had significantly higher JH titres than did non-aggressive bees within each colony. Inter-colonial variation in aggressiveness was maintained during summer and winter, suggesting a genetic basis for these differences.

Reproduction in workers. Division of labour between queens and workers is fundamental to the organization of complex insect societies. Queens lay eggs and workers perform tasks related to colony growth and development. The evolution of division of labour between queens and workers has been studied extensively. However, the proximal mechanisms underlying the regulation of worker reproduction are known for only a few species of social Hymenoptera. Workers in many species of social Hymenoptera are able to produce haploid male offspring, a consequence of haplodiploidy. Regulation of worker reproduction appears to be mediated by JH. JH treatment results in ovary development in the sweat bee, *Lasioglossum zephyrurn* (Bell, 1973). *Polistes gallicus* queen wasps inhibit the ovary development and egg laying behaviour of subordinate workers via behavioural interactions. Queen bumble bees inhibit worker reproduction via pheromones that also depress the JH titre. Treatment of bumble bee workers with exogenous JH resulted in ovary development and oviposition. In the honey bee, *Apis mellifera,* queen pheromones, together with brood pheromones also inhibit ovary development and egg-laying behaviour in workers. Workers usually lay eggs in the presence of a queen only at extremely low frequencies. In queenless colonies; however, worker oviposition is more extensive and may result in the production of large number of males. It has been demonstrated that the effects of JH on worker reproduction are more apparent in the absence of a queen. Haemolymph levels of JH increase in workers with age. Low titres of JH are associated with brood care (nursing behaviour) while a higher titre at about 3 weeks of age induces foraging (Robinson *et al.*, 1989). Robinson *et al.* (1992) determined the role of JH in worker honey bee reproduction in queenless colonies and found that:

(i) Egg-laying workers had low haemolymph titres of JH, as did bees engaged in brood care, while foragers had significantly higher titres.

(ii) Egg-laying workers have significantly lower rates of JH biosynthesis than foragers do.

(iii) Ovary development was inhibited slightly by application of the JH analogue methoprene to l-day-old bees, but was not affected by application to older bees, at least some already displaying egg-laying behaviour.

17.10 Pheromone production

The pheromones are biosynthesized in the pheromone glands from fatty acid precursors. Attempts have been made to show the role of endocrine system in pheromonal production. Pheromone production in insects is usually age related and coincides with the maturation of gonads, and in some cases with feeding. The observation that females of certain species have repeated reproductive cycles and that mating occurs only during defined periods of each cycle led to the proposal that pheromone production might be under hormonal control (Barth, 1965). Early work on cockroaches established that females require the presence of functional corpora allata in order to produce sex pheromones and the role of JH in regulating pheromone production was established. In female house fly, *Musca domestica* ovarian produced 20-hydroxyecdysone was shown to be the key hormone inducing sex pheromone production (Adams *et al.*, 1984). In Lepidoptera pheromone production was reported to be governed by a brain factor termed as 'pheromone biosynthesis activating neuropeptide' (PBAN) and its structure was elucidated by Raina *et al.*, 1989.

Tillman *et al.* (1999) reviewed the biosynthetic pathways and endocrine regulation of insect pheromone production and suggested that insects have evolved one or a few tissue-specific auxiliary or modified enzymes that transform the products of normal metabolism to pheromone compounds. Insects utilize at least three hormonal messengers to regulate pheromone biosynthesis. In Dictyoptera and Coleoptera pheromone production is induced by JH III. In the female *Musca domestica* and possibly other Diptera, it appears that during hydrocarbon sex pheromone biosynthesis, ovarian-produced ecdysteroids regulate pheromone synthesis. Lepidopteran sex pheromone biosynthesis is often mediated by PBAN through alteration of enzyme activities at one or more steps prior to or during fatty acid synthesis or during modification of the carbonyl group.

In the female German cockroach, *Blattella germanica* pheromone production is correlated with oocyte growth, and is slow in those females carrying ootheca (Schal *et al.*, 1990). Application of JH analogue to allatectomized insects stimulates pheromone production. Implantation of ootheca in virgin females inhibits oocyte growth and pheromone production.

The reproductive behaviour in moths is dependent on chemical communication and in most species, females produce and release an attractive blend of pheromone to which conspecific males orient. Production of the pheromone blend is under the regulation of a neuropeptide termed PBAN (Raina, 1993). The direct action of PBAN has been demonstrated by studies *in vitro* that showed synthetic PBAN stimulated pheromone-gland tissue for pheromone production. PBAN has also been detected in *Bombyx mori, Helicoverpa zea* and *H. armigera*. The exact tissue involved in PBAN production was determined to be the intersegmental membrane that is situated between the eighth and ninth abdominal segments (Rafaeli and Gileadi, 1996). Female moths call conspecific males during specific periods when they emit their pheromone. Zhu *et al.* (1995) found that the pheromone production in the female turnip moth, *Agrotis segetum* is under the control of a brain factor, which was a PBAN-like substance. Decapitation of a female resulted in a significant decrease in pheromone production. Pheromone production was restored by injection of homogenates of either male or female brain-suboesophageal ganglion or the corpora cardiaca alone. The major pheromone component of the Indian meal moth, *Plodia interpunctella* was identified as (Z,E)-9,12-tetradecadienyl acetate [ZETA] (Brady *et al.*, 1971). Some octopamine agonists were found to suppress *in vitro* biosynthesis of the pheromone production and calling behaviour in *P. interpunctella* (Hirashima *et al.*, 2001). Biogenic amines, such as octapamine (OA), play a key role as neurotransmitters, neurohormones and neuromodulators in insect systems with a physiological role analogous to adrenaline in vertebrates (Evans, 1980). In *H. armigera,* OA and OA agonists significantly inhibit pheromone production caused by PBAN in intact and decapitated moths as well as in pheromone-gland incubations *in vitro* (Rafaeli and Gileadi, 1996).

17.11 Colour change

Colour change in insects results either from the concentration or dispersion of pigment cells or from contraction and expansion of melanophores. Such a change in body colour is short-term and periodic and is termed physiological colour change. By contrast, morphological colour change is long-term or irreversible change in the colour of integument. Such changes are produced during development, in response to conditions of light, temperature, humidity or other conditions.

Physiological colour change in stick insect, *Carausius morosus* occurs due to movement of pigment granules in epidermal cells. It becomes black at night and resumes its brown colour during day. During day the dark brown pigment is found at the internal end of epidermal cell while orange pigment is concentrated around the nucleus. This leads to light colouration of the insect during day. At night dark brown pigment moves towards the outer end of the cell and orange granules spread evenly, and resulting into darkening of the integument. The movement of pigments is related to light conditions. Janda (1934) concluded that a substance that is released from the brain and transported by the haemolymph initiates this colour change. Dupont-Raabe (1957) showed that a secretion from the neurosecretory cells in the tritocerebrum controls the colour change. The corpora cardiaca also releases a substance causing moderate pigment migration with the development of intermediate colouration.

In transparent aquatic larvae of a phantom midge, *Chaoborus crystallinus* colour change is restricted to the surface of air-sacs which are covered with melanophores containing black pigment and capable of expansion and contraction. Movements of melanophores cause the colour change, and against a dark background the chromatophores are distended covering the entire surface of air-sacs while in light background chromatophores contracts. Gersch (1956) shown that at least three neurohormones with partial antagonistic effects are involved in colour change.

Novák (1955) showed that the size of ventral black spots on abdomen of large milkweed bug, *Oncopeltus fasciatus* is governed by JH activity. The size of spots decreased in insects kept at high temperature.

Locust nymphs exist in a green or yellow solitary form while gregarious form is extensive black with a yellow background. Carlisle and Ellis (1962) showed that corpora allata and prothoracic glands regulate these colour differences. Since locusts change colour from the solitarious light green to the darker pigments of black or brown in the gregarious phase, enormous studies have taken place to explain this colour change. There was great excitement when an albino strain of the *Locusta migratoria* was discovered in Okinawa, Japan. Rahman *et al.* (2003) discovered that a peptide [His7]-corazonin produced in the brain, then stored and released from the corpora cardiaca promotes the black pigmentation characteristic of gregarious locusts. To date, the function of this peptide is not fully understood; however, it has been found that [His7]-corazonin itself cannot induce gregarious behaviour. Breuer *et al.* (2002) summarized the findings relating to the underlying mechanism of phase transition in locusts, from the non-swarming solitarious phase to the swarming gregarious phase. These phases differ in many traits, such as colouration, morphometrics and behaviour. The most comprehensive theory at present to explain the switch from the non-swarming to the swarming form is that the locusts are brought together by the heterogeneity of the environment. JH induces certain solitarious characteristics, such as green colouration, but is not the primary causal factor. The discovery of [His7]-corazonin, made possible by using a novel assay system (the Okinawa albino mutant of *Locusta migratoria*), which was known to be deficient in this hormone. This peptide, which is produced in the brain and is most likely released via the corpora cardiaca, promotes the gregarious black pigmentation. It also plays a role in morphometrical phase change as well as in behavioural alterations. Tanaka (2004) examined the possible involvement of this peptide in the induction of dark colour in nymphs of American grasshopper, *Schistocerca americana*. [His7]-corazonin showed a dose-dependent induction of black patterns, but it did not affect the green-brown polyphenism. Injection of JH III into reddish nymphs obtained at a low temperature caused them to turn green in a dose-dependent manner, but it had no effect on the expression of black patterns. The results suggested that [His7]-corazonin and JH are involved in the control of black pattern expression and green-brown polyphenism, respectively in this grasshopper.

Chapter 18

Pheromones

Chemical odours are important for communication and in insects many behavioural patterns rely on sense of smell. The chemicals used for intraspecific communication are termed 'pheromones'. These substances were defined as "substances that are secreted to the outside by one individual and received by a second individual of the same species in which they release a specific reaction, for example a definite behaviour or developmental process." The term pheromone was first used by Karlson and Butenandt (1959), and Karlson and Lüscher (1959). Insects use pheromones for different purposes; the most common ones are sex and alarm pheromones. Sex pheromones allow organisms to lure other members of opposite sex of the same species so they can mate while alarm pheromones are used to warn other insects. There are also pheromones that are used by insects to find their nest and for other purposes. Besides, some pheromones cause permanent physiological changes in insects. The queen of a beehive, for example, releases chemicals that suppress the fertility of the female workers in the hive. The complete behaviour of an insect can be determined by the use of chemicals and in some cases species also make use of the 'pheromonal infrastructure' of others. Wilson and Bossert (1963) divided pheromones into 'releasers' and 'primers'. The releaser pheromone induces an immediate behavioural change while the primer pheromone trigger a chain of physiological/ developmental events, e.g. sexual maturation, and the change in behaviour is not immediate (Fig. 18.1). Hundreds of releaser pheromones are known; in contrast,

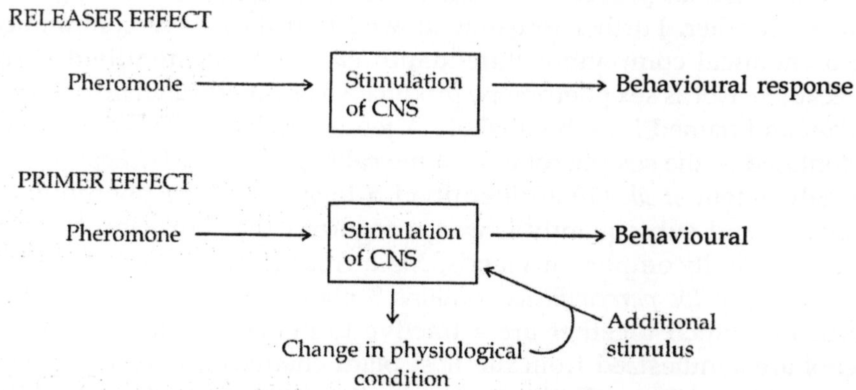

Fig. 18.1. Schematic representation of the releaser and primer effects of pheromones on behaviour of insects (modified after Wilson and Bossert, 1963).

very few primer pheromones have been identified, primarily because they are much more difficult to assay. A major function of primer pheromones is to coordinate the timing of physiological and behavioural development (Wyatt, 2003).

Primer pheromones of social insects have been identified in: the honey bee queen substance inhibiting queen cell construction by workers; a pheromone that delays reproductive development in workers of the bumble bee, *Bombus terrestris*; and a secretion that suppresses the differentiation of new soldiers in the termite, *Nasutitermes lujae*. In red imported fire ant, *Solenopsis invicta* queen-produced primer pheromones play important roles in regulating many basic features of ant societies, e.g. inhibition of dealation (wing shedding), ovary development in winged virgin queens, inhibition of the production of winged male and female sexuals, and mutual inhibition of egg production among egg-laying queens in polygyne (multiple-queen) colonies. Vargo (1997) reported that queen poison gland is the source of the pheromone inhibiting dealation and ovary development in virgin queens of *S. invicta*.

The chemicals that are involved in communication are now termed 'semiochemicals', and such chemicals are further divided into 'allelochemicals' and 'pheromones'. Allelochemicals are mainly involved in interspecific communication while pheromones are cues for intraspecific communication. Secretion of a pheromone may release different behavioural patterns by the receiving individual and pheromones are considered multicomponent for this reason. Some early workers compared these chemicals with hormones and as pheromones act externally, terms like 'exohormones' and 'ectohormones' were also used for these chemicals. Some chemicals not secreted by an insect and act as pheromone, are often referred as 'parapheromones', e.g. methyl eugenol that attract males of some fruit flies, *Bactrocera* species.

In the world of insects, pheromones are very common and the first pheromones were discovered in the 1930's during research on silkworm, *Bombyx mori*. Female moths of this species proved to be able to excite males, while the two couldn't see or hear each other. Further research showed that they were communicating by using a chemical compound. Butendandt *et al.* (1959) identified (*E*)10-(Z)12-hexadecadien-1-ol as sex pheromone produced from abdominal tip of virgin female of *B. mori* and named it as 'bombykol'. A few years later, (Z)-7-dodecenyl acetate was identified as the sex pheromone of the cabbage looper, *Trichoplusia ni* (Berger, 1966). Silverstein *et al.* (1966) identified a blend of three pheromones for *Ips paraconfusus* and subsequently it was recognized that most insect pheromones consisted of multicomponent blends. Male frass (fragments of host tissue and faecal pellets) of *Ips paraconfusus* contains 3 pheromones: ipsenol, ipsdienol, and *cis*-verbenol, which together are attractive to males and females. Ipsenol and ipsdienol are synthesized from the host plant chemical, myrcene, in males only while *cis*-verbenol is synthesized from another host plant chemical (-)α-pinene, in both sexes. These pheromones accumulate in the hindgut upon exposure to vapours of the host plant chemicals.

Over the past four decades, extensive researches on insect pheromones have resulted in the chemical and/or behavioural elucidation of pheromone components from several thousand species of insects, with much of the work concentrating on sex pheromones from economically important pests. Significant contributions on pheromones and their role in insects' behaviour have been published by: Wilson (1963), Jacobson (1965, 1972), Butler (1967), Blum (1969, 1985), Beroza (1970), Ebling and Highnam (1970), Law and Regnier (1971), Shorey (1973, 1976), Birch (1974a,b), Roelofs (1975), Shorey and McKelvey (1977), Birch and Haynes (1982), Bell and Cardé (1984), Baker (1985), Haynes and Birch (1985), Tamaki (1985), Borden (1985), Payne *et al.* (1986), Mayer and McLaughlin (1991), Agosta (1992), Cardé and Minks (1997), Howse *et al.* (1998), Meer *et al.* (1998), Blomquist and Vogt (2003a), and Wyatt (2003).

18.1 Location of pheromone production or release

Pheromone producing gland cells can be individual cells or clusters of cells forming gland, and are located externally on different body parts of insects, including antennae, head, thorax, legs and abdomen. Percy-Cunningham and MacDonald (1987) reviewed the anatomical location of the cells or tissues involved in pheromone biosynthesis, accumulation, and release. Though such locations have a wide range, but the abdomen appears to be the most common location in the species of Dictyoptera, Coleoptera, and Lepidoptera. In Hymenoptera pheromone producing glands are mainly located in head and abdominal regions.

In American cockroach, *Periplaneta americana* the pheromone seems to be produced in the gut and is released from faecal pellets. In the female German cockroach, *Blattella germanica* the sex pheromone is produced in a gland located on the anterior of the 10th abdominal tergite. The contents of secretory vesicles from cells in the gland are transported through long ducts to the cuticular surface for release. Many coleopterans have been examined to verify the sites of sex or aggregation pheromone synthesis, accumulation, and/or release, and in most cases diversified abdominal glands have been found to release pheromones. In other coleopteran taxa, sex or aggregation pheromones are synthesized, accumulated, or released in other sites. In male red flour beetle, *Tribolium castaneum* a setiferous patch is located in the prothoracic femora. In Dutch elm beetle, *Scolytus multistriatus* and the large European elm bark beetle, *Scolytus scolytus* pheromone production, storage, and release is correlated with accessory glands associated with vaginal palpi; however, in other scolytid beetles, alimentary canal and Malpighian tubules are associated with the accumulation of aggregation pheromone. Flight muscles, hindgut and fat body tissue are other structures involved in pheromone synthesis in coleopterans.

In Diptera pheromones are synthesized mainly in specialized subcuticular abdominal epidermal cells (oenocytes) and deposited onto the cuticular surface.

Among Hymenoptera, the glands are located in the head and also in the abdomen of queen honey bee; and in abdomen of pine sawflies, *Diprion similis* and *Neodiprion pratti*. Locations of exocrine glands in worker honey bee are depicted in figure 18.2. Mandibular and Nasanov (Nassanoff's) glands in honey bees are located in head and abdomen, respectively. These glands are well-developed in queen and workers, but greatly reduced in the drone bees. In ants poison and Dufour's glands are associated with sting while. Pavan's gland opens on the ventral surface of the abdomen above VI abdominal sternite.

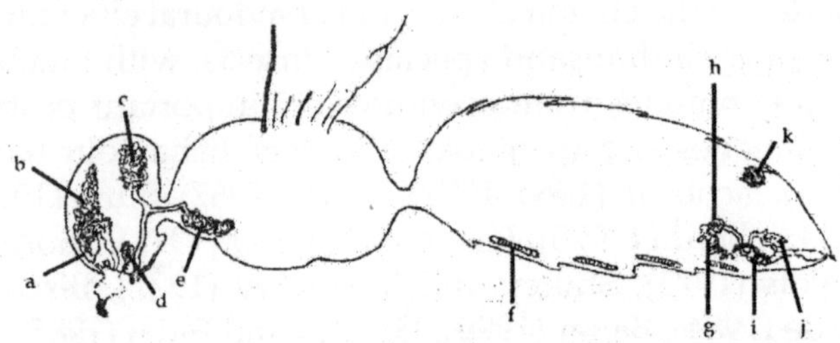

Fig. 18.2. Exocrine glands of honey bee worker. a- mandibular gland, b- hypopharyngeal (maxillary) gland, c- head labial gland, d- postgenal gland, e- thorax labial gland, f- wax gland, g- poison gland, h- vesicle of poison gland, i- Dufour's gland, j- Koschevnikov's gland, k- Nasanov (Nassanoff's gland) (modified after Wilson, 1971).

In the aphid, *Schizaphis borealis* the pheromone emitting organ is probably located on the hind tibiae of oviparous female. Pheromones producing abdominal glands are also found in termites and fleas.

Most lepidopteran females produce and release sex pheromone components from bulbous extrudable glands located between the 8th and 9th abdominal segments and occasionally between 7th and 8th abdominal segments. These glands have secretory cells that are hypertrophied and modified epidermal cells. Some lepidopterans have paired tubular glands in the abdomen. Male Lepidoptera often produce aphrodisiacs scents from the glands associated with wing scales, termed 'androconia' (Fig. 18.3). These scales may be scattered or in groups and terminate in a row of minute processes called 'fimbriae'. Glandular cells are considered to be connected with scales. Scales may be located on wings, legs and abdomen. The scents play a part during courtship when both the male and female fly together. The gentle flapping of the wings by the male, drift the pheromones from the scales towards the female, encouraging her to settle down and mate. Some male Lepidoptera has a yellow

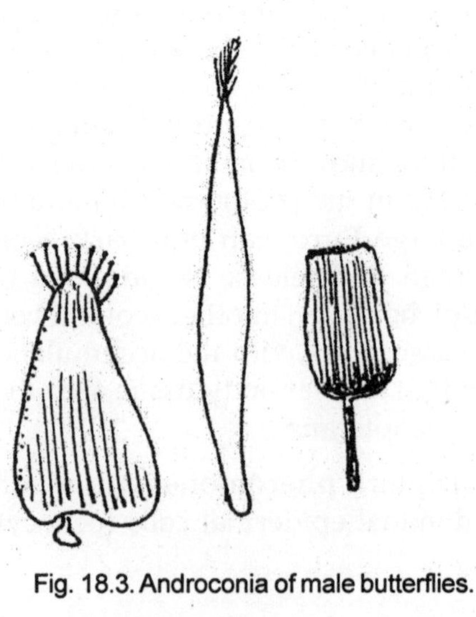
Fig. 18.3. Androconia of male butterflies.

brush-like organ that is usually tucked inside the last segment of the abdomen. When extruded during courtship, it forms a beautiful globular structure with very fine hairs. These hairs are called 'hair pencils' and sometimes rubbed or brushed against the antennae of the female to invoke a cooperative response. Males of many species of nocturnal Lepidoptera have special organs associated with scent glands which may be rather large and protruding bifurcate formations (coremata) on genital segments, wing glands and brushes, and genital hair-pencils. Specialized scent organs are located on tarsi in Noctuidae, and on abdomen in Sphingidae and diurnal moths.

18.2 Pheromone biosynthesis

It has been recognized that most insect pheromones consist of multicomponent blends and single-component pheromones are rare. A question asked in several systems was whether pheromone components were derived from dietary components and altered only minimally or synthesized *de novo*. Bjostad *et al.* (1987) demonstrated that in the Lepidoptera, most of the sex pheromone components were synthesized *de novo*, with some exceptions. Early studies in the cotton boll weevil, *Anthonomus grandis* gave evidence that the monoterpenoid pheromone components could arise both from modification of dietary precursors (Thompson and Mitlin, 1979) and from *de novo* biosynthesis (Mitlin and Hedin, 1974). Studies in bark beetles yielded convincing evidence that aggregation pheromone components were synthesized by the slight modification of host tree derived monoterpenoid precursors (Hughes, 1974). Seybold *et al.* (1995) later demonstrated pheromone synthesis in bark beetles was more complicated, and some pheromone components are synthesized *de novo*. The biosynthetic pathways of pheromones for many insect species have been determined and it has now become clear that some pheromone-specific enzymes are responsible for producing the myriad of pheromone molecules.

It became apparent that the products of normal metabolism, particularly those of the fatty acid and isoprenoid pathways, were modified by a few pheromone gland-specific enzymes to produce the myriad of pheromone molecules. Now a clear understanding of the biosynthetic pathways for many of the lepidopteran pheromones is known. Chain shortening of fatty acids is also involved in producing the queen pheromone in honey bees (Plettner *et al.*, 1996). In some insects, fatty acid elongation followed by decarboxylation produce the hydrocarbon pheromones, e.g. in lepidopterans, dipterans, German cockroach, and social insects (Blomquist and Vogt, 2003b).

18.3 Pheromone detection

A variety of chemicals are released from exocrine glands in the form of liquid, thin films, aerosols, or gases. Certain nonvolatile chemicals are used by insects to mark food resources, routes, territories, home range, etc. while volatile chemicals travel long distances in air or water. Two different receiving systems, i.e. taste and smell are used by insects to receive the chemical stimuli. The taste stimuli are sensed mainly in aquatic medium and the signals are water-borne. However, certain aquatic insects also use olfactory signals for communication. The air-borne signals vary widely, and are received by chemoreceptors (olfactory receptors) which trap odour molecules and transfer them to specific recognition site. The odour signal depolarizes the neural axon and generates a nerve impulse. The water-borne compounds, on the other hand, had a limited range of qualities.

Vogt (2005) reviewed the molecular basis of pheromone detection in insects. Insect chemoreceptor cells are surrounded by characteristic cuticular specializations forming the sensilla, and occur in the form of bristles, hairs, pegs, plates and pits. Insect pheromone receivers are complex structures, and mainly the trichoid sensillum (Fig. 3.4) that acts as highly efficient filter and receives the odour molecules. A sensillum may be uniporous (with one pore) or multiporous (with many pores). Each antenna of male silkworm, *Bombyx mori* has about 17,000 sensilla that respond to sex-signalling chemicals produced by the female, and each sensillum has up to 3,000 pores. Insect antennae (Fig. 18.4) are by far the most common organs where sensilla are located. The sensilla are also located on tarsi, mouthparts, ovipositior and other body parts. Studies of whole antennal responses frequently utilize the electroantennogram (EAG), and give a measurement of the summed receptor potentials of olfactory receptors responding to a stimulus. In moth the exocrine gland compound is liberated in the air and detected by specific sensilla located on the antennae of moth of opposite sex. The odour molecules diffuse through the pore openings connected to the fine tubules and initiate an electrical change in the membrane of the receptor cell, which is transmitted through the nerve axons. While communicating with the use of pheromone, a message is transmitted as a signal in coded form through a medium to a receiver. The sensory receptors recognize and interpret the code and provide a particular meaning. The signals are distinctive and species-specific to avoid ambiguity. These signals have the advantages, as they can be used in dark and can travel around obstacles.

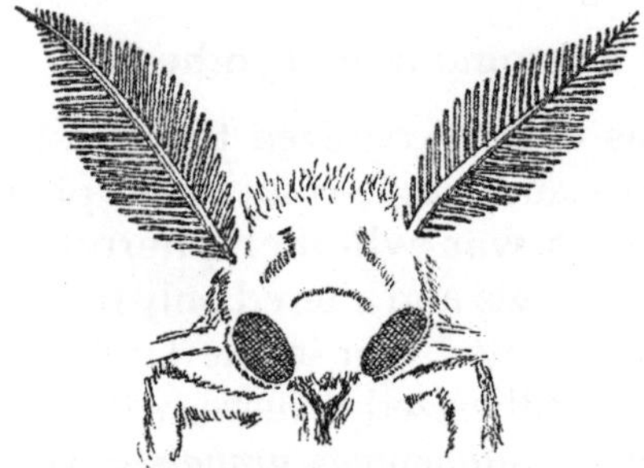

Fig. 18.4. The feathery antennae of a male moth can detect scent of a female from a considerable distance.

The olfactory system in male moths is so sensitive that males detect pheromone plumes many hundreds of metres downwind from a female. Key components of this highly specific chemosensory system are the pheromone binding proteins (PBPs), which transport the pheromone to the sensory neuron. The receptors trigger a cascade, which converts the chemical message to neural impulses. The impulses are then analysed in a specialized part of the brain, the olfactory bulb. In order to 'reset' the sensory cell, catabolic enzymes rapidly degrade pheromone not bound by the PBP. Other scavenger proteins may be involved in removing PBP-pheromone complex by endocytosis. Pheromone removal from the antennae is extremely important because it ensures that the antennae are resensitized after a stimulus.

Detailed studies on the pheromone olfactory systems have been made in male moths, which are housed in antennal sensory hairs (sensilla trichodea). Two or three olfactory neurons project their dendrites into the hollow space of one hair. Electrophysiological recordings indicate that antennae of male *Bombyx mori* can detect molecules of its sex pheromone, bombykol with high degree of specificity. Such a high level of olfactory function is provided by the complex biochemical system of cooperating olfactory proteins in the antenna, in which odorant binding proteins (OBPs) play an important, yet poorly understood role. Kaissling (1974) proposed a model for pheromone detection in *B. mori* in which pheromone molecules were transported to neuronal receptor proteins via pore-tubules and were subsequently inactivated by some rapid but non-enzymatic process. Vogt and Riddiford (1981) identified the pheromone binding protein (PBP) and sensilla esterase (SE) of polyphemus moth, *Antheraea polyphemus*. Vogt *et al.* (1985) proposed a new model for pheromone detection, and in this model, PBPs transported pheromone to receptor proteins and SE rapidly inactivated pheromone by enzymatic degradation. The SE of *A. polyphemus* is known to be only one type of odour degrading enzyme (ODE). Now PBPs have become established as only a subclass of a much larger family of insect odorant binding proteins (OBPs) that have been identified and/or cloned from various insect orders, such as Dictyoptera, Orthoptera, Isoptera, Lepidoptera, Coleoptera, Diptera, Hymenoptera, and Hemiptera (Vogt *et al.*, 1999). Earlier it was considered that the OBPs are restricted to Endopterygota and Paraneoptera. Riviere *et al.* (2003) described the first OBP from a dictyopteran insect, Madeira cockroach, *Rhyparobia maderae*. The PBP of *R. maderae* (PBPRma) shares all the hallmarks of the OBP family and is expressed specifically in the female adult antennae, the sex that perceives the sexual pheromone. The identification of OBPs from a primitive termite, *Zootermopsis nevadensis* (Ishida *et al.*, 2002) suggests that the PBP-gene family is also distributed throughout the neopteran orders.

Although genetic evidence strongly suggests that insect OBPs play a role in odorant discrimination, it is not known how strongly OBPs discriminate in binding to ligands, nor it is clear what features of an odorant molecule (for example size, functionality, or the locations and conformations of double bonds) are involved in that stage of discrimination by the olfactory system. The hydrophobic odorants (pheromones) are solubilized by OBPs and transported through an aqueous environment (sensillar lymph) to the olfactory receptors, where the signal transduction starts (Kaissling, 2001). OBPs occur at high concentrations in the lymph of insect olfactory sensilla, and are subdivided into pheromone-binding proteins (PBPs) and general odorant binding proteins (GOBPs). A third class of small proteins have also been identified in several sensorial organs from a number of insect orders. Although they may function as carrier proteins, they have been separated into a group of chemosensory specific proteins (CSPs), characterized by four cysteine residues and with low sequence similarity to OBPs. The PBPs are not the only proteins that function in the transport of hydrophobic ligands through aqueous media. The GOBPs, which are related in sequence to the PBPs and the CSPs, which are in a different protein family are either associated with chemosensory structures or have been shown to bind small ligands.

Male moths of *Bombyx mori* are attracted to the female-produced sex pheromone, bombykol. The *B. mori* pheromone binding protein (BmPBP) is known to undergo a pH-dependent conformational change from a 'closed' to an 'open' conformation when the environment pH drops below 5.8. Experiments with model membranes have indicated that the pH drop near a cell membrane is sufficient to trigger this change, and it has been suggested that the PBP could switch to an open conformation as it approaches the neuron, thus releasing pheromone in the immediate vicinity of waiting olfactory receptor proteins (Wojtasek and Leal, 1999).

In *Drosophila melanogaster* an OBP named LUSH is essential for the recognition of a water-soluble semiochemical (ethanol) (Kim and Smith, 2001). On the basis of their binding abilities, OBPs have been considered to play a significant part in the remarkable selectivity of the insect olfactory system, but the literature is also dichotomous (Campanacci *et al.*, 2001) with data suggesting that some OBPs are specific, whereas others may be more 'sloppy'.

18.4 Classification of pheromones

Most pheromones are naturally occurring glandular secretions or body metabolites. Though pheromones have been classified on the basis of their chemical structures, but their classification on the basis of behavioural responses elicited by insects is of importance for behavioural studies, and these chemicals have been broadly classified as:

- Alarm pheromones
- Aggregation pheromones
- Epideictic pheromones
- Trail-marking pheromones
- Sex pheromones

18.4.1 Alarm pheromones

Alarm behaviour is provoked by the presence of a predator or a threat to the nest as in many social insects. The alarm behaviour of many eusocial insects is well-known when nest disturbance induces many bees, ants, termites, or wasps to an aggressive behaviour. Some characteristic alarm behaviours are: raised heads or antennae, running, movement to the source of the pheromone, biting and stinging, and firing repellent chemicals. Factors like pheromone concentration, length of exposure, and situation also play a crucial role in the reaction to alarm pheromones. The function of alarm pheromones is to warn colony members of an enemy, not to ward off intruders.

In social insects, alarm behaviour mostly occurs as the result of an external threat to a colony or as a result of a signal produced by nestmates to protect the colony and its brood (Landolt *et al.*, 1998). Alarm pheromones occur in all groups of social insects and are employed in or near the nest. However, in ants and termites these pheromones are also emitted outside of the nest when enemies attack workers that are following a food trail. The alarm pheromones may have evolved from defensive compounds, as in social Hymenoptera these are usually produced in defensive glands (Hölldobler and Wilson, 1990). Alarm pheromones have been shown to be highly volatile, and are normally produced in large amounts. These pheromones represent a wide range of chemical structures including terpenoids, alcohols, aldehydes, ketones, esters, nitrogen heterocycles, and sulphur containing compounds.

In ants alarm pheromones are secreted usually in the mandibular gland but have also been found in the Dufour's, pygidial, venom, and supra-anal glands. Alarm pheromones in ants have been proven to be the least species-specific of all pheromone classes; the same alarm pheromones can be found in a number of ant species. Despite the common use of alarm pheromones, the responses can differ drastically among ant species (Wilson, 1971). One of the simplest examples of alarm communication is that of *Acanthomyops* colonies. These are large, subterranean ants, and releases alarm pheromones from its mandibles as well as the Dufour's gland. They respond to alarm pheromones by the 'fight' reaction. The general response is raised antennae followed by opened mandibles and then running in the direction of the pheromone source. If more alarm pheromones are not released, the signal fades out within a few minutes (Wilson, 1971). The slave-making ant, *Formica pergandei* and *F. subintegra* release pheromones which contain decyl, dodecyl and tetradecyl acetates from Dufour's glands. These are also components of alarm pheromone of other formicine ants. When the slave-making ants enter a host nest, they release large amounts of pheromone, which causes communication disturbance among defenders. These pheromones have also been termed as 'propaganda pheromone' (Regnier and Wilson, 1971). The imported fire ant, *Solenopsis saevissima* is known to be aggressive and also has a complex alarm system. When in combat, the workers release two substances: one from head that acts as an alarm pheromone and another from the Dufour's gland that acts as a trail pheromone to orient other workers to the enemy (Wilson, 1971). In the ant, *Pheidole biconstricta* the pygidial gland secretes an alarm pheromone that increases running and aggressive behaviours. Worker ants of this species show varying degrees of alarm behaviour to pygidial gland secretions but their response are not so strong to mandibular, poison, or Dufour's glands secretions. Leaf cutting ants release an alarm pheromone from their mandibular glands and respond aggressively to them. *Crematogaster* ant species that live symbiotically with the acacia tree produce potent alarm pheromones from their mandibles.

In termites, defence reactions can be released by various signals from nestmates that include vibratory movements and head banging, or by chemical compounds, i.e., trail and alarm pheromones. The occurrence of alarm pheromones in termites was first demonstrated by Ernst (1959) in a *Nasutitermes* species. They are mostly produced in the frontal glands of soldiers and consist of volatile monoterpenes. Soldiers and workers have a complementary role in defence and respond differently during defence recruitment. Soldiers are first recruited by alarm pheromones that act as short-range attractants. Workers are recruited later by trail pheromones or soldier defensive secretions and thereafter participate in defence. Workers bite enemies as well as covering them with faecal material. These chemicals also act as toxicants and irritants to enemies, but their main function is to invoke alarm behaviour.

Honey bees, *Apis* species are well-known among beekeepers that once you have been stung by a honey bee, others will attack shortly after. The alarm pheromone is only found in guard workers and foragers do not respond to alarm pheromones. The alarm pheromone is located in the membranes surrounding the sting but the bee does not have to sting the enemy in order to release the pheromone. The first response after perceiving a potential enemy is for the bee to open its sting chamber and release the alarm pheromone to recruit other bees from the colony. If the intruder is a big enough threat, the bee will sting its enemy. Typically the barbed sting is detached from the bee along with the poison gland and Dufour's gland that are still attached to the sting. There have been many studies to determine factors that elicit defence in honey bees. The alarm pheromone is released from autotomized stings embedded into attacked animals, thereby creating a chemical signal which guides other guards to the enemy. Alarm pheromones of honey bees and stingless bees can mobilize masses of individuals and orchestrate probably the best insect defence on earth (Schmidt, 1998).

Pheromonal communication of alarm is widespread in the social Vespidae; however, it is not universal. Alarm pheromones have been demonstrated for a number of species of social Vespidae including several hornets and yellowjackets. The normal response to an enemy is a mass stinging attack. Maschwitz (1964) first demonstrated alarm pheromone responses in the yellowjackets, *Vespula vulgaris* and *V. germanica* in response to crushed wasps and body parts. Alarm behaviour in the *Vespa* species occurred in response to the squashed sting apparatus, sting sac, and solvent extract of the sting sac. Pheromone-mediated alarm has since been observed in many other vespines. Veith *et al.* (1984) identified 2-methyl-

3-butene-2-ol as a component of the alarm pheromone of *Vespa crabro*, and stimulated the hornet with squashed venom sacs or venom. The source of alarm pheromones in social wasps generally is the venom, although the head is implicated as an additional source of alarm pheromone in some *Vespa* species. It has been considered that alarm pheromone in venom is released when wasps spray venom or venom is deposited when wasps sting. An alarm pheromone originating in the head of workers may be released at the mouthparts and applied or evaporated from the mandibles. Reed and Landolt (2000) gave experimental evidence that an alarm pheromone is deposited on a substrate or target when attacked by southern yellowjackets, *Vespula squamosa*. They demonstrated that alarmed southern yellowjacket apply chemicals to targets that they attack and that such attacked targets may subsequently remain active in eliciting alarm and attack responses from yellowjackets. Alarm pheromone production is also reported in the other wasp species of the genera *Polybia* and *Polistes*.

Aphids are important pest insects, damaging plants by sucking nutrients from the plant tissues and by transmitting plant viruses. Predators like ladybirds, lacewings and hoverfly larvae have been shown to strongly influence the growth and persistence of aphid colonies. Under predator attack, aphids secrete droplets from the cornicles (a pair of tubular structures located on the posterior abdomen). Cornicle fluid was originally thought to serve a direct defensive function by gluing predator mouthparts together; however, this fluid is now known to contain an additional alarm pheromone. The main and sometimes only component of this pheromone is the sesquiterpene, (*E*)-β-farnesene (EBF) (Bowers *et al.*, 1972). EBF triggers various behavioural reactions in con- and heterospecific aphids such as increased alertness, withdrawal of the stylets from the plant, walking behaviour and dropping off the host plant. Though the aphids that emit EBF are mostly killed by predators, but other colony members may escape in time, and therefore, get benefit from the release of the alarm pheromone (Montgomery and Nault, 1977). The evolution of this altruistic alarm communication has only been possible because of the peculiarities of the life cycle of many aphid species, as they reproduce asexually and all aphids descending from a female that hatches are genetically identical. This clonal structure of aphid colonies makes the evolution of alarm pheromone an extreme case of kin selection, because most receivers are genetically identical to the signaller, and the escape benefit enhances the survival of same genotype (McAllister and Roitberg, 1987). Aphids exhibit polyphenism and may be either winged or wingless. The winged morphs are produced in response to adverse environmental conditions such as crowding or poor plant quality. These morphs disperse and colonize new hosts while the wingless morphs stay on the plant on which they are born. Dixon and Agarwala (1999) reported that the presence of natural enemies increase the proportion of winged morphs in pea aphids, *Acyrthosiphon pisum*. Kunert *et al.* (2005) showed that the exposure to alarm pheromone also induces aphids to give birth to winged dispersal morphs that leave their host plants. They demonstrated that the alarm pheromone is the only volatile compound emitted from aphid colonies under predator attack and its emission is proportional to predator activity. The artificial alarm pheromone induces groups of aphids but not single individuals to produce a higher proportion of winged morphs among their offspring. Furthermore, aphids react more strongly to the frequency of pheromone release than the amount of pheromone delivered. The EBF leads to a 'pseudo crowding' effect whereby alarm pheromone perception causes increased walking behaviour in aphids resulting in an increase in the number of physical contacts between individuals, similar to what happens when aphids are crowded. Both nymphs and adults of bed bug, *Cimex lectularius* discharge alarm pheromone from thoracic glands in response to irritation and alerted individuals show a response similar to aphids (Levinson *et al.*, 1974).

18.4.2 Aggregation pheromones

Many insect species aggregate for feeding, mating, hibernation, defence, overcoming the resistance of host, or in response to environmental conditions. For aggregation insects also use pheromones to communicate. A pheromone that releases behaviour in conspecifics leading to an increase in their density in the vicinity of pheromone source is termed an aggregation pheromone. These pheromones are distinct from sex pheromones which are produced by one sex and affect the other, and their function is more profound. Such pheromones are manly found in Coleoptera, especially the bark and ambrosia beetle. Besides insects of the orders: Orthoptera, Dictyoptera, Hemiptera, Diptera, Lepidoptera, Mecoptera, and Hymenoptera also release aggregation pheromones. Aggregation pheromones play a key role in mating and other behaviour. Many chemicals, including hydrocarbons, alcohols, aldehydes, ketones, esters, acids, anhydrides, amines, and nitriles have been identified as insect aggregation pheromones, and released by eggs, larvae and adults. Aggregation pheromones of the same species in different stages of the life cycle can be distinct, and the same pheromone may induce different behaviour in different developmental stages or physical states.

Aggregation in bark beetles may be induced by a complex of secondary attractants (host odours plus beetle produced compounds). These may include one or more host monoterpenes, one or more beetle produced compounds that induce aggregation, but may have differential effect on sexes, and one or more compounds which vary in effect from aggregation to anti-aggregation depending on concentration (Borden, 1985). Thus, a pheromone has an anti-aggregation effect at higher concentration, and it becomes a multifunctional pheromone at varying concentrations. Bark beetles of the genera *Ips, Dendroctonus, Scolytus, Tomicus, Pityogenes, Pityophthorus,* and *Orthotomicus* belong to an insect group that as adults bore into the host plant for the purpose of laying eggs. Bark beetle adults and larvae in northern temperate climates generally feed on phloem/cambium of conifers. In the most tropical zones, the majority of species feed on wood (xylem) and on phloem of broad-leave trees and shrubs. Semiochemicals from both the tree and the beetle have many functions during the life cycle of a bark beetle. Once an individual or pair begins to release an aggregation pheromone, the likelihood that the tree is colonized depends on: (i) the population level of beetles available for recruitment to the attack, and (ii) the resistance (health) of the tree and its ability to produce defensive resin. Successful colonization and reproduction by a bark beetle in a living tree requires release of enough aggregation pheromone to ensure the attraction of sufficient conspecifics to overwhelm the host tree defences, but after killing the tree and securing mates, pheromone should not be released any longer in order to avoid further competition for bark areas (Berryman *et al.*, 1985). Semiochemicals play a role in cooperation among beetles when killing the tree and in their avoidance of competition. Pitman *et al.* (1966) reported that male of Californian five-spined engraver, *Ips confusus* produce the pheromone after 2 hours of infestation and an increase in pheromone production occurs during next 96 hours followed by a decrease; and by day 14 the pheromone is hardly detectable. The bark and wood of ponderosa pine (the host) become attractive only 4-6 hours after males are introduced into entrance tunnels and maximum production of pheromone occurs during 18-144 hours. Attractancy is initiated by males feeding in phloem-cambial tissues of the host and is delayed until food is passed through the gut. As pheromone production is initiated only after feeding, either a precursor is ingested and metabolized or digestive metabolism initiates secretory activity in specialized cells (Wood *et al.*, 1966). The mountain pine beetle, *Dendroctonus ponderosae* after dispersing, searches for either by cues emitted from the host or by following the scent trail of the aggregation pheromone emitted by another member of its species that has already found a host. It has been theorized that the beetles need to aggregate in sufficient numbers to overcome the host resistance, to utilize the host optimally, and establish themselves successfully so that they can rear their broods (Borden, 1982). However, they do not overpopulate the resource by emitting an anti-aggregation pheromone once the appropriate numbers of beetles have established in the tree.

In Colorado potato beetle, *Leptinotarsa decemlineata* individuals of both sexes produce pheromones. Dickens *et al.* (2002) identified a male-produced aggregation pheromone for the beetle. The male beetles produced only minor amounts of the pheromone; however, its production could be enhanced by topical application of JH III and antennectomy. The pheromone was identified as (*S*)-3,7-dimethyl-2-oxo-oct-6-ene-1,3-diol [(*S*)-CPB I], a unique structure for an insect pheromone. Antennal receptors of both sexes responded selectively to the (*S*)-enantiomer. As both JH III and antennectomy increase levels of the pheromone, it indicates the existence of a feedback system involving antennal input, and this system may be under hormonal control. This pheromone differs from the paradigm of a female-produced pheromone for this beetle.

Males of confused flour beetle, *Tribolium confusum* produce an aggregation pheromone while females produce a sex pheromone. The adults of *T. confusum* aggregate at low densities; at intermediate densities they distribute randomly; and at high densities, they distribute uniformly. The uniformity in spacing is achieved by secretion of quinones from thoracic and abdominal glands in the beetles. The chemicals also act as repellents above a certain concentration (Naylor, 1959). Similarly in cotton boll weevil, *Anthonomus grandis* aggregation pheromones are produced by one sex and sex pheromones by another. Sap beetles, *Carpophilus* species are cosmopolitan pests which occur on a wide variety of hosts, including fruits and grains, both before and after harvest. These beetles are known to respond to male-produced aggregation pheromones as well as host-related volatiles. Walgenbach *et al.* (1983) showed evidence for the existence of male-produced aggregation pheromones in three stored product weevils, *Sitophilus zeamais*, *S. granarius* and *S. oryzae*. Interspecific attractions between weevil species suggest that their pheromones are closely related.

Many springtails aggregate in moist places to get protection against desiccation and their aggregation is induced by pheromones. In large milkweed bug, *Oncopeltus fasciatus* pheromone-mediated aggregation assists in: (i) mutual benefit from aposematic colouration, (ii) protection of newly eclosed adults, and (iii) cumulative effect of lytic salivary enzymes on the host (Aller and Caldwell, 1979).

Ishii (1970) reported that abdominal extracts of nymphs of German cockroach, *Blattella germanica* caused strong aggregation and the aggregation pheromone was also found in the faeces. He found that the aggregation pheromones do not appear to be species-specific and the nymphs of several correlated species aggregated in response to each others' pheromones. Aggregation pheromones in part serve a function of sex pheromone, if the insects of both sexes tend to aggregate for the purpose of mating. Conversely, sex pheromone may also act as aggregation pheromone. Female of American cockroach, *Periplaneta americana* produce a sex pheromone that attract not only male, but also nymphs to a protected habitat.

In the desert locust, *Schistocerca gregaria* a number of behaviourally active aggregation pheromone components have been identified from adult male, and from third, fourth and fifth instar nymphs. The aggregation pheromone system of gregarious *S. gregaria* is derived from the insects themselves and their faeces. It has been demonstrated that two distinct sets of releaser pheromones modulate the aggregation behaviour of the desert locust, a juvenile aggregation pheromone specific to the nymphs and an adult aggregation pheromone specific to the adult stages. Young adults (immature adults) of either sex did not produce a stimulus with significant activity. Of the older adults (sexually mature adults), only the males produced the aggregation pheromone to which both sexes were equally responsive. Mature solitary desert locust adults were also responsive to the components of the adult pheromone system. Faecal volatiles are part of the aggregation pheromone complex. The nymphs aggregate in response to faecal volatiles of nymphs and young adults but were indifferent to volatiles emitted from adult faeces. The nymphal pheromone system consists of three sets of compounds: aliphatic C_6C_8-C_{10} aldehydes and their corresponding acids produced by the insects themselves, augmented by faecal phenols and guaiacol. The adult aggregation pheromone system consists of 2-phenylacetonitrile, guaiacol, phenol and benzaldehyde (Obeng-Ofori *et al.*, 1998).

18.4.3 Epideictic pheromones

The pheromones stimulating aggregation or attraction are common in insects; however, studies demonstrate that some insects also release pheromones which influence their population density, and elicit their dispersal or spacing. Such chemicals are termed 'epideictic pheromones'. Thus, epideictic or spacing pheromones are those which stimulate dispersal from a resource once an optimal density has been reached, and limit intraspecific competition. Both immature and adult insects secrete such olfactory and tactile chemicals. Prokopy (1981) reviewed and gave many examples for the occurrence of epideictic pheromones in phytophagous insects.

Many phytophagous or parasitoid insects produced active compounds or sets of compounds termed as 'marking pheromones' (MPs). Evidence for the use of actively-produced MPs has been found in five holometabolous insect orders including Coleoptera, Diptera, Hymenoptera, Lepidoptera, and Neuroptera (Nufio and Papaj, 2001). MPs are distributed among at least 20 families of phytophagous insects (Prokopy, 1981; Roitberg and Prokopy, 1987). In hymenopteran parasitoids, evidence of an ability to discriminate between parasitized and unparasitized hosts has been gathered for 150 to 200 species and in nearly every family (Lenteren, 1981). Oviposition behaviour in phytophagous and parasitoid insects is often modified by the presence of conspecific brood (eggs and larvae). Often, females avoid laying eggs on or in hosts bearing brood, a behaviour that acts to reduce the level of competition suffered by their offspring. Avoidance of occupied hosts is typically mediated by cues and/or signals associated with brood.

Host marking behaviour is a conspicuous and well-studied aspect of the oviposition behaviour of many tephritids, particularly frugivorous species. In tephritids, host marking behaviour generally involves dragging the aculeus about the fruit or other part, resulting in deposition of a substance termed a host marking pheromone (HMP). HMPs exert numerous and complex effects on behaviour in female and male flies, including suppression of oviposition attempts, interruption of oviposition once initiated, stimulation of emigration from areas of high infestation, reduction of clutch size and number of clutches laid per visit, and arrestment of males on fruits in cases where the mating system is resource based. Marking of an oviposition site has been reported in many frugivorous species of the genera *Anastrepha, Ceratitis, Rhagoletis* and *Paraceratitiella*. Prokopy (1972a) reported that polyphagous female of apple maggot fly, *Rhagoletis pomonella* oviposits in fruits of many hosts. A single fruit can sustain only a limited number of larvae and in case of hawthorn, only one larva can develop to maturity in one fruit. After ovipositing the female deposit a pheromone on the host surface which deters other females who may subsequently deposit their eggs. The oviposition deterring pheromone (ODP) is secreted from posterior half of the midgut. Females who repeatedly contact oviposition deterring pheromone show erratic flight patterns and may also leave the tree to find fresh oviposition sites. Similar oviposition deterring pheromones have also been described in other fruit flies, viz. European cherry fruit fly, *R. cerasi*; Mediterranean fruit fly, *Ceratitis capitata*; South American fruit fly, *Anastrepha fraterculus*; and West Indian fruit fly, *A. obliqua*.

In large white cabbage butterfly, *Pieris brassicae* it was long believed that an epideictic pheromone was employed by the females on the eggs and leaf surface when they oviposit batches of eggs on cabbage and other hosts (Rothschild and Schoonhoven, 1977). The pheromone deters oviposition of further eggs for about 2 days. Besides, the presence of feeding larvae also deters oviposition and involves another pheromone. Recent work has shown that the host plant itself responds to the presence of *P. brassicae* eggs and that systemic changes in leaf surface chemistry may be the primary source of an oviposition deterring effect (Blaakmeer *et al.*, 1994). Pheromones released by the mandibular glands of larvae of the Mediterranean flour moth, *Ephestia kuehniella* deter oviposition by adult females as long as the quantity present at the site is relatively large (Corbet, 1973). In corn earworm, *Helicoverpa zea* the frass left by feeding larvae and scales left by the female during oviposition deter further oviposition (Prokopy, 1981).

In most phytophagous insects host marking is usually obligatory; however, deployment of MP is occasionally facultative. Presumably, facultative marking reduces the overall costs of marking by deploying a signal only when it is beneficial to do so. The females of a *Delia* sp. (Fam. Anthomyiidae) mark one of their host species but not the other. On developing flower bud of *Polemonium foliosissimum*, female deposits a single egg and an associated MP (Zimmerman, 1979). The same *Delia* sp. also utilizes *Ipomopsis aggregata* as a larval resource but, in contrast to its first host, females often lay more than one egg on this host, and do not deposit MP on this second host. Zimmerman (1980) suggested that females fail to mark *I. aggregata* plants not because these hosts support relatively more offspring to maturity, but because egg mortality on *I. aggregata* is so severe that females experience little fitness losses if consecutive females reuse an infested host. In *P. foliossimum*, by contrast, egg mortality is low and female reuse leads to relatively intense larval competition between clutches.

In cases where more than one larva can develop in a given resource or where larvae are mobile enough to find new hosts, MPs may allow females to assess not just presence versus absence of brood but also the amount of brood. In such cases, females may make a graded assessment of the level to which a host has been previously utilized. Females of the cowpea seed beetle, *Callosobruchus maculatus* use chemical and tactile cues associated with eggs to assess not only whether or not a host bean has been utilized but also the number of eggs associated with that host. Females have been found to selectively reuse hosts that bear a lower than average number of eggs (Wilson, 1988).

Scolytid beetles of the genera *Ips*, *Dendroctonus* and *Scolytus* aggregate on their host trees for feeding and breeding. They release aggregation pheromones to maintain optimal densities of attack. However, the distribution of attacks over a host is often very regular and highly specific. This indicates that some spacing mechanism is involved in their host colonization. Female of Douglas fir beetle, *Dendroctonus pseudotsugae* produces pheromones frontalin and sendenol, which stimulate aggregation and mass arrival at a host tree. Once the male arrives and stridulates outside a female's entrance hole, both sexes release 3,2-methy-cyclohexenone (3,2-MCH) and the male also releases 3,3-MCH, both of which reduce further arrivals (St. Clair *et al.*, 1977).

Some social insects maintain definite territories for foraging and intruders in theses territories are attacked. Many social colonies have colony-specific odours which appear to delineate the social space of a given colony. Hölldobler and Wilson (1990) suggested that in weaver ant, *Oecophylla longinoda* even distribution of colonies appears to be maintained by pheromone, laid down by foraging workers.

Density dependent spacing chemicals have also been determined in grain aphid, *Rhopalosiphum padi*, and at high densities these arrest further colonization. The active spacing chemicals were identified as: MHO (6-methyl-5-hepten-2-one), (-) and (+) MHOH (6-methyl-5-hepten-2-ol), and 2-T (2-tridecanone). It remains unclear whether the active volatiles are released by aphids or by the leaves, but it is likely that the MHO and MHOH are of aphid origin and the 2-T is released by the plant (Quiroz *et al.*, 1997). MHO and MHOH act as semiochemicals in various insects, and 2-T is a known insect repellent, found naturally both in insects and plants.

In entomophagous Hymenoptera, females may use internal or external or both marks to indicate which hosts have been previously exploited. Whether MPs are deposited on the inside or the outside of the host depends on the life-stage attacked. Egg parasitoids tend to mark hosts externally while parasitoids utilizing other host stages tend to mark hosts internally.

18.4.4 Trail-marking pheromones

The trail-marking pheromones are used as navigational aids in directing other members of the colony to a distant location, which vary from a few metres to hundreds of metres, to find mates or to maintain social integrity during migration of colonies. In most cases, the trails are laid by conspecific foraging workers as they return home from a food source. These trails are used by other foragers to find food sources. Terrestrial trails are laid in continuous or intermittent lines on solid substrates. In contrast, in aerial trails pheromones plumes are released in water or air from a fixed point and the signals are generated in the medium.

The chemical foraging trails of termites and ants are the most widely studied and recognized foraging signals used by eusocial insects. These trails are composed of pheromones, which signal to other individuals through smell or taste. Increasingly complex communication in foraging is correlated with colony size. Ants deposit a certain amount of pheromone while walking, and each ant prefers to follow a direction rich in pheromone (Fig. 18.5). This enables the ant colony to quickly find the shortest route. In ants glandular sources of trail-marking pheromones are more diverse, including: Dufours's gland, poison gland, pygidial gland, rectal gland, Pavan's gland, tibial glands and dorsal glands. In termites the pheromone originates in sternal gland and the trail is marked by dragging or periodically touching the sternum of the abdomen to the surface. Ants following an odour trail walk in a zigzag fashion in which each antenna appears to be moved alternately in and out of the odour field. It has been presumed that if both the antennae would remain continuously in the odour field, habituation might quickly occur. By passing in and out of the threshold concentration level, maximal sensitivity to the stimulus would persist. In addition, such a system imparts a clear orientation axis to the single path. This also indicates that intact insect probably make simultaneous comparison between the stimuli perceived by each antenna. The type of orientation is chemotropotaxis (Haynes and Birch, 1985). Evidence for this mechanism is supplied by experiments with ant species that show that if one antenna is amputated, the ant repeatedly overcorrects to the other side, and if the antennae are glued into a crossed position the ant has difficulty following the trail. In order for such a system to work, there must be a steep concentration gradient at the trail boundary and such a steep gradient will only be obtained by a rapidly diffusing substance. The excitatory component(s) of the trail pheromone in most ants is quite volatile; however, several termite species of *Nasutitermes* release a highly stable, nonvolatile pheromone.

Fig. 18.5. Scent trail laid by a forging worker ant.

Wilson (1959) investigation into the trailing responses of red imported fire ants, *Solenopsis invicta* lead to the discovery that Dufour's gland secretions function both as releaser and orienter in trail following. When a foraging worker discovers a novel food source it needs help in securing a food source, and it lays a chemical trail down in the direction of the home nest. The chemical trail markers are laid on the ground through the sting. Ant pheromone trail consists of a series of smears made by periodic extension, contact and withdrawal of the sting with the ground substrate. Food quality is conveyed by the strength of the chemical trail and the strength of a trail varies depending upon the level of trail reinforcement and recruitment. Wilson (1962) showed that the imported fire ant, *Solenopsis saevissima* secretes a trail pheromone from the Dufour's gland. The secretion is exuded through the sting and the trail is highly species-specific. When a worker finds a food source, she returns to the nest laying a trail of highly volatile pheromone. Wilson showed that volatility is essential for the simple and elegant method of controlling recruitment. The recruitment in fire ants also provides a good example of insect 'mass communication' in which a group of individuals share the information.

In red imported fire ant, *S. invicta* a trail formed of Dufour's gland secretion acts as guide to workers already foraging, but does not assist in recruitment. The pheromone compound is (*Z,E*)-α-farnesene. Two other compounds, a sesquiterpene and an orientation inducer pheromone are also secreted, of these the former acts as an attractant and the latter stimulates trail-following. A forager returning to the nest releases the pheromone in high concentration and stimulates some of the reserve foragers to follow her return trail, or to join in the traffic along an existing trail (Meer *et al.*, 1989). Reinhard and Kaib (2001) investigated trail-laying and trail-following behaviours of French subterranean termite, *Reticulitermes santonensis*. During foraging for food, termites walk slowly (on average, 2.3 mm/s) and lay a dotted trail by dabbing the abdomen at intervals on the ground. When food is discovered they return at a quick pace (on average, 8.9 mm/s) to the nest, laying a trail for recruiting nestmates to the food source. While laying this recruitment trail, the workers drag the abdomen continuously on the ground.

Meliponine stingless bees exploit food resources with strategies ranging from solitary foraging to the mass recruitment of nestmates. These bees employ a matching diversity of recruitment communication strategies, ranging from odour communication to the potential communication of resource location through sounds produced inside the nest. The aggressive defence and attack of good food sources is a key element in their foraging competition. Several species of stingless bees readily attack and displace conspecifics and interspecifics on rich food sources. A forager of *Trigona postica* that finds food makes several direct flights between the nest and food source, and deposits her mandibular gland secretion on vegetation, stones, or earth at a distance of 2-3 m. She alerts her nestmates with an acoustic signal and flies back in front of the alerted bees to the food source (Linaduer and Kerr, 1960). Similarly, workers of *Trigona subterranea* also produce a trail-following pheromone from mandibular glands. At high concentration the same pheromone releases alarm and attack behaviour in *T. subterranea*. Nieh *et al.* (2003a) showed that *Trigona hyalinata*, an aggressive stingless bee, uses a pulsed mass-recruitment system that is highly focussed in time and space. Foragers produced an ephemeral, polarized, odour trail consisting of mandibular gland secretions. Surprisingly, the odour trail extended only a short distance away from the food source, instead of providing a complete trail between the nest and the food source (as has been described for other stingless bees). This abbreviated trail may represent an intermediate strategy between full trail-marking, found in some stingless bees, and odour marking of the food alone, which is found in stingless bees and honey bees.

Larvae of Madrone butterfly, *Eucheira socialis* secrete a trail pheromone from the ventral surface of the posterior tip of the abdomen. They mark trails by bringing the secretory site into brief contact with the substrate during a locomotive cycle. Most marking by foragers is carried out when they move to new branches and little when they move over established trails or return to the communal shelter after feeding. Caterpillars make careful comparisons of alternative pathways at choice points and select newer and stronger trails over older and weaker trails (Fitzgerald and Underwood, 1998). Some other insects also lay down terrestrial trails. The eastern tent caterpillar, *Malacosoma americanum* form chemical trail, which stimulates recruitment of foraging caterpillars, in a way similar to recruitment in foraging Hymenoptera. American cockroach, *Periplaneta americana* lays down a series of spots trail pheromone to which both adults and immature respond apparently.

In contrast to the terrestrial trail pheromones aerial trail pheromones are generally released by stationary insects and are then dependent on air movements to generate the trail. These pheromones are known to exist in a number of insects of the orders: Coleoptera, Lepidoptera, Hemiptera, Mecoptera, Diptera, Hymenoptera, and Orthoptera. Bossert and Wilson (1963) suggested that the effective chemical communication for such chemicals is influenced by the wind velocity, pheromone emission rate, and the behavioural threshold of responding insect. All the components of an aerial trail are not necessarily perceived simultaneously, and there are active regions where the perception of an additional component might alter the behaviour of a responding insect. Vorhis-Key and Baker (1982) documented that the argentine ant, *Iridomyrmex humilis* can orient upwind to a point source of its terrestrial trail pheromone, (Z)-9-hexadecenal. In addition, the ants also orient upwind in a uniformly permeated pheromone field. It appears that the ants were using anemotaxis or wind-steering to orient in these pheromone fields.

18.4.5 Sex pheromones

Sex pheromones are most familiar and dramatic among insects' pheromones, since these occur commonly and the behavioural patterns they stimulate are easy to observe. These pheromones bring the individuals of both sexes together, thereby increasing the chances of successful mating (Fig. 18.6). Studies on these compounds and the behaviours they mediate have indicated that these pheromones generally consist of more than one chemical component. Sex pheromones are mainly released by insects of the order Lepidoptera. Besides, insects of the orders: Coleoptera, Orthoptera, Hemiptera, Diptera, Dictyoptera, and Hymenoptera also release such chemicals. The sex pheromones are mainly released by female insects; however, in some cases males or individual of both sexes release these pheromones. Individuals of opposite sex are usually attracted after receiving the chemical stimuli, but in some beetles, both sexes may be attracted by the same scent at some appropriate location where they aggregate and mate. In most cases the pheromone releasing insects exhibit certain stereotyped postures termed 'calling postures'. The sex pheromone may attract an insect of opposite sex in low concentrations and the same chemical elicits courtship behaviour at the higher concentrations encountered at close range. In some insects separate substances are involved in attraction and courtship. Birch (1974b) used the term 'aphrodisiacs' for the substances produced by either sex (usually by male and often as only part of a complex pattern of courtship behaviour) that facilitates courtship or prepare the opposite sex for copulation after the pair have been brought together. Often, attractants are produced by one sex and aphrodisiacs by the other.

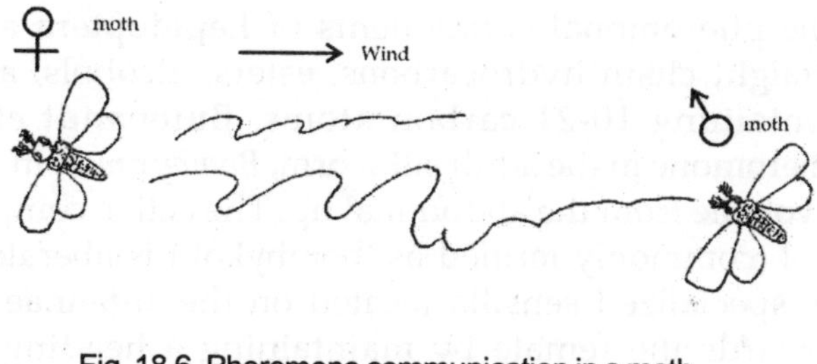

Fig. 18.6. Pheromone communication in a moth.

The sex pheromone of the German cockroach, *Blattella germanica* is a contact pheromone produced by females and perceived by males through antennal contact. The pheromone is made of four components, each of which produces the complete programme of mating behaviour in males. The pheromone components include: 3,11-dimethylnonacosan-2-one, 20-hydroxy-3,11-dimethylnonacosan-2-one, 29-oxo-3,11-dimethylnonacosan-2-one, and 3,11-dimethylheptacosan-2-one (Schal *et al.*, 1991).

Pettersson (1970) first published an evidence of sex pheromone in the aphid *Schizaphis arrhenatheri*. He concluded that the active substance was emitted by special organs, so called pseudorhinaria (consisting of pheromone producing glandular epidermis cells covered by scent plaques), on the hind tibiae of oviparous females and detected by males. Males were clearly attracted to females and the activity of the pheromone was found to fade rapidly. Dawson *et al.* (1987) identified sex pheromone of vetch aphid, *Megoura viciae* and it consisted of two synergistic monoterpenoid components: (-)-(1R,4aS,7S,7aR)-nepetalactol and (+)-(4aS,7S,7aR)-nepetalactone. The same compounds, in varying ratios, are in fact released by sexual females of a number of other aphids, such as the blackberry aphid, *Sitobion fragariae*; grain aphid, *Sitobion avenae*; black bean aphid, *Aphis fabae*; pea aphid, *Acyrthosiphon pisum*; cabbage aphid, *Myzus persicae*; and *Cryptomyzus* species while the nepetalactol alone is released by the spring green aphid, *Schizaphis graminum* and grain aphid, *Rhopalosiphum padi*.

The pheromonal components of Lepidoptera appear generally to be aliphatic straight chain hydrocarbons, esters, alcohols, acetates, aldehydes and ketones containing 10-21 carbon atoms. Butenandt *et al.* (1959) identified first sex pheromone in the adult silkworm, *Bombyx mori* in which the virgin female produces a volatile from the abdominal tip. The active compound (*E*)10-(*Z*)12-hexadecadien-1-ol (commonly termed as 'bombykol') is liberated into the wind and is detected by specialized sensilla located on the antennae of male. The male propel itself towards the female by maintaining a heading into the air stream because of pheromone-laden air stream. The antennal receptor is innervated by two nerves; one is sensitive to bombykol and the other to the minor component of the pheromone. In this case, the message is about receptivity for mating which is transmitted as a simple coded signal.

The female of oriental fruit moth, *Grapholita molesta* dispenses a pheromone mixture which contains: the *Z* and *E* isomers of 8-dodecenyl acetate, (*Z*)-8-dodecenyl alcohol and dodecyl alcohol. The scent is carried in air currents and sexually mature males orient upwind (Cardé *et al.*, 1975).

The virgin female of the tobacco hornworm, *Manduca sexta* releases a blend of sex pheromone from a gland near the tip of her abdomen. Twelve pheromone-like compounds have been identified in solvent rinses of the female sex pheromone gland. Behavioural observations in the wind tunnel and in the field revealed that eight of them, namely hexadecanal (16:Ald), three isomeric hexadecenals (Z9, E11 and Z11), two isomeric hexadecadienals (EE and EZ) and two isomeric hexadecatrienals (EEE and EEZ), play a role in the attraction of males to conspecific females (Tumlinson *et al.*, 1989). The pheromone is carried away by the gentle evening breezes making an invisible plume of pheromone molecules. When a male moth encounters a plume of pheromone his behaviour changes in a few very characteristic ways: (i) he turns into the wind and begins flying slowly upwind (the wind direction is critical information for any insect tracking an odour plume since moving into the wind while smelling the odour should usually take the insect in the direction of the odour source), (ii) he begins to execute a continuous series of regular turns back-and-forth across the wind. It has been suggested that upon encountering pheromone a timing circuit in the central nervous system of the moth is activated, which serves to initiate the turns, and (iii) while flying the male maintains his altitude near that of the pheromone plume.

Males of many fruit flies (Fam. Tephritidae) are known to release chemical signals to attract females and influence mating success. The virgin females are more attracted by sex pheromones than mated ones. A considerable number of chemical components may function as pheromone of male Mediterranean fruit fly, *Ceratitis capitata*. Baker *et al.* (1985) identified nine volatile components emitted and/or extracted from sexually mature male *C. capitata* which include: ethyl-(*E*)-3-octenoate, genranyl acetate, (3*E*,6*E*)-1,3,6,10-dodecatetraene (*E*,*E*-α-farnesene), 3,4-dihydro-2H-pyrole (delta^1-pyroline), *E*-2-hexenoic acid, dihydro-3-

methylfuran-2(3H)-one, 2-ethyl-3,5-dimethylpyrazine, linalool, and ethyl acetate. Jang *et al.* (1989) detected 69 compounds using gas chromatographic analysis from sexually mature laboratory reared males of *C. capitata* and identified 56 of these compounds. The sexually active calling male rest immobile, and everts the ballon-like rectal epithelial sac, thus presumably releasing pheromone. Receptive females respond to male odour by flying upwind, zigzagging back and forth within the plume of pheromone in wind tunnel tests. The females in the field are originally attracted from a long-range to males in a lek by chemical stimuli. Sex pheromones produced by males that are attractive to females have also been reported in oriental fruit fly, *Bactrocera dorsalis;* melon fly, *B. cucurbitae* and Queensland fruit fly, *B. tryoni.*

The sex pheromone of the house fly, *Musca domestica* has four components namely, (Z)-9-tricosene, *cis*-9,10-eposytricosane, (Z)-14-tricosen-10-one, and methylalkanes. These components are produced by females and influence male courtship behaviour. The synthesis of these pheromone components is related to the age of females. Newly emerged females do not have detectable amounts of pheromone, and sexually mature females have large amounts. Pheromone biosynthesis was related to ovarian development and pheromone synthesis did not begin until the ovaries reached stage 4, the early vitellogenic stage.

18.5 Honey bee pheromones

In eusocial insects many categories of pheromones have been described which regulate different behaviours. The entire social coordination of insect colonies is largely mediated by pheromones. Several behaviours of social insects are coordinated by simultaneous or sequential release of pheromones. The coordination of behaviours by pheromones is made possible by great diversity of exocrine glands which exist in social insects. In many social insects a pheromone or pheromone blend plays a diverse role in a variety of social behaviours and this is termed 'pheromone parsimony' (Blum, 1974). The honey bee queen and its role its role in sexual and social interaction is a most astonishing example of pheromonal parsimony.

Honey bees have one of the most complex pheromonal communication systems, possessing 15 known glands that produce an array of compounds (Free, 1987). In *Apis mellifera* both releaser and primer pheromones are secreted by individuals of all castes. Alarm pheromone is released by the Koschevnikov's gland, near the sting shaft, and consists of more than 40 chemical compounds, including isopentyl acetate (IPA), butyl acetate, 1-hexanol, 1-butanol, 1-octanol, hexyl acetate, octyl acetate, n-pentyl acetate and 2-nonanol. Alarm pheromone is released by worker bees to alert other bees of danger and is released when a bee stings another animal. This pheromone attracts other bees to the location and causes the other bees to behave defensively, i.e. sting or charge.

Both larvae and pupae emit a 'brood recognition' pheromone. This inhibits the ovarian development in workers and helps nurse bees to distinguish worker larvae from drone larvae and pupae. Honey bee larvae produce a ten-component blend of fatty-acid esters called brood pheromone. Drones produce a pheromone that attracts other flying drones to promote drone aggregations at sites suitable for mating with virgin queens.

The Dufour's gland secretion is a complex of chemicals and it allows worker bees to distinguish between eggs laid by the queen (which are attractive), and those laid by workers.

Nasanov gland is located on upper surface of abdomen, between the last and next-to-last segments. When a bee raises its abdomen and flexes the last segment down a fragrance exudes from a pouch opened by the motion. Nasanov pheromone is released by workers to orient returning forager bees back to the colony. Bees use this pheromone to find the entrance to their colony or hive, and they release them on flowers so other bees know which flowers have nectar. The scent gland or Nasanov (Nassanoff's) gland secretion of worker honey bee consists of: nerol, (*E,E*)-farnesol, (*Z*)- and (*E*)-citral, geraniol, nerolic acid and geranic acid (Pickett *et al.*, 1980).

The queen honey bee is known to regulate a number of behaviours of colony workers and remains constantly surrounded by workers. In case the queen dies, the workers construct queen cells and the ovaries of adult workers begin to enlarge. A multicomponent pheromone termed as 'queen substance' is secreted from the mandibular glands of queen bee, coats her body and is transferred to workers by trophallaxis. The honey bee queen produces pheromones that function in both releaser and primer roles such as attracting a retinue of workers around her, attracting drones on mating flights, preventing workers from reproducing at the individual (worker egg-laying) and colony (swarming) level, and regulating several other aspects of colony functioning. Queen mandibular pheromone (QMP) is emitted by the queen bee and affect social behaviour, maintenance of the hive, swarming, mating behaviour, and inhibition of ovary development in workers. Slessor *et al.* (1988) reported the main compounds of queen substance as: 9-oxo-(*E*)-2-decenoic acid (9-ODA) and 9-hydroxy-(*E*)-2-decenoic acid (9-HDA). It has also been reported that both enantiomers, the (+) and (-) of 9-HDA are present along with two other substances namely, methyl *p*-hydroxybenzoate (HOB), and 4-hydroxy-3-methoxyphenylethanol (HVA). These compounds elicit 'retinue behaviour', in which workers cluster around the queen, making frequent contact with her. The queen substance is distributed among the workers and a licker bee plays a role in the distribution of pheromone. The substance checks the ovarian development in workers and makes them aware about the presence of queen in the hive.

The Queen retinue pheromone (QRP) is formed of 5 QMP compounds and the 4 other compounds listed below. All these compounds are important for the retinue attraction of worker bees around their queen (Keeling *et al.*, 2003).

- Methyl (Z)-octadec-9-enoate (methyl oleate)
- (*E*)-3-(4-hydroxy-3-methoxyphenyl)-prop-2-en-1-ol (coniferyl alcohol)
- Hexadecan-1-ol
- (Z9,Z12,Z15)-octadeca-9,12,15-trienoic acid (linolenic acid)

Another key aspect of colony social organization in honey bees is the division of labour among workers. Adult worker honey bees perform tasks within the nest such as brood care 'nursing' during the first 2-3 weeks of life and then switch to foraging and colony defence. But division of labour in honey bee colonies is not rigid, because bees are sensitive to changes in their social environment, especially colony age structure. One response to changes in colony age structure is a change in the typical pattern of behavioural maturation. For example, in colonies lacking older bees (foragers), some bees initiate foraging even at young stage. Huang and Robinson (1992) suggested that the age at onset of foraging in honey bee colonies is regulated by worker-worker interactions. Old bees inhibit the maturation of younger bees and when a portion of a colony's foragers is removed, young bees develop faster. In addition, two other honey bee primer pheromones, queen mandibular pheromone (QMP) and brood pheromone (BP), also play a role in regulating workers' behavioural maturation.

Ethyl oleate (EO) is a 'forager pheromone' released by older forager bees to slow the maturing of nurse bees (Leoncini *et al.*, 2004). This pheromone acts as a chemical inhibitory factor, delaying age at onset of foraging. EO is present in highest concentrations in the honey crop, suggesting that it is transmitted via trophallaxis. These results suggest that worker's behavioural maturation is modulated via trophallaxis, a form of food exchange that also serves as a prominent communication channel in insect societies.

The queen releases a 'footprint pheromone' from tarsal glands that is deposited on the comb as she walks across it. This inhibits queen cell construction. Besides, the bees also secrete rectal gland pheromone, tarsal (Arnhart) gland pheromone, tergite gland pheromone, wax gland and comb pheromone, and laying worker pheromone.

Pheromones analogous to honey bee queen substance have been reported in ants, *Solenopsis invicta, Myrmica rubra*; halictine bee, *Halictus duplex*; and oriental hornet, *Vespa orientalis*.

18.6 Interspecific pheromonal communication

Interspecific pheromonal communication in insects may be circumstantial or in some species it has functionally evolved. Usually parasites, predators, commensals and competitors show their response to pheromones of other species. Many insects exploit host odours in searching food or oviposition sites. Exploitation of chemical signals by parasites and predators is quite common. Hosts' pheromone form a component of communication system and exploitation of such signals may confer a selective advantage to natural enemies. Such chemicals are termed kairomones. Some predators have broken the chemical code of social insects while in some other cases it appears that the predator has managed to insert its own chemical code into that of social insect nest to invade.

The alarm pheromone (decyl acetate plus dodecyl acetate) of western flower thrips, *Frankliniella occidentalis* is present in the anal fluid produced by 2nd-instar nymphs, and is used in defence against arthropod predators. However, the predators like minute pirate bug, *Orius tristicolor* and phytoseiid mite, *Neoseiulus cucumeris* use the pheromone as a prey-finding kairomone (Teerling *et al.*, 1993).

Epideictic pheromones of larvae of Mediterranean flour moth, *Ephestia kuehniella* can be exploited by ichneumon wasp, *Venturia canescens*. Braconid parasitoid, *Opius lectus* locates its host, the maggot of *Rhagoletis pomonella* by responding to the pheromone that the female deposits on the fruit following oviposition.

The female parasitoids learn to associate the pheromonal clues in finding the hosts. Female of *Bracon mellitor*, a parasitoid of larvae of cotton boll weevil, *Anthonomus grandis* probe the frass with her ovipositor. The braconid parasitoid, *Orgilus lepidus* is attracted by the frass of potato tuber moth, *Phthorimaea operculella*. Tricosane and other hydrocarbons associated with eggs of corn earworm, *Helicoverpa zea* cause host-seeking by the egg parasitoid, *Trichogramma evanescens* and other species.

Aggregation pheromones of bark beetles are used by their predators to locate the hosts. *Telenomus remus*, a scelionid egg-parasitoid respond to sex pheromones of fall armyworm, *Spodoptera frugiperda*. The pheromones of social insects like ants and termites are commonly exploited by their symbiotic guests. Many myrmecophile ants are able to follow trail pheromones laid down by host army ants while some ants and their guests called myrmecophiles live in ants' nests, where they are groomed and cared. The slave-making ant, *Formica pergandei* and *F. subintegra* release propaganda pheromone which causes communication disturbance among defenders (section 18.4.1). Some beetles are also able to follow the scent trail laid by ants. Hölldobler (1971a) discussed a well-known example of interspecific communication between ants and European staphylinid beetle, *Atemeles pubicollis* (see section 11.5). As cited by Hölldobler the guests beetles have broken the ants' code, that is, attained the ability to 'speak' ants' language, which involves a diversity of contact, visual and chemical cues.

The staphylinid termitophile, *Trichopsenius frosti* releases cuticular hydrocarbons identical to those of its host, eastern subterranean termite, *Reticulitermes flavipes*. Howard *et al.* (1980) suggested this as a form of 'chemical mimicry' which allows the beetles to be accepted by termites in their nest.

Some insects release pheromones which assist in mutualistic communication. *Ips pini* and *I. paraconfusus* have overlapping distributional range and infest common host, ponderosa pine. Both beetle species secrete pheromones which attract conspecifics; however, each species produces a chemical that interrupts the attraction of the other. This assists in reproductive isolation and avoidance of interspecific competition (Birch and Wood, 1975). Many lepidopterans interrupt the pheromonal blend released by other closely related species. Presence of female Indian meal moths, *Plodia interpunctella* inhibit the response to male almond moths, *Cadra cautella.*

In the well-known mutualistic relation between aphids and ants, the aphids gain 'protection service' provided by ants. The tending ants respond to aphid alarm pheromones by raising their antennae.

Chapter 19

Behavioural Genetics

A number of behavioural adaptations have contributed significantly for the dominance of insects over the globe. These traits of insects are programmed genetically to arise stereotypically. The behavioural traits also add another dimension, which leads to evolution of structural features to overcome the environmental constraints. However, many behavioural traits are modified environmentally and physiologically, e.g. virgin and mated females respond differently to identical stimuli of sex pheromone. The behaviour of higher animals is partly under genetic control and partly learned. Consequently, it becomes difficult to study the genetic basis of behavioural traits in these species. In insects behaviour is under a strict genetic control, which makes them good 'first' research objects in behavioural studies. Natural selection causes the introduction of variability, which may be inherited and the favoured variations increase among individuals of a population. The morphological and behavioural variability may be complementary to each other. Though in the life span of any individual morphological characters are fixed at a particular developmental stage, but they may change with age or as a result of learning. Learning is of course of little significance in insects, but there are experimental evidences that support learning can modify innate behaviour, e.g. some insects can learn to run through a maze. Taking insects as test animals a number of studies on behavioural genetics have been conducted, as the behaviour is stereotyped and easy to study. Single gene or multiple genes may govern these traits.

The texts of Hirsch and McGuire (1982), Ehrman and Parsons (1976, 1981), Fuller and Thompson (1978), Hall (1985), and Hoy (1994) are valuable for studying behavioural genetics. Behavioural genetics is concerned with the genetic analysis of individual differences in behavioural traits. These differences may occur naturally or may be induced through mutations. Naturally occurring variations form the basis of traditional behavioural genetic approaches aimed at demonstrating a genetic basis for individual differences. However, induced behavioural mutations, which block or alter the normal pattern of behaviour, provide a useful tool for understanding how genes influence behaviour. Thus, behavioural genetics is the study of the hereditary control of an organism's action. The action of gene can affect:

(i) The sensory organs, thus changing information input.

(ii) The effector organ(s), thus altering the response.

(iii) An intermediary system (nervous, endocrine), thus altering coordination and perceptual capacities.

The behavioural genetics can also play an important role in finding the nature and causes of behavioural patterns shown by insects. This may happen in three ways: (i) through the impact of behavioural genetic research, (ii) by providing valuable experimental preparations, e.g. genetic mosaics, strains, etc., and (iii) behavioural genetic research provide important cautions regarding the orderliness we can expect in formulating the laws of behaviour.

19.1 Single gene and behaviour

It is rather easy to trace the behavioural patterns, which are governed by single gene. Sometimes the effect of one gene may affect more than one change, e.g. 'pleiotropy'. If the changes are both behavioural and morphological it becomes simple to analyse them. A great deal of work has been done by taking *Drosophila* species as test insects. A number of mutants of *Drosophila* have been produced by alteration of a single gene and it has been possible to account for particular behaviour patterns. These mutants display differences in behavioural patterns, e.g. perturbed sexual behaviour, less sexual vigour, sluggishness, hyperkinetic (consuming oxygen at an exaggerated rate and dying earlier), taxis, unable to fly, under stress, etc. Many mutants are temperature sensitive, and display abnormal trait only above and below a certain temperature. Bastock (1956) crossed sex-linked yellow mutant with wild stock of *Drosophila melanogaster* for seven generations, which ensured that both were genetically similar, except in the region of yellow gene. She observed that males of yellow mutant were less successful in mating with wild females.

Rothenbuler (1964) analysed nest sanitization behaviour of European honey bee, *Apis mellifera*. The larvae infected with American foul brood eventually die and their cadavers contaminate the hive. The bee's races differ in their response to diseased larvae. The hygienic strains removed diseased larvae, whereas non-hygienic strains were unable to do so. The difference in behaviour is determined by two independently segregating loci; one involve in uncapping of cells with diseased larvae and the other with removal of cadavers from the cells. All hybrids of two strains were observed non-hygienic, thereby indicating the recessive nature of gene governing the behaviour. The genes that determine this behaviour do not induce structural or physiological changes. This is one of the most famous experiments in behaviour genetics, as it deals with behavioural pattern of adaptive value and shows that the complex behavioural traits may be governed by presence of a single allele in an animal's genotype.

UURR x uurr

(Non-hygine) (Hygienic)

↓

UuRr

(Hybrid, non-hygienic)

Rothenbuler backcrossed (UuRr x uurr) the hybrids with hygienic strain and obtained four groups in equal ratio.

(i) Bees only uncap the cell with dead larvae (uuRr).

(ii) Bees failed to uncap, but remove cadavers (Uurr).

(iii) Non-hygienic bees- neither uncapped nor removed cadavers (UuRr).

(iv) Hygienic bees- uncapped cells and removed cadavers (uurr).

Gilliam *et al.* (1983) discussed the natural defence against chalkbrood. Hygienic bees detect, uncap, and remove diseased brood from the combs before the disease becomes infectious. Arathi and Spivak (2001) analysed the behaviour of hygienic *Apis mellifera* bees in colonies and observed that the performance of hygienic behaviour depended on the proportion of hygienic bees in the colony. There was a significant increase in the percentage of cells cleaned with an increase in the percentage of hygienic bees in the colony. The number of times an individual bee performed hygienic behaviour also differed depending on the genotypic composition of the colony. A comparison of the ages of the bees performing hygienic behaviour in the different colonies indicated that the age at task performance varied significantly depending on the composition of the colony the bees cohabited and their genotype.

Ross and Keller (1998) found a single genomic element marked by the Gp-9 gene to be responsible for the existence of two distinct forms of social organization, polygyne (many queens rule the colony) and monogyne (only one queen lays eggs), in the red imported fire ant, *Solenopsis invicta.* This genetic factor influenced the reproductive phenotypes and behavioural strategies of queens and determined whether workers tolerated a single fertile queen or multiple queens per colony. This factor also affected worker tolerance of queens with alternate genotypes, explaining the dramatic differences in Gp-9 allele frequencies observed between monogyne and polygyne forms in the wild. These findings reveal how a single genetic factor can have major effects on complex social behaviour and influence the nature of social organization. Krieger and Ross (2002) reported that there were different proteins that corresponded to the two types of social behaviour in the ants. The protein was reported to be an odorant binding protein, which is crucial in allowing insects to perceive chemicals signals. In fire ant societies, workers recognize the queens, and then make the choice to accept the queens as egg layers or not. Since the gene Gp-9 encodes the odorant binding protein, they realized that the differences in fire ant society must come from a deviation within the gene.

19.2 Polygenes and behaviour

Most behavioural patterns are under polygenic control and directional selection experiments were found useful in establishing whether a particular behaviour has a genetic basis or not (Ehrman and Parsons, 1976). In these experiments individuals are selected at extreme low or high behavioural pattern distribution with the view to form distinct high or low lines in subsequent generations. If a trait does not have a genetic basis, then attempt to select extremes would fail. Hirsch and Erlenmeyer-Kimling (1962) showed that selection of *Drosophila melanogaster* that has previously been assumed to be negatively geotactic could be separated into both positive and negative geotactic lines. Three major chromosomes were found to exert an influence on the level of geotaxis in *Drosophila*. Several other behaviours, e.g. locomotor activity, eclosion behaviour, chemotaxis, duration of copulation, optomotor response, phototaxis, etc. are under polygenic control in *Drosophila*.

Experiments on the circadian system of *Drosophila melanogaster* established the endogenous nature of the eclosion rhythm. It was observed that period length of this rhythm under constant darkness (DD) is dependent upon the genetic make-up of the fly strain, thus suggesting involvement of specific genes (see section 15.8). Konopka and Benzer (1971) isolated three period mutants in eclosion rhythm, long (per^L), short (per^S) and arrhythmic (per^0), which mapped to a single genetic locus. These mutant flies also exhibited similar defects in locomotor activity rhythms. The pleiotropic effect of each *per* mutation on both eclosion and locomotor activity rhythm established the existence of a single oscillator underlying different rhythms in different developmental stages of the fly. Restoration of a short-period activity rhythm in per^0 flies by transplantation of adult brain from per^S flies on the abdomen of the recipient demonstrated the presence of this oscillator in the fly brain (Handler and Konopka, 1979). In transgenic experiments, expression of a wild-type copy of this gene restored normal behavioural rhythms in arrhythmic per^0 flies. Several early studies of these mutants shed light on the role of *per* in maintenance of circadian rhythms. Both *per* mRNA and protein levels exhibited rhythmic abundance, which reflected the behavioural rhythm: a near 24-h molecular rhythm in the wild-type flies, no rhythm in per^0 flies, a short period rhythm in per^S flies, and a long period rhythm in per^L flies. Subsequent characterization of these flies using genetic mosaic and transgenic approaches defined a group of 20–30 lateral neurons in the adult fly brain as the anatomic site controlling activity rhythm.

Bennet-Clark and Ewing (1970) studied the love song of *Drosophila*. A courting male extends one or both wings and generate a characteristic sound as he approaches nearer to female (Fig. 11.5). The female receives this sound through the aristae and if she is receptive, mating ensues. If the female is unreceptive due to immaturity, already mated or of a different species, she produces her own song resulting into male to turn away. Each *Drosophila* species produce unique love song, which constitute a powerful isolation mechanism and act as barrier to interbreeding. Thus, the love songs of *Drosophila* maintain the genetic integrity of each species. Though *Drosophila obscura* sings no song, but it mates only in daytime and uses visual cues to recognize the conspecifics. Other *Drosophila* species breed at night and use acoustic cues. Males of *Drosophila melanogaster* also use tactile cues to recognize the females. Recorded oscillograms of *Drosophila* species also exhibit differences in pulse patterns. The closely related *Drosophila pseudoobscura* and *D. persimilis* can be easily distinguished on the basis of their different love songs. Dobzhansky *et al.* (1964) studied *D. paulistorum*, a superspecies consisting of closely related races, which are believed to be in the process of splitting into distinct species. The individuals of these races differ behaviourally and their hybrids are sterile. Of the six races, love songs of five are quite similar and that of sixth is unique. This suggests that the original barrier to interbreeding of these races is some other mechanism and the song pattern evolved later. Hybridization experiments suggested that the inheritance of song pattern is sex-linked and genes governing this pattern are located on X chromosomes. The inheritance of pulse interval is rather more complex and may be partly controlled by genes on chromosomes other than sex chromosomes.

Hoikkala and Lumme (1987) observed that in *Drosophila virilis* group, the males of six species keep their wings extended while producing a train of successive sound pulses with very short intervals (20 ms) between pulses. The males of the remaining five species produce only one sound pulse during each vibration, which results in species-specific songs with long intervals (in *D. littoralis* up to 300 ms) between successive sound pulses. Gene (or genes) that makes *D. littoralis* males to set their wings in a resting position after each sound pulse is located at the proximal end of the X chromosome. Hoikkala and Lumme (1990) genetically analysed a strain of *D. littoralis* with an unusual low wing-beat frequency (<250 Hz) using marker strains, the anomalous frequency was found to be caused by genes on all major autosomes, the strongest factors being on the second chromosome.

Drosophila ananassae has a unique status because of certain peculiarities in its genetic behaviour. The most unusual feature of this species is its relatively high frequency of spontaneous male recombination. The results of studies on non-sexual behaviour, such as phototactic responses, eclosion rhythm, and preferences for oviposition and pupation sites, suggested that these behaviours are under polygenic control, with a substantial amount of additive genetic variation (Singh and Singh, 2003). Markow and Smith (1979) suggested that the phototactic behaviour of *D. ananassae* is under polygenic control and influenced by additive genetic variation. Pupation site preference is an important step in *Drosophila* pre-adult (larval) behaviour because the site selected by larvae can have a decisive influence on their subsequent survival. The positive response to selection for high and low pupation height, and the intermediate pupation height of F_1 hybrids produced by reciprocal crosses between high and low lines, showed that pupation height in *D. ananassae* is under polygenic control (Singh and Pandey, 1993). Spieth (1966a) studied the mating behaviour of light and dark forms of *D. ananassae* and found that various laboratory stocks developed sexual isolation, the extent of which varied from weak to marked. The males court females by slightly spreading and vibrating both wings, and produce courtship songs by vibrating their wings intermittently during the sequence of courtship behaviour. These forms were later found to be different sibling species (*D. ananassae* and *D. pallidosa*) of the *D. ananassae* complex. Although potentially interfertile, these sibling species are genetically distinct and their isolation is maintained by strong mating preferences (Futch, 1966). *D. ananassae* and *D. pallidosa* are useful for studying sexual isolation because of their sympatric distribution and because of the lack of post-mating isolation between them. Yamada *et al.* (2002a,b) reported that a very narrow region on the second chromosome was involved in controlling the female's discriminatory behaviour among courting males in *D. ananassae*. They also recorded and analysed male courtship songs in several strains of *D. ananassae* and *D. pallidosa* and observed species-specificity in the courtship song parameters. These parameters were suggested to play a role in mate recognition that enforced their sexual isolation. Because mate discrimination is well-developed and is the only known mechanism that prevents gene flow between species, these loci may have played crucial roles in the evolution of reproductive isolation, and therefore, in the speciation of these two species (Doi *et al.*, 2001). Singh and Chatterjee, (1987) tested mating propensity using mutant and wild-type isofemale strains of *D. ananassae* and the results confirm the genetic basis of this phenomenon. There is a positive correlation between mating propensity and fertility in *D. ananassae*.

Mating of *Drosophila melanogaster* is a stereotypically patterned behaviour consisting of a fixed sequence of actions that are primarily under genetic control. Mutations that disrupt specific aspects of mating activities offer a starting point for exploring the molecular machineries underlying sexual behaviour. Several genes, identified as causing aberrant sexual behaviour when mutated, have been isolated and cloned, providing molecular probes for expression and mosaic analyses that can be used in specifying the cells responsible for the behaviour (Yamamoto *et al.*, 1997).

The calling songs of crickets are species-specific and produced by males to attract the females. The females can recognize conspecifics songs and do not waste time and energy by responding to others. The sound frequency of song and temporal pattern of procession of pulse is termed 'phase structure'. Bentley and Hoy (1972) analysed songs of field crickets, *Teleogryllus commodus*, *T. oceanicus* and their hybrids. The oscillograms revealed differences in all three cases. In song of *T. oceanicus* intertrill interval was absent, but it was present in *T. commodus*. The hybridization experiments suggested that intertrill interval is associated with the X chromosomes received by male from the mother. The hybrid females preferred the songs of hybrid males only. Hoy (1974) concluded that X chromosome and unknown number of autosomes both control the temporal pattern of the song. Hill *et al.* (1972) showed that crickets *T. commodus* and *T. oceanicus* can discriminate between the songs of their own species and that of another species.

In honey bees queens and workers are alternative forms of the adult female and represent one of the best-known examples of insect polyphenism. Evans and Wheeler (1999) studied queen- and worker-specific gene expression in the development of *Apis mellifera*. Numerous genes appear to be differentially expressed between the two castes. Two are particularly promising as potential regulators of caste differentiation. The results make it clear that the process involves specific activation of genes in workers as well as in queens. Hughes *et al.* (2003) showed that worker caste development in the leaf cutting ant, *Acromyrmex echinatior* has a significant genetic component. Individuals of different patrilines within the same colony differ in their propensities to develop into minor or major workers. They also examined the relative paternal contributions made to the two worker castes in individual colonies. The study eliminated environmental and other factors by using ants from the same colony, and later sequenced genes from several groups of ants in order to determine which ants were related by the same patriline.

19.3 Gynandromorphs

These organisms express both male and female characteristics by virtue of being a mosaic of male and female cells and are termed gynandromorphs (Fig. 19.1). These individuals can be referred to as being 'genetic mosaics' and 'chimeras'. A gynandromorphy may be bilateral (half male and half female), anterior-posterior (one sex on anterior and other on posterior side), or mixed with patches of male and female tissues. In a number of hymenopterans unmated females can produce haploid males from unfertilized eggs. In some others the fertilized female can lay both type of eggs, resulting into development of diploid females and haploid males. Sometimes, an egg with two nuclei is laid. Only one of which is fertilized. The resultant insect is a mixture of male and female characters. These gynandromorphs can be produced in the laboratory and assist in the study of nature of relationships between structural changes and stereotyped behaviour. In a classical study, Whiting (1932) demonstrated the potential of gynandromorphs for behavioural genetics. He observed the behaviour of gynandromorphs of the parasitic wasp, *Bracon hebetor* and obtained evidence, which suggested that the organ controlling sex-specific behavioural patterns lies in the brain of this wasp.

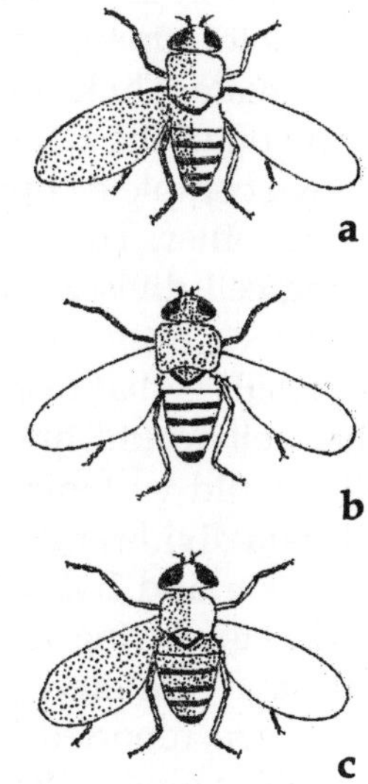

Fig. 19.1. Different gynandromorphs in a fly, (a)- Bilateral gynandromorph, (b)- Anterior-posterior gynandromorphy, (c)- Sex-piebald.

Szabad and Fajszi (1982) studied the sexual behaviour of *Drosophila melanogaster* gynandromorphs. In some gynandromorphs, certain female functions were missing because the corresponding control sites (foci) were either composed of male tissue or did not develop. A given gynandromorph can also show elements of both male and female reproductive pathways. Many of the gynandromorphs did not lay eggs, while some nonovipositing gynandromorphs laid eggs. Some of the noninseminated gynandromorphs laid eggs at the rate characteristic for inseminated females, suggesting that an oviposition focus (mapping in the head region) suppresses oviposition in virgin females, but not in gynandromorphs. Some of the inseminated gynandromorphs oviposited eggs at a low rate, possibly because the focus responsible for detection of insemination could not function properly. Some of the inseminated gynandromorphs laid unfertilized eggs. Thus, sexual behaviour in *Drosophila* does not depend on the presence (or absence) of the ovary or germ line.

In insects cell division of zygote is completely determinate; this means that decisions about what a cell will become are made with each division. In Lepidoptera sex is determined, by X and Y chromosomes, and XX is male and XY (or only one X) is female. Thus, the number of X chromosomes is important. When cells divide, a copy of all the DNA is passed to both cells. Therefore, before the division there are two complete copies of DNA in each of the chromosomes, which are attached to each other. These copies of the DNA in each chromosome are then separated during cell division. Occasionally the two copies of a particular chromosome do not detach from each other during cell division (called non-disjunction), and the end result is that one of the cells will end up missing an entire chromosome. The other cell would have two or three X chromosomes. Thus, one of the resulting cells would be female, while the other cell would be male. Consequently, these cells will divide and produce tissues of different sexes. The resultant insect appears half male and half female 'bilateral gynandromorph'. The non-disjunction can also occur during later divisions, and the formed insect is termed 'mosaic' and will have scattered patches of male and female tissues. Such gynandromorphs have been reported in tiger swallowtail, *Papilio glaucus* and adults of forest tent caterpillar moth, *Malacosoma disstria.*

19.4 Altruism

Altruism is a self-defensive behaviour, commonly observed in social insects. The youngones toward their parents or conspecific sisters selflessly show it. Altruism is of interest from evolutionary point of view, because it implies the existence of group selection. The soldiers of many eusocial insects perform the only function of colony defence. The workers of social groups without soldier caste have either barbed stings or use other structural modifications for protecting the colony. When a worker honey bee stings, the sting remain embedded in the victim's body and the bee is fatally injured. Many wasps and ants also exhibit similar defensive manoeuvre. Female workers of most social insects sacrifice their egg-laying ability in presence of queen. In stingless bees and ants trophic eggs are laid, which are fed to larvae and queen. The honey ants filled their crops with liquid food and simply act as storage system. These altruistic behavioural patterns have evolved through natural selection.

The Darwinian theory of natural selection says competitive selfishness is the driving force of survival. Altruism or unselfishness is inconsistent with the theory as it lowers the survival chance. Hamilton (1964a,b) provided an explanation for altruistic behaviour and suggested that altruism can evolve by natural selection if directed towards close genetic relatives. Thus, copies of genes lost due to personal sacrifice by the altruist are compensated by the survival and reproduction of other copies of the same genes in beneficiaries of altruism. Richard Dawkins (1989), author of the famous book *'The Selfish Gene'*, first proposed the idea for the recognition of genes in the bodies of other individuals, when he used the phrase 'green beard gene'. He hypothesized a gene that gives its bearer some conspicuously visible-external label (say a 'green beard'), and also makes such individuals to be nice to other 'green-bearded' individuals. The idea of green beard has remained a matter of considerable curiosity and interest.

Keller and Ross (1998) gave first possible example of a green beard gene in red imported fire ant, *Solenopsis invicta*. They were looking for variable genes because this permits them to estimate the genetic proximity among different members of the ant colony and identified a peculiar gene called Gp-9, which exists in two forms, B and b. Each female ant inherits one copy of the gene from its mother and another from its father, and has anyone of the three possible (BB, Bb or bb) genotypic compositions. They found that both workers and queens possessing the bb gene die early and such genes are probably recessive lethals. However, the real peculiarity that they found was that while both BB and Bb types exist among the workers, only Bb types are to be found among the queens. This happens because the workers with Bb gene combination kill all BB queens attempting to reproduce. Thus, the workers having a copy of b gene (Bb workers) can recognize the same b gene in queens (Bb queens). Keller and Ross concluded that b is a green beard gene and permits its bearers to recognize other bearers. It appears that BB queens and Bb queens possess different smell. Besides, this smell is somewhat related with reproduction, as BB queens are not killed until they begin to reproduce. They also suggested that BB queens are so strong that if allowed to reproduce, they will eliminate other queens, whereas Bb queens are reasonable and permit coexistence of many queens in the colony. The workers prefer to have many queens in the colony.

19.5 Kin selection

Hamilton (1964a,b) proposed well-known genetic theory termed as 'kin selection' (see section 14.5.4). Hamilton's ideas were widely appreciated and extensively applied. The ability to discriminate kin from non-kin has been reported in many social insects. In social Hymenoptera with a system of haplodiploidy a female is more closely related genetically to her full sister than to her daughter. Many social hymenopteran queens are known to mate, multiply and produce different patrilines of daughters. Thus, the workers may have to rear half sisters as often or probably more often than they can rear full sisters. If workers had the ability to distinguish between their full and half sisters then of course they could give preferential treatment to their full sisters. For this reason the mechanism of kin recognition is of particular interest in social Hymenoptera. Greenberg (1988) demonstrated the first example of kin recognition in a primitive eusocial sweat bee, *Lasioglossum zephyrum*. The guard bee sitting at the entrance to the burrow prevents entry of both parasites as well as non-nestmate conspecifics. He showed that the probability of acceptance of an intruder depends on how closely the intruder is genetically related to the guard bee. The guard bees also use the odour signals to recognize their nestmates.

Apis mellifera workers recognize queens as well as each other on the basis of genetic relatedness. Recent findings also suggest that honey bee workers are capable of giving differential treatment to their full and half sister larvae.

Ants have also been used for studies on kin recognition. The acacia ant, *Pseudomyrmex ferruginea* uses genetically determined recognition labels (probably pheromones) produced by each individual. In *Camponotus* workers can discriminate between their full and half sisters. In other genera, e.g. *Odontomachus, Acromyrmex* and *Cataglyphis* there is an ability to discriminate nastemates but different levels of genetic relatedness cannot be discriminated.

19.6 Natural selection

The example of peppered moth, *Biston betularia* has provided a natural example that has been intensively studied (see section 13.2.1.3). The phenomenon of industrial melanism involves natural selection and has affected over 70 species of moths in Britain. Prior to 1800, dark-coloured or melanic moths were rare and were therefore collectors' items. During the industrial revolution, soot and other industrial wastes darkened tree trunks and the light-coloured morph of the moth became rare and the dark morph became abundant. In 1819, the first melanic morph was seen; by 1886, it was far more common, illustrating rapid evolutionary change. The cause of this change was thought to be selective predation by birds, which favoured camouflage colouration in the moth. Kettlewell (1958) did release-recapture experiments using both morphs (Table 19.1). By observing bird predation from blinds, he could confirm that conspicuousness of moth greatly influenced the chance it would be eaten.

Table 19.1. Recapture success of different morphs of *Biston betularia* in release-recapture experiments.

	Light moth (%)	Dark moth (%)
Non-industrial woods	14.6	4.7
Industrial woods	13	27.5

19.7 Genetic determination of behaviour

To what extent is animal behaviour genetically determined? This question has been mainly investigated since the time Sir Francis Galton who might be considered as the Father of Behavioural Genetics. Interest in the genetic control of behaviour has become even stronger with the advent of genetic theories for the evolution of behaviour. The influence of genes on behaviour is obvious but it seems to be a complex phenomenon. Some behavioural patterns are regulated by a single gene but in most cases several genes probably involved to bring about any kind of behaviour.

Drosophila species exhibit distinct differences in their mating behaviours and the signals emitted by the flies during mating. This assists in the recognition of individuals of their own species. Variation within the species also gives the females a chance to exercise mate choice between the males of same species. Genetic basis of variation within and between the species can be studied:

(i) By using hybridization methods, using visible mutant genes or molecular markers.

(ii) With the aid of biometrical methods by studying the quality of variation in the studied traits and by tracing the selection pressures, which have affected these traits during their evolution.

(iii) By sequencing genes found to affect behavioural traits and studying variation in their structure within and between the species.

Selection pressures affecting behavioural traits can be studied by letting the females to show what kind of males they would accept as courtship partners and what kind of males are most favoured courters. Many *Drosophila* species have been found to exhibit genetic variation in traits affecting mate choice. This variation forms a basis for sexual selection exercised by females within the species and it may also give rise to sexual selection between geographically isolated populations. Analysing the genetic basis of variation in traits affecting mate choice is essential for understanding how species-specific courtship behaviours have evolved. It is also very important to find out whether variation within the species is or is not qualitatively similar to that between the species.

Dozens of loci affecting different aspects of behaviour have been already identified and cloned in *Drosophila*. Some of these genes determine variation in such ethologically complex phenotypes as the male 'love song' that is produced during courtship and the locomotor 'sleep-wake' activity cycles that are controlled by the circadian clock. Although the evolutionary analysis of most behavioural genes in *Drosophila* is relatively new, it has already given important insights into the forces shaping the molecular variation at these loci and their functional consequences (Peixoto, 2002).

The genetic basis of a complex behavioural phenomenon namely 'polyethism' has also been determined in *Apis mellifera*. It is well-known that honey bee workers share amongst themselves a variety of complex tasks such as nest building, cleaning, nursing brood, food storage, guarding, temperature regulation, foraging, etc. Robinson and Page (1988); Frumhoff and Baker (1988) demonstrated that workers change their behavioural roles as they grow older and such behavioural variability among worker honey bees has a genetic basis. Robinson and Page (1988) established experimental honey bee colonies by artificially inseminating queen bees such that she produced electrophoretically distinguishable patrilines of daughter. Then they collected three samples of about 40 bees each from such colonies: the first sample consisted of 'control' bees picked randomly irrespective of what behaviour they were performing; the second sample consisted of bees which were guarding the nest entrance; and the third sample of bees that were performing the task of cleaning the cells of dead bees. They determined the genetic composition of bees in each group and observed that the worker bees fathered by some males were more likely to guard the nest while those fathered by a different male were more likely to remove dead bees.

Frumhoff and Baker (1988) set up *Apis mellifera* colonies by artificially inseminating queen bees by a method that the drones used would produce daughters not only distinguishable by electrophoresis but by body colour itself. They found that bees of one patriline are significantly more likely to specialize in the task of grooming nestmates, but common behaviours as feeding nestmates are performed almost equally by both patrilines. The queen of *Apis mellifera* mates with several males that create genetic heterogeneity among her daughters. Such variation may be an important component to division of labour within genetically heterogeneous colonies. Thus, different genotypes of bees have different propensities for performing different behaviours, and such multiple mating by the queen may be a better way of ensuring efficient colony organization.

The jewel wasp, *Nasonia* (Fam. Pteromalidae) is a superior organism for genetic research. It is a parasitoid that kills and lays eggs upon the pupae of various flies. A key feature of *Nasonia* is its interesting form of sex determination called 'haplodiploidy', which affords the insect an ability to control the sex of its offspring. The important features of wasp are: (i) short generation time, (ii) easy detection and mapping of visible mutations, (iii) availability of visible and molecular markers, (iv) ability to inbreed and produce healthy inbred isogenic lines, (v) presence of three closely related and interfertile species, and (vi) ability to produce hundreds of genetically identical (clonal) recombinant genotypes in the F_3 generation. These features make *Nasonia* an excellent organism for basic studies in genetics, including developmental genetics, evolutionary genetics, molecular evolution and comparative genomic research. Three closely related species in the genus *Nasonia*, viz. *N. vitripennis*, *N. longicornis*, and *N. giraulti* provide an abundance of molecular markers for comparative genomic and evolutionary studies. Differences between the species in behaviour (e.g. courtship and mate preference), development and morphology provide excellent opportunities for behavioural genetic research.

19.8 Behavioural genetics and evolution

Evolution consists of changes in the genetic constitution of populations, and can occur if there is genetic variability in the population. The amount of genetic variation in the population measures its evolutionary potential. In recent years it has become possible to estimate genetic variation by application of innovated techniques. One fundamental step in the process of evolution is the formation of new species.

Darwin's theory of evolution of organisms by means of natural selection leads to the spread of favourable variations and suggests that the organisms with favourable variations are more likely to survive and reproduce. Though natural selection remains the fundamental process directing evolutionary changes, but its behaviour depends on genetic variability (Fig. 19.2).

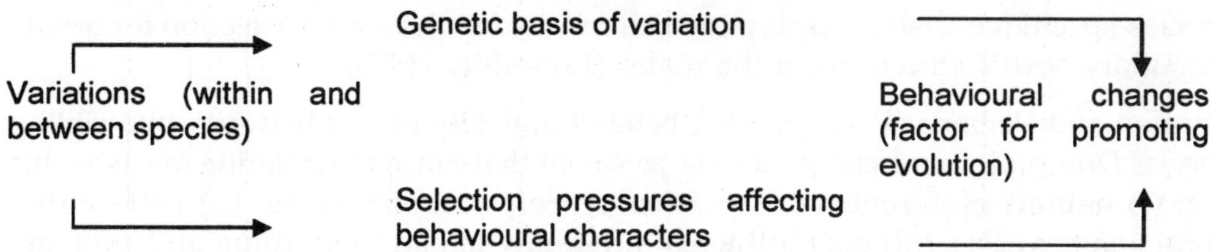

Fig. 19.2. Genetically determined variations and their impact on behaviour.

The occurrence of variations is a pervasive phenomenon of natural populations. Insects too exhibit conspicuous variations, e.g. colour pattern in butterflies, ladybird beetles, grasshoppers, etc. Occurrences of discontinuous variants called morphs (castes of social insects) often have striking differences. The genes not only determine morphological characters, but also patterns of behaviour that can be transmitted from parent to offspring. Many such behavioural patterns governed by genes are discussed in this section. Though direct evidences of evolution of behaviour are not much available, but a number of indirect evidences allow us to state the way behaviour has evolved. The behavioural changes also occur slowly and in successive stages of adaptations. The examples of camouflage and eyespots in sphinx moth are glistening examples of evolution of body colour and markings. Caspari (1963) proposed that the genetic basis of behavioural characters is important because genes are the basic unit that is transmitted and reshuffled during evolution and are arranged by selective forces into patterns of adaptive action.

Species in the family Drosophilidae, especially *Drosophila melanogaster* have been subject of intense genetic, evolutionary, developmental, molecular and behavioural researches. Tremendous amount of work has been done by taking these flies as test insects. In extraordinary Hawaiian fauna of *Drosophila* many closely related species exhibit slight differences in their food preferences and resting places. Dobzhansky *et al.* (1956) noted differential attraction of *Drosophila* species to different yeasts. Spieth (1966b) suggested that secondary sexual characters in males of Hawaiian *Drosophila* species are manifestations of the complex courtship behaviour. Some of these traits usually define genetic groupings. Studies on genitalia, behaviour, and genetics provided evidence that these characters were species group differences and used primarily by males to display courtship behaviour. The sexual selection in Hawaiian *Drosophila* was an extremely important part of evolutionary process in the Hawaiian fauna and it was possible to determine the 'direction of evolution' of a group by analysing data obtained from mate preference studies (Kaneshiro, 1999). More than 500 endemic species of Drosophilidae have been described from Hawaiian Islands and many more are yet to be added. The evidences based on morphological, behavioural, genetic (including DNA sequencing) and developmental data indicate that the group is monophyletic and formed from a single founder. The tremendous morphological diversity observed in the Hawaiian *Drosophila* is a manifestation of the elaborate, species-specific courtship displays, which resulted in the sexual selection for ornate secondary sexual structures in the males (Kaneshiro, 1999).

Ehrman and Parsons (1976) pointed behavioural differences between two sibling species *Drosophila pseudoobscura* and *D. persimilis* that can act as isolating mechanisms are: (i) habitat preferences- *D. persimilis* prefers cooler areas, (ii) differential attraction to yeasts, (iii) both attracted differently to bait in morning and evening, (iv) *D. persimilis* has more affinity to light, and (v) differences in courtship songs. Of these, the differences in courtship songs mainly divided the two populations in distinct groups.

Rice (1985), Rice and Salt (1990) presented experimental evidence for the possibility of sympatric speciation in *Drosophila melanogaster*. They established a stock population of *D. melanogaster* and pupae from the culture were placed into a habitat maze. To find food, newly emerged flies had to negotiate the maze with different environmental gradients, viz. light and dark (phototaxis), up and down (geotaxis), and scent of acetaldehyde and ethanol (chemotaxis). They cultured two strains (with different eye colour) of flies that had chosen opposite habitats. One strain emerged early, flew upward and was attracted to dark and acetaldehyde. The other emerged late, flew downward and was attracted to light and ethanol. Pupae from these two strains were placed together in the maze and the adults were allowed to mate at the food site and were collected. A selective penalty was imposed on flies and females that switched habitats were destroyed. None of their gametes passed into the next generation. Males that switched habitats received no penalty. After 25 generations of this mating tests reproductive isolation between the two strains emerged and habitat specialization was also produced. In case without the penalty against habitat switching the result was the same, and thus, the results indicated the possibility of sympatric speciation.

The *Drosophila willistoni* complex of species consists of a cluster of closely related species and subspecies. Their phylogeny has recently been resolved using DNA sequencing. Each morphological species has a unique courtship song, despite significant differences between strains within species (Ritchie and Gleason, 1995).

It has been proposed that many host-specific phytophagous insects originated by shifting and adapting to new host plant species. Sympatric speciation is the splitting of one evolutionary lineage into two without the occurrence of geographic isolation. Thus, cases of sympatric speciation have tremendous implications for understanding of species and their origins. Berlocher and Feder (2002) reviewed the evidence for sympatric speciation via host shifting for phytophagous insects and propose a set of testable predictions for distinguishing geographic mode (allopatric versus sympatric) of divergence. They concluded that sympatric speciation is a viable hypothesis.

Bush (1969) developed a model for sympatric speciation in American hawthorn fly, *Rhagoletis pomonella* (Fam. Tephritidae) based on host preference behaviour as a reproductive isolating mechanism. Sometime during the 19th century it began to infest apple trees and then cherries, roses, pears and possibly other members of the Rosaceae. This shift has led the formation of 'host races', an intermediate step in speciation. The apple host race of *R. pomonella* is different from the hawthorn race in several ways: (i) eclosion of apple flies is temporally earlier in the year, because of earlier fruiting phenology of apple trees, (ii) allozyme data show frequency differences between the races at several sites, and (iii) apple and hawthorn flies respond positively to the odour of chemical volatiles emanating from the surface of their natal fruit but not to non-natal volatiles. It has also been observed that female offspring collected from hawthorn (*Crataegus* species) and apple are more likely to select that particular host for oviposition. Genetic differences between flies on these two hosts have been found at 6 out of 13 allozyme loci (Feder *et al.*, 1988). Laboratory studies have shown an asynchrony in emergence time of adults between these two host races (Smith, 1988). Flies from apple trees take about 40 days to mature, whereas flies from hawthorn trees take 54-60 days to mature. Hybridization studies show that host preferences are inherited, and it may represent the early stages of a sympatric speciation in *R. pomonella*. Bush (1975) suggested that formation of new species involves allopatric, parapatric or sympatric reproductive isolation. In sympatric speciation changes in behaviour may be a factor for reproductive isolation. Thus, in phyto- and zoophagous insects host selection mechanisms and their behavioural preferences to hosts have played a major role in the formation of new species.

Feder and Filchak (1999) reviewed the ecological and genetic evidence for host-plant mediated selection in *R. pomonella* and reached the conclusions: (i) differences in the fruiting phenologies of apple and hawthorn trees exert different selection pressures on the diapause and eclosion time characteristics of the host races, (ii) host-related fitness trade-offs for *R. pomonella* may not be due to disruptive selection affecting any one specific life history stage, rather it is the sum total of directional selection pressures acting across different life-stages that generates divergent selection on apple and hawthorn flies, and (iii) natural *R. pomonella* populations harbour a good deal of genetic variation for development-related traits. This variation allows fly populations to rapidly respond to temporal vagaries in local environmental conditions across years, as well as to broad-scale geographic differences that exist across the range of the species. Perhaps most importantly, this variation gives *R. pomonella* the flexibility to explore and adapt to novel plants.

Bush (1966) on the basis of cytogenetic data established five taxonomic species group of North American *Rhagoletis*, viz. *R. pomonella, R. tabellaria, R. ribicola, R. cingulata* and *R. suavis* that contain most of Nearctic *Rhagoletis* species. Phylogenetic analysis of mitochondrial cytochrome oxidase II nucleotide sequences supported monophyly of these groups (Smith and Bush, 1997). Monophyly of these groups was also supported by mitochondrial 16S rDNA sequences analysis (McPheron and Han, 1997) and electrophoretic enzyme alleles (Berlocher and Bush, 1982).

Eurosta solidaginis (Fam. Tephritidae) is a gall-forming fly that is associated with goldenrod plants. Mainly it lays eggs in *Solidago altissima*, but in some areas it uses *S. gigantea*. Craig *et al.* (1993) compared reproductive behaviour of the flies associated with the two hosts. In host choice experiments, each fly strain frequently laid eggs in its own host. They also performed several mating experiments and observed that in absence of hosts, females mated with males from either strain. When males of both strains were present, statistically significant positive assortative mating was seen. When both hosts and flies from both populations were present, females waited on the buds of the host that they are normally associated with and the males fly to the host to mate. Like the *Rhagoletis* case, this may represent the beginning of a sympatric speciation.

Abrahamson *et al.* (2001) studied the interactions of goldenrod host plants, *Solidago*, the gall fly, *Eurosta solidaginis* and the stem- and gall-boring beetle, *Mordellistena convicta* (Fam. Mordellidae). *E. solidaginis* has developed genetically differentiated and reproductively isolated host races that are associated with the ancestral host *Solidago altissima* and the derived host *S. gigantea*. The galling of *Solidago* by *Eurosta* created a new resource that has led to a subsequent host range expansion by *M. convicta*. The beetles from stems and galls are genetically distinct and likely shifted from stems to galls. Beetles from *S. altissima* versus *S. gigantea* galls exhibit assortative mating and higher preference for and/or performance on their natal host.

Gray and Cade (2000) attempted to clarify the role of sexual selection in speciation in North American field crickets, *Gryllus texensis* and *G. rubens*. These species are cryptic sister with extensive areas of both sympatry and allopatry and occur throughout much of the south-central and southeastern United States. Males of these species can be separated only on the basis of calling songs. *G. texensis* produce trills with a pulse rate of about 80 pulses per second (p/s), and *G. rubens* trill at about 56 p/s (both at 25°C). Combined with previous work their results suggested that divergent sexual selection might have caused speciation between these taxa. Gray (2005) experimentally manipulated the courtship interactions of *G. rubens* and *G. texensis* to examine the potential of close-range courtship interactions to limit interspecific gene flow. Responses of males to females and of females to male courtship song and males *per se* were examined for four paired sympatric and allopatric populations. *G. rubens* males were more likely to court conspecific females, but male *G. texensis* courted females of both species equally. If paired with conspecific males, female *G. rubens* and *G. texensis* both preferred conspecific courtship song. Thus, males *G. rubens* were more discriminating than male *G. texensis*, whereas female *G. texensis* were more discriminating than female *G. rubens*. The net effect of courtship interactions would appear to limit interspecific gene flow only weakly.

It is widely accepted that most behavioural patterns of insects are genetically programmed and in terms of evolution all individuals of a population behave to increase their own 'genetic fitness'. The genetic fitness is a measure of the number of copies of genes an animal contributes towards the genetic composition of the generation. Any animal behaves to increase its own fitness and it should behave in such a way so as to maximize its reproductive success. Thus, all individuals should be selfish and concerned with them only.

However, the phenomenon of altruism and mutual cooperation as found in many social insect groups cannot be explained by considering genetic fitness. The altruistic behaviour could be explained on the basis of kin selection in which any individual would sacrifice its own reproduction for the sake of its kin who share same proportion of its genes.

Chapter **20**

Behavioural Systematics

Systematics is commonly defined as the 'study of biological diversity and the relationships among organisms'. Taxonomy is a component of systematics specifically focussing on the theory and practice of classification. The two terms are not clearly separable and both are frequently used interchangeably by biologists. Howksworth and Bisby (1988) suggested that taxonomy includes a range of different areas from the description and naming of new taxa (nomenclature), the arrangement of organisms into a convenient classificatory system (classification), and the construction of identification systems for particular groups of organisms. Systematics includes taxonomy with the addition of theoretical and practical aspects of evolution, genetics and speciation. Systematics plays a central role in biology by providing the means for characterizing the organisms that we study. Through the production of classifications that reflect evolutionary relationships it also allows predictions and testable hypotheses. By allowing taxa to be correctly identified classifications provide a key to the literature and a means for organizing information (Danks, 1988).

20.1 Characters and their use in systematics

Taxonomic characters are defined as attributes of a taxon that allow its differentiation or potential differentiation from others. It is assumed that the characters used in taxonomy are fixed as being under genetic control. Differences between taxa may be insignificant in some cases; however, more often they are extensive and numerous. Taxonomic characters can be suitably categorized as morphological, physiological, biochemical, numerical, embryological, molecular, ecological, and behavioural (ethological). Most taxonomic classifications and keys for identification are based on morphological characters as these characters can be easily observed and evaluated for variation. The other kinds of characters often require expensive equipments, live materials and are more difficult to study. Biological characters, e.g. embryological, ecological, and behavioural have frequently been used at all levels in classification as well as for hypothesizing phylogenetic relationships among recognized species and groups of species. Of these, behavioural and ecological characters have been particularly popular and interesting. Taxonomic characters have a dual function, first these are indicators of differences and second they function as indicators of relationships. Early taxonomists used almost exclusively morphological characters to distinguish taxa and as a basis of classification. Though morphological characters still form a source for classification, they are being supplemented to an over increasing extent by other kinds of characters.

Insect behaviour is plastic, but it also often has a basis that it is largely genetically determined. The plasticity in behaviours makes them poor taxonomic characters; however, a number of behaviours are stereotyped and provide good characters. It is a general tendency among systematists to pay less attention to behavioural characters than morphological characters. This is partly due to the fact that behaviour requires long period of study. An observer must take his observations at the right season and at the right place, and often all sorts of difficulties prevent or delay the observations. Another reason for paying less attention to behavioural characters is that many considered them to be variable. However, certain behaviours are fixed and often characterize the whole group. Important works on behavioural systematics appeared in the contributions made by Spieth (1952, 1966b), Mayr (1958), Cullen (1959), Evans (1957, 1966b), Wenzel, (1992), Queiroz and Wimberger (1993), Martins (1996), Wimberger and Queiroz (1996) and Aluja and Norrbom (1999).

Taxonomists generally believe that the best classification of living organisms will eventually take into account a wide range of attributes, functional as well as morphological. They should also make efforts to accumulate different kinds of evidences. However, taxonomists frequently view behavioural characters with suspicion as:

(i) These characters can not be studied in preserved specimens.
(ii) These characters are intermittent even in living specimens as certain behavioural patterns are exhibited only during a particular period of the day or during breeding season.

Ernst Mayr (1963) cited "indeed, behavioural characters are often clearly superior to morphological characters in the study of closely related species, particularly sibling species." The comparative ethology of related species has been mainly used for identification/separation of taxa of various groups like bees, wasps, flies, orthopterans, hemipterans, etc. These studies have also been found important in cladistic and phylogenetic analysis. Behavioural characters as shown by individuals of closely related taxa are obviously important as they assist in recognition of conspecifics and in isolating mechanisms.

20.1.1 Sound production

Behavioural characters are often fast evolving and the acoustic behaviour of insects has proven to be important in insect speciation. Recent advancements in the study of behaviour and development of modern sound recording devices and conversion of these sounds into graphic patterns have recognized to be fairly valuable in systematic studies. Grasshoppers, katydids and crickets are well-known for their abilities to jump and particularly for singing by males (females are typically silent).

The calls of grasshoppers (usually diurnal), katydids and crickets (usually nocturnal) can be heard during their breeding seasons.

The Orthoptera may be broadly subdivided into two suborders: Ensifera (crickets) and Caelifera (grasshoppers). Members of the suborder Ensifera often display acoustic communication and the acoustic signals are typically produced by adult males to attract the females. The calling songs produced by the males are highly species-specific, with each cricket species producing a unique song. The Ensifera includes the families: Gryllidae (true crickets) and Tettigoniidae (bush crickets), and both gryllids and tettigoniids produce sound by rubbing the fore wings together and also possess tympanal ears located on fore legs. Crickets provide excellent systems of long and short-range acoustic communication in the context of mate finding and provide great opportunities for behavioural studies. Field crickets (Subfam. Gryllinae) could be located or collected by using acoustic signals (section 21.6). The acoustic sampling of these crickets may prove a reliable, efficient and non-invasive method of sampling field crickets. Female crickets are tuned to the calling songs of conspecific males and tend to ignore heterospecific males that may be calling at the same time. The role of calling song as a pre-zygotic mechanism of species isolation has been investigated in sympatric cricket species.

Crickets are among the few insect groups that can be identified to species, metres away, in the field. Males of most cricket species advertise to females with a distinctive call, enabling the females to find appropriate males and biologists to learn what species occur in an area. Easy identification of species in the field makes crickets of special value to systematists, ecologists and behaviourists. By tuning in to cricket calls, a systematist can discover morphologically poorly defined species (= sibling species), discover rare or hard-to-collect species, and easily and accurately map geographical and ecological distributions (Walker and Masaki, 1989).

Fulton (1932) first used calling songs to distinguish cricket species, and found that in many cases a supposed species produces more than one calling song. He recognized four songs of *Gryllus assimilis* in North Carolina and showed that these populations had different seasonal cycles, occurred in different (though overlapping) habitats, and were reproductively incompatible in laboratory crosses (Fulton, 1952). Subsequently more cryptic species of *G. assimilis* were recorded in the eastern United States on the basis of their calling songs. Later these species were also distinguished by using usual morphological features.

Alexander (1962b) taxonomic experience with three families of insects, Gryllidae (crickets), Tettigoniidae (katydids and long-horned grasshoppers) and Cicadidae (cicadas) gave strong support to the view that sound signals produced by the insects belonging to these groups can be well utilized for their classification, and a number of orthopterans have been discovered or rediscovered by careful analysis of their songs. Songs of more than 40 species of North American crickets were analysed and described (Walker, 1964; Alexander, 1967). Otte and Alexnader (1983) published a monograph on Australian crickets which adequately illustrates the effectiveness of using cricket songs to facilitate a faunal study. Of the 376 species, about 250 were first recognized by songs during a single year of fieldwork.

Species of North American crickets of the genus *Gryllus* have been used as models to investigate many important topics in behavioural studies which include the acoustical communication. Alexander (1968) derived a 'probable phylogeny' of *Gryllus* species based on a phenetic analysis of a limited number of behavioural and ecological characters. This phylogeny was used to infer the pattern and mode of evolution of life cycle (i.e. developmental stage at which diapause occurs), calling song and habitat associations. Li and Zheng (1999) analysed the characteristics of songs of six species of *Loxoblemmus* (Fam. Gryllidae) and showed that the sound structure of crickets can be used as an important criterion for their identification. Songs of different species differed in sound structure, while those of the same species were alike. Lian and Li (2002) made comparative studies on the characteristics of the common sound types, including call song, courtship song and rival song of three species of crickets (*Dianemobius furumagiensis, Tartarogryllus minusculus, Loxoblemmus equestris*) and showed that different sound types of same species have certain similar sound characteristics although there exists some difference in the frequency domain characteristics and the time domain characteristics. Comparatively, call song was most useful for taxonomy among the three sound types of crickets.

Grasshoppers recognize conspecifics partner by acoustic communication at large distances. Grasshoppers of the subfamily Gomphocerinae have acquired the most elaborate songs due to the peculiarities of the structure of sound production apparatus and mating strategy (Otte, 1970). A male of Gomphocerinae produces a 'calling song' when being alone, and the female that is ready to mate produces a 'response song'. Usually the sound has a broad frequency spectrum, therefore that the specificity of the songs lies not in their frequency band but in the pattern of amplitude over time.

Helversen and Helversen (1994) conducted numerous behavioural and electrophysiological studies and showed that the grasshoppers mainly use temporal parameters (phrase duration, the number of phrases, syllable/pause ratio, etc.) for the song recognition. They studied whether the females of two closely related species *Chorthippus albomarginatus* and *C. oschei* distinguish between con- and heterospecific courtship songs. Although these courtship songs differ in many temporal parameters, the two species are known to hybridize in nature. In choice mating experiments, the females were found to mate rather assortatively with conspecific males, but in 10-20 per cent of cases they mated with heterospecifics. The acoustic component was the most important for the recognition.

Courtship songs in *Drosophila* can vary largely between closely related species. In species of *D. virilis* group, the male courtship songs play an important role both in species recognition and in sexual selection within the species in the wild. These songs can easily be distinguished from each other by the length of the sound pulses and interpulse intervals. Hoikkala and Lumme (1987) observed that in *D. virilis* group, the males of six species keep their wings extended while producing a train of successive sound pulses with very short intervals while the males of the remaining five species produce only one sound pulse during each vibration (see section 19.2). *Drosophila pseudoobscura* and *D. persimilis* can be easily distinguished on the basis of their different love songs (Dobzhansky *et al.*, 1964).

Many host-specific phytophagous insects produce acoustic signals during mating. The acoustic calls are specific to each species and intermediate characteristics are evident in hybrid offspring. In Hemiptera both aerial- and substrate-transmitted vibrations results in sound production. Claridge (1985) reviewed sound production and acoustic behaviour in the Hemiptera like cicadellids, delphacids and other smaller Auchenorrhyncha with respect to their taxonomic and evolutionary significance for speciation. Acoustic studies in the Hemiptera have shown how rapidly prezygotic reproductive barriers involving acoustic signals may become established (Claridge and Vrijer, 1993). The differences in sounds produced by closely related, even morphologically indistinct taxa, indicate the potential contribution of sound characters, in conjunction with morphological and molecular evidence, for resolving evolutionary relationships among psyllids. Both male and female psyllids produce species-specific and sex-specific acoustic signals during mating. Their acoustic signals, distributional data, ecology, and host plant habitat can be used to elucidate the processes and mechanisms of speciation. A combined approach of this sort has enormous potential to explore the relative roles of natural selection related to ecology and habitat, versus sexual selection and competition related to interactions within and between species. Tishechkin (2003) recorded oscillograms and described vibrational signals of 50 species of leafhoppers belonging to different families. He discussed the taxonomic status and relationships of various taxa on the basis of bioacoustic characters.

Green lacewings of the order Neuroptera are a good case study to look at the role of courtship songs in reproductive isolation and at evolutionary changes in song features among closely related species. Male and female green lacewings establish duetting behaviour through a low frequency tremulation courtship song that always precedes copulation. A combination of playback experiments, laboratory hybridization, electrophoretic and mitochondrial DNA studies has shown that many species of green lacewings are really groups of previously unknown cryptic sibling biological species (Wells and Henry, 1992).

20.1.2 Light production

The mating flashes of male fireflies have long been considered a valuable aid in taxonomy. The study of firefly systematics was started in 1910 by Frank A.

McDermott, a chemist with a hobby in fireflies. McDermott (1917) reported that in fireflies flashed mating signals were useful for species identification. McDermott (1964) published a catalogue entitled "*Taxonomy of the Lampyridae*" including nearly 2000 described species of the Lamypridae of the world. McDermott's taxonomic work thus began with locating, recognizing, discriminating and expermineting with mating signals, and he thought about species and reproductive isolating mechanisms in flireflies. He put his conclusions into practical applications by making what apperas to be the first formal taxonomic decision based on mating signals. Modern firefly systematics began when it was realized that the luminescent pattern emitted by flying, mate-seeking males provided an invaluable clue for recognizing species in nature. An approach in systematics giving special attention to these signals 'semios' could be called 'semiosystematics', the focus thus being on the coded patterns that the insects are using for interbreeding (gene flow), and/or non-interbreeding (isolation) (Lloyd, 1990).

Herbert Barber (1951) showed that the flash patterns of *Photuris* fireflies can be used as taxonomic characters for the species identification. Indeed, many taxonomists have discovered the presence of morphologically cryptic species of fireflies using these signals. Mating flashes have been found to be very important in terms of reproductive isolation, and these are generally consistent and simple signals repeated at fairly regular intervals (Lloyd, 1969). Ohba (1983) proposed a classification scheme for the various mating systems found in Japanese fireflies, designated according to representative taxa.

In fireflies a 'flash dialogue' of male ensues to help him in finding the mate and to make sure that she's the right species. Most of the time, male emits a single type of flash pattern that is species-specific while the female emits a stereotyped response. The response latency, the delay between call and response, is a critical factor for species recognition (Vencl *et al.*, 1994). This has frequently been used as a useful taxonomic allusion.

It is relatively easy to recognize fireflies of different types even by their flash patterns. For example in *Photinus pyralis* the males have a dancing 'J' pattern flight, flashing about each 4-5 seconds, pausing after each flash and hovering to look for the female response which occurs about 2 seconds later. *Photuris* is a slightly larger firefly the adults of which have a distinctive arched body in contrast to *Photinus* which has a flat body. *Photuris* often may be seen flashing in rapid succession, especially when alarmed or captured, and has the ability to mimic the flash response of a number of different species of *Photinus*.

20.1.3 Host associations

The gall-making habit is species-specific in many insects and found in six orders of insects: Coleoptera, Lepidoptera, Hemiptera, Thysanoptera, Diptera, and Hymenoptera. Most gall-inducing species diversity is clustered in only a few insect groups, such as the gall midges (Diptera: Cecidomyiidae) and the gall wasps

(Hymenoptera: Cynipidae). Gall insects concentrate on all parts of the plant and the part of the plant attacked helps in their identification. The gall serves both as a shelter and a food source and the gall insect is partially protected from parasites and predators. Many inquilines live in insect galls as associates in this microecosystem. Galls are basically of two types, open and closed. The open galls are produced by piercing mouthpart forms like aphids, psyllids, and coccids. Closed galls are made by larvae with mandibles such as Coleoptera, Lepidoptera, Diptera, and Hymenoptera. Most galls are simple (monothalamous) and contain a single larva. Fewer galls are compound (polythalamous) and have several larvae in separate chambers. Galls can also be classified according to their form and texture. The gall-making habit has been evolved independently in different insect groups.

Phylogenetic clustering of gall-inducing species is also apparent among the scale insects. The gall-inducers are known from ten coccoid families, the highest proportion belongs to the Eriococcidae. The gall shapes are recognizably species-specific and structural complexity is often considerable. The shape of the gall appears to be independent of the host plant species and the host plant organ on which the gall grows is determined by the species and sex of the scale insect (Gullan *et al.*, 2004). Sexual dimorphism of gall morphology is apparent among most eriococcid scale species in which males are known to induce galls.

In many insects, leaf mining behaviours are also important from taxonomic point of view. Insects of four orders have evolved the leaf mining habit: Coleoptera, Lepidoptera, Diptera, and Hymenoptera. There is little difference between a borer and a miner except that the borer feeds deeper in the tissues of plant. Bark beetles that mine under the bark in the phloem and cambium layer could be considered miners or borers. The larvae of flies belonging to family Agromyzidae also show habits of both leaves mining and boring. Some leaf mining species also make galls. Some Chironomidae dig grooves in leaves, removing the upper epidermis and packing the channels with frass. Many larvae of the order Lepidoptera have similar habits. The host associations in many taxa are species-specific. Drew and Hancock (1994) placed 52 species of fruit flies (including 40 new species) found in Asia in *Bacterocera dorsalis* complex. For discussing these species, in addition to standard morphological characters they also used chemical analysis of male pheromones, tissue enzyme electrophoresis and host plant data (both commercial and wild hosts).

The bark beetles are associated with pine trees and can be easily distinguished on the basis of their behavioural characters. The size, shape, and location of the galleries made by adults and larvae as they move through the inner bark can also be used to identify these bark beetles. The parental and larval galleries made by different species of bark beetles differ in shapes and patterns. Besides, the oviposition behaviour in these galleries also varies in different species. Each species of bark beetles also exhibit a definite pattern of infestation.

The host associations of some parasitoids are also important for taxonomic studies. *Nasonia* is a cosmopolitan parasitoid of pupae of cyclorrhaphous Diptera. The genus was earlier regarded as monotypic that including only one species *N. vitripennis*. Darling and Werren (1990) described two more sibling species in this genus namely *N. longicornis* and *N. giraulti* that parasitize blow fly pupae. They summarized behavioural and genetic data for separation of three *Nasonia* species that corroborate morphological analysis.

20.1.4 Nest construction

Many insect species construct nests/cases for rearing of offspring, shelter, and protection. These differ in architecture, location, and materials used for construction. Such behaviours afford diagnostic characters for separation of closely related taxa and phylogenetic analysis. The two bee genera *Anthidium* and *Dianthidium* were slow to be recognized on morphological basis, yet all known species of the former construct their nests of cottony plant fibres, while those of latter from resinous exudations and sand or small pebbles (Mayr *et al.*, 1953). Similarly various species of the termite genus *Apicotermes* are separated on the basis of structures of nests (Schmidt, 1955).

The bees of subfamily Halictinae form a fascinating group for the studies of social structure in insects. They exhibit one of the richest social spectra among insects, ranging from the strictly solitary life to the fairly developed eusociality, with inclusion of communal and semisocial life (Michener, 1974). The variable structure of nests of these bees may indicate to some extent the systematic relationships among the numerous species. Their nest architecture is quite variable among species or supraspecific taxa (Sakagami and Michener, 1962). Batra (1966) described nests of Indian halictines which vary in architecture in different species.

The use of extraneous materials in the construction of nests or larval or pupal cases provides characters at various levels in the classification of caddisworms and bagworms. Weaver and Morse (1986) reviewed phylogeny of the families of Trichoptera to provide a basis for understanding the probable evolution of feeding tactics, and case or retreat constructions by larvae. This phylogeny implies that the ancestral caddisfly larva was probably a tube-dwelling detritivore, inhabiting humus and detrital mats near the shores of lentic or lotic-depositional habitats. This ancestor evolved into a tube-case-making detritivore and scraper in the ancestor of Integripalpia and into a retreat-making collector-gatherer in the ancestor of Annulipalpia. All other larval feeding and case-making tactics evolved from these ancestral habits.

The vespines (yellowjackets and hornets) construct well-designed prominent nests which can be found in different places. The nests are usually surrounded by a masticated paper envelope. The envelope surrounds tiers of round combs. When tiers are added as the colony grows, the outer envelope is expanded. The architecture of nests has been used in phylogenetic studies by using behavioural characters (Greene, 1979) or a combination of behavioural and morphological characters (Carpenter, 1987). The behavioural characters include nest architecture and the behaviour of the colony and the individuals. Greene (1979) traced the phylogeny of yellowjackets of the genera *Vespula* and *Dolichovespula* by using these two kinds of behaviours. In *Dolichovespula* there is no morphological difference between the castes while in *Vespula* there is differentiation. The *Dolichovespula* queen maintains dominance over the workers in the colony by eating the eggs laid by workers while in *Vespula* the queen maintains control over the workers through pheromones. The nest colour, paper texture, and the way the tiers of comb are supported were considered as important taxonomic characters in nest architecture. Greene (1979) concluded that *Vespula* arose from *Dolichovespula* which arose from *Vespa*, which implies that *Dolichovespula* is more closely related to *Vespula*.

Evans and Matthews (1971) studied nesting behaviour and larval structure of several closely resembling nyssonine sand wasps in eastern Australia. Of these, *Austrogorytes bellicosus* makes multicellular nests in coarse sandy soil and provisions them with eurymelid bugs, laying the egg longitudinally on the topmost bug in the cell. *Clitemnestva plomleyi* regarded as a structurally primitive sand wasp, preys upon leafhoppers and is believed to utilize pre-existing vertical holes in hard soil. *Sphecius pectoralis* digs a long burrow in friable sand and makes up to 12 cells, provisioning them with cicadas. Four species of the genus *Bembecinus* also prey on homopterous insects but lay eggs in the empty cells on a pedestal of sand grains and provisioning is progressive.

20.1.5 Defence secretions

Termites constitute ideal taxa for studying behavioural systematics. Many features in termites are chemical based and much information is available on morphological, chemical, physiological and behavioural aspects of termites' secretions. The soldier defence system in higher termites is chemical based and has been used in assessing the relationship among taxa. The frontal and salivary gland secretions have been analysed in a number of termite species, which provide a data base for systematic study. Prestwich (1983) traced phylogeny of Isoptera based on mechanical and chemical defences.

Among lower termites only Mastotermitidae has chemical defence mechanism (Moore, 1968). In Rhinotermitidae members of subfamilies Coptotermitinae, Termitogetoninae, Psammotermitinae, Heterotermitinae, Prorhinotermitinae, and Rhinotermitinae have well-developed frontal glands and the phylogeny of the family was discussed by (Quennedey and Deligne, 1975). Among higher termites (Termitidae) taxa of the subfamily Apicotermitinae are without soldier and with dehiscent workers. The members of Termitinae, Macrotermitinae, and Nasutitermitinae utilize soldier secretions (Prestwich, 1979). The nasute soldiers were earlier hypothesized to be diphyletic (Sands, 1957); however, subsequently considered as monophyletic (Prestwich and Collins, 1981). The defence gland secretions of pentatomid, coreid, and lygaeid bugs also have species-specific composition (Aldrich *et al.*, 1979).

20.1.6 Pheromones

Pheromones are chemicals produced as messengers that affect the behaviour of other individuals of same species. They are usually wind borne but may be placed on soil, vegetation or various items. Tom Eisner, a foremost authority in the science of chemical use by insects, claims that each species of insect relies on some one hundred chemicals in its life, to engage in such routine activities as finding food and mates, aggregating to take advantage of food resources, protecting sites of oviposition, and escaping predation. It has been found that pheromones may convey different signals when presented in combinations or concentrations. Pheromones are species-specific in many groups of insects. Wood *et al.* (2002) analysed extracts of the alarm pheromone found in mandibular glands of three coexisting *Crematogaster* species, viz. *C. mimosae*, *C. nigriceps* and *C. sjostedti* and observed distinct differences in the 28 volatile compounds that were identified. The variations of gland components in these ant species may facilitate species identification and lead to species-specific alarm and defence responses that influence their competitive interactions. 3-Hexanol, 3-methylbutanoic acid, 2-methylbutanoic acid, 3-octanone, 3-octanol, phenylacetaldehyde, 2-phenylpropenal and 3-decanone were found in all three species. The mandibular glands of *C. nigriceps* contain 7 compounds not detected in the two other species; 3-methyl-2-pentanone, 3-methyl-2-pentanol, 3-methyl-2-hexanol, 3-methyl-2-heptanone, 3-methyl-2-heptanol, 2-phenylethanol and 2-methylheptanoic acid. *C. sjostedti* mandibular gland secretion also contains 7 compounds not detected in the other two species; butanoic acid, 3-heptanol, 2-methylpentanoic acid, glycerol, 6-methyl-3-octanol, 2-pyrrolidinone and 3-nonanol. Nonanal was the only compound detected solely in extracts of *C. mimosae*. Urbanova *et al.* (2004) analysed components of marking pheromones produced by labial gland of males of six cuckoo bumble bee species, viz. *Bombus vestalis*, *B. rupestris*, *B. campestris*, *B. sylvestris*, *B. bohemicus*, and *B. maxillosus*. The specimens of *B. rupestris* and *B. campestris* had a significantly different composition of the labial gland secretions. Populations of three other species studied, viz. *B. vestalis*, *B. sylvestris* and *B. bohemicus* produced similar components. A different marking pheromone of *B. maxillosus* was reported for the first time.

20.1.7 Reproductive behaviour

Recent studies on animal behaviour, as well as the revised interpretation of many earlier observations, indicate that differences in reproductive behaviour among insects are the most important factor in restricting random mating between closely related forms. Behavioural studies have contributed significantly in phylogenetic studies and differences in behaviours between closely related taxa are also important, especially if these differences include possible isolating mechanisms (Michener, 1953).

Mechanisms which produce reproductive isolation fall into two broad categories, pre- and postmating mechanisms. Premating isolating mechanisms operate to keep species separate before mating occurs. Behavioural isolating mechanisms rely on organisms making a choice of whether to mate and a choice of mate. Differences in courtship behaviour, for instance, may be sufficient to prevent mating from occurring. A behavioural isolating mechanism should result in some sort of positive assortative mating.

The sexual or behavioural isolating mechanisms that act before fertilization (Mayr, 1942) are thought to precede the evolution of postzygotic isolating mechanisms. We need to understand the genetic basis of sexual isolation to know insight about the early stages of speciation. However, mating behaviours are complex traits, with variation attributable to multiple interacting loci with individually small effects, whose expression depends on the environment. Thus, understanding the genetic architecture of sexual isolation requires that we overcome the twin obstacles of mapping genes causing differences between organisms that, by definition, do not interbreed (Orr, 2005) and solving the problem of genetically dissecting complex behavioural traits (Anholt and Mackay, 2004).

Behavioural differences represent a rich character source for insect systematics at all levels but particularly for species discrimination. Courtship is often species-specific and sometimes is very useful in separating closely related species of parasitic Hymenoptera. Excellent examples of courtship differences occur among cryptic species of parasitic Hymenoptera. In *Aphytis* (Fam. Aphelinidae) obvious behavioural acts are coupled with species-specific sex pheromones released by both sexes. Besides, other behaviours such as host searching and host finding may also differ among close relatives. Behavioural traits may also have some utility at higher taxonomic levels. For example, in blister beetles, genera or species groups can often be characterized by the position the male takes during courtship delivery.

Drosophila species present an ideal model system for investigating the genetic basis of sexual isolation. Several species pairs are only partially reproductively isolated, producing fertile hybrids that can be backcrossed to one of the parental species. Courtship behaviour of *Drosophila* is composed of sequential actions that exchange auditory, visual, and chemosensory signals between males and females, allowing for individual components of the behaviour to be quantified and separated. Courtship is initiated when the male aligns himself with the female, using visual and olfactory signals for orientation. Male also uses pheromonal cues for sex and species recognition, followed by wing vibration to produce a species-specific courtship song. After courtship initiation, the male again uses pheromonal cues by licking the female's genitalia, after which he will attempt to copulate. The female can accept the male or reject him by moving away. Successful copulation is accompanied by the transfer of sperm and seminal fluids.

Spieth (1947) was the first to observe behaviour of species belonging to *Drosophila willistoni* group. Spieth (1966a) studied the mating behaviour of light and dark forms of *Drosophila ananassae* and found that various laboratory stocks developed sexual isolation, the extent of which varied from weak to marked. These forms were later found to be different sibling species (*D. ananassae* and *D. pallidosa*) of the *D. ananassae* complex. Although potentially interfertile, these sibling species are genetically distinct and their isolation is maintained by strong mating preferences (Futch, 1966). *D. ananassae* and *D. pallidosa* are useful for studying sexual isolation because of their sympatric distribution and because of the lack of post-mating isolation between them. Courtship songs play a crucial role in sexual isolation between these two species. Yamada *et al.* (2002b) recorded and analysed male courtship songs in several strains of *D. ananassae* and *D. pallidosa* and observed species-specificity in the courtship song parameters. Doi *et al.* (2001) demonstrated that male courtship songs play a dominant role in female mate discrimination.

Dobzhansky and Pavlovsky (1971) reported a speciation event that occurred in a laboratory culture of *Drosophila paulistorum* sometime between 1958 and 1963. The culture was descended from a single inseminated female that was captured in the Llanos of Colombia. In 1958 this strain produced fertile hybrids when crossed with conspecifics of different strains from Orinocan. From 1963 onward crosses with Orinocan strains produced only sterile males. Initially no assortative mating or behavioural isolation was seen between the Llanos strain and the Orinocan strains.

Rice (1985), Rice and Salt (1990) presented experimental evidence for the possibility of sympatric speciation in *Drosophila melanogaster*. They established a stock population of flies and pupae from the culture were placed into a habitat maze. Newly emerged flies had to negotiate the maze with several environmental gradients to find food. The flies had to make three choices of which way to go. The first was between light and dark (phototaxis), second was between up and down (geotaxis), and the last was between the scent of acetaldehyde and the scent of ethanol (chemotaxis). This divided the flies among eight habitats. The flies were further divided by the time of day of emergence. In total the flies were divided among 24 spatio-temporal habitats. They then cultured two strains of flies that had chosen opposite habitats. One strain emerged early, flew upward and was attracted to dark and acetaldehyde while the other emerged late, flew downward and was attracted to light and ethanol. Pupae from these two strains were placed together in the maze. Flies emerged were allowed to mate at the food site and were collected. The two strains differed in eye colour. A selective penalty was imposed on flies that switched habitats. Females that switched habitats were destroyed. None of their gametes passed into the next generation. Males that switched habitats received no penalty. The two strains showed reproductive isolation after 25 generations of this mating test. Habitat specialization was also produced. The experiment was repeated without the penalty against habitat switching. The result was the same- reproductive isolation was produced. Their results, they stated, show the possibility of sympatric speciation.

Headrick and Goeden (1994) described behaviours, viz. wing displays, courtship, copulation, grooming, feeding, and waste elimination of 48 species of California fruit flies (Fam. Tephritidae). The majority of behaviours exhibited by fruit fly adults were sexual and males displayed majority of these. Male sexual displays were divided into two categories: aggregation displays and courtship displays. The behavioural data-base was used for developing hypotheses about plesiomorphic versus apomorphic behaviours and how they relate to known phylogenies. Headrick and Goeden (1999) characterized some genera of subfamily Tephritinae in a phylogenetic context by discussing various behavioural patterns like wing display, courtship, copulatory induction behaviour, copulation, aggression, territoriality, and resource guarding. They reported that sexual behaviours in these flies are highly diverse and typically overt in their presentation. These patterns assist to promote comparison and will perhaps find utility in defining phylogenetic relationships. Flies with antler-like head projections are known in males of genera of tribe Phytalmiini (Fam. Tephritidae) and vary from the small, disk-shaped protrusions on the lower half of the face to the massive branched structures (Dodson, 1997). These cuticular projections mainly arise from distinct notches at the margins of the eyes and have been studied with respect to the mating behaviours like defence of territories by males and male-male encounters. *Phytalmia* species though exhibit interesting similarities in these behaviours but their behavioural patterns also differ from each other. In *Anastrepha* species (Fam. Tephritidae) daily rhythms in male calling vary sharply and their onset fluctuates from before sunrise to after sunset, depending on the species. Aluja *et al.* (1999) reported that the daily pattern of calling rhythms appears to be one of the most hard-wired behaviours in *Anastrepha* species; but it is also influenced by environmental conditions. They illustrated distribution of calling activities in 20 *Anastrepha* species. Drew (1974) showed that males of each fruit fly species of Dacini responded to either cue-lure or methyl eugenol while some species did not respond to either. White (1999) showed the differences in percentage species belonging to different subgenera of *Bactrocera* and *Dacus* responding to synthetic male lures. Sivinski *et al.* (1999) presented a catalogue of sexual and agonistic behaviours in tephritid fruit flies and listed distribution of sexual behaviours in various taxa.

Hybridization experiments have been a frequent source of data for taxonomic decisions at the species level, particularly in parasitic Hymenoptera (Pinto and Stouthamer, 1994). Species decisions in groups such as the Eulophidae, Aphelinidae, Trichogrammatidae, and some Braconidae have often utilized crossing data. Generally, the weight given to these data in taxonomic decision making depends on whether the results show compatibility or incompatibility. However, the compatibility of laboratory cultures must be used with caution as the laboratory is an obviously artificial habitat which brings individuals together that may not co-occur in nature. In nature there may be several isolating mechanism which prevent the hybridization, e.g. *Trichogramma minutum* and *T. platneri* can be differentiated only on the basis of crossing incompatibility (Pinto and Stouthamer, 1994).

In minute featherwing beetles (Fam. Ptiliidae) spermatozoa are morphologically distinctive and large. In all eight species of *Bambara invisibilis* group sperm and spermathecae show diagnostic differences and are highly coadapted morphologically. Only a small number of sperm are accommodated in the spermatheca, and only a single large egg is matured at a time. Dybas and Dybas (1981) hypothesized that the limited number of male and female gametes in these beetles confers high adaptive value and selective advantage to biomechanical improvements that increase precision and control in transfer and storage of sperm as well as in fertilization of the egg. Natural selection for economy of gametes and intrasexual reproductive competition in the male sex for sperm precedence are postulated as driving forces in the evolution of the coadapted sperm-spermatheca system. Accompanying the strong stabilizing selection and coadaptation within species that is inferred from the morphological data, there is a consistent pattern of interspecific morphological and taxonomic variation in sperm and spermathecae among all eight closely-related sympatric species, suggesting that these complementary structures may function as reproductive isolating barriers. Fielding and Knisley (1995) reported detailed observations and quantifications of the specific copulatory behaviours of two tiger beetles, *Cicindela dorsalis* and *C. puritana* and suggested that differences in such behaviour between cicindelid species and genera may have important functional and phylogenetic significance.

20.2 Behaviour, speciation and evolution

The use of behavioural traits as characters in reconstructing phylogenies is a relatively recent phenomenon. Most early evolutionary studies have been restricted to derive phylogenies from non-behavioural characters. However, in recent years attempts have been made for cladistic analyses by using behavioural traits in many arthropods, and in many cases behavioural traits typically show better phylogenetic resolution than morphological ones.

Darwin (1859) published his famous text '*The Origin of Species by Means of Natural Selection, or the Preservation of Favoured Races in the Struggle for Life*', which led the taxonomists to understand that the similarity among organisms is largely due to propinquity of descent. It was also recognized that similarities in behavioural traits among organisms were probably due to the sharing of common evolutionary histories. As a result efforts were made to trace the phylogenies by combining sciences of genetics, palaeontology, evolutionary biology, and behavioural traits, which resulted into modern synthesis in systematic biology. These kinds of studies were motivated by the desire to learn evolutionary pathways of different kinds of behaviour and not to use the behavioural traits as part of the evidence for actually making evolutionary trees. Using this method, behavioural traits are aligned with the corresponding taxa on dendrograms to see if the behavioural traits had evolved in any sort of non-random array among the various taxa involved. Evans (1966b) discussed biology of sand wasps genera and constructed a hypothetical evolutionary tree from the morphological data. Various behavioural traits were then superimposed on to the tree. Evans suggests that the provisioning in sand wasps proceeds through an evolutionary process beginning with a primitive behaviour of mandibular prey carriage, followed by construction of cocoons, followed by progressive stages of nest provisioning, to the most advanced stage where wasps oviposit in empty cells.

In recent years more sophisticated and repeatable phylogeny-producing algorithms are available and efforts have been made to analyse phylogenies on the basis of behavioural homology. Behavioural characteristics have also been used in the construction of cladograms in some groups. Several kinds of behavioural characters have been used in phylogenetic reconstructions. These characters include: feeding, locomotion, nest architecture, nest location, social systems, territorial behaviour, courtship, parental care, and reproductive behaviour (Wimberger and Quieroz, 1996). Stuart and Hunter (1998) generated cladograms on the basis of cocoon construction behaviour in black flies, and these cladograms were compared with a cladogram generated from morphological and cytological data. The cladogram generated from behavioural patterns showed nearly complete analogy with the morphological/cytological tree, with only minor inconsistencies.

Drosophila species have been subject of intense genetic, evolutionary, developmental, molecular and behavioual studies. The Hawaiian Islands have been recognized as one of the best places in the world for conducting evolutionary research on *Drosophila* species. In Hawaii, flies of the genus *Drosophila* have undergone spectacular adaptive radiation, resulting in the evolution of more than 500 species of *Drosophila* that are found nowhere else on earth. This taxonomic uniqueness is reflected in behaviour and morphology. Hawaiian *Drosophila* sing songs, as do continental *Drosophila;* however, the Hawaiian songs have diverged strongly in form and mechanism of production. These mechanisms are radical departures from the continental wing song mechanism and are further examples of the remarkable behavioural innovation that has occurred in the *Drosophila* of Hawaii during their evolutionary transit through these islands (Hoy *et al.*, 1988).

More than a quarter of the known species of the family Drosophilidae live in the Hawaii archipelago. Most Hawaiian *Drosophila* are endemic and all of the endemic species may be descendants of a single gravid female arrived at the oldest island Kauai about six million years ago. Despite the huge morphological diversity, there is little genetic diversity among the endemic species. It appears that courtship behaviour has driven speciation in Hawaiian *Drosophila*. Even closely related sympatric species have differences in their reproductive behaviour. This is an example of sexual selection causing speciation probably due to the effects of genetic drift in small founder population. Spieth (1966b) was the first who suggested that elaborate, oftentimes bizarre, secondary sexual structures found in males of most Hawaiian *Drosophila* species are manifestations of complex courtship behaviour displayed by the males. The sexual selection was considered as an extremely important part of evolutionary process in the Hawaiian fauna. Carson (1997) presented a discussion of how sexual selection in "local genetically rich populations can either maintain a cohesive status quo or forge a disruptive evolutionary novelty." He suggested that evolution of novelty in Hawaiian *Drosophila* is even more evident when the population is subjected to bottlenecks

Rhagoletis pomonella is a model for incipient sympatric speciation (divergence without geographic isolation) by host-plant shifts. Host-specific mating is a key feature of *Rhagoletis* biology, as it is for many phytophagous insect specialists (Bush, 1966). Because these flies overlap broadly in their geographic distributions across North America and appear largely interfertile in crosses, Bush (1969) hypothesized that the six or more members of the *R. pomonella* group radiated sympatrically in the absence of geographic isolation by means of host-plant shifts. The apple infesting population of *R. pomonella* originated by a shift from hawthorn in the mid-1800s represents an example of host race formation in action. In particular, the shift of the species *R. pomonella* from its native ancestral host hawthorn, *Crataegus* species to introduced domesticated apple, *Malus pumila* is often cited as an example of incipient sympatric speciation (Berlocher and Feder, 2002).

Feder and Bush (1989) suggested the divergence of apple maggot fly, *Rhagoletis pomonella* into two sympatric races because of the host fruit trees, the two emerging races choose to inhabit. One race lives and breeds on hawthorn trees, while the divergent race prefers apple trees. This was because apple trees were only introduced to North America around 200 years ago and the fruiting in two hosts occur at different times. Natural selection has adapted the 'apple-based' fly race to this difference in timing by matching their emergence to when the apples mature. Consequently the two races never really come into contact with each other and thus they are reproductively isolated. Hybrids between the two races have been documented, but they were found to have reduced fitness (see section 19.8). Linn *et al.* (2003) showed that *R. pomonella* flies use volatile compounds emitted from the surface of ripening fruit as important chemosensory cues to recognize and distinguish among their host plants. Linn *et al.* (2004) further reported that host-specific mating may play a dual role as an important postzygotic as well as a premating reproductive barrier to isolate sympatric *Rhagoletis* flies.

The European cherry fruit fly, *Rhagoletis cerasi* shows strong ecological similarities to *R. pomonella*, which are: (i) infestation of two different host plants, viz. European fly honeysuckle, *Lonicera xylosteum* and sweet cherry, *Prunus avium*, and (ii)

divergent phenological and behavioural adaptations of flies on different hosts (Schwarz *et al.*, 2003). Flies on these two hosts show a marked difference in emergence time that appears to be an adaptation to the fruiting time of their respective host plants (Boller and Bush, 1974). The two forms show a differentiation in their host choice behaviour and their response to host marking pheromone. For these reasons the two forms of *R. cerasi* are generally regarded as host races.

The biology and behaviour of dacine fruit flies (Fam. Tephritidae) are such that species have a close-knit relationship with their host plants. The reproductive biology of species is dependent upon the factors within the host plant habitat. Consequently, understanding the patterns of distribution of their endemic host plants and speciation of their host plant ecosystems will contribute to elucidating the phylogenetic relationships of the Dacini. Evidence indicates that there has been considerable evolution of dacine species (particularly of genus *Bactrocera*) in association with their rain forest host plant species (Drew and Hancock, 1999).

Fireflies (Fam. Lampyridae) present one of the best systems available for studying sexual selection and phylogeny. The origin of photic organs is known to precede their use in the context of sexual signalling, originally playing a role as a nocturnal warning signal for unpalatable larvae (Branham and Wenzel, 2001). Sivinski (1981) hypothesized that lampyrid larvae use their photic emissions as aposematic displays. The photic signals in some species are used in sexual display and females discriminate between males on the basis of such signal parameters. Lloyd (1997) offered the possibility that certain behaviours, such as males flashing their distinctive signals, were likely convergent among many species. The luminous displays serve to communicate an individual's sex, species, and exact location (Barber, 1951). Thus, the primary function of luminous behaviour in adults is to facilitate pair formation (Lloyd, 1966). Some fireflies do not produce light and rely on pheromones for pair formation while in some other species the sedentary female produces a pheromone that acts as a long-range sexual signal and emits a glow that males use to locate the source of the pheromone at close range. Lloyd (1971) termed this type of communication system as 'Signal System I'. In most cases males fly while broadcasting their species-specific flash pattern and females are sedentary, responding to the male signal with a species-specific signal following a species-specific delay. A short flash dialogue commonly ensues until the male either departs or lands near the female, whereupon mating may occur. This type of photic communication system has been termed 'Signal System II'. Lloyd (1971) argued that signal system II was derived from signal system I.

Carpenter (1987) used cladistics to analyse the relationships of the genera of Vespinae (*Vespula, Dolichovespula, Provespa* and *Vespa*), and the major species groups of *Vespula*. In this analysis, seventeen morphological and seven behavioural characters were analysed separately as well as together. The morphological characters alone resulted in the phylogenetic tree. Within the genus *Vespula*, the subgenera *Rugovespula* and *Paravespula* (=*V. vulgaris* group) form a clade. The sister group to this is a group containing *Vespula* (*S. Str.*) (=*V. rufa* group) and the *V. squamosa* species group. *Dolichovespula* is a sister group to *Vespula* (*S. Lat.*); *Provespa* is the sister group to this clade, and *Vespa* is the sister group to all the other vespines.

Alexander (1991) provided an analysis of *Apis* phylogeny using both morphological and behavioural characters. A total of 20 (including two behavioural) characters were used in the phylogenetic reconstruction of the six species. Wenzel (1993) used a total of 21 behavioural characters in a study of paper wasp evolution. Comparisons were made of a phylogeny based on morphological characters with one derived from behavioural characters. Of the two cladograms presented, neither showed much difference in the branching patterns of individual taxa. Crespi *et al.* (1998) compared several types of data (including mitochondrial DNA, morphology, behaviour, and gall morphology) to determine the evolution of the soldier caste in gall-forming thrips. Of these, all data except gall morphology provided a fairly consistent phylogeny. Gwynne (1995) discussed Orthoptera phylogeny and created a cladogram using 67 characters, of which two were behavioural. The main behavioural characters included male's position during copulation and facing of both sexes during copulation. The morphological characters like presence of a harp, folding of the tegmina, and others pertaining to singing were used. It was observed that behaviour has significant impact on the evolution of particular body structures.

Bosch *et al.* (2001) used cladistic analysis to study the evolution of 29 characters related to nesting behaviour, nest provisioning, oviposition, cocoon spinning, defecation, and life cycle (nest characters) in 11 species of *Osmia* bees (Fam. Megachilidae). The phylogenetic trees based on molecular and behavioural characters supported the monophyly of the genus *Osmia*.

Chapter 21

Behavioural Manipulations and Insect Pest Management

Pests evolved prior to human on this earth; however, the term came into existence when the interest of man clashed with them. An overview of the world of pest animals reveals their great diversity and constancy. Efforts made by man to manage the problems posed by pests also have a long history and human civilization has been competing with insects, rodents, diseases, and weeds for survival throughout its history. Man through his technology has attempted and created tools to combat these pests. A pest is any injurious or nuisance species, the control of which is felt to be necessary either for economic or for social reasons. A pest is considered to be anything that: (i) injures humans, animals, crops, structures, or possessions, (ii) competes with humans, domestic animals or crops for food, feed, or water, and (iii) spreads disease to humans, domestic animals, or crops.

Presently we are faced with acute problems of saving our crops, trees and many other items from the attack of insect pests. Further, with an increase in global population and development of new scientific innovations, there has been considerable advancement in the pest management practices. This had also resulted in the discoveries of many new pesticides. With the advent of synthetic organic insecticide DDT in 1939, followed by a number of organophosphates, terpenes, carbamates, pyrethroids, etc. and their indiscriminate use throughout the world led to hazards and disturbances to the 'balance of nature'. Taking these facts a number of dreadful pesticides were banned in different countries and concept of safe, economic and judicious application of pesticides came into existence (Yazdani and Agarwal, 1997). A new era started in the field of pest management when Rachel Carson (1962) published her famous manuscript '*Silent Spring*'. She stressed on the preservation of natural fauna and flora and on the need of conserving natural resources. Consequently the concept of pest management was postulated where emphasis was given to bring a pest population below economic injury level rather than eliminating the pest. The practice of using insect behaviour to protect valued resources has been known from centuries; however, in the last 35 years there has been an increased desire to reduce the reliance on pesticides and considerable efforts have also been made to manipulate pests behaviour to protect various resources and there has been increased interest in the management of insect pests by behavioural means. The behaviour of insect results from the integration by its central nervous system of a variety of inputs that derive from stimuli acting on different receptors. To manipulate behaviour one must change either the inputs or the processing of those inputs by the central nervous system. The insect behaviour is usually manipulated through inputs to the behaviour and more specifically through the stimuli that generate these inputs (Harris and Foster, 1995). Foster and Harris (1997) discussed methods using stimuli to manipulate behaviour of a pest for the purpose of protecting a valued resource. They defined 'manipulation' as the use of stimuli that either stimulate or inhibit behaviour and thereby change its expression. Three principals are involved in behavioural manipulation method:

(i) Behaviour of the pest.
(ii) A means by which the behaviour is manipulated appropriately.
(iii) A method that utilizes the behavioural manipulation for protection of a resource from a pest.

The methods were divided into two categories: those that manipulate behaviour over a long-distance (finding-type behaviour) and those that manipulate behaviour at a short-distance (acceptance-type behaviour). To describe the type of stimuli, mainly chemicals, they used the terms attractant and repellents for long-distance stimuli and stimulant and deterrent for short-distance stimuli.

21.1 Chemical control of behaviour

Although conventional insecticides are still a basis of chemical control, they are gradually being superseded by less toxic compounds that disrupt insect development or modify behaviour. The use of behaviour-modifying chemicals has found wide application, particularly with insect pheromones, as a tool of great importance in a number of pest management systems. Some of these new chemical weapons are much safer for the environment and more species-specific than most conventional insecticides. Such chemicals in its environment mediate much of an insect's behaviour and can be used to attract pests to traps or baits, or repel them from residence, crops, or domestic animals. Behavioural messages are conveyed by a wide array of chemical compounds, which are collectively known as 'semiochemicals' (a term proposed by Law and Regnier, 1971). These chemicals may facilitate communication between the members of a single species (pheromones) or between members of different species (allelochemicals) (Whittaker and Feeny, 1971). Functionally, semiochemicals may have a wide range of activity. They may serve as attractants or repellents, stimulate or inhibit feeding, provoke flight or inhibit it, or simply elicit behaviour patterns at inappropriate times.

To recognize the potential of semiochemicals in crop protection, a greater understanding of insect/insect and insect/plant interactions and insect chemical ecology is essential. Semiochemicals, when employed alone, often give ineffective or insufficiently robust pest control. Use of semiochemicals should, therefore, be combined with other approaches in integrated management strategies. The main components of such strategies are pest monitoring, to allow accurate timing of pesticide treatments; combined use of semiochemicals, host plant resistance and trap crops, to manipulate pest behaviour; and use of selective insecticides or biological control agents, to reduce pest populations. The objective is to draw together these approaches into a 'push-pull' or 'stimulo-deterrent diversionary strategy' (SDDS). In SDDS, the harvestable crop is protected by host-masking agents, repellents, antifeedants or oviposition deterrents (Pickette *et al.*, 1997).

Allelochemicals are chemicals that are significant to individuals of a species different from its source species. Allelochemicals are subdivided into several groups depending on whether the physiological or behavioural response of the receiver is adaptively favourable to the emitter but not the receiver (allomone); response is favourable to the receiver but not the emitter (kairomone); or response is favourable to both emitter and receiver (synomone) (section 11.4). Allomones are mostly defensive chemicals, producing negative response in insects and reducing chances of contact and utilization. They act as repellents, flight arrestants, oviposition and feeding deterrents, and toxicants. Kairomones on the other hand, are advantageous to an insect, promoting host finding, oviposition and feeding. They include attractants, excitants and stimulants.

Pheromones are released by member of a species and cause a specific physiological or behavioural response in another member of the same species. A pheromone may be 'releaser', which induces an immediate behavioural change, or 'primer' which does not result in immediate behavioural change. Pheromones may be further classified on the basis of the interaction mediated, such as alarm, aggregation, trail-marking, epideictic, social, and sex pheromone. It is the sex pheromones of insects that are of particular interest to agricultural integrated pest management (IPM) practitioners. The methods using pheromones for pest management, therefore, depend on utilizing the normal behavioural response of an insect to a synthetic pheromone source/blend and destroying it subsequently. As sex pheromones elicit behavioural responses from their conspecific counterparts at very low doses, low application rates can be used to influence the natural communication between insects in all pheromone-mediated control techniques.

The concept of IPM is based on the recognition that no single approach to pest control offers a universal solution, and that the best crop protection can be provided by a fusion of various tactics and practices based on sound ecological principles. Insects for performing daily routine activities as finding food and mates, aggregating, protecting sites of oviposition, and escaping predation use such chemicals. It has been observed that pheromones may convey different signals when presented in combinations or concentrations. They travel slowly, do not fade quickly, and are effective over a long-range.

Important texts enumerating the role of behaviour-modifying chemicals in insect pest management are published by: Shorey and McKelvey (1977), Mitchell (1981), Nordlund *et al.* (1981), Bell and Cardé (1984), Kydonieus and Beroza (1986), Jutsum and Gordon (1989), Ridgway *et al.* (1990), Metcalf and Metcalf (1992), Arn, *et al.* (1992), Cardé and Minks (1997), Howse *et al.* (1998)), Hardie and Minks (1999), Srivastava and Bhattacharya (2003), and Gut *et al.*, (2004).

21.2 Pheromones

Karlson and Lüscher (1959) proposed the term 'pheromone' to represent chemical(s) used for communication between individuals of a given species. Sex pheromones are among the most powerful of chemical attractants. Ever since Butenandt *et al.* (1959) first discovered them in adult silkworms, *Bombyx mori*, these chemicals have aroused great interest because of their potential as pest management agents. With the rapid growth in the identification of insect pheromones in the early 1970s, great expectations were evoked as to the potential of these chemicals for use as effective and ecofriendly methods of insect control. It was believed that large-scale replacement of insecticides would be possible through the use of pheromones. However, such claims have remained largely unfulfilled with the exception of certain Lepidoptera and bark beetles (Dhaliwal *et al.*, 1994). Thousands of insect pheromones have been identified and today we have a much clearer view of the limitations and possibilities associated with use of insect pheromones in IPM programmes.

The primary potential applications of insect pheromones in a sophisticated and environmentally friendly IPM system are for detection and monitoring of populations, mass trapping, attract and kill, and for mating disruption, etc. These uses take advantage of sex pheromones on which a vast majority of insect pests rely to mediate reproduction. Long-lasting pheromones allow marking of territorial boundaries or food sources. Other signals are very short-lived, and are intended to provide an immediate message, such as a short-term warning of danger or a brief period of reproductive readiness. Over the last 40 years, scientists have identified pheromones from over 1,500 different species of insects and have found wide application in the fields of agriculture, forestry, and urban pest management.

The use of pheromones has been instrumental in increasing the effectiveness of both monitoring insect populations and in providing adequate information to enable implementation of cost effective control. In an insect pest management the population dynamics of a given species may not be precisely known, but long term monitoring can help to elucidate patterns in population life cycles. These patterns can then be combined with the known biology of the insect to define the parameters of an effective control campaign (Norton and Mumford, 1993). Over the last two or so decades information collected from insect monitoring has led to the production of pest forecasting systems that allow agriculturists and livestock farmers to begin spraying or chemotherapy at precise time intervals for maximum impact.

21.2.1 Exploiting pheromones

The specificity of a pheromone and incredible ability of an insect to detect its minute quantities, have offered much potential to use these chemicals as a pest management tool. Once the structure of the behavioural chemical is determined it can be produced synthetically and then efficient systems are devised to utilize the same for a range of management purposes. Besides, consumer demand for pesticide-free produce is increasing steadily, pushing pheromone use forward. The availability of synthetic pheromone 'gossyplure' formulated to disrupt mating of pink bollworm, *Pectinophora gossypiella* has fostered alternative management strategies and efforts have been made to control this pest by using synthetic commercial formulation in many countries. For the control of pink bollworm, Baker *et al.* (1990) used Shin-Etsu plastic 'rope' dispenser at the beginning of flowering (pin-square stage) at the base of the cotton plant at a rate of 1,000/ha with a total application of 75 g of pheromone per hectare in southwestern United States. The sex pheromone released by female diamondback moth, *Plutella xylostella* was found to be most attractive to males. The components of pheromone of the female moth are: (Z)-11-hexadecenal and (Z)-11-hexadecenyl acetate. These two compounds have the greatest attracting powers when in a ratio of 7:3 respectively (Tamaki *et al.*, 1977). With the discovery of dispalure, (Z)-7,8-epoxy-2-methyl octadecanol, the female sex pheromone of gypsy moth, *Lymantria dispar* (Bierl *et al.*, 1970), an extra tool for its monitoring and control was offered. The female sex pheromone of codling moth, *Cydia pomonella* is commonly called codlemone and its main component (*E,E*)-8,10-dodecadien-1-ol (*E,E*-8,10-12:OH) is thought to camouflage the female's natural emission, thus pushing the males into false trail following (Bartell, 1982). The control of codling moth through mating disruption is being adopted in area-wide programmes in many countries.

The cotton boll weevil, *Anthonomus grandis* is a good example of an important pest where mass trapping has been applied successfully. Traps for the weevils are baited with grandlure, which acts both as sex pheromone attracting females and as an aggregation pheromone in early and late season when it attracts both sexes. Bark beetles of the genera *Dendroctonus, Ips,* and *Scolytus* are the most destructive pests of forests in the northern hemisphere and cause heavy damage to forest trees. The southern pine beetle (SPB), *Dendroctonus frontalis* is considered to be the most important cause of damage and mortality to pine trees. The chemical ecology of SPB involves a complex of chemical compounds produced by beetle as well as host plants. These chemicals serve both intra- and interspecific communication functions. Upon landing on a tree, females bite into the bark and if the host is found suitable, they release the primary aggregation pheromone, frontalin, and the synergist *trans*-verbenol. Besides, females also release small quantities of verbenone and *endo*-brevicomin, both serving as synergists to enhance the attractiveness of frontalin. These compounds, in combination with host volatiles, primarily α-pinene, stimulate mass aggregation of conspecifics (predominantly males) to the host (Renwick and Vité, 1969). The concentration of verbenone and *endo*-brevicomin also increases with an increase in beetle population. However, at some unknown threshold, these compounds begin to inhibit beetles response to the aggregation pheromones and cause arriving beetles to switch their attack to neighbouring trees (Payne *et al.*, 1978). The switching of mass attack from one host tree to a neighbouring tree may be the result of both the cessation of release of attractive compounds (frontalin and α-pinene) and the increased concentration of inhibitor pheromones, (verbenone and *endo*-brevicomin).

Several natural enemies of SPB also respond to kairomones emitted from the beetles or associated fungi, including: the bark beetle destroyer clerid, *Thanasimus dubius* (to the aggregating pheromone, frontalin); the predatory fly, *Medetera bistriata* (to the semiochemicals released by infested logs); the parasitoid, *Dinotiscus dendroctoni* (to the compounds collected from beetle-infested trees); and some other species. The use of semiochemicals as management tools shows considerable promise in reducing damage and mortality by bark beetles. Some of these compounds have already been successfully used to monitor population trends or as a mass trapping and/or disruption tactic against forest insect pests (Borden, 1993).

The banana weevil, *Cosmopolites sordidus* produces an aggregation pheromone that attracts both males and females. This pheromone has been identified and synthesized and is being recommended as an effective agent in the trapping and control of the weevil. The synergism between banana plant extract (kairomones) and the synthetic pheromones in attracting *C. sordidus* need to be exploited (Tinzaara *et al.*, 2002).

Other instances in which efforts have been made to monitor/manage populations of pests or successful population reduction/suppression were achieved in case of insect pests like: beet armyworm, *Spodoptera exigua;* tobacco cutworm moth, *S. litura;* Indian meal moth, *Plodia interpunctella;* Mediterranean flour moth, *Ephestia kuehniella;* almond moth, *Cadra cautella;* tobacco budworm, *Heliothis virescens;* tomato pinworm, *Keiferia lycopersicella;* European corn borer, *Ostrinia nubilalis;* yellow stem borer, *Scirpophaga incertulus;* rice stem borer, *Chilo suppressalis;* cocoa moth, *Conopomorpha cramerella;* corn earworm, *Helicoverpa zea;* rough bollworm, *Earias vitella;* oriental fruit moth, *Grapholita molesta;* grapevine moth, *Eupoecilia ambiguella;* grape fruit moth, *Lobesia botrana;* leopard moth, *Zeuzera pyrina;* tea tussock moth, *Euproctis pseudoconspersa;* brinjal fruit borer, *Leucinodes orbonalis;* pink maize stalk borer, *Sesamia nonagriodes;* diamondback moth, *Plutella xylostella;* currant clearwing moth, *Synanthedon tipuliformis;* Asiatic rhinoceros beetle, *Oryctes rhinoceros;* Asian palm weevil, *Rhynchophorus ferrugineus;* palm weevil, *Rhynchophorus palmarum;* sweet potato weevil, *Cylas formicarius;* West Indian cane weevil, *Metamasius hemipterus;* spruce bark beetle, *Ips typographus;* confused flour beetle, *Tribolium confusum;* cigarette beetle, *Lasioderma serricorne;* citrus mealybug, *Planococcus citri,* etc.

Presently, pheromones have been used or perceived for use in the ways as detailed below:

Detection and monitoring. Pheromones are species-specific and every species has an exclusive sex pheromone. These can be exploited to determine the pest population size, temporal distribution of pest, detection of outbreaks, and establishment of emergence time of adults. Traps containing the appropriate sex pheromone are placed in the fields or orchards and reliable information concerning the size of the pest population can be acquired. Trap design is also critical for effective use of traps in monitoring insect populations. Traps vary in design and size, depending on the behaviour of the target insects (Fig. 21.1). Consistent trapping protocols are essential for population evaluations, determination of spray thresholds, and year-to-year comparisons. Pest cyclic occurrence can also be determined and subsequently one can decide when to spray pesticides. This approach does not eliminate the use of standard pesticides, but it does reduce the

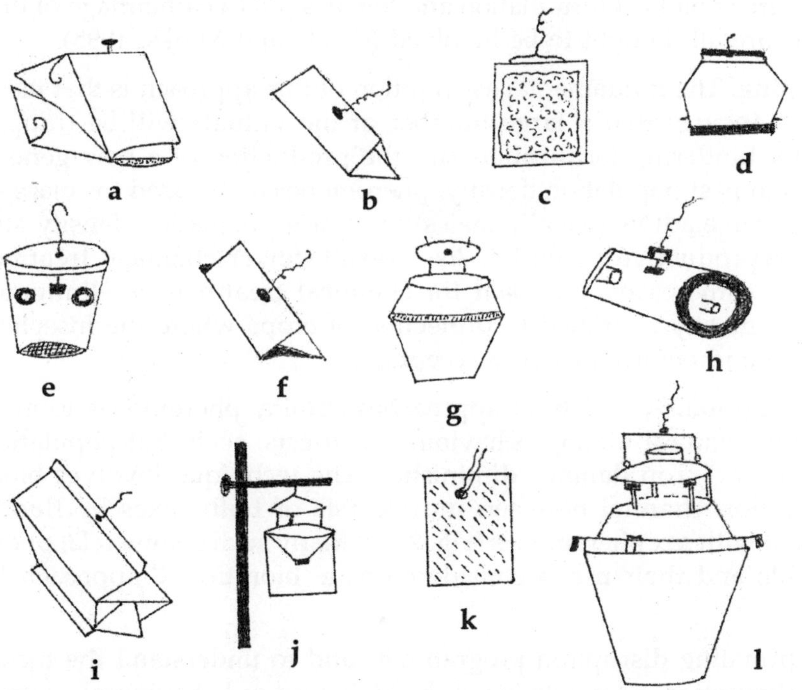

Fig. 21.1. Some commercially available traps used for monitoring/reducing insect pest population. Pheromones/attractants are used to lure insects. (a)- Wing trap, (b)- Delta trap, (c)- Vertical yellow sticky trap, (d)- Pherocon II trap, (e)- Nadel trap, (f)- Jackson trap, (g)- Funnel trap, (h)- Steiner trap, (i)- Tent trap, (j)- Funnel trap, (k)- Sticky card trap, (l)- Bucket trap.

quantity of pesticide applied to the field. Efficient monitoring makes it possible to time treatments to coincide with the start of the pest's active season or with massive attacks and thus, the pheromone serves as a sensitive detector that may be exploited to make intelligent decisions about pesticide application. The economic threshold (ETLs) levels of insect pests can also be determined in terms of pheromone trap (Plate 21.1) catches to decide about the timing of insecticide applications (Jutsum and Gordon, 1989).

The development of active pheromones for many insect pests and their formulation in lures for deployment in custom-designed traps is increasingly providing information required to complement and sustain rational pest management systems. Thus, monitoring traps offer benefits, viz. (i) early detection and location of infestation, (ii) indication of the severity of infestation, (iii) determination of the most favourable time for control measures to be applied, and (iv) a quality control aid.

Pheromones have been effectively used for monitoring and control of lepidopterous pests. The exact mechanism of how the synthetic pheromone disrupts the orientation of males is not precisely known, although habituation, competition between source points of formulation and females plus a camouflage of the female pheromone are all thought to be involved (Cardé and Minks, 1995).

Mass trapping. The fundamental assumption of this approach is that use of traps in the field ensures that a large number of individuals will be trapped, thus significantly hindering fertilization and the production of a new generation of pests. At high pest population density, pheromone can be used for mass trapping of sexually active adults (usually males) to reduce population density and lower the pest's reproductive potential. As a result, fewer chemical treatments are required; in some cases, the need for chemical treatment is eliminated. This technique is mostly restricted to protection of crops where the insecticides are not effective and environmentally acceptable.

Mating disruption. A variety of approaches employ pheromones to manipulate or disrupt the natural mating behaviours of insects, such that population levels are reduced and crop damage diminishes. The technique involves blocking of communication channel between individuals of both sexes by flooding the environment with sex pheromones (Fig. 21.2). Many such commercial pheromones are available and their use is considered as a 'biorational' approach to insect control.

To develop mating disruption programmes and to understand the mechanisms of mating disruption, a knowledge of the biological or behavioural characteristics of the target pest, and how pheromones move and distribute themselves within agricultural systems is essential (Cardé and Minks, 1995). In has been shown that the pheromone plume is not a uniform cloud, but rather a series of filaments of pheromone interspersed with pockets of zero to low concentrations due to natural turbulence in the air (Sanders, 1997). The male is exposed to a series of rapidly changing concentrations, which requires both the interpretation of the odour and a resetting of the antennal receptors within milliseconds to perceive the next pheromone molecule (Leal, 2003). The male proceeds to fly upwind in a series of surges interspersed with crosswind casting when it detects pockets of air with reduced concentrations of pheromone. For most species, a turbulent plume with pockets of high and low concentrations is required for proper upwind flight; if a column of air is filled uniformly with pheromone, then the sensory structures may fatigue and no longer respond to the pheromone. Cardé *et al.* (1998) demonstrated that the mating disruption of the pink bollworm, *Pectinophora gossypiella* occurs from a combination of mechanisms including camouflage, competition between pheromone dispensers and females, habituation, and some advancement of the rhythm of the male's response to the female's pheromone, which may result in asynchrony in sexual behaviours.

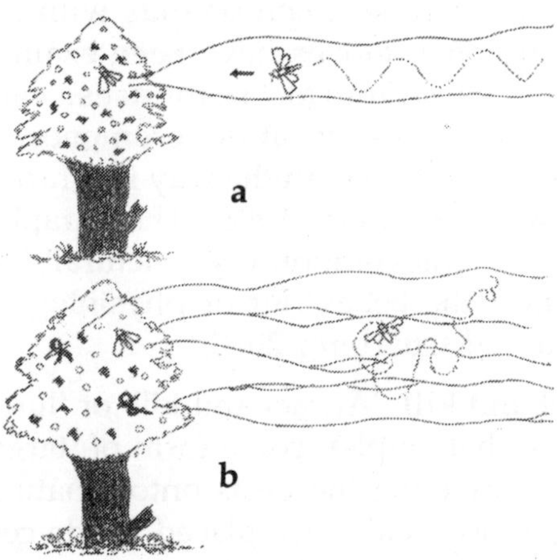

Fig. 21.2. Mating disruption in which the moth uses its own communication system to its detriment. (a)- Male moth follows the pheromone to its source- the calling female and mating commences. (b)- Male moth confused by the synthetic pheromone released from the dispenser and fails to locate the female.

Mating disruption may be accomplished in two ways, false trail following or confusion. The formation of false trails from dispensers that act as female (or male) mimics, greatly outnumber the calling insects. False trail following results due to placing of more point sources of pheromone in a crop than the anticipated numbers of females in that crop. Emission of pheromone is relatively low from each source such that a downwind trail is created and not lost in a background of released pheromone. Males following these trails are thought to spend their mating energies in pursuit of artificial pheromone sources. Male confusion is considered to be the result of ambient pheromone concentrations sufficient to hide the trails of calling females. It may occur by habituation of responsiveness of the nervous system as a result of exposure to the pheromone. Specific receptor sites on the antennae respond to the pheromone molecules and mating disruption can also be achieved by blocking of antennal receptors through the use of modified pheromones. When a receptor site is continually activated by high ambient concentrations of pheromones, the receptor site becomes unresponsive and the insect becomes navigationally sightless.

Numerous synthetic compounds with molecular structures similar to naturally occurring pheromones have been examined and tested for their ability to elicit similar behavioural responses. Successful compounds are termed 'parapheromones' or 'antipheromones' depending on whether they attract the insects or interfere with the way in which insects perceive the natural pheromones (Dhaliwal and Arora, 1996). The parapheromones have been loosely defined as "anthropogenic compounds structurally related to natural pheromone compounds that affect the behaviour or physiology of the insect's communication system" (Renou and Guerrero, 2000).

Attract and kill. 'Attract and kill' or 'lure and kill' strategies fall into two groups: (i) those that employ some form of 'target device', e.g. a trap, and (ii) those that rely on attracting the pests onto a natural surface, e.g. host tree foliage. In first strategy insecticides are placed inside centrally located containers or other point sources that are baited with sex or aggregation pheromones, eliminating the need to scatter the insecticide in the field. The insect is attracted to the dispenser loaded with the pheromone and comes into contact with the insecticide. The efficiency of chemical pest control is increased, and environmental pollution is minimized.

Inhibition of oviposition. 'Anti-oviposition' or 'oviposition deterring' pheromones have been reported in some Lepidoptera and Diptera. Such pheromones act as epideictic pheromones with respect to their effects of reducing intraspecific competition. It was long believed that females of large white cabbage butterflies, *Pieris brassicae* add an anti-oviposition pheromone to the eggs, which inhibit egg-laying by conspecific females (Rothschild and Schoonhoven, 1977). Recent work; however, has shown that the host plant itself responds to the presence of *P. brassicae* eggs and that systemic changes in leaf surface chemistry may be the primary source of an oviposition-deterring effect (Blaakmeer *et al.*, 1994). Females of European cherry fruit fly, *Rhagoletis cerasi* and apple maggot fly, *R. pomonella* mark the host fruits with oviposition deterring pheromones (ODP) around the oviposition punctures (section 18.4.3). The pheromone deters other females of the apple maggot fly from ovipositing on the same host (Prokopy, 1972a). Aluja and Boller (1992) reported that application of synthetic host marking anti-oviposition pheromone in a cherry orchard significantly reduced the infestation of *R. cerasi*. Oviposition deterrents for females of other fruit flies, viz. *Bactrocera oleae, Ceratitis capitata, Anastrepha fraterculus* have also been reported. Other preliminary trials using such pheromones in IPM programmes are in progress. In the mustard beetle, *Phaedon cochleariae* adults are known to be deterred from feeding and oviposition by the exocrine glandular secretion of conspecific co-occurring larvae. The exocrine larval secretion released by feeding larvae may adsorb to the surface of Chinese cabbage leaves, and thus, convey the information about their former or actual presence (Rostás and Hilker, 2002).

Alarm pheromone. Many aphid species release a linear sequiterpenoid, (*E*)-β-farnesene (EBF), which acts as alarm pheromone upon the approach or attack of a predator or parasitoid (Bowers *et al.*, 1972). This volatile chemical is released through the cornicles as part of a defence secretion and can be detected on the antennae of neighbouring aphids. When aphids detect EBF they stop feeding and are stimulated to disperse rapidly by walking away or dropping off. Sensitivity to EBF varies among aphid species and alates of responsive species seem to be particularly sensitive. EBF is also found in some plants like common hop, *Humulus lupulus* and wild potato, *Solanum berthaultii* that repels the cabbage aphid, *Myzus persicae* (Pickett *et al.*, 1992). This pheromone has been used to mobilize feeding aphids so that insecticidal sprays more easily target them.

Francis *et al.* (2004) studied the attraction of three aphids, *Myzus persicae, Acyrthosiphon pisum*, and *Brevicoryne brassicae* and three plants, *Vicia faba, Brassica napus*, and *Sinapis alba* by the predatory two-spotted ladybird beetle, *Adalia bipunctata* larvae and adults. Both predatory larvae and adults were attracted only by *A. pisum* and *M. persicae* on all hosts. (*E*)-β-farnesene, was the effective semiochemical for the ladybird. Plant leaves alone or in association with nonstressed aphids did not have any attraction for the predator in all cases.

21.2.2 Pheromone-dispensing technologies

Many pheromone-dispensing technologies have shown promise and success under commercial conditions and are now available to growers. These include:

Microencapsulated pheromones (MECs). These are small droplets of pheromone enclosed within a polymer capsule, which controls the pheromone release rate. The capsules can be applied as suspensions by using a sprayer. The effective field longevity of these materials generally ranges from days to weeks depending on climatic conditions, capsule size and properties, and the pheromone components.

Hand-applied dispensers. Large, hand-applied dispensers include systems with an impermeable reservoir fitted with a permeable membrane for regulating pheromone release, and analogous laminate dispensers that consist of a central pheromone-containing core sandwiched between two polymer films. These dispensers (Fig. 21.3) may be of various shapes and sizes and can be applied by specialized ground or air applicators. Pheromones can be mixed into paraffin

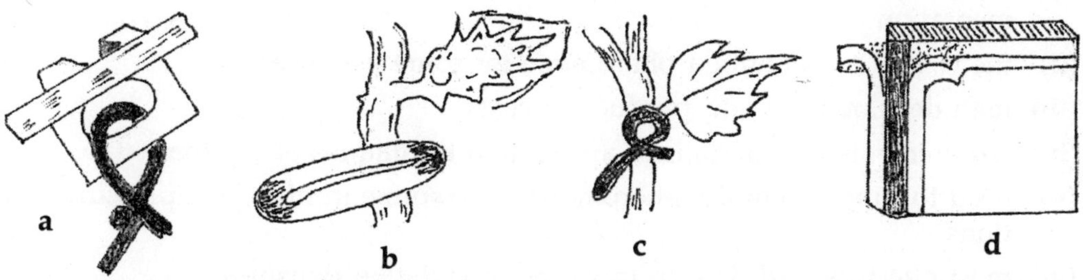

Fig. 21.3. (a)- Pheromone dispenser, (b)- Selibate ring, (c)- Twist-tie, (d)- Hercon's plastic laminate lure.

wax or aqueous paraffin emulsions and applied directly to the tree using hand-dispensers. These dispensers also include pheromone-impregnated polymer spirals, and twist-tie 'ropes' (Plate 21.2) that consist of a pheromone-filled plastic tube with a wire spline along one side.

Hollow fibres. These consist of a short, impermeable plastic tube that is sealed at one end and filled with pheromone. The fibres are often combined with an adhesive material during application and may require specialized ground or air equipment to apply.

High-emission dispensers. These dispensers were developed to emit larger quantities of pheromone and use fewer dispensers. The pheromone if filled in a pressurized aerosol can, which dispenses puffs of pheromone at fixed time intervals.

21.2.3 Advantages and limitations of pheromone-based pest management systems

Many advantages of pheromone-based pest management systems are common to other biologically based management approaches, including:

(i) highly selective
(ii) no detectable residues
(iii) negligible health risks
(iv) a more rapid registration process
(v) no accumulation in wildlife or groundwater
(vi) reduced worker re-entry or pre-harvest intervals
(vii) limited impacts on other management practices
(viii) strong tool for managing insecticide resistance pest species
(ix) only the primary target species responds to the pheromone
(x) no effect on non-target or biological control agents
(xi) does not result in outbreaks of secondary pests or pest resurgence

The limitations of such systems are:

(i) high degree of selectivity may be a significant obstacle to large-scale implementation
(ii) the lack of an identified pheromone for some pest species
(iii) high development and production costs
(iv) requirements for specialized application techniques or equipment
(v) need to supplement the pheromone programme in high pest-pressure situations
(vi) most effective with low to moderate population densities
(vii) monitoring the target pest with pheromone traps also can become problem-

atic, because the pheromones used to disrupt mating will also prevent pest insects from locating traps

21.3 Allelochemicals

Plant secondary compounds acting as allelochemicals are the key regulating factors of the coevolutionary processes and these relationships between plant chemicals and insects are seminal in understanding behaviour, ecology, distribution and speciation of insects. The potential application of allelochemicals comprise of development of new, safer and effective agrochemicals as pesticides and growth promoters. Being natural products allelochemicals are safer to use and easily biodegradable (Yazdani and Agarwal, 2005). Two types of allelochemicals are important for the pest management programmes, viz. 'allomones' and 'kairomones.'

The allelochemicals not only affect organisms primarily through sensory modalities like olfaction and gustation, but also through tactile senses and via direct physiological pathways when ingested or absorbed. Allelochemical web links the various trophic levels in any community and an understanding of interactions between trophic levels provides greater potential for the manipulations of their communication systems in pest management programmes. Host plant selection by phytophagous insects is a complex process involving the sequential effects of plant produced allelochemicals on insect behaviour through host finding, feeding, oviposition, and growth and development. In this process, plant chemicals influence insect behaviour (Kogan, 1976) by acting as:

Allomones

(i) *Repellents.* Orient insects away from plants.

(ii) *Excitants.* Initiate or accelerate movements.

(iii) *Suppressants.* Deter feeding or oviposition.

(iv) *Antibiotics.* Disrupt normal growth and development.

(v) *Antixenotics.* Disrupt normal host selection.

Kairomones

(i) *Attractants.* Orient insects toward plants.

(ii) *Arrestants.* Slow down or stop insect movement.

(iii) *Excitants.* Elicit biting, piercing, or oviposition.

Takahashi (1993) suggested six categories of semiochemicals, which include most behaviour modification active compounds:

(a) *Locomotor stimulants.* Cause kinetic reaction in an insect by increasing locomotion.

(b) *Arrestants.* Cause kinetic reaction in an insect by decreasing locomotion or slowing the turning rate.

(c) *Attractants.* Cause an insect to make movements towards them.

(d) *Repellents.* Cause an insect to make movements away from them.

(e) *Feeding, mating and oviposition stimulants.* Elicit any of these reactions.

(f) *Feeding, mating and oviposition deterrents.* Inhibit/suppress one of the behaviour reactions.

Some of the best-documented allomones include the cyclic hydroxamic acid (DIMBOA) in corn, gossypol and related compounds in cotton, steroidal glucosides in potato and saponins in alfalfa. Pentadecanal in rice exhibits allomonic properties against brown planthopper, *Nilaparvata lugens*. The *Brassica* species with low isothiocynates are more readily consumed by herbivores, but species with high concentration of these chemicals are avoided.

The crop or animal volatiles that are used by pests or parasites to detect its host are kairomones. It is generally acknowledged that presence of plant extracts can improve the attraction of pheromones and trap catches. If the crop kairomones are identified, then the opportunity exists to breed crop plants with lower levels of such chemicals and thereby decreasing crop attractiveness for the pests (Pickett, 1985).

21.3.1 Attractants

Many chemicals attract insects and can be used for monitoring or reducing pest populations. Because these attractants are used in ways that do not injure other animals or humans, and no residue is left on foods or feeds, they can be used in an ecofriendly manner in pest management programmes. Chemicals that act as attractants or carry other messages across distances are volatile compounds and when released into the air, they can be detected by insects having receptors to detect a specific compound. The most common use of chemical attractants is in traps to monitor insect populations. For population monitoring, chemical attractants usually are impregnated or encased in a paper, rubber or plastic lure that slowly releases the active component(s) over a period of several days or weeks. Most traps use an adhesive-coated surface or a funnel-shaped entrance to capture the target insect. These traps are very sensitive and may capture pest insects that are present at densities too low to detect. This attribute can be extremely important when the goal of a sampling programme is to detect 'foreign' or 'exotic' pests as soon as they enter an area so that control measures can be initiated immediately. Besides, traps baited with chemical attractants capture only one species or a narrow range of species. Attractant-baited traps are also useful to determine first appearance and peak activity period of a pest in a given area. Sensitivity and specificity make attractant-baited traps efficient, labour-saving tools.

Traps may employ attractants in combination with toxicants, sterilants, or pathogens to reduce pest densities. Trapping is generally directed at adult stage and it may target either or both sexes. The use of traps has been demonstrated to detect exotic pests such as the gypsy moth, *Lymantria dispar;* Mediterranean fruit fly, *Ceratitis capitata; Bactrocera* species; and many other pest species whenever infestations are detected in new areas.

Mass trapping programmes using chemical attractants have targeted such important pests as fruit flies, Japanese beetle, *Vespula* species, European chafer, etc.; however, field-scale successes have been limited. Mass trapping of the Japanese beetle, *Popillia japonica* has been effective, because the lure is a combination of kairomone and sex pheromone. Phenethyl propionate + eugenol + geraniol (3: 7: 3) and japonilure are used for trapping both sexes of the beetle (Ladd and Klein, 1986). Amlure (propyl 1,4-benzodioxan-2-carboxylate) has been used for trapping of European chafer, *Amphimallon majalis* (McGovern *et al.*, 1970).

A number of fruit flies, *Bactrocera* species are attracted by attractants, and these can be used in traps. Most works pertaining to the study of population dynamics of fruit flies have been done by using attractants in traps and different types of traps and attractants have been used for this purpose. The first sets of attractants were essentially food lures based on natural plant products, flavouring essence and ammonia. These attractants were advantageous as both sexes of many species respond to them. Protein hydrolysates, for example, serve as feeding attractants for fruit flies of the genera *Anastrepha* and *Rhagoletis*. These chemicals can be applied to sticky traps to improve catch, or combined with an insecticide and sprayed on fruit crops to suppress active infestations. Howlett (1912) observed that oil of citronella (*Cymbopogon nardus*, Fam. Andropogonaceae) was attractive to the males of *B. diversa* and *B. zonata*. Further investigation (Howlett, 1915) showed that the attractive component was the phenyl propanoid methyl eugenol, or 3-4 dimethoxy -1 allylbenzene, which was also attractive to males of *B. dorsalis*. The discovery was almost forgotten until the *B. dorsalis* was discovered in Hawaii in 1946, and then sincere efforts were made to search lures, which resulted into rediscovery of effectiveness of methyl eugenol (Steiner, 1952). During the outbreak of *B. dorsalis* in Hawaii, oil of citronella was evaluated by USDA and found very effective. Oil of citronella contains about 8 per cent methyl eugenol, and it was soon found that the purified chemical was far superior (Steiner, 1952). By using methyl eugenol Steiner collected more than four million males of *B. dorsalis* in Hawaii. Traps (Plate 21.3) containing 1g of methyl eugenol trapped as many as 2600-7300 males of *B. dorsalis* per day and became a standard monitoring device to detect incipient oriental fruit fly invasions (Steiner, 1957). The technique was refined by using 2.5 x 2.5 x 0.37" cane fibre blocks saturated with 97 per cent methyl eugenol and 3 per cent naled insecticide. Such blocks were dropped from aircraft over the Island of Rota in the Marianas at about two weeks intervals and the population reduction of *B. dorsalis* was at least 99.6 per cent in seven months (Steiner *et al.*, 1965).

Efforts were also made to screen other chemicals as lure for fruit flies and anisyl acetone, or 4 (*p*-methoxyphenyl)-2-butanone was discovered to be an effective attractant for the melon fly *B. cucurbitae* (Barthel *et al.*, 1957). A derivative cue-lure or 4-(*p*-acetoxyphenyl)-2-butanone, was found to more effective (Beroza *et al.*, 1960). Cue-lure is rapidly hydrolyzed to form 4-(*p*-hydroxyphenyl)-2-butanone or raspberry ketone that is very effective lure for melon fly. Metcalf (1990) reported at least 176 species of *Bactrocera* are attracted to cue-lure, and 58 to methyl eugenol. Trimedlure, a synthetic chemical is produced commercially as an attractant for the Mediterranean fruit fly, *Ceratitis capitata*. The synthetic product is a mixture of isomeric *tert*-butyl 4- and 5-chloro-*cis*- and *trans*-2-methylcyclohexane-1-carboxylates, of these the most active component is the (1*S*,2*S*,4*R*) isomer (Metcalf and Metcalf, 1992). Ceralure, a recently developed highly attractive and persistent analogue of trimedlure has improved persistency and appears to be a promising tool for male annihilation technique.

Such chemicals can be used in traps and are highly effective because of their specificity and environmental compatibility. The technique is based on the use of attractant/insecticide system, in which males are annihilated from the environment and called as 'male annihilation technique' (MAT). In MAT continuous use of attractant leads to annihilation of male flies, thus their chances of mating with the females become less and the same reduces production of further progeny. The MAT has been used successfully to control *B. cucurbitae* on the Island of Hawaii using fibre blocks treated with a mixture of 95 per cent cue-lure and 5 per cent naled insecticide. The treated blocks were tied above the ground on trees @ 585 blocks per mi^2 and were replaced each month. The melon fly population declined to 99.9 per cent after the fourth treatment (Cunningham and Steiner, 1972). By using attractants successful eradication of fruit flies have been achieved from many small Islands and in other cases their populations were suppressed to a great extent by continuous mass trapping in some countries.

The use of olfactory feeding attractants or baits in combination with insecticides has been a well-established practice since the 1950s in case of tephritid fruit flies such as *Ceratitis capitata* and *Bactrocera oleae*. The use of a protein hydrolysate-insecticide mixture in a 'lure and kill' strategy allows the operator to spray a reduced surface area of crop relying on the attractant to bring the flies to the treated areas. Protein hydrolysates are used as bait in many fruit-growing areas of the world for fruit fly control and have undoubtedly reduced environmental contamination by insecticides (Jones and Casagrande, 2000). The use of bait/insecticide system is termed 'bait application technique' (BAT).

Boller (1983) suggested simultaneous use of bait and attractant mixed with insecticide (double attack method) and reported that the technique is consistently more effective than either technique alone. Presently, the most powerful technique for eradication/suppression of tephritids populations is the simultaneous use of bait/malathion system and male annihilation. In the experiments on isolated populations of Queensland fruit fly, *B. tryoni* in Western NSW, Australia this method was observed to be more effective (Bateman *et al.*, 1966). Kumar and Agarwal (1998) and Agarwal and Kumar (1999) found mango pulp mixed with methyl eugenol as most effective formulation against *B. dorsalis* and *B. zonata*, respectively. Kumar and Agarwal (2005) reported that a formulation consisting of soya powder, cue lure and malathion was effective in annihilation of male flies of *B. cucurbitae*. Hancock *et al.* (2000) discussed the results of an eradication programme of *B. papayae* in Queensland, Australia by combining MAT and BAT. Firbeboard blocks impregnated with methyl eugenol (13.5-15 ml) and malathion (4.5-5 ml) were used @ 400-600 blocks/km^2 and almost 3 million blocks were used during 1995-1998. Besides, protein bait applications (5 L protein, 2 L malathion and 100 L water) were carried out as spot leaf treatments (100 ml/tree). The flies were mostly trapped in first five months of campaign and by October, 1996 trap catches declined by 99.9 per cent.

In olive fruit fly, *Bactrocera oleae* olfactory attractants have been described for both males and females. The possibility of combining visual attraction of adult flies to yellow sticky traps, with the response of males to the female produced sex pheromone (1,7 dioxaspiro [5,5] undecane), and the female's response to ammonium salts, has been investigated as a population suppression mechanism (Broumas *et al.*, 1983).

Kairomones have also been used for controlling disease-causing insects of man and livestock. Tsetse flies, *Glossina* species are the principal vectors of the African trypanosomiasis diseases of humans and domestic animals. One of the environmentally benign tsetse control methods relies on baits containing kairomonal attractants. The odours of host chemicals contain several chemicals, which are now widely used in traps. For *G. morsitans* and *G. pallidipes*, the attractive components of ox breath are CO_2, acetone and 1-octen-3-ol (Vale and Hall, 1985). The attractivity of buffalo and ox urine to *Glossina* species is due to a phenolic fraction that contains 3-n-propylphenol as one of the active principals (Bursell *et al.*, 1988).

Artificial honeydews have been used experimentally to attract, arrest and feed aphidophagous insects, with subsequent reduction in aphid population (Hagen and Bishop, 1979). The green lacewings, *Chrysoperla carnea* can be attracted by applying artificial honeydew to plants. This promotes egg laying in desired location and potentially increases predation of plant pests.

The screwworm, *Cochliomyia hominivorax* adult suppression system (SWASS) consists of a palletized formulation containing the chemical lure, food (dried blood), a feeding stimulant (sugar) and an insecticide. This combination attracts the flies and induces them to feed, thereby acquiring a lethal dose of insecticide (Coppedge *et al.*, 1980).

21.3.2 Deterrents

Deterrent is a chemical that inhibits behaviour such as feeding or oviposition, when applied to a site where such behaviour normally occurs. In pest management, a deterrent is directly applied to a resource to reduce or prevent the effects of pest behaviour like feeding. Chemicals that inhibit feeding of insects on a treated surface without necessarily killing or repelling them are called 'antifeedants'. These can result in cessation of feeding, either temporarily or permanently, when tested. These represent a different approach to crop protection in that the elimination of the pest is not the prime objective rather the pest individuals are inhibited to feed upon the crop or commodity. They are also termed 'gustatory repellents' or 'feeding deterrents'. Insects possess highly adapted behaviours for feeding that are elicited by specific chemical signal stimuli. These signals are detected by a number of chemosensilla located primarily on the mouthparts. The chemosensilla of Lepidoptera in particular that respond to taste stimuli, are among the most studied and have been selected for a biorational design of antifeedants (Koul, 1998). Antifeedants inhibit the taste receptors of mouth region and in the absence of gustatony stimulus, the insect fails to recognize the treated host as food. A large variety of plant allomones protect plants from herbivory and offer an opportunity for new approaches in pest management. The success of pest management tactics using such chemicals would be more with compounds that produce both behavioural and physiological effects (Jermy, 1990). A series of compounds have been synthesized as active feeding inhibitors. Glycosides of some steroidal alkaloids like demissine and tomatine found in Solanaceae reduce feeding by Colorado potato beetle, *Leptinotarsa decemlineata* (Koul, 1982). Azadirachtin and related analogues, isolated from Indian neem tree, *Azadirachta indica* are potential antifeedants against a number of pest insects. Azadirachtin has deterrent effects on behaviour of locusts, aphids, jassids, caterpillars, etc. and modify feeding and oviposition. Antifeedants should be systemic otherwise as the plant grows, the insect will soon find a place (area) on which the chemical has not being applied. The systemic antifeedants also destroy the internal feeders and may provide resistance to plant after application. Azadirachtin has the systemic property and protects newly growing leaves from feeding damage.

The larvae of mustard beetle, *Phaedon cochleariae* were deterred from feeding by application of extracts of plants in the genus *Ajuga* to the young leaves at the top of mustard plants; beetles moving down to feed on older leaves were controlled by an insect growth regulator (Griffiths *et al.*, 1991).

Pyrethrum and many other plant products/extracts are found to act as antifeedants against many pests. Some synthetic organic insecticides also act as antifeedants, e.g. arprocarb (a carbamate compound) is a systemic antifeedant against cotton boll weevil, *Anthonomus grandis*. Wood preservatives, such as pentachlorophenol, act as feeding deterrents to termites and other wood-dwelling insects.

Oviposition deterrents are limited and host marking pheromone of the European cherry fruit fly, *Rhagoletis cerasi* deters oviposition by conspecifics. An extract of pheromone sprayed on cherry trees reduced fruit infestation by wild fly larvae tenfold (Katsoyannos and Boller, (1980).

21.3.3 Repellents

The strategy for using repellents is generally to stop a pest from finding a valued resource. Use of repellents to repel pest insects is the oldest and most widely used method. Repellents are 'chemicals that cause insects to move away from their sources'. The plant-derived volatile allomones such as 6-carbon aliphatic aldehydes and alcohols, the so-called 'green leaf volatiles' (GLVs), have been suggested as both repellents and deterrents for tephritid flies. The GLV compounds are botanically ubiquitous, being the chief aroma contributors to the common green odour of leaves and immature unripe 'green' fruits (Visser, 1986). The release of GLVs is also indicative of plant tissue damage; as these are direct simple, cleavage products of fatty acid breakdown. The damage to plant tissues caused by mechanical means and insect feeding and oviposition. The GLVs have strong repellency against females of olive fruit fly, *Bactrocera oleae*.

Repellents also include a wide range of chemicals from volatile substances active in vapour form to protect man and animals from biting and blood-sucking insects. When these insects are attracted to hosts by skin odours and carbon dioxide from their breath the repellent jam the insects' sensors and confuse them so they are unable to land and bite the host successfully. Mosquitoes are the best studied of the biting insects, and they are known to use visual, thermal, and olfactory stimuli to locate a blood meal. Visual stimuli appear to be important for in-flight orientation, particularly over long-ranges, whereas olfactory stimuli become more important when the mosquito is close to its host. Carbon dioxide and lactic acid from the skin and the breath appear to be the main insect attractants. Skin warmth and moisture also serve as attractants at close range. Repellents are commonly used to protect humans from insects like mosquitoes and black flies, which are vectors of diseases; however, they are effective only at short distances from the treated surfaces. DEET (N,N-diethyl-m-toluamide) and dimethyl phthalate are versatile and effective insect repellent and commonly used to repel mosquitoes, ticks, fleas, biting flies and chiggers. Mothballs and p-dicholorobenzene are placed in drawers and closets to prevent infestation by a variety of insects that feed on stored products and natural fibres.

Insect repellents fall into two categories: synthetic chemical repellents and natural (plant-derived) repellents. A large number of plants have been tested as sources of insect repellents. Plants whose essential oils reportedly have repellent activity include: citronella, cedar, verbena, geranium, lavender, Indian neem, pine, catnip, cinnamon, rosemary, soyabean, eucalyptus, basil, garlic, peppermint, pyrethrum, etc.

21.3.4 Stimulants

Most known stimulants are involved in either feeding or oviposition, mainly the former. A number of secondary plant compounds that have no nutritional value to either plant or insect have been the subjects of considerable study. The plant as a chemical defence against herbivores may manufacture these substances but they unwittingly serve as token feeding stimulants to a selected group of specially adapted species. Many appear to have no effects at all on insect herbivores, while others stimulate feeding and/or growth. Use of feeding stimulants (phagostimulants) allows the application of reduced dosages of insecticides assuaging some of the environmental and health concerns associated with traditional insect control tactics. Sugars are one of the best-documented groups of substances because of their feeding stimulant potential. Several other substances that may have feeding stimulant properties are amino acids, lipids, ascorbic acid, glucosinolates, and cucurbitacins. Besides, a variety of commercial feeding stimulants have been developed which elicit strong feeding responses. Many feeding stimulants are allelochemical-based and attract the pests. The addition of feeding stimulants to entomopathogens has in some cases increased the effectiveness of the control.

Sugar responses have also been tested on several major insect pests belonging to the orders: Coleoptera, Lepidoptera, Diptera and Hemiptera. Protein hydrolysate has been used in the eradication of tephritid fruit flies and sugars for the control of flies around animal rearing facilities. Many other substances have also been tested for phagostimulant properties on a variety of insect pests. The cucurbitacins (a class of tetracyclic terpenes) produced by cucurbits act as attractants and feeding incitants for spotted cucumber beetles, *Diabrotica undecimpunctata* and banded cucumber beetle, *D. balteata*. In small trials, such substances have been found effective in reducing populations of several species of cucumber beetles. Dried and ground bitter squash roots and fruits, containing cucurbitacins, induced several cucumber beetles, to feed on poisoned baits, increasing the kill proportionately (Metcalf *et al.*, 1987). Allelochemical-based toxic baits are potential tool for greatly reducing the amount of insecticide used to protect crops. Fresh and dried garbanzo bean extracts were positive phagostimulants for the tobacco budworm, *Heliothis virescens* (Potter and Watson 1983).

The herbivore-induced plant volatiles (HIPVs) also include compounds that attract pollinating insects and natural enemies of insect pests. Identification of key volatile compounds in HIPVs that attract natural enemies would be useful for establishing effective pest management by behavioural manipulation of the natural enemies. The attraction of parasitoids to plant volatiles could be of use to monitor population of natural enemies or potentially to retain them in crop to attack colonizing pests.

Oviposition stimulants may induce oviposition by a pest on nonhost plant; however, such chemicals have not been commercially used in pest management. Unnithan and Saxena (1990) reported that sorghum shoot fly, *Atherigona soccata* can be stimulated to lay eggs on non-host (corn) by applying acetone extract of sorghum. In small field trails, the extract diverted oviposition from sorghum.

21.3.5 Arrestants

In the feeding sequence, once the insect has located the food plant an arrestant may prevent further searching behaviour. The arrestants may be perceived as odours or tastes and only a few have been isolated and identified. The coumarin produced by sweetclover arrests flight by the sweetclover weevil, *Sitona cylindricollis* (Heidweg and Thorsteinson 1961). The ultimate effect of such a chemical is aggregation of the insects on the host plant.

21.4 Semiochemicals and augmentation of natural enemies

Recent discoveries of tritrophic level interactions among plants, herbivores, and parasitoids/predators have demonstrated how tightly interwoven these components are and illustrate the importance of multitrophic perspectives for effective and sustainable pest management strategies. Plants have long been known to possess toxins and other chemicals that serve to discourage herbivore feeding. Allelochemicals tend to serve as messengers mediating interactions among host plants, insects and their natural enemies. Both plant chemicals (allomones) and host chemicals (kairomones) play important role in long-range and short-range location of hosts by natural enemies (Vet and Dicke, 1992). Besides, they also use volatiles emanating from scales, faeces, glands, culticular components; and by-products of feeding, excretion, etc. The composition of the chemical signal often differs with different species of hosts or with different plants. The behaviour and success of natural enemies undergo considerable variation when the insects colonize different host plants and modify the behavioural sequence of natural enemies. Successful parasitism involves a series of interactions often mediated by allelochemicals, including host habitat location, host location and host acceptance (Vinson, 1976). Thus, tritrophic interactions essentially focus on the abilities of natural enemies to use plant traits to locate insects, besides emphasizing on the ecological and evolutionary implications of the food chain (Price *et al.*, 1980). The natural enemies use chemical, tactile or visual cues to distinguish between infested and non-infested plants.

The allelochemicals that may be plant odorants or host chemicals dominate the atmosphere and form 'aerial bouquets', which stimulate the behavioural patterns of natural enemies by providing vital signals. Besides, volatiles released by damaged plants (also termed as infochemicals) after insect feeding tend to increase the efficiency of natural enemies and plants producing such alarm responses attract more parasitoids and predators. Allelochemicals may be useful in pest management programmes for manipulation of densities of natural enemies and the behavioural response of a natural enemy to a host's kairomones or plant allomones could be utilized to recognize promising species in classical biological control. Insect behaviour and factors that mediate behaviour must be understood to assure the success of such techniques. According to Tumlinson *et al.* (1993a), regardless of the approach used to employ parasitoid insects for pest control, ecological aspects involving their foraging behaviour are important issues to be investigated. The most common manipulation of allelochemicals in augmenting biological control is through exposing the parasitoids to these chemicals prior to release (Vet and Groenwold, 1990). These allochemicals may also be sprayed on the crop to attract the parasitoids. Tumlinson *et al.* (1993b) suggested that hymenopterous parasitoids exploit allelochemicals emitted by plants in response to insect herbivore feeding, and learn to respond to different blends of chemicals that indicate the location of their hosts; they can be trained to respond to a specific odour blend. Thus, it may be possible to increase their effectiveness for biological control by conditioning them, prior to their release, to search for a target pest in a particular crop.

The terpenes produced by damaged pines are kairomones for pest beetles, but if these chemicals are used by parasitoids to locate the bark beetles, the terpenes act as synomones. The α-pinene and myrcene produced by damaged pines are kairomones for *Dendroctonus* species, but synomones for pteromelid parasitoids that attack pine beetles. Thus, allelochemical in the tritrophic interaction could act as:

- An allomone that deters non-specialist phytophages
- A kairomone that attracts specialist phytophages
- A synomone that attracts the parasitoid of the phytophage

Noldus and Lenteren (1985) have shown that the sex pheromone of cabbage moth, *Mamestra brassicae* females were absorbed on to the leaf surface of Brussel sprouts to such an extent that it was capable of eliciting behavioural responses from *Trichogramma evanescens*. The spotted stem borer, *Chilo partellus* produces volatiles attracting the parasitoid, *Cotesia flavipes* and an exogenous elicitor of this plant response has been identified in the regurgitate of the larva (Potting *et al.*, 1995). A comparison of volatiles of the frass of cotton bollworm, *Helicoverpa armigera* indicates that larvae fed on artificial diet had no significant kairomonal compounds, while the frass of *H. armigera* fed on *Abutilon* and *Gossypium* plants indicated the presence of several host-seeking kairomonal compounds that considerably influenced parasitoid behaviour. Therefore the behaviour and success of natural enemies undergo considerable variations when the herbivores colonize different host plant species in response to different volatile compounds that modify the behavioural sequence of natural enemies (Lewis and Martin, 1990). Kairomones from moth scales tend to influence the parasitic potential of *Trichogramma chilonis*. Volatile compounds emanating from the scales of *Helicoverpa armigera* (its natural host) and *Corcyra cephalonica* (a laboratory host) moths were identified as hexatriacontane, pentacosane, heptadecene, docosane, and 2,6,10-dodecatrienal-3,7-11-trimethyl which increased the activity of *Trichogramma chilonis* (Ananthakrishnan *et al.*, 1991). They discussed the significance of an array of compounds from moth scales with kairomonal activity for manipulating entomophagous insects in biological control programmes. The abdominal tip exudates of *Spodoptera litura* females were identified as complex compounds containing dodecane, heptadecane, octadecane and pentadecenoic acid which influenced the activity of the egg parasitoid, *Telenomus remus* (Ananthakrishnan and Senrayan, 1992). McCall *et al.* (1994) studied infochemicals released by cotton plants, which they named herbivore-induced volatile emissions. They focussed on investigating the collected volatiles from undamaged, freshly damaged (0-2 hours after feeding) and old damaged plants (16-19 hours after feeding), attacked by corn earworms, *Helicoverpa zea.* According to them, healthy plants produce specific odours that are recognized by some herbivore insects. They also pointed that damaged plants, on the other hand, would produce antifeedants (attempting to control the damage caused by the herbivore insects), and moreover, blend of volatile compounds that would attract the natural enemies. Paré and Tumlinson (1998) suggested that these chemical signals are mixtures of volatile terpenoids among other compounds that are released by plants, leading insect to distinguish between infested and non-infested plants. Röse *et al.* (1998) reported that cotton plants under herbivore attack release volatile semiochemicals that attract natural enemies of the herbivores to the damaged plant. The volatiles released in response to herbivory are not only released from the damaged leaves but from the entire cotton plant. Cotton leaves damaged by larvae were attractive to the generalist parasitoid, *Cotesia marginiventris* and the specialist parasitoid, *Glabromicroplitis croceipes.* Artificially damaged cotton plants that released green leafy volatiles and constitutive terpenoids were less attractive for *G. croceipes* and *C. marginiventris.*

Turlings and Tumlinson (1992) reported that maize seedlings respond to insect herbivore injury by releasing relatively large amounts of several characteristic terpenoids and as a result, become highly attractive to parasitoids. Chemical evidence showed that the induced emission of volatiles was not limited to the sites of damage but occurred throughout the plant. The systemic release of volatiles by injured maize coincided with attractiveness to the parasitoid, *C. marginiventris.* Volicitin, (N-[17-hydroxylinolenoyl]-L-glutamine, a key compound that induces plants to release these signals when attacked by herbivore insects, was isolated from oral secretions of beet armyworm caterpillars, *Spodoptera exigua* (Alborn *et al.*, 1997). Volicitin induced the maize seedlings to emit volatile compounds that attracted females of the parasitoid, *C. marginiventris.* Mechanical damage of the leaves, without application of this compound, did not trigger release of the same blend of volatiles. They concluded that volicitin is a key component in a chain of chemical signals and biochemical processes that regulate tritrophic interactions. Moraes *et al.* (1998) demonstrated that tobacco, cotton, and corn plants release volatile blends that are specific to damage caused by the corn earworm, *Helicoverpa zea* and the tobacco budworm, *Heliothis virescens.* They further demonstrated that the specialist parasitoid, *Cardiochiles nigriceps* uses these volatile blends to distinguish a plant with its host, *H. virescens* as well as to distinguish a plant with a non-host, *H. zea.* Fatouros *et al.* (2005) demonstrated that the parasitic wasp, *Trichogramma brassicae* uses an anti-aphrodisiac produced by its host large white cabbage butterfly, *Pieris brassicae* which is passed from male to female during mating. The compound benzyl cyanide is used to make female butterflies less attractive to other conspecific males. However, this parasitic wasp has evolved the ability to exploit this intraspecific communication of its insect host and use the information obtained as a foraging cue to locate its host, and will eventually parasitize the newly laid butterfly eggs.

Some predators also use kairomones for finding the prey. Adults of anthocorid bugs utilize alarm pheromones of aphids and thrips for finding their prey. Larvae and adult coccinellids detect their prey both by olfactory and visual cues. Lewis *et al.* (1977) demonstrated the response of larvae of common lacewing, *Chrysoperla carnea* to kairomones present in scales of corn earworm, *Helicoverpa zea*. The kairomones act as stimulant for predation on eggs of corn earworm.

21.5 Visual lures

Many pest insects are positively phototactic and attracted by light and its component colours. Visual lures used in insect management fall into three categories: (i) lights (incandescent, fluorescent, and ultraviolet) that attract insects; (ii) coloured objects that are attractive because of specific reflectance; and (iii) shapes or silhouettes that stand out against a contrasting background. Light traps (Fig. 21.4) are commonly used as pest management strategy as they provide an easy-to-use non-pesticide alternative tool for reducing and suppressing a number of pests, particularly moths.

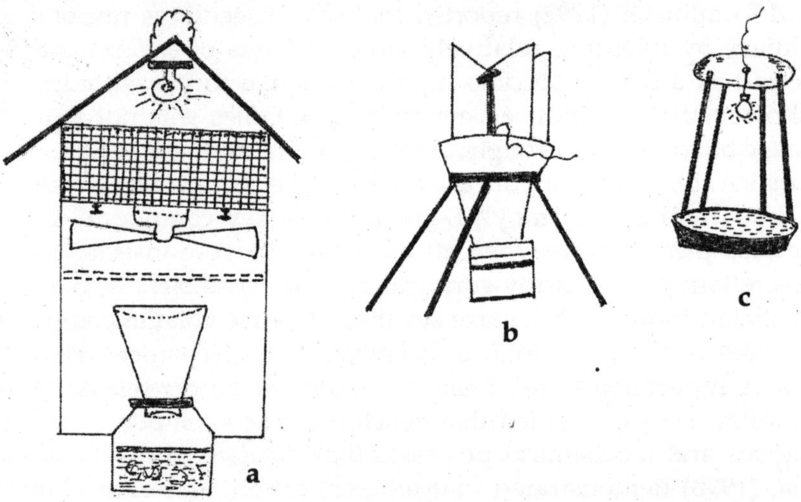

Fig. 21.4a-c. Light traps used to attract photo positive insects.

A great number of insect species are attracted to light of various wavelengths. Although different species respond uniquely to specific portions of the visible and nonvisible spectrum, most traps or other devices that rely on light to attract insects use fluorescent bulbs or bulbs that emit ultraviolet wavelengths (black lights). Studies have shown that ultraviolet light is the most effective part of the light spectrum for attracting night-flying insects. Some important pests that are attracted to light include the tobacco cutworm moth, *Spodoptera litura*; yellow stem borer, *Scirpophaga incertulus*; spotted stem borer, *Chilo partellus*; European corn borer, *Ostrinia nubilalis*; codling moth, *Cydia pomonella*; cabbage looper, *Trichoplusia ni*; tiger moth, *Amsacta moorei*; diamondback moth, *Plutella xylostella*; peach twig borer, *Anarsia lineatella*; rice leaf roller moths, *Cnaphalocrocis medinalis*; rice green leafhopper, *Nephotettix nigropictus*; green leafhopper, *N. virescens*; potato leafhopper, *Empoasca fabae*; brown planthopper, *Nilaparvata lugens*; rice gall midge, *Orseolia oryzae*; many cutworms and armyworms, many beetles, termites and some grasshoppers. Lights and light traps are used with varying degrees of success in monitoring populations and in mass trapping. These traps also provide useful information about the timing of appearance and relative abundance of many pests. In many light traps a low voltage 'electrocutor' or 'bug zappers' surrounds the lamps. These light electrocutors attract and kill a wide variety of insects, the great majority of which are not pests.

Some day-flying insects also show their behavioural response to specific coloured objects, e.g. yellow objects attract many insects and are often used in traps designed to capture winged aphids and adult whiteflies. Aphids respond to different wavelengths of light differently depending upon their developmental stage, population density, suitability of their host plant and environmental conditions such as temperature and wind speed. The migratory aphids alight when they encounter an object of the appropriate colour (usually yellow) and size (usually the largest). Yellow traps or yellow polyethylene sheets covered with transparent glue are used for aphid control. Yellow, sticky-coated cards or plastic cups are widely used in mass trapping programmes to control whiteflies in greenhouses.

Daylight fluorescent yellow colour reflects little light below 500 nm but a great deal between 500-600 nm, and this colour attracts a wide range of tephritid flies (Prokopy, 1972b). The adults locate vegetation evidently primarily by vision. Green colour of foliage and/or the darkness of tree silhouette are the main stimuli eliciting arrival of apple maggot flies, *Rhagoletis pomonella*. Red spheres and yellow cards also attract *R. pomonella*. Small sticky-coated daylight fluorescent yellow rectangles are widely used for mass trapping and monitoring tephritid populations, especially temperate species (Prokopy, 1977).

21.6 Acoustic traps

The acoustic signals produced by insects can be effectively used in tracking and trapping pest individuals and therefore have potential for use in pest management programmes. Insects of the orders Orthoptera and Hemiptera usually disseminate their sound signals through air for the purpose of reproduction or defence. Sound-baited traps have been developed for trapping of crickets and a few other insects; however, the use of such traps is not so common when compared with light, chemical, or bait traps. Though in recent years advances in electronics have greatly increased the technologies in which sound signals can be detected, analysed, synthesized, amplified, reproduced and broadcasted but these have certain limitations because most insects do not respond to sound signals and the attraction to sound signals is restricted to only a few groups of individuals. Besides these, their use and operation is rather costly.

Acoustic traps are most likely to be used for species that exhibit positive phonotaxis under natural environments. The traps generally catch individuals of one or only a few species as the sound signals in insects are species-specific. Sound-baited traps were first used against mosquitoes. Ulagaraj and Walker (1973) proved that flying mole crickets land at electronic reproductions of their calling songs and they developed basic techniques for using sound to trap mole crickets. The technique involves use of a sound source, a catching device and a controller. Walker (1982) improved the application of technique and collected a large number of southern mole cricket, *Scapteriscus acletus* and West Indian mole cricket, *S. vicinus* at various sites in Florida. The record catches of *S. acletus* and *S. vicinus* for one station in one night was 3297 and 2174, respectively. Walker (1988) discussed the use of acoustic traps for agricultural insects. Insects like crickets (Gryllotalpidae), field crickets (Gryllidae) and ormiine flies (Tachinidae) have been mainly trapped by using such traps. The crickets are attracted to conspecifics calling songs while ormiine flies respond to calling songs of their hosts. These traps are also valuable in studying behaviour, ecology, collection of specimens and population monitoring.

Tachinid flies of the tribe Ormiini are parasitoids of crickets and katydids and their females find their hosts by homing on the hosts' calling songs. Cade (1975, 1981) trapped gravid females of *Euphasiopteryx ochracea* at broadcasts of recorded songs of *Gryllus integer*. Walker (1989) developed a trap for *E. ochracea* and trapped as many female individuals as in sticky trap. Webb *et al.* (1983) collected females of Caribbean fruit flies, *Anastrepha suspensa* with sound-baited traps within field-caged guava trees. Field crickets are another group of pest insects that have been caught in sound-baited traps. Campbell and Shipp (1979) developed traps for common field cricket, *Teleogryllus commodus*, an important pest of pastures in Australia and New Zealand. Cade (1981) trapped *Gryllus* species in acoustic traps in North America. Sanjayan and Prakash (2006) discussed the role of acoustic communication in assessment, monitoring and management of insect pests.

21.7 Insect growth regulators (IGRs)

IGRs are chemicals that alter normal growth and development by obstructing the function of insect endocrine system. These chemicals are also called 'biorational' or 'third-generation insecticides', and disrupt the normal activity of the endocrine system of insects, affecting the development, reproduction, or metamorphosis of the target insect. These compounds can be used to stimulate development at inappropriate times or inhibit it at other times. IGRs may be defined as "materials that are natural biochemicals or exogenously applied chemicals that cause morphological and physiological changes during the growth and development of insects". Synthetic compounds possessing activities of juvenile hormone and moulting hormone of insects, often termed as 'mimics' or IGRs. Chitin synthesis inhibitors (CSIs) are also included in the group of IGRs. They are highly effective against specific stages of many insect pests, with a short environmental persistence and low level of mammalian toxicity. Most IGRs do not affect adults; however, for many pests, it is the larval stages that are the most destructive and affected by IGRs. They have a much slower mode of action than synthetic chemical insecticides. Besides their influence on metamorphosis and reproductive functions of insects, these chemicals have influences on diapause, behaviour and communication (through pheromone production). In social insects they are known to regulate caste differentiation and morphogenesis.

21.7.1 Juvenile hormone analogues

Under the influence of neurohormone, the corpora allata releases one or a mixture of hormones that control metamorphosis and regulate reproductive development in insects. In adult female insects, JH stimulates the deposition of yolk in the eggs and affects accessory gland activity and pheromone production. In social insects they are known to regulate caste differentiation and have special influence on morphogenesis. Chemically the JH is a non-sterolic lipid of the category of terpenes.

In some insects, the final moult produces an adult that looks pretty much like the earlier nymphal stages. In others, the final moult of the larva produces a pupa. After a period of metamorphosis, the adult emerges. The hormone ecdysone triggers larva-to-larva moults as long as JH is present. In the absence of JH, ecdysone promotes the pupa-to-adult moult. Thus, normal metamorphosis seems to occur when the output of JH diminishes spontaneously in the mature larva. When solutions of JH are sprayed on mature larvae, or on the foliage upon which they are feeding, their normal development is upset. This raises the possibility of using JH as an insecticide. A number of 'JH analogues (JHAs)' or 'JH mimics' or 'juvenoids' have been synthesized and used for the control of a range of pests of public health and stored products, viz. mosquitoes, pharaoh ants, flies, cockroaches, fleas, etc. There are numerous reports on the occurrence of plant metabolites having JH activity, e.g. phytojuvenoids, farnesol, juvocimene-I and II, sesamin, juvadecene, sesamolin, etc. The first JHA was discovered by chance from paper and termed as 'paper factor'. It was named as juvabione and chemically identified as the methyl ester of todomatuic acid. Along with its analogue dehydrojuvabione, it is found in balsam fir tree, *Abies balsamea*.

Since the early 1970s, numerous analogues of JH have been tested for insecticidal activity. Most of the early analogues resemble JH in their basic terpenoid structure. Many plants also possess JHAs where they may function as defensive compounds against herbivorous insects. Two JH analogues, methoprene and hydroprene are used in controlling many dipteran and household pests, *e.g.* flea larvae, hemipterans, lice, some beetles, flies, mosquitoes, and pharaoh ants. A few active compounds with less apparent similarity to JH (aromatic non-terpenoidal JHAs) like fenoxycarb, pyriproxifen have also been synthesized (Dhadialla *et al.*, 1998). Other closely related naturally occurring JHAs are JH-0, JH-I, JH-II, JH-III and 4-methyl JH-I. Different formulations of fenoxycarb have been used for the control of a number of coleopterous and lepidopterous pests. Pyriproxifen is active against mosquitoes, other dipterans and whiteflies. These chemicals interfere with normal development so that the insect fails to reach the adult stage or the formed adult is sterile or malformed. Exposure to these compounds at moulting results in the production of insects containing mixed larval/pupal or larval/adult morphologies. When exposed to a JHA, the insect remains in an immature state, and is rendered unable to successfully moult into the adult stage or become reproductively viable. The efficacy of these compounds is greatest when normal juvenile hormone titres are low, as in the last larval or early pupal stages. Thus, timing of application of JHA is important for successful control. Another useful property of these compounds is that, in adults, they disrupt normal reproductive physiology and act as a method of birth control. JHAs are not specific for any pest under attack, as the molecules based on the JH system are more or less common to all insect species. The reason for the increasing popularity of JHAs includes their short residual effects in the field and the few and short sensitive phases of the target insects. However, since JHAs act only on certain stages, their application must be timed precisely and, owing to their low field stability, it should to be repeated several times.

21.7.2 Anti-juvenile hormone analogues

Bowers *et al.* (1976) discovered two plant products in the blue billygoatweed, *Ageratum houstonianum* and named precocene I and II. These are also called 'JH antagonists' or 'anti-juvenile hormone analogues' (AJHAs). Precocenes appear to exert a cytotoxic action on the corpora allata, mediated by an oxidative bioactivation *in vivo*. These compounds damage the corpora allata thus preventing JH from doing its normal function. In immature insects, exposure to anti-JH compounds may result in premature (precocious) development of adult structures or behaviours. The adults of many species became sterilized after exposure, and diminished feeding was also observed. AJHs are found in many plants and several derivatives have also been synthesized. Anti-JH compounds seem to be most effective against hemipterans. Despite their unique mode of action, these IGRs have not been developed into commercial products because they break down too rapidly in presence of oxygen. Although JH-based insecticides were found to be effective against certain groups of insect pests, their shortcomings soon became apparent; however, current JH research provides new opportunities and ideas for the development of innovative integrated pest management (IPM) tools and strategies aimed at disrupting JH functions.

21.7.3 Moulting hormone analogues

The steroidal compounds like ecdysone, ecdysterone and other ecdysteroids are secreted by prothoracic glands are responsible for normal moulting, growth and maturation of insects. Ecdysteroids are synthesized from cholesterol or phytosteroids in the diet, because insects cannot synthesize the steroid nucleus. Ecdysone or ecdysone-like compounds or moulting hormone analogues (MHAs) occur in numerous plant species especially those belonging to Pteridophyta, Gymnospermae and Angiospermae, where they evidently serve as a defence against insect herbivores. The exogenous application of MHAs leads to an increased titre of ecdysone in insects, which cannot be metabolized or excreted rapidly enough to prevent hormonal imbalance resulting in moulting promotion and death of insects. The search for ecdysteroid agonists or MHAs has been very successful and RH-5992 (tebufenozide) is an important compound. Tebufenozide is used as safe and selective insecticide acting as ecdysone agonists and is most effective when applied in egg stage but does not affect adults. RH-2485, the newest member of the bisacylhydrazine class, seems to be even more active than tebufenozide, acting against a wide range of lepidopteran pests.

21.7.4 Anti-moulting hormone analogues

Certain compounds like azosterols and non-steroidal secondary and tertiary amines antagonize the conversion of phytosterol to cholesterol required as a base for the synthesis of ecdysone in insects. These compounds are termed anti-moulting hormone analogues (AMHAs) and cause marked reduction in moulting hormone titre of insects; as a result the moulting cycle is delayed. Such compounds also occur in some plants like Indian neem, *Azadirachta indica* and African medicinal shrub, *Plumbago articulata*. Azadirachtin is structurally similar to insect hormones called 'ecdysones' and its major action is to modify insect haemolymph ecdysteroid titres due to a blockage of release of prothoracicotropic hormone from the corpus cardiacum complex. Azadirachtin-containing products have been formulated and are being marketed as botanical insecticides. Some azadirachtin-induced morphogenic effects in insects are: (i) prolonged larval development, (ii) death before or during ecdysis, (iii) formation of larval-pupal intermediates or pupal-adult intermediates, (iv) inability to remove exuviae completely, and (v) moulted adults with crippled wings and legs (Peter and Ananthakrishnan, 1996). Brassinosteroids represent the first true antiecdysteroids observed. They are a family of growth-promoting hormones found in plants and have striking structural similarities with the ecdysteroids. The effect of brassinosteroids on insects may be explained by their competition with moulting hormones at the binding site of the hormone receptor, which results, for example, in delayed moulting (Richter and Koolman, 1991). Two triterpenoids, cucurbitacins B and D isolated from seeds of a cruciferous candytuft plant, *Iberis umbellata* were demonstrated to be insect steroid hormone antagonists acting at the ecdysteroid receptor (Dinan *et al.*, 1997).

21.7.5 Chitin synthesis inhibitors

The insect cuticle serves as a first line of defence against physical and biotic factors. It also provides a site for attachment of muscles and helps in locomotion. Chitin is a major constituent of cuticle. Chitin synthesis inhibitors (CSIs), e.g. diflubenzuron, triflumuron, flufenoxuron, novaluron, buprofezin, lufenuron, etc. interfere with the synthesis of chitin. These chemicals inhibit the moulting process (apolysis) by blocking the activity of chitin synthetase, an enzyme needed by epidermal cells when constructing a new exoskeleton. They act rather slowly (after 2-5 days), but eventually disrupt any process that involves construction of new cuticle (e.g., moulting, hatching, pupation) and are most effective when used against the immature stages of a pest. Diflubenzuron (a benzoylurea compound) is registered under the trade name Dimilin and used for controlling gypsy moths, codling moths, grasshoppers, leafhoppers, whiteflies, scale insects, boll weevils, flies, lice, and various other pests. Plant products like azadirachtin and plumbagin were also found to possess CSI effects.

21.8 Other behavioural manipulations in pest management

The range of other behavioural options available for the management of a pest may vary with the pest species, pest intensity or severity, and effectiveness of the option. For effective use of these methods, an understanding of pest biology, ecology, and behaviour is essential. Some pest management strategies involving pests' behaviour are given below.

Sanitation has generally pertained to the destruction of breeding sites or other pest reservoirs. This is due to a potential relationship between the number of insects eliminated and decrease in the population's reproductive potential that occurs when breeding or overwintering sites are destroyed. The southern pine beetle, *Dendroctonus frontalis* produce aggregation pheromones when colonizing insects attack the host trees. The pheromones rapidly attract high number of beetles to a concentrated cluster of trees, and expedient salvage of infested timber can often remove a substantial population from a forest stand and prevent proliferation of infestation (Belanger *et al.*, 1993).

Black cutworms, *Agrotis ipsilon* are negatively phototactic and active during the night. They tend to avoid light and hide in dark locales. Small heaps of grass and plant debris are placed in the fields, where they assemble in large numbers. Such cutworms can be collected and destroyed.

Disruption of the continuity of pest biological rhythms by manipulation of crop phenology is an effective control strategy for many pests. Producing phonological asynchrony in pest/crop interactions can be accomplished by alternating the time of planting (or harvest) or by selecting a cultivar with desirable developmental traits that avoid pest peaks.

Some insects have very specific requirements and will die or move away in the absence of their required hosts. Host selectivity may occur in either the egg-laying behaviour of the female or the feeding behaviour of the larva. Mobility of an insect species is also important because it influences how far an insect can travel to search out an acceptable host plant when it is presented with a less preferred plant host. Depending on the species and life-stage, the degree of mobility varies from a few inches to several miles. Many pest management recommendations focus on selecting rotations that decrease certain pest populations because of unavailability of feeding and oviposition sites. Insect pests like white grubs, wireworms and corn rootworms, whose major feeding stage occurs in the soil are controlled by crop rotations.

Trap cropping is the establishment of a preferred host of a specified pest within or adjacent to the main crop. The intention is to attract a high percentage of the pest into the trap crop, where it can be destroyed more efficiently than under usual field conditions. The trap crop must be highly attractive and the target insect must be mobile enough to aggregate readily in the trap areas.

The strategy 'perimeter trapping' is based on the premise that pests like fruit flies invade commercial orchards from neighbouring vegetation and abandoned orchards. If effective traps are placed in the perimeter of the orchard, a high proportion of the invading individuals can be intercepted before entering the orchard. Such a method has proven effective in reducing populations of apple maggot fly, *Rhagoletis pomonella*.

A strategy termed as 'push-pull' involves the behavioural manipulations of stem borers infesting maize and sorghum. It involves 'pushing' the insects away from the harvestable, economic crops, and 'pulling' them onto a trap crop where their population is reduced. Thus, antifeedants, non-host volatiles, parasitoid attractants and compounds associated with plant defence can be used to achieve the 'push', while the sex pheromone and host volatiles can be used to 'pull' the insects onto the trap crop. Khan *et al.* (1997) used the 'push-pull' strategy by intercropping maize and sorghum crops with molasses grass, *Melinis minutiflora*. The stem-borers were 'pushed' away from the main crop, significantly reducing the level of infestation. In addition, volatile compounds produced by the molasses grass increased larval parasitism of stem-borers by the parasitoid, *Cotesia sesamiae*. The 'pull' was provided by planting yet another plant around the maize- the forageable crop Napier grass, *Pennisetum purpureum* which is much more attractive to the stem-borers than the cereal host, but interferes with the normal development of the stem-borer larvae. Khan *et al.* (2001) developed an intercropping and trap crop system, using a 'push-pull' strategy for the control of stem borers in small-scale maize farming systems. Napier grass, *P. purpureum* and Sudan grass, *Sorghum sudanense* were used as trap plants, whereas molasses grass, *Melinis minutiflora* and two species of desmodium, *Desmodium uncinatum* and *D. intortum* were grown to repel ovipositing stem borers. The integrated 'push-pull' strategies were shown to increase parasitism of stem borers through attraction of parasitoids to one of the intercrops, molasses grass.

Nalyanya *et al.* (2000) evaluated the efficacy of insecticide bait in combination with attractants (pull), repellents (push), or both (push-pull), using a hydramethylnon-based bait, faeces-contaminated surfaces as an attractant, and methyl neodecanamide treated surfaces to repel German cockroaches, *Blattella germanica*. Both adult males and first-instar cockroaches chose shelters nearest the attractant-treated surfaces and the results indicated that the strategy could be used to displace pests from resources or commodities that are to be protected, and simultaneously lure the pest to an attractant source coupled with a pest control agent. Concentrating cockroaches into a limited area facilitate the precision-targeting of the pest population and promises to reduce insecticide use.

Insects like caterpillars, ants and mealy bug nymphs crawl up the tree trunk for feeding from the soil to the shoots. The banding material like grease or 'Ostico paint' smeared around the trunk arrests the crawling insects below the banded part where either they die of starvation due to lack of feeding sites or by getting entangled with the sticky banding material.

Application of bait spray to non-host vegetation surrounding host plants of melon flies is a common practice for melon fly control in Hawaii. The intent is to attract (with bait spray droplets) immigrating melon fly females to sprayed sites, where they ingest feeding stimulant and insecticide before entering cultivated fields of hosts (Nishida,1958). Most gravid melon fly females overnight on favoured non-host plants in border areas adjacent to cultivated hosts before entering cultivated fields during the day to oviposit. Prokopy *et al.* (2003) in Hawaii, investigated the effectiveness of GF-120 fruit fly bait spray applied to border area plants (*Sorghum* sp.) that completely surrounded (on all four sides) patches of cucumbers, a favoured host of melon flies. They found that GF-120 fruit fly bait spray applied to a broad and dense swath of sorghum (50 cm deep) was very effective in preventing 4-week-old released protein-deprived (since eclosion) melon fly females from entering cucumber patches.

The interactions between migration and other aspects of pest biology are also critical in developing pest management strategies to deal with mobile pests. For example, in the case of aphids and whiteflies, dispersal of only a few individuals that are insecticide resistant or virus vectors into a crop may give rise to offspring that will cause a significant impact on crop yields. Exploitation of such behaviours of pest insects may assist in their management.

Some insects, such as grasshoppers, locusts, moths and lacewings can detect ultrasounds. Although it is implied that ultrasound has special properties that make it more repellent than audible sound, there is no evidence to support this; or its possible use in pest control is yet to be explored.

References

Abrahamson, W.G., Eubanks, M.D., Blair, C.P. & Whipple, A.V. 2001. Gall flies, inquilines, and goldenrods: a model for host-race formation and sympatric speciation. *Amer. Zool.*, 41: 928-938.

Abrams, P.H. 2000. The evolution of predator-prey interactions: theory and evidence. *Ann. Rev. Ecol. Syst.*, 31: 79-105.

Abramson, C.I. 1994. *A Primer of Invertebrate Learning: The Behavioural Perspective*. Amer. Psychol. Assoc., Washington.

Acharya, L. & McNeil, J.N. 1998. Predation risk and mating behaviour: the responses of moths to bat-like ultrasound. *Behav. Ecol.*, 9: 552-558.

Adams, T.S., Dillwith, J.W. & Blomquist, G.J. 1984. The role of 20-hydroxyecdysone in house fly sex pheromone biosynthesis. *J. Insect Physiol.*, 30: 287-294.

Adler, H.E. 1971. Orientation: sensory basis. *Ann. N.Y. Acad. Sci.*, 188: 3-4.

Agarwal, M.L. & Kapoor, V.C. 1986. Indian Dacini (Diptera: Tephritidae) and their host plant relationships. In: *Fruit Flies of Economic Importance*. R. Cavalloro (ed.). A.A. Balkema, Rotterdam, pp. 51-60.

Agarwal, M.L. & Kumar. P. 1999. Relative efficacy of bait and attractant combinations against peach fruit fly, *Bactrocera zonata* (Saunders). *Pestology*, 23 (1): 23-26.

Agarwal, M.L. & Singh, R. 2005. *Amazing World of Honey bees*. RAU Publ., Samastipur, Bihar.

Agarwal, M.L. & Yazdani, S.S. 1991. A note on growth potential of melon fruit fly, *Dacus* (*Zeugodacus*) *cucurbitae* Coquillett in relation to certain host plants in north Bihar. *J. Res.* (*BAU*), 3(1): 61-62.

Agosta, W.C. 1992. *Chemical Communication: The Language of Pheromones*. W.H. Freeman & Co., N.Y.

Ahmad, F., Aslam, M. & Razaq, M. 2004. Chemical ecology of insects and tritrophic interactions. *J. Res. Sci.*, 15(2): 181-190.

Aidley, D.J. (ed.) 1981. *Animal Migration*. Cambridge Univ. Press, Cambridge.

Alborn, H.T., Turlings, T.C.J., Jones, T.H., Stenhagen, G., Loughrin, J.H. & Tumlinson, J.H. 1997. An elicitor of plant volatiles from beet armyworm oral secretion. *Science*, 276(5314): 945-948.

Alcock, J. 1971. Interspecific differences in avian feeding behaviour and the evolution of Batesian mimicry. *Behaviour*, 40: 1-9.

Alcock, J. 1979. Selective male choice by female of *Harpobittacus australis* (Mecoptera: Bittacidae). *Psyche*, 86: 213-217.

Alcock, J. 1984. *Animal Behaviour- An Evolutionary Approach*. (3rd ed.). Sinauer Assoc. Inc. Publ. Sunderland, Massachusetts.

Alcock, J. 1994. Postinsemination associations between males and females in insects: the mate-guarding hypothesis. *Ann. Rev. Entomol.*, 39: 1-21.

Alcock, J. 1996. The relation between male body size, fighting, and mating success in Dawson's burrowing bee, *Amegilla dawsoni* (Apidae, Apinae, Anthophorini). *J. Zool.*, 239(4): 663-674.

Alcock, J. & Gwynne, D. 1988. The mating system of *Vanessa kershawi*: males defend landmark territories as mate encounter sites. *J. Res. Lepidoptera,* 26(1-4): 116-124.

Alcock, J. & Gwynne, D. 1991. Evolution of insect mating system: the impact of individual selectionist thinking. In: *Reproductive Behaviour of Insects Individuals and Populations.* W.J. Bailey & J. Ridsdill-Smith (eds.). Chapman & Hall, London, pp. 10-41.

Aldrich, J.R., Blum, M.S. & Fales, H.M. 1979. Species specific natural products of adult male leaf-footed bugs (Hemiptera: Heteroptera). *J. Chem. Ecol.,* 5: 53-62.

Alexander, B.A. 1991. Phylogenetic analysis of the genus *Apis* (Hymenoptera: Apidae). *Ann. Entomol. Soc. Am.,* 84(2): 137-149.

Alexander, R.D. 1961. Aggressiveness, territoriality and sexual behaviour in field crickets (Orthoptera: Gryllidae). *Behaviour,* 17: 130-223.

Alexander, R.D. 1962a. Evolutionary change in cricket acoustical communication. *Evolution,* 16: 443-467.

Alexander, R.D. 1962b. The role of behavioural study in cricket classification. *Syst. Zool.,* 11: 53-72.

Alexander, R.D. 1964. The evolution of mating behaviour in arthropods. *Symp. Roy. Entomol. Soc. Lond.,* 2: 78-94.

Alexander, R.D. 1967. Acoustical communication in arthropods. *Ann. Rev. Entomol.,* 12: 495-526.

Alexander, R.D. 1968. Life cycle origins, speciation and related phenomena in crickets. *Quart. Rev. Biol.,* 43: 1-41.

Alexander, R.D., Marshall, D.C. & Cooley. J.R. 1997. Evolutionary perspectives on insect mating. In: *The Evolution of Mating Systems in Insects and Arachnids.* Vol. 1. J.C. Choe, & B.J. Crespi (eds.). Cambridge Univ. Press, Cambridge, pp. 4-31.

Alexander, R.M. 1981. Flight. In: *Locomotion Energetics in Arthropods.* C.F. Herreid & C.R. Foutner (eds.). Plenum Press, N.Y., pp. 247-278.

Alexander, R.M. 2003. *Principles of Animal Locomotion.* Princeton Univ. Press, Princeton, N.J.

Aller, T. & Caldwell, R.L. 1979. An investigation of the possible presence of an aggregation pheromone in the milkweed bugs, *Oncopeltus fasciatus* and *Lygaeus kalmii. Physiol. Entomol.,* 4: 287-290.

Alloway, T.M. 1969. Effect of low temperature upon acquisition and retention in the grain beetle *Tenebrio molitor. J. Comp. Physiol. Psychol.,* 69: 1-8.

Alloway, T.M. 1972. Learning and memory in insects. *Ann. Rev. Entomol.,* 17: 43-56.

Alpatov, A.M., Zotov, V.A., Tshernyshev, W.B. & Rietveld, W.J. 1999. Endogenous circadian rhythm is a crucial tool for survival of the sand-desert tenebrionid beetle *Trigonoscelis gigas* Reitter. *Biol. Rhythm Res.,* 30(1): 104-109.

Altner, H. & Loftus, R. 1985. Ultrastructure and function of insect thermo- and hygroreceptors. *Ann. Rev. Entomol.,* 30: 273-295.

Altner, H. & Prinninger, L. 1980. Ultrastructure of invertebrate chemo-, thermo- and hygroreceptors and its functional significance. *Int. Rev. Cytol.,* 67: 69-139.

Aluja, M. & Boller, E.F. 1992. Host marking pheromone of *Rhagoletis cerasi*: field deployment of synthetic pheromone as a novel cherry fruit fly management strategy. *Entomol. Exp. Appl.,* 65: 141-147.

Aluja, M. & Norrbom, A.L. (eds.) 1999. *Fruit Flies (Tephritidae): Phylogeny and Evolution of Behaviour.* CRC Press, Boca Raton.

Aluja, M., PinÞero, J., Jácome, I., Díaz-Fleischer, F. & Sivinski, J. 1999. Behaviour of flies in the genus *Anastrepha* (Trypetinae: Toxotrypanini). In: *Fruit Flies (Tephritidae): Phylogeny and Evolution of Behaviour*. M. Aluja & A.L. Norrbom (eds.). CRC Press, Boca Raton, pp. 375-406.

Ananthakrishnan, T.N. 1992. *Dimensions of Insect-Plant Interactions*. Oxford & IBH Publ. Co. Pvt. Ltd., New Delhi.

Ananthakrishnan, T.N. & Senrayan, R. 1992. Phytochemical-induced responses, a vital factor governing insect host-parasite/ predator interactions. *Phytophaga*, 4(2): 87-94.

Ananthakrishnan, T.N., Senrayan, R., Murugesan, S. & Annadurai. R.S. 1991. Kairomones of *Helicoverpa armigera* and *Corcyra cephalonica* and their influence on the parasite potential of *Trichogramma chilonis* (Trichogrammatidae: Hymenoptera). *J. Biosci.*, 16(3): 111-119.

Andersen, N.M. 1976. A comparative study of locomotion on the water surface in semiaquatic bugs (Insecta, Hemiptera, Gerromorpha). *Vidensk. Meddr dansk naturh. Foren.*, 139: 337-396.

Anderson, J. 1985. Low thigmo-kinesis, a key mechanism in habitat selection by riparian *Bembidion* (Carabidae) species. *Oikos*, 44(3): 499-505.

Andersson, M. 1984. The evolutions of eusociality. *Ann. Rev. Ecol. Syst.*, 15: 165-189.

Andersson, M. 1994. *Sexual Selection*. Princeton Univ. Press, Princeton, N.J.

Aneshansley, D., Eisner, T., Widom, J.M. & Widom, B. 1969. Biochemistry at 100°C: the explosive discharge of bombardier beetle (*Brachinus*). *Science*, 165: 61-63.

Anholt, R.R.H. & Mackay, T.F.C. 2004. Quantitative genetic analyses of complex behaviours in *Drosophila*. *Nat. Rev. Genet.*, 5: 838-849.

Aoki, S. 1977. *Colophina clematis* (Homoptera, Pemphigidae), an aphid species with soldiers. *Kontyu*, 45: 276-282.

Aprille, J.R., Lagace, C.J., Modica-Napolitano J. & Trimmer, B.A. 2004. Role of nitric oxide and mitochondria in control of firefly flash. *Integr. Comp. Biol.*, 44(3): 213-219.

Arathi, H.S. & Spivak, M. 2001. Influence of colony genotypic composition on the performance of hygienic behaviour in the honey bee, *Apis mellifera* L. *Anim. Behav.*, 62(1): 57-66.

Aristotle (c. mid 4th century B.C.). *History of Animals (Historia Animalium)*. (Trans. A.L. Peck). Loeb Classical Library, Cambridge; Harvard Univ. Press, Cambridge, Massachusetts; and Heinemann, London, 1965-91, 3 vols.

Arms, K., Feeny, P. & Lederhouse, R.C. 1973. Sodium: stimulus for puddling behaviour by tiger swallowtail butterflies. *Science*, 185: 372-374.

Arn, H., Tóth, M. & Priesner, E. 1992. *List of Sex Pheromones of Lepidoptera and Related Attractants*. OILB-SROP, Paris.

Arnqvist, G., Jones, T.M. & Elgar, M.A. 2003. Insect: reversal of sex roles in nuptial feeding. *Nature*, 424(6947): 387.

Arnqvist, G. & Nilsson, T. 2000. The evolution of polyandry: multiple mating and female fitness in insects. *Anim. Behav.*, 60: 145-164.

Arora, M.P. 1992. *Animal Behaviour*. Himalaya Publ. House, Mumbai.

Aschoff, J. (ed.) 1965. *Circadian Clocks*. North-Holland Publ., Amsterdam.

Aschoff, J. 1984. Circadian timing. *Ann. N.Y. Acad. Sci.*, 423: 442-468.

Askew, R.R. 1971. *Parasitic Insects*. Heinemann Educational Books, London.

Athanas, M.M., Schroder, R.F.W. & Sidor, A.M. 1996. Parental care in *Erixestus winnemana* (Hymenoptera: Pteromalidae), an egg parasite of *Calligrapha* (Coleoptera: Chrysomelidae). *Entomol. News*, 107(3): 161-165.

Atkins, G., Atkins, S., Schoun, D. & Stout, J.F. 1987. Scototaxis and shape discrimination in the female cricket *Acheta domestius* in an arena and on a compensatory treadmill. *Physiol. Entomol.*, 12(2): 125-133.

Atkins, M.D. 1978. *Insects in Perspectives*. Macmillan, N.Y.

Atkins, M.D. 1980. *Introduction to Insect Behaviour*. Macmillan, N.Y.

Attygalle, A.B. & Morgan, E.D. 1985. Ant trail pheromones. *Adv. Insect Physiol.*, 18: 1-30.

Atwal, A.S. 1955. Influence of temperature, photoperiod, and food on the speed of development, longevity, fecundity, and other qualities of the diamondback moth, *Plutella maculipennis* (Curtis) (Tineidae, Lepidoptera). *Aust. J. Zool.*, 3: 185-221.

Atwal, A.S. & Bains, S.S. 1974. *Applied Animal Ecology*. Kalyani Publ., Ludhiana.

Autrum, H. 1963. Anatomy and physiology of sound receptors in invertebrates. In: *Acoustic Behaviour of Animals*. R.-G. Busnel (ed.). Elsevier Publ. Co., Amsterdam.

Averill, A.L. & Prokopy, R.J. 1987. Intraspecific competition in the tephritid fruit fly *Rhagoletis pomonella*. *Ecology*, 68(4): 878-886.

Averof, M. & Cohen, S.M. 1997. Evolutionary origin of insect wings from ancestral gills. *Nature*, 385: 627-630.

Awasthi, V.B. 2001. *Principles of Insect Behaviour*. Scientific Publ., Jodhpur.

Awasthi, V.B. 2005. *Insect Neuroendocrines*. Scientific Publ., Jodhpur.

Baader, A.P. 1996. The significance of visual landmarks for navigation of the giant tropical ant, *Paraponera clavata* (Formicidae, Ponerinae). *Insectes Sociaux*, 43(4): 435-450.

Baer, B., Morgan, E.D. & Schmid-Hempel, P. 2001. A nonspecific fatty acid within the bumble bee mating plug prevents females from remating. *Proc. Natl. Acad. Sci., USA*, 98(7): 3926-3928.

Baerends, G.P. 1941. Fortpflanzungsverhalten und Orientierung der Grabwespe *Ammophila campestris*. *Tijdschr. Entomol.*, 84: 68-75.

Baerends, G.P. 1950. Specializations in organs and movements with a releasing function. *Symp. Soc. Exp. Biol.*, 4: 337-360.

Bailey, W.J. 1991a. *Acoustic Behaviour of Insects: An Evolutionary Perspective*. Chapman & Hall, London.

Bailey, W.J. 1991b. Mate finding: selection on sensory cues. In: *Reproductive Behaviour of Insects Individuals and Populations*. W.J. Bailey & J. Ridsdill-Smith (eds.). Chapman & Hall, London, pp. 42-74.

Bailey, W.J. & Cunninghum, R.J. 1990. Song power, spectral distribution and female phonotaxis in the bush cricket *Requena verticalis* (Tettigoniidae: Orthoptera): active female choice or passive attraction. *Anim. Behav.*, 40(1): 33-42.

Bailey, W.J. & Ridsdill-Smith, J. (eds.) 1991. *Reproductive Behaviour of Insects Individuals and Populations*. Chapman & Hall, London.

Baker, H.G. 1961. *Ficus* and *Blastophaga*. *Ecology*, 15: 375-379.

Baker, R. (ed.) 1981. *The Mystery of Migration*. Viking Press, N.Y.

Baker, R., Herbert, R.H. & Grant, G.G. 1985. Isolation and identification of the sex pheromone of the Mediterranean fruit fly *Ceratitis capitata* (Wied.). *J. Chem. Soc. Chem. Commun.*, 1985: 824-825.

Baker, R.R. 1978. *The Evolutionary Ecology of Animal Migration*. Hodder & Stoughton, London.

Baker, R.R. 1983. Insect territoriality. *Ann. Rev. Entomol.*, 28: 65-89.

Baker, R.R. 1987. Integrated use of moon and magnetic compasses by the heart-and-dart moth, *Agrotis exclamationis*. *Anim. Behav.*, 35(1): 94-101.

Baker, R.R. & Mather, J.G. 1982. Magnetic compass sense in the large yellow underwing moth, *Noctua pronuba* L. *Anim. Behav.*, 30(2): 543-548.

Baker, T.C. 1985. Chemical control of behaviour. In: *Comprehensive Insect Physiology, Biochemistry and Pharmacology. Behaviour*. Vol. 9. G.A. Kerkut & L.I. Gilbert (eds.). Pergamon Press, Oxford, pp. 621-672.

Baker, T.C. 1989. Pheromones and flight behaviour. In: *Insect Flight*. G.J. Goldsworthy & C.H. Wheeler (eds.). CRC Press, Boca Raton, pp. 231-255.

Baker, T.C., Staten, R.T. & Flint, H.M. 1990. Use of the pink bollworm pheromone in the southwestern United States. In: *Behaviour-Modifying Chemicals for Insect Management*. R.L. Ridgway, R.M. Silverstein & M.N. Inscoe (eds.). Marcel Dekker, N.Y., pp. 417-436.

Balakrishnan, R. & Pollack, G.S. 1996. Recognition of courtship song in the field cricket, *Teleogryllus oceanicus*. *Anim. Behav.*, 51: 353-366.

Banks, A.N. & Srygley, R.B. 2003. Orientation by magnetic field in leaf-cutter ants, *Atta colombica* (Hymenoptera: Formicidae). *Ethology*, 109: 835-846.

Barber, H.S. 1951. North American fireflies of the genus *Photuris*. *Smithson. Inst. Misc. Coll.*, 117: 1-58.

Bargen, H., Saudhof, K. & Poehling, H.M. 1998. Prey finding by larvae and adult females of *Episyrphus balteatus*. *Entomol. Exp. Appl.*, 87(3): 245-254.

Barnard, C.J. 1983. *Animal Behaviour: Ecology and Evolution*. Croom Helm., London.

Barnes, T.C. & Kohn, H.I. 1932. The effect of temperature on the leg posture and speed of creeping in the ants. *Biol. Bull.*, 62: 306-312.

Barrozo, R.B. & Lazzari, C.R. 2004. The response of the blood-sucking bug *Triatoma infestans* to carbon dioxide and other host odours. *Chem. Senses*, 29(4): 319-329.

Bartell, R.J. 1982. Mechanism of communication disruption by pheromones in the control of Lepidoptera: a review. *Physiol. Entomol.*, 7: 353-364.

Barth, R.H. 1965. Insect mating behaviour: endocrine control of a chemical communication system. *Science*, 149: 882-883.

Barth, R.H. & Laster, L.J. 1973. Neuro-hormonal control of sexual behaviour in insects. *Ann. Rev. Entomol.*, 18: 445-472.

Barthel, W.F., Green, N., Keiser, I. & Steiner, L.F. 1957. Anisyl acetone, synthetic attractant for male melon fly. *Science*, 126: 654.

Barton Browne, L. 1974. *The Experimental Analysis of Insect Behaviour*. Springer-Verlag, N.Y.

Barton Browne, L. 1975. Regulatory mechanism in insect feeding. *Adv. Insect Physiol.*, 11: 1-116.

Barton Browne, L. 1995. Ontogenic changes in feeding behaviour. In: *Regulatory Mechanisms in Insect Feeding*. R.F. Chapman & G. Boer, de (ed.). Chapman & Hall, N.Y., pp. 307-342.

Bastock, M. 1956. A gene mutation which changes a behaviour pattern. *Evolution*, 10: 421-439.

Bateman, M.A., Friend, A.H. & Hampshire, F. 1966. Population suppression in the Queensland fruit fly, *Dacus (Strumeta) tryoni*. II Experiments on isolated populations in Western New South Wales. *Aust. J. Agric. Res.*, 17: 699-718.

Bateman, P.W. 1998. Mate choice for novel partners in the cricket *Gryllus bimaculatus*. *Ecol. Entomol.*, 23: 473-475.

Bateman, P.W. & Ferguson, J.W.H. 2004. Male mate choice in the Botswana armoured ground cricket *Acanthoplus discoidalis* (Orthoptera: Tettigoniidae: Hetrodinae). Can, and how, do males judge female mating history? *J. Zool., Lond.*, 262: 305-309.

Bates, H.W. 1862. Contribution to the insect fauna of Amazon valley. *Trans. Linn. Soc. Lond.*, 23: 495-566.

Batra, L.R. & Batra, S.W.T. 1979. Termite-fungus mutualism. In: *Insect Fungus Symbiosis, Nutrition, Mutualism and Commensalism*. L.R. Batra (ed.). Allanheld & Osmun, Montclair, N.J., pp. 117-163.

Batra, S.W.T. 1966. Nests and social behaviour of halictine bees of India (Hymenoptera: Halictidae). *Indian J. Entomol.*, 28(3): 375-393.

Batra, S.W.T. 1975. Termites (Isoptera) eat and manipulate symbiotic fungi. *J. Kansas Entomol. Soc.*, 48(1): 89-92.

Batra, S.W.T. 1977. Bees of India (Apoidea), their behaviour, management and a key to the genera. *Oriental Ins.*, 11(3): 289-324.

Batra, S.W.T. & Batra, L.R. 1966. The fungal gardens of insects. *Sci. Amer.*, 217: 112-120.

Bays, M. & Pyza, E. 2001. Circadian rhythms in the first neuropile of the optic lobe in the house fly. *J. Exp. Biol.*, 204: 3303-3310.

Beard, R.L. 1963. Insect toxins and venoms. *Ann. Rev. Entomol.*, 8: 1-18.

Beck, S.D. 1963. Physiology and ecology of photoperiodism. *Bull. Entomol. Soc. Am.*, 9: 8-16.

Beck, S.D. 1965. Resistance of plants to insects. *Ann. Rev. Entomol.*, 10: 207-232.

Beck, S.D. 1980. *Insect Photoperiodism*. (2nd ed.). Academic Press, N.Y.

Beebe, W. 1947. Notes on the Hercules beetle, *Dynastes hercules* (Linn.) at Rancho Grande, Venezuela, with special reference to combat behaviour. *Zoologica*, 32: 109-116.

Beier, W. 1968. Beeinflussung der inneren Uhr der Bienen durch Phasenverschiebung des Licht-Dunkel-Zeitgebers. *Z. Bienenforsch.*, 9: 356-378.

Beier, W. & Lindauer, M. 1970. Der Sonnestand als Zeitgeber für die Biene. *Apidologie*, 1: 5-28.

Belanger, R.P., Hedden, R.L. & Lorio, P.L. 1993. Management strategies to reduce losses from the southern pine beetle. *South. J. Appl. Forest.*, 17: 150-154.

Beling, I. 1929. Über das Zeitgedächtnis der Bienen. *Z. vergl. Physiol.*, 9: 259-338.

Bell, W.J. 1973. Factors controlling initiation of vitellogenesis in a primitively social bee, *Lasioglossum zephyrum* (Hymenoptera: Halictidae). *Insectes Sociaux*, 20: 253-260.

Bell, W.J. & Barth, R.H. Jr. 1971. Initiation of yolk deposition by juvenile hormone. *Nature*, 230: 220-222.

Bell, W.J. & Cardé, R.T. 1984. *Chemical Ecology of Insects*. Chapman & Hall, London.

Bell, W.J. & Kramer, E. 1979. Search and anemotactic behaviour of cockroaches. *J. Insect Physiol.*, 25(8): 631-640.

Bennet-Clark, H.C. 1984. Insect hearing: acoustic and transduction. In: *Insect Communication*. T. Lewis (ed.). Academic Press, London, pp. 49-82.

Bennet-Clark, H.C. 1989. Cricket songs and the physics of sound production. In: *Cricket Behaviour and Neurobiology*. F. Huber, T.E. Moore & W. Loher (eds.). Cornell Univ. Press, Ithaca, N.Y., pp. 227-261.

Bennet-Clark, H.C. & Ewing, A.W. 1970. The love song of fruit fly. *Sci. Amer.*, 233: 84-92.

Bennet-Clark, H.C. & Lucey, E.C.A. 1967. The jump of the flea: a study of the energetics and a model of the mechanism. *J. Exp. Biol.*, 45: 151-164.

Bennett, M.F. & Renner, M. 1963. The collecting performance of honey bees under laboratory conditions. *Biol. Bull.*, 125: 416-430.

Bentley, B.L. 1977. Extra-floral nectarines and protection by pugnacious body guards. *Ann. Rev. Ecol. Syst.*, 8: 407-427.

Bentley, D.R. & Hoy, R.R. 1972. Genetic control of the neuronal network generating cricket (*Teleogryllus gryllus*) song patterns. *Anim. Behav.*, 20: 478-492.

Berger, R.S. 1966. Isolation, identification and synthesis of the sex attractant of the cabbage looper, *Trichoplusia ni. Ann. Entomol. Soc. Am.*, 59: 767-771.

Berlocher, S.H. & Bush, G.L. 1982. An electrophoretic analysis of *Rhagoletis* (Diptera: Tephritidae) phylogeny. *Syst. Zool.*, 31: 136-155.

Berlocher, S.H. & Feder, J.L. 2002. Sympatric speciation in phytophagous insects: moving beyond controversy? *Ann. Rev. Entomol.*, 47: 773-815.

Bernard, G.D. 1979. Red-absorbing visual pigment of butterflies. *Science*, 203: 1125-1127.

Bernays, E.A. 1985. Regulation of feeding behaviour. In: *Comprehensive Insect Physiology, Biochemistry and Pharmacology. Regulation: Digestion, Nutrition, Excretion* Vol. 4. G.A. Kerkut & L.I. Gilbert (eds.). Pergamon Press, Oxford, pp. 1-32.

Bernays, E.A. 1986. Diet-induced head allometry among foliage chewing insects and its importance for graminivores. *Science*, 4737: 495-497.

Bernays, E.A. & Chapman, R.F. 1994. *Host-Plant Selection by Phytophagous Insects*. Chapman & Hall, N.Y.

Bernays, E.A., Oppenheim, S., Chapman, R.F., Kwon, H. & Gould, F. 2000. Taste sensitivity of insect herbivores to deterrents is greater in specialists than in generalists: a behavioural test of the hypothesis with two closely related caterpillars. *J. Chem. Ecol.*, 26(2): 547-563.

Bernays, E.A. & Wrubel, R.P. 1985. Learning by grasshoppers: association of colour/light intensity with food. *Physiol. Entomol.*, 10(4): 359-369.

Bernhard, C.G. (ed.) 1966. *The Functional Organization of the Compound Eye*. Pergamon, Oxford.

Beroza, M. (ed.) 1970. *Chemicals Controlling Insect Behaviour*. Academic Press, N.Y.

Beroza, M., Alexander, B.H., Steiner, L.F., Mitchell, W.C. & Miyashita, D.H. 1960. New synthetic lures for the male melon flies. *Science*, 131: 1044-1045.

Berryman, A.A., Dennis, B., Raffa, K.F. & Stenseth, N.C. 1985. Evolution of optimal group attack with particular reference to bark beetles (Coleoptera: Scolytidae). *Ecology*, 66: 898-903.

Bess, H.A. & Haramoto, F.H. 1961. Contribution to the biology and ecology of oriental fruit fly, *Dacus dorsalis* in Hawaii. *Hawaii Agr. Exp. Sta. Tech. Bull.* 44: 1-30.

Beugnon, G. 1987. Time-compensated celestial orientation in the cricket, *Pteronemobius lineolatus. Behav. Processes*, 15(1): 109-116.

Bhandari, R. & Pathak, S.C. 1997. Sexual aggressiveness in relation to reproduction, among males of *Blattella germanica* Linn. (Blattoidea: Blattellidae). *Insect Environ.*, 3(3): 78-79.

Bierl, B.A., Beroza, M. & Collier, C.W. 1970. A potent sex attractant of the gypsy moth: its isolation, identification and synthesis. *Science*, 170: 87-89.

Billard, G. & Bruyant, C. 1905. Sue un mode particulier de locomotion de certains *Stenus. C. r. Séane. Soc. Biol.*, 59: 102-103.

Billen, J. (ed.) 1992. *Biology and Evolution of Social Insects*. Leuven Univ. Press, Leuven.

Birch, M.C. (ed.) 1974a. *Pheromones*. North-Holland, Amsterdam.

Birch, M.C. 1974b. Aphrodisiacs pheromones in insects. In: *Pheromones*. Birch, M.C. (ed.). North-

Holland, Amsterdam, pp. 115-134.

Birch, M.C. & Haynes, K.F. 1982. *Insect Pheromones*. The Institute of Biology's, Studies in Biology, No. 147. Edward Arnold, London.

Birch, M.C. & Wood, D.L. 1975. Mutual inhibition of the attractant pheromone response of two species of *Ips* (Coleoptera: Scolytidae). *J. Chem. Ecol.*, 1: 101-113.

Biron, K. 2003. Fireflies, dead fish and a glowing bunny: a primer on bioluminescence. *BioTeach J.*, 1: 19-26.

Birukow, G. 1953. Menotaxis im polarisierten Licht bei *Geotrupes sylvaticus* Panz. *Naturwissenschaften*, 40: 611.

Birukow, G. 1956. Lichtkompassorientierung beim Wasserlaufer *Velia currens* F. am Tage und zur Nachtzeit. I. Herbst-und Winterversuche. *Z. Tierpsychol.*, 13: 463-484.

Birukow, G. 1960. Innate type of chronometry in insect orientation. *Cold Spring Harb. Symp. Quant. Biol.*, 25: 403-412.

Birukow, G. 1966. Orientation behaviour in insects and factors which influence it. In: *Insect Behaviour*. P.T. Haskell (ed.). *Symp. Roy. Entomol. Soc. Lond.*, No. 3, pp. 1-12.

Bishop, J.A. 1972. An experimental study of the cline of industrial melanism in *Biston betularia* (L.) (Lepidoptera) between urban and rural North Wales. *J. Anim. Ecol.*, 41: 209-243.

Bishop, J.A. & Cook, L.M. 1975. Moths, melanism and clean air. *Sci. Amer.*, 232: 90-99.

Bjostad, L.B., Wolf, W.A. & Roelofs, W.L. 1987. Pheromone biosynthesis in lepidopterans: desaturation and chain shortening. In: *Pheromon Biochemistry*. G.D. Prestwich & G.J. Blomquist (eds.). Academic Press, N.Y., pp. 77-120.

Blaakmeer, A., Hagenbeek, D., Beek, T.A. van, Groot, A.E. de, Schoonhoven, L.M. & Loon, J.J.A. van, 1994. Plant response to eggs vs. host marking pheromone as factors inhibiting oviposition by *Pieris brassicae*. *J. Chem. Ecol.*, 20: 1657-1665.

Blake, G.M. 1958. Diapause and the regulation of development in *Anthrenus verbasci* (L.) (Col. Dermestidae). *Bull. Entomol. Res.*, 49: 751-775.

Blake, G.M. 1960. Decreasing photoperiod inhibiting metamorphosis in an insect. *Nature*, 188: 168-169.

Blake, G.M. 1963. Shortening of a diapause-controlled life cycle by means of increasing photoperiod. *Nature*, 198: 462-463.

Blake, R.W. 1986. Hydrodynamics of swimming in the water boatman *Cenocorixa bifida*. *Can. J. Zool.*, 64(8): 1606-1613.

Blest, A.D. 1957. The function of eyespot patterns in Lepidoptera. *Behaviour*, 11: 209-226.

Blest, A.D. 1964. Protective display and sound production in some new world arctiid and ctenuchid moths. *Zoologica*, 49: 161-181.

Blest, A.D., Collett, T.S. & Pye, J.D. 1963. The generation of ultrasonic signals by a New World arctiid moth. *Proc. Roy. Soc. Lond. Ser. B, Biol. Sci.*, 158: 196-207.

Blomquist, G.J. & Vogt, R.G. (eds.) 2003a. *Insect Pheromone Biochemistry and Molecular Biology: The Biosynthesis and Detection of Pheromones and Plant Volatiles*. Elsevier Academic Press, London.

Blomquist, G.J. & Vogt, R.G. 2003b. Biosynthesis and detection of pheromones and plant volatiles-introduction and overview. In: *Insect Pheromone Biochemistry and Molecular Biology: The Biosynthesis and Detection of Pheromones and Plant Volatiles*. Elsevier Academic Press, London, pp. 3-18.

Blum, M.S. 1969. Alarm pheromones. *Ann. Rev. Entomol.*, 14: 57-80.

Blum, M.S. 1974. Pheromonal sociality in the Hymenoptera. In: *Pheromones*. Birch, M.C. (ed.). North-Holland, Amsterdam, pp. 222-249.

Blum, M.S. 1978. Biochemical defences of insects. In: *Biochemistry of Insects*. M. Rockstein (ed.). Academic Press, N.Y., pp. 465-513.

Blum, M.S. 1981. *Chemical Defence of Arthropods*. Academic Press, N.Y.

Blum, M.S. 1985. Alarm Pheromones. In: *Comprehensive Insect Physiology, Biochemistry and Pharmacology. Behaviour*. Vol. 9. G.A. Kerkut & L.I. Gilbert (eds.). Pergamon Press, Oxford, pp. 193-224.

Blum, M.S. & Blum, N.A. (eds.) 1979. *Sexual Selection and Reproductive Competition in Insects*. Academic Press, N.Y.

Blum, M.S., Crewe, R.M., Kerr, W.E., Keith, L.H., Garrison, A.W. & Walker, M.M. 1970. Citral in stingless bees: isolation and functions in trail-laying and robbing. *J. Insect Physiol.*, 16: 1637-1648.

Blum, M.S. & Ross, G.N. 1965. Chemical releasers of social behaviour: V. Source, specificity and properties of the odour trail pheromone of *Tetramorium guineense* (F.) (Formicidae: Myrmieinae). *J. Insect Physiol.*, 11: 857-868.

Boake, C.R.B. & Capranica, R.R. 1982. Aggressive signals in "courtship" chirps of a gregarious cricket. *Science*, 218: 580-582.

Boeckh, J. 1980. Neural basis of coding of chemosensory quality at the receptor cell level. In: *Olfaction and Taste VII*. Starre, H. van der (ed.). IRL Press, London, pp. 113-122.

Bollenbacher, W.E., Granger, N.A., Katahira, E.J. & O'Brien, M.A. 1987. Developmental endocrinology of larval moulting in the tobacco hornworm, *Manduca sexta*. *J. Exp. Biol.*, 128: 175-192.

Boller, E.F. 1983. Biotechnical methods for the management of fruit fly populations. In: *Fruit Flies of Economic Importance*. R. Cavalloro (ed.). *CEC/IOBC Symp. Athens*, 1982. Balkema, Rotterdam, pp. 342-352.

Boller, E.F. & Bush, G.L. 1974. Evidence for genetic variation in populations of the European cherry fruit fly, *Rhagoletis cerasi* (Diptera: Tephritidae) based on physiological parameters and hybridization experiments. *Entomol. Exp. Appl.*, 17: 279-293.

Bolwig, N. 1946. Senses and sense-organs of the anterior end of the house fly larva. *Vidensk. Medd. Dansk naturh. Foren.*, 109: 81-217.

Bonduriansky, R. 2001. The evolution of male mate choice in insects: a synthesis of ideas and evidence. *Biol. Rev.*, 76: 305-339.

Boppré, M. 1984. Redefining 'pharmacophagy'. *J. Chem. Ecol.*, 10: 1151-1154.

Borden, J.H. 1982. Aggregation Pheromones. In: *Bark Beetles in North American Conifers. Ecology and Evolution*. J.B. Mitton & K.B. Sturgeon (eds.). Univ. Texas Press, Austin, pp. 74-139.

Borden, J.H. 1985. Aggregation pheromone. In: *Comprehensive Insect Physiology, Biochemistry and Pharmacology. Behaviour*. Vol. 9. G.A. Kerkut & L.I. Gilbert (eds.). Pergamon Press, Oxford, pp. 257-285.

Borden, J.H. 1993. Strategies and tactics for the use of semiochemicals against forest insect pests in North America. In: *Pest Management: Biologically Based Technologies*. R.D. Lumsden & J.L. Vaughn (eds.). *ACS Conf. Proc., Amer. Chem. Soc., Wash. D.C.*, pp. 265-279.

Bosch, J., Maeta, Y. & Rust, R. 2001. A phylogenetic analysis of nesting behaviour in the genus *Osmia* (Hymenoptera: Megachilidae). *Ann. Entomol. Soc. Am.*, 94(4): 617-627.

Bossert, W.H. & Wilson, E.O. 1963. The analysis of olfactory communication among animals. *J. Theor. Biol.*, 5: 443-469.

Bowen, M.F., Saunders, D.S., Bollenbacher, W.E. & Gilbert, L.I. 1984. *In vitro* reprogramming of the photoperiodic clock in an insect brain-retrocerebral complex. *Proc. Natl. Acad. Sci., USA*, 81(18): 5881-5884.

Bowerman, R.F. 1977. The control of arthropod walking. *Comp. Biochem. Physiol.*, 56A: 231-247.

Bowers, W.S., Nault, L.R., Webb, R.E. & Dutky, S.R. 1972. Aphid alarm pheromone: isolation, identification, synthesis. *Science*, 177: 1121-1122.

Bowers, W.S., Ohta, T. Cleere, J.S. & Marsella, P.A. 1976. Discovery of insect anti-juvenile hormones in plants. *Science*, 193: 542-547.

Brackenbury, J. 1999. Fast locomotion in caterpillars. *J. Insect Physiol.*, 45(6): 525-533.

Brackenbury, J. 2000. Locomotory modes in the larva and pupa of *Chironomus plumosus* (Diptera, Chironomidae). *J. Insect Physiol.*, 46(12): 1517-1527.

Brady, J. 1972. Spontaneous, circadian components of tsetse fly activity. *J. Insect Physiol.*, 18: 471-484.

Brady, J. 1974. The physiology of insect circadian rhythms. *Adv. Insect Physiol.*, 11: 1-115.

Brady, J. & Crump, A.J. 1978. The control of circadian activity rhythm in tsetse flies: environment or physiological clock? *Physiol. Entomol.*, 3: 179-190.

Brady, U.E., Tumlinson, J.H., Brownlee, R.G. & Silverstein, R.M. 1971. Sex stimulant and attractant in the Indian meal moth and in the almond moth. *Science*, 171: 802-804.

Brakefield, P.M. 1985. Polymorphic Müellerian mimicry and interactions with thermal melanism in ladybirds and a soldier beetle: a hypothesis. *Biol. J. Linn. Soc.*, 26: 243-267.

Brandt, H. 1934. Die Lichtorientierung der Mehlmotte *Ephestia kuehniella* Zeler. *Z. vergl. Physiol.*, 20: 646-673.

Branham, M.A. & Greenfield, M.D. 1996. Flashing males win mate success. *Nature*, 381: 745-746.

Branham, M.A. & Wenzel, J.W. 2001. The evolution of bioluminescence in cantharoids (Coleoptera: Elateriodea). *Florida Entomol.*, 84: 565-586.

Braun, G. & Bicker, G. 1992. Habituation of an appetitive reflex in the honey bee. *J. Neurophysiol.*, 67(3): 588-598.

Breed, M.D. 1983. Correlations between juvenile hormone and aggressive behaviour in worker honey bees. *Insectes Sociaux*, 30: 482-495.

Breed, M.D., Garry, M.F., Pearce, A.N., Bjostad, L., Hibbard, B. & Page, R.E. 1995. The role of wax comb in honey bee nestmate recognition: genetic effects on comb discrimination, acquisition of comb cues by bees, and passage of cues among individuals. *Anim. Behav.*, 50: 489-496.

Breed, M.D., Michener, C.D. & Evans, H.E. (eds.) 1982. *The Biology of Social Insects*. Westview Press, Boulder, Colorado.

Breland, O.P & Simmons, E.S. 1970. Preliminary studies of the spermatozoa and the male reproductive system of some whirligig beetles (Coleoptera: Gyrinidae). *Entomol. News*, 81: 101-110.

Breuer, M., Hoste, B. & Loof, A. de, 2003. The endocrine control of phase transition: some new aspects. *Physiol. Entomol.*, 28(1): 3-10.

Brian, M.V. 1974. Caste differentiation in *Myrmica rubra*: the role of hormones. *J. Insect Physiol.*, 20(7): 1351-1365.

Broadley, R.A. & Stringer, I.A.N. 2001. Prey attraction by larvae of the New Zealand glowworm, *Arachnocampa luminosa* (Diptera: Mycetophilidae). *Invert. Biol.*, 120(2): 170-177.

References

Brodsky, A.K. 1994. *The Evolution of Insect Flight*. Oxford Univ. Press, Oxford.

Broumas, Th., Katsoyannos, P., Yamvrias, C., Liaropoulos, C., Haniotakis, G. & Strong, F. 1983. Control of the olive fruit fly in a pest management trail in olive culture. In: *Fruit Flies of Economic Importance*. R. Cavalloro (ed.). *CEC/IOBC Symp. Athens*, 1982. Balkema, Rotterdam, pp. 584-592.

Brower, L.P. 1958. Experimental studies of mimicry in some North American butterflies. I. The monarch *Danaus plexippus* and viceroy, *Limenitis archippus*. *Evolution*, 12: 32-47.

Brower, L.P. 1969. Ecological chemistry. *Sci. Amer.*, 220: 22-29.

Brower, L.P. 1977. Monarch migration. *Nat. Hist.*, 86: 40-53.

Brower, L.P. 1996. Monarch butterfly orientation: missing pieces of a magnificent puzzle. *J. Exp. Biol.*, 199: 93-103.

Brower, L.P. Alcock, J. & Brower, J.V.Z. 1971. Avian feeding behaviour and the selective advantage of incipient mimicry. In: *Ecological Genetics and Evolution*. E.R. Creed (ed.). Blackwell, Oxford, pp. 261-274.

Brower, L.P., Brower, J.V.Z. & Collins, C.T. 1963. Experimental studies of mimicry. 7. Relative palatability and Müllerian mimicry among neotropical butterflies of the subfamily Helioconiinae. *Zoologica*, 48: 65-83.

Brower, L.P., Brower, J.V.Z., Cranston, F.P. 1965. Courtship behaviour of the queen butterfly, *Danaus gilippus berenice* (Cramer). *Zoologica*, 50: 1-39.

Brower, L.P., Cook, L.M. & Cronz, H.J. 1967. Predator responses to artificial Batesian mimics released in a neotropical environment. *Evolution*, 21: 11-23.

Brown, A.W.A. 1966. The attraction of mosquitoes to hosts. *J. Amer. Med. Assoc.*, 196: 159-162.

Brown, F.A. 1965. A unified theory for biological rhythms. In: *Circadian Clocks*. J. Aschoff (ed.). North-Holland, Amsterdam, pp. 231-261.

Brown, F.A., Hastings, J.W. & Palmer, J.D. 1970. *The biological clock: Two Views*. Academic Press, N.Y.

Brown, J.L. 1975. *The Evolution of Behaviour*. W.W. Norton, N.Y.

Brown, W.D. 1999. Mate choice in tree crickets and their kin. *Ann. Rev. Entomol.*, 44: 371-396.

Brues, C.T. 1946. *Insect Dietary. An Account for the Food Habits of Insects*. Harvard Univ. Press, Cambridge, Massachusetts.

Brun, R. 1914. *Die Raumorientierung der Ameisen und das Orientierungsproblem im Allgemeinen*. Gustav Fischer, Jena, Austria.

Buck, J.B. 1937. Studied on the firefly. I. The effects of light and other agents on flashing in *Photinus pyralis*, with special reference to periodicity and diurnal rhythm. *Physiol. Zool.*, 10: 45-58.

Buck, J.B. 1948. The anatomy and physiology of light organ in fireflies. *Ann. N.Y. Acad. Sci.*, 49: 397-483.

Buck, J.B. 1978. Functions and evolutions of bioluminescence. In: *Bioluminescence in Action*. P.J. Herring (ed.). Academic Press, London, pp. 419-460.

Buck, J.B. & Buck, E. 1978. Toward a functional interpretation of synchronous flashing by fireflies. *Amer. Nat.*, 112: 471-492.

Buddenbrock, W. von, 1917. Die Lichtkompassbewegungen bei den Insekten, insbesondere den Schmetterlingsraupen. *Sitz. Ber. Heidelb. Akad. Wiss., Math.-Naturwiss. Kl.* 8B, 1-26.

Buddenbrock, W. von, 1921. Der Rhythmus der Schreitbewegungen der Stabheuschrecke *Dixippus*. *Biol. Zentralbl.*, 41: 41-48.

Buddenbrock, W. von & Schulz, E. 1933. Beiträge zur Kenntnis der Lichtkompassbewegung und der Adaptation des Insektenauges. *Zool. Jahrb. Physiol.*, 52: 513-536.

Buitenhuis, R., McNeil, J.N., Boivin, G. & Brodeur, J. 2004. The role of honeydew in host searching of aphid hyperparasitoids. *J. Chem. Ecol.*, 30(2): 273-285.

Bullock, T.H. & Horridge, G.H. 1965. *Structure and Function of Nervous System of Invertebrates*. W.H. Freeman & Co., San Francisco.

Bünning, E. 1973. *The Physiological Clock*. (Rev. 3rd ed.). Springer-Verlag, N.Y.

Burk, T. 1983. Male aggression and female choice in a field cricket (*Teleogryllus oceanicus*): the importance of courtship song. In: *Orthopteran Mating Systems: Sexual Competition in a Diverse Group of Insects*. D.T. Gwynne & G.K. Morris (eds.). Westview Press, Boulder, Colorado, pp. 97-119.

Burk, T. & Webb, J.C. 1983. Effect of size on calling propensity, song parameters, and mating success in Caribbean fruit flies, *Anastrepha suspensa* (Loew) (Diptera: Tephritidae). *Ann. Entomol. Soc. Am.*, 76: 678-682.

Burkhardt, D.1964. Colour discrimination in insects. *Adv. Insect Physiol.*, 2: 131-173.

Burmeister, H. 1836. *A Manual of Entomology*. (Trans. W.E. Shuckard). Edward Churton, London.

Burns, M.D. 1973. The control of walking in Orthoptera. I. Leg movements in normal walking. *J. Exp. Biol.*, 58: 45-48.

Bursell, E., Gough, A.J.E., Beevor, P.S., Cork, A., Hall, D.R. & Vale, G.A. 1988. Identification of components of cattle urine attractive to tsetse flies, *Glossina* spp. (Diptera: Glossinidae). *Bull. Entomol. Res.*, 78: 281-291.

Burtt, E.T. 1951. The ability of adults grasshoppers to change colour on burnt ground. *Proc. Roy. Entomol. Soc. Lond. A, Gen. Entomol.*, 26: 45-48.

Burtt, E.T. & Catton, W.T. 1966. Image formation and sensory transmission in the compound eye. *Adv. Insect Physiol.*, 3: 2-46.

Bush, G.L. 1966. The taxonomy, cytology, and evolution of the genus *Rhagoletis* in North America (Diptera: Tephritidae). *Bull. Mus. Comp. Zool.*, 134: 431-562.

Bush, G.L. 1969. Sympatric host race formation and speciation in frugivorous flies of the genus *Rhagoletis*. *Ann. N.Y. Acad. Sci.*, 49: 397-482.

Bush, G.L. 1975. Modes of animal speciation. *Ann. Rev. Ecol. Syst.*, 6: 339-364.

Busnel, R.-G. 1963. *Acoustic Behaviour of Animals*. Elsevier Publ. Co., Amsterdam.

Butenandt, A., Beckmann, R., Stamm, D. & Hecker., E. 1959. Über den Sexual-Lockstoff des Seidenspinners *Bombyx mori*, Reindarstellung und konstitution. *Z. Naturforsch.*, 14: 283-384.

Butler, C.G. 1967. Insect pheromones. *Biol. Rev.*, 42: 42-87.

Butler, C.G., Fletcher, D.J.C. & Walter, D. 1969. Nest-entrance marking with pheromones by the honey bee- *Apis mellifera* L. and by a wasp *Vespula vulgaris* L. *Anim. Behav.*, 17: 142-147.

Buttel-Reepen, H. von, 1900. Sind die Bienen 'Reflexmaschinen'? *Biol. Zbl.*, 20: 97.

Byers, J.A. 1988. Novel diffusion-dilution method for release of semiochemicals: testing pheromone component ratios on western pine beetle. *J. Chem. Ecol.*, 14(1): 199-212.

Byers, J.A. 1991. Simulation of mate finding behaviour in pine shoot beetles, *Tomicus piniperda*. *Anim. Behav.*, 41: 649-660.

Byers, J.A. 1993. Avoidance of competition by spruce bark beetles, *Ips typographus* and *Pityogenes chalcographus*. *Experientia*, 49: 272-275.

Byers, J.A. 1995. Host tree chemistry affecting colonization in bark beetles. In: *Chemical Ecology of Insects* 2. R.T. Cardé & W.J. Bell (eds.). Chapman & Hall, N.Y., pp. 154-213.

Cade, W.H. 1975. Acoustically orienting parasitoids: fly phonotaxis to cricket song. *Science*, 190: 1312-1313.

Cade, W.H. 1978. Of cricket song and sex. *Nat. Hist.*, 87: 64-73.

Cade, W.H. 1981. Field cricket spacing, and the phonotaxis of crickets and parasitoid flies to clumped and isolated songs. *Z. Tierpsychol.*, 55: 365-375.

Cade, W.H. 1985. Insect mating and courtship behaviour. In: *Comprehensive Insect Physiology, Biochemistry and Pharmacology. Behaviour*. Vol. 9. G.A. Kerkut & L.I. Gilbert (eds.). Pergamon Press, Oxford, pp. 591-619.

Caldwell, R.I. & Rankin, M.A. 1972. Effect of juvenile hormone mimic on flight in the milkweed bug *Oncopeltus fasciatus*. *J. Comp. Endocr.*, 19: 601-604.

Callahan, P.S. 1975. Insect antennae with special reference to the mechanism of scent detection and the evolution of sensilla. *Int. J. Insect Morphol. Embryol.*, 4: 381-430.

Camhi, J.M. 1974. Neural mechanisms of response modification in insects. In: *Experimental Analysis of Insect Behaviour*. L. Barton Browne (ed.). Springer-Verlag, Berlin, pp.60-86.

Campanacci, V., Krieger, J., Bette, S., Sturgis, J.N., Lartigue, A., Cambillau, C., Breer, H. & Tegoni, M. 2001. Revisiting the specificity of *Mamestra brassicae* and *Antheraea polyphemus* pheromone-binding proteins with a fluorescence binding assay. *J. Biol. Chem.*, 276: 20078-20084.

Campbell, B.C. & Duffey, S. 1981. Alleviation of α-tomatine-induced toxicity to the parasitoid, *Hyposoter exiguae*, by phytosteroles in the diet of host, *Heliothis zea*. *J. Chem. Ecol.*, 7: 927-946.

Campbell, D.J. & Shipp, E. 1979. Regulation of spatial pattern in populations of the field cricket *Teleogryllus commodus* (Walker). *Z. Tierpsychol.*, 51: 260-268.

Carayon, J. 1964. Un cas d'offrande nuptiale chez les Hétéroptères. *Comp. Rend. Acad. Sci. Paris*, 259: 4815-4818.

Carazo, P., Sanchez, E., Font, E. & Desfilis, E. 2004. Chemosensory cues allow male *Tenebrio molitor* beetles to assess the reproductive status of potential mates. *Anim. Behav.*, 68(1): 123-129.

Cardé, R.T., Baker, T.C. & Roelofs, W.L. 1975. Ethological function of components of a sex attractant system for oriental fruit moth, *Grapholitha molesta* (Lepidoptera: Tortricidae). *J. Chem. Ecol.*, 1: 475-491.

Cardé, R.T. & Minks, A.K. 1995. Control of moth pests by mating disruption: successes and constraints. *Ann. Rev. Entomol.*, 40: 559-585.

Cardé, R.T. & Minks, A.K. (eds.) 1997. *Insect Pheromone Research: New Directions*. Chapman & Hall, N.Y.

Cardé, R.T., Staten, R.T. & Mafra-Neto, A. 1998. Behaviour of pink bollworm near high-dose, point sources of pheromone in field wind tunnels: insights into mechanisms of mating disruption. *Entomol. Exp. Appl.*, 89: 35-46.

Carlisle, D.B. & Ellis, P.E. 1962. Endocrine glands and phase in locusts. *Symp. Genet.*, 10: 219-224.

Carlisle, D.B. & Ellis, P.E. 1963. Prothoracic gland and gregarious behaviour in Locusts. *Nature*, 200: 603-604.

Carlson, A.D. & Copeland, J. 1985. Communication in insects: I. Flash communication in fireflies. *Quart. Rev. Biol.*, 60: 415-436.

Carlosn, S.D. & Chi, C. 1979. The functional morphology of insect photoreceptor. *Ann. Rev. Entomol.*, 24: 379-416.

Carpenter, G.D.H. 1946. Mimetic polymorphism. *Nature*, 158: 277-279.

Carpenter, G.D.H. & Ford, E.B. 1933. *Mimicry*. Methuen & Co. Ltd., London.

Carpenter, J.M. 1987. Phylogenetic relationships and classification of the Vespinae (Hymenoptera: Vespidae). *Syst. Entomol.*, 12: 413-431.

Carpenter, J.M. 1991. Phylogenetic relationships and the origin of social behaviour in Vespidae. In: *The Social Biology of Wasps*. K.C. Ross & R.W. Matthews (eds.). Cornell Univ. Press, Ithaca, N.Y., pp. 7-32.

Carroll, M.J. & Berenbaum, M.R. 2002. Behavioural responses of the parsnip webworm to host plant volatiles. *J. Chem. Ecol.*, 28: 2191-2200.

Carroll, S.B., Weatherbee, S.D. & Langeland, J.A. 1995. Homeotic genes and the regulation and evolution of insect wing number. *Nature*, 375: 58-61.

Carson, H.L. 1997. Sexual selection: a driver of genetic change in Hawaiian *Drosophila*. *J. Hered.*, 88: 343-352.

Carson, R. 1962. *Silent Spring*. Houghton Mifflin, Boston.

Carthy, J.D. 1952. The return of ants to their nests. *Int. Congr. Entomol. 9th Trans.*, 1: 365-369.

Carthy, J.D. 1958. *An Introduction to the Behaviour of Invertebrates*. Allen & Unwin, London.

Carthy, J.D. 1965. *The Behaviour of Arthropods*. W.H. Freeman & Co., San Francisco.

Caspari, E. 1963. Genes and the study of behaviour. *Amer. Zool.*, 3: 97-100.

Cayre, M., Strambi, C. & Strambi, A. 1994. Neurogenesis in an adult insect brain and its hormonal control. *Nature*, 368: 57-59.

Chadwick, L.E. 1953a. The motion of the wings. In: *Insect Physiology*. K.D. Roeder (ed.). John Wiley & Sons, N.Y., pp. 577-614.

Chadwick, L.E. 1953b. Aerodynamics and flight metabolism. In: *Insect Physiology*. K.D. Roeder (ed.). John Wiley & Sons, N.Y., pp. 615-636.

Chapman, R.F. 1959. Field observations on the behaviour of hoppers of the red locust (*Nomadacris septemfasciata* Serville). *Anti-Locust Bull.*, no. 33, 51 pp.

Chapman, R.F. 1965. The behaviour of nymphs of *Schistocerca gregaria* (Forskål) (Orthoptera: Acrididae) in a temperature gradient, with special reference to temperature preference. *Behaviour*, 24: 283-317.

Chapman, R.F. 1972. *The Insects: Structure and Function*. (ESLB ed.). English Univ. Press Ltd., London.

Chapman, R.F. 1974. The chemical inhibition of feeding by phytophagous insects: a review. *Bull. Entomol. Res.*, 64: 339-363.

Chapman, R.F. 1982. Chemoreception: the significance of receptor numbers. *Adv. Insect Physiol.*, 16: 247-356.

Chapman, R.F. 1995. Mechanism of food handling by chewing insects. In: *Regulatory Mechanisms in Insect Feeding*. R.F. Chapman & G. Boer, de (ed.). Chapman & Hall, N.Y., pp. 3-31.

Chapman, R.F. 2003. Contact chemoreception in feeding by phytophagous insects. *Ann. Rev. Entomol.*, 48: 455-484.

Chapman, R.F., Bernays, E.A. & Stoffolano, J.G. Jr. 1987. *Perspectives in Chemoreception and Behaviour*. Springer-Verlag, N.Y.

Chapman, R.F. & Boer, G. de (ed.) 1995. *Regulatory Mechanisms in Insect Feeding*. Chapman & Hall, N.Y.

Chattopadhyay, A.K. & Sukul, N.C. 1994. Anti-predator strategy of larval aggregation pattern in *Aspidomorpha miliaris* (Chrysomelidae: Coleoptera). *Entomon*, 19(3-4): 125-130.

Chevrier, C. & Bressac, C. 2002. Sperm storage and use after multiple mating in *Dinarmus basalis* (Hymenoptera: Pteromalidae). *J. Insect Behav.*, 15(3): 385-398.

Chittka, L. 1996. Optimal sets of colour receptors and opponent processes for coding of natural objects in insect vision. *J. Theor. Biol.*, 181: 179-196.

Chittka, L. & Tautz, J. 2003. The spectral input to honey bee visual odometry. *J. Exp. Biol.*, 206: 2393-2397.

Chittka, L., Williams, N.M., Rasmussen, H. & Thomson, J.D. 1999. Navigation without vision: bumble bee orientation in complete darkness. *Proc. Roy. Soc. Lond. Ser. B, Biol. Sci.*, 266: 45-50.

Choe, J.C. & Crespi, B.J. (eds.) 1997a. *Social Behaviour in Insects and Arachnids*. Cambridge Univ. Press, Cambridge.

Choe, J.C. & Crespi, B.J. (eds.) 1997b. *The Evolution of Mating Systems in Insects and Arachnids*. Cambridge Univ. Press, Cambridge.

Claridge, M.F. 1985. Acoustic signals in the Homoptera: behaviour, taxonomy, and evolution. *Ann. Rev. Entomol.*, 30: 297-317.

Claridge, M.F. & Vrijer, P.W.F. de, 1993. Reproductive behaviour: the role of acoustic signals in species recognition and speciation. In: *Planthoppers: Their Ecology and Management*. R.F. Denno & T.J. Perfect (eds.). Chapman & Hall, N.Y., pp. 216-233.

Clark, B.R. & Faeth, S.H. 1997. The consequences of larval aggregation in the butterfly *Chlosyne lacinia*. *Ecol. Entomol.*, 22(4): 408-415.

Clarke, C.A. & Sheppard, P.M. 1960. The evolution of mimicry in the butterfly *Papilio dardanus*. *Heredity*, 14: 163-173.

Clarke, C.A. & Sheppard, P.M. 1972. Genetic and environmental factors influencing pupal colour in swallowtail butterflies *Battus philenor* (L.) and *Papilio polytes* L. *J. Entomol. A*, 46: 123-133.

Cock, R. de & Matthysen, E. 2003. Glow-worm larvae bioluminescence (Coleoptera: Lampyridae) operates as an aposematic signal upon toads (*Bufo bufo*). *Behav. Ecol.*, 14(1): 103-108.

Čokl, A. & Virant-Doberlet, M. 2003. Communication with substrate-borne signals in small plant-dwelling insects. *Ann. Rev. Entomol.*, 48: 29-50.

Colazza, S., Fucarino, A., Peri, E., Salerno, G., Conti, E. & Bin, F. 2004. Insect oviposition induces volatile emission in herbaceous plants that attracts egg parasitoids. *J. Exp. Biol.*, 207: 47-53.

Cole, C.L. & Adkisson, P.L. 1964. Daily rhythm in the susceptibility of an insect to a toxic substance. *Science*, 144: 1148-1149.

Coleman, R.A., Barker, A.M. & Fenner, M. 1999. Parasitism of the herbivore *Pieris brassicae* L. (Lep., Pieridae) by *Cotesia glomerata* L. (Hym., Braconidae) does not benefit the host plant by reduction of herbivory. *J. Appl. Entomol.*, 123(3): 171-177.

Collett, M., Collett, T.S. & Wehner, R. 1999. Calibration of vector navigation in desert ants. *Curr. Biol.*, 9(18): 1031-1034.

Collett, M., Harland, D. & Collett, T.S. 2002. The use of landmarks and panoramic context in the performance of local vectors by navigating honey bees. *J. Exp. Biol.*, 205: 807-814.

Cook, D. 1988. Sexual selection in dung beetles. II. Female fecundity as an estimate of male reproductive success in relation to horn size and alternative behavioural strategies in *Onthophagus binodis* Thunberg (Scarabaeidae: Onthophagini). *Aust. J. Zool.*, 36: 521-532.

Cook, L.M. 2000. Changing views on melanic moths. *Biol. J. Linn. Soc.*, 69: 431-441.

Cook, M.J.W. 1986. *Bemisia tabaci*, a literature survey on cotton whitefly with an annotated bibliography. *FAO, United Nations, C.A.B. Inst. Biol. Contl.*, pp. 1-121.

Coppedge, J.R., Brown, H.E., Goodenough, J.L., Tannahill, F.H., Snow, J.W., Petersen, H.D. & Hofman, H.D. 1980. Field suppression of new formulation of the screwworm adult suppression system. *J. Econ. Entomol.*, 73: 411-414.

Corbet, P.S. 1958. Lunar periodicity of aquatic insects in Lake Victoria. *Nature*, 182: 330-331.

Corbet, P.S. 1964. Temporal patterns of emergence in aquatic insects. *Can. Entomologist*, 96: 264-279.

Corbet, S.A. 1973. Oviposition pheromone in larval mandibular glands of *Anagasta kuehniella*. *Nature*, 243: 537-538.

Corbet, T.P. 1962. *A Biology of Dragonflies*. Witherby Ltd., London.

Cornell, H.V. & Hawkins, B.A. 2003. Herbivore responses to plant secondary compounds: a test of phytochemical coevolution theory. *Amer. Nat.*, 161(4): 507-522.

Cortesero, A.M., Stapel, J.O. & Lewis, W.J. 2000. Understanding and manipulating plant attributes to enhance biological control. *Biol. Cont.*, 17(1): 35-49.

Cott, H.B. 1940. *Adaptive Colouration in Animals*. Methuen & Co. Ltd., London.

Couturier, A. & Robert, P. 1958. Recherches sur les migrations du hanneton commun (*Melolontha melolontha* L.). *Annals Epiphyt.*, 9: 257-329.

Craig, T.P., Itami, J.K., Abrahamson, W.G. & Horner, J.D. 1993. Behavioural evidence for host-race formation in *Eurosta solidaginis*. *Evolution*, 47: 1696-1710.

Crane, J. 1952. A comparative study of innate defensive behaviour in Trinidad mantids. *Zoologica*, 37: 259-293.

Crankshaw, O.S. 1979. Female choice in relation to calling and courtship songs in *Acheta domesticus*. *Anim. Behav.*, 27: 1274-1275.

Cratsley, C.K. & Lewis, S.M. 2003. Female preference for male courtship flashes in *Photinus ignitus* fireflies. *Behav. Ecol.*, 14(1): 135-140.

Cratsley, C.K. & Lewis, S.M. 2005. Seasonal variation in mate choice of *Photinus ignitus* fireflies. *Ethology*, 111: 89-100.

Cratsley, C.K., Rooney, J.A. & Lewis, S.M. 2003. Limits to nuptial gift production by male fireflies. *J. Insect Behav.*, 16: 361-370.

Creed, E.R. 1966. Geographic variation in the two-spot ladybird in England and Wales. *Heredity*, 21: 57-72.

Creed, E.R. 1971. Melanism in the two-spot ladybird, *Adalia bipunctata* in Great Britain. In: *Ecological Genetics and Evolution*. E.R. Creed (ed.). Blackwell, Oxford, pp. 134-151.

Crespi, B.J. 1992. Eusociality in Australian gall thrips. *Nature*, 359: 724-726.

Crespi, B.J., Carmean, D.A., Mound, L.A., Worobey, M. & Morris, D. 1998. Phylogenetics of social behaviour in Australian gall-forming thrips: evidence from mitochondrial DNA sequence, adult morphology and behaviour, and gall morphology. *Mol. Phylogenet. Evol.*, 9: 163-180.

Crespi, B.J., Morris, D.C. & Mound, L.A. 2004. *Evolution of Ecological and Behavioral Diversity: Australian Acacia Thrips as Model Organisms*. Australian Biol. Resources Study & Australian Natl. Insect Coll., CSIRO: Canberra, Australia.

Crespi B.J. & Mound, L.A. 1997. Ecology and evolution of social behaviour among Australian gall thrips and their allies. In: *Evolution of Social Behaviour in Insects and Arachnids*. J. Choe & B.J. Crespi (eds.). Cambridge Univ. Press, pp. 166-180.

Crocker, R.L. & Hamon, A.B. 1983. Factors affecting orientation of *Dialeurodes citri* (Homoptera: Aleyrodidae) on persimmon leaves. *Florida Entomol.*, 66(1): 208-210.

Crozier, R.H. & Pamilo, P. 1996. *Evolution of Social Insect Colonies: Sex Allocation and Kin Selection*. Oxford Univ. Press, Oxford.

Crumpacker, D.W. & Williams, J.S. 1973. Density, dispersion and population structure in *Drosophila pseudoobscura*. *Ecol. Monogr.*, 43: 499-538.

Cruse, H. 1990. What mechanisms coordinate leg movement in walking arthropods? *Trends Neurosci.*, 13(1): 15-21.

Cruse, H. & Graham, D. 1984. Models for the analysis of walking in arthropods. In: *Coordination of Motor Behaviour*. B.M.H. Bush & F. Clarac (eds.). *Soc. Exp. Biol. Seminar Ser.* 24. Cambridge Univ. Press, Cambridge, pp. 283-301.

Cullen, J.M. 1959. Behaviour as a help in taxonomy. *Syst. Assoc. Publ.*, 3: 131-140.

Cunningham, R.T. & Steiner, L.F. 1972. Field trials of cue-lure + naled on saturated fibre block for control of the melon fly by male annihilation technique. *J. Econ. Entomol.*, 65: 505-509.

Cuoturier, A. & Robert, P. 1955. Maintain de la direction de vol chez *Melolontha melolontha* L. *Comp. Rend.*, 240: 2561-2563.

Dacke, M., Nordström, P. & Scholtz, C.H. 2003. Twilight orientation to polarized light in the crepuscular dung beetle *Scarabaeus zambesianus*. *J. Exp. Biol.*, 206: 1535-1543.

Dadd, R.H. 1963. Feeding behaviour and nutrition in grasshoppers and locusts. *Adv. Insect Physiol.*, 1: 47-111.

Dalton, S. 1975. *Borne on the Wind. The Extraordinary World of Insects in Flight*. E.P. Dutton, N.Y.

Daly, K.C. & Figueredo, A.J. 2000. Habituation of sexual response in male *Heliothis* moths. *Physiol. Entomol.*, 25(2): 180-190.

Damman, H. 1986. The osmaterial glands of the swallowtail butterfly *Eurytides marcellus* as a defence against natural enemies. *Ecol. Entomol.*, 11(2): 261-265.

Danks, H.V. 1988. Systematics in support of entomology. *Ann. Rev. Entomol.*, 33: 271-296.

Danthanarayana, W. 1976. Diel and lunar flight periodicities in the light brown apple moth, *Epiphyas postvittana* (Walker) (Tortricidae) and their possible adaptive significance. *Aust. J. Zool.*, 24(1): 65-73.

Danthanarayana, W. (ed.) 1986. *Insect Flight: Dispersal and Migration*. Springer-Verlag, Berlin.

Darling, D.C. & Werren, J.H. 1990. Biosystematics of *Nasonia* (Hymenoptera: Pteromalidae): two new species reared from bird's nests in North America. *Ann. Entomol. Soc. Am.*, 83(3): 352-370.

Darwin, C. 1859. *The Origin of Species by Means of Natural Selection, or the Preservation of Favoured Races in the Struggle for Life*. John Murray, London.

Darwin, C. 1871. *The Descent of Man and Selection in Relation to Sex*. John Murray, London.

Darwin, C. 1872. *The Expression of the Emotions in Man and the Animals*. John Murray, London.

Davey, K.G. 1965. *Reproduction in the Insects*. Oliver & Boyd, Edinburgh.

Davey, K.G. & Huebner, E.1974. The response of the follicle cells of *Rhodnius prolixus* to juvenile hormone and antigonadotropin *in vitro*. *Can. J. Zool.*, 52: 1407-1412.

Davies, N.B. 1978. Territorial defence in the speckled wood butterfly (*Pararge aegeria*): the resident always wins. *Anim. Behav.*, 26: 138-147.

Davis, N.T., Homberg, U., Dircksen, H., Levine, R.B. & Hildebrand, J.G. 1993. Crustacean cardioactive peptide-immunoreactive neurons in the hawk moth *Manduca sexta* and changes in their immunoreactivity during postembryonic development. *J. Comp. Neurol.*, 338: 612-627.

Dawkins, M.S. 1986. *Unraveling Animal Behaviour*. Longman, London.

Dawkins, R. 1989. *The Selfish Gene*. Oxford Univ. Press, Oxford.

Dawson, G.W., Griffiths, D.C., Janes, N.F., Mudd, A., Pickett, J.A., Wadhams, L.J., and Woodcock, C.M. 1987. Identification of an aphid sex pheromone. *Nature*, 325: 614-616.

Delcomyn, F. 1971a. The locomotion of the cockroach *Periplaneta*. *J. Exp. Biol.*, 54: 443-452.

Delcomyn, F. 1971b. The effect of limb amputation on locomotion in the cockroach *Periplaneta americana*. *J. Exp. Biol.*, 54: 453-469.

Delcomyn, F. 1981. Insect locomotion on land. In: *Locomotion Energetics in Arthropods*. C.F. Herreid & C.R. Foutner (eds.). Plenum Press, N.Y., pp. 103-125.

Delcomyn, F. 1985a. Insect locomotion: past, present and future. In: *Insect Locomotion*. M. Gewecke & G. Wendler (eds.). Paul Parey Sci. Publ., N.Y., pp. 1-18.

Delcomyn, F. 1985b. Walking and running. In: *Comprehensive Insect Physiology, Biochemistry and Pharmacology. Nervous System: Structure and Motor Function*. Vol. 5. G.A. Kerkut & L.I. Gilbert (eds.). Pergamon Press, Oxford, pp. 439-466.

Delcomyn, F. 1985c. Factors regulating insect walking. *Ann. Rev. Entomol.*, 30: 239-256.

Delcomyn, F. 1999. Walking robots and the central and peripheral control of locomotion in insects. *Autonomous Robots*, 7: 259-270.

Delcomyn, F. 2004. Insect walking and robotics. *Ann. Rev. Entomol.*, 49: 51-70.

Delcomyn, F. & Usherwood, P.N.R. 1973. Motor activity during walking in the cockroach *Periplaneta americana*. *J. Exp. Biol.*, 59: 629-642.

DeLuca, M. & McElroy, W.D. (eds.) 1981. *Bioluminescence and Chemiluminescence*. Academic press, N.Y.

Demoll, R. 1918. *Der Flug der Insekten und der Vögel*. Fisher, Jena.

Denlinger, D.L. 1985. Hormonal control of diapause. In: *Comprehensive Insect Physiology, Biochemistry and Pharmacology. Endocrinology II*. Vol. 8. G.A. Kerkut & L.I. Gilbert (eds.). Pergamon Press, Oxford, pp. 353-412.

Denlinger, D.L., Giebultowicz, J. & Saunders, D.S. (eds.) 2001. *Insect Timing: Circadian Rhythmicity to Seasonality*. Elsevier Publ. Co., Amsterdam.

Deonier, D.L. 1974. Ultraviolet-reflective surfaces on *Ochthera mantis mantis* (DeGeer) (Diptera: Ephydridae). Preliminary report. *Zoologica*, 34: 159-214.

Dethier, V.G. 1953a. Mechanoreception. In: *Insect Physiology*. K.D. Roeder (ed.). John Wiley & Sons, N.Y., pp. 523-543.

Dethier, V.G. 1953b. Chemoreception. In: *Insect Physiology*. K.D. Roeder (ed.). John Wiley & Sons, N.Y., pp. 544-576.

Dethier, V.G. 1953c. Vision. In: *Insect Physiology*. K.D. Roeder (ed.). John Wiley & Sons, N.Y., pp. 488-522.

Dethier, V.G. 1957. Chemoreception and behaviour of insects. *Survey of Biological Progress*, 3. Academic Press, N.Y., pp. 149-183.

Dethier, V.G. 1963. *The Physiology of Insect Senses*. John Wiley & Sons, N.Y.

Dethier, V.G. 1966. Feeding behaviour. In: *Insect Behaviour*. P.T. Haskell (ed.). *Symp. Roy. Entomol. Soc. Lond.*, 3: 46-58.

Dethier, V.G. 1976. *The Hungry Fly*. Harvard Univ. Press, Cambridge, Massachusetts.

Dethier, V.G. 1982. Mechanisms of host plant recognition. *Entomol. Exp. Appl.*, 31: 49-56.

Dethier, V.G., Barton Browne, L. & Smith, C.N. 1960. The designation of chemicals in terms of responses they elicit from insects. *J. Econ. Entomol.*, 53: 134-136.

Dethier, V.G. & Chadwick, L.E. 1948. Chemoreception in insects. *Physiol. Rev.*, 28: 220-254.

Dethier, V.G. & Stellar, E. 1963. *Animal Behaviour- Its Evolutionary and Neurological Basis*. Prentice-Hall of India Pvt. Ltd., N. Delhi.

Devi, P.K., Yadav, D.N. & Jha, A. 2003. Predatory behaviour and feeding potential of *Paederus fuscipes*. *Indian J. Entomol.*, 65(3): 319-323.

Dhadialla, T.S., Carlson, G.R. & Le, D.P. 1998. New insecticides with ecdysteroidal and juvenile hormone activity. *Ann. Rev. Entomol.*, 43: 545-569.

Dhadialla, T.S. & Raikhel, A.S. 1994. Endocrinology of mosquito vitellogenesis. In: *Perspectives in Comparative Endocrinology*. K.G. Davey, R.E. Peter, & S.S. Tobe (eds.). Natl. Res. Coun. Canada, Ottawa, pp. 275-281.

Dhaliwal, G.S. & Arora, R. 1996. *Principles of Insect Pest Management*. Natl. Agril. Tech. Inf. Centre, Ludhiana.

Dhaliwal, G.S., Bhathal, J.S. & Bhathal, S.S. 1994. Novel approaches in insect pest management: an overview. In: *Trends in Agricultural Insect Pest Management*. G.S. Dhaliwal & R. Arora (eds.). Commonwealth Publ., N. Delhi, pp. 469-511.

Dicke, M., Baarlen, P. van, Wessels, R. & Dijkman, H. 1993. Herbivory induces systemic production of plant volatiles that attract herbivore predators: extraction of endogenous elicitor. *J. Chem. Ecol.*, 19: 581-599.

Dicke, M., Sabelis, M.W., Takabayashi, J., Bruin, J. and Posthumus, M.A. 1990. Plant strategies of manipulating predator-prey interactions through allelochemicals: prospects for application in pest control. *J. Chem. Ecol.*, 16(11): 3091-3118.

Dickens, J.C., James, E., Oliver, J.E., Hollister, B., Davis, J.C. & Klun, J.A. 2002. Breaking a paradigm: male-produced aggregation pheromone for the Colorado potato beetle. *J. Exp. Biol.*, 205: 1925-1933.

Dickinson, M.H., Lehmann, F.-O. & Sane, S.P. 1999. Wing rotation and the aerodynamic basis of insect flight. *Science*, 284: 1954-1960.

Dinan, L., Whiting, P., Girault, J.P., Lafont, R., Dhadialla, T.S., Cress, D.E., Mugat, B., Antoniewski, C. & Lepesant, J.A. 1997. Cucurbitacins are insect steroid hormone antagonists acting at the ecdysteroid receptor. *Biochem. J.*, 327(3): 643-650.

Ding, D., Swedenborg, P.D. & Jones, R.L. 1989. Plant odour preferences and learning in *Macrocentrus grandii* (Hymenoptera: Braconidae), a larval parasitoid of the European corn borer, *Ostrinia nubilalis* (Lepidoptera: Pyralidae). *J. Kansas Entomol. Soc.*, 62(2): 164-176.

Dingle, H. 1972. Migration strategies of insects. *Science*, 175: 1327-1335.

Dingle, H. 1978. *Evolution of Insect Migration and Diapause*. Springer-Verlag, N.Y.

Dingle, H. 1985. Migration. In: *Comprehensive Insect Physiology, Biochemistry and Pharmacology. Behaviour*. Vol. 9. G.A. Kerkut & L.I. Gilbert (eds.). Pergamon Press, Oxford, pp. 375-415.

Dingle, H. 1996. *Migration: The Biology of Life on the Move*. Oxford Univ. Press, Oxford.

Dingle, H. & Arora, G. 1973. Experimental studies of migration in bugs of the genus *Dysdercus*. *Oecologia*, 12: 119-140.

Dixon, A.F.G. & Agarwala, B.K. 1999. Ladybird-induced life history changes in aphids. *Proc. Roy. Soc. Lond. Ser. B, Biol. Sci.*, 266: 1549-1553.

Dobzhansky, T., Cooper, D.M., Phaff, H.J., Knapp, E.P. & Carson, H.L. 1956. Studies on the ecology of *Drosophila* in the Yosemite region of California. 4. Differential attraction of species of *Drosophila* to different species of yeasts. *Ecology*, 37: 544-550.

Dobzhansky, T., Ehrman, L., Pavlovsky, O. & Spassky, B. 1964. The superspecies *Drosophila paulistorum*. *Proc. Natl. Acad. Sci., USA*, 51(1): 3-9.

Dobzhansky, T. & Pavlovsky, O. 1971. Experimentally created incipient species of *Drosophila*. *Nature*, 230: 289-292.

Dodson, G.N. 1997. Resource defence mating system in antlered flies, *Phytalmia* spp. (Diptera: Tephritidae). *Ann. Entomol. Soc. Am.*, 90: 496-504.

Doi, M., Matsuda, M., Tomaru, M., Matsubayashi, H. & Oguma, Y. 2001. A locus for female discrimination behaviour causing sexual isolation in *Drosophila*. *Proc. Natl. Acad. Sci., USA*, 98(12): 6714-6719.

Dolley, W.L. Jr. & White, J.D. 1951. The effect of illuminance on the reversal temperature in the drone fly *Eristalis tenax*. *Biol. Bull.*, 100: 84-89.

Dongen, S. van, Matthysen, E., Sprengers, E. & Dhondt, A.A. 1998. Mate selection by male winter moths *Operophtera brumata* (Lepidoptera, Geometridae): adaptive male choice or female control? *Behaviour*, 135(1): 29-42.

Doolan, J.M. & Young, D. 1989. Relative importance of song parameters during flight phonotaxis and courtship in the bladder cicada *Cystosoma saundersii*. *J. Exp. Biol.*, 141: 113-131.

Downes, J.A. 1969. The swarming and mating flights of Diptera. *Ann. Rev. Entomol.*, 14: 271-298.

Drake, V.A. & Gatehouse, A.G. (eds.) 1995. *Insect Migration Tracking Resources Through Space and Time*. Cambridge Univ. Press, Cambridge.

Dreisig, H. 1975. Environmental control of the daily onset of luminescent activity in glowworms and fireflies (Coleoptera: Lampyridae). *Oecologia*, 18(2): 85-99.

Dreisig, H. 1978. The circadian rhythm of bio-luminescence in the glow-worm, *Lampyris noctiluca* L. (Coleoptera, Lampyridae). *Behav. Ecol. Sociobiol.*, 3: 1-18.

Drew, R.A.I. 1974. The responses of fruit fly species (Diptera: Tephritidae) in the South Pacific area to male attractants. *J. Aust. Entomol. Soc.*, 13: 267-270.

Drew, R.A.I. & Hancock, D.L. 1994. The *Bactrocera dorsalis* complex of fruit flies (Diptera: Tephritidae: Dacinae) in Asia. *Bull. Entomol. Res.* (*Suppl. 2*), pp. 1-68.

Drew, R.A.I. & Hancock, D.L.1999. Phylogeny of the tribe Dacini (Dacinae) based on morphological, distributional and biological data. In: *Fruit Flies (Tephritidae): Phylogeny and Evolution of Behaviour*. M. Aluja & A.L. Norrbom (eds.). CRC Press, Boca Raton, pp. 491-504.

Drosopoulos, S. & Claridge, M.F. (eds.) 2006. *Insect Sounds and Communication: Physiology, Behaviour, Ecology, and Evolution*. CRC Press, Boca Raton.

Dubois, R. 1885. Note sur la physiologie des pyrophores. *C. R. Soc. Biol.*, 2: 559.

Dudley, R. 2000. *The Biomechanics of Insect Flight. Form, Function, Evolution*. Princeton Univ. Press, Princeton, N.J.

Duffey, S.S. 1977. Arthropods allomones: chemical effronteries and antagonists. *Proc. XV Int. Cong. Entomol. Wash.*, pp. 323-394.

Duffey, S.S. 1980. Sequestration of plant natural product by insects. *Ann. Rev. Entomol.*, 25: 447-477.

DuMortier, B. 1963. Morphology of sound emission apparatus in Arthropoda. In: *Acoustic Behaviour of Animals*. R.-G. Busnel (ed.). Elsevier Publ. Co., Amsterdam, pp. 277-345.

Dunlap, J.C., Loros, J.J. & DeCoursey, P.J. (eds.) 2003. *Chronobiology: Biological Timekeeping*. Sinauer Assoc. Inc. Publ. Sunderland, Massachusetts.

Dupont-Raabe, M. 1957. Les mécanismes de l'adaptation chromatique chez les insectes. *Arch. Zool. Exp. Gén.*, 94: 61-294.

Dybas, L.K. & Dybas, H.S. 1981. Coadaptation and taxonomic differentiation of sperm and spermathecae in featherwing beetles. *Evolution*, 35(1): 168-174.

Dyer, F.C. 2002. The biology of the dance language. *Ann. Rev. Entomol.*, 47: 917-949.

Dyer, F.C., Berry, N.A. & Richard, A.S. 1993. Honey bee spatial memory- use of route-based memories after displacement. *Anim. Behav.*, 45(5): 1028-1030.

Eberhard, M.J.W. 1978. Polygyny and the evolution of social behaviour in wasps. *J. Kansas Entomol. Soc.*, 51: 832-856.

Eberhard, W.G. 1983. Behaviour of the adult bottle brush weevils (*Rhinostomus barbirostris*) (Coleoptera: Curculionidae). *Rev. Biol. Trop.*, 31: 233-244.

Eberhard, W.G. 1996. *Female Control: Sexual Selection by Cryptic Female Choice*. Princeton Univ. Press, Princeton, N.J.

Ebling, J. & Highnam, K.C. 1970. *Chemical Communication. Studies in Biology*. No. 19. Crane, Russak & Co., N.Y.

Edery, I., Zwiebel, L.J., Dembinska, M.E. & Rosbash, M. 1994. Temporal phosphorylation of the *Drosophila* period protein. *Proc. Natl. Acad. Sci., USA*, 91(6): 2260-2264.

Edmunds, L.N. 1988. *Cellular and Molecular Bases of Biological Clocks*. Springer-Verlag, N.Y.

Edmunds, M. 1972. Defensive in Ghanaian praying mantids. *Zool. J. Linn. Soc.*, 51: 1-32.

Edmunds, M. 1974. *Defence in Animals. A Survey of Anti-Predator Defences*. Longman, N.Y.

Edwards, J.S. 1961. The action and composition of the saliva of an assassin bug, *Platymerus rhadamanthus* Gaerst (Hemiptera, Reduviidae). *J. Exp. Biol.*, 38: 61-77.

Eggert, A.-K. 1992. Alternative male mate-finding tactics in burying beetles. *Behav. Ecol.*, 3(3): 243-254.

Eggert, A.-K. & Sakaluk, S.K. 1994. Sexual cannibalism and its relation to male mating success in sagebrush cricket *Cyphoderris strepitans* (Haglidae: Orthoptera). *Anim. Behav.*, 47: 1171-1177.

Eggert, A.-K. & Sakaluk, S.K. 1995. Female-coerced monogamy in burying beetles. *Behav. Eco. Sociobiol.*, 37: 147-153.

Ehrlich, P.R. & Raven, P.H. 1964. Butterflies and plants, a study in coevolution. *Evolution*, 18: 586-608.

Ehrlich, P.R. & Raven, P.H. 1967. Butterflies and plants. *Sci. Amer.*, 216: 104-113.

Ehrman, L. & Parsons, P.A. 1976. *The Genetics of Behaviour*. Sinauer Assoc. Inc. Publ. Sunderland, Massachusetts.

Ehrman, L. & Parsons, P.A. 1981. *Behaviour Genetics and Evolution*. McGraw-Hill Book Co. Inc., N.Y.

Eibl-Eibesfeldt, I. 1970. *Ethology. The Biology of Behaviour*. Holt, Rinehart & Winston, N.Y.

Eisemann, C.H. & Rice, M.J. 1989. Behavioural evidence for hygro- and mechanoreception by ovipositor sensilla of *Dacus tryoni* (Diptera: Tephritidae). *Physiol. Entomol.*, 14: 273-277.

Eisenstein, E.M. & Cohen, M.J. 1966. Learning in an isolated insect ganglia. *Anim. Behav.*, 13: 104-108.

Eisner, T. 1964. Catnip: its raison d'etre. *Science*, 146: 1318-1320.

Eisner, T. 1965. Defensive spray of a phasmid. *Science*, 148: 966-968.

Eisner, T. 1970. Chemical defence against predation in arthropods. In: *Chemical Ecology*. E. Sondheimer & J.B. Simone (eds.). Academic Press, N.Y., pp. 157-217.

Eisner, T. & Aneshansley, D.J. 1999. Spray aiming in the bombardier beetle: photographic evidence. *Proc. Natl. Acad. Sci., USA*, 96: 9705-9709.

Eisner, T., Aneshansley, D.J., Eisner, M., Attygalle, A.B., Alsop, D.W. & Meinwald, J. 2000. Spray mechanism of the most primitive bombardier beetle (*Metrius contractus*). *J. Exp. Biol.*, 203, 1265-1275.

Eisner, T., Goetz, M.A., Hill, D.E., Smedley, S.R. & Meinwald, J. 1997. Firefly "femmes fatales" acquire defensive steroids (lucibufagins) from their firefly prey. *Proc. Natl. Acad. Sci., USA*, 94(18): 9723-9728.

Eisner, T. & Meinwald, J. 1966. Defensive secretions of arthropods. *Science*, 153: 1341-1350.

Eisner, T., Silberglied, R.E., Aneshansley, D., Carrel, J.E. & Howland, H.C. 1969. Ultraviolet video-viewing: the television camera as an insect eye. *Science*, 166: 1172-1174.

Eisner, T., Tassell, E. van & Carrel, J.E. 1967. Defensive use of a faecal shield by a beetle larva. *Science*, 158: 1471-1473.

El-Helaly, M.S., Rawash, I.A. & Ibrahim, E.G. 1981. Phototaxis of the adult whitefly, *Bemisia tabaci* Gennadius to the visible light. I. Effect of the exposure period on the insect's response to different wavelengths of the visible light spectrum using a devised simple technique. *Acta Phytopathol. Acad. Sci. Hung.*, 16(1-2): 181-188.

Elsner, N. & Popov, A.V. 1978. Neuroethology of acoustic communication. *Adv. Insect Physiol.*, 13: 229-355.

Emerson, A.E. 1959. Social Insects. *Encyclopedia Britannica*, 20: 871-878.

Emmel, T.C. 1965. A new mimetic assemblage of lycid and cerambycid beetles in central Chiapas. *South-western Naturalist*, 10: 14-16.

Emsley, M.G. 1965. Speciation in *Heliconius*: morphology and geographical distribution. *Zoologica*, 50: 191-254.

Endo, N., Abe, M., Sekine, T. & Matsuda, K. 2004. Feeding stimulants of Solanaceae-feeding lady beetle, *Epilachna vigintioctomaculata* (Coleoptera: Coccinellidae) from potato leaves. *Appl. Entomol. Zool.*, 39(3): 411-416.

Engelmann, F. 1979. Insect vitellogenin: identification, biosynthesis and role in vitellogenesis. *Adv. Insect Physiol.*, 14: 49-108.

Engelmann, F. 1983. Vitellogenesis controlled by juvenile hormone. In: *Endocrinology of Insects*. R.G.H. Downer & H. Laufer (eds.). Alan R. Liss, Inc. N.Y., pp. 259-270.

Engelmann, F. 2002. Ecdysteroids, juvenile hormone and vitellogenesis in the cockroach *Leucophaea maderae*. *J. Insect Sci.*, 2(20): 1-8.

Ernst, E. 1959. Beobachtungen beim Spritzakt der *Nasutitermes* Soldaten. *Rev. Suisse Zool.*, 66: 289-295.

Esch, H. 1967. The evolution of bee language. *Sci. Amer.*, 216: 96-104.

Esch, H.E. & Burns, J.E. 1995. Honey bees use optic flow to measure the distance of a food source. *Naturwissenschaften*, 82: 38-40.

Esch, H.E., Zhang, S., Srinivasan, M.V. & Tautz, J. 2001. Honey bee dances communicate distances measured by optic flow. *Nature*, 411(6837): 581-583.

Etheredge, J.A., Perez, S.M., Taylor, O.R. & Jander, R. 1999. Monarch butterflies (*Danaus plexippus* L.) use a magnetic compass for navigation. *Proc. Natl. Acad. Sci., USA*, 96(24): 13845-13846.

Evans, D.L. & Schmidt, J.O. (eds.) 1990. *Insect Defence. Adaptive Mechanisms and Strategies of Prey and Predators*. State Univ. of N.Y. Press, Albany, N.Y.

Evans, H.E. 1957. *Studied on the Comparative Ethology of Digger Wasps of the Genus Bembix*. Cornell Studied in Entomology, Cornell Univ. Press, Ithaca, N.Y.

Evans, H.E. 1966a. *Life on a Little-Known Planet*. E.P. Dutton, N.Y.

Evans, H.E. 1966b. *The Comparative Ethology and Evolution of Sand Wasps*. Harvard Univ. Press, Cambridge, Massachusetts.

Evans, H.E. 1969. Phoretic copulation in Hymenoptera. *Entomol. News*, 80: 113-124.

Evans, H.E. 1984. *Insect- A Textbook of Entomology*. Addison-Wesley Publ. Co., Massachusetts.

Evans, H.E. & Eberhard, M.J.W. 1970. *The Wasps*. Univ. Michigan Press, Ann Arbor, Michigan.

Evans, H.E. & Matthews, R.W. 1971. Nesting behaviour and larval stages of some Australian nyssonine sand wasps (Hymenoptera: Sphecidae). *Aust. J. Zool.*, 19(3): 293-310.

Evans, J.D. & Wheeler, D.E. 1999. Differential gene expression between developing queens and workers in the honey bee, *Apis mellifera*. *Proc. Natl. Acad. Sci., USA*, 96(10): 5575-5580.

Evans, M.E.G. 1977. Locomotion in the Coleoptera Adephaga, especially Carabidae. *J. Zool. Lond.*, 181: 189-226.

Evans, P.D. 1980. Biogenic amines in the insect nervous system. *Adv. Insect Physiol.*, 15: 317-473.

Evans, W.G. 1964. Infra-red receptors in *Melanophila acuminata* DeGeer. *Nature*, 202: 211.

Evans, W.G. 1976. Circadian and circatidal locomotory rhythms in intertidal beetle, *Thalassotrechus barbarae* (Horn): Carabidae. *J. Exp. Mar. Biol. Ecol.*, 22: 79-90.

Ewer, J., Gammie, S.C. & Truman, J.W. 1997. Control of insect ecdysis by a positive-feedback endocrine system: roles of eclosion hormone and ecdysis triggering hormone. *J. Exp. Biol.*, 200: 869-881.

Ewing, A.W. 1983. Functional aspects of *Drosophila* courtship. *Biol. Rev.*, 58: 275-292.

Ewing, A.W. 1984. Acoustic signals in insect sexual behaviour. In: *Insect Communication*. T. Lewis (ed.). Academic Press, London, pp. 223-240.

Ewing, A.W. 1989. *Arthropod Bioacoustics. Neurobiology and Behaviour*. Cornell Univ. Press, Ithaca, N.Y.

Exner, S. 1891. *Die Physiologie der Facetierten Augen von Krebsen und Insecten*. Franz Deuticke, Leipzig and Wien.

Fabre, J.H. 1879-1908. *Souvenirs Entomologiques*. 10 vols. Delagrave, Paris.

Farner, D.S.1985. Annual rhythms. *Ann. Rev. Entomol.*, 47: 65-82.

Farrell, B.D.1998. "Inordinate fondness" explained: why are there so many beetles? *Science*, 281: 555-559.

Fatouros, N.E., Huigens, M.E., Loon. J.J.A. van, Dicke, M. & Hilker, M. 2005. Butterfly anti-aphrodisiac lures parasitic wasps. *Nature*, 433(7027): 704.

Feder, J.L. & Bush, G.L. 1989. A field test of differential host-plant usage between two sibling species of *Rhagoletis pomonella* fruit flies (Diptera: Tephritidae) and its consequences for sympatric models of speciation. *Evolution*, 43: 1813-1819.

Feder, J.L., Chilcote, C.A. & Bush. G.L. 1988. Genetic differentiation between sympatric host races of the apple maggot fly, *Rhagoletis pomonella*. *Nature*, 336: 61-64.

Feder, J.L. & Filchak, K.E. 1999. It's about time: the evidence for host plant-mediated selection in the apple maggot fly, *Rhagoletis pomonella*, and its implications for fitness trade-offs in phytophagous insects. *Proc. 10th Int. Symp. Insect Plant Relationship*. Kluwer Acad. Publ., Netherlands, pp. 211-225.

Fent, K. 1986. Polarized skylight orientation in the desert ant *Cataglyphis*. *J. Comp. Physiol. A*, 158(2): 145-150.

Ferenz, H.J. 1975. Photoperiodic and hormonal control of reproduction in male beetles, *Pterostichus nigrita*. *J. Insect Physiol.*, 21: 331-341.

Fernandez, A.T. & Randolph, N.M. 1967. A photoperiodic effect on the daily susceptibility of the house fly to trichlorfon. *J. Econ. Entomol.*, 60: 1633-1636.

Fielding, K. & Knisley, C.B. 1995. Mating behaviour in two tiger beetles, *Cicindela dorsalis* and *C. puritana* (Coleoptera: Cicindelidae). *Entomol. News*, 106(2): 61-67.

Filippi, L., Hironaka, M. & Nomakuchi, S. 2001. A review of the ecological parameters and implications of subsociality in *Parastrachia japonensis* (Hemiptera: Cydnidae), a semelparous species that specializes on a poor resource. *Popul. Ecol.*, 43(1): 41-50.

Filippi, L., Nomakuchi, S., Kuki, K. & Tojo, S. 1995. Adaptiveness of parental care in *Parastrachia japonensis* (Hemiptera: Cydnidae). *Ann. Entomol. Soc. Am.*, 88: 374-383.

Finch, S. 1978. Volatile plant chemicals and their effect on host plant finding by the cabbage root fly (*Delia brassicae*). *Entomol. Exp. Appl.*, 24: 350-359.

Finch, S. 1980. Chemical attraction of plant-feeding insects to plants. *Appl. Biol.*, 5: 67-143.

Finks, L.S. & Brower, L.P. 1981. Birds can overcome the cardenolide defence of monarch butterflies in Mexico. *Nature*, 291: 67-70.

Fitzgerald, T.D. & Underwood, D.L.A. 1998. Trail marking by the larva of the Madrone butterfly *Eucheira socialis* and the role of the trail pheromone in communal foraging behaviour. *J. Insect Behav.*, 11(2): 247-263.

Fjerdingstad, E.J. & Boomsma, J.J. 1998. Multiple mating increases the sperm stores of *Atta colombica* leafcutter ant queens. *Behav. Ecol. Sociobiol.*, 42(4): 257-261.

Fletcher, B.S. 1974. The ecology of a natural population of the Queensland fruit fly, *Dacus tryoni* V. The dispersal of adults. *Aust. J. Zool.*, 22: 189-202.

Fleury, F., Allemand, R., Vavre, F., Fouillet, P. & Boulétreau, M. 2000. Adaptive significance of a circadian clock: temporal segregation of activities reduces intrinsic competitive inferiority in *Drosophila* parasitoids. *Proc. Roy. Soc. Lond. Ser. B, Biol. Sci.*, 267: 1005-1010.

Fondacaro J.D. & Butz, A. 1970. Circadian rhythm of locomotor activity and susceptibility to methyl parathion of adult *Tenebrio molitor* (Coleoptera: Tenebrionidae). *Ann. Entomol. Soc. Am.*, 63: 952-955.

Fontaine, A.R., Olsen, N., Ring, R.A. & Singla, C.L. 1991. Cuticular metal hardening of mouthparts and claws of some forest insects of British Columbia. *J. Entomol. Soc. Brit. Columbia*, 88: 45-55.

Ford. E.B. 1940. Polymorphism and taxonomy. In: *The New Systematics*. J.S. Huxley (ed.). Oxford Univ. Press, Oxford, pp. 493-513.

Ford, E.B. 1953. The genetic polymorphism in the Lepidoptera. *Adv. Genet.*, 5: 43-87.

Ford, E.B. 1955. Polymorphism and taxonomy. *Heredity*, 9: 255-264.

Ford, E.B. 1965. *Genetic Polymorphism*. Fabre & Fabre, London.

Ford, E.B. 1975. *Ecological Genetics*. (4th ed). Chapman & Hall, London.

Forel, A. 1886-1888. Experience et remarques critiques sur les sensations des insects. *Recucil Zool. Suisse*, 2: 1-76.

Forel, A. 1908. *The Sense of Insects*. (Transl. M. Yearsley), Methuen & Co., Ltd., London.

Forel, A. 1910. *Das Sinnesleben der Insekten*. Ernst Reinhardt, München.

Foster, S.P. & Harris, M.O. 1997. Behavioural manipulation methods for insect pest management. *Ann. Rev. Entomol.*, 42: 123-146.

Fox, L.R. 1988. Diffuse coevolution within complex communities. *Ecology*, 69: 906-907.

Fraenkel, G.S. 1932. Die Wanderungen der Insekten. *Ergebn. Biol.*, 9: 1-238.

Fraenkel, G.S. & Gunn. D.L. 1961. *The Orientation of Animals, Kineses, Taxes and Compass Reactions*. (Rev. ed.). Dover Publ., N.Y.

Francis, F., Lognay, G. & Haubruge, E. 2004. Olfactory responses to aphid and host plant volatile releases: (*E*)-beta-farnesene an effective kairomone for the predator *Adalia bipunctata*. *J. Chem. Ecol.*, 30(4): 741-755.

Fraser, J.F.D. & Rothschild, M. 1960. Defence mechanisms in warningly coloured moths and other insects. *Proc. 11th Int. Cong. Entomol., BIII (Symp. 4)*: 249-256.

Free, J.B. 1987. *Pheromones of Social Bees*. Comstock, Ithaca, N.Y.

French, R.A. 1967. Long distance of movement of two migrant Lepidoptera in relation to synoptic weather conditions. *Biometeorol.*, 2: 565-569.

Friend, W.G. & Smith, J.J.B. 1977. Factors affecting feeding by blood-sucking insects. *Ann. Rev. Entomol.*, 22: 309-331.

Frisch, K. von, 1937. The language of bees. *Sci. Progr.*, 32: 29-37.

Frisch, K. von, 1946. Die Tänze der Bienen. *Oesterr. Zool. Z.*, 1: 1-48.

Frisch, K. von, 1947. The dances of the honey bee. *Bull. Anim. Behav.*, 5: 1-32.

Frisch, K. von, 1949. Die Polarisation des Himmelslichtes als Orientierender faktor bei den Tänzan der Bienen. *Experientia*, 5(4): 142-148.

Frisch, K. von, 1950. Die Sonne als Kompass im Leben der Bienen. *Experientia*, 6: 210-221.

Frisch, K. von, 1954. *The Dancing Bees*. Methuen & Co. Ltd., London.

Frisch, K. von, 1962. Dialects in the language of the bees. *Sci. Amer.*, 207: 78-87.

Frisch, K. von, 1967. *The Dance, Language and Orientation of Bees*. (Trans. L.E. Chadwick). Harvard Univ. Press, Cambridge, Massachusetts.

Frisch, K. von, 1971. *Bees-Their Vision, Chemical Senses and Language*. (Rev. ed.). Cornell Univ. Press, Ithaca, N.Y.

Frisch, K. von & Lindauer, M. 1954. Himmel und Erde im Konkurrenz bei der Orientierung Beinen. *Naturwissenschaften*, 41: 245-253.

Frisch, K. von, Lindauer, M. & Daumer, K. 1960. Über die Wahrnehmung polarisierten Lichtes durch das Bienenauge. *Experientia*, 16: 289-301.

Froy, O., Gotter, A.L., Casselman, A.L. & Reppert, S.M. 2003. Illuminating the circadian clock in monarch butterfly migration. *Science*, 300: 1303-1305.

Frumhoff, P.C. & Baker, J. 1988. A genetic component to division of labour within honey bee colonies. *Nature*, 333: 358-361.

Frye, M.A. & Dickinson, M.H. 2001. Fly flight: a model for the neural control of complex behaviour. *Neuron*, 32(3): 385-388.

Fu, X.-H., Ohba, N. & Lei, C.L. 2004. Morphological and biological observations on aquatic firefly *Luciola substriata* (Gorham) (Coleoptera: Lampyridae) in China. *Acta Entomol. Sinica*, 47(3): 372-378.

Fukuda, S. 1952. Function of the pupal brain and suboesophageal ganglion in the production of non-diapause and diapause eggs in the silkworm. *Annot. Zool. Japan*, 25: 149-155.

Fukushi, T. 1989. Learning and discrimination of coloured papers in the walking blowflies, *Lucilia cuprina*. *J. Comp. Physiol. A*, 166(1): 57-64.

Fukushi, T. & Wehner, R. 2004. Navigation in wood ants *Formica japonica*: context dependent use of landmarks. *J. Exp. Biol.*, 207(19): 3431-3439.

Fullard J.H. 1987. Sensory ecology and neuroethology of moths and bats: interactions in a global perspective. In: *Recent Advances in the Study of Bats*. M.B. Fenton, P.A. Racey & J.M.V. Rayner (eds.). Cambridge Univ. Press, Cambridge, pp. 244-272.

Fuller, J.L. & Thompson, W.R. 1978. *Foundations of Behaviour Genetics*. The CV Mosby Co., N.Y.

Fulton, B.B. 1932. North Carolina's singing Orthoptera. *J. Elisha Mitchell Sci. Soc.*, 47: 55-69.

Fulton, B.B. 1952. Speciation in field cricket. *Evolution*, 6: 283-295.

Futch, D.G. 1966. A study of speciation in South Pacific populations of *Drosophila ananassae*. *Univ. Tex. Publ.*, 6615: 79-120.

Gadagkar, R. 1990. Origin and evolution of eusociality: a perspective from studying primitively eusocial wasps. *J. Genet.*, 69(2): 113-125.

Gadagkar, R. 2001. Division of labour and organization of work in the primitively eusocial wasp *Ropalidia marginata*. *Proc. Indian Natl. Sci. Acad. Part B*, 67(6): 397-422.

Gade, G., Hoffmann, K.H. & Spring, J.H. 1997. Hormonal regulation in insects: facts, gaps, and future directions. *Physiol. Rev.*, 77(4): 963-1032.

Galil, J. & Eisikowitch, D. 1969. On the pollination ecology of *Ficus sycomorus* in East Africa. *Ecology*, 49: 259-269.

Gammie, S.C. & Truman, J.W. 1997. Neuropeptide hierarchies and the activation of sequential motor behaviours in the hawk moth, *Manduca sexta*. *J. Neurosci.*, 17: 4389-4397.

Gary, N.E. & Martson, J. 1971. Mating behaviour of drone honey bees with queen models (*Apis mellifera* L.). *Anim. Behav.*, 19: 299-304.

Gatehouse, A.G. 1997. Behaviour and ecological genetics of wind-borne migration by insects. *Ann. Rev. Entomol.*, 42: 475-502.

Gaul, A.T. 1951. Addition to vespine biology. VII. Orientation flight. *Bull. Brooklyn Entomol. Soc.*, 46: 54-56.

Gersch, M. 1956. Untersuchungen zur Frage der hormonalen Beeinflussung der Melanophoren bei der *Corethra*-larve. *Z. vergl. Physiol.*, 39: 190-208.

Gewecke, M. & Wendler, G. (ed.) 1985. *Insect Locomotion*. Paul Parey Sci. Publ., N.Y.

Ghent, A.W. 1960. A study of group feeding behaviour of larvae of the jack pine sawfly, *Neodiprion pratti* Roh. *Behaviour*, 16: 110-148.

Ghiradella, H. 1998. The anatomy of light production: the fine structure of the firefly lantern. In: *Microscopic Anatomy of Invertebrates*. M. Locke & F.W. Harrison (eds.). John Wiley & Sons, N.Y., pp. 363-381.

Gibbs, M., Lace, L.A., Jones, M.J. & Moore, A.J. 2004. Intraspecific competition in the speckled wood butterfly *Pararge aegeria*: effect of rearing density and gender on larval life history. *J. Insect Sci.*, 4: 1-6.

Giebultowicz, J.M. 2000. Molecular mechanism and cellular distribution of insect circadian clocks. *Ann. Rev. Entomol.*, 45: 769-793.

Gilbert, L.I. & King, D.S. 1973. Physiology of growth and development: endocrine aspects. In: *The Physiology of Insecta*. Vol. I. (2nd ed.). M. Rockstein (ed.). Academic Press, N.Y., pp. 249-379.

Gill, J.S. & Lewis, C.T. 1971. Systemic action of an insect feeding deterrent. *Nature*, 232: 402-403.

Gilliam, M., Taber, S. III & Richardson. G.V. 1983. Hygienic behaviour of honey bees in relation to chalkbrood disease. *Apidologie*, 14: 29-39.

Gillott, C. 1996. Male insect accessory glands: functions and control of secretory activity. *Invert. Repro. Develop.*, 30(1-3): 199-205.

Giray, T., Giovanetti, M. & West-Eberhard, M.J. 2005. Juvenile hormone, reproduction, and worker behaviour in the neotropical social wasp *Polistes canadensis*. *Proc. Natl. Acad. Sci., USA.*, 102(9): 3330-3335.

Givois, V. & Pollack, G.S. 2000. Sensory habituation of auditory receptor neurons: implications for sound localization. *J. Exp. Biol.*, 203(17): 2529-2537.

Godden, D.H.1973. A re-examination of circadian rhythmicity in *Carausius morosus*. *J. Insect Physiol.*, 19: 1377-1386.

Goldsmith, T.H. & Bernard, G.D. 1974. The visual system of insects. In: *The Physiology of Insecta*. Vol. 2. M. Rockstein (ed.). Academic Press, N.Y., pp. 166-272.

Gonzalez, A., Hare, J.F. & Eisner, T. 1999. Chemical egg defence in *Photuris* firefly 'femmes fatales'. *Chemoecol.*, 9(4): 177-185.

Goodenough, J., McGuire, B. & Wallace, R.A. 1993. *Perspectives on Animal Behaviour*. John Wiley & Sons, N.Y.

Goodman, L.J. 1965. The role of certain optometre reactions in regulating stability in the rolling plane during flight in the desert locust, *Schistocerca gregaria*. *J. Exp. Biol.*, 42: 385-408.

Görner, P. 1973. Beispiele einer Orientierung ohne richtende Aussenreize. *Fortsch. Zool.*, 21: 21-45.

Gotwald, W. 1995. *Army Ants: The Biology of Social Predation*. Cornell Univ. Press, Ithaca, N.Y.

Gould, J.L. 1975. Honey bee recruitment: the dance language controversy. *Science*, 189: 685-693.

Gould, J.L. 1976. The dance-language controversy. *Quart. Rev. Biol.*, 51: 211-244.

Gould, J.L. 1982. *Ethology*. Norton, N.Y.

Gould, J.L. 1986a. Pattern learning by honey bees. *Anim. Behav.*, 34(4): 990-997.

Gould, J.L. 1986b. The locale map of honey bees: do insects have cognitive maps. *Science*, 232: 861-863.

Gould, J.L. 1987. Landmark learning by honey bees. *Anim. Behav.*, 35(1): 26-34.

Gould, J.L., Kirschvink, J.L. & Deffeyes, K.S. 1978. Bees have magnetic remanence. *Science*, 201: 1026-1028.

Graber, V. 1877-1879. *Die Insekten*. 2 bd. Oldenbourg, München.

Graham, D. 1972. A behavioural analysis of the temporal organization of walking movement in the 1st instar and adult stick insect *Carausius morosus*. *J. Comp. Physiol.*, 81: 23-52.

Graham, D. 1985. Pattern and control of walking in insects. *Adv. Insect Physiol.*, 18: 31-140.

Gray, D.A. 2005. Does courtship behaviour contribute to species-level reproductive isolation in field crickets? *Behav. Ecol.*, 16(1): 210-206.

Gray, D.A. & Cade, W.H. 2000. Sexual selection and speciation in field crickets. *Proc. Natl. Acad. Sci., USA*, 97(26): 14449-14454.

Gray, J. 1953. Undulatory propulsion. *Q. J. Microsc. Sci.*, 94: 551-578.

Gray, J. 1968. *Animal Locomotion*. Weidenfeld & Nicolson, London.

Greenberg, L. 1988. Kin recognition in the sweat bee, *Lasioglossum zephyrum*. *Behav. Genet.*, 18(4): 425-438.

Greene, A. 1979. Behavioural characters as indicators of yellowjacket phylogeny (Hymenoptera: Vespidae). *Ann. Entomol. Soc. Am.*, 72: 614-619.

Greene, E., Orsak, L.J. & Whitman, D.W. 1987. A tephritid fly mimics the territorial displays of its jumping spider predators. *Science*, 236: 310-311.

Griffiths, D.C., Maniar, S.P., Merritt, L.A., Mudd, A., Pickett, J.A., Pye, B.J., Smart, L.E. & Wadhams, L.J. 1991. Laboratory evaluation of pest management strategies combining antifeedants with insect growth regulator insecticides. *Crop Prot.*, 10(2): 145-151.

Gullan, P.J. & Cranston, P.S. 2000. *The Insects: An Outline of Entomology*. (2nd ed.). Blackwell Sci., Oxford.

Gullan P.J., Miller, D.R. & Cook, L.G. 2004. Gall-inducing scale insects (Hemiptera: Sternorrhyncha: Coccoidea). In: *Biology, Ecology, and Evolution of Gall-Inducing Arthropods*. A. Raman, C.W. Schaefer & T.M. Whithers (eds.). Oxford & IBH Publ. Co. Pvt. Ltd., N. Delhi, pp. 159-229.

Gunn, D.L. & Cosway, C.A. 1938. The temperature and humidity relations of the cockroach. V. Humidity preference. *J. Exp. Zool.*, 15: 555-563.

Gupta, K.K. & Sehgal, S.S. 1997. Mechanism of mate finding and recognition in mating behaviour of red cotton bug. *Bull. Entomol.*, 38(1/2): 18-22.

Gut, L.J., Stelinski, L.L., Thomson, D.R. & Miller, J.R. 2004. Behaviour-modifying chemicals: prospects and constraints in IPM. In: *Integrated Pest Management: Potential, Constraints and Challenges*. O. Koul, G.S. Dhaliwal & G.W. Cuperus (eds.). CABI Press, N.Y., pp. 73-122.

Gwynne, D.T. 1984. Courtship feeding increases female reproductive success in bush crickets. *Nature*, 307: 361-363.

Gwynne, D.T. 1990. The katytid spermatophore: evolution of a parental investment. In: *The Tettigoniidae: Biology, Systematics and Evolution*. W.J. Bailey & D.C.F. Rentz (eds.). Crawford House Press, Bathurst, pp. 27-40.

Gwynne, D.T. 1995. Phylogeny of the Ensifera (Orthoptera): a hypothesis supporting multiple origins of acoustical signalling, complex spermatophores and maternal care in crickets, katydids, and weta. *J. Orth. Res.*, 4: 203-218.

Haddow, A.J. & Gillett, J.D. 1957. Observation on the oviposition cycle of *Aedes* (*Stegomyia*) *aegypti* (Linnaeus). *Ann. Trop. Med. Prasitol.*, 51: 159-169.

Hadlington, P. 1987. *Australian Termites and Other Common Timber Pests*. NSW Univ. Press. Kensington, NSW.

Hafez, M. 1950. On the behaviour and sensory physiology of the house fly larva, *Musca domestica* L. I. Feeding stage. *Parasitology*, 40: 215-236.

Hagedorn, H.H. 1974. The control of vitellogenesis in the mosquito, *Aedes aegypti*. *Amer. Zool.*, 14: 1207-1217.

Hagedorn, H.H. 1985. The role of ecdysteroid in reproduction. In: *Comprehensive Insect Physiology, Biochemistry and Pharmacology. Endocrinology II.* Vol. 8. G.A. Kerkut & L.I. Gilbert (eds.). Pergamon Press, Oxford, pp. 205-262.

Hagen, H.A. 1858. Monographie der termiten. *Limaea Entomol.*, 12.

Hagen, K.S. 1987. Nutritional ecology of terrestrial insect predators. In: *Nutritional Ecology of Insects, Mites, Spiders, and Related Invertebrates.* F. Slansky Jr. & J.G. Rodriquez (eds.). John Wiley & Sons, N.Y., pp. 533-577.

Hagen, K.S. & Bishop, G.W. 1979. Use of supplemental foods and behavioural chemicals to increase the effectiveness of natural enemies. In: *Biological Control and Insect management.* D.W. Davis, J.A. McMartry, & S.C. Hoyt (eds.). *Calif. Agril. Exp. Stn. Publ.*, 4096, pp. 49-60.

Hagvar, S. 2001. Occurrence and migration on snow, and phenology of egg-laying in the winter-active insect *Boreus* sp. (Mecoptera). *Norwegian J. Entomol.*, 48(1): 51-60.

Hall, J.C. 1985. Genetic analysis of behaviour. In: *Comprehensive Insect Physiology, Biochemistry and Pharmacology. Behaviour.* Vol. 9. G.A. Kerkut & L.I. Gilbert (eds.). Pergamon Press, Oxford, pp. 287-373.

Hall, J.C. 2003. Genetics and molecular biology of rhythms in *Drosophila* and other insects. *Adv. Genet.*, 48: 1-286.

Hallberg, E., Pers, J.N.C. van der, & Haniotakis, G.E. 1984. Functional sensilla of *Dacus oleae*: fine structure characteristics. *Entomologia Hellenica*, 2: 41-46.

Halliday, T.R. & Slater, P.J.B. (eds.) 1983. *Animal Behaviour.* (3 vols.). Blackwell Scientific Publ., Oxford.

Hamilton, W.D. 1964a. The genetical evolution of social behaviour I. *J. Theor. Biol.*, 7: 1-16.

Hamilton, W.D. 1964b. The genetical evolution of social behaviour II. *J. Theor. Biol.*, 7: 17-52.

Hamner, K.C., Flinn, J.C., Sirohi, G.S., Hoshizaki, T. & Carpenter, B.H. 1962. Studies on the biological clock at the south pole. *Nature*, 195: 476-480.

Hancock, D.L., Osborne, R., Broughton, S. & Gleeson, P. 2000. Eradication of *Bactrocera papayae* (Diptera: Tephritidae) by male annihilation and protein baiting in Queensland, Australia. In: *Area-Wide Control of Fruit Flies and Other Insect Pests.* K.H. Tan (ed.). Penerbit Universiti Sains, Malaysia, pp. 381-388.

Handler, A.M. & Konopka, R.J. 1979. Transplantation of a circadian pacemaker in *Drosophila. Nature*, 279: 236-238.

Hangartner, W. 1969a. Trail laying in the subterranean ant, *Acanthomyops interjectus. J. Insect Physiol.*, 15: 1-4.

Hangartner, W. 1969b. Structure and variability of the individual odour trail in *Solenopsis* (Formicidae). *Z. vergl. Physiol.*, 62: 111-120.

Hans, H. & Thorsteinson, A.J. 1961. The influence of physical factors and host plant odour on the induction and termination of dispersal flights in *Sitona cylindricollis* Fahr. *Entomol. Exp. Appl.*, 4: 165-177.

Hansell, M. 1985. Ethology. In: *Comprehensive Insect Physiology, Biochemistry and Pharmacology. Behaviour.* Vol. 9. G.A. Kerkut & L.I. Gilbert (eds.). Pergamon Press, Oxford, pp. 3-93.

Harari, A.R., Brockmann, H.J. & Landolt, P.J. 2000. Intrasexual mounting in the beetle *Diaprepes abbreviatus* (L.). *Proc. Roy. Soc. Lond. Ser. B, Biol. Sci.*, 267(1457): 2071-2079.

Harborne, J.B. 1993. *Introduction to Ecological Biochemistry.* (4th ed.). Academic Press, London.

Hardie, J. & Lees, A.D. 1985. Endocrine control of polymorphism and polyphenism. In:

Comprehensive Insect Physiology, Biochemistry and Pharmacology. Endocrinology II. Vol. 8. G.A. Kerkut & L.I. Gilbert (eds.). Pergamon Press, Oxford, pp. 441-490.

Hardie, J. & Minks, A.K. (ed.) 1999. *Pheromones of Non-Lepidopteran Insects Associated with Agricultural Plants*. CABI Publ., Wallingford.

Harker, J.E. 1954. Diurnal rhythm in *Periplaneta americana* L. *Nature*, 173: 689-690.

Harker, J.E. 1956. Factors controlling the diurnal rhythm of activity in *Periplaneta americana* L. *J. Exp. Biol.*, 33: 224-234.

Harris, M.O. & Foster, S.P. 1995. Behaviour and integration. In: *Chemical Ecology of Insects* 2. R.T. Cardé & W.J. Bell (eds.). Chapman & Hall, N.Y., pp. 1-46.

Harrison, J.M. & Breed, M.D. 1987. Temporal learning in the giant tropical ant, *Paraponera clavata*. *Physiol. Entomol.*, 12(3): 317-320.

Hartfelder, K. 2000. Insect juvenile hormone: from "status quo" to high society. *Braz. J. Med. Biol. Res.*, 33(2): 157-177.

Hartfelder, K. & Engels, W. 1998. Social insect polymorphism: hormonal regulation of plasticity in development and reproduction in the honey bee. *Curr. Topics Develop. Biol.*, 40: 45-77.

Hartland-Rowe, R. 1955. Lunar rhythm in the emergence of an ephemeropteran. *Nature*, 176: 657.

Harvey, E.N. 1920. *The Nature of Animal Light*. J.P. Lippincott Co., Philadelphia.

Harvey, E.N. 1952. *Bioluminescence*. Academic Press, N.Y.

Harvey, E.N. & Stevens, K.P. 1928. The brightness of the light of the West Indian elaterid beetle, *Pyrophorus*. *J. Gen. Physiol.*, 12: 269-272.

Hashimoto, H. 1966. Discovery of *Clunio takahashii* from Japan. *Jap. J. Zool.*, 14: 13-29.

Haskell, P.T. 1961. *Insect Sounds*. Quadrangle Book, Chicago.

Haskell, P.T. (ed.) 1966. *Insect Behaviour. Symp. Roy. Entomol. Soc. Lond.*, No. 3.

Haskell, P.T. 1974. Sound production. In: *The Physiology of Insecta*. Vol. 2. (2nd ed.). M. Rockstein (ed.). Academic Press, N.Y., pp. 353-410.

Havukkala, I. 1980. Klinokinetic and klinotactic humidity reactions of the beetle *Hylobius abietis* and *Tenebrio molitor*. *Physiol. Entomol.*, 5(2): 133-140.

Haynes, K.F. & Birch, M.C. 1985. The role of other pheromones, allomones and kairomones in the behavioural responses of insects. In: *Comprehensive Insect Physiology, Biochemistry and Pharmacology. Behaviour*. Vol. 9. G.A. Kerkut & L.I. Gilbert (eds.). Pergamon Press, Oxford, pp. 225-255.

Headrick, D.H. & Goeden, R.D. 1994. Reproductive behaviour of California fruit flies and the classification and evolution of Tephritidae (Diptera) mating systems. *Studia Dipterologica*, 1: 194-252.

Headrick, D.H. & Goeden, R.D. 1999. Behaviour of flies in the subfamily. Tephritinae. In: *Fruit Flies (Tephritidae): Phylogeny and Evolution of Behaviour*. M. Aluja & A.L. Norrbom (eds.). CRC Press, Boca Raton, pp. 671-707.

Hebb, D.O. 1953. Heredity and environment in mammalian behaviour. *Brit. J. Anim. Behav.*, 1: 43-47.

Heidweg, H. & Thorsteinson, A.J. 1961. The influence of physical factors and host plant odour on the induction and termination of dispersal flights in *Sitona cylindricollis* Fabr. *Entomol. Exp. Appl.*, 4: 165-177.

Heinrich, B. 1979. *Bumble Bee Economics*. Harvard Univ. Press, Cambridge, Massachusetts.

Heitler, W.J. 1974. The locust jump. Specializations of the metathoracic femoral-tibial joint. *J. Comp. Physiol.*, 89: 93-104.

Heitler, W.J. & Burrows, M. 1977. The locust jump. I. The motor programme. *J. Exp. Biol.*, 66: 203-219.

Helfrich, C., Cymborowski, B. & Engelmann, W. 1985. Circadian activity rhythm of the house fly continues after optic tract severance and lobectomy. *Chronobiol. Int.*, 2: 19-32.

Helmy, O. & Jander, R. 2003. Topochemical learning in black carpenter ants (*Camponotus pennsylvanicus*). *Insectes Sociaux*, 50(1): 32-37.

Helversen O. von, & Helversen D. von, 1994. Forces driving coevolution of song and song recognition in grasshoppers. *Fortsch. Zool.*, (39): 253-284.

Heran, H. 1956. Ein Beitrag zur Frage nach der Wahrnehmungsgrundlage der Entfernungsweisung der Bienen. *Z. vergl. Physiol.*, 42: 103-163.

Heran, H. & Lindauer, M. 1963. Windcompensation und Seitenwindkorrektur der Bienen beim Flug über Wasser. *Z. vergl. Physiol.*, 47: 39-55.

Herman, W.S. 1973. The endocrine basis of reproductive inactivity in monarch butterflies overwintering in Central California. *J. Insect Physiol.*, 19: 1883-1887.

Herman, W.S. 1975. Endocrine regulation of posteclosion enlargement of the male and female reproductive glands in monarch butterflies. *Gen. Comp. Endocr.*, 26: 534-540.

Herman, W.S. 1985. Hormonally mediated events in adult monarch butterflies. In: *Migration: Mechanisms and Adaptive Significance*. M.A. Rankin (ed.). *Univ. Texas, Marine Sci.Inst. Cont.*, 27 (Suppl.), Austin, USA, pp. 799-815.

Herman, W.S. & Tatar, M. 2001. Juvenile hormone regulation of longevity in the migratory monarch butterfly. *Proc. Roy. Soc. Lond. Ser. B, Biol. Sci.*, 268(1485): 2509-2514.

Hermann, H.R. (eds.) 1979-1982. *Social Insects*. Vols. I-IV. Academic Press, N.Y.

Hickling, R, & Brown, R.L. 2000. Analysis of acoustic communication by ants. *J. Acoust. Soc. Am.*, 108(4): 1920-1929.

Hill, K.G., Loftus-Hills, J.J. & Gartside, D.F. 1972. Pre-mating isolation between Australian field crickets *Teleogryllus commodus* and *T. oceanicus* (Orthoptera: Gryllidae). *Aust. J. Zool.*, 20: 153-163.

Hinde, R.A. 1970. *Animal Behaviour, A Synthesis of Ethology and Comparative Psychology*. (2nd ed.). McGraw-Hill Book Co. Inc., N.Y.

Hinde, R.A. 1982. *Ethology*. Oxford Univ. Press, Oxford.

Hinton, H.E. 1951. A new chironomid from Africa, the larva of which can be dehydrated without injury. *Proc. Zool. Soc. Lond.*, 121: 371-380.

Hinton, H.E. 1955. Protective devices of endopterygote pupae. *Trans. Soc. Brit. Entomol.*, 12: 49-92.

Hinton, H.E. 1964. Sperm transfer in insects and the evolution of haemocoelic insemination. *Symp. Roy. Entomol. Soc. Lond.*, 2: 95-107.

Hirashima, A., Eiraku, T., Watanabe, Y., Kuwano, E., Taniguchi, E. & Eto, M. 2001. Identification of novel inhibitors of calling and *in vitro* [^{14}C] acetate incorporation by pheromone glands of *Plodia interpunctella*. *Pest Manag. Sci.*, 57(8): 713-720.

Hirsch, J. & McGuire, T.R. (eds.) (1982). *Behaviour-Genetic Analysis*. Hutchinson Ross & Co., Stroudsburg.

Hirsch, J. and Erlenmeyer-Kimling, L. 1962. Studies in experimental behavioural genetics. IV. Chromosomes analyses for geotaxis. *J. Comp. Physiol. Psychol.*, 55: 732-739.

Hochuli, D.F. 2001. Insect herbivory and ontogeny: how do growth and development influence feeding behaviour, morphology and host use? *Austral Ecol.*, 26(5): 563-570.

Hocking, B. 1971. Blood-sucking behaviour of terrestrial arthropods. *Ann. Rev. Entomol.*, 16: 1-26.

Hodgson, E.S. 1958. Chemoreception in arthropods. *Ann. Rev. Entomol.*, 3: 19-36.

Hodgson, E.S. 1974. Chemoreception. In: *The Physiology of Insecta.* Vol. 2. M. Rockstein (ed.). Academic Press, N.Y., pp. 127-165.

Hoikkala, A. & Lumme, J. 1987. The genetic basis of evolution of the male courtship sounds in the *Drosophila virilis* group. *Evolution,* 41: 827-845.

Hoikkala, A. & Lumme, J. 1990. Inheritance of male courtship sound characteristics in *Drosophila littoralis. Behav. Genet.*, 20(3): 423-435.

Hölldobler, B. 1971a. Communication between ants and their guests. *Sci. Amer.*, 224: 86-93.

Hölldobler, B. 1971b. Homing in the harvester ant *Pogonomyrmex badius. Science,* 171: 1149-1151.

Hölldobler, B. 1978. Ethological aspects of chemical communication in ants. *Adv. Study Behav.*, 8: 75-115.

Hölldobler, B. 1980. Canopy orientation: a new kind of orientation in ants. *Science,* 210(4465): 86-88.

Hölldobler, B. & Wilson, E.O. 1990. *The Ants.* Harvard Univ. Press, Cambridge, Massachusetts.

Hooper, J. 2002. *Of Moths and Men: The Untold Story of Science and the Peppered Moth.* W.W. Norton, N.Y.

Horridge, G.A. 1962. Learning of leg position by the ventral nerve cord of headless insects. *Proc. Roy. Soc. Lond. Ser. B, Biol. Sci.*, 157: 33-52.

Horridge, G.A.1975. *The Compound Eye and Vision of Insects.* Clarendon Press, Oxford.

Howard, R.W., McDaniel, C.A. & Blomquist, G.J. 1980. Chemical mimicry as an integrating mechanism; cuticular hydrocarbons of a termitophile ant and its host. *Science,* 210: 431-433.

Howksworth, D.L. & Bisby, F.A. 1988. Systematics: the keystone of biology. In: *Prospects in Systematics. Systematics Association Special Volume. No. 36.* D.L. Howksworth, (ed.). Clarendon Press, Oxford, pp. 3-30.

Howlett, F.M. 1912. The effect of oil of citronella on two species of *Dacus. Trans. Entomol. Soc. Lond.*, 60: 412-418.

Howlett, F.M. 1915. Chemical reactions of fruit flies. *Bull. Entomol. Res.*, 6: 297-305.

Howse, P.E. 1964. The significance of the sound produced by the termite *Zootermopsis angusticollis* (Hagen). *Anim. Behav.*, 12: 284-300.

Howse, P.E. 1970. *Termites: A Study in Social Behaviour.* Hutchinson Univ. Library, London.

Howse, P.E. 1974. Design and function in the insect brain. In: *Experimental Analysis of Insect Behaviour.* L. Barton Browne (ed.). Springer-Verlag, Berlin, pp. 180-194.

Howse, P.E. 1975. Brain structure and behaviour in insects. *Ann. Rev. Entomol.*, 20: 359-379.

Howse, P.E., Stevens, I.D.R. & Jones, O.T. 1998. *Insect Pheromones and Their Use in Pest Management.* Chapman & Hall, London.

Hoy, M.A. 1994. *Insect Molecular Genetics: An Introduction to Principles and Applications.* Academic Press, California.

Hoy, R.R. 1974. Genetic control of acoustic behaviour in crickets. *Amer. Zool.*, 14: 1067-1080.

Hoy, R.R., Hoikkala, A. & Kaneshiro, K. 1988. Hawaiian courtship songs: evolutionary innovation in communication signals of *Drosophila. Science*, 240(4849): 217-219.

Hoyle, G. 1955. Neuromuscular mechanisms of a locust skeletal muscle. *Proc. Roy. Soc. Lond. Ser. B, Biol. Sci.*, 143: 343-367.

Hoyle, G. 1964. Exploration of neuronal mechanisms underlying behaviour in insects. In: *Neural Theory and Modelling*. R. Reiss (ed.). Stanford Univ. Press, Stanford, pp. 346-376.

Hoyle, G. 1970. Cellular mechanisms underlying behaviour-neurobiology. *Adv. Insect Physiol.*, 7: 349-444.

Hoyle, G. 1974. Neural control of skeletal muscles. In: *The Physiology of Insecta*. Vol. 4. M. Rockstein (ed.). Academic Press, N.Y., pp. 175-236.

Hoyle, G. 1976. Arthropod walking. In: *Neural Control of Locomotion*. R.M. Herman, S. Grillner, P.S.G. Stein, & D.G. Stuart (eds.). Plenum Press, N.Y., pp. 137-179.

Hsiao, T.H. 1985. Feeding behaviour. In: *Comprehensive Insect Physiology, Biochemistry and Pharmacology. Behaviour*. Vol. 9. G.A. Kerkut & L.I. Gilbert (eds.). Pergamon Press, Oxford, pp. 471-512.

Hsiao, T.H. 1986. Specificity of certain chrysomelid beetles for Solanaceae. In: *Solanaceae: Biology and Systematics*. W.G. D'Arcy (ed.). Columbia Univ. Press, N.Y., pp. 345-363.

Huang, Z.-Y., Kuang, H.O., Kuang, B.Y. & Qin, Y. 2001. Juvenile hormone and division of labour in *Apis cerana*. *Proc. 7th IBRA Conf. on Tropical Bees and 5th AAA Conf.*, Thailand, 2000, *J. Apic. Res. (Suppl.)*, 40: 3-6.

Huang, Z.-Y., Robinson, G.E., 1992. Honey bee colony integration: worker-worker interactions mediated hormonally regulated plasticity in division of labour. *Proc. Natl. Acad. Sci., USA*, 89: 11726-11729.

Huang, Z.-Y., Robinson, G.E., Tobe, S.S., Yagi, K.J., Strambi, C., Strambi, A. & Stay, B. 1991. Hormonal regulation of behavioral development in the honey bee is based on changes in the rate of juvenile hormone biosynthesis. *J. Insect Physiol.*, 37(10): 733-741.

Huber, F. 1974. Neural integration (central nervous system). In: *The Physiology of Insecta*. Vol. 4. M. Rockstein (ed.). Academic Press, N.Y., pp. 3-100.

Huber, J.P. 1810. *Recherches sur les Moeurs des Fourmis Indigènes*. J.J. Paschoud, Paris.

Hughes, G.M. 1952. The coordination of insect movement. I. The walking movements of insects. *J. Exp. Biol.*, 29: 267-284.

Hughes, G.M. 1957. The coordination of insect movements. II. The effect of limb amputation and the cutting of commissures in the cockroach *Blatta orientalis*. *J. Exp. Biol.*, 34: 306-333.

Hughes, G.M. 1958. The coordination of insect movements. III. Swimming in *Dytiscus*, *Hydrophilus* and a dragonfly nymph. *J. Exp. Biol.*, 35: 567-583.

Hughes, G.M. & Mill, P.J. 1974. Locomotion. Terrestrial. In: *The Physiology of Insecta*. Vol. 3. (2nd ed.). M. Rockstein (ed.). Academic Press, N.Y., pp. 335-379.

Hughes, P.R. 1974. Myrcene: a precursor of pheromones in *Ips* beetles. *J. Insect Physiol.*, 20: 1271-1275.

Hughes, W.O.H., Sumner, S., Borm, S. van & Boomsma, J.J. 2003. Worker caste polymorphism has a genetic basis in *Acromyrmex* leaf-cutting ants. *Proc. Natl. Acad. Sci., USA*, 100(16): 9394-9397.

Huntingford, F.A. 1984. *The Study of Animal Behaviour*. Chapman & Hall, London.

Imms, A.D. 1947. *Insect Natural History*. Collins, London.

Imms, A.D. 1957. *A General Text Book of Entomology*. (9th ed.). (Revised by O.W. Richards & R.G.

Davies). Methuen & Co., London.

Injeyan, H.S. & Tobe, S.S. 1981. Phase polymorphism in *Schistocerca gregaria*: assessment of juvenile hormone synthesis in relation to vitellogenesis. *J. Insect Physiol.*, 27: 203-210.

Inouye, D.W. 1978. Resource partitioning in bumble bees: experimental studies of foraging behaviour. *Ecology*, 59(4): 672-678.

Ishida, Y., Chiang, V.P., Haverty, M.I. & Leal, W.S. 2002. Odorant-binding proteins from a primitive termite. *J. Chem. Ecol.*, 28: 1887-1893.

Ishii, S. 1970. An aggregation pheromone of the German cockroach, *Blattella germanica* L. 2. Species specificity of the pheromone. *Appl. Entomol. Zool.*, 5: 53-41.

Iyengar, V.K. & Eisner, T. 1999. Female choice increases offspring fitness in an arctiid moth (*Utetheisa ornatrix*). *Proc. Natl. Acad. Sci., USA*, 96(26): 15013-15016.

Jacobs-Jessen, U.F. 1959. Zur Orientierung der Hummel und einiger anderer Hymenopteran. *Z. vergl. Physiol.*, 41: 597-641.

Jacobson, M. 1965. *Insect Sex Attractants*. John Wiley & Sons, N.Y.

Jacobson, M. 1972. *Insect Sex Pheromones*. Academic Press, N.Y.

Jaffe, K., Ramos, C., Lagall, C. & Parra, L. 1990. Orientation cues using by ants. *Insectes Sociaux*, 37(2): 101-115.

Janda, V. 1934. Contribution to the knowledge of periodic colour change in *Dixippus morosus*. *Mém. Soc. Sci. Bohème*, pp. 1-30.

Jander, R. 1957. Die Optische Richtungsorientierung der roten Waldameise (*Formica rufa* L.). *Z. vergl. Physiol.*, 40: 162-238.

Jander, R. 1963. Insect orientation. *Ann. Rev. Entomol.*, 8: 95-114.

Jander, R. 1975. Ecological aspects of spatial orientation. *Ann. Rev. Ecol. Syst.*, 6: 171-188.

Jang, E.G., Light, D.M., Flath, R.A., Nagata, J.T., Mon, T.R. 1989. Eectroantennogram responses of the Mediterranean fruit fly, *Ceratitis capitata* to identified volatile constituents from calling moths. *Entomol. Exp. Appl.*, 50: 7-19, 248-251.

Janz, N., Bergström, A. & Sjögren, A. 2005. The role of nectar sources for oviposition decisions of the common blue butterfly *Polyommatus icarus*. *Oikos*, 109: 535-538.

Janzen, D.H. 1966. Coevolution of mutualism between ants and acacias in Central America. *Evolution*, 20: 249-275.

Janzen, D.H. 1967. Interaction of the bull's-horn acacia (*Acacia cornigera* L.) with an ant inhabitant (*Pseudomyrmex ferruginea* F. Smith) in eastern Mexico. *Univ. Kansas Sci. Bull.*, 67: 315-558.

Janzen, D.H. 1980. When is it coevolution? *Evolution*, 34: 611-612.

Järvilehto, M.1985. The eye: vision and perception. In: *Comprehensive Insect Physiology, Biochemistry and Pharmacology. Nervous System: Sensory*. Vol. 6. G.A. Kerkut & L.I. Gilbert (eds.). Pergamon Press, Oxford, pp. 355-429.

Jaycox, E.R. 1976. Behavioural changes in worker honey bees (*Apis mellifera* L.) after injection with synthetic juvenile hormone (Hymenoptera: Apidae). *J. Kansas Entomol. Soc.*, 49: 165-170.

Jermy, T. 1976. Insect-host plant relationships- coevolution or sequential evolution? *Symp. Biol. Hung.*, 16: 109-113.

Jermy, T. 1990. Prospects of antifeedant approach to pest control- a critical review. *J. Chem. Ecol.*, 16: 3151-3166.

Johnson, C.G. 1960. A basis for a general system of insect migration and dispersal by flight. *Nature*, 186: 348-350.

Johnson, C.G. 1963. The aerial migration of insects. *Sci. Amer.*, 209: 132-138.

Johnson, C.G. 1966. A functional system of adaptive dispersal by flight. *Ann. Rev. Entomol.*, 11: 233-260.

Johnson, C.G. 1969. *Migration and Dispersal of Insects by Flight*. Metheun & Co. Ltd., London.

Johnson, C.G. 1974. Insect migration. In: *The Physiology of Insecta*. Vol. 3. M. Rockstein (ed.). Academic Press, N.Y., pp. 280-334.

Johnson, C.G. 1976. *Insect Migration*. Carolina Biology Reader No. 84. Carolina Biol. Supply Co., Burlington, NC.

Johnson, C.H. & Golden, S.S. 1999. Circadian programs in cyanobacteria: adaptiveness and mechanism. *Ann. Rev. Microbiol.*, 53: 389-409.

Jolivet, P. 1998. *Interrelationship Between Insects and Plants*. CRC Press, Boca Raton.

Joly, P. & Joly, L. 1953. Résultat de greffes de corpora allata chez *Locusta migratoria migratorioides* L. *Ann. Sci. Nat. Zool.*, 15: 331-345.

Jones, O.T. & Casagrande, E.D. 2000. The use of semiochemical-based devices and formulation in area-wide programmes: a commercial perspective. In: *Area-Wide Control of Fruit Flies and Other Insect Pests*. K.H. Tan (ed.). Penerbit Universiti Sains, Malaysia, pp. 439-448.

Jones, R.E. 1991. Host location and oviposition on plants. In: *Reproductive Behaviour of Insects Individuals and Populations*. W.J. Bailey & J. Ridsdill-Smith (eds.). Chapman & Hall, London, pp. 107-138.

Joron, M. 2003. Aposematic colouration. In: *Encyclopedia of Insects*. R.T. Cadré & V.H. Resh (eds.). Academic Press, N.Y., pp. 39-45.

Joseph, K.J. 1998. Biology and breeding behaviour of the elephant dung beetle, *Heliocopris dominus* Bates (Coprinae: Scarabaeidae). *Entomon*, 23(4): 325-329.

Jutsum, A.R. & Goldsworthy, G.J. 1976. Fuel for flight in *Locusta*. *J. Insect Physiol.*, 22: 243-249.

Jutsum, A.R. & Gordon, R.F.S. 1989. *Insect Pheromones in Plant protection*. John Wiley & Sons, N.Y.

Kaiser, E. & Michl, H. 1958. *Die Biochemie der Tierischen Gifte*. Franz Deuticke, Austria.

Kaissling, K.-E. 1971. Insect olfaction. In: *Handbook of Sensory Physiology*. L. Beidler (ed.). Vol.4. *Chemical Senses*. Springer-Verlag, N.Y., pp. 351-431.

Kaissling, K.-E. 1974. Sensory transduction in insect olfactory receptors. In: *Biochemistry of Sensory Function*. L. Jaenicke (ed.). Springer-Verlag, Heidelberg, pp. 243-273.

Kaissling, K.-E. 2001. Olfactory perireceptor and receptor events in moths: a kinetic model. *Chem. Senses*, 26:125-150.

Kaitala, A. & Dingle, H. 1992. Spatial and temporal variation in wing dimorphism of California populations of the water strider *Aquarius remigis* (Heteroptera: Gerridae). *Ann. Entomol. Soc. Am.*, 85(5): 590-595.

Kaitala, A. & Wiklund, C. 1995. Female mate choice and mating costs in the polyandrous butterfly *Pieris napi* (Lepidoptera: Pieridae). *J. Insect behav.*, 8(3): 355-363.

Kakimoto, T., Fujisaki, K. & Miyatake, T. 2003. Egg laying preference, larval dispersion, and cannibalism in *Helicoverpa armigera* (Lepidoptera: Noctuidae). *Ann. Entomol. Soc. Am.*, 96(6): 793-798.

Kalmring, K. & Elsner, N. 1985. *Acoustic and Vibrational Communication in Insects*. Parey Verlag, Hamburg.

Kambysellis, M.P. & Williams, C.M. 1971. *In vitro* development of insect tissues. II. The role of ecdysone in the spermatogenesis of silkworms. *Biol. Bull.*, 141: 541-552.

Kammer, A.E. 1985. Flying. In: *Comprehensive Insect Physiology, Biochemistry and Pharmacology. Nervous System: Structure and Motor Function.* Vol. 5. G.A. Kerkut & L.I. Gilbert (eds.). Pergamon Press, Oxford, pp. 491-552.

Kaneshiro, K.Y. 1999. Sexual selection and speciation in Hawaiian *Drosophila* (Drosophilidae): a model system for research in Tephritidae. In: *Fruit Flies (Tephritidae): Phylogeny and Evolution of Behaviour.* M. Aluja & A.L. Norrbom (eds.). CRC Press, Boca Raton, pp. 861-877.

Kanuck, M.J. & Sullivan, D.J. 1992. Ovipositional behaviour and larval development of *Aphidencyrtus aphidivorus* (Hymenoptera: Encyrtidae), an aphid hyperparasitoid. *J. N.Y. Entomol. Soc.*, 100: 527-532.

Karban, R. & Baldwin, I.T. 1997. *Induced Responses to Herbivory.* Univ. Chicago Press, Chicago.

Karlson, P. & Butenandt, A. 1959. Pheromones (ectohormones) in insects. *Ann. Rev. Entomol.*, 4: 39-58.

Karlson, P. & Lüscher, M. 1959. "Pheromones": a new term for a class of biologically active substances. *Nature*, 183: 55-56.

Kataoka, H., Toschi, A., Li, J.P., Carney, R.L., Schooley, D.A. & Kramer, S.J. 1989. Identification of an allatotropin from adult *Manduca sexta. Science*, 243: 1481-1483.

Katsoyannos, B.I. 1988. Response to shape, size and colour. In: *Fruit Flies: Their Biology, Natural Enemies and Control.* A.S. Robinson & G. Hooper (eds.). In: *World Crop Pests.* W. Helle (ed.). Vol. 3A. Elsevier Science Publ., Amsterdam, pp. 307-324.

Katsoyannos, B.I. & Boller, E.F. 1980. Second field application of oviposition-deterring pheromone of the European cherry fruit fly *Rhagoletis cerasi* L. (Diptera: Tephritidae). *Z. Angew. Entomol.*, 89: 278-281.

Keasar, T., Ney-Nifle, M., Mangel, M. & Swezey, S. 2001. Early oviposition experience affects patch residence time in a foraging parasitoid. *Entomol. Exp. Appl.*, 98(2): 123-132.

Keeling, C.I., Slessor, K.N., Higo, H.A. & Winston, M.L. 2003. New components of the honey bee (*Apis mellifera* L.) queen retinue pheromone. *Proc. Natl. Acad. Sci., USA*, 100(8): 4486-4491.

Keil, T.A. 1997. Comparative morphogenesis of sensilla: a review. *Int. J. Insect Morphol. Embryol.*, 26: 151-160.

Keiser, I., Kobayashi, R.M., Miyashita, D.H. Harris, E.J., Schneider, E.I. & Chambers, D.L. 1974. Suppression of Mediterranean fruit flies by oriental fruit flies in mixed infestation in guava. *J. Econ. Entomol.*, 67: 353-360.

Keller, L. & Ross, K.G. 1998. Selfish genes: a green beard in the red fire ant. *Nature*, 394: 573-575.

Kellogg, F.E. 1970. Water vapour and carbon dioxide receptors in *Aedes aegypti. J. Insect Physiol.*, 16: 99-108.

Kennedy, J.S. 1939. The behaviour of the desert locust (*Schistocerca gregaria* (Forsk.)) (Orthoptera) in an outbreak centre. *Trans Roy. Entomol. Soc. Lond.*, 89: 385-542.

Kennedy, J.S. 1951. The migration of the desert locust (*Schistocerca gregaria* Forsk.). *Phil. Trans. Roy. Soc. B*, 235: 163-290.

Kennedy, J.S. 1961. A turning point in the study of insect migration. *Nature*, 189: 785-791.

Kennedy, J.S. 1975. Insect dispersal. In: *Insects, Science and Society.* D. Pimental (ed.). Academic Press, N.Y., pp. 103-119.

Kennedy, J.S. 1983. Zigzagging and casting as a response to windborne odour: a review. *Physiol. Entomol.* 8:109-120.

Kennedy, J.S. 1985. Migration, behavioural and ecological. In: *Migration: Mechanisms and Adaptive Significance.* M.A. Rankin (ed.). Contributions in Marine Science Suppl. Vol. 27, pp. 7-26.

Kennedy, J.S. & Marsh, D. 1974. Pheromone-regulated anemotaxis in flying moths. *Science*, 184: 999-1001.

Kennedy, J.S. & Moorehouse, J.E. 1969. Laboratory observations on locust responses to windborne grass odour. *Entomol. Exp. Appl.*, 12: 489-503.

Kerkut, G.A. & Gilbert, L.I. (ed.) 1985. *Comprehensive Insect Physiology and Pharmacology*. Vol. 9. *Behaviour*. Pergamon Press, Oxford.

Kesten, U. 1969. Zur Morphologic und Biologic von *Anatis ocellata* (L.) (Coleoptera: Coccinellidae). *Z. Angew. Entomol.*, 63: 412-445.

Kester, K.M. & Barbosa, P. 1991. Behavioural and biological constraints imposed by plants on insect parasitoids: implications for biological control. *Biol. Cont. Theor. Appl.*, 1: 94-106.

Kettlewell, H.B.D. 1955a. Recognition of appropriate backgrounds by the pale and black phase of Lepidoptera. *Nature*, 175: 934.

Kettlewell, H.B.D. 1955b. Selection experiments on industrial melanism in the Lepidoptera. *Heredity*, 9: 323-342.

Kettlewell, H.B.D. 1958. A survey of the frequencies of *Biston betularia* (L.) (Lepidoptera) and its melanic forms in Great Britain. *Heredity*, 12: 51-72.

Kettlewell, H.D.B. 1961. The phenomenon of industrial melanism in Lepidoptera. *Ann. Rev. Entomol.*, 6: 245-262.

Kettlewell, H.D.B. 1973. *The Evolution of Melanism. The Study of a Recurring Necessity with Special Reference to Industrial Melanism in Lepidoptera*. Clarendon, N.Y.

Key, K.H.L. & Day, M.F. 1954. A temperature-controlled physiological colour response in the grasshopper *Kosciuscola tristis* Sjost. (Orthoptera: Acrididae). *Aust. J. Zool.*, 2: 309-339.

Khan, Z.R., Ampong-Nyarko, K., Chiliswa, P., Hassanali, A., Kimani, S., Lwande, W., Overholt, W.A., Pickett, J.A., Smart, L.E., Wadhams, L.J. & Woodcock, C.M. 1997. Intercropping increases parasitism of pests. *Nature*, 388(6643): 631-632.

Khan, Z.R., Pickett, J.A., Wadhams, L. & Muyekho, F. 2001. Habitat management strategies for the control of cereal stem borers and striga in maize in Kenya. *Insect Sci. Appl.*, 21(4): 375-380.

Kim, M.-S. & Smith, D.P. 2001. The invertebrate odorant-binding protein LUSH is required for normal olfactory behaviour in *Drosophila*. *Chem. Senses*, 26: 195-199.

Kimble, G.A. 1961. *Hilgard and Marquis' Conditioning and Learning*. Appleton-Century-Crofts, N.Y.

Kimsey, L.S. 1980. The behaviour of male orchid bees (Apidae, Hymenoptera, Insecta) and the question of leks. *Anim. Behav.*, 28: 996-1004.

Kingsolver, J.G. & Daniel, T.L. 1995. Mechanism of food handling by fluid-feeding insects. In: *Regulatory Mechanisms in Insect Feeding*. R.F. Chapman & G. Boer, de (ed.). Chapman & Hall, N.Y., pp. 32-73.

Kirby, W. & Spence, W. 1815-1828. *Introduction to Entomology*. 4 vols. Longman, London.

Kirchner, W.H. 1993. Acoustical communication in honey bees. *Apidologie*, 24(3): 297-307.

Kline, D.L., Takken, W., Wood, J.F. & Carlson, D.A. 1990. Field studies on the butanone, carbon dioxide, honey extract, 1-octen-3-ol, L-lactic acid and phenols as attractants for mosquitoes. *Med. Vet. Entomol.*, 4: 383-391.

Kloot, W.G. van der & Williams, C.M. 1953. Cocoon construction by the cecropia silkworm. *Behaviour*, 5: 141-174.

Klopfer, P.H. & Hailman, J.P. 1967. *An Introduction to Animal Behaviour. Ethology's First Century.* Prentice-Hall, Englewood Cliffs, N.J.

Knuttel, H. & Fiedler, K. 2001. Host-plant-derived variation in ultraviolet wing patterns influences mate selection by male butterflies. *J. Exp. Biol.*, 204(14): 2447-2459.

Koeppe, J.K., Fuchs, M., Chen, T.T., Hunt, L.M., Kovalick, G.E. & Briers, T. 1985. The role of juvenile hormone in reproduction. In: *Comprehensive Insect Physiology, Biochemistry and Pharmacology. Endocrinology II.* Vol. 8. G.A. Kerkut & L.I. Gilbert (eds.). Pergamon Press, Oxford, pp. 165-203.

Koga, T. & Hayashi, K. 1993. Territorial behaviour of both sexes in the water strider *Metrocoris histrio* (Hemiptera: Gerridae) during the mating season. *J. Insect Behav.*, 6(1): 65-77.

Kogan, M. 1976. The role of chemical factors in insect/plant interrelationships. *Proc. XV Int. Congr. Entomol. Wash.*, 211-227.

Kokkinn, M.J. & William, W.D. 1989. An experimental study of phototactic responses of *Tanytarsus barbitarsis* Freeman (Diptera: Chironomidae). *J. Mar. Freshwat. Res.*, 40(6): 693-702.

Koltermann, R. 1971. Std-Periodik in der langzeiterinnerung an Duft-und Fabrsignale bei der Honigbiene. *Z. vergl. Physiol.*, 75: 49-68.

Konopka, R.J. & Benzer, S. 1971. Clock mutants of *Drosophila melanogaster. Proc. Natl. Acad. Sci., USA*, 68: 2112–2116.

Korb, J. & Heinze, J. 2004. Multilevel selection and social evolution of insect societies. *Naturwissenschaften*, 91(6): 291-304.

Korb, J. & Linsenmair, K.E. 2000. Thermoregulation of termite mounds: what role does ambient temperature and metabolism of the colony play? *Insectes Sociaux*, 47(4): 357-363.

Koul, O. 1982. Insect feeding deterrent in plants. *Indian Rev. Life Sci.*, 2: 97-125.

Koul, O. 1998. Behaviour and development modifying approaches from plant defence mechanisms. In: *Critical Issues in Insect Pest Management.* G.S. Dhaliwal & E.A. Heinrichs (eds.). Commonwealth Publ. New Delhi, pp. 221-249.

Koyama, J. 1989. Mating pheromones. In: *Fruit Flies: Their Biology, Natural Enemies and Control.* A.S. Robinson & G. Hooper (eds.). In: *World Crop Pests.* W. Helle (ed.). Vol. 3A. Elsevier Science Publ., Amsterdam, pp. 165-168.

Kozhanova, N.I. 2000. Hormonal control of insect gametogenesis. *Tsitologiya*, 42(2): 115-127.

Kozocik, J.J. 1981. Stepping patterns in cockroach *Periplaneta americana. J. Exp. Biol.*, 90: 357-360.

Kramer, M.G. & Marden, J.H. 1997. Almost airborne. *Nature*, 385: 403-404.

Kranz, B.B., Schwaz, M.P., Mound, L.A. & Crespi, B.J. 1999. Social biology and sex ratios of the eusocial gall-inducing thrips *Kladothrips hamiltoni. Ecol. Entomol.*, 24(4): 432-442.

Krebs, J.R. & Davies, N.B. 1997. *Behavioural Ecology: An Evolutionary Approach.* (4th ed.). Blackwell Scientific Publ., Oxford.

Krieger, M.J.B. & Ross, K.G. 2002. Identification of a major gene regulating complex social behaviour. *Science*, 295: 328-332.

Krishna, K. & Weesner, F.M. (eds.) 1969. *Biology of Termites.* Vol.1. Academic Press, N.Y.

Krishna, K. & Weesner, F.M. (eds.) 1970. *Biology of Termites.* Vol.2. Academic Press, N.Y.

Kudo, S., Ishibashi, E. & Makino, S. 1995. Reproductive and subsocial behaviour in the ovoviviparous leaf beetle *Gonioctena sibirica* (Coleoptera: Chrysomelidae). *Ecol. Entomol.*, 20: 367-373.

Kühn, A. 1919. *Die Orientierung der Tiere im Raum.* Fischer, Jena.

Kumar, B. & Agarwal, M.L. 2005. Efficacy of different attractant and bait combinations against *Bactrocera cucurbitae* (Coquillett). *Indian J. Plant Prot.*, 33(2): 194-196.

Kumar, V. & Agarwal, M.L. 1998. Efficacies of different bait combinations against oriental fruit fly, *Bactrocera dorsalis* (Hendel). *J. Res. (BAU)*, 10 (1): 83-86.

Kumashiro, M. & Sakai, M. 2001. Reproductive behaviour in the male cricket *Gryllus bimaculatus* DeGeer. II. Neural control of the genitalia. *J. Exp. Biol.*, 204(6): 1139-1152.

Kunert, G., Otto, S., Röse, U.S.R., Gershenzon, J. & Weisser, W.W. 2005. Alarm pheromone mediates production of winged dispersal morphs in aphids. *Ecology Letters*, 8: 596-603.

Kupier, J.W. 1962. The optics of the compound eye. *Symp. Soc. Exp. Biol.*, 16: 58-71.

Kutsch, W. 1974. The development of flight pattern in locusts. In: *Experimental Analysis of Insect Behaviour*. L. Barton Browne (ed.). Springer-Verlag, Berlin, pp.149-158.

Kutter, H. 1950. Über eine neue, extrem parasitische Ameise 1. Mitteilung. *Mitt. Schweiz. Entomol. Ges.*, 23: 81-94.

Kuwahara, Y. & Casida, J.E. 1973. Quantitative analysis of the sex pheromone of several phycitid moths by electron-capture gas chromatography. *Agril. Biol. Chem.*, 37: 681-684.

Kydonieus, A.F. & Beroza, M. 1986. *Insect Suppression with Controlled Release Pheromone System*. Vols. I & II. CRC Press, Boca Raton.

La Greca, M. 1947. Su una particolare maniera di deambulazione di un Acridid: *Tropidopola cylindrica* (Marsch.). *Boll. Zool. Torino*, 14: 83-104.

Ladd, T.L. & Klein, M.G. 1986. Japanese beetle (Coleoptera: Scarabaeidae) response to colour traps with phenethyl propionate + eugenol + geraniol (3: 7: 3) and japonilure. *J. Econ. Entomol.*, 79: 84-86.

Lall, A.B. & Worthy, K.M. 2000. Action spectra of the female's response in the firefly *Photinus pyralis* (Coleoptera: Lampyridae): evidence for an achromatic detection of the bioluminescent optical signal. *J. Insect physiol.*, 46(6): 965-968.

Laloi, D., Gallois, M., Roger, B. & Pham-Delegue, M.H. 2001. Changes with age in olfactory conditioning performance of worker honey bees (*Apis mellifera*). *Apidologie*, 32(3): 231-242.

Laloi, D., Sandoz, J.C., Picard-Nizou, A.L., Marchesi, A., Pouvreau, A., Tasei, J.N., Poppy, G. & Pham-Delegue, M.H. 1999. Olfactory conditioning of the proboscis extension in bumble bees. *Entomol. Exp. Appl.*, 90(2): 123-129.

Lambin, M. 1987. A method for identifying the nearby spatial cues used by animals during transverse orientation. *Behav. Process*, 14(1): 1-10.

Lamunyon, C.W. 1997. Increased fecundity, as a function of multiple mating, in an arctiid moth, *Utetheisa ornatrix*. *Ecol. Entomol.*, 22(1): 69-73.

Lance, D.R. 1983. Host-seeking behaviour of the gypsy moth, the influence of polyphagy and highly apparent host plants. In: *Herbivorous Insects, Host Seeking Behaviour and Mechanism*. S. Ahmad (ed.). Academic Press, N.Y., pp. 201-224.

Land, M.F. 1985. The eye: optics. In: *Comprehensive Insect Physiology, Biochemistry and Pharmacology. Nervous System: Sensory*. Vol. 6. G.A. Kerkut & L.I. Gilbert (eds.). Pergamon Press, Oxford, pp. 225-275.

Landolt, P.J., Jeanne, R.L. & Reed, H.C. 1998. Chemical communication in social wasps. In: *Pheromone Communication in Social Insects: Ants, Wasps, Bees, and Termites*. R.K. Meer vander, M. Breed, K.E. Espelie & M.L. Winston (eds.). Westview Press, Boulder, Colorado, pp. 216-235.

Law, J.M. & Regnier, F.E. 1971. Pheromones. *Ann. Rev. Biochem.*, 40: 533-548.

Leal, W.S. 2003. Proteins that make sense. In: *Insect Pheromone Biochemistry and Molecular Biology*. G.J. Blomquist & R.G. Vogt (eds.). Elsevier Publ. Co., Amsterdam, pp. 447-476.

Lee, J.C., Bernays, E.A. & Wrubel, R.P. 1987. Does learning play a role in host location and selection by grasshoppers? In: *Insect-Plants. Proc. VI Int. Symp. on Insect Plant Relationships* (1986). V. Labeyrie, G. Fabres & D. Lachaise (eds.). Dr W. Junk, Netherland, pp. 125-127.

Lees, A.D. 1959. The role of photoperiod and temperature in the determination of parthenogenetic and sexual forms in the aphid *Megoura viciae* Buckton-I. The influence of these factors on apterous virginoparae and their progeny. *J. Insect Physiol.*, 3: 92-117.

Lees, A.D. 1960. The role of photoperiod and temperature in the determination of parthenogenetic and sexual forms in the aphid *Megoura viciae* Buckton-II. The operation of 'interval timer' in young clones. *J. Insect Physiol.*, 4: 154-175.

Lees, A.D. 1964. The location of the photoperiodic receptors in the aphid *Megoua viciae*. *J. Exp. Biol.*, 41: 119-133.

Lehane, M.J. 2005, *The Biology of Blood-Sucking in Insects* (2nd ed.). Cambridge Univ. Press, Cambridge.

Lehmann, H. 1923. Biologische Beobachtungen an *Notonecta gluca*. *Zool. Jahrb. Syst.*, 46: 121-156.

Lehner, P.N. 1979. *Handbook of Ethological Methods*. Garland STPM Press, N.Y.

Lehrman, D.S. 1953. A critique of Konrad Lorenz's theory of instinctive behaviour. *Ouart. Rev. Biol.*, 28: 337-363.

Lenoir, A., D'Ettorre, P. & Errard, C. 2001. Chemical ecology and social parasitism in ants. *Ann. Rev. Entomol.*, 46: 573-599.

Lenteren, J.C. van, 1981. Host discrimination by parasitoids. In: *Semiochemicals: Their Role in Pest Control*. D.A. Nordlund, R.L. Jones, & W. J. Lewis (eds.). John Wiley & Sons, N.Y., pp. 153-173.

Leoncini, I., Conte, Y. Le, Costagliola, G., Plettner, E., Toth, A.L., Wang, M., Huang, Z., Bécard, J.-M., Crauser, D., Slessor, K.N. & Robinson, G.E. 2004. Regulation of behavioural maturation by a primer pheromone produced by adult worker honey bees. *Proc. Natl. Acad. Sci., USA*, 101(50): 17559-17564.

Less, D.R. 1971. The distribution of melanism in the pale brindled beauty moth, *Phigalia pedaria*, in Great Britain. In: *Ecological Genetics and Evolution*. E.R. Creed (ed.). Blackwell, Oxford, pp. 152-174.

Lessman, C.A. & Herman, W.S. 1983 Seasonal variation in haemolymph juvenile hormone of adult monarchs *Danaus plexippus* (Lepidoptera). *Can. J. Zool.*, 61: 88-94.

Lessman, C.A., Rollins, L. & Herman, W.S. 1982. Effects of juvenile hormones I, II, and III on reproductive tract growth in male and female monarch butterflies. *Comp. Biochem. Physiol.*, 71A: 141-144.

Leuthold, R.H. 1968. A tibial gland scent-trail and trail-lying behaviour in the ant *Crematogaster ashmeadi* Mayr. *Psyche*, 75: 233-248.

Levinson, H.Z., Levinson, A.R. & Maschwitz, U. 1974. Action and composition of the alarm pheromone of the bed bug, *Cimex lectularius* L. *Naturwissenschaften*, 61: 684-685.

Lewis, S.M. 2004. Multiple mating and repeated copulations: effects on male reproductive success in red flour beetles. *Anim. Behav.*, 67(4): 799-804.

Lewis, S.M. & Jutkiewicz, E. 1998. Sperm precedence and sperm storage in multiply mated red flour beetles. *Behav. Ecol. Sociobiol.*, 43(6): 365-369.

Lewis, T. (ed.) 1984. *Insect Communication*. Academic Press, London.

Lewis, W.J., Jones, R.L. & Sparks, A.N. 1972. A host-seeking stimulant for the egg parasitoid *Trichogramma evenescens*: its source and a demonstration of its laboratory and field activity. *Ann. Entomol. Soc. Am.*, 65: 1087-1089.

Lewis, W.J. & Martin, W.R. 1990. Semiochemicals for use with parasitoids: status and future. *J. Chem. Ecol.*, 16: 3067-3090.

Lewis, W.J., Nordlund, D.A., Gross, H.R. Jr., Jones, R.L. & Jones, S.L. 1977. Kairomones and their use for the management of entomophagous insects. V. Moth scales as a stimulant for predation of *Heliothis zea* eggs by *Chrysoperla carnea* Stephens larvae. *J. Chem. Ecol.*, 3: 483-487.

Li, K. & Zheng, Z.M. 1999. Sound characteristics analysis and the identification of six species of *Loxoblemmus* (Orthoptera: Grylloidea). *Entomotaxonomia*, 21(1): 17-21.

Li, Y.H. & Liu, S.S. 2004. Learning in phytophagous insects. *Acta Entomol. Sinica*, 47(1): 106-116.

Lian, Z.M. & Li, K. 2002. The comparative studies on the common sound types of crickets (Orthoptera: Grylloidea). *Entomotaxonomia*, 24(1): 45-51.

Liimatainen, J., Hoikkala, A. & Shelly, T. 1997. Courtship in *Ceratitis capitata* (Diptera: Tephritidae): comparison of wild and mass-reared males. *Ann. Entomol. Soc. Am.*, 90 (6): 836-843.

Lin, N. 1963. Territorial behaviour in the cicada killer wasp *Sphecius speciosus* (Drury) (Hymenoptera: Sphecidae). I. *Behaviour*, 20: 115-133.

Lin, N. & Michener, C.D. 1972. Evolution of sociality in insects. *Quart. Rev. Biol.*, 47: 131-159.

Lindauer, M. 1960. Time-compensated sun orientation in bees. *Cold Spring Harb. Symp. Quant. Biol.*, 25: 371-377.

Lindauer, M. 1970. Lernen und Gedächtnis, Versuche an der Honigbiene. *Naturwissenschaften*, 57: 463-467.

Lindauer, M. 1971. *Communication Among Social Bees*. (Rev ed.). Harvard Univ. Press, Cambridge, Massachusetts.

Lindauer, M. 1976. Foraging and homing flight of the honey bee: some general problems of orientation. In: *Insect Flight*. R.C. Rainey (ed.). *Roy. Entomol. Soc. (Lond.) Symp.* 7, pp. 199-216.

Linaduer, M. & Kerr, W.E. 1960. Communication between workers of stingless bees. *Bee World*, 41: 29-41 & 65-71.

Lindauer, M. & Martin, H. 1968. Die Schwereorientierung der Bienen unter dem Einfluss des Erdmagnetfeldes. *Z. vergl. Physiol.*, 60: 219-243.

Linn, C. Jr., Dambroski, H.R., Feder, J.L., Berlocher, S.H., Nojima, S. & Roelofs, W.L. 2004. Postzygotic isolating factor in sympatric speciation in *Rhagoletis* flies: reduced response of hybrids to parental host-fruit odours. *Proc. Natl. Acad. Sci., USA*, 101(51): 17753-17758.

Linn, C. Jr., Feder, J.L., Nojima, S., Dambroski, H.R., Stewart H., Berlocher, S.H. & Roelofs, W. 2003. Fruit odour discrimination and sympatric host race formation in *Rhagoletis*. *Proc. Natl. Acad. Sci., USA*, 100(20): 11490-11493.

Linsley, E.G. 1958. The ecology of solitary bees. *Hilgardia*, 27(19): 543-599.

Linsley, E.G., Eisner, T. & Klots, A.B. 1961. Mimetic assemblage of sibling species of lycid beetles. *Evolution*, 15: 15-29.

Little, A.E.F., Murakami, T., Mueller, U.G. & Currie, C.R. 2003. The infrabuccal pellet piles of fungus-growing ants. *Naturwissenschaften*, 90(12): 558-562.

Lloyd, J.E. 1965. Aggressive mimicry in *Photuris* firefly femmes fatalis. *Science*, 149: 653-654.

Lloyd, J.E. 1966. Studies on the flash communication system in *Photinus* fireflies. *Univ. Michigan Mus. Zool. Misc. Pub. No.* 130, 95 pp.

Lloyd, J.E. 1969. Flashes of *Photuris* fireflies: their value and use in recognizing species. *Florida Entomol.*, 52: 29-35.

Lloyd, J.E. 1971. Bioluminescent communication in insects. *Ann. Rev. Entomol.*, 16: 97-122.

Lloyd, J.E. 1977. Bioluminescence and communication. In: *How Animals Communicate*. T.A. Sebeok (ed.). Indiana Univ. Press, Bloomington, pp. 164-183.

Lloyd, J.E. 1978. Insect bioluminescence. In: *Bioluminescence in Action*. P.J. Herring (ed.). Academic Press, London, pp. 241-272.

Lloyd, J.E. 1979. Sexual selection in luminescent beetles. In: *Sexual Selection and Reproductive Competition in Insects*. M.S. Blum & N.A. Blum (eds.). Academic Press, N.Y., pp. 293-342.

Lloyd, J.E. 1981. Mimicry in the sexual signals of fireflies. *Sci. Amer.*, 245(1): 138-145.

Lloyd, J.E. 1983. Bioluminescence and communication in insects. *Ann. Rev. Entomol.*, 28: 131-160.

Lloyd, J.E. 1990. Firefly semiosystematics and predation: a history. *Florida Entomol.*, 73(1): 51-66.

Lloyd, J.E. 1991. Lampyridae (Cantharoidea). In: *Immature Insects*. Vol.2. F.W. Stehr (ed.). Kendall-Hunt Publ. Co., Dubuque, IA., pp. 427-428.

Lloyd, J.E. 1997. Firefly mating ecology, selection and evolution. In: *The Evolution of Mating Systems in Insects and Arachnids*. Vol. 1. J.C. Choe, & B.J. Crespi (eds.). Cambridge Univ. Press, Cambridge, pp. 184-191.

Lo, N., Tokuda, G., Watanabe, H., Rose, H., Slaytor, M., Maekawa, K., Bandi, C. & Noda, H. 2000. Evidence from multiple gene sequences indicates that termites evolved from wood-feeding cockroaches. *Curr. Biol.*, 10(13): 801-804.

Loeb, J. 1918. *Forced Movements, Tropism and Animal Conduct*. J.B. Lippincott Co., Philadelphia.

Loher, W. 1972. Circadian control of stridulation in the cricket *Teleogryllus commodus* Walker. *J. Comp. Physiol.*, 79: 173-190.

Loher, W. 1974. Circadian control of spermatophore formation in the cricket *Teleogryllus commodus* Walker. *J. Insect Physiol.*, 20: 1155-1172.

Longo, N. 1964. Probability learning and habit reversal in the cockroach. *Amer. J. Psychol.*, 77: 29-41.

Lorenz, K. 1937. Über die bildung des Instinkbegriffs. *Naturwissenschaften*, 25: 289-300.

Lorenz, K. 1939. Vergleichende Verhaltens-forschung. *Zool. Anz. Suppl. Bd.*, 12: 69-102.

Lorenz, K. 1950. The comparative method of studying innate behavioural patterns. *Symp. Soc. Exp. Biol.*, 4: 221-268.

Lorenz, K. 1952. *King Solomon's Ring: New Light on Animal Ways*. Thomas Y. Crowell Co., N.Y.

Lorenz, K. 1965. *Evolution and Modification of Behaviour*. Univ. Chicago Press, Chicago.

Lorenz, K. 1970, 1971. *Studies on Animal and Human Behaviour*. (Vol. 1 & 2). Harvard Univ. Press, Cambridge, Massachusetts.

Lubbock, J. 1883. *Fourmis, Abeilles et Guêpes*. Paris.

Lund, E.J. 1911. On the structure, physiology and use of photogenic organs, with special reference to the Lampyridae. *J. Exp. Zool.*, 11: 415-467.

Lüscher, M. 1960. Hormonal control of caste determination in termites. *Ann. N.Y. Acad. Sci.*, 89: 549-563.

Lüscher, M. 1961. Air-conditioned termite nests. *Sci. Amer.*, 205(1): 138-145.

MacArthur, R.H. 1962. Some generalized theorems of natural selection. *Proc. Natul. Acad. Sci.*, 48: 1893-1897.

Magnus, D. 1958. Experimentalle Untersuchunger zur Bionomie und Ethologie des Kaisermantels *Argynnis paphia* L. (Lep. Nymphalidae). *Z. Tierpsychol.*, 15: 397-426.

Mairan, J.J. de, 1729. Observation Botanique. Histoire de L'Académie Royale des Sciences de Paris, pp. 35-36.

Majerus, M.E.N. 1998. *Melanism: Evolution in Action.* Oxford Univ. Press, Oxford.

Majerus, M.E.N., Brunton, C.F.A. & Stalker. J. 2000. A bird's eye view of the peppered moth. *J. Env. Biol.*, 13(2): 155-159.

Maleszka, R. & Helliwell, P. 2001. Effect of juvenile hormone on short-term olfactory memory in young honey bees (*Apis mellifera*). *Horm. Behav.*, 40: 403-408.

Mamood, A.N. & Waller, G.D. 1990. Recovery of learning responses by honey bees following a sublethal exposure to permethrin. *Physiol. Entomol.*, 15(1): 55-60.

Manning, A. 1966. Sexual behaviour. In: *Insect Behaviour.* P.T. Haskell (ed.). *Symp. Roy. Entomol. Soc. Lond.*, 3: 59-68.

Manning, A. & Dawkins, M.S. 1992. *An Introduction to Animal Behaviour.* (4th ed.). Cambridge Univ. Press, Cambridge.

Manton, S.M. 1972. The evolution of arthropod locomotory mechanisms. Part 10. Locomotory habits, morphology and evolution of hexapod classes. *Zool. J. Linn. Soc.*, 51: 203-400.

Mappes, J. & Kaitala, A. 1994. Experiments with *Elasmucha grisea* L. (Heteroptera: Acanthosomatidae): does a female parent bug lay as many eggs as she can defend? *Behav. Ecol.*, 5(3): 314-317.

Marden, J.H., O'-Donnell, B.C., Thomas, M.A. & Bye, J.Y. 2000. Surface-skimming stoneflies and mayflies: the taxonomic and mechanical diversity of two-dimensional aerodynamic locomotion. *Physiol. Biochem. Zool.*, 73(6): 751-764.

Markl, H. 1962. Borstenfelder an den Gelenken als Schwaresinnesorgane bei Ameisen und anderen Hymenopteren. *Z. vergl. Physiol.*, 45: 475-569.

Markl, H. 1974. Insect behaviour: functions and mechanisms. In: *The Physiology of Insecta.* Vol. 3. M. Rockstein (ed.). Academic Press, N.Y., pp. 3-148.

Markl, H. & Lindauer, M. 1965. Physiology of insect behaviour. In: *The Physiology of Insecta.* Vol. 2. M. Rockstein (ed.). Academic Press, N.Y., pp.3-122.

Markow, T.A. & Smith, L.D. 1979. Genetics of phototactic behaviour in *Drosophila ananassae*, a member of the *melanogaster* species group. *Behav. Genet.*, 9: 61-67.

Marler, P. & Hamilton, W.J. 1966. *Mechanisms of Animal Behaviour.* John Wiley & Sons, N.Y.

Martin, U., Martin, H. & Lindauer, M. 1978. Transplantation of a time-signal in honey bees. *J. Comp. Physiol.*, 124: 193-201.

Martins, E.P. (ed.) 1996. *Phylogenies and the Comparative Method in Animal Behaviour.* Oxford Univ. Press, Oxford.

Maschwitz, U. 1964. Alarm substances and alarm behaviour in social Hymenoptera. *Nature*, 204: 324-327.

Mast, S.O. 1911. *Light and the Behaviour of Organisms.* John Wiley & Sons, N.Y.

Mast, S.O. 1938. Factors involved in the orientation of lower organisms to light. *Biol. Rev.*, 13: 186-224.

Masters, A.R., Malcolm, S.B. & Brower, L.P. 1988. Monarch butterfly (*Danaus plexippus*) thermoregulatory behaviour and adaptations for overwintering in Mexico. *Ecology*, 69: 458-467.

Mather, M.H. & Roitberg, B.D. 1987. A sheep in wolf's clothing: tephritid flies mimic spider predators. *Science*, 236: 308-309.

Mathur, R. 1994. *Animal Behaviour*. Rastogi & Co., Meerut.

Matsumoto, Y. 1970. Volatile organic sulphur compounds as insect attractants with special reference to host selection. In: *Control of Insect Behaviour by Natural Products*. D.L. Wood, R.M. Silverstein & M. Nakajima (eds.). Academic Press, N.Y., pp. 133-160.

Matthews, R.W. & Matthews, J.R. 1978. *Insect Behaviour*. John Wiley & Sons, N.Y.

Mayer, M.S. & McLaughlin, J.R. 1991. *Handbook of Insect Pheromones and Sex Attractants*. CRC Press, Boca Raton.

Mayer, R.J. & Candy, D.J. 1969. Control of haemolymph concentration during locust flight. An adipokinetic hormone from the corpora cardiaca. *J. Insect Physiol.*, 15: 598-611.

Mayr, E. 1942. *Systematics and the Origin of Species*. Columbia Univ. Press, N.Y.

Mayr, E. 1958. Behaviour and systematics. In: *Behaviour and Evolution*. A. Roe & G.G. Simpson (eds.). Yale Univ. Press, New Haven, pp. 341-362.

Mayr, E. 1963. *Animal Species and Evolution*. Belknap Press, Harvard Univ. Press, Cambridge, Massachusetts.

Mayr, E., Linsley, E.G. & Usinger, R.L. 1953. *Methods and Principles of Systematic Zoology*. McGraw-Hill Book. Co. Inc., N.Y.

Mazokhin-Porschnyakov, G.A. 1969. *Insect Vision*. Plenum Press, N.Y.

Mbata, G.N., Shu, S. & Ramaswamy, S.B. 1997. Rhythmicity of mating and oviposition in *Callosobruchus subinnotatus* (Pic.) (Coleoptera: Bruchidae). *J. Insect Behav.*, 10(3): 409-423.

McAllister, M.K. & Roitberg, B.D. 1987. Adaptive suicidal behaviour in pea aphids. *Nature*, 328(6133): 797-799.

McCaffery, A.R. & Page, W.W. 1978. Factors influencing the production of long-winged *Zonocerus variegatus*. *J. Insect Physiol.*, 24: 465-472.

McCall, P.J., Turlings, T.C.J., Loughrin, J., Proveaux, A.T. & Tumlinson, J.H. 1994. Herbivore-induced volatiles emissions from cotton (*Gossypium hirsutum* L.) seedlings. *J. Chem. Ecol.*, 20: 3039-3050.

McClelland, G.H.A. & Green, C.A. 1970. Subtle periodicity of pupation in rapidly developing mosquitoes. *Bull. Wld. Hlth. Org.*, 42: 951-955.

McConnell, E. & Richards, A.G. 1955. How fast can a cockroach run? *Bull. Brooklyn Entomol. Soc.*, 50: 36-43.

McDermott, F.A. 1917. Observations on the light emission of American Lampyridae: the photogenic function as a mating adaptation. *Fifth Paper Canadian Entomol.*, 49: 53-61.

McDermott, F.A. 1964. The taxonomy of the Lampyridae. *Trans. Amer. Entomol. Soc.*, 90: 1-72.

McElroy, W.D. 1951. Properties of the reaction utilizing adenosine triphosphate for bioluminescence. *J. Biol. Chem.*, 191: 547-557.

McElroy, W.D. & DeLuca, M. 1973. Chemical and enzymatic mechanisms of firefly luminescence. In: *Chemiluminescence and Bioluminescence*. M.J. Cormier, D.M. Hercules & J. Lee (eds.). Plenum Press, N.Y., pp. 285-309.

McElroy, W.D. & DeLuca, M. 1985. Biochemistry of insect luminescence. In: *Comprehensive Insect*

Physiology, Biochemistry and Pharmacology. Regulation: Digestion, Nutrition, Excretion Vol. 4. G.A. Kerkut & L.I. Gilbert (eds.). Pergamon Press, Oxford, pp. 553-563.

McElroy, W.D. & Hastings, J.W. 1955. Biochemistry of firefly luminescence. In: *The Luminescence of Biological System*. F.H. Johnson (ed.). Amer. Assoc. Adv. Sci. Press, Washington, DC, pp. 161-198.

McElroy, W.D. & Seliger, H.H. 1962. Biological luminescence. *Sci. Amer.*, 207: 2-14.

McElroy, W.D., Seliger, H.H. & DeLuca, M. 1974. Insect bioluminescence. In: *The Physiology of Insecta*. Vol. 2 (2nd ed.). M. Rockstein (ed.). Academic Press, N.Y., pp. 411-460.

McFarland, D. 1985. *Animal Behaviour*. Pitman Publ. Ltd., London.

McGovern, T.P., Fiori, B., Beroza, M. & Ingangi, J.C. 1970. Propyl 1,4-benzoxidan-2-carboxylate, a potent attractant for the European chafer. *J. Econ. Entomol.*, 63: 168.

McIver, S.B. 1975. Cuticular mechanorecptors of arthropods. *Ann. Rev. Entomol.*, 20: 381-397.

McIver, S.B. 1985. Mechanoreception. In: *Comprehensive Insect Physiology, Biochemistry and Pharmacology. Nervous System: Sensory*. Vol. 6. G.A. Kerkut & L.I. Gilbert (eds.). Pergamon Press, Oxford, pp. 71-132.

McMichael, B., Foote, B. & Bowker, B. 1990. Biology of *Rivellia melliginis* (Diptera: Platystomatidae): a consumer of the nitrogen-fixing root nodules of black locust (Leguminosae). *Ann. Entomol. Soc. Am.*, 83: 241-243.

McMillan, V.E. 1991. Variable mate-guarding behaviour in the dragonfly *Plathemis lydia* (Odonata: Libellulidae). *Anim. Behav.*, 41: 979-987.

McMullen, L.H. & Atkins, M.D. 1962. On the flight and host selection of the Douglas-fir beetle *Dendroctonus pseudotsugae* Hopk. (Coleoptera: Soclytidae). *Can. Entomol.*, 94: 1309-1325.

McNeil, J.N., Laforge, M., Bédard, C. & Cusson, M. 1996. Juvenile hormone production and sexual maturation in true armyworm, *Pseudaletia unipuncta* (Haw.) (Lepidoptera: Noctuidae): a comparison of migratory and non-migratory populations. *Arch. Insect Biochem. Physiol.*, 32(3/4): 575-584.

McPheron, B.A. & Han, H.-Y. 1997. Phylogenetic analysis of North American *Rhagoletis* (Diptera: Tephritidae) and related genera using mitochondrial DNA sequence data. *Mol. Phylogenet. Evol.*, 7: 1-16.

Meats, A. & Hartland, C.L. 1999. Upwind anemotaxis in response to cue-lure by the Queensland fruit fly, *Bactrocera tryoni*. *Physiol. Entomol.*, 24(1): 90-97.

Meder, E. 1958. Uber die Einberechnung der Sonnenwanderung bei der Orientierung der Honigbiene. *Z. vergl. Physiol.*, 40: 610-641.

Medugorac, I. & Lindauer, M. 1967. Das Zeitgedächtnis der Bienen unter dem Einfluss von Narkose und von sozialen Zeitgebern. *Z. vergl. Physiol.*, 55: 450-474.

Meer, R.K. vander, Breed, M., Espelie, K.E. & Winston, M.L. 1998. *Pheromone Communication in Social Insects: Ants, Wasps, Bees, and Termites*. Westview Press, Boulder, Colorado.

Meer, R.K. vander, Lofgren, C.S. & Alvarez, F.M. 1989. The orientation inducer pheromone of the fire ant, *Solenospis invicta*. *Physiol. Entomol.*, 15: 483-488.

Mellor, H.E. & Hamilton, J.G.C. 2003. Navigation of *Lutzomyia longipalpis* (Diptera: Psychodidae) under dusk or starlight conditions. *Bull. Entomol. Res.*, 93(4): 315-322.

Menzel, R. 1993. Associative learning in honey bees. *Apidologie*, 24(3): 157-168.

Menzel, R., Brandt, R., Gumbert, A., Komischke, B. & Kunze, J. 2000. Two spatial memories for honey bee navigation. *Proc. Roy. Soc. Lond. Ser. B, Biol. Sci.*, 267(1447): 961-968.

Menzel, R., Geiger, K., Joerges, J., Müller, U. & Chittka, L. 1998. Bees travel novel homeward routes by integrating separately acquired vector memories. *Anim. Behav.*, 55: 139-152.

Metcalf, R.L. 1990. Chemical ecology of dacine fruit flies (Diptera: Tephritidae). *Ann. Entomol. Soc. Am.*, 83 (6): 1017-1030.

Metcalf, R.L., Ferguson, J.E., Lampman, R. & Andersen, J.E. 1987. Dry cucurbitacin-containing baits for controlling diabroticite beetles (Coleoptera: Chrysomelidae). *J. Econ. Entomol.*, 80: 870-875.

Metcalf, R.L. & Metcalf, E.R. 1992. *Plant Kairomones in Insect Ecology and Control*. Chapman & Hall, N.Y.

Michelsen, A. & Larsen, O.N. 1985. Hearing and sound. In: *Comprehensive Insect Physiology, Biochemistry and Pharmacology. Nervous System: Sensory*. Vol. 6. G.A. Kerkut & L.I. Gilbert (eds.). Pergamon Press, Oxford, pp. 495-556.

Michelsen, A. & Nocke, H. 1974. Biophysical aspects of sound communication in insects. *Adv. Insect Physiol.*, 10: 247-296.

Michener, C.D. 1953. Life history studies in insect systematics. *Syst. Zool.*, 2: (3): 112-118.

Michener, C.D. 1969. Comparative social behaviour of bees. *Ann. Rev. Entomol.*, 14: 299-342.

Michener, C.D. 1974. *The Social Behaviour of the Bees: A Comparative Study*. Belknap Press, Harvard Univ. Press, Cambridge, Massachusetts.

Michener, C.D. & Lange, R.B. 1958. Distinctive type of primitive social behaviour among bees. *Science*, 127: 1046-1047.

Michener, C.D. & Michener, M.H. 1951. *American Social Insects*. Van Nostrand, N.Y.

Miklas, N., Lasnier, T. & Renou, M. 2003. Male bugs modulate pheromone emission in response to vibratory signals from conspecifics. *J. Chem. Ecol.*, 29(3): 561-574.

Miller, P.L. 1974a. The neural basis of behaviour. In: *Insect Neurobiology*. J.E. Treherne (ed.). American Elsevier, N.Y., pp. 359-430.

Miller, P.L. 1974b. Rhythmic activities and the insect nervous system. In: *Experimental Analysis of Insect Behaviour*. L. Barton Browne (ed.). Springer-Verlag, Berlin, pp.114-138.

Milne, A. 1960. Biology and ecology of the garden chafer *Phyllopertha horticola* (L.) VII. The flight season: male and female behaviour and concluding discussion. *Bull. Entomol. Res.*, 51: 353-378.

Min, K.M., Jones, N., Borst, D.W. & Rankin, M.A. 2004. Increased juvenile hormone levels after long-duration flight in the grasshopper, *Melanoplus sanguinipes*. *J. Insect Physiol.*, 50(6): 531-537.

Minis, D.H. 1965. Parallel peculiarities in the entrainment of a circadian rhythm and photoperiodic induction in the pink bollworm (*Pectinophora gossypiella*). In: *Circadian Clocks*. J. Aschoff (ed.). North-Holland, Amsterdam, pp. 333-343.

Minis, D.H. & Pittendrigh, C.S. 1968. Circadian oscillation controlling hatching: its ontogeny during embryogenesis of a moth. *Science*, 159: 534-536.

Mitchell, E.R. 1981. *Management of Insect Pests with Semiochemicals: Concepts and Practice*. Plenum Press, N.Y.

Mitlin, N. & Hedin, P.A. 1974. Biosynthesis of grandlure, the pheromone of the boll weevil, *Anthonomus grandis*, from acetate, mevalonate, and glucose. *J. Insect Physiol.*, 20: 1825-1831.

Mittelstaedt, H. 1950. Physiologie des Gleichgewichtssinnes bei fliegenden Libellen. *Z. vergl. Physiol.*, 32: 422-463.

Mittelstaedt, H. 1954. Regelung und Steuerung bei der Orientierung der Lebewesen. *Regelungstechnik*, 2: 226-232.

Mittelstaedt, H. 1962. Control systems of orientation in insects. *Ann. Rev. Entomol.*, 7: 177-198.

Mitter, C. & Brooks, D.R. 1983. Phylogenetic aspects of coevolution. In: *Coevolution*. D.J. Futuyma & M. Stalkin (eds.). Sinauer Assoc. Inc. Publ. Sunderland, Massachusetts, pp. 65-98.

Miyata, A. 1983. *Handbook of the Moth Ecology- Moth as an Indicator of the Environment*. (In Japanese). Showado Printing Publ. Div., Nagasaki, Japan.

Monteith, L.G. 1960. Influence of plants other than the food plants of their hosts on host-finding by tachinid parasites. *Can. Entomol.*, 92: 641-652.

Montgomery, M.E. & Nault, L.R. 1977. Comparative response of aphids to the alarm pheromone, (*E*)-beta-farnesene. *Entomol. Exp. Appl.*, 22: 236-242.

Moore, B.P. 1968. Studies on the chemical composition and function of the cephalic gland secretion in Australian termites. *J. Insect Physiol.*, 14: 33-39.

Moore, D., Angel, J.E., Cheeseman, I.M., Fahrbach, S.E. & Robinson, G.E. 1998. Timekeeping in the honey bee colony: integration of circadian rhythms and division of labour. *Behav. Ecol. Sociobiol.*, 43(3): 147-160.

Moore, T.E. 1993. Acoustic signals and speciation in cicadas (Insecta: Homoptera: Cicadidae). In: *Evolutionary Patterns and Processes, Linnaean Society Symposium No 14*. D.R. Lees & D. Edwards (eds.). Academic Press, London, pp. 269-284.

Moraes, C.M. de, Lewis, W.J., Pare, P.W., Alborn, H.T. & Tumlinson, J.H. 1998. Herbivore-infested plants selectively attract parasitoids. *Nature*, 393(6685): 570-573.

Moraes, C.M. de, & Mescher, M.C. 2004. Biochemical crypsis in the avoidance of natural enemies by an insect herbivore. *Proc. Natl. Acad. Sci., USA*, 101(24): 8993-8997.

Morris, R.F., Miller, C.A., Greenbank, D.D. & Mott, D.G. 1956. The population dynamics of the spruce budworm in Canada. *Proc. 10th Intern. Cong. Entomol., Montreal*, 4: 137-149.

Mueller, U.G. & Gerardo, N. 2002. Fungus-farming insects: multiple origins and diverse evolutionary histories. *Proc. Natl. Acad. Sci., USA*, 99(24): 15247-15249.

Mueller, U.G., Rehner, S.A. & Schultz, T.R. 1998. The evolution of agriculture in ants. *Science*, 281: 2034-2038.

Mueller, U.G., Schultz, T.R., Currie, C.R., Adams, R.M.M. & Malloch, D. 2001. The origin of the attine ant-fungus symbiosis. *Quart. Rev. Biol.*, 76: 169-197.

Muller, D., Gerber, B., Hellstern, F., Hammer, M. & Menzel, R. 2000. Sensory preconditioning in honey bees. *J. Exp. Biol.*, 203(8): 1351-1364.

Muller, F. 1873. Beiträge zur kenntniss der termiten. *Jenaische Z. Nat.*, pp. 75-87.

Muller, F. 1874. Recent trends on termites and honey bees. *Nat. Febr.*, 19.B.9.

Müller, H.J. 1955. Die Saisonformenbildung von *Araschnia levana*, ein photoperiodisch gesteuerter Diapause-effekt. *Naturwissenschaften*, 42: 134-135.

Müller, J. 1826. *Zur vergleichenden Physiologie Gesichtssinnes des Menschen und Tiere*. Cnobloch, Leipzig.

Müller, J.F.T. 1879. *Ituna* and *Thyridia*: a remarkable case of mimicry in butterflies. *Trans. Entomol. Soc. (Proc.)*, 1879: 20-28.

Mumcuoglu, Y., Galun, R. & Ikan, R. 1986. The aggregation response of human body louse (*Pediculus humanus*) (Insecta: Anoplura) to its excretory products. *Insect Sci. Appl.*, 7(5): 629-632.

Munyiri, F.N. & Ishikawa, Y. 2004. Endocrine changes associated with metamorphosis and diapause induction in the yellow-spotted longicorn beetle, *Psacothea hilaris*. *J. Insect Physiol.*, 50(11): 1075-1081.

Myers, M.P., Wager-Smith, K., Wesley, C.S., Young, M.W. & Sehgal, A. 1995. Positional cloning and sequence analysis of the *Drosophila* clock gene, *timeless*. *Science*, 270: 805-808.

Nachtigall, W. 1962a. Funktionelle Morphologie, Kinematic und Hydromechanik des Ruderapparates von *Gyrinus*. *Z. vergl. Physiol.*, 45: 193-226.

Nachtigall, W. 1962b. Die Mechanik der Schwimmbewegungen bei der *Acilius*- Larve (Coleoptera, Dytiscidae). *Int. Revue Ges. Hydrobiol.*, 47: 413-430.

Nachtigall, W. 1963. Zur Lokomotionsmechanik Schwimmender Dipterenlarven, *Z. vergl. Physiol.*, 46: 449-466.

Nachtigall, W. 1964. Die Schwimmechanik der Wasserinsekten. *Ergebn. Biol.*, 271: 39-78.

Nachtigall, W. 1974a. *Insects in Flight*. George Allen & Unwin Ltd., London.

Nachtigall, W. 1974b. Locomotion: mechanics and hydrodynamics of swimming in aquatic insects. In: *The Physiology of Insecta*. Vol. 3. M. Rockstein (ed.). Academic Press, N.Y., pp. 381-432.

Nachtigall, W. 1981. Insect flight aerodynamics. In: *Locomotion Energetics in Arthropods*. C.F. Herreid & C.R. Foutner (eds.). Plenum Press, N.Y., pp. 127-162.

Nachtigall, W. 1985. Swimming in aquatic insects. In: *Comprehensive Insect Physiology, Biochemistry and Pharmacology. Nervous System: Structure and Motor Function*. Vol. 5. G.A. Kerkut & L.I. Gilbert (eds.). Pergamon Press, Oxford, pp. 467-490.

Nalepa, C.A. 1984. Colony composition, protozoan transfer and some life history characteristics of the wood roach *Cryptocercus punctulatus* Scudder (Dictyoptera: Cryptocercidae). *Behav. Ecol. Sociobiol.*, 14: 273-279.

Nalyanya, G., Moore, C.B. & Schal, C. 2000. Integration of repellents, attractants, and insecticides in a "push-pull" strategy for managing German cockroach (Dictyoptera: Blattellidae) populations. *J. Med. Entomol.*, 37(3): 427-434.

Narain, A.S. 1979. A study of the behavioural response of adults of *Nymphula* (Lepidoptera: Pyralidae) to different wavelengths in the visible spectrum. *Z. Angew. Zool.*, 66(1): 27-34.

Nayar, J. 1967. Endogenous diurnal rhythm of pupation in a mosquito population. *Nature*, 214: 828-829.

Naylor, A.F. 1959. An experimental analysis of dispersal in the flour beetle, *Trobolium confusum*. *Ecology*, 40: 453-465.

Neilsen, E.T. 1961. On the habits of the migratory butterfly *Ascia monuste* L. *Biol. Meddr.*, 23: 1-81.

Neumann, D. 1963. Über die Steuerung lunaren Schwärmperiodik der Mücke *Clunio marinus*. *Verh. dt. Zool. Ges. Jena*, 1962, pp.275-285.

Neumann, D. 1966. Die intraspezifische Variabilität der lunaren und täglichen Schlüpfzeiten von *Clunio marinus* (Diptera; Chironomidae). *Verh. dt. Zool. Ges. Jena*, 1965, pp.223-233.

Neumann, D. 1976. Adaptations of chironomids to intertidal environments. *Ann. Rev. Entomol.*, 21: 387-414.

Neumann, D. 1989. Circadian components of semilunar and lunar timing mechanisms. *J. Biol. Rhythms*, 4(2): 285-294.

Neville, A.C. 1965. Circadian organization of chitin in some insect skeletons. *Q. J. Microsc. Sci.*, 106: 315-325.

Neville, A.C. 1970. Cuticular ultrastructure in relation to the whole insect. *Symp. Roy. Entomol. Soc. Lond.*, 5: 17-39.

Ng, S.M. 1986. The geotactic and phototactic responses of four species of aphidophagous coccinellid larvae. In: *Ecology of Aphidophaga. Proc. II Symp. Zvikovske Podhardi.* I. Hodek (ed.). Dr W. Junk, Dordrecht (Netherlands). *Ser. Entomol.* 35: 57-68.

Nicol, J.A.C. 1978. Bioluminescence and vision. In: *Bioluminescence in Action.* P.J. Herring (ed.). Academic Press, London, pp. 367-398.

Nieh, J.C. 2004. Recruitment communication in stingless bees (Hymenoptera, Apidae, Meliponini). *Apidologie*, 35: 159-182.

Nieh, J.C., Barreto, L.S., Contrera, F.A.L. & Imperatriz-Fonseca, V.L. 2004. Olfactory eavesdropping by a competitively foraging stingless bee, *Trigona spinipes. Proc. Roy. Soc. Lond. Ser. B, Biol. Sci.*, 271: 1633-1640.

Nieh, J.C., Contrera, F.A.L. & Nogueira-Neto, P. 2003a. Pulsed mass recruitment by a stingless bee, *Trigona hyalinata. Proc. Roy. Soc. Lond. Ser. B, Biol. Sci.*, 270: 2191-2196.

Nieh, J.C., Contrera, F.A.L., Ramírez, S. & Imperatriz-Fonseca, V.L. 2003b. Variation in the ability to communicate 3-D resource location by stingless bees from different habitats. *Anim. Behav.*, 66: 1129-1139.

Nieh, J.C. & Roubik, D.W. 1998. Potential mechanisms for the communication of height and distance by a stingless bee, *Melipona panamica. Behav. Ecol. Sociobiol.*, 43: 387-399.

Nijhout, H.F. 1994. *Insect Hormones.* Princeton Univ. Press, Princeton, N.J.

Nishida, T. 1958. The influence of yeast hydrolysate on the behaviour of the melon fly. *J. Econ. Entomol.*, 51: 140-143.

Nishiitsutsuji-Uwo, J. & Pittendrigh, C.S. 1968. Central nervous system control of circadian rhythmicity in the cockroach. III. The optic lobes, locus of the driving oscillator? *Z. vergl. Physiol.*, 58: 14-46.

Nisimura, T. & Numata, H. 2001. Endogenous timing mechanism controlling the circannual pupation rhythm of the varied carpet beetle *Anthrenus verbasci. J. Comp. Physiol. A*, 187(6): 433-440.

Noldus, L.P.J.J. & Lenteren, J.C. van, 1985. Kairomones for the egg parasite *Trichogramma evanescens* Westwood. I. Effect of volatile substances released by two of its hosts, *Pieris brassicae* L. and *Mamestra brassicae*, L. *J. Chem. Ecol.*, 11(6): 781-791.

Nordlander, G., Eidmann, H.H., Jacobsson, U., Nordenhen, H. & Sjoedin, K. 1986. Orientation of the pine weevil *Hylobius abietis* to underground sources of host volatiles. *Entomol. Exp. Appl.*, 41(1): 91-100.

Nordlund, D.A. 1981. Semiochemicals: a review of the terminology. In: *Semiochemicals: Their Role in Pest Control.* D.A. Nordlund, R.L. Jones & W.J. Lewis (eds.). John Wiley & Sons, N.Y., pp. 13-28.

Nordlund, D.A., Jones, R.L. & Lewis, W.J. 1981. *Semiochemicals: Their Role in Pest Control.* John Wiley & Sons, N.Y.

Nordlund, D.A., Strand, M.R., Lewis, W.J. & Vinson, S.B. 1987. Role of kairomones from host accessory gland secretion in host recognition by *Telenomus remus* and *Trichogramma pretiosum*, with partial characterization. *Entomol. Exp. Appl.*, 44(1): 37-43.

Norton, G.A. & Mumford, J.D. 1993. *Decision Tools for Pest Management.* CAB Int., Wallingford.

Novák, V.J.A. 1955. To the knowledge of the mechanism conditioning the development of the ventral black pattern in the abdomen of the bug *Oncopeltus fasciatus* Dal. *Acta Soc. Zool. Èsl.*,

19: 233-246.

Novák, V.J.A. 1975. *Insect Hormones.* (2nd ed.). Chapman & Hall, London.

Nowosielski, J.W., Patton, R.L & Naegele, J.A. 1964. Daily rhythm of narcotic sensitivity in the house cricket, *Gryllus domesticus* L., and the two-spotted spider mite, *Tetranychus urticae* Koch. *J. Cell Physiol.*, 63: 393-398.

Nufio, C.R. & Papaj, D.R. 2001. Host marking behaviour in phytophagous insects and parasitoids. *Entomol. Exp. Appl.*, 99: 273-293.

O'Donnell, S. & Foster, R.L. 2001. Thresholds of response in nest thermoregulation by worker bumble bees, *Bombus bifarius nearcticus* (Hymenoptera: Apidae). *Ethology*, 107: 387-399.

Obeng-Ofori, D., Torto, B., Njagi, P.G.N. & Hassanali, A. 1998. Aggregation pheromone complex of the desert locust, *Schistocerca gregaria* (Forskål) (Orthoptera: Acrididae): current Status. *J. Ghana Sci. Assoc.*, 1(1): 69-83.

Ohba, N. 1983. Studies on the communication system of Japanese fireflies. *Sci. Rpt. Yokosuka City* Mus., No. 30, 62 pp.

Oldfield, B.P. 1980. Accuracy of orientation in female crickets, *Teleogryllus oceanicus* (Gryllidae): dependence on song spectrum. *J. Comp. Physiol. A*, 141: 93-99.

Oliveira, P.S. & Hölldobler, B. 1989. Orientation and communication in the neotropican ant *Odontomachus bauri* Emey (Hymenoptera, Formicidae, Ponerinae). *Ethology*, 83(2): 154-166.

Onyeocha, F.A. & Fuzeau-Braesch, S. 1991. Circadian rhythm changes in toxicity of the insecticide dieldrin on larvae of the migratory locust *Locusta migratoria migratorioides. Chronobiol. Int.*, 8(2): 103-109.

Orphanidis, P.S. & Kalmoukos, P.E. 1970. Negative chemotropism of *Dacus oleae* (Gmel.) adults against essential oils. *Annals Inst. Phytopath. Benaki, N.S.*, 9: 288-294.

Orr, H.A. 2005. The genetic basis of reproductive isolation: insights from *Drosophila. Proc. Natl. Acad. Sci., USA*, 102 (Suppl. 1): 6522-6526.

Otte, D. 1970. A comparative study of communicative behaviour in grasshoppers. *Misc. Publ. Mus. Zool. Univ. Michigan*, 141: 1-168.

Otte, D. 1972. Simple versus elaborate behaviour in grasshoppers: an analysis of communication in the genus *Syrbula. Behaviour*, 42: 291-322.

Otte, D. 1974. Effects and functions in the evolution of signalling systems. *Ann. Rev. Ecol. Syst.*, 5: 385-417.

Otte, D. 1977.Communication in Orthoptera. In: *How Animals Communicate.* T.A. Sebeok (ed.). Indiana Univ. Press, Bloomington, pp. 334-361.

Otte, D. & Alexander, R.D. 1983. The Australian crickets (Orthoptera: Gryllidae). *Acad. Nat. Sci. Philadelphia, Monograph*, 22: 1-477.

Otte, D. & Joern, A. 1975. Insect territoriality and its evolution: population studies of desert grasshoppers on creosote bushes. *J. Anim. Ecol.*, 44: 29-54.

Page, T.L. 1978. Interaction between bilaterally paired components of the cockroach circadian system. *J. Comp. Physiol.*, 124: 225-236.

Page, T.L. 1985. Clocks and circadian rhythms. In: *Comprehensive Insect Physiology, Biochemistry and Pharmacology. Nervous System: Sensory.* Vol. 6. G.A. Kerkut & L.I. Gilbert (eds.). Pergamon Press, Oxford, pp. 577-652.

Page, T.L. & Block, G.D. 1980. Circadian rhythmicity in cockroaches: effects of early post-embryonic development and ageing. *Physiol. Entomol.*, 5: 271-281.

Palmer, J.D. 1976. *An Introduction to Biological Rhythms.* Academic Press, N.Y.

Palmer, J.D. 1995. The *Biological Rhythms and Clocks of Intertidal Animals.* Oxford Univ. Press, Oxford.

Papaj, D.R. & Lewis, A.C. (eds.) 1993. *Insect Learning: Ecological and Evolutionary Perspectives.* Chapman & Hall, London.

Papi, F. 1992. General aspects. In: *Animal Homing.* F. Papi (ed.). Chapman & Hall, London, pp. 1-18.

Papov, A.V. & Shuvalov, V.F. 1977. Phonotactic behaviour of crickets. *J. Comp. Physiol.,* 119(1): 111-126.

Pardi, L. 1955. Orientamento solare in un Tenebrionide alofilo: *Phaleria provincialis* Fauv. (Coleopt.). *Boll Ist Zool. Torino,* 5, 1.

Paré, P.W. & Tumlinson, J.H. 1998. Cotton volatiles synthesized and released distal to the site of insect damage. *Phytochemistry,* 47(4): 521-526.

Park, T. 1948. Experimental studied of interspecific competition. I. Competition between populations of the flour beetle, *T. confusum* Duval and *T. castaneum* Herbst. *Ecol. Monogr.,* 18: 265-308.

Parker, G.A. 1970. Sperm competitiveness and its evolutionary consequences in insects. *Biol. Rev.,* 45: 525-567.

Parker, G.A. 1974. Courtship persistence and female-guarding as male time investment strategies. *Behaviour,* 48: 157-184.

Parker, G.A. 1979. Sexual selection and sexual conflict. In: *Sexual Selection and Reproductive Competition in Insects.* M. Blum & N. Blum (eds.). Academic Press, N.Y., pp. 123-166.

Parry, K. & Morgan, E.D. 1979. Pheromones of ants: a review. *Physiol. Entomol.,* 4: 161-189.

Pasteur, G. 1982. A classificatory review of mimicry systems. *Ann. Rev. Ecol. Syst.,* 13: 169-199.

Pavan, M. 1959. Biochemical aspects of insect poisons. *Intrn. Cong. Biochem.* 4th, Vienna, 12: 15-36.

Pavlov, I.P. 1927. *Conditional Reflexes.* Oxford Univ. Press, London.

Payne, T.L., Birch, M.C. & Kennedy, C.E.J. (eds.) 1986. *Mechanisms in Insect Olfaction.* Clarendon Press, Oxford.

Payne, T.L., Coster, J.E., Richerson, J.V., Edson, L.J. & Hart, E.R. 1978. Field response of the southern pine beetle to behavioural chemicals. *Environ. Entomol.,* 7(4): 578-582.

Pearce, A.N., Huang, Z.Y. & Breed, M.D. 2001. Juvenile hormone and aggression in honey bees. *J. Insect Physiol.,* 47: 1243-1247.

Peckham, G.W. & Peckham, E.G. 1898. *On the Instincts and Habits of Solitary Wasps.* Houghton-Mifflin, Boston.

Peixoto, A.A. 2002. Evolutionary behavioural genetics in *Drosophila. Adv. Genet.,* 47: 117-50.

Pellmyr, O. & Krenn, H.W. 2002. Origin of a complex key innovation in an obligate insect-plant mutualism. *Proc. Natl. Acad. Sci., USA,* 99(8): 5498-5502.

Pemberton, R.W. & Lee, J.H. 1996. The influence of extrafloral nectaries on parasitism of an insect herbivore. *Amer. J. Bot.,* 83(9): 1187-1194.

Pener, M.P. 1985. Hormonal effects on flight and migration. In: *Comprehensive Insect Physiology, Biochemistry and Pharmacology. Endocrinology II.* Vol. 8. G.A. Kerkut & L.I. Gilbert (eds.). Pergamon Press, Oxford, pp. 491-550.

Pener, M.P. 1991. Locust phase polymorphism and its endocrine relations. *Adv. Insect Physiol.,*

23: 1-79.

Pener, M.P. 1992. Environmental cues, endocrine factors, and reproductive diapause in male insects. *Chronobiol. Int.*, 9(2): 102-113.

Percy-Cunningham, J.E. & MacDonald, J.A. 1987. Biology and ultrastructure of sex pheromone-producing glands. In: *Pheromone Biochemistry*. G.D. Prestwich & G.J. Blomquist (eds.). Academic Press, N.Y., pp. 27-75.

Perdeck, A.C. 1958. The isolating value of specific song patterns in two sibling species of grasshoppers (*Chorthippus brunneus* Thunb. and *C. biguttulus* L.). *Behaviour*, 12: 1-75.

Perez, Y., Verdier, M. & Pener, M.P. 1971. The effect of photoperiod on male sexual behaviour in a north Adriatic strain of the migratory locust. *Entomol. Exp. Appl.*, 14: 245-250.

Perttunen, V. 1951. The humidity preferences of various carabid species (Col., Carabidae) of wet and dry habitats. *Ann. Entomol. Finnica*, 17: 72-84.

Peter, A.J. & Ananthakrishnan, T.N. 1996. Neuroendocrine implications of azadirachtin-treated cotton grasshopper, *Cyrtacanthacris tatarica* Linn. (Acrididae: Orthoptera). In: *Biotechnical Perspectives in Chemical Ecology of Insects*. T.N. Ananthakrishnan (ed.). Oxford & IBH Publ. Co. Pvt. Ltd., N. Delhi, pp. 89-101.

Petit, C. & Ehrman, L. 1969. Sexual selection in *Drosophila*. *Evol. Biol.*, 2: 157-191.

Pettersson, J. 1970. An aphid sex attractant I. Biological studies. *Entomol. Scand.*, 1: 63-73.

Phelan, P.L. & Baker, T.C. 1986. Male-size-related courtship success and intersexual selection in the tobacco moth, *Ephestia elutella*. *Experientia*, 42: 1291-1293.

Phillips, L.H. & Konishi, M. 1973. Control of aggression by singing in crickets. *Nature*, 241: 64-65.

Pickett, J.A. 1985. Production of behaviour-controlling chemicals by crop plants. *Phil. Trans. Roy. Soc. Lond.*, 318: 203-211.

Pickett, J.A., Wadhams, L.J. & Woodcock, C.M. 1997. Developing sustainable pest control from chemical ecology. *Agril. Ecosyst. Environ.*, 64(2): 149-156.

Pickett, J.A., Wadhams, L.J., Woodcock, C.M. & Hardie, J. 1992. The chemical ecology of aphids. *Ann. Rev. Entomol.*, 37: 67-90.

Pickett, J.A., Williams, I.H., Martin, A.P. & Smith, M.C. 1980. Nasanov pheromone of the honey bee, *Apis mellifera* L. (Hymenoptera: Apidae). Part I. Chemical characterization. *J. Chem. Ecol.*, 6: 425-434.

Pierce, G.W. 1948. *The Songs of Insects*. Harvard Univ. Press, Cambridge, Massachusetts.

Pinto, J.D. & Stouthamer, R. 1994. Systematics of the Trichogrammatidae with emphasis on *Trichogramma*. In: *Biological Control with Egg Parasitoids*. E. Wajnberg & S.A. Hassan (eds.). CAB International, Wallingford, pp. 1-36.

Pitman, G.B., Renwick, J.A.A. & Vité, J.P. 1966. Studies on the pheromone of *Ips confusus* (LeConte). IV. Isolation of the attractive substance by gas-liquid chromatography. *Contrib. Boyce Thompson Inst.*, 23: 243-250.

Pittendrigh, C.S. 1954. On temperature independence in the clock controlling emergence time in *Drosophila*. *Proc. Natl. Acad. Sci., USA*, 40: 1018-1029.

Pittendrigh, C.S. 1958. Perspectives in the study of biological clocks. In: *Perspective in Marine Biology*. A.A. Buzzati-Traverso (ed.). Univ. California Press, California, pp. 239-268.

Pittendrigh, C.S. 1960. Circadian rhythms and circadian organization of living systems. *Cold Spring Harb. Symp. Quant. Biol.*, 25: 159-184.

Pittendrigh, C.S. 1967. Circadian systems. I. The driving oscillation and its assay in *Drosophila pseudoobscura*. *Proc. Natl. Acad. Sci., USA*, 58: 1762-1767.

Pittendrigh, C.S. 1981. Circadian systems: entrainment. In: *Biological Rhythms. Handbook of Behavioural Neurobiology*. Vol. 4. J. Aschoff (ed.). Plenum Press, N.Y., pp. 95-124.

Pittendrigh, C.S. 1993. Temporal organization: reflections of a Darwinian clockwatcher. *Ann. Rev. Physiol.*, 55: 17-54.

Pittendrigh, C.S. & Minis, D.H. 1964. The entrainment of circadian oscillations by light and their role as photoperiodic clocks. *Amer. Nat.*, 98: 261-294.

Pittendrigh, C.S. & Takamura, T. 1989. Latitudinal clines in the circadian properties of a pacemaker. *J. Biol. Rhythms*, 4: 217-235.

Plettner E., Slessor K.N., Winston M.L. & Oliver J.E. 1996. Caste-selective pheromone biosynthesis in honey bees. *Science*, 271: 1851-1853.

Pollack, G.S. & Hoy, R.R. 1979. Temporal pattern as a cue for species-specific calling song recognition in crickets. *Science*, 204: 429-432.

Pollard, E., Hall, M.L. & Bibby, T.J. 1984. The clouded yellow butterfly migration of 1983. *Entomol. Gaz.*, 35(4): 227-234.

Popham, E.J. & Vickere, H. 1979. Tsetse fly reactions to light and humidity. *Expererentia*, 35(2): 194-196.

Potter, M.F. & Watson, T.F. 1983. Garbanzo bean as a potential feeding stimulant for use with a nuclear polyhedrosis virus of the tobacco budworm (Lepdioptera: Noctuidae). *J. Econ. Entomol.*, 76: 449-451.

Potting, R.P.J., Vet, L.E.M. & Dicke, M. 1995. Host microhabitat location by stem borer parasitoid *Cotesia flavipes*: the role of herbivore volatiles and locally and systemically induced plant volatiles. *J. Chem. Ecol.*, 21: 525-539.

Poulton, E.B. 1890. *The Colours of Animals*. Trübner & Co Ltd., London.

Pradhan S. 1965. New biotic theory of the periodicity of locust cycle. *Indian J. Entomol.*, 27: 95-101.

Prestwich, G.D. 1979. Termite chemical defence: new natural products and chemosystematics. *Sociobiology*, 4: 127-140.

Prestwich, G.D. 1983. Chemical systematics of termite exocrine secretions. *Ann. Rev. Ecol. Syst.*, 14: 287-311.

Prestwich, G.D. & Collins, M.S. 1981. Chemotaxonomy of *Subulitermes* and *Nasutitermes* termite soldier defence secretions. Evidence against the hypothesis of diphyletic evolution of the Nasutitermitinae. *Biochem. Syst. Ecol.*, 9: 83-88.

Price, P.W. 1984. *Insect Ecology* (2nd ed.). John Wiley & Sons, N.Y.

Price, P.W., Bouton, C.E., Gross, P., McPheron, B.A., Thompson, J.N. & Weis, A.E. 1980. Interactions among three trophic levels: influence of plants on interactions between insect herbivores and natural enemies. *Ann. Rev. Ecol. Syst.*, 11: 41-65.

Price, T., Schluter, D. & Heckman, N.E. 1993 Sexual selection when the female directly benefits. *Biol. J. Linn. Soc.*, 48: 187-211.

Pringle, J.W.S. 1938a. Proprioception in insects. I. A new type of mechanical receptor from the palps of the cockroach. *J. Exp. Biol.*, 15: 101-113.

Pringle, J.W.S. 1938b. Proprioception in insects. II. The action of campaniform sensilla on the legs. *J. Exp. Biol.*, 15: 114-131.

Pringle, J.W.S. 1940. The reflex mechanism of the insect leg. *J. Exp. Biol.*, 17: 8-17.

Pringle, J.W.S. 1957. *Insect Flight*. Cambridge Univ. Press, Cambridge.

Pringle, J.W.S. 1968. Comparative physiology of the flight motor. *Adv. Insect Physiol.*, 5: 163-228.

Pringle, J.W.S. 1974. Locomotion: flight. In: *The Physiology of Insecta*. Vol. 3. (2nd ed.). M. Rockstein (ed.). Academic Press, N.Y., pp. 433-476.

Pritchatt, D. 1968. Avoidance of electric shock by the cockroach *Periplaneta americana*. *Anim. Behav.*, 16: 178-185.

Prokopy, R.J. 1972a. Evidence for a marking pheromone deterring repeated oviposition in apple maggot flies. *Environ. Entomol.*, 1: 326-332.

Prokopy, R.J. 1972b. Response of apple maggot flies to rectangles of different colours and shades. *Environ. Entomol.*, 1: 720-726.

Prokopy, R.J. 1977. Stimuli influencing trophic relations in Tephritidae. *Colloq. Int. CNRS*, 265: 305-336.

Prokopy, R.J. 1981. Epideictic pheromones that influence spacing patterns of phytophagous insects. In: *Semiochemicals: Their Role in Pest Control*. D.A. Nordlund, R.L. Jones, & W. J. Lewis (eds.). John Wiley & Sons, N.Y., pp. 181-213.

Prokopy, R.J. 1986. Visual and olfactory stimulus interaction in resource finding by insects. In: *Mechanisms in Insect Olfaction*. T.L. Payne, M.C. Birch, & C.E.J. Kennedy (eds.). Clarendon Press, Oxford, pp. 81-89.

Prokopy, R.J. & Boller, E.F. 1971. Response of European cherry fruit flies to coloured rectangles. *J. Econ. Entomol.*, 64(6): 144-1147.

Prokopy, R.J., Duan, J.J. & Vargas, R.I. 1996. Potential for host range expansion in *Ceratitis capitata* flies: impact of proximity of adult food to egg-laying sites. *Ecol. Entomol.*, 21(3): 295-299.

Prokopy, R.J. & Economopoulos, A.P. 1975. Attraction of laboratory-cultured and wild *Dacus oleae* flies to sticky-coated McPhail traps of different colours and odours. *Environ. Entomol.*, 4(2): 187-192.

Prokopy, R.J., Miller, N.W., Piñero, J.C., Barry, J.D., Tran, L.C., Oride, L. & Vargas, R.I. 2003. Effectiveness of GF-120 fruit fly bait spray applied to border area plants for control of melon flies. *J. Econ. Entomol.*, 96: 1485-1493.

Prokopy, R.J. & Owens, E.D. 1978. Visual generalist visual specialist phytophagous insects: host selection behaviour an application to management. *Entomol. Exp. Appl.*, 24: 609-620.

Prokopy, R.J. & Owens, E.D. 1983. Visual detections of plants by herbivorous insects. *Ann. Rev. Entomol.*, 28: 337-364.

Prokopy, R.J. & Webster, R.P. 1978. Oviposition-detrerring pheromone of *Rhagoletis pomonella*: a kairomone for its parasitoid *Opius lectus*. *J. Chem. Ecol.*, 4(4): 481-494.

Prop, N. 1962. Protection against birds and parasites in some species of tenthredinid larvae. *Arch. Neer. Zool.*, 13: 380-447.

Pszczolkowski, M.A. & Dobrowolski, M. 1999. Circadian dynamics of locomotor activity and deltamethrin susceptibility in the pineweevil, *Hylobius abietis*. *Phytoparasitica*, 27(1): 19-25.

Pureswaran, D.S. & Borden, J.H. 2003. Is bigger better? Size and pheromone production in the mountain pine beetle, *Dendroctonus ponderosae* Hopkins (Coleoptera: Scolytidae). *J. Insect Behav.*, 16(6): 765-782.

Pyza, E. & Meinertzhagen, I.A. 1995. Monopolar cell axons in the first optic neuropile of the house fly, *Musca domestica* L., undergo daily fluctuations in diameter that have a circadian basis. *J. Neurosci.*, 15: 407-418.

Queiroz, A. de & Wimberger, P.H. 1993. The usefulness of behaviour for phylogeny estimation: levels of homoplasy in behavioural and morphological characters. *Evolution*, 47: 46-60.

Quennedey, A. & Deligne, J. 1975. L'arme frontale des soldats de termites. I. Rhinotermitidae. *Insectes Sociaux*, 22: 243-267.

Quiroz, A., Pettersson, J., Pickett, J.A., Wadhams, L.J., Niemeyer, H.M. 1997. Semiochemicals mediating spacing behaviour of bird cherry-oat aphid, *Rhopalosiphum padi* feeding on cereals. *J. Chem. Ecol.*, 23(11): 2599-2607.

Raabe, M. 1982. Neuroendocrine control of reproduction in insects. *J. Physiol.*, 78(6): 553-557.

Rabe, W. 1953. Beitrage zum Orientierungsproblem der Wasserwanzen. *Z. vergl. Physiol.*, 35: 300-325.

Rafaeli, A. & Gileadi, C. 1996. Down regulation of pheromone biosynthesis: cellular mechanisms of pheromonostatic responses. *Insect Biochem. Molec. Biol.*, 26: 797-807.

Raghu, S., Drew, R.A.I. & Clarke, A.R. 2004. Influence of host plant structure and microclimate on the abundance and behaviour of a tephritid fruit fly. *J. Insect Behav.*, 17(2): 179-190.

Rahman, M.M., Baggerman, G., Schoofs, L., Loof, A. de, & Breuer, M. 2003. Presence of [His7]-corazonin in the central nervous system of a newly isolated albino strain of *Schistocerca gregaria* (Orthoptera: Acrididae)- mass spectrometric and immunocytochemical evidence. *European J. Entomol.*, 100(3): 455-458.

Raina, A.K. 1993. Neuroendocrine control of sex pheromone biosynthesis in Lepidoptera. *Ann. Rev. Entomol.*, 38: 329-349.

Raina, A.K., Jaffe, H., Kempe, T.G., Keim, P., Blacher R.W., Fales, H.M., Riley, C.T., Klun, J.A., Ridgway, R.L. & Hayes, D.K. 1989. Identification of a neuropeptide hormone that regulates sex pheromone production in female moths. *Science*, 244: 796-798.

Rainey, R.C. 1958. Some observations on flying locusts and atmospheric turbulence in eastern Africa. *Quart. J. Roy. Meteorol. Soc.*, 84: 334-354.

Rainey, R.C. 1974. Biometry and insect flight: some aspects of energy exchange. *Ann. Rev. Entomol.*, 19: 407-439.

Rainey, R.C. (ed.) 1976. *Insect flight*. Blackwell Scientific Publ., Oxford.

Ramamurthy, B.N. 1966. Adaptive colouration in *Labidura riparia* (Pallas) (Insecta- Dermaptera) to desert habitat. *Proc. 2nd All India Cong. Zool.*, Varanasi, 1962, pt. II.

Rankin, M.A. 1974. The hormonal control of flight in the milkweed bug *Oncopeltus fasciatus*. In: *Experimental Analysis of Insect Behaviour*. L. Barton Browne (ed.). Springer-Verlag, Berlin, pp.317-328.

Rankin, M.A., Caldwell, R.L. & Dingle, H. 1972. An analysis of a circadian rhythm of oviposition in *Oncopeltus fasciatus*. *J. Exp. Biol.*, 56: 353-359.

Read, D.P., Feeney, P.P. & Root, R.B. 1970. Habitat selection by the aphid parasite *Diaeretiella rapae* (Hymenoptera: Braconidae) and hyperparasite *Charips brassicae* (Hymenoptera: Cynipidae). *Can. Entomol.*, 102: 1567-1578.

Readio, J., Chen, M.-H. & Meola, R. 1999. Juvenile hormone biosynthesis in diapausing and nondiapausing *Culex pipiens* (Diptera: Culicidae). *J. Med. Entomol.*, 36(3): 355-360.

Réaumer, R.A.F. de, 1734-1742. *Memoires pour Servir a l'Histoire des Insectes*. (6 vols.). Imprimerie Royale, Paris.

Reavey, D. 1993. Why body size matters to caterpillars. In: *Caterpillars: Ecological and Evolutionary Constraints on Foraging*. N.E. Stamp & T.M. Casey (eds.). Chapman & Hall, N.Y., pp. 170-202.

Reed, H.C. & Landolt, P.J. 2000. Application of alarm pheromone to targets by southern yellowjackets (Hymenoptera: Vespidae). *Florida Entomol.*, 83(2): 193-196.

Reeve, H.K. 1991. *Polistes*. In: *The Social Biology of Wasps*. K.G. Ross & R.W. Matthews (eds.). Cornell Univ. Press, Ithaca, N.Y., pp. 99-148.

Regnier. F.E. & Wilson, E.O. 1971. Chemical communication and "propaganda" in slave-marker ants. *Science*, 172: 267-269.

Reguera, P. & Gomendio, M. 2002. Flexible oviposition behaviour in the golden egg bug (*Phyllomorpha laciniata*) and its implications for offspring survival. *Behav. Ecol.*, 13(1): 70-74.

Reinhard, J. & Kaib, M. 2001. Trail communication during foraging and recruitment in the subterranean termite *Reticulitermes santonensis* De Feytaud (Isoptera, Rhinotermitidae). *J. Insect Behav.*, 14(2): 157-171.

Renner, M. 1957. Neue Versuche über den Zeitsinn der Honigbiene. *Z. vergl. Physiol.*, 40: 85-118.

Renner, M. 1959. Über ein weiteres Versetzungs-experiment zur analyse des Zeitsinnes und der Sonnenorientierung der Honigbiene. *Z. vergl. Physiol.*, 42: 449-483.

Renou, M. & Guerrero, A. 2000. Insect parapheromones in olfaction research and semiochemical based pest control strategies. *Ann. Rev. Entomol.*, 45: 605-630.

Renwick, J.A.A. 2001. Variable diets and changing taste in plant-insect relationships. *J. Chem. Ecol.*, 27(6): 1063-1076.

Renwick, J.A.A. & Vité, J.P. 1969. Bark beetle attractants: mechanisms of colonization by *Dendroctonus frontalis*. *Nature*, 224: 1222-1223.

Reppert, S.M., Zhu, H.S. & White, R.H. 2004. Polarized light helps monarch butterflies navigate. *Current Biol.*, 14(2): 155-158.

Rettenmeyer, C.W. 1963. Behavioual studies of army ants. *Kansas Univ. Sci. Bull.*, 44(9): 281-465.

Rettenmeyer, C.W. 1970. Insect mimicry. *Ann. Rev. Entomol.*, 15: 43-74.

Ribeiro, S.T. 1989. Group effects and aposematism in *Jadera haematoloma* (Hemiptera: Rhopalidae). *Ann. Entomol. Soc. Am.*, 82(4): 466-475.

Rice, M.J. 1975. Insect mechanoreceptor mechanisms. In: *Sensory Physiology and Behaviour*. R. Galun, P. Hillman, I. Parnas & R. Werman (eds.). Plenum Press, N.Y., pp. 135-165.

Rice, W.R. 1985. Disruptive selection on habitat preference and the evolution of reproductive isolation: an exploratory experiment. *Evolution*, 39: 645-646.

Rice, W.R. & Salt, G.W. 1990. The evolution of reproductive isolation as a correlated character under sympatric conditions: experimental evidence. *Evolution*, 44: 1140-1152.

Richards, A.M. 1960. Observations on the New Zealand glowworm *Arachnocampa luminosa* (Skuse) 1890. *Trans. Roy. Soc. New Zealand*, 88: 559-574.

Richards, G. 1973. The historical development of nineteenth and twentieth century studies on the behaviour of insects. In: *History of Entomology*. R.F. Smith, T.E. Mittler & C.N. Smith (eds.). Annual Reviews, Palo Alto, California, pp. 477-502.

Richards, G. 1981. Insect hormones in development. *Biol. Rev. Camb. Philos. Soc.*, 56: 501-549.

Richards, M.H., Wettberg, E.J. von & Rutgers, A.C. 2003. A novel social polymorphism in a primitively eusocial bee. *Proc. Natl. Acad. Sci., USA*, 100(12): 7175-7180.

Richards, O.W. 1927. Sexual selection and allied problems in insects. *Biol. Rev.*, 2: 298-364.

Richards, O.W. 1953. *The social Insects*. Macdonald, London.

Richards, O.W. 1965. Concluding remarks on the social organization of insect communities. *Symp. Zool. Soc. Lond.*, 14: 169-172.

Richards, O.W. & Richards, M.O. 1951. Observations of the social wasps of South America (Hymenoptera, Vespidae). *Trans. Roy. Entomol. Soc. Lond.*, 102(1): 1-170.

Richter, K. & Koolman, J. 1991. Antiecdysteroid effects of brassinosteroids in insects. In: *Brassinosteroids: Chemistry, Bioactivity, and Application.* H.G. Cutler, T. Yokota, & G. Adam (eds.). *Amer. Chem. Soc., Wash., D.C.*, pp. 265-278.

Ricker, J.P., Brzorad, J.N. & Hirsch, J. 1986. A demonstration of discriminative conditioning in the blow fly, *Phormia regina. Bull. Psychon Soc.*, 24(3): 240-243.

Riddiford, L.M. 1982. Changes in translatable mRNA's during the larval-pupal transformation of the epidermis of the tobacco hornworm. *Devel. Biol.*, 92: 330-342.

Riddiford, L.M. 1991. Hormonal control of sequential gene expression in insect epidermis. In: *Physiology of the Insect Epidermis.* K. Binnington & A. Retnakaran (eds.). CSIRO Publ., Melbourne, pp. 46-54.

Riddiford, L.M. & Truman, J.W. 1978. Biochemistry of insect hormones and insect growth regulators. In: *Biochemistry of Insects.* M. Rockstein (ed.). Academic Press, N.Y., pp. 308-357.

Ridgway, R.L., Silverstein, R.M. & Inscoe, M.N. (eds.) 1990. *Behaviour-Modifying Chemicals for Insect Management.* Marcel Dekker, N.Y.

Riley, J.R., Greggers, U., Smith, A.D., Reynolds, D.R. & Menzel, R. 2005. The flight paths of honey bees recruited by the waggle dance. *Nature*, 435: 205-207.

Ritchie, M.G. & Gleason, J.M. 1995. Rapid evolution of courtship song pattern in *Drosophila willistoni* sibling species. *J. Evol. Biol.*, 8: 463-479.

Ritchie, M.G., Sunter, D. & Hockham, L.R. 1998. Behavioural components of sex role reversal in the tettigoniid bush cricket *Ephippiger ephippiger. J. Insect Behav.*, 11(4): 481-491.

Ritz, M.S. & Sakaluk, S.K. 2002. The role of the male's cerci in copulation and mate guarding in decorated crickets (*Gryllodes sigillatus*). *J. Zool. Lond.*, 257: 519-523.

Riviere, S., Lartigue, A., Quennedey, B., Campanacci, V., Farine, J.P., Tegoni, M., Cambillau, C. & Brossut, R. 2003. A pheromone-binding protein from the cockroach *Leucophaea maderae*: cloning, expression and pheromone binding. *Biochem. J.*, 371(2): 573-579.

Roberts, S.K. 1960. Circadian activity in cockroaches. I. The freerunning rhythm in steady state. *J. Cell. Comp. Physiol.*, 55: 99-110.

Roberts, S.K. 1962. Circadian activity in cockroaches. II. Entrainment and phase-shifting. *J. Cell. Comp. Physiol.*, 59: 175-186.

Roberts, S.K. 1966. Circadian activity rhythms in cockroaches. III. The role of endocrine and neural factors. *J. Cell Physiol.*, 67: 473-486.

Roberts, S.K. 1974. Circadian rhythms in cockroaches. Effect of optic lobe lesions. *J. Comp. Physiol.*, 88: 21-30.

Robinson, G.E. 2002. Genomics and integrative analyses of division of labour in honey bee colonies. *Amer. Nat.*, 160: S160-S172.

Robinson, G.E. & Page, R.E. 1988. Genetic determination of guarding and undertaking in honey bee colonies. *Nature*, 333: 356-358.

Robinson, G.E., Page, R.E., Strambi, C. & Strambi, A. 1989. Hormonal and genetic control of behavioural integration in honey bee colonies. *Science*, 246: 109-112.

Robinson, G.E., Strambi, C., Strambi, A. & Huang, Z.-Y. 1992. Reproduction in worker honey bees is associated with low juvenile hormone titres and rates of biosynthesis. *J. Comp. Endocrinol.*, 87: 471-480.

Robinson, M.H. 1969. Defence against visually hunting predators. In: *Evolutionary Biology*. T. Dobzhansky, M.K. Hecht & W.C. Steere (eds.). Meredith Corp., N.Y., pp. 225-259.

Roces, F. 1990. Olfactory conditioning during the recruitment process in a leaf-cutting ant. *Oecologia*, 83(2): 261-262.

Roces, F. & Tautz, J. 2001. Ants are deaf. *J. Acoust. Soc. Am.*, 109(6): 3080-3082.

Roeder, K.D. 1937. The control of tonus and locomotor activity in the praying mantis (*Mantis religiosa* L.). *J. Exp. Biol.*, 76: 353-374.

Roeder, K.D. 1965. Moths and ultrasound. *Sci. Amer.*, 212: 94-102.

Roeder, K.D. 1967. *Nerve Cells and Insect Behaviour*. (Rev. ed.). Harvard Univ. Press, Cambridge, Massachusetts.

Roeder, K.D. 1970. Episodes in insect brains. *Amer. Sci.*, 58: 378-389.

Roeder, K.D. 1974. Some neurological mechanisms of simple behaviour. *Adv. Study Behav.*, 5: 2-46.

Roeder, K.D. & Treat, A.E. 1957. Ultrasonic reception by the tympanic organs of noctuid moths. *J. Exp. Zool.*, 134: 127-158.

Roelofs, W.L. 1975. Insect communication- chemical. In: *Insect, Science, and Society*. D. Pimentel (ed.). Academic Press, N.Y., pp. 79-99.

Roenneberg, T., Daan, S. & Merrow, M. 2003. The art of entrainment. *J. Biol. Rhythms*, 18: 183-194.

Roitberg, B.D. & Prokopy, R.J. 1987. Insects that mark host plants. *Biosci.*, 37: 400-406.

Rojas, J.C. & Wyatt, T.D. 1999. Role of visual cues with host odour during the host-finding behaviour of the cabbage moth. *Entomol. Exp. Appl.*, 91(1): 59-65.

Romoser, W.S. 1981. *The Science of Entomology*. (2nd ed.). Macmillan, N.Y.

Rose, U., Ferber, M. & Hustert, R. 2001. Maturation of muscle properties and its hormonal control in an adult insect. *J. Exp. Biol.*, 204: 3531-3545.

Röse, U.S.R., Lewis, W.J. & Tumlinson, J.H. 1998. Specificity of systemically released cotton volatiles as attractants for specialist and generalist parasitic wasps. *J. Chem. Ecol.*, 24(2): 303-319.

Rosén, W.Q., Han, G.-B. & Löfstedt, C. 2003. The circadian rhythm of the sex-pheromone-mediated behavioural response in the turnip moth, *Agrotis segetum*, is not controlled at the peripheral level. *J. Biol. Rhythms*, 18(5): 402-408.

Rosin, R. 1988. Do honey bees still have a "dance language"? *Amer. Bee J.*, 128: 267-268.

Ross, K.G. & Keller, L. 1998. Genetic control of social organization in an ant. *Proc. Natl. Acad. Sci., USA*, 95(24): 14232-14237.

Ross, K.G. & Matthews, R.W. (eds.) 1991. *The Social Biology of Wasps*. Cornell Univ. Press, Ithaca, N.Y.

Rossel, S. & Wehner, R. 1982. The bee's map of the E-vector pattern in the sky. *Proc. Natl. Acad. Sci., USA*, 79(14): 4451-4455.

Rossel, S. & Wehner, R. 1986. Polarization vision in bees. *Nature*, 323: 128-131.

Rostás, M. & Hilker, M. 2002. Feeding damage by larvae of the mustard leaf beetle deters conspecific females from oviposition and feeding. *Entomol. Exp. Appl.*, 103(3): 267-277.

Roth, L.M. 1948. A study of mosquito behaviour. An experimental laboratory study of the sexual behaviour of *Aedes aegypti*. *Amer. Midl. Natur.*, 40: 265-352.

Roth, L.M. & Eisner, T. 1962. Chemical defences of arthropods. *Ann. Rev. Entomol.*, 7: 107-134.

Roth, L.M. & Willis, E.R. 1957. The medical and veterinary importance of cockroaches. *Smithson. Inst. Misc. Coll.*, 134: 1-147.

Rothenbuler, N. 1964. Behaviour genetics of nest cleaning in honey bees. 4. Responses of F_1 and backcross generations to disease-killed brood. *Amer. Zool.*, 4: 111-123.

Rothschild, M. 1961. Defensive colours and Müllerian mimicry among insects. *Trans. Roy. Entomol. Soc. Lond.*, 113: 101-121.

Rothschild, M. 1967. Mimicry, the deceptive way of life. *Nat. Hist.*, 76: 44-51.

Rothschild, M. 1971. Speculation about mimicry. In: *Ecological Genetics and Evolution*. E.R. Creed (ed.). Blackwell, Oxford, pp. 202-223.

Rothschild, M. 1989. Moths and memory. *Endeavour*, 13(1): 15-19.

Rothschild, M. & Schoonhoven, L.M. 1977. Assessment of egg load by *Pieris brassicae* (Lepidoptera: Pieridae). *Nature*, 266: 352-355.

Ruck, P. 1964. Retinal structure and photoreception. *Ann. Rev. Entomol.*, 9: 83-102.

Ruiter, L. de, 1956. Countershading in caterpillars. An analysis of its adaptive significance. *Arch. Néerl. Zool.*, 11: 285-341.

Ruiter, L. de, 1958. Some remarks on problems of the ecology and evolution of mimicry. *Arch. Neerl. Zool.*, 13.1, Suppl., pp. 351-368.

Ruther, J., Reinecke, A. & Hilker, M. 2002. Chemical orientation in *Melolontha* cockchafers. *Bull.-OILB/SROP*, 25(7): 15-20.

Ruther, J., Reinecke, A., Tolasch, T. & Hilker, M. 2001. Make love not war: a common arthropod defence compound as sex pheromone in the forest cockchafer *Melolontha hippocastani*. *Oecologia*, 128(1): 44-47.

Rutowski, R.L. 1978. The courtship behaviour of the small sulphur butterfly, *Eurema lisa* (Lepidoptera: Pieridae). *Anim. Behav.*, 26: 892-903.

Rutz, W., Gerig, L., Wille, H. & Lüscher, M. 1976. The function of juvenile hormone in adult worker honey bees, *Apis mellifera*. *J. Insect Physiol.*, 22: 1485-1491.

Ruxton, G.D., Sherreatt, T.N. & Michel, P.S. 2005. *Avoiding Attack: The Evolutionary Ecology of Crypsis, Warning Signals and Mimicry*. Oxford Univ. Press, Oxford.

Safranek, L. & Williams, C.M. 1984. Critical weights for metamorphosis in the tobacco hornworm, *Manduca sexta*. *Biol. Bull.* 167(3): 555-567.

Sakagami, S.F. & Michener, C.D. 1962. *The Nest Architecture of the Sweat Bees* (*Halictinae*). Univ. Kansas Press, Lawrence.

Sakai, S., Kato, M. & Nagamasu, H. 2000. *Artocarpus* (Moraceae)- gall midge pollination mutualism mediated by a male-flower parasitic fungus. *Amer. J. Bot.*, 87(3): 440-445.

Sakai, T. & Ishida, N. 2001. Circadian rhythms of female mating activity governed by clock genes in *Drosophila*. *Proc. Natl. Acad. Sci., USA*, 98(16): 9221-9225.

Sakaluk, S.K. 1991. Post-copulatory mate guarding in decorated crickets. *Anim. Behav.*, 41: 207-216.

Sakaluk, S.K. & Cade, W.H. 1980. Female mating frequency and progeny production in singly and doubly mated house and field crickets. *Can. J. Zool.*, 58: 404-411.

Sanders, C.J. 1997. Mechanisms of mating disruption in moths. In: *Insect Pheromone Research: New Directions*. R.T. Cardé & A.K. Minks (eds.). Chapman & Hall, N.Y., pp. 333-358.

Sands, W.A. 1957. Soldier mandibles of the Nasutitermitinae (Isoptera: Termitidae). *Insectes Sociaux*, 4: 13-24.

Sanjayan, K.P. & Prakash, R. 2006. Acoustic communication in insects: a paradigm for assessment, monitoring and management of pest and their natural enemies. In: *Integrated Pest Management and Biocontrol*. S.C. Dwivedi & Nalini Dwivedi (eds.). Pointer Publ., Jaipur, pp. 287-306.

Santschi, F. 1911. Observations et remarques critiques sur le méchanisme de l'orietation chez les fourmis. *Revue Suisse Zool.*, 19: 303-338.

Santschi, F. 1913. A propos de l'orientation virtuelle chez les fourmis. *Bull. Soc. Hist. Nat. Afr. Nord.*, 4-6: 231-235.

Santschi, F. 1923. L'orientation sideral de fourmis, et quelques considérations sur leurs differentes possibilités d'orientation. *Mem. Soc. Vaudoise Sci. Natur.*, 1: 137-176.

Sargent, T.D. 1968. Cryptic moths: effects on background selection of painting the circumocular scales. *Science*, 159: 100-101.

Saunders, D.S. 1974. Circadian rhythms and photoperiodism in insects. In: *The Physiology of Insecta*. Vol. 2. M. Rockstein (ed.). Academic Press, N.Y., pp. 461-533.

Saunders, D.S. 1982. *Insect clocks*. (2nd ed.). Pergamon Press, Oxford.

Saunders, D.S. 1997. Insect circadian rhythms and photoperiodism. *Invert. Neurosci.*, 3(2-3): 155-164.

Schal, C., Burns, E.L. & Blomquist, G.J. 1990. Endocrine regulation of female contact sex pheromone in the German cockroach, *Blattella germanica*. *Physiol. Entomol.*, 15: 81-91.

Schal, C., Burns, E.L., Gadot, M., Chase, J. & Blomquist, G.J. 1991. Biochemistry and regulation of pheromone production in *Blattella germanica* (L.) (Dictyoptera, Blattellidae). *Insect Biochem.*, 21(1): 73-79.

Schliep, W. 1910. Der Farbenwechsel von *Dixippus morosus* (Phasmidae). *Zool. Jahrb. Physiol.*, (*Physiol.*), 30: 45-132.

Schmidt, E.L. & Williams, C.M. 1953. Physiology of insect diapause. 5. Assay of the growth and differentiation hormone of Lepidoptera by the method of tissue culture. *Biol. Bull.*, 105(1): 174-187.

Schmidt, J.O. 1982. Biochemistry of insect venoms. *Ann. Rev. Entomol.*, 27: 339-368.

Schmidt, J.O. 1998. Mass action in honey bees: alarm, swarming and the role of releaser pheromones. In: *Pheromone Communication in Social Insects: Ants, Wasps, Bees, and Termites*. R.K. Meer vander, M. Breed, K.E. Espelie & M.L. Winston (eds.). Westview Press, Boulder, Colorado, pp. 257-313.

Schmidt, R.S. 1955. Termite (*Apicotermes*) nests- important ethological material. *Behaviour*, 8: 344-356.

Schneider, F. 1962. Dispersal and migration. *Ann. Rev. Entomol.*, 7: 223-242.

Schneider, S.S., Stamps, J.A. & Gary, N.E. 1986. The vibration dance of the honey bee. II. The effects of foraging success on daily patterns of vibration activity. *Anim. Behav.*, 34: 386-391.

Schneirla, T.C. 1953a. Modifiability in insect behaviour. In: *Insect Physiology*. K.D. Roeder (ed.). John Wiley & Sons, N.Y., pp. 723-747.

Schneirla, T.C. 1953b. Insect behaviour in relation to its setting. In: *Insect Physiology*. K.D. Roeder (ed.). John Wiley & Sons, N.Y., pp. 685-722.

Schneirla, T.C. 1957a. A comparison of species and genera in the ant subfamily Dorylinae with respect to functional pattern. *Insectes Sociaux*, 4(3): 259-298.

Schneirla, T.C. 1957b. Theoretical consideration of cyclic processes in doryline ants. *Proc. Amer.*

Phil. Soc., 101: 106-133.

Schneirla, T.C., Brown, R.Z. & Brown. F.C. 1954. The bivouac or temporary nest as an adaptive factor in certain terrestrial species of army ants. *Ecol. Monogr.*, 24: 269-296.

Schöne, H. 1951. Die Lichtorientierung der larvaen von *Acilius sulcatus* L. and *Dytiscus marginalis* L. *Z. vergl. Physiol.*, 33: 63-98.

Schoonhoven, L.M. 1967. Chemoreception of mustard oil glucosides in larvae of *Pieris brassicae*. *Proc. K. Ned. Akad. Wet.*, Ser. C, 70: 556-568.

Schoonhoven, L.M. 1968. Chemosensory bases of host plant selection. *Ann. Rev. Entomol.*, 13: 115-136.

Schoonhoven, L.M. 1982. Biological aspects of antifeedants. *Entomol. Exp. Appl.*, 31: 57-69.

Schoonhoven, L.M., Jermy, T. & Loon. J.J.A. van, 1998. *Insect-Plant Biology: From Physiology to Evolution*. Chapman & Hall, N.Y.

Schoonhoven, L.M. & Loon, J.J.A. van, 2002. An inventory of taste in caterpillars: each species its own key. *Acta Zool. Acad. Scient. Hungaricae*, 48 (Suppl.1): 215-263.

Schroder, R.F.W., Sidor, A.M. & Athanas, M.M. 1996. Parental care in *Erixestus winnemana* (Hymenoptera: Pteromalidae), an egg parasite of *Calligrapha* (Coleoptera: Chrysomelidae). *Entomol. News*, 107(3): 161-165.

Schulz, D.J. & Robinson, G.E. 2001. Octopamine influences division of labour in honey bee colonies. *J. Comp. Physiol. A*, 187: 53-61.

Schwartzkopff, J. 1974. Mechanoreception. In: *The Physiology of Insecta*. Vol. 2. M. Rockstein (ed.). Academic Press, N.Y., pp. 273-352.

Schwarz, D., McPheron, B.A., Hartl, G.B., Boller, E.F. & Hoffmeister, T.S. 2003. A second case of genetic host races in Rhagoletis? A population genetic comparison of sympatric host populations in the European cherry fruit fly, *Rhagoletis cerasi*. *Entomol. Exp. Appl.*, 108: 11-17.

Schwind, R. 1989. A variety of insects are attracted to water by reflected polarized light. *Naturwissenschaften*, 76(8): 377-378.

Scott, M.P. 1989. Male parental care and reproductive success in the burying beetle, *Nicrophorus orbicollis*. *J. Insect Behav.*, 2 : 133-137.

Scott, M.P. 1998. The ecology and behaviour of burying beetles. *Ann. Rev. Entomol.*, 43: 595-618.

Seabrook, W.D. 1977. Insect chemosensory responses to other insects. In: *Chemical Control of Insect Behaviour*. H.H. Shorey & J.J. McKelvey Jr. (eds.). John Wiley & Sons, N.Y., pp. 15-43.

Seevers, C.H. 1965. The systematics, evolution and zoogeography of staphylinid beetles associated with army ants. *Fieldiana Zool.*, 47: 137-351.

Seidelmann, K. & Ferenz, H.J. 2002. Courtship inhibition pheromone in desert locusts, *Schistocerca gregaria*. *J. Insect Physiol.*, 48(11): 991-996.

Sen-Sarma, P.K. 1988. Homeostasis in termite society. *Presidential address, 75th Indian Sci, Cong, Pune, (Section Zool., Entomol. and Fish.)*, 38 pp.

Sen-Sarma, P.K. & Chatterjee, P.N. 1966. The effect of preconditioning on the humidity reactions of the workers of *Microcerotermes beesoni* (Isoptera-Termitidae). *Insectes Sociaux*, 13: 267-276.

Seth, R.K., Kaur, J.J., Rao, D.K. & Reynolds, S.E. 2002. Sperm transfer during mating, movement of sperm in the female reproductive tract, and sperm precedence in the common cutworm *Spodoptera litura*. *Physiol. Entomol.*, 27(1): 1-14.

Seybold, S.J., Quilici, D.R., Tillman, J.A., Vanderwel, D., Wood, D.L. & Blomquist G.J. 1995. *De novo* biosynthesis of the aggregation pheromone components ipsenol and ipsdienol by the pine bark beetles *Ips paraconfusus* Lanier and *Ips pini* (Say) (Coleoptera: Scolytidae). *Proc. Natl. Acad. Sci., USA*, 92(18): 8393-8397.

Sharma, V.K. 2003. Adaptive significance of circadian clocks. *Chronobiol. Int.*, 20(6): 901-919.

Sharma, V.K., Lone, S.R. & Goel, A. 2004a. Clocks for sex: loss of circadian rhythms in ants after mating? *Naturwissenschaften*, 91(7): 334-337.

Sharma, V.K., Lone, S.R., Goel, A. & Chandrashekaran, M.K. 2004b. Circadian consequences of social organization in the ant species *Camponotus compressus*. *Naturwissenschaften*, 91(8): 386-390.

Sheeba, V., Sharma, V.K. Chandrashekaran, M.K. & Joshi, A. 1999a. Effect of different light regimes on pre-adult fitness in *Drosophila melanogaster* populations reared in constant light for over six hundred generations. *Biol. Rhythm Res.*, 30(4): 424-433.

Sheeba, V., Sharma, V.K., Chandrashekaran, M.K. & Joshi, A. 1999b. Persistence of eclosion rhythm in *Drosophila melanogaster* after 600 generations in an aperiodic environment. *Naturwissenshaften*, 86: 448-449.

Shen, J.X., Xu, Z.M. & Hankes, E. 1998: Direct homing behaviour in the ant *Tetramorium caespitum* (Formicidae, Myrmicinae). *Anim. Behav.*, 55: 1443-1450.

Sheppard, P.M. 1962. Some aspects of the geography, genetics and taxonomy of a butterfly. In: *Taxonomy and Geography*. D. Nichols (ed.). Syst. Assoc. Publ. 4, pp. 135-152.

Sheppard, P.M. 1975. *Natural Selection and Heredity*. (4th ed.). Hutchinson Univ. Library, London.

Shipp, E. & Otton, J. 1976a.Circadian rhythms of sensitivity to insecticides in *Musca domestica* (Diptera: Muscidae). *Entomol. Exp. Appl.*, 19: 163-171.

Shipp, E. & Otton, J. 1976b. Diel changes in DDT absorption and breakdown rates and respiratory rhythm in the house fly, *Musca domestica*. *Entomol. Exp. Appl.*, 19: 235-242.

Shorey, H.H. 1962. Nature of the sound produced by *Drosophila melanogaster* during courtship. *Science*, 137: 677-678.

Shorey, H.H. 1973. Behavioural responses to insect pheromones. *Ann. Rev. Entomol.*, 18: 349-380.

Shorey, H.H. 1976. *Animal Communication by Pheromones*. Academic Press, N.Y.

Shorey, H.H. & Gaston, L.K. 1965. Sex pheromones of noctuid moths. V. Circadian rhythm of pheromone-responsiveness in males of *Autographa californica, Heliothis virescence, Spodoptera exigua* and *Trichoplusia ni* (Lepidoptera: Noctuidae). *Ann. Entomol. Soc. Am.*, 58: 597-600.

Shorey, H.H. & McKelvey, J.J. Jr. (eds.) 1977. *Chemical Control of Insect Behaviour: Theory and Application*. John Wiley & Sons, N.Y.

Silverstein, R.M., Rodin, J.O. & Wood, D.L. 1966. Sex attractants in frass produced by male *Ips confusus* in ponderosa pine. *Science*, 154: 509-510.

Simpson, S.J. 1990. The pattern of feeding. In: *Biology of Grasshoppers*. R.F. Chapman, & A. Joern (eds.). John Wiley & Sons, N.Y., pp. 73-103.

Simpson, S.J. & Raubenheimer, D. 1993a. A multi-level analysis of feeding behaviour: the geometry of nutritional decisions. *Philos. Trans. Roy. Soc. London, Ser. B*, 342: 381-402.

Simpson, S.J. & Raubenheimer, D. 1993b. The central role of the haemolymph in the regulation of nutrient intake in insects. *Physiol. Entomol.*, 18(4): 395-403.

Simpson, S.J. & Simpson, C.L. 1990. Mechanisms of nutritional compensation by phytophagous insects. In: *Plant Insect Interactions*. Vol. II. E.A. Bernays, (ed.). CRC Press, Boca Raton, pp. 111-160.

Singarajah, K.V. 1988. Spectral sensitivity of motion-sensitive units of the butterfly ventral nerve cord. *J. Insect Physiol.*, 34(11):1005-1012.

Singh, B.N. & Chatterjee, S. 1987. Variation in mating propensity and fertility in isofemale strains of *Drosophila ananassae*. *Genetica*, 73: 237-242.

Singh, B.N. & Pandey, M.B. 1993. Selection for high and low pupation height in *Drosophila ananassae*. *Behav. Genet.*, 23: 239-243.

Singh, H. 1995. *A Textbook of Animal Behaviour*. Anmol Publ. Pvt. Ltd., N. Delhi.

Singh, M.P. 1949. On the possibility of *Agrotis ipsilon* Rott. passing the summer months in the plains of Bihar breeding on alternate host plants. *Indian J. Entomol.*, 11(1): 103-105.

Singh, S.R. & Singh, B.N. 2003. Behavioural genetics of *Drosophila ananassae*. *Genet. Mol. Res.*, 2 (4): 394-409.

Sivinski, J., Aluja, M., Dodson, G., Friedberg, A., Headrick, D., Kaneshiro, K. & Landolt, P. 1999. Topics in the evolution of sexual behaviour in the Tephritidae. In: *Fruit Flies (Tephritidae): Phylogeny and Evolution of Behaviour*. M. Aluja & A.L. Norrbom (eds.). CRC Press, Boca Raton, pp. 751-792.

Sivinski, J. & Pereira, R. 2005. Do wing markings in fruit flies (Diptera: Tephritidae) have sexual significance? *Florida Entomol.*, 88: 321-324.

Sivinski, J.M. 1981. The nature and possible functions of luminescence in Coleoptera larvae. *Coleopt. Bull.*, 35(2): 167-179.

Sivinski, J.M. 1998. Phototropism, bioluminescence, and the Diptera. *Florida Entomol.*, 81(3): 282-292.

Sivinski, J.M. & Wing, S.R. 2004. Visual mating signals. In: *Encyclopedia of Entomology*. Vol. 3. J.L. Capinera (ed.). Kluwer Academic Publ., Dordrecht, Netherlands, pp. 2470-2481.

Skals, N., Plepys, D. & Lofstedt, C. 2003. Foraging and mate-finding in the silver-Y moth, *Autographa gamma* (Lepidoptera: Noctuidae) under the risk of predation. *Oikos*, 102(2): 351-357.

Slansky, F.1993. Nutritional ecology: the fundamental quest for nutrients. In: *Caterpillars: Ecological and Evolutionary Constraints on Foraging*. N.E. Stamp & T.M. Casey (eds.). Chapman & Hall, N.Y., pp. 29-91.

Slater, P.J.B. 1985. *An Introduction to Ethology*. Cambridge Univ. Press, Cambridge.

Slater, P.J.B. 1999. *Essentials of Animal Behaviour*. Cambridge Univ. Press, Cambridge.

Slessor, K.N., Kaminski, L.-A., King, G.G.S., Borden, J.H. & Winston, M.L. 1988. Semiochemical basis of the retinue response to queen honey bees. *Nature*, 332: 354-356.

Slifer, E.H.1970. The structure of arthropod chemoreceptors. *Ann. Rev. Entomol.*, 15: 121-142.

Smedley, S.R., Lafleur, K.A., Gibbons, L., Arce, J.E., Brown, J.T. & Lozier, M.L. 2002. Glandular hairs: pupal chemical defence in a non-native ladybird beetle (Coleoptera: Coccinellidae). *Northeastern-Naturalist*, 9(3): 253-266.

Smith, B.H. & Breed, M.D. 1995. The chemical basis for nestmate recognition and mate discrimination in social insects. In: *Chemical Ecology of Insects* 2. R.T. Cardé & W.J. Bell (eds.). Chapman & Hall, N.Y., pp. 287-317.

Smith, D.C. 1988. Heritable divergence of *Rhagoletis pomonella* host races by seasonal asynchrony. *Nature*, 336: 66-67.

Smith, D.S. 1969. The fine structure of haltere sensilla in the blow fly, *Calliphora erythrocephala* (Meig.) with scanning electron microscopic observations on the haltere surface. *Tissue Cell*, 1: 443-484.

Smith, D.S. & Treherne, J.E. 1963. Functional aspects of the organization of insect nervous system. *Adv. Insect Physiol.*, 1: 401-484.

Smith, J.J. & Bush, G.L.1997. Phylogeny of genus *Rhagoletis* (Diptera: Tephritidae) inferred from DNA sequences of mitochondrial cytochrome oxidase II. *Mol. Phylogenet. Evol.*, 7: 33-43.

Smith, J.J.B. 1985. Feeding mechanisms. In: *Comprehensive Insect Physiology, Biochemistry and Pharmacology. Regulation: Digestion, Nutrition, Excretion.* Vol. 4. G.A. Kerkut & L.I. Gilbert (eds.). Pergamon Press, Oxford, pp. 33-86.

Smith, R.L. 1976a. Male brooding behaviour of the water bug *Abedus herberti* (Hemiptera: Belostomatidae). *Ann. Entomol. Soc. Am.*, 69: 740-747.

Smith, R.L. 1976b. Brooding behaviour of a male water bug *Belostoma flumineum* (Hemiptera: Belostomatidae). *J. Kansas Entomol. Soc.*, 49: 333-343.

Smith, R.L. 1979. Paternity assurance and altered roles in the mating behaviour of a giant water bug, *Abedus herberti* (Heteroptera: Belostomatidae). *Anim. Behav.*, 35: 1200-1209.

Snodgrass, R.E. 1935. *Principles of Insect Morphology*. McGraw-Hill Book Co. Inc., N.Y.

Snodgrass, R.E. 1954. The dragonfly larva. *Smithson. Inst. Misc. Coll.*, 123(2): 1-36.

Sokolove, P.G. 1975. Locomotory and stridulatory circadian rhythms in the cricket, *Teleogryllus commodus*. *J. Insect Physiol.*, 21: 537-558.

Sokolove, P.G. & Loher, W. 1975. Role of eyes, optic lobes, and pars intercerebralis in locomotory and stridulatory circadian rhythms of *Teleogryllus commodus*. *J. Insect Physiol.*, 21: 785-799.

Sorge, D., Nauen, R., Range, S. & Hoffmann, K.H. 2000. Regulation of vitellogenesis in the fall armyworm, *Spodoptera frugiperda* (Lepidoptera: Noctuidae). *J. Insect Physiol.*, 46(6): 969-976.

Sotavalta, O. 1953. Recordings of high wing-stroke and thorax vibration frequency in some midges. *Biol. Bull.*, 104: 439-444.

Sotthibandhu, S. & Baker, R.R. 1979. Celestial orientation by the large yellow underwing moth, *Noctua pronuba* L. *Anim. Behav.*, 27: 786-800.

Southwick, E.E. & Moritz, R.F.A. 1987. Social synchronization of circadian rhythms of metabolism in honey bees (*Apis mellifera*). *Physiol. Entomol.*, 12: 209-212.

Southwood, T.R.E. 1962. Migration of terrestrial arthropods in relation to habitat. *Biol. Rev.*, 37: 171-214.

Southwood, T.R.E. 1973. The insect/plant relationship- an evolutionary perspective. In: *Insect Plant Relationship*. H.F. Emden, van (ed.). *Symp. Roy. Entomol. Soc. Lond.*, 6: 3-30.

Southwood, T.R.E. 1981. Ecological aspects of insect migration. In: *Animal Migration*. D.J. Aidley (ed.) Cambridge Univ. Press, Cambridge, pp. 196-208.

Sowig, P. 1996. Duration and benefits of biparental brood care in the dung beetle *Onthophagus vacca* (Coleoptera: Scarabaediae). *Ecol. Entomol.*, 21(1): 81-86.

Spieth, H.T. 1947. Sexual behaviour and isolation in *Drosophila*. I. The mating behaviour of species of *willistoni* group. *Evolution*, 1: 17-31.

Spieth, H.T. 1952. The mating behaviour of the genus *Drosophila* (Diptera). *Bull. Amer. Mus. Nat. Hist.*, 99: 399-474.

Spieth, H.T. 1966a. Mating behaviour of *Drosophila ananassae* and *ananassae*-like flies from the Pacific. *Univ. Tex. Publ.*, 3: 133-146.

Spieth, H.T. 1966b. Courtship behaviour of endemic Hawaiian *Drosophila: Univ. Texas Publ.*, 6615, pp. 245-313.

Spieth, H.T. 1974. Courtship in *Drosophila*. *Ann. Rev. Entomol.*, 19: 385-405.

Spradberg, J.P. 1973. *Wasps. An Account of the Biology and Natural History of Solitary and Social Wasps with Particular Reference to Those of the British Isles*. Sidgwick & Jackson, London.

Sreenivasulu, V., Subramanyam, G. & Rajarami Reddy, G. 1996. Cardiotoxicity of the grasshopper *Poecilocerus pictus* defensive secretion in the garden lizard, *Calotes nemoricola*. *Entomon*, 21(1): 19-21.

Srinivasan, M.V., Zhang, S., Altwein, M. & Tautz, J. 2000. Honey bee navigation: nature and calibration of the "odometer". *Science*, 287(5454): 851-853.

Srivastava, R.P. & Bhatttacharya, A.K. 2003. Sex pheromone in integrated pest management. In: *Biopesticides and Bioagents in Integrated Pest Management*. R.P. Srivastava (ed.). International Book Dist. Co. Lucknow, pp. 721-809.

St. Clair, A., Goulding, R.L. & Rudinsky, J.A. 1977. Controlled-release dispensing system for 3,2-MCH, antiaggregative pheromone of the Douglas-fir beetle. *Z. Angew. Entomol.*, 83: 297-300.

St. Hilaire, I.G. 1859. Histoire Générale des Régnes Organiques.

Städler, E. 1974. Host plant stimuli affecting oviposition behaviour of the eastern spruce budworm. *Entomol. Exp. Appl.*, 17: 176-188.

Städler, E. 1984. Contact chemoreception. In: *Chemical Ecology of Insects*. W.J. Bell & R.T. Cardé (eds.). Chapman & Hall, N.Y., pp. 3-35.

Steinberg, S., Dicke, M. & Vet, L.E.M. 1993. Relative importance of infochemicals from first and second trophic level in long-range host location by the larval parasitoid *Cotesia glomerata*. *J. Chem. Ecol.*, 19: 47-59.

Steiner, L.F. 1952. Methyl eugenol as an attractant for the oriental fruit fly. *J. Econ. Entomol.*, 45: 241-248.

Steiner, L.F. 1957. Low-cost plastic fruit fly trap. *J. Econ. Entomol.*, 50: 508-509.

Steiner, L.F., Mitchell, W.C., Harris, E.J., Kozuma, T.T. & Fujimoto, M.S. 1965. Oriental fruit fly eradication by male annihilation. *J. Econ. Entomol.*, 58 (5): 961-964.

Stengel, M.M.D. 1974. Migratory behaviour of the female of the common cockchafer *Melolontha melolontha* L. and its neuroendocrine regulation. In: *Experimental Analysis of Insect Behaviour*. L. Barton Browne (ed.). Springer-Verlag, Berlin, pp. 297-303.

Stern, D.L. & Foster, W.A. 1996. The evolution of sterile soldiers in aphids. *Biol. Rev.*, 71: 27-79.

Stiedl, O. & Kalmring, K. 1989. The importance of song and vibratory signals in the behaviour of the bush cricket *Ephippiger ephippiger* Fiebig (Orthoptera, Tettigoniidae): taxis by females. *Oecologica*, 80(1): 142-144.

Stinner, R.E., Barfield, C.S., Stimae, J.L. & Dohse, L. 1983. Dispersal and movement of insect pests. *Ann. Rev. Entomol.*, 28: 319-335.

Stoffolano, J.G. Jr. & Yin, L.R.S. 1987. Structure and function of the ovipositor and associated sensilla of the apple maggot, *Rhagoletis pomonella* (Walsh) (Diptera: Tephritidae). *Int. J. Insect Morphol. Embryol.*, 16: 41-70.

Stoks, R., Bruyn, L.D. & Matthysen, E. 1997. The adaptiveness of intense contact mate guarding by males of the emerald damselfly, *Lestes sponsa* (Odonata, Lestidae): the male's perspective. *J. Insect Behav.*, 10: 289-298.

Stolz, U., Velez, S., Wood, K.V., Wood, M. & Feder, J.L. 2003. Darwinian natural selection for orange bioluminescent colour in a Jamaican click beetle. *Proc. Natl. Acad. Sci., USA*, 100(25): 14955-14959.

Stowe, M.K. 1988. Chemical mimicry. In: *Chemical Mediation of Coevolution*. K. Spencer (ed.). Academic Press, N.Y., pp. 513-580.

Straw, N.A. 1989. Evidence for an oviposition-deterring pheromone in *Tephritis bardanae* (Diptera: Tephritidae). *Oecologia*, 78: 121-130.

Struwe, G. 1972. Spectral sensitivity of the compound eye in butterflies (*Heliconius*). *J. Comp. Physiol.*, 79: 191-196.

Stuart, A.E. & Hunter, F.F. 1998. End-products of behaviour versus behavioural characters: a phylogenetic investigation of pupal cocoon construction and form in some North American black flies (Diptera: Simuliidae). *Syst. Entomol.*, 23(4): 387-398.

Subramanyan, M.V.V. & Krishnamoorthy, R.V. 1990. Gravity orientation in the silkworm *Bombyx mori*. *J. Ecobiol.*, 2(1): 24-32.

Sueur, J. & Aubin, T. 2004. Acoustic signals in cicada courtship (order Hemiptera, genus *Tibicina*). *J. Zool., Lond.*, 262: 217-224.

Sullivan, J.P., Fahrbach, S.E., Harrison, J.F., Capaldi, E.A., Fewell, J.H. & Robinson, G.E. 2003. Juvenile hormone and division of labour in honey bee colonies: effects of allatectomy on flight behaviour and metabolism. *J. Exp. Biol.*, 206: 2287-2296.

Sullivan, J.P., Jassim, O., Fahrbach, S.E. & Robinson, G.E. 2000. Juvenile hormone paces behavioural development in the adult worker honey bee. *Horm. Behav.*, 37: 1-14.

Sullivan, W.N., Cawley, B., Hayes, D.K., Rosenthal, J. & Halberg, F. 1970. Circadian rhythm in susceptibility of house files and Madeira cockroaches to pyrethrum. *J. Econ. Entomol.*, 63: 159-163.

Sumana, A. & Gadagkar, R. 2003. *Ropalidia marginata*- a primitively eusocial wasp society headed by behaviourally non-dominant queens. *Curr. Sci.*, 84(11): 1464-1468.

Sutcliffe, J.F. 1987. Distance orientation of biting flies to their hosts. *Insect Sci. Appl.*, 8: 611-616.

Sutcliffe, J.F & McIver, S.B. 1984. Mechanics of blood-feeding in black flies (Diptera, Simuliidae). *J. Morphol.*, 180(2): 125-144.

Sutherland, O.R.W. 1977. Plant chemicals influencing insect behaviour. *New Zealand Entomol.*, 6(3): 222-228.

Svensson, B.G. & Petersson, E. 1987. Sex-role reversed courtship behaviour, sexual dimorphism and nuptial gifts in the dance fly, *Empis borealis* (L.). *Ann. Zool. Fenn.*, 24: 323-334.

Svensson, G.P., Lofstedt, C. & Skals, N. 2004. The odour makes the difference: male moths attracted by sex pheromones ignore the threat by predatory bats. *Oikos*, 104(1): 91-97.

Szabad, J. & Fajszi, C. 1982. Control of female reproduction in *Drosophila*: genetic dissection using gynandromorphs. *Genetics*, 100(1): 61-78.

Tachon, G., Murray, A.M., Gray, D.A. & Cade, W.H. 1999. Agonistic displays and the benefits of fighting in the field cricket, *Gryllus bimaculatus*. *J. Insect Behav.*, 12(4): 533-543.

Takahashi, S. 1993. Insect behaviour modifiers. In: *Bioactive Natural Products- Detection, Isolation and Structural Determination*. S.M. Colegate & R.J. Molyeux (eds.). CRC Press, Boca Raton, pp. 241-263.

Takeda, K. 1961. Classical conditioned response in the honey bee. *J. Insect Physiol.*, 6: 168-179.

Takeda, M. & Skopik, S.D. 1997. Photoperiodic time measurement and related physiological mechanisms in insects and mites. *Ann. Rev. Entomol.*, 42: 323-349.

Tallamy, D.W., Stull, J., Ehresman, N.P., Gorski, P.M. & Mason, C.E. 1997. Cucurbitacins as feeding and oviposition deterrents to insects. *Environ. Entomol.*, 26(3): 678-783.

Tamaki, Y. 1985. Sex pheromones. In: *Comprehensive Insect Physiology, Biochemistry and Pharmacology. Behaviour*. Vol. 9. G.A. Kerkut & L.I. Gilbert (eds.). Pergamon Press, Oxford, pp. 145-191.

Tamaki, Y., Kawasaki, K., Yamada, H., Koshihara, T., Osaki, N., Ando, T., Yoshida, S. & Kakinohana, H. 1977. (Z)-11-Hexadecenal and (Z)-11-hexadecenyl acetates: sex-pheromone components of the diamondback moth (Lepidoptera: Plutellidae). *Appl. Entomol. Zool.*, 12: 208-210.

Tanaka, S. 2001. Endocrine mechanisms controlling body-colour polymorphism in locusts. *Arch. Insect Biochem. Physiol.*, 47(3): 139-149.

Tanaka, S. 2004. Hormonal control of body-colour polyphenism in the American grasshopper, *Schistocerca americana*: a function of [His7]-corazonin. *Ann. Entomol. Soc. Am.*, 97(2): 302-309.

Tauber, E., Cohen, D., Greenfield, M.D. & Pener, M.P. 2001. Duet singing and female choice in the bush cricket *Phaneroptera nana*. *Behaviour*, 138(4): 411-430.

Tauber, M.J., Tauber, C.A. & Masaki, S. 1986. *Seasonal Adaptations of Insects*. Oxford Univ. Press, Oxford.

Taub-Montemayor, T.E., Palmer, J.O. & Rankin, M.A. 1997. Endocrine regulation of reproduction and diapause in the boll weevil, *Anthonomus grandis* Boheman. *Arch. Insect Biochem. Physiol.*, 35: 455-477.

Taylor, B. & Jones, M.D.R. 1969. The circadian rhythm of flight activity in the mosquito *Aedes aegypti* (L.): the phase-setting effects of light-on and light-off. *J. Exp. Biol.*, 51: 59-70.

Taylor, L.R. 1986. The four kinds of migration. In: *Insect Flight: Dispersal and Migration*. W. Danthanarayan (ed.). Springer-Verlag, Berlin, pp. 265-280.

Taylor, S.J. 1988. Skototaxis in three species of flat bugs (Heteroptera: Aradidae). *Ann. Entomol. Soc. Am.*, 81(3): 537-538.

Teerling, C.R., Gillespie, D.R. & Borden, J.H. 1993. Utilization of western flower thrips alarm pheromone as a prey-finding kairomone by predators. *Can. Entomol.*, 125(3): 431-437.

Teimann, D. L. 1970. Nature's toy train, the railroad worm. *Natl. Geogr.*, 138: 56-57.

Telfer, W.H. 1954. Immunological studies of insect metamorphosis. II. The role of a sex-limited protein in egg formation by the cecropia silkworm. *J. Gen. Physiol.*, 37: 539-558.

Thiery, D. & Visser, J.H. 1986. Masking of host plant odour in the olfactory orientation of the Colorado potato beetle. *Entomol. Exp. Appl.*, 41(2): 165-172.

Thomas, L.K. & Manica, A. 2003. Filial cannibalism in an assassin bug. *Anim. Behav.*, 66(2): 205-210.

Thompson, A.C. & Mitlin, N. 1979. Biosynthesis of the sex pheromone of the male boll weevil from monoterpene precursors. *Insect Biochem.*, 9: 293-294.

Thompson, J.N. 1994. *The Coevolutionary Process*. Univ. Chicago Press, Chicago.

Thompson, J.N. 1999. Specific hypotheses on the geographic mosaic of coevolution. *Amer. Nat.*, 153 (Suppl.): 1-14.

Thompson, J.N. & Pellmyr, O. 1991. Evolution of oviposition behaviour and host preference in Lepidoptera. *Ann. Rev. Entomol.*, 36: 65-89.

Thorne, B.L. 1997. Evolution of eusociality in termites. *Ann. Rev. Ecol. Syst.*, 28: 27-54.

Thorne, B.L., Breisch, N.L. & Muscedere, M.L. 2003. Evolution of eusociality and the soldier caste in termites: influence of intraspecific competition and accelerated inheritance. *Proc. Natl. Acad. Sci., USA*, 22: 12808-12813.

Thornhill, R. 1976. Sexual selection and nuptial feeding behaviour in *Bittacus apicalis* (Insecta: Mecoptera). *Amer. Nat.*, 110: 529-548.

Thornhill, R. & Alcock, J. 1983. *The Evolution of Insect Mating Systems*. Harvard Univ. Press, Cambridge, Massachusetts.

Thorsteinson, A.J. 1960. Host selection in phytophagous insects. *Ann. Rev. Entomol.*, 5: 193-218.

Thorpe, W.H. 1939a. Further studies on the olfactory conditioning in a parasitic insect: the nature of conditioning process. *Proc. Roy. Soc. Lond. Ser. B, Biol. Sci.*, 126: 379-397.

Thorpe, W.H. 1939b. Further studies of pre-imaginal olfactory conditioning in insects. *Proc. Roy. Soc. Lond. Ser. B, Biol. Sci.*, 127: 424-433.

Thorpe, W.H. 1950. A note on detour behaviour with *Ammophila pubescens* Curt. *Behaviour*, 2: 257-264.

Thorpe, W.H. 1963. *Learning and Instincts in Animals*. (2nd ed.). Methuen & Co. Ltd., London.

Thorpe, W.H. 1974. *Animal Nature and Human Nature*. Methuen & Co. Ltd., London.

Thorpe, W.H. 1979. *The Origin and Rise of Ethology*. Heinemann, London.

Thorpe, W.H. & Jones, F.G.W. 1937. Olfactory conditioning and its relation to the problem of host selection. *Proc. Roy. Soc. Lond. Ser. B, Biol. Sci.*, 124: 56-81.

Tillman, J.A., Seybold, S.J., Jurenka, R.A. & Blomquist, G.J. 1999. Insect pheromones- an overview of biosynthesis and endocrine regulation. *Insect Biochem. Mol. Biol.*, 29(6): 481-514.

Timmins, G.S., Robb, F.J., Wilmot, C.M., Jackson, S.K. & Swartz, H.M. 2001. Firefly flashing is controlled by gating oxygen to light-emitting cells. *J. Exp. Biol.*, 204: 2795-2810.

Tinbergen, N. 1942. An objective study of the innate behaviour of animals. *Biblioth. Biother.*, 1: 39-98.

Tinbergen, N. 1951. *The Study of Instinct*. Oxford Univ. Press, Oxford.

Tinbergen, N. 1953. *Social Behaviour in Animals*. Methuen & Co. Ltd., London.

Tinbergen, N. 1955. Some aspects of ethology, the biological study of animal behaviour. *Advan. Sci.*, 12: 17-27.

Tinbergen, N. 1958. *Curious Naturalists*. Doubleday Garden City, N.Y.

Tinbergen, N. 1963. On aims and methods of ethology. *Z. Tierpsychol.*, 20: 404-433.

Tinbergen, N. 1965. *Animal Behaviour*. Time-Life Books, N.Y.

Tinbergen, N. 1972. *The Animal and its World. Exploration of an Ethologist. 1932-1972. Vol. I. Field Studies*. Harvard Univ. Press, Cambridge, Massachusetts.

Tinbergen, N. 1973. *The Animals in its World: Explorations of an Ethologist*. (Vol. 1 & 2). Harvard Univ. Press, Cambridge, Massachusetts.

Tinbergen, N. & Kruyt, W. 1938. Über die Orientierung des Bienenwolfes (*Philanthus triangulum* Fabr.). III. Die Bevorzugung bestimmter Wegmarken. *Z. vergl. Physiol.*, 25: 292-334.

Tinzaara, W., Dicke, M., Huis, A. van, & Gold, C.S. 2002. Use of infochemicals in pest management with special reference to the banana weevil, *Cosmopolites sordidus* (Germar) (Coleoptera: Curculionidae). *Insect Sci. Appl.*, 22(4): 241-261.

Tipping, C. 1995. The longest migration. In: *University of Florida Book of Insect Records*. 1997. T.J. Walker (ed.), pp. 23-25.

Tishechkin, D. Yu. 2003. Vibrational communication in leafhoppers from Ulopides subfamilies group (Homoptera: Cicadellidae) and Membracidae, with notes on the classification of higher taxa. *Russian Entomol. J.*, 12(1): 10-58.

Tomioka, K. & Chiba, Y. 1992. Characterization of an optic lobe circadian pacemaker by *in situ* and *in vitro* recording of neural activity in the cricket, *Gryllus bimaculatus*. *J. Comp. Physiol. A*, 171: 1-7.

Tooker, J.F. & Hanks, L.M. 2004. Stereochemistry of host plant monoterpenes as mate location cues for the gall wasp *Antistrophus rufus*. *J. Chem. Ecol.*, 30(2): 473-477.

Torr, S.J. 1989. The host-orientated behaviour of tsetse flies (*Glossina*): the interaction of visual and olfactory stimuli. *Physiol. Entomol.*, 14(3): 325-340.

Trimmer, B.A., Aprille, J.R., Dudzinski, D.M., Lagace, C.J., Lewis, S.M., Michel, T., Qazi, S. & Zayas, R.M. 2001. Nitric oxide and the control of firefly flashing. *Science*, 292(5526): 2486-2488.

Trivers, R.L. 1972. Parental investment and sexual selection. In: *Sexual Selection and the Descent of Man, 1871-1971*. B. Campbell (ed.). Aldine, Chicago, pp. 136-179.

Trivers, R.L. & Hare, H. 1976. Haplodiploidy and the evolution of social insects. *Science*, 191: 249-263.

Trujillo-Cenoz, O. 1972. The structural organization of the compound eye in insects. In: *Physiology of Photoreceptor Organs*. M.G.E. Fuortes (ed.) (vol. III). Springer-Verlag, Berlin, pp. 5-62.

Truman, J.W. 1971a. Circadian rhythm and physiology with special reference to neuroendocrine processes in insects. *Proc. Int. Symp. Circadian Rhythmicity* (Wageningen, 1971), pp. 111-135.

Truman, J.W. 1971b. Hour-glass behaviour of the circadian clock controlling eclosion of the silk moth *Antheraea pernyi*. *Proc. Natl. Acad. Sci., USA*, 68: 595-599.

Truman, J.W. 1972a. Physiology of insect rhythms. I. Circadian organization of the endocrine events underlying the moulting cycle of larval tobacco hornworm. *J. Exp. Biol.*, 57: 805-820.

Truman, J.W. 1972b. Physiology of insect rhythms. II. The silkworm brain as the location of the biological clock controlling eclosion. *J. Comp. Physiol.*, 81: 99-114.

Truman, J.W. & Riddiford, L.M. 1970. Neuroendocrine control of ecdysis in silk moths. *Science*, 167: 1624-1626.

Truman, J.W. & Riddiford, L.M. 1974. Hormonal mechanisms underlying insect behaviour. *Adv. Insect Physiol.*, 10: 297-352.

Trumbo, S.T. & Eggert, A.-K. 1994. Beyond monogamy: territory quality influences sexual advertisement in male burying beetles. *Anim. Behav.*, 48: 1043-1047.

Trumen, J. 1976. Extraretinal photoreception in insects. *Photochem. Photobiol.*, 23: 215-225.

Tumlinson, J.H., Brennan, M.M., Doolittle, R.E., Mitchell, E.R., Brabham, A., Mazomenos, B.E., Baumhover, A.H. & Jackson, D.M. 1989. Identification of a pheromone blend attractive to *Manduca sexta* (L.) males in a wind tunnel. *Arch. Insect Biochem. Physiol.*, 10: 255-271.

Tumlinson, J.H., Lewis, W.J. & Vet, L.E.M. 1993a. How beneficial parasitic wasps find their hosts. *Sci. Amer.*, 268(3): 100-106.

Tumlinson, J.H., Turlings, T.C.J. & Lewis, W.J. 1993b. Semiochemically mediated foraging behaviour in beneficial parasitic insects. *Arch. Insect Biochem. Physiol.*, 22(3-4): 385-391.

Turlings, T.C.J. & Tumlinson, J.H. 1992. Systemic release of chemical signals by herbivore-injured corn. *Proc. Natl. Acad. Sci., USA*, 89(17): 8399-8402.

Turner, C.H. 1913. Behaviour of the common roach (*Periplaneta orientalis* L.) on an open maze. *Biol. Bull.*, 25: 348-365.

Tutt, J.W. 1902. *Migration and Dispersal of Insects*. Elliot Stock, London.

Tyschen, P.H. 1977. Mating behaviour of the Queensland fruit fly, *Bactrocera tryoni* in field cages. *J. Aust. Entomol. Soc.*, 16: 459-465.

Ugolini, A. 1985. Initial orientation and homing in workers of *Polistes gallicus* (L.) (Hymenoptera, Vespidae). *Z. Tierpsychol.*, 69(2): 133-140.

Ugolini, A. 1986. Homing ability in *Polistes gallicus* (L.) (Hymenoptera, Vespidae). *Monit. Zool. Ital.*, 20(1): 1-15.

Ulagaraj, S.M. & Walker, T.J. 1973. Phonotaxis of crickets in flight: attraction of male and female crickets to male calling songs. *Science*, 182: 1278-1279.

Underwood, T.J., Tallamy, D.W. & Pesek, J.D. 1997. Bioluminescence in firefly larvae: a test of the aposematic display hypothesis (Coleoptera: Lampyridae). *J. Insect Behav.*, 10(3): 365-370.

Unnithan, G.C. & Saxena, K.N. 1990. Diversion of ovposition by *Atherigona soccata* (Diptera: Muscidae) to nonhost maize with sorghum seedling extract. *Environ. Entomol.*, 19: 1432-1437.

Urbanova, K., Halik, J., Hovorka, O., Kindl, J. & Valterova, I. 2004. Marking pheromones of the cuckoo bumble bee males (Hymenoptera, Apoidea, *Bombus* Latreille): compositions of labial gland secretions of six species found in the Czech Republic. *Biochem. Syst. Ecol.*, 32(11): 1025-1045.

Urquhart, F.A. 1960. *The Monarch Butterfly*. Univ. Toronto Press, Toronto.

Urquhart, F.A. 1976. Found at last: the monarch winter home. *Natl. Geogr.*, 150: 161-173.

Uvarov, B.P. 1966. *Grasshoppers and Locusts. A Handbook of General Acridology*. Vol. 1. Cambridge Univ. Press, Cambridge.

Uvarov, B.P. 1977. *A Handbook of General Acridology*. Vol.2. Centre for Overseas Pest Research, London.

Vafopoulou, X. & Steel, C.G.H. 1991. Circadian regulation of synthesis of ecdysteroids by prothoracic glands of the insect *Rhodnius prolixus*: evidence of a dual oscillator system. *Gen. Comp. Endocrinol.*, 83(1): 27-34.

Vafopoulou, X. & Steel, C.G.H. 1996. The insect neuropeptide prothoracicotropic hormone is released with a daily rhythm: re-evaluation of its role in development. *Proc. Natl. Acad. Sci., USA.*, 93(8): 3368-3372.

Vafopoulou, X. & Steel, C.G.H. 1999. Daily rhythm of responsiveness to prothoracicotropic hormone in prothoracic glands of *Rhodnius prolixus*. *Arch. Insect Biochem. Physiol.*, 41(3): 117-123.

Vahed, K. 1998. The function of nuptial feeding in insects: a review of empirical studies. *Biol. Rev.*, 73(1): 43-78.

Vale, G.A. & Hall, D.R. 1985. The role of 1-octen-3-ol, acetone and carbon dioxide in the attraction of tsetse flies, *Glossina* spp. (Diptera: Glossinidae), to ox odour. *Bull. Entomol. Res.*, 75: 209-217.

Valen, L. van, 1973. A new evolutionary theory. *Evol. Theory*, 1: 1-30.

Varalakshmi, Y. & Mohan, P.M. 2000. Neural control of spinning and pupation in the silkworm *Bombyx mori* L. *J. Adv. Zool.*, 21(2): 104-107.

Vargo, E.L. 1997. Poison gland of queen fire ants (*Solenopsis invicta*) is the source of a primer pheromone. *Naturwissenschaften*, 84: 507-510.

Vea, E.V., Cutkomp, L.K. & Halberg F. 1977. Interrelationships of oxygen consumption and insecticide sensitivity. *Chronobiologia*, 4: 313-323.

Vecht, J. van der, 1961. Evolution in a group of Indo-Australian *Eumenes*. *Evolution*, 15: 468-477.

Veith, H.J., Koeniger, N. & Maschwitz, U. 1984. 2-methyl-3-butene-2-ol, a major component of the alarm pheromone of the hornet, *Vespa crabro*. *Naturwissenschaften*, 71: 328-329.

Vencl, F.V., Blasko, B.J. & Carlson, A.D. 1994. Flash Behaviour of female *Photuris versicolor* fireflies in simulated courtship and predatory dialogues. *J. Insect Behav.*, 7(6): 843-858.

Venkatraman, T.V. 1954. A note on the possibility of *Agrotis ipsilon* Rott., migrating from the plains to the hills during summer months in parts of India. *Indian J. Entomol.*, 16(2): 197-198.

Vet, L.E.M. & Dicke, M. 1992. Ecology of infochemical use by natural enemies in a tritrophic context. *Ann. Rev. Entomol.*, 37: 141-172.

Vet, L.E.M. & Groenwold, A.W. 1990. Semiochemicals and learning in parasitoids. *J. Chem. Ecol.*, 16: 3119-3135.

Villet, M.H. 1987. Sound pressure levels of some African cicadas (Homoptera: Cicadoidea). *J. Entomol. Soc. Sth. Afr.*, 50: 269-273.

Vinson, S.B. 1976. Host selection by insect parasitoids. *Ann. Rev. Entomol.*, 21: 109-133.

Vinson, S.B. 1984a. Parasitoid-host relationship. In: *Chemical Ecology of Insects*. W.J. Bell & R.T. Cardé (eds.). Chapman & Hall, N.Y., pp. 205-233.

Vinson, S.B. 1984b. How parasitoids locate their hosts: a case of insect espionage. In: *Insect Communications*. T. Lewis (ed.). Academic press, London, pp. 325-348.

Vinson, S.B. 1985. The behaviour of parasitoids. In: *Comprehensive Insect Physiology, Biochemistry and Pharmacology. Behaviour*. Vol. 9. G.A. Kerkut & L.I. Gilbert (eds.). Pergamon Press, Oxford, pp. 417-469.

Virant-Doberlet, M. & Čokl, A. 2004. Vibrational communication in insects. *Neotrop. Entomol.*, 33(2): 121-134.

Visser, J.H. 1986. Host odour perception in phytophagous insects. *Ann. Rev. Entomol.*, 31: 121-144.

Viviani, V.R. 2001. Fireflies (Coleoptera: Lampyridae) from Southeastern Brazil: habitats, life history, and bioluminescence. *Ann. Entomol. Soc. Am.*, 94: 129-145.

Viviani, V.R. 2002. The origin, diversity, and structure, function relationships of insect luciferases. *Cell. Mol. Life Sci.*, 59(11): 1833-1850.

Viviani, V.R. & Bechara, E.J.H. 1997. Bioluminescence and biological aspects of Brazilian railroad-worms (Coleoptera: Phengodidae). *Ann. Entomol. Soc. Am.*, 90(3): 389-398.

Vogt, R.G. 2005. Molecular basis of pheromone detection in insects. In: *Comprehensive Insect Physiology, Biochemistry, Pharmacology and Molecular Biology. Vol. 3. Endocrinology*. L.I. Gilbert, K. Iatro & S. Gill (eds.). Elsevier Academic Press, London, pp. 753-804.

Vogt, R.G., Callahan, F.E., Rogers, M.E. & Dickens J.C. 1999. Odorant binding protein diversity and distribution among the insect orders, as indicated by LAP, an OBP-related protein of the true bug *Lygus lineolaris* (Hemiptera, Heteroptera). *Chem. Senses*, 24: 481-495.

Vogt, R.G. & Riddiford, L.M. 1981. Pheromone binding and inactivation by moth antennae. *Nature*, 293: 161-163.

Vogt, R.G., Riddiford, L.M. & Prestwich, G.D. 1985. Kinetic properties of a sex pheromone-degrading enzyme: the sensillar esterase of *Antheraea polyphemus*. *Proc. Natl. Acad. Sci., USA*, 82: 8827-8831.

Vorhis-Key, S.E. van & Baker, T.C. 1982. Trail Pheromone conditioned anemotaxis by the Argentine ant. I. *Iridomyrmex humilis*. *J. Insect Physiol.*, 27: 363-370.

Vuline, K. 1987. Swimming in whirligig beetle (Col. Gyrinidae): a possible role of the pygidial gland secretion. *Coleop. Bull.*, 41(2): 151-153.

Waage, J.K. 1979. Dual function of the damselfly: sperm removal and transfer. *Science*, 203: 916-

918.

Wada, K. & Munakata, K. 1968. Naturally occurring insect control chemicals. Isoboldine, a feeding inhibitor and cocculolidine, an insecticide, in the leaves of *Cocculus trilobus* D.C. *J. Agril. Food Chem.*, 16: 471-474.

Wahl, O. 1932. Neue Untersuchungen über das Zeitgedächtnis der Bienen. *Z. vergl. Physiol.*, 16: 529-589.

Walgenbach, C.A., Phillips, J.K., Faustini, D.L. & Burkholder, W.E. 1983. Male-produced aggregation pheromone of the maize weevil, *Sitophilus zeamais*, and interspecific attraction between three *Sitophilus* species. *J. Chem. Ecol.*, 9: 831-841.

Walker, T.J. 1964. Cryptic species among sound producing ensiferan Orthoptera (Gryllidae and Tettigoniidae). *Quart. Rev. Biol.*, 39: 345-355.

Walker, T.J. 1982. Sound traps for sampling mole cricket flights (Orthoptera: Gryllotalpidae: *Scapteriscus*). *Florida Entomol.*, 65(1): 105-110.

Walker, T.J. 1988. Acoustic traps for agriculturally important insects. *Florida Entomol.*, 71(4): 484-492.

Walker, T.J. 1989. A live trap for monitoring *Euphasiopteryx* and tests with *E. ochracea* (Diptera: Tachinidae). *Florida Entomol.*, 72(2): 314-319.

Walker, T.J. & Masaki, S. 1989. Natural history. In: *Cricket Behaviour and Neurobiology*. F. Huber, T.E. Moore & W. Loher (eds.). Cornell Univ. Press, Ithaca, N.Y., pp. 1-42.

Wallace, E.K., Albert, P.J. & McNeil, J.N. 2004. Oviposition behaviour of the eastern spruce budworm *Choristoneura fumiferana* (Clemens) (Lepidoptera: Tortricidae). *J. Insect Behav.*, 17(2): 145-154.

Ware, G.W. & McComb, M. 1970. Circadian susceptibility of pink bollworm moths to azinphosmethyl. *J. Econ. Entomol.*, 63: 1941-1942.

Waters, D.A. 2003. Bats and moths: what is there left to learn? *Physiol. Entomol.*, 28 (4): 237-250.

Watson, J.A.L. & Abbey, H.M. 1985. Seasonal cycles in *Naustitermes exitiosus* (Hill) (Isoptera, Termitidae). 1. Caste development. *Sociobiology*, 10(1): 73-92.

Wearing, C.H. & Hutchins, R.F.N. 1973. α-farnesene a naturally occurring oviposition stimulant for the codling moth, *Laspeyresia pornonella*. *J. Insect Physiol.*, 19: 1251-1256.

Weatherston, J. 1967. The chemistry of arthropod defensive substances. *Ouart. Rev.*, 21: 287-313.

Weatherston, J. & Percy, J.L. 1970. Arthropod defensive secretions. In: *Chemicals Controlling Insect Behaviour*. M. Beroza (ed.). Academic Press, N.Y., pp. 95-144.

Weaver, J.S. III & Morse, J.C. 1986. Evolution of feeding and case-making behaviour in Trichoptera. *J. North Amer. Benthol. Soc.*, 5(2): 150-158.

Webb, J.C., Burk, T. & Sivinski, J. 1983. Attraction of female Caribbean fruit flies, *Anastrepha suspensa* (Diptera: Tephritidae), to the presence of males and male-produced stimuli in field cages. *Ann. Entomol. Soc. Am.*, 76: 996-998.

Weber, N.A. 1972. *Gardening Ants: The Attines*. Memoir 92, Amer. Philos. Soc., Philadelphia.

Wedell, N. & Karlsson, B. 2003. Paternal investment directly affects female reproductive effort in an insect. *Proc. Roy. Soc. Lond. Ser. B, Biol. Sci.*, 270: 2065-2071.

Wehner, R. 1971 *Information Processing in the Visual Systems of Arthropods*. Springer-Verlag, N.Y.

Wehner, R. 1976. Polarized-light navigation by insects. *Sci. Amer.*, 235: 106-115.

Wehner, R., Bernard, G.D. & Geiger, E. 1975. Twisted and non-twisted rhabdoms and their significance for polarization detection in the bees. *J. Comp. Physiol.*, 104(3): 225-245.

Wehner, R., Michel, B. & Antonsen, P. 1996. Visual navigation in insects: coupling of egocentric and geocentric information. *J. Exp. Biol.*, 199: 129-140.

Wehner, R. & Rossel, S. 1985. The bee's celestial compass: a case study in behaviour neurobiology. *Fortschr. Zool.*, 31: 11-53.

Weis-Fogh, T. 1964. Control of basic movements in flying insects. *Symp. Roy. Soc. Exp. Biol.*, 18: 343-361.

Wellings, P.W. 1991. Host location and oviposition on animals. In: *Reproductive Behaviour of Insects Individuals and Populations*. W.J. Bailey & J. Ridsdill-Smith (eds.). Chapman & Hall, London, pp. 75-107.

Wellington, W.G. 1949. The effect of temperature and moisture upon the behaviour of the spruce budworm, *Choristoneura fumiferana* Clemens (Lepidoptera: Tortricidae). *Sci. Agric.*, 29: 210-229.

Wellington, W.G. 1953. Motor responses evoked by dorsal ocelli of *Sarcophaga aldrichi* Parker, and the orientation of the fly to plane polarized light. *Nature*, 172: 1177-1179.

Wellington, W.G. 1974. Bumble bee ocelli and navigation at dusk. *Science*, 183: 550-551.

Wells, H., Wells, P.H. & Cook, P. 1990. The Importance of overwinter aggregation for reproductive success of monarch butterflies (*Danaus plexippus*). *J. Theor. Biol.*, 147: 115-131.

Wells, M.M. & Henry, C.S. 1992. Behavioural responses of green lacewings (Neuroptera: Chrysopidae: *Chrysoperla*) to synthetic mating songs. *Anim. Behav.*, 44(4): 641-652.

Wendler, G. 1966. The coordination of walking movements in arthropods. *Sym. Soc. Exp. Biol.*, 20: 229-250.

Wenner, A.M. 1964. Sound communication in honey bees. *Sci. Amer.*, 210: 116-124.

Wenner, A.M. 1971. *The Bee Language Controversy: An Experiment in Science*. Educational Programmes Improvement Corp., Boulder, Co.

Wenner, A.M. & Johnson, D.L. 1967. Honey bees: do they use direction and distance information provided by their dancers? *Science*, 158: 1076-1077.

Wenner, A.M. & Wells, P.H. 1990. *Anatomy of a Controversy: The Question of a "Language" Among Bees*. Columbia Univ. Press, N.Y.

Wenzel, J.W. 1992. Behavioural homology and phylogeny. *Ann. Rev. Ecol. Syst.*, 23: 361-381.

Wenzel, J.W. 1993. Application of the biogenetic law to behavioural ontogeny: a test using nest architecture in paper wasps. *J. Evol. Biol.*, 6: 229-247.

Wetterer, J.K. 1999. The ecology and evolution of worker size-distribution in leaf-cutting ants (Hymenoptera: Formicidae). *Sociobiology*, 34(1): 119-144.

Wheeler, W.M. 1910. *Ants, Their Structure, Development, and Behaviour.* Columbia Univ. Press, N.Y.

Wheeler, W.M. 1918. A study of some ant larvae with a consideration of the origin and meaning of social habits among insects. *Proc. Amer. Phil. Soc.*, 57: 293-343.

Wheeler, W.M. 1923. *Social Life Among the Insects*. Harcourt, Brace, N.Y.

Wheeler, W.M. 1928. *The Social Insects: Their Origin and Evolution*. Kegan Paul, Trench Trubner & Co., Ltd., London.

White, I.M. 1999. Morphological features of the tribe Dacini (Dacinae): their significance to behaviour and classification. In: *Fruit Flies (Tephritidae): Phylogeny and Evolution of Behaviour*. M. Aluja & A.L. Norrbom (eds.). CRC Press, Boca Raton, pp. 505-533.

White, R.H. 1985. Insect visual pigments and colour vision. In: *Comprehensive Insect Physiology, Biochemistry and Pharmacology. Nervous System: Sensory*. Vol. 6. G.A. Kerkut & L.I. Gilbert (eds.). Pergamon Press, Oxford, pp. 431-493.

Whiting, P.W. 1932. Reproductive reactions of sex mosaics of a parasitic wasp, *Habrobracon juglandis*. *J. Comp. Psychol.*, 14: 345-363.

Whittaker, R.H. & Feeny, P. 1971. Allelochemicals: chemical interactions between species. *Science*, 171: 757.

Wickler, W. 1968. *Mimicry in Plants and Animals*. Weidenfeld & Nicolson, London.

Wigglesworth, V.B. 1934. The physiology of ecdysis in *Rhodnius prolixus* (Hemiptera). II. Factors controlling moulting and metamorphosis. *Quart. J. Mic. Sci.*, 77: 191-222.

Wigglesworth, V.B. 1936. The function of corpus allatum in the growth and reproduction of *Rhodnius prolixus* (Hemiptera). *Quart. J. Mic. Sci.*, 79: 91-121.

Wigglesworth, V.B. 1941. The sensory physiology of the human louse, *Pediculus humanus* DeGeer (Anoplura). *Parasitology*, 33: 67-109.

Wigglesworth, V.B. 1964a. *The Life of Insects*. The New American Library, N.Y.

Wigglesworth, V.B. 1964b. The hormonal regulation of growth and reproduction in insects. *Adv. Insect Physiol.*, 2: 247-336.

Wigglesworth, V.B. 1970. *Insect Hormones*. W.H. Freeman & Co., San Francisco.

Wigglesworth, V.B. 1972. *The Principles of Insect Physiology*. (7th Ed.). ELBS and Chapman & Hall, London.

Wigglesworth, V.B. 1973. The evolution of insect wings and flight. *Nature*, 246: 127-129.

Wigglesworth, V.B. & Gillett, J.D. 1934. The function of the antennae in *Rhodnius prolixus* (Hemiptera) and the mechanism of orientation to the host. *J. Exp. Biol.*, 11: 120-139.

Wilcox, S. 1979. Sex discrimination in *Gerris remigis*: role of a surface wave signals. *Science*, 206: 1325-1327.

Wilde, J. de, & Boer, J.A. de, 1961. Physiology of diapause in the adult Colorado beetle. II. Diapause as a case of pseudoallatectomy. *J. Insect Physiol.*, 6: 152-161.

Wilde, J. de & Loof, A. de, 1973a. Reproduction. In: *The Physiology of Insecta*. Vol. I. (2nd ed.). M. Rockstein (ed.). Academic Press, N.Y., pp. 11-95.

Wilde, J. de & Loof, A. de, 1973b. Reproduction- endocrine control. In: *The Physiology of Insecta*. Vol. I. (2nd ed.). M. Rockstein (ed.). Academic Press, N.Y., pp. 97-157.

Wilkin, P.J. 1985. Aerodynamics. In: *Comprehensive Insect Physiology, Biochemistry and Pharmacology. Nervous System: Structure and Motor Function*. Vol. 5. G.A. Kerkut & L.I. Gilbert (eds.). Pergamon Press, Oxford, pp. 553-570.

Wille, J. 1920. Biologie und Bekamptung der deutschen Schabe. *Monogr. Angew. Entomon*, 5: 1-141.

Williams, C.B. 1930. *The Migration of Butterflies*. Oliver and Boyd, Edinburgh.

Williams, C.B. 1957. Insect migration. *Ann. Rev. Entomol.*, 2: 163-180.

Williams, C.B. 1958. *Insect Migration*. Collins, London.

Williams, C.M. 1952. Physiology of insect diapause. IV. The brain and prothoracic glands as an endocrine system in the cecropia silkworm. *Biol. Bull.*, 103: 120-138.

Williams, C.M. & Adkisson, P.L. 1964. Physiology of insect diapause. XIV. An endocrine mechanism for the photoperiodic control of pupal diapause in the oak silkworm *Antheraea pernyi*. *Biol. Bull.*, 127: 511-525.

Williams, G.C. 1966. *Adaptation and Natural Selection*. Princeton Univ. Press, Princeton, N.J.

Willis, E.R. & Roth, L.M. 1950. Humidity reactions of *Tribolium castaneum* (Herbst.). *J. Exp. Zool.*, 115: 561-588.

Willmer, P.G., Gilbert, F., Ghazoul, J., Zalat, S. & Semida, F.M. 1994. A novel form of territoriality: daily paternal investment in an anrtophorid bee. *Anim. Behav.*, 48: 535-549.

Wilsbacher, L.D. & Takahashi, J. 1998. Circadian rhythms: molecular basis of the clock. *Curr. Opin. Genet. Dev.*, 8: 595-602.

Wilson, D.M. 1966. Insect walking. *Ann. Rev. Entomol.*, 11: 103-122.

Wilson, D.M. 1968. The nervous control of insect flight and related behaviour. *Adv. Insect Physiol.*, 5: 289-334.

Wilson, D.M. 1971. Stabilizing mechanisms in insect flight. In: *Gravity and the Organization*. S.A. Gordon & M.J. Cohen (eds.). Univ. Chicago Press, Chicago, pp. 169-176.

Wilson, E.O. 1959. Source and possible nature of the odour trail of fire ants. *Science*, 129: 643-644.

Wilson, E.O. 1962. Chemical communication among workers of fire ant *Solenopsis saevissima* (Fr. Smith). 1. The organization of mass foraging. 2. An information analysis of the odour trail. 3. The experimental introduction of social responses. *Anim. Behav.*, 10: 134-164.

Wilson, E.O. 1963. Pheromones. *Sci. Amer.*, 208: 100-114.

Wilson, E.O. 1971. *The Insect Societies*. Belknap Press, Harvard Univ. Press, Cambridge, Massachusetts.

Wilson, E.O. 1975a. Slavery in ants. *Sci. Amer.*, 232: 32-36.

Wilson, E.O. 1975b. *Sociobiology: The New Synthesis*. Belknap Press, Harvard Univ. Press, Cambridge, Massachusetts.

Wilson, E.O. & Bossert, W.H. 1963. Chemical communication among animals. *Rec. Prog. Hormone Res.*, 19: 673-716.

Wilson, K. 1988. Egg laying decisions by the bean weevil *Callosobruchus maculatus*. *Ecol. Entomol.*, 13: 107-118.

Wimberger, P.H. & Queiroz, A. de, 1996. Comparing behavioural and morphological characters as indicators of phylogeny, In: *Phylogenies and the Comparative Method in Animal Behaviour*. E.P. Martins (ed.). Oxford Univ. Press, Oxford, pp. 206-233.

Winfree, A.T. 1986. The *Timing of Biological Clocks*. Scientific American Lib., No. 19, N.Y.

Winston, M.L. 1987. *The Biology of the Honey Bee*. Harvard Univ. Press, Cambridge, Massachusetts.

Woiwod, I.P., Reynolds, D.R. & Thomas, C.D. (eds.) 2001. *Insect Migration: Mechanisms and Consequences*. CABI Publ., Wallingford, U.K.

Wojtasek, H. & Leal, W.S. 1999. Conformational change in the pheromone-binding protein from *Bombyx mori* induced by pH and by interaction with membranes. *J. Biol. Chem.*, 274: 30950-30956.

Wolf, E. 1927. Über das Heimkehrvermögen der Bienen. *Z. vergl. Physiol.*, 6: 221-254.

Wood, D.L. 1982. The role of pheromones, kairomones and allomones in the host selection and colonization behaviour of bark beetles. *Ann. Rev. Entomol.*, 27: 411-446.

Wood, D.L., Browne, L.E., Silverstein, R.M. & Rodin, J.O. 1966. Sex pheromones of bark beetles-I. mass production, bioassay, source, and isolation of the sex pheromone of *Ips confusus* (LeC.). *J. Insect Physiol.* 12: 523-536.

Wood, W.F., Palmer, T.M. & Stanton, M.L. 2002. A comparison of volatiles in mandibular glands from three *Crematogaster* ant symbionts of the whistling thorn acacia. *Biochem. Syst. Ecol.*, 30(3): 217-222.

Woodhouse, E.J. & Dutt, H.L. 1913. Work on the caterpillar pest of mokameh. *Agril. J. India*, 8(4): 372.

Wright, S. 1978. *Evolution and the Genetics of Populations. Vol. 4: Variability Within and Among Natural Populations*. Univ. Chicago Press, Chicago.

Wyatt, G.R. & Davey, K.G. 1996. Cellular and molecular actions of juvenile hormone. II. Roles of juvenile hormones in adult insects. *Adv. Insect Physiol.*, 26: 1-155.

Wyatt, T.D. 2003. *Pheromones and Animal Behaviour: Communication by Smell and Taste*. Cambridge Univ. Press, Cambridge.

Wynn, H. & Vahed, K. 2004. Male *Gryllus bimaculatus* guard females to delay them from mating with rival males and to obtain repeated copulations. *J. Insect Behav.*, 17(1): 53-66.

Yamada, H., Matsuda, M. & Oguma, Y. 2002a. Genetics of sexual isolation based on courtship song between two sympatric species: *Drosophila ananassae* and *D. pallidosa*. *Genetica*, 116: 225-237.

Yamada, H., Sakai, T., Tomaru, M., Doi, M., Matsuda, M. & Oguma, Y. 2002b. Search for species-specific mating signal in courtship songs of sympatric sibling species, *Drosophila ananassae* and *D. pallidosa*. *Genes Genet. Syst.*, 77(2): 97-106.

Yamamoto, D., Jallon, J.-M. & Komatsu, A. 1997. *Genetic dissection of sexual behaviour in Drosophila melanogaster*. *Ann. Rev. Entomol.*, 42: 551-585.

Yamamura, N. 1986. An evolutionarily stable strategy (ESS) model of postcopulatory guarding in insects. *Theor. Popul. Biol.*, 29: 438-455.

Yang, E., Lee, D.W. & Wu, W.N. 2003. Action spectra of phototactic responses of the flea beetle, *Phyllotreta striolata*. *Physiol. Entomol.*, 28(4): 362-367.

Yazdani, S.S. & Agarwal, M.L. 1997. *Elements of Insect Ecology*. Narosa Publ. House, N. Delhi.

Yazdani, S.S. & Agarwal, M.L. 2005. *Essentials of Scientific Crop Protection*. Kalyani Publ., Ludhiana.

Yin, C., Qin, W. & Stoffolano, J.G. Jr. 1999. Regulation of mating behaviour by nutrition and the corpus allatum in both male and female *Phormia regina* (Meigen). *J. Insect Physiol.*, 45(9): 815-822.

Young, D. 1989. *Nerve Cells and Animal Behaviour*. Cambridge Univ. Press, Cambridge.

Young, M. 1998. The molecular control of circadian behavioural rhythms and their entrainment in *Drosophila*. *Ann. Rev. Biochem.*, 67: 135-152.

Young, M. & Musgrave, I. 2005. "Moonshine: why the peppered moth remains an icon of evolution," *Skeptical Inquirer*, March-April, pp. 23-28.

Youthed, G.J. & Moran, V.C. 1969. The lunar-day activity rhythm of myrmeleontid larvae. *J. Insect Physiol.*, 15: 1259-1271.

Yuval, B. & Galun, R. 1987. Aspects of compound-conditioning to gustatory stimuli in the house fly *Musca domestica* L. *J. Insect Physiol.*, 33(3): 159-165.

Zacharuk, R.Y. 1980. Ultrastructure and function of insect chemosensilla. *Ann. Rev. Entomol.*, 25: 27-47.

Zeh, D.W. & Smith, R.L. 1985. Paternal investment by terrestrial arthropods. *Amer. Zool.*, 25: 785-805.

Zera, A.J. & Denno, R.F. 1997. Physiology and ecology of dispersal polymorphism in insects. *Ann. Rev. Entomol.*, 42: 207-231.

Zerrahn, G. 1933. Formdressur und Formunterscheidung bei der Honigbiene. *Z. vergl. Physiol.*, 20: 117-150.

Zhao, Z.W. & Zera, A.J. 2004. The haemolymph JH titre exhibits a large-amplitude, morph-dependent, diurnal cycle in the wing-polymorphic cricket, *Gryllus firmus*. *J. Insect Physiol.*, 50(1): 93-102.

Zhu, J., Millar, J. & Löfstedt, C. 1995. Hormonal regulation of sex pheromone biosynthesis in the turnip moth, *Agrotis segetum*. *Arch. Insect Biochem. Physiol.*, 30(1): 41-59.

Zimmerman, M. 1979. Oviposition behaviour and the existence of an oviposition-deterring pheromone in *Hylemya*. *Environ. Entomol.*, 8: 277-279.

Zimmerman, M. 1980. Selective deposition of an oviposition deterring pheromone by *Hylemya*. *Environ. Entomol.*, 9: 321-324.

Zink, A.G. 2003. Quantifying the costs and benefits of parental care in female treehoppers. *Behav. Ecol.*, 14(5): 687-693.

Zitnan, D., Kingan, T.G., Hermesman, J.L. & Adams, M.E. 1996. Identification of ecdysis-triggering hormone from an epitracheal endocrine system. *Science*, 271(5245): 88-91.

Zitnan, D., Ross, L.S., Zitnanova, I., Hermesman, J.L., Gill, S.S. & Adams, M.E. 1999. Steroid induction of a peptide hormone gene leads to orchestration of a defined behavioural sequence. *Neuron*, 23: 523-535.

Index of Scientific Names

C

Y

Z

Subject Index

B

C

G

H

Q

R

Y

Z

Plate 1.1. A caterpillar coil itself and becomes apparently feign death if disturbed (Photo courtsey S.K. Jain).

Plate 6.1. Pond-skater (Fam. Gerridae) (Photo courtsey and with permission from John Moore). © John Moore.

Plate 6.2. Water stick insect, *Ranatra linearis* (Photo courtsey and with permission from John Tyler). © John Tyler.

Plate 6.3. Giant water bug, *Belostoma* sp. (Photo M.L. Agarwal).

Plate 7.1. Dung beetles rolling a dung ball (Photo courtsey and with permission from John Moore). © John Moore.

Plate 8.1. The migratory painted lady, *Cynthia cardui* (Photo courtsey and with permission from David L. Mohn). © David L. Mohn.

Plate 8.2. The migratory red admiral butterfly, *Vanessa atalanta* (Photo courtsey and with permission from John Tyler). © John Tyler.

Plate 8.3. Indian tortoiseshell, *Aglais cashmirensis*. The butterfly *cashmirensis* was reported at 18,000 feet in Himalayas (Photo courtsey and with permission from Ramana Athreya). © Ramana Athreya.

Plate 9.1. Cephalopharyngeal skeleton of fruit fly, *Bactrocera* larva. The fruit fly larva has fluid feeding mouthparts and inhabits the bacterial soup in the decaying areas of infested fruit. It bores into the interior of the host, macerates the tissues and ingests the broken-down tissues and associated bacteria (Photo M.L. Agarwal).

Plate 9.2. Leaf miner feeds inside the blade of a leaf, between the epidermal layers (Photo M.L. Agarwal).

Plate 9.3a. Mango shoot gall formed by the feeding of gall forming mango shoot gall psyllid (Photo M.L. Agarwal).

Plate 9.3b. Galls on a leaf (Photo M.L. Agarwal).

Plate 9.4. Litchi fruit and shoot borer, *Conopomorpha cramerella* bores into and feed on litchi leaves and fruits (Photo M.L. Agarwal).

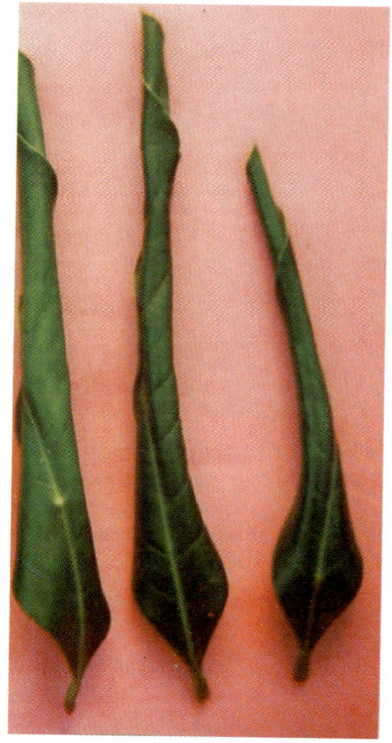

Plate 9.5. The larva of litchi leaf roller, *Statherotis discana* hides and feed on the leaf of litchi, rolling the leaf while feeding on it (Photo M.L. Agarwal).

Plate 9.6. Feeding by a leaf skeletonizer on leaf tissues (Photo M.L. Agarwal).

Plate 10.1. Male hercules beetles. The spectacular clash between two male beetles for dominance and access to a nearby female is well-known. A minor male stands little chance against a major male. (Photo courtesy and with permission from Alex Wild). ©Alex Wild.

Plate 10.2. Mating pair of dragonflies forming the characteristic 'wheel' position (Photo courtsey and with permission from John Moore). © John Moore.

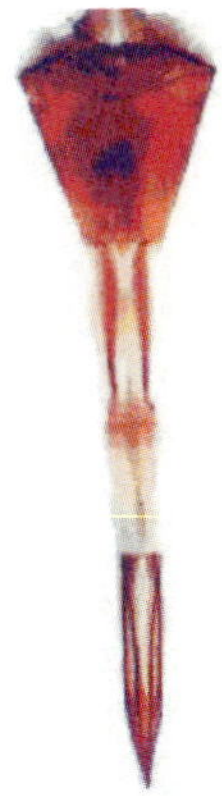

Plate 10.3. Ovipositor of a fruit fly, *Bactrocera* sp. The last three abdominal segments are modified to form a tube-like substitutional ovipositor which facilitates the act of egg deposition inside the host (Photo M.L. Agarwal).

Plate 10.4. Male giant water bug, *Abedus* sp. carrying their mate's eggs on his back. Many of the eggs have already hatched (Photo courtesy and with permission from Alex Wild). © Alex Wild.

Plate 11.1. A caterpillars looking like bird dropping and resemble with something that a bird would never eat (Photo courtsey and with permission from David L. Mohn). © David L. Mohn.

Plate 11.2. A moth resembling bird dropping (Photo courtsey and with permission from John Tyler). © John Tyler.

Plate 12.1. Trophallaxis between two members of an ant colony (Photo courtsey and with permission from John Moore). © John Moore.

Plate 12.2. Mutualistic relationship between ants and sap-sucking hemipterans which excrete sugary substance 'honeydew' (Photo M.L. Agarwal).

Plate 12.3. Ant, *Formica* workers tending aphids (Photo courtesy and with permission from John R. Conway). © John R. Conway.

Plate 12.4a. *Apis mellifera* bee collecting pollens from a flower. The mutualistic relationship between the flower and bee is beneficial for both (Photo M.L. Agarwal).

Plate 12.4b. European honey bee, *Apis mellifera* collecting nectar from mustard flower (Photo courtesy Neeraj Kumar).

Plate 12.5. A hummingbird hawk moth, *Macroglossum* sp. The moth has exceptionally powerful flight and hover over flowers as it feed on wing (Photo courtsey and with permission from John Moore). © John Moore.

Plate 12.6. Fungal garden of leaf cutting ant, *Atta cephalotes* (Photo Courtesy and with permission from A. San Juan). © A. San Juan.

Plate 13.1. Praying mantis, the raptorial forelegs are used for defence and catching prey (Photo courtsey and with permission from David L. Mohn). © David L. Mohn.

Plate 13.2. Protective colouration as shown by a katydid with its background (Photo M.L. Agarwal).

Plate 13.3. A katydid resembling in colouration with a leaf (Photo courtsey S.K. Jain).

Plate 13.4. The orchid mantid, *Hymenopus coronatus* blends beautifully with the pink flower spike of an orchid (Photo courtsey and with permission from John Moore). © John Moore.

Plate 13.5a. The stick insect resembles with colour of foliage and twigs on which it rests. The insect is in harmony among the grasses (Photo M.L. Agarwal).

Plate 13.5b. A stick insect resembling a green twig with thorns (Photo courtsey and with permission from Ramana Athreya). © Ramana Athreya.

Plate 13.6. The leaf insect, *Phyllium* resembling a leaf (Photo courtsey and with permission from John Moore). © John Moore.

Plate 13.7. A mantid. It is difficult to find it in the dead litter (Photo courtsey and with permission from Ramana Athreya). © Ramana Athreya.

Plate 13.8. The female lappet moth, *Paralebeda plagifera* in typical rest position and resembling a dead leaf (Photo courtsey and with permission from David L. Mohn). © David L. Mohn.

Plate 13.9. *Microblepsis leucosticta*, a moth resembling a feather (Photo courtsey and with permission from John Moore). © John Moore.

Plate 13.10. A caterpillar resembling a twig both in colour and form (Photo M.L. Agarwal).

Plate 13.11. Deadleaf *Kallima* butterfly has extraordinary resemblance to leaves, which is exhibited by undersurface of the closed wings.

Plate 13.12. Autumn leaf butterfly, *Doleschalia bisaltide* resembling a dry leaf (Photo courtsey and with permission from Ramana Athreya). © Ramana Athreya.

Plate 13.13. Tailed sulphur butterfly, *Dercas verhuelli* resembling a green leaf (Photo courtsey and with permission from Ramana Athreya). © Ramana Athreya.

Plate 13.14. The peppered moth, *Biston betularia* (normal and melanic forms). The melanic and normal forms get advantages under different environmental conditions and provide an excellent example of impact of natural selection on biological adaptation, playing a vital role in evolution (from the experiment of H.D.B. Kettlewell).

Plate 13.15. Viceroy butterfly, *Basilarchia archippus*.

Plate 13.16. Monarch butterfly, *Danaus plexippus*.

Plate 13.17. A caterpillar exhibiting dangerous or unpleasant attributes by means of characteristic structures like eyespots, colour pattern and horn (Photo courtesy P.P. Singh).

Plate 13.18. Bihar hairy caterpillar, *Spilarctia obliqua* with urticating hairs, containing irritating substances secreted by glands located at their bases (Photo M.L. Agarwal).

Plate 13.19. The spiny Bihar hairy caterpillars of *Spilarctia obliqua* live in aggregated condition and remain protected up to 2nd instar, and when become somewhat larger they disperse and feed apart (Photo courtesy P.P. Singh).

Plate 13.20. A silverline butterfly, *Spindasis* sp. The posterior end of body has appendages similar to those found on anterior end (Photo M.L. Agarwal).

Plate 13.21. The defensive display of eyespots on wings of a moth deflects the predator's attack (Photo M.L. Agarwal).

Plate 13.22. Defence strategy of a caterpillar, when disturbed it jumps with a silk thread and hangs itself in the mid-air (Photo courtesy I.P. Singh).

Plate 13.23. Caterpillar of oleander hawk moth, *Daphnis nerii* (Photo courtsey and with permission from David L. Mohn). © David L. Mohn.

Plate 14.1. Nest of a mason wasp constructed from mud (Photo courtesy S.K. Jain).

Plate 14.2. Nest of a solitary wasp constructed from mud and divided into cells (Photo courtesy S.K. Jain).

Plate 14.3. Workers of primitive subterranean termite in nest below ground. They form an extensive underground tunnelling system and damage plant parts (Photo courtesy P.P. Singh).

Plate 14.4. Mound formed by termites above their subterranean nest (Photo courtesy Binay Kumar).

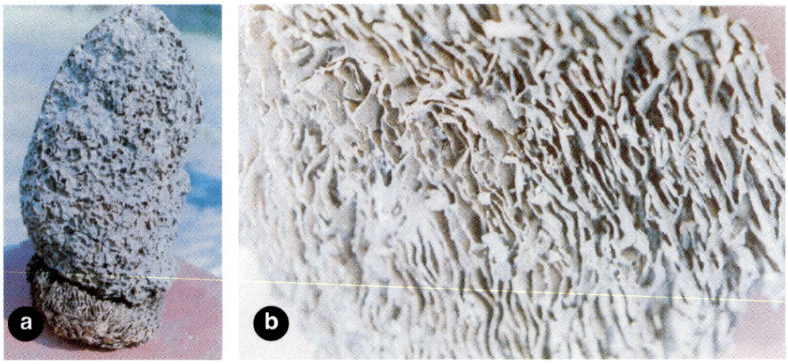

Plate 14.5a,b. Fungal garden from the termite nest. Termites are most conspicuous fungus-growing social insects. Each fungal garden is enclosed in a close fitting cavity lined with a mixture of saliva and dirt (Photo M.L. Agarwal).

Plate 14.6a-e. The nests of paper wasps. The nest is constructed above the ground hanging with a petiole from the tree. The nest material is a water-resistant gray paper made up of a mixture of masticated wood and the salivary secretion of the female wasps (Photo 14a-c M.L. Agarwal and 14d,e courtesy Neeraj Kumar).

Plate 14.7. An apiary of giant honey bee, *Apis dorsata* in nature. A large number of bee colonies exist on the same tree (Photo courtesy Neeraj Kumar).

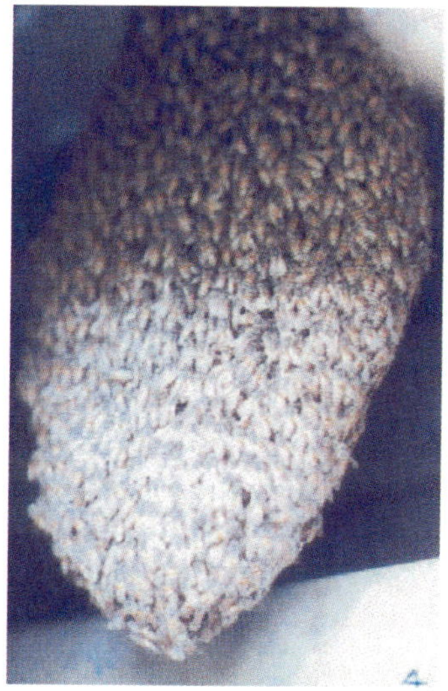

Plate 14.8. Indian honey bee, *Apis cerana* colony on a tree trunk after swarming (Photo courtesy R. Singh).

Plate 14.9. European honey bee, *Apis mellifera* colony with queen, drones and workers (Photo courtesy R. Singh).

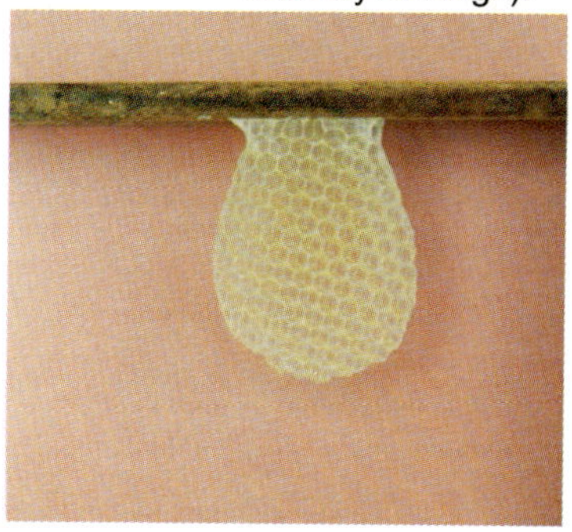

Plate 14.10. Wax comb of European honey bee, *Apis mellifera* suspended vertically from a bee hive frame (Photo M.L. Agarwal).

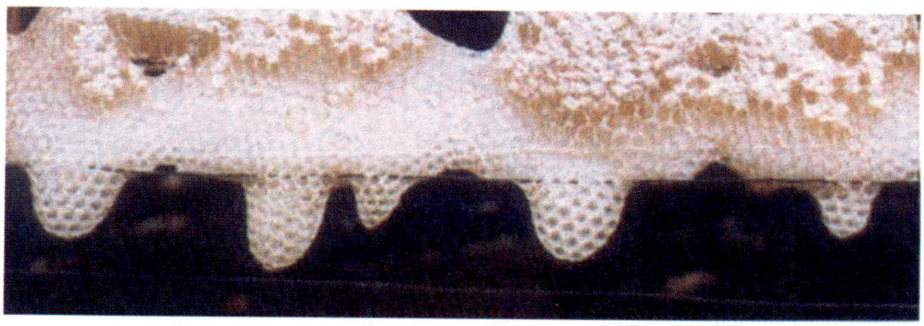

Plate 14.11. *Apis mellifera* comb showing mature queen cells, packed honey cells and empty cells (Photo courtesy R. Singh).

Plate 14.12. Caribbean pine leaf cutting ant, *Atta laevigata*. It is one of the largest leafcutter ants (Photo courtesy and with permission from Alex Wild). © Alex Wild.

Plate 14.13. Leaf cutting ant, *Atta cephalotes*. Notice the piles of forage that coat the upper surface of the nest mound (Photo Courtesy and with permission from A. San Juan). © A. San Juan.

Plate 14.14. The foraging workers of leaf cutting ants, *Atta cephalotes* returning back to the nest (Photo Courtesy and with permission from A. San Juan). © A. San Juan.

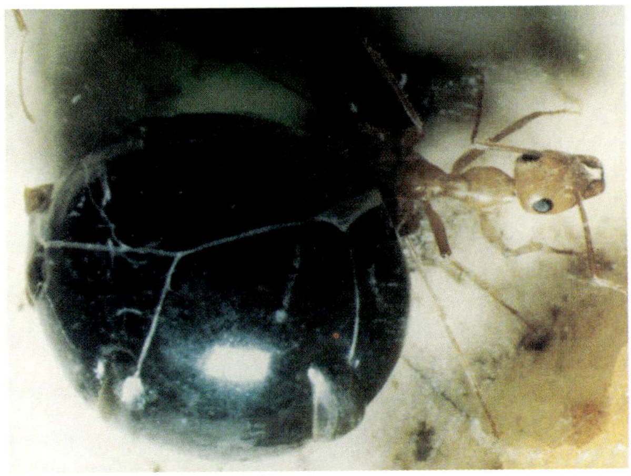

Plate 14.15. Mexican honeypot ant, *Myrmecocystus mexicanus*. Honeypot ants have an unusual food storage system. Some members of each colony act as living receptacles known as 'repletes'. The abdomens of honeypot ant workers are so distended that they have difficulty in moving and are forced to remain permanently in the nest as living carbohydrate stores (Photo courtesy and with permission from John R. Conway). © John R. Conway.

Plate 14.16. The nest of weaver ant, *Oecophylla* sp. formed by binding fresh leaves with silk. The ants pull the leaves into just the right position, holding them until workers bring larvae, which then proceed to secrete a silky substance from their well-developed silk glands, which acts as a 'glue', binding the leaves together. The ants take up residence within the newly established leaf nest. In the nest construction a remarkable collaborative effort follows (Photo M.L. Agarwal).

Plate 14.17. Green tree ants, *Oecophylla smaragdina* repairing nest by using silky substance (Photo courtsey and with permission from John Tyler). © John Tyler.

Plate 14.18. A major worker of army ant, *Eciton burchellii.* The mandibles of ants are usually large and powerful, and able to give painful bites (Photo courtesy and with permission from Alex Wild). © Alex Wild.

Plate 16.1. A firefly, *Pyractomena* sp. (Photo courtesy and with permission from J.E. Lloyd). © J.E. Lloyd.

Plate 16.2. A female *Photuris sp.* ready to devour a prey male. These *femmes fatales* use flash signals to lure in males of other species by imitating their females' flash patterns (Photo courtesy and with permission from J.E. Lloyd). © J.E. Lloyd.

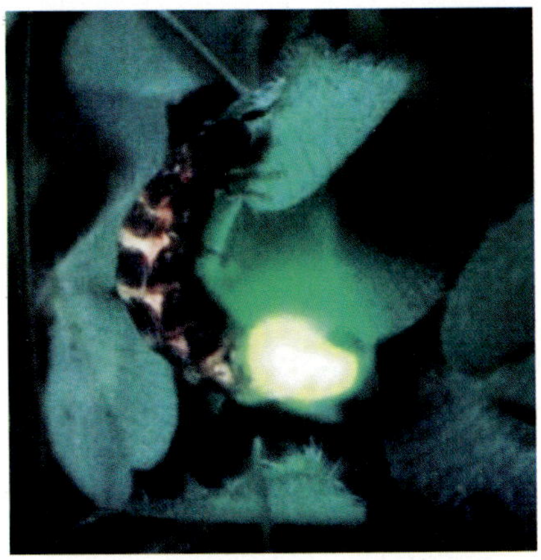

Plate 16.3. Glowworm, *Lampyris noctiluca* (Photo courtsey and with permission from Robin Scagell). © Robin Scagell.

Plate 16.4. European Glowworm, *Lampyris noctiluca* glowing female on grass stem (Photo courtsey and with permission from John Tyler). © John Tyler.

Plate 16.5. *Lampyris noctiluca* males: the white one is newly hatched (Photo courtsey and with permission from John Tyler). © John Tyler.

Plate 16.6. Glowworms, *Arachnocampa* lying in wait for prey (Photo Courtesy and with permission from Garry Maguire). © Garry Maguire.

Plate 16.7. Glowworm, *Arachnocampa* in final pupal stage (Photo Courtesy and with permission from Garry Maguire). © Garry Maguire.

Plate 16.8. Male and female glowworm flies, *Arachnocampa* mating adjacent to a pupal casing from which the female fly has emerged (Photo Courtesy and with permission from Garry Maguire). © Garry Maguire.

Plate 16.9. Luminous lanternfly, *Fulgora lanternaria* gives light when male and female are placed together.

Plate 16.10. A firefly, *Photinus pyralis* (Photo courtesy and with permission from J.E. Lloyd). © J.E. Lloyd.

Plate 21.1. A pheromone trap for catching the adult moths.

Plate 21.2. Twist-tie rope, a pheromone dispenser.

Plate 21.3. Steiner trap used for trapping adult males of *Bactrocera* spp. The attractant and insecticide mixture is suspended in the trap to annihilate the male flies (Photo M.L. Agarwal).